FACTORS FOR UNIT CONVERSIONS

Quantity	Equivalent Values
Mass	$1 \text{ kg} = 1000 \text{ g} = 0.001 \text{ metric ton} = 2.20462 \text{ lb}_m = 35.27392 \text{ oz}$ $1 \text{ lb}_m = 16 \text{ oz} = 5 \times 10^{-4} \text{ ton} = 453.593 \text{ g} = 0.453593 \text{ kg}$
Length	$1 \text{ m} = 100 \text{ cm} = 1000 \text{ mm} = 10^6 \text{ microns } (\mu m) = 10^{10} \text{ angstroms (Å)}$ $= 39.37 \text{ in.} = 3.2808 \text{ ft} = 1.0936 \text{ yd} = 0.0006214 \text{ mile}$ $1 \text{ ft} = 12 \text{ in.} = 1/3 \text{ yd} = 0.3048 \text{ m} = 30.48 \text{ cm}$
Volume	$1 \text{ m}^3 = 1000 \text{ liters} = 10^6 \text{ cm}^3 = 10^6 \text{ ml}$ $= 35.3145 \text{ ft}^3 = 220.83 \text{ imperial gallons} = 264.17 \text{ gal}$ $= 1056.68 \text{ qt}$ $1 \text{ ft}^3 = 1728 \text{ in.}^3 = 7.4805 \text{ gal} = 0.028317 \text{ m}^3 = 28.317 \text{ liters}$ $= 28\,317 \text{ cm}^3$
Force	$1 \text{ N} = 1 \text{ kg·m/s}^2 = 10^5 \text{ dynes} = 10^5 \text{ g·cm/s}^2 = 0.22481 \text{ lb}_f$ $1 \text{ lb}_f = 32.174 \text{ lb}_m \text{·ft/s}^2 = 4.4482 \text{ N} = 4.4482 \times 10^5 \text{ dynes}$
Pressure	$1 \text{ atm} = 1.01325 \times 10^5 \text{ N/m}^2 \text{ (Pa)} = 101.325 \text{ kPa} = 1.01325 \text{ bars}$ $= 1.01325 \times 10^6 \text{ dynes/cm}^2$ $= 760 \text{ mm Hg at } 0°C \text{ (torr)} = 10.333 \text{ m H}_2\text{O at } 4°C$ $= 14.696 \text{ lb}_f\text{/in.}^2 \text{ (psi)} = 33.9 \text{ ft H}_2\text{O at } 4°C$ $= 29.921 \text{ in Hg at } 0°C$
Energy	$1 \text{ J} = 1 \text{ N·m} = 10^7 \text{ ergs} = 10^7 \text{ dyne·cm}$ $= 2.778 \times 10^{-7} \text{ kW·h} = 0.23901 \text{ cal}$ $= 0.7376 \text{ ft-lb}_f = 9.486 \times 10^{-4} \text{ Btu}$
Power	$1 \text{ W} = 1 \text{ J/s} = 0.23901 \text{ cal/s} = 0.7376 \text{ ft·lb}_f\text{/s} = 9.486 \times 10^{-4} \text{ Btu/s}$ $= 1.341 \times 10^{-3} \text{ hp}$

Example: The factor to convert grams to lb_m is $\left(\dfrac{2.20462 \text{ lb}_m}{1000 \text{ g}} \right)$

WILEY SERIES IN
CHEMICAL ENGINEERING

Bird, Armstrong and Hassager: DYNAMICS OF POLYMERIC LIQUIDS, Vol. I FLUID MECHANICS

Bird, Hassagar, Armstrong and Curtiss: DYNAMICS OF POLYMERIC LIQUIDS, Vol. II KINETIC THEORY

Bird, Stewart, and Lightfoot: TRANSPORT PHENOMENA

Brownell and Young: PROCESS EQUIPMENT DESIGN: VESSEL DESIGN

Davis: NUMERICAL METHODS AND MODELING FOR CHEMICAL ENGINEERS

Doraiswamy and Sharma: HETEROGENEOUS REACTIONS: ANALYSIS, EXAMPLES AND REACTOR DESIGN, Vol. 1, Vol. 2

Felder and Rousseau: ELEMENTARY PRINCIPLES OF CHEMICAL PROCESSES, 2nd Edition

Foust, Wenzel, Clump, Maus and Andersen: PRINCIPLES OF UNIT OPERATIONS, 2nd Edition

Froment and Bischoff: CHEMICAL REACTOR ANALYSIS AND DESIGN

Franks: MODELING AND SIMULATION IN CHEMICAL ENGINEERING

Henley and Seader: EQUILIBRIUM—STAGE SEPARATION OPERATIONS IN CHEMICAL ENGINEERING

Hill: AN INTRODUCTION TO CHEMICAL ENGINEERING KINETICS AND REACTOR DESIGN

Jawad and Farr: STRUCTURAL ANALYSIS AND DESIGN OF PROCESS EQUIPMENT

Kellogg Company: DESIGN OF PIPING SYSTEMS, Revised 2nd Edition

Levenspiel: CHEMICAL REACTION ENGINEERING, 2nd Edition

Nauman and Buffham: MIXING IN CONTINUOUS FLOW SYSTEMS

Rase: CHEMICAL REACTOR DESIGN FOR PROCESS PLANTS PRINCIPLES AND TECHNIQUES, Vol. 1

Rase and Barrow: PIPING DESIGN FOR PROCESS PLANTS

Reklaitis: INTRODUCTION TO MATERIAL AND ENERGY BALANCES

Rudd, Fathi-Afshar, Trevino and Stadtherr: PETROCHEMICAL TECHNOLOGY ASSESSMENT

Rudd and Watson: STRATEGY OF PROCESS ENGINEERING

Sandler: CHEMICAL AND ENGINEERING THERMODYNAMICS

Smith and Corripio: PRINCIPLES AND PRACTICE OF AUTOMATIC PROCESS CONTROL

Smith and Missen: CHEMICAL REACTION EQUILIBRIUM ANALYSIS

Ulrich: A GUIDE TO CHEMICAL ENGINEERING PROCESS DESIGN AND ECONOMICS

Welty, Wicks and Wilson: FUNDAMENTALS OF MOMENTUM, HEAT AND MASS TRANSFER, 3rd Edition

ELEMENTARY PRINCIPLES OF CHEMICAL PROCESSES

Elementary Principles of Chemical Processes

Second Edition

Richard M. Felder
Ronald W. Rousseau

Department of Chemical Engineering
North Carolina State University
Raleigh, North Carolina

John Wiley & Sons

New York Chichester
Brisbane Toronto
Singapore

Library of Congress Cataloging in Publication Data:

Felder, Richard M., 1939–
 Elementary principles of chemical processes.

 Includes index.
 1. Chemical processes. I. Rousseau,
Ronald W., 1943- II. Title.
TP155.7.F44 1986 660.2′8 85-12058
ISBN 0-471-87324-1

Printed in the United States of America

10 9 8 7 6 5 4 3 2 1

About the Authors

Richard M. Felder, is Professor of Chemical Engineering at North Carolina State University, Raleigh, North Carolina. Dr. Felder received a BChE degree from the City College of New York in 1962 and a PhD in chemical engineering from Princeton University in 1966. He worked for the Atomic Energy Research Establishment (Harwell, England) and Brookhaven National Laboratory before joining the North Carolina State faculty in 1969. At North Carolina State he has won the Sigma Xi Faculty Research Achievement Award and has twice been designated a University Outstanding Teacher. He won the R. J. Reynolds Award for Excellence in Teaching, Research, and Extension in 1982, was designated an Alumni Distinguished Professor in 1983, and won the AT&T Foundation Award for Excellence in Engineering Education in 1985. Dr. Felder's research interests include engineering process simulation and optimization, coal gasification technology, membrane transport processes, and radioisotope applications in process engineering. He has authored or coauthored over 75 papers on chemical process engineering and engineering education, and has presented seminars and short courses on process analysis to industrial and research institutions in the United States, Europe, and South America.

Ronald W. Rousseau is Professor of Chemical Engineering at North Carolina State University, Raleigh, North Carolina. He received his BS, MS, and PhD at Louisiana State University and has been a member of the faculty at North Carolina State since 1969. In 1982–1983 he was Visiting Professor of Chemical Engineering at Princeton University. Dr. Rousseau has received the American Institute of Chemical Engineers Forest Products Division Award and, at North Carolina State University, the Sigma Xi Faculty Research Achievement Award. He has been a consultant to chemical companies and given short courses on chemical process technology to industrial personnel in the United States, Canada, Europe, and the Middle East. He has served on the Editorial Advisory Board of the *AIChE Journal* and on the Advisory Committee to the Chemical and Process Engineering Division of the National Science Foundation. Dr Rousseau's research interests are separation processes (especially crystallization), absorption and stripping, distillation, adsorption, and phase equilibria. He has presented results of his research in over 75 publications and at professional meetings and seminars in the United States, Asia, and Europe.

v

Preface
to the First Edition

An introductory stoichiometry course traditionally plays several important roles in the chemical engineering curriculum. On the most obvious level, it prepares the student to formulate and solve material and energy balances on chemical process systems and lays the foundation for subsequent courses in thermodynamics, unit operations, kinetics, and process dynamics. More fundamentally, it introduces the engineering approach to solving process-related problems: breaking a process down into its components, establishing the relations between known and unknown process variables, assembling the information needed to solve for the unknowns using a combination of experimentation, empiricism, and the application of natural laws, and, finally, putting the pieces together to obtain the desired problem solution.

We have tried in this book to fulfill each of these functions. Moreover, recognizing that the stoichiometry course is often the students' first real encounter with what they think may be their chosen profession, we have attempted to provide in the text a realistic, informative, and positive introduction to the practice of chemical engineering.

We begin the book with a qualitative discussion of the kinds of problems engineers must face in connection with several apparently dissimilar processes and, after a brief introduction to fundamental techniques of engineering calculations, systematically develop the structure of elementary process analysis: what process variables are and how they are expressed, measured, and calculated; laws of nature that govern the performance of processes; and physical properties of process materials that must be determined in order to design a new process or analyze an existing one.

The chemical process constitutes the structural and motivational framework for the presentation of all of the text material. When we bring in concepts from physical chemistry—for example, vapor pressure, solubility, and compressibility—we introduce them as quantities whose values are required to determine process variables or to perform material and/or energy balance calculations on a process. When we discuss computational techniques such as curve-fitting, root-finding methods, and numerical integration, we present them on the same need-to-know basis in the context of process analysis.

An important feature of the book is a set of industrial process case studies, which demonstrate the role of single-unit calculations in the analysis of multiple-unit processes. We have designed the case studies to be worked on as term projects by individuals or (preferably) small teams of students, beginning after the students have completed the introductory chapter on material balances (Chapter 4). In each study, the students are asked to produce a flowchart of a moderately complex process from a given description and to perform material and energy balance calculations on the process; in addition, they are called on to answer questions that require them to think about how the overall process is structured and why it might be structured that way.

Knowing the problems associated with the case study, the students tend to be on the lookout for information in the formal course material that will help them obtain the required solutions. The case study thus provides both a motivation for learning the text material and a feeling for the contextual significance of this material; moreover, it introduces the common engineering discipline of starting with a large, multifaceted problem and systematically building up the blend of information and technique needed to solve it.

In writing the book we have tried to avoid pedanticism without sacrificing thoroughness or rigor. The writing style is largely informal; all solution techniques are illustrated by examples; and short "Test Yourself" questions of the type used effectively in the self-paced approach to education help focus the students' attention and reinforce the main points in each section. Most of the problems concern real processes and contain realistic data; they are designed to provide practice in all of the methods discussed in the chapters they follow and, in addition, illustrate the range of activities encompassed by chemical engineering, both in the traditional areas of chemical processing and in such fields as environmental science and technology and biomedicine.

The SI system of units is used widely but not exclusively throughout the text, and extensive SI data tables, including steam tables, are contained in appendices. Computer programming is not covered explicitly, but applications of computers in process analysis are discussed, and problems for which computer solutions are appropriate are given.

We acknowledge with gratitude the many contributions of colleagues and friends who have helped us in the preparation of the book. Proceeding in no particular order, our thanks go to Professors David Marsland, of North Carolina State University, and Richard Seagrave, John Stevens, and George Burnet, of Iowa State University, who read the manuscript and offered many helpful suggestions for its improvement; Russ O'Dell, who worked out solutions and corrected several dozen mistakes in the manuscript after the authors had convinced themselves there were none left; Jim Ferrell, our beloved leader, who gave us much moral and financial support throughout the seemingly unending years in which the book was being written; Bobbie and Tess, for more than we could summarize here; Kenneth Felder, who performed with great diligence the horrible job of proofreading data tables; Ron Jr., David, Brett, Elena, and Gary for just being there; and Thurman Poston, the engaging engineering editor, who provided us with many broad perspectives, deep insights, and free lunches. We also thank students who served as guinea pigs during the formative stages of the book's development, and we sincerely apologize for all those problems they involuntarily and painfully debugged for us; in particular, we raise our glasses to the Lost Generation of Fall 1973, who had the miserable fortune to get the first draft as a course text.

We are especially indebted to Dr. James Fair of the Monsanto Company, Mr. Norman Kaplan of the U.S. Environmental Protection Agency, and Dr. Ray E. Harrison of Westvaco, Inc., who reviewed draft copies of the case studies and suggested changes to bring the process descriptions closer in line with industrial practice.

In some cases we chose to retain deviations from practice for pedagogic reasons; any such deviations should be attributed to us and not the reviewers.

Last, and most of all, we thank Magnificent Mary Wade, who uncomplainingly and with great good humor typed revision after revision, until the authors, unable to stand any more, declared the book done.

Richard M. Felder
Ronald W. Rousseau

Preface
to the Second Edition

Contemplating the warm reception given to the first edition of this book by both professors and students, we have for the most part elected to retain the overall text structure and philosophy in the second edition. The principal additions are three new case studies, hundreds of new and modified problems, a new chapter on computer-aided balances, and related appended material on iterative solution of nonlinear equations. In the paragraphs that follow, we summarize these and other new features of the text.

We have combined the introductory material on engineering calculations, including unit systems and conversions, dimensional homogeneity, and process data handling (Chapters 2 and 4 of the first edition), into a single introductory chapter. In the first chapter on material balances, we have expanded the discussion of recycle and purge calculations, and we introduce the concept of extent of reaction, thereby setting the stage for subsequent treatments of generalized stoichiometry in thermodynamics and kinetics courses. In the same chapter, we introduce the concept of chemical equilibrium, showing—without treading too heavily on the rightful domain of the thermodynamics course—how equilibrium compositions can be calculated from given values of equilibrium constants.

The section on equations of state now includes cubic equations—in particular, the Soave modification of the Redlich-Kwong equation—along with virial equations and the compressibility factor equation of state. The discussion of vapor-liquid equilibrium has been expanded to include bubble and dew point calculations and boiling point (Txy and Pxy) diagrams, and a section on colligative properties of solutions has been added.

If there has been a significant change in recent years in the way material and energy balance calculations are performed, it is the increased extent to which computers are used for these calculations. We recognize the inevitability of this trend and applaud it; however, we feel strongly that prospective chemical engineers should understand the principles of balance calculations before they are turned loose on ASPEN or its equivalent. Consequently, although we have substantially expanded our coverage of computer-aided balance calculations in this edition, we introduce the topic only after providing a solid foundation in the conservation laws.

In Chapter 10 on computer-aided balances, we introduce the sequential modular and simultaneous equation approaches to flowsheet simulation and illustrate how these approaches are implemented in practice. In presenting this material, we have tried to provide enough information so that the student understands how process flowsheet simulators work and what they can accomplish, but not so much detail that the perspective of the introductory text is lost. The material on numerical methods in Appendix A has been similarly expanded to provide a more extensive coverage of iterative techniques for solving nonlinear algebraic equations, material fundamental to both approaches to flowsheet simulation.

We have added a number of complex problems to the text, some of which have been designated as appropriate for computer solution. We have retained and in some cases expanded the "Test Yourself" questions sprinkled throughout the book, intended to help students make sure they have grasped the important concepts in each section. In addition, we have included several "Creativity Exercises," in which the student is asked to brainstorm imaginative applications of the techniques presented in the text and to come up with innovative approaches to solving problems that arise in the context of the text material.

It is impossible to thank individually all of our faculty colleagues and students who contributed the corrections and suggestions that we have used in preparing this edition. With apologies to those whom we might inadvertently omit, we express our gratitude to Ray Anthony, Albert Babb, J. B. Bredenberg, Bill Clements, Cameron Crowe, Sam Davis, Ehud Finkelstein, Dave Guinnup, Richard Hanks, Curt Johnson, R. S. Larson, Terry Lenz, Octave Levenspiel, P. K. Lim, Carl Locke, Jerry Meldon, Robert Romig, Donald Sundberg, Edward Tefft, Charles Walker, and Gordon Youngquist, and most particularly to Victor Agreda, Tom Canning, and Raphael Katzen for assistance in the development of the case studies. We thank our colleagues in the chemical engineering departments of the University of Colorado and Princeton University, without whose hospitality during our sabbatical leaves we would probably still be struggling to finish Chapter 2. We also thank those of our students who took the trouble to point out errors in early printings; we like to think they did it out of a sense of cooperation and common purpose, and not just to collect the quarters.

Finally, we thank Sandra and Linda for their moral support during the years of preparation of this edition, and we salute Merrill Floyd and Don Ford, whose perpetual good spirits and patience in the face of one missed deadline after another made the preparation a much less painful experience than it might otherwise have been.

<div align="right">

RMF
RWR

</div>

To the Instructor

The organization of this text has been planned to provide enough flexibility to accommodate classes with diverse backgrounds within the scope of a one-semester or two-quarter course. We anticipate that classes starting with a minimal background in engineering analysis will cover the first nine chapters and one case study. Classes of students who have been exposed to dimensional analysis and elementary data correlation can skip or skim Chapter 2; and classes whose freshman chemistry courses cover the systematic use of units to describe and analyze reactive systems may omit portions of Chapter 3. The time gained as a result of these omissions may be used to cover additional sections in Parts 1 through 3, such as Chapter 10 on computer-aided balances or Chapter 11 on transient balances, or to cover appended material on numerical analysis, or to work through an additional case study. Covering all of Chapters 2 and 3 should, in any event, take no more than six lecture hours. A one-quarter course should cover Chapters 1 through 6.

In our presentation of material balance procedures we place greater emphasis on algebraic solution techniques and less on such devices as tie elements than do most authors. We have adopted this approach for several reasons. The algebraic technique is perfectly general, applying to processes both with and without tie elements, and is better adapted to computer-aided balancing. When tie elements are stressed, we find that students tend to use them as crutches, and when (as in most real industrial processes) there is no tie element, the students often tend to flounder unless they are accustomed to the algebraic approach. Furthermore, we believe that much of the computational efficiency attributed to the use of tie elements is illusory. If students are accustomed to labeling unknown variables on a flowchart, *and if they are grounded in the principle that the balance involving the fewest unknowns should be written first*, they can solve most material balance problems as quickly without explicitly using the notion of a tie element as they could using this notion.

A criticism sometimes leveled at the stoichiometry course is its emphasis on drill and routine solution methods, an emphasis that gives the student little idea of the wide range of problems likely to be encountered by the practicing engineer and the imaginative and creative abilities needed to solve some of them. Unfortunately, much of the drill-like aspect of the course is necessary: The only way we have found to teach students the engineering approach to process analysis is to have them practice it, again and again, until they get it.

In an effort to show the other side of the coin, we have included several "Creativity Exercises," in which the students are asked to come up with as many ideas as they can in response to questions related to the text material. The creativity exercises can be used in a variety of ways: as focal points for in-class brainstorming sessions, as individual extra-credit assignments, or as individual or group projects with rewards (e.g. bonus points on subsequent tests) for the solutions exhibiting the greatest fluency (quantity of solutions, practical or not), flexibility (range of approaches to the solution), and originality (novelty of solutions, cleverness, humor). Use of the exercises

provides variety and challenge for the students—especially the more creative ones. The student-generated solutions also provide valuable anecdotal material for the faculty lounge, and occasionally turn up some really good ideas.

We have discussed in the Preface the motivational aspects of the case studies and the way the studies complement the formal text material and vice versa. An additional benefit occurs if the assignments are made to groups, an approach we regularly use in our classes. We invariably see the groups starting out in a state of semianarchy, with each group member doing much the same thing as every other member, and then developing coherence as the weeks go by. By the end of the term the students have learned how to divide the labor appropriately and to learn from one another (since they know they are liable to be tested on any part of the project, not just the part for which they were responsible). This is the part of the course the students usually say they enjoyed most. We have also found that periodic conferences between the groups and the instructor to discuss the case studies provide educational benefits to all parties concerned.

RMF
RWR

Contents

PART ONE
ENGINEERING PROBLEM ANALYSIS

Chapter One

What Chemical Engineers Sometimes Do for a Living

Since you bought this book, you are probably thinking of spending the next 3 years learning to be a chemical engineer and then going to work as one for the next 40; even so, it is a fairly safe bet that, like most people in your position, you have only a limited idea of what chemical engineering is. A logical way for us to begin this book might therefore be with a definition of chemical engineering, but, unfortunately, no universally accepted definition has ever been formulated. Instead, we will cite some examples of problems that recent chemical engineering graduates have been called on to solve. You might ponder these examples to see if any of them seem to present the sort of intellectual challenge you can see yourself taking on and enjoying.

EXAMPLE 1

A chemist in your company's research and development division has discovered that if he mixes two reactants in a certain proportion at an elevated temperature, he obtains a product more valuable than both reactants. The company contemplates manufacturing the product using a process based on this reaction. At this point the matter becomes an engineering problem or, more precisely, hundreds of engineering problems.

1. What should the reaction be carried out in? A long pipe? A large tank? Several smaller tanks? An extremely large test tube? How large? Made of what? Does it have to be heated? If so, how much and how? With an electrical heater inside or outside the reactor? By passing a hot fluid through a heating coil in the reactor? By heating the reactants before they get into the reactor? Does the reaction supply its own heat, so that heating is needed only for startup? If so, can it "run away" and possibly explode? Should control measures be introduced to prevent this? What kind?

2. Where should the reactants be obtained? Buy them, or make them? In what proportions should they be fed to the reactor?

3. Should the reactor effluent, which contains the product and unconsumed reactants, be sold as is, or should the product be separated from the reactants and the latter be sent back to the reactor? If separation is desirable, how can it be accomplished? Heat the mixture and draw off and condense the vapor, which will be richer in the more volatile substances than the original mixture? Add another substance that extracts the product and is immiscible with the reactants, and then separate the two phases mechanically? If all of the process materials are gases at the reaction temperature, can the mixture be cooled to a

temperature at which the product condenses but the reactants do not, or vice versa, or if they are liquids can the mixture be cooled to a temperature at which the product crystallizes? If one of these alternatives is chosen, what kind of equipment is needed? What size? What materials? What are the heating or cooling requirements? Are controls needed to keep the operation of the process within rigid limits? What kind of controls? Should they be manual or automatic?

4. How should the reactant and product streams be moved to and from the reactor and any heating, cooling, and separation equipment involved in the process? By gravity from a raised feed tank? With pumps, or blowers, or compressors, or conveyor belts? What kinds? How big? In pipes made of what?

5. Is enough known about the reaction system to be able to answer all of these questions, or should additional laboratory studies be carried out? What studies? Can the laboratory data be used directly to design the industrial plant, or should a smaller pilot plant be constructed first to test the design? How much smaller?

6. What can possibly go wrong with the process, and what can be done if and when it does?

7. Are there waste products that result from the process? In what quantities? Are they potentially harmful if released untreated into the environment? If so, in what way? What should be done to reduce pollution hazards? Chemical treatment of the wastes? Dump liquid and solid wastes into containers, seal and cart out to sea? Disperse gases in the atmosphere with a high stack? Precipitate solids electrostatically from gas exhausts?

8. How much of the process should be automated, and how should it be done?

9. How much does all of this cost? How much can the product be sold for, and to whom? How much money will the process net each year? Is it enough to make it worthwhile? If so, where should the plant be built?

10. Once the plant has been built, what procedure should be followed for "startup"?

11. Six months later when startup has been achieved, why is the product not coming out the way it did in the laboratory? Is it an equipment malfunction or a change in conditions somewhere between the laboratory and industrial process? How can you find out? What can be done to correct the problem? Is it necessary to shut down the operation for modifications?

12. Is it significant or just a coincidental series of bad breaks that there have been three explosions and four fires within 6 months in the reactor unit? In either case, how do you stop them from recurring?

13. All sorts of other things are going wrong with the process operation. Why weren't they on the list of things that could possibly go wrong? What can be done about them?

14. When the process finally starts working perfectly and the next day an order comes down to change the product specifications, how can it be done without redesigning the entire process? Why didn't they think of this before they built the plant?

EXAMPLE 2

Electricity is generated by burning a fuel in a boiler furnace, using the heat of combustion to generate steam at a high pressure, ejecting the steam through an expansion nozzle, and impinging the resulting high velocity steam jet on the blades of a turbine rotor that drives an electrical generator.

Your company is designing a process plant and plans to include an on-site power-generating facility. Until recently, the boiler would have been routinely designed to burn a fuel oil or a natural gas; however, dwindling domestic reserves of petroleum and natural gas and potential dramatic increases in the price of foreign crude oil have raised serious doubts about the long-range availability of these fuels. It is consequently necessary to consider a variety of alternative energy sources. Principal among them is coal, which exists in the United States in amounts that can satisfy the nation's energy requirements for the next 100 to 500 years, by current estimates.

Once again, a battery of widely varying engineering problems is involved in selecting and implementing a design.

1. How much do gas and oil cost now? Will they be available for the expected lifetime of the power plant? If so, how much will their cost increase? If supplies of these fuels should run out after the plant has been designed and constructed, will it be possible to switch to an alternative fuel without major modifications, or will a new boiler be required?

2. How much does coal cost now, and how rapidly is *its* cost likely to increase?

3. The stack gas from a coal-fired boiler furnace is bound to contain pollutants such as carbon monoxide, sulfur dioxide, soot, ash, and tars, in addition to the normal nonhazardous combustion products (CO_2 and H_2O). What are the expected emission rates of these substances? To what extent are other hazardous trace components of coal, such as lead and arsenic, likely to be emitted? Might the anticipated pollutant emission rates exceed the limits on emissions prescribed by federal regulations? Could they lead to health hazards in the vicinity of the plant? How do the emissions compare with those obtained when gas and oil are burned?

4. Which of the many available filtration, centrifugal separation, wet scrubbing, and electrostatic precipitation processes can be used to reduce emissions to acceptably low levels? Of the possible processes, which are the most economical?

5. Should an attempt be made to recover the sulfur in the stack gas and convert it to a marketable form (such as sulfuric acid), or is it more economical simply to convert it to a disposable solid waste such as calcium sulfate? If the latter, how do you get rid of the mountain of waste that will rapidly accumulate? What about the ash collected in the furnace grate and recovered from the stack gas—is it purely a waste material, or can it be used economically for landfill, as a fertilizer component, or as a filler for construction materials?

6. Instead of cleaning the stack gas following the combustion of coal, would it be better to pretreat the coal and remove potential pollutants prior to burning?

How? Pulverize the coal and put in a liquid whose density is such that the heavy sulfur-containing minerals sink and the combustible carbon floats? Gasify the coal, reacting it with steam and oxygen to form a producer gas of CO and H_2, and then scrubbing this gas to remove impurities (CO_2, H_2S, and particulates)? What should then be done with the producer gas? Burn it on the spot? React it further to form methane (synthetic natural gas) or a higher molecular weight liquid fuel?

7. Are any alternative energy sources feasible, either in the short or the long run, for meeting some or all of the plant's energy needs? Solar collectors? Wind-driven turbines? Burning wood, or municipal refuse? Is it possible that an economical way will be found to extract oil from the millions of tons of shale reserves in the western United States? How can the plant be designed now so that maximum advantage may be taken of these alternative sources should any of them become practicable?

8. At what rate must the chosen fuel be burned to obtain the energy required to meet the anticipated plant needs?

9. How much air should be fed to burn the required amount of fuel? Just enough to provide the quantity of oxygen needed to react with the fuel, or more than this amount to raise the likelihood of complete fuel consumption? How much more? Should the air be preheated to increase the efficiency of the steam-generation operation? If so, can the hot stack gases be used to accomplish this preheating?

10. What temperatures and pressures will the furnace have to withstand? What materials of construction should be used? How should the fuel and air be introduced? How should the ash and stack gases be withdrawn?

11. Should the steam tubes run through the combustion zone, or should they be embedded in the furnace walls? How many tubes should there be, and of what size, and of what material? At what pressure and temperature should the steam be generated?

12. What type of process variable measurement and control instrumentation should be installed to handle the routine operation of the power plant? What can possibly go wrong in the plant operation? How can the system be designed to reduce the likelihood of any malfunction? Do any of the anticipated problems represent potential hazards to anyone working in the area? If so, can the system be designed to fail safe? What kinds of warning devices should be installed to alert people in the area that something is wrong? Should an automated shutdown procedure be built into the system so that a minor malfunction cannot turn into a major catastrophe before anyone has time to react? If so, what type of procedure?

13. After all is said and done, is building and running the power plant preferable to buying electricity from the local utility, or would it be better to accept the

higher cost of purchased electricity in order to avoid the capital costs, fuel handling and solid waste disposal problems, and potential environmental hazards associated with on-site power generation?

EXAMPLE 3

In the course of its metabolic activities the human body produces certain waste products, which are picked up by the blood from the cells and tissues in which they are produced, and are then filtered out of the blood in the kidneys, eventually being excreted in urine. Occasionally the kidneys stop functioning, and the wastes, having no way to get out, build up in the blood. When they reach a high enough level, the patient dies.

Two remedial procedures can be followed once the kidneys malfunction. The first is to transplant a good kidney, which requires that someone willing to donate a kidney be found; this procedure is extraordinarily expensive and frequently fails when the recipient's body rejects the transplanted organ. The second is to use an artificial kidney, which is an out-of-the-body device to remove substances from blood that are normally removed by the kidneys. The normal procedure is to construct a circuit in which blood flows from an artery in the patient's arm through the purifying device and back into a vein.

In the operation of an artificial kidney several substances are removed from a flowing stream, an operation that chemical engineers have dealt with (in other contexts) for years; in fact, chemical engineers working in collaboration with physicians have been in the forefront of the development of artificial organs. The particular process used in most artificial kidneys is called dialysis: the blood flows on one side of a semipermeable membrane, a fluid that initially contains no waste metabolites flows on the other, and the wastes diffuse out of the blood, through the membrane and into the other fluid, which is discarded once it leaves the device.

The engineering problems involved in the design of such a device are numerous. What membrane material will allow wastes to pass through but will keep proteins in the blood? How thick must it be? Typical membranes are several thousandths of an inch thick—how do you make something that thin that won't tear or puncture if you look at it the wrong way? Should the membranes be flat and stacked in layers with blood and dialysate flowing between alternate layers, or should they be thin-walled fibers with blood on the inside and dialysate on the outside, or vice versa? Is the patient's heart adequate to pump the blood through the dialyzer, or should an outside pump be added to the circuit? How can the system be designed to be fail safe if a membrane springs a leak? What should the flow rate of the dialysate be? Can the artificial kidney be made disposable so that the bother of cleaning can be avoided? Can it be made wearable so that the patient doesn't have to spend 24 hours a week in bed getting dialyzed? What can be done to minimize the likelihood of blood clotting? Are there any alternatives to dialysis (such as passing the blood through a bed of material that adsorbs the wastes) that might be advantageous in any respect? Is there enough information to evaluate them, or should laboratory

studies be carried out to acquire the necessary data? What studies—what data? Can a sufficiently precise description (model) be obtained for the artificial kidney so that proper unit and operating conditions can be prescribed based on a preliminary diagnosis of the patient?

The different problems suggested in these examples are clearly too diverse to fall into a single category, ranging as they do from mathematical analysis to plumbing. Also, they are not all pure chemical engineering problems: mechanical engineers design boilers and heat exchangers as often as chemical engineers do, and design of instrumentation is more often the province of electrical engineers than chemical engineers. For that matter, they are not all pure engineering problems: certainly few engineers would attempt to design an artificial kidney without collaborating or consulting with a physician, and fewer would attempt by themselves to see whether their design works, even if they could manage to find a willing patient.

The single feature that all of these problems have in common is that people who call themselves chemical engineers work to solve them. Some of the specific knowledge needed to solve these problems will be presented later in the curriculum, a small amount of it is contained in this text, and most of it must be learned after graduation. There are, however, fundamental techniques that have been developed for setting up and attacking engineering problems that are independent of the particular problem under consideration. What some of these techniques are and how and when to use them are the subject of this book.

Chapter Two

Introduction to Engineering Calculations

The examples of Chapter 1 suggest the range of problems encompassed by chemical engineering, both in traditional areas of chemical processing and in such relatively new fields as environmental science and technology and biomedical engineering. The differences between the systems described in these examples—a chemical manufacturing process, an electrical power generator and pollution control facility, and an artificial organ—are obvious; in this chapter we examine the similarities.

One similarity is that all of the systems described involve **processes,** or operations designed to transform raw materials (A and B, fuel and air, uremic blood) into desired products (C, combustion products, purified blood). Many of the problems that arise in connection with the design of a new process or the analysis of an existing one are of a certain type: Given amounts and properties of the raw materials, calculate amounts and properties of the products, or vice versa.

In this book we outline a systematic approach to the solution of problems of this type. This chapter presents basic techniques for expressing the values of system variables and for setting up and solving equations that relate these variables. In Chapter 3 we discuss the variables of specific concern in process analysis—temperatures, pressures, chemical compositions, and amounts or flow rates of raw material and product streams—describing how they are defined, calculated, and, in some cases, measured. In the same chapter we outline several elementary techniques for the extraction of desired information from process variable measurement data.

Parts 2 and 3 of the book deal with the laws of conservation of mass and energy, which relate the inputs and outputs of chemical processes, electrical power plants, the human body, nuclear reactors, and the like. These laws of nature constitute the underlying structure of all of process design and analysis; in the same way, the techniques that we present beginning in the next section underlie all applications of the laws.

2.1 UNITS AND DIMENSIONS

A measured or counted quantity has a numerical **value** (2.47) and a **unit** (whatever there are 2.47 of). It is useful in most engineering calculations—and essential in many—to write both the value and the unit of each quantity appearing in an equation:

$$2 \text{ meters}, \quad \tfrac{1}{3} \text{ second}, \quad 4.29 \text{ kilograms}, \quad 5 \text{ gold rings}$$

A **dimension** is a property that can be measured, such as length, time, mass, or temperature, or calculated by multiplying or dividing other dimensions, such as

length/time (velocity), length³ (volume), or mass/length³ (density). Measurable units (as opposed to countable units) are specific values of dimensions that have been defined by convention, custom, or law, such as grams for mass, seconds for time, and centimeters or feet for length.

Units can be treated like algebraic variables when quantities are added, subtracted, multiplied, or divided. *The numerical values of two quantities may be added or subtracted only if the units are the same.*

$$3 \text{ cm} - 1 \text{ cm} = 2 \text{ cm} \qquad (3x - x = 2x)$$

but

$$3 \text{ cm} - 1 \text{ mm (or 1 s)} = ? \qquad (3x - y = ?)$$

On the other hand: *Numerical values and their corresponding units may always be combined by multiplication or division.*

$$3 \text{ N} \times 4 \text{ m} = 12 \text{ N} \cdot \text{m}$$

$$\frac{5.0 \text{ km}}{2.0 \text{ h}} = 2.5 \text{ km/h}$$

$$7.0 \frac{\text{km}}{\text{h}} \times 4 \text{ h} = 28 \text{ km}$$

$$3 \text{ m} \times 4 \text{ m} = 12 \text{ m}^2$$

$$6 \text{ cm} \times 5 \frac{\text{cm}}{\text{s}} = 30 \text{ cm}^2/\text{s}$$

$$\frac{6 \text{ g}}{2 \text{ g}} = 3 \qquad (3 \text{ is a } \textbf{dimensionless} \text{ quantity})$$

2.2 CONVERSION OF UNITS

A measured quantity can be expressed in terms of any units having the appropriate dimension. A particular velocity, for instance, may be expressed in ft/s, miles/h, cm/yr, or any other ratio of a length unit to a time unit. The numerical value of the velocity naturally depends on the unit chosen.

The equivalence between two expressions of a given quantity may be defined in terms of a ratio:

$$\frac{1 \text{ cm}}{10 \text{ mm}} \qquad \text{(1 centimeter per 10 millimeters)} \qquad \textbf{(2.2-1)}$$

$$\frac{10 \text{ mm}}{1 \text{ cm}} \qquad \text{(10 millimeters per centimeter)} \qquad \textbf{(2.2-2)}$$

$$\left[\frac{10 \text{ mm}}{1 \text{ cm}}\right]^2 = \frac{100 \text{ mm}^2}{1 \text{ cm}^2} \qquad \textbf{(2.2-3)}$$

Ratios of the form of Equations 2.2-1, 2, and 3 are known as **conversion factors.**

To convert a quantity expressed in terms of one unit to its equivalent in terms of another unit, multiply the given quantity by the conversion factor (new unit/old unit). For example, to convert 36 mg to its equivalent in grams, write

$$(36 \text{ mg}) \times \frac{1 \text{ g}}{1000 \text{ mg}} = 0.036 \text{ g} \qquad (2.2\text{-}4)$$

(Note how the old units cancel, leaving the desired unit.) An alternative way to write this equation is to use a vertical line instead of the multiplication symbol

$$\frac{36 \text{ mg}}{} \left| \frac{1 \text{ g}}{1000 \text{ mg}} = 0.036 \text{ g} \right.$$

Carrying along units in calculations of this type is the best way of avoiding the common mistake of multiplying when you mean to divide and vice versa. In the given example, the result is known to be correct because milligrams cancel out leaving only grams on the left side, whereas

$$\frac{36 \text{ mg}}{} \left| \frac{1000 \text{ mg}}{1 \text{ g}} = 36\,000 \text{ mg}^2/\text{g} \right.$$

is clearly wrong. (More precisely, it is not what you intended to calculate.)

If you are given a quantity having a compound unit (e.g., miles/h, cal/g·°C), and you wish to convert it to its equivalent in terms of another set of units, set up a **dimensional equation:** write the given quantity and its units on the left, write the units of conversion factors that cancel the old units and replace them with the desired ones, then fill in the values of the conversion factors, and carry out the indicated arithmetic operations to find the desired value. (See Example 2.2-1.)

TEST YOURSELF

1. What is a conversion factor?
2. What is the conversion factor for s/min (s = second)?
3. What is the conversion factor for min^2/s^2? (See Equation 2.2-3.)
4. What is the conversion factor for m^3/cm^3?

EXAMPLE 2.2-1 Conversion of Units

Convert an acceleration of 1 cm/s² to its equivalent in km/y².

Solution

$$\frac{1 \text{ cm}}{\text{s}^2} \left| \frac{3600 \text{ s}^2}{\text{h}^2} \right| \frac{24^2 \text{ h}^2}{1 \text{ day}^2} \left| \frac{365^2 \text{ day}^2}{1 \text{ yr}^2} \right| \frac{1 \text{ m}}{10^2 \text{ cm}} \left| \frac{1 \text{ km}}{10^3 \text{ m}} \right.$$

$$= \frac{(3600 \times 24 \times 365)^2}{10^2 \times 10^3} \frac{\text{km}}{\text{yr}^2} = \boxed{9.95 \times 10^9 \text{ km/yr}^2}$$

A principle illustrated in this example is that raising a quantity (in particular, a conversion factor) to a power raises its units to the same power. The conversion factor for h^2/day^2 is therefore the square of the factor for h/day

$$\left(\frac{24 \text{ h}}{1 \text{ day}}\right)^2 = 24^2 \frac{h^2}{day^2}$$

2.3 SYSTEMS OF UNITS

A system of units has the following components:

1. *Base units,* or units for the dimensions of mass, length, time, temperature, electrical current, and light intensity.

2. *Multiple units,* which are defined as multiples or fractions of base units such as minutes, hours, and milliseconds, all of which are defined in terms of the base unit of a second. Multiple units are defined for convenience rather than necessity: It is simply more convenient to refer to 3 yr than to 94 608 000 s.

3. *Derived units,* obtained in one of two ways:
 (a) By multiplying and dividing base or multiple units (cm^2, ft/min, $kg \cdot m/s^2$, etc.). Derived units of this type are referred to as **compound units.**
 (b) As defined equivalents of compound units (e.g., 1 erg = 1 $g \cdot cm/s^2$, 1 lb_f = 32.174 $lb_m \cdot ft/s^2$).

In 1960 an international conference formulated a system of metric units that has gained widespread acceptance in the scientific and engineering community. It is known as the "Systeme Internationale d'Unites," or the **SI system** for short. Two of the base units of the SI system—the ampere for electrical current and the candela for luminous intensity—will not concern us in this book. A third, the degree Kelvin for temperature, will be discussed later. The others are the meter (m) for length, the kilogram (kg) for mass, and the second (s) for time.

Unit prefixes are used in the SI system to indicate powers of ten. The most common of these prefixes and their abbreviations are mega (M) for 10^6 (1 megawatt = 1 MW = 10^6 watts), kilo (k) for 10^3, centi (c) for 10^{-2}, milli (m) for 10^{-3}, micro (μ) for 10^{-6}, and nano (n) for 10^{-9}. The conversion factors between, say, centimeters and meters are therefore 10^{-2} m/cm and 10^2 cm/m. The principal SI units and prefixes are summarized in Table 2.3-1.

An SI convention is to use blanks rather than commas when writing large numbers: for example, 34 607 for 34,607. We will use this notation throughout the text.

The **CGS system** is almost identical to the SI system, the principal difference being that grams (g) and centimeters (cm) are used instead of kilograms and meters as the base units of mass and length. The principal units of the CGS system are shown in Table 2.3-1.

The base units of the **American engineering system** are the foot (ft) for length, the pound-mass (lb_m) for mass, and the second (s) for time. This system has two principal

TABLE 2.3-1

SI and CGS Units

Base Units		
Quantity	Unit	Symbol
Length	meter (SI)	m
	centimeter (CGS)	cm
Mass	kilogram (SI)	kg
	gram (CGS)	g
Moles	gram-mole	mol or g-mole
Time	second	s
Temperature	kelvin	K
Electric current	ampere	A
Light intensity	candela	cd

Multiple Unit Prefixes

Mega (M) = 10^6
Kilo (k) = 10^3
Centi (c) = 10^{-2}
Milli (m) = 10^{-3}
Micro (μ) = 10^{-6}
Nano (n) = 10^{-9}

Derived Units

Quantity	Unit	Symbol	Equivalent in Terms of Base Units
Volume	liter	l or L	0.001 m^3
			1000 cm^3
Force	newton (SI)	N	1 kg·m/s^2
	dyne (CGS)		1 g·cm/s^2
Pressure	pascal (SI)	Pa	1 N/m^2
Energy, work	joule (SI)	J	$1 \text{ N·m} = 1 \text{ kg·m}^2/\text{s}^2$
	erg (CGS)		$1 \text{ dyne·cm} = 1 \text{ g·cm}^2/\text{s}^2$
	gram-calorie	cal	$4.184 \text{ J} = 4.184 \text{ kg·m}^2/\text{s}^2$
Power	watt	W	$1 \text{ J/s} = 1 \text{ kg·m}^2/\text{s}^3$

difficulties associated with it. The first is the occurrence of conversion factors (such as 1 ft/12 in.) which, unlike those in metric systems, are not multiples of ten; the second, which has to do with the unit of force, is discussed in the next section.

Factors for converting from one system of units to another may be obtained by taking ratios of quantities listed in the table on the inside front cover of this book. A larger table of conversion factors is given on the inside back cover of *Perry's Chemical Engineers' Handbook*,[1] and still another table is given on pp. 1-4 through 1-18 of the *Handbook*.

[1] R. H. Perry and D. W. Green, Eds., *Perry's Chemical Engineers' Handbook*, 6th Edition, McGraw-Hill, New York, 1984.

TEST YOURSELF

1. What are the factors (numerical values and units) needed to convert
 (a) meters to millimeters?
 (b) nanoseconds to seconds?
 (c) square centimeters to square meters?
 (d) cubic feet to cubic meters (use the conversion factor table on the inside front cover)?
 (e) horsepower to British thermal units per second?
2. What is the derived unit for velocity in the SI system? The CGS system? The American engineering system?

EXAMPLE 2.3-1 Conversion Between Systems of Units

Convert $23 \ \text{lb}_m \cdot \text{ft/min}^2$ to its equivalent in $\text{kg} \cdot \text{cm/s}^2$.

Solution

As before, begin by writing the dimensional equation, fill in the units of conversion factors (new/old) and then the numerical values of these factors, and then do the arithmetic. The result is

$$\frac{23 \ \text{lb}_m \cdot \text{ft}}{\text{min}^2} \left| \frac{0.454 \ \text{kg}}{1 \ \text{lb}_m} \right| \frac{100 \ \text{cm}}{3.281 \ \text{ft}} \left| \frac{1 \quad \text{min}^2}{(60)^2 \ \text{s}^2} \right. \qquad \begin{array}{l} \text{(Cancellation of units} \\ \text{leaves } \text{kg} \cdot \text{cm/s}^2) \end{array}$$

$$= \frac{(23)(0.454)(100) \ \text{kg} \cdot \text{cm}}{(3.281)(3600) \ \text{s}^2} = \boxed{0.088 \ \frac{\text{kg} \cdot \text{cm}}{\text{s}^2}}$$

2.4 FORCE AND WEIGHT

According to Newton's second law of motion, **force** is proportional to the product of mass and acceleration (length/time²). Natural units of force are, therefore, $\text{kg} \cdot \text{m/s}^2$ (SI), $\text{g} \cdot \text{cm/s}^2$ (CGS), and $\text{lb}_m \cdot \text{ft/s}^2$ (American engineering). Force occurs very frequently in scientific and engineering problems, however, and it would be awkward to have to carry around these complex units in all calculations. To avoid this necessity, derived force units have been defined in each system. In the metric systems, the derived force units (the **newton** in the SI system, the **dyne** in the CGS system) are defined to equal the natural units:

$$1 \ \text{newton (N)} \equiv 1 \ \text{kg} \cdot \text{m/s}^2 \qquad (2.4\text{-}1)$$

$$1 \ \text{dyne} \equiv 1 \ \text{g} \cdot \text{cm/s}^2 \qquad (2.4\text{-}2)$$

In the American engineering system, the derived force unit—called a **pound-force** (lb_f)—is defined as the product of a unit mass (1 lb_m) and the acceleration of gravity

at sea level and 45° latitude, which is 32.174 ft/s²:

$$1 \text{ lb}_f \equiv 32.174 \text{ lb}_m \cdot \text{ft/s}^2 \qquad (2.4\text{-}3)$$

The fact that the units of mass and force in the American system are both called pounds is a common source of confusion. Make sure you keep them distinct in your mind: a lb_f is no more the same as a lb_m than a kick is the same as a foot.

To convert a force from a defined unit (such as newtons) to a natural unit ($\text{kg} \cdot \text{m/s}^2$), a conversion factor—usually denoted by the symbol g_c—must be used. From the defining equations for the former units—(2.4-1) for the newton, (2.4-2) for the dyne, and (2.4-3) for the pound-force—the conversion factor is

$$g_c \equiv 1 \frac{\text{kg} \cdot \text{m/s}^2}{\text{N}} = 1 \frac{\text{g} \cdot \text{cm/s}^2}{\text{dyne}} = 32.174 \frac{\text{lb}_m \cdot \text{ft/s}^2}{\text{lb}_f} \qquad (2.4\text{-}4)$$

The equation that relates force in defined units to mass and acceleration is

$$F = ma/g_c \qquad (2.4\text{-}5)$$

The **weight** of an object is the force exerted on the object by gravitational attraction. Suppose that an object of mass m is subjected to a gravitational force W (W is by definition the weight of the object) and that if this object were falling freely its acceleration would be g. The weight, mass, and free-fall acceleration of the object are related by Equation 2.4-6

$$W = mg/g_c \qquad (2.4\text{-}6)$$

The acceleration of gravity varies directly with the mass of the attracting body (the earth, in most problems) and inversely with the square of the distance between the centers of mass of the attracting body and the object being attracted. The value of g at sea level and 45° latitude and the corresponding value of g/g_c are given below in each system of units; g does not vary much with position on the earth's surface and (within moderate limits) altitude, however, and the given values may accordingly be used for most conversions between mass and weight.

$$g = 9.8066 \text{ m/s}^2 \Rightarrow g/g_c = 9.8066 \text{ N/kg}$$
$$g = 980.66 \text{ cm/s}^2 \Rightarrow g/g_c = 980.66 \text{ dyne/g} \qquad (2.4\text{-}7)$$
$$g = 32.174 \text{ ft/s}^2 \Rightarrow g/g_c = 1 \text{ lb}_f/\text{lb}_m$$

Note: Here g is an acceleration and g_c is a factor for converting from one force unit to another. This is not a source of confusion in the SI and CGS systems, but it is in the American engineering system, since g and g_c are numerically almost equal. Remember that g *varies with position*—it is different at different points on the earth's surface, and different on the moon than on the earth—*but g_c is a constant* and always has the value given by Equation 2.4-4.

The factors needed to convert from one force unit to another in a different system are given in the table on the inside front cover.

TEST YOURSELF

1. How is g_c defined, and what are its numerical values in the SI, CGS, and American engineering systems?
2. What is a force of $2\ \text{kg·m/s}^2$ equivalent to in newtons? What is a force of $2\ \text{lb}_m \cdot \text{ft/s}^2$ equivalent to in lb_f?
3. If the acceleration of gravity at a point is $g = 9.8\ \text{m/s}^2$ and an object is resting on the ground at this point, is this object accelerating at a rate of $9.8\ \text{m/s}^2$?
4. Suppose an object weighs 9.8 N at sea level. What is its mass? Would its mass be greater, less, or the same on the moon? How about its weight?
5. Suppose an object weighs $2\ \text{lb}_f$ at sea level. What is its mass? Would its mass be greater, less, or the same at the center of the earth? How about its weight? (Careful!)

EXAMPLE 2.4-1 Weight and Mass

Water has a density of $62.4\ \text{lb}_m/\text{ft}^3$. How much does $2.000\ \text{ft}^3$ of water weigh (1) at sea level and 45° latitude (2) in Denver, Colorado, where the altitude is 5374 ft and the acceleration of gravity is $32.139\ \text{ft/s}^2$?

Solution

The mass of the water is

$$M = \left(62.4\ \frac{\text{lb}_m}{\text{ft}^3}\right)(2\ \text{ft}^3) = 124.8\ \text{lb}_m$$

The weight of the water is

$$W(\text{lb}_f) = (124.8\ \text{lb}_m)\frac{g}{g_c}\left(\frac{\text{lb}_f}{\text{lb}_m}\right)$$

1. At sea level $g/g_c = 1\ \text{lb}_f/\text{lb}_m$, so that $W = 124.8\ \text{lb}_f$.
2. In Denver, $g/g_c = 32.139/32.174 = 0.9989$, and $W = 124.7\ \text{lb}_f$.

As this example illustrates, the error incurred by assuming that $g/g_c = 1\ \text{lb}_f/\text{lb}_m$ is normally quite small as long as you remain on the earth's surface—in a satellite or on another planet it would be a different story.

2.5 DIMENSIONAL HOMOGENEITY AND DIMENSIONLESS QUANTITIES

We began our discussion of units and dimensions by saying that quantities can be added and subtracted only if their units are the same. If the units are the same, it follows that the dimensions of each term must be the same. For example, if two quantities can be expressed in terms of grams/second, both must have the dimension

(mass/time). This suggests the following rule:

Every valid equation must be dimensionally homogeneous: That is, all additive terms on both sides of the equation must have the same dimensions.

Consider the equation

$$V(m/\dot{s}) = V_0(m/s) + g(m/s^2)t(s) \qquad (2.4\text{-}8)$$

This equation is dimensionally homogeneous, since each of the terms V, V_0, and gt has the same dimensions (length/time). On the other hand, the equation $V = V_0 + g$ is not dimensionally homogeneous (why not?), and therefore cannot possibly be valid.

Equation 2.4-8 is both dimensionally homogeneous *and* consistent in its units, in that each additive term has the units m/s. If values of V_0, g, and t with the indicated units are substituted into the equation, the addition may be carried out to determine the value of V. If an equation is dimensionally homogeneous but its additive terms have inconsistent units, the terms (and hence the equation) may be made consistent simply by applying the appropriate conversion factors.

For example, suppose that in the dimensionally homogeneous equation $V = V_0 + gt$ it is desired to express the time (t) in minutes and the other quantities in the units given above. The equation can be written as

$$V(m/s) = V_0(m/s) + g(m/s^2)t(min)(60 \text{ s/min})$$
$$= V_0 + 60gt$$

Each additive term again has units of m/s (verify this), so the equation is consistent.

The converse of the given rule is not necessarily true—an equation may be dimensionally homogeneous and invalid. For example, if M is the mass of an object, then the equation $M = 2M$ is dimensionally homogeneous, but it is also obviously incorrect.

EXAMPLE 2.5-1 Dimensional Homogeneity

Consider the equation

$$D(ft) = 3t(s) + 4$$

1. If the equation is valid, what are the dimensions of the constants 3 and 4?
2. If the equation is consistent in its units, what are the units of 3 and 4?
3. Derive an equation for distance in meters in terms of time in minutes.

Solution

1. For the equation to be valid, it must be dimensionally homogeneous, so that each term must have the dimension of length. The constant 3 must therefore have the dimension $\boxed{\text{length/time}}$, and 4 must have the dimension $\boxed{\text{length}}$.
2. For consistency of units, the constants must be $\boxed{3 \text{ ft/s}}$ and $\boxed{4 \text{ ft}}$.

3. Define new variables $D'(m)$ and $t'(min)$. The equivalence relations between the old and new variables are

$$D(\text{ft}) = \frac{D'(\text{m})}{} \left| \frac{3.2808 \text{ ft}}{1 \text{ m}} \right. = 3.28D'$$

$$t(s) = \frac{t'(\text{min})}{} \left| \frac{60 \text{ s}}{1 \text{ min}} \right. = 60t'$$

Substitute these expressions in the given equation

$$3.28D' = (3)(60t') + 4$$

and simplify by dividing through by 3.28

$$\boxed{D'(\text{m}) = 55t'(\text{min}) + 1.22}$$

Exercise: What are the units of 55 and 1.22?

Example 2.5-1 illustrates a general procedure for rewriting an equation in terms of new variables having the same dimensions but different units:

1. Define new variables (e.g., by affixing primes to the old variable names) that have the desired units.
2. Write expressions for each old variable in terms of the corresponding new variable.
3. Substitute these expressions in the original equation and simplify.

A **dimensionless quantity** can be a pure number (2, 1.3, $\frac{5}{2}$) or a multiplicative combination of variables with no net dimensions.

$$\frac{M(\text{g})}{M_o(\text{g})} \qquad \frac{D(\text{cm})V(\text{cm/s})\rho(\text{g/cm}^3)}{\mu(\text{g/cm}\cdot\text{s})}$$

A quantity such as M/M_o or $DV\rho/\mu$ is also called a **dimensionless group**.

Exponents (such as the 2 in X^2), transcendental functions (such as log, exp $\equiv e$, sin, ...) *and arguments of transcendental functions (such as the X in* sin *X) must be dimensionless quantities.* For example, 10^2 makes perfect sense, but $10^{2\text{ ft}}$ is meaningless, as is log (20 s) or sin (3 dynes).

EXAMPLE 2.5-2 Dimensional Homogeneity and Dimensionless Groups

A quantity k depends on the temperature T in the following manner:

$$k\left(\frac{\text{mol}}{\text{cm}^3\cdot\text{s}}\right) = 1.2 \times 10^5 \exp\left(-\frac{20\,000}{1.987T}\right)$$

The units of the quantity 20 000 are cal/mol, and T is in K (degrees Kelvin). What are the units of 1.2×10^5 and 1.987?

Solution

Since the equation must be dimensionally homogeneous and exp is dimensionless, 1.2×10^5 should have the same units as k, mol/cm$^3 \cdot$s. Moreover, since the argument of exp must be dimensionless, we can write

$$\frac{20\,000\text{ cal}}{\text{mol}} \left| \frac{1}{T(\text{K})} \right| \frac{\text{mol} \cdot \text{K}}{1.987\text{ cal}} \qquad \text{(All units cancel)}$$

The answers are thus

$$\boxed{1.2 \times 10^5 \text{ mol/cm}^3 \cdot \text{s} \quad \text{and} \quad 1.987 \text{ cal/mol} \cdot \text{K}}$$

TEST YOURSELF

1. What is a dimensionally homogeneous equation? If an equation is dimensionally homogeneous, is it necessarily valid? If an equation is valid, is it necessarily dimensionally homogeneous?
2. If $y(\text{m/s}^2) = az(\text{m}^3)$, what are the units of a?
3. What is a dimensionless group? What multiplicative combination of $r(m)$, $s(\text{m/s}^2)$, and $t(s)$ would constitute a dimensionless group?
4. If $z(\text{lb}_f) = a \sin(Q)$, what are the units of a and Q?

2.6 ARITHMETIC CALCULATIONS: SCIENTIFIC NOTATION, SIGNIFICANT FIGURES, AND PRECISION

Both very large and very small numbers are commonly encountered in process calculations. A convenient way to represent such numbers is to use **scientific notation,** in which a number is expressed as the product of another number (usually between 0.1 and 10) and a power of 10. Examples:

$$123\,000\,000 = 1.23 \times 10^8 \text{ (or } 0.123 \times 10^9)$$
$$0.000028 = 2.8 \times 10^{-5} \text{ (or } 0.28 \times 10^{-4})$$

The **significant figures** of a number are the digits from the first nonzero digit on the left to either (a) the last digit (zero or nonzero) on the right if there is a decimal point, or (b) the last nonzero digit of the number if there is no decimal point. For example,

2300 or 2.3×10^3 has two significant figures.

2300. or 2.300×10^3 has four significant figures.

2300.0 or 2.3000×10^3 has five significant figures.

23 040 or 2.304×10^4 has four significant figures.

0.035 or 3.5×10^{-2} has two significant figures.

0.03500 or 3.500×10^{-2} has four significant figures.

(*Note:* The number of significant figures is easily shown and seen if scientific notation is used.)

The number of significant figures in the reported value of a measured or calculated quantity provides an indication of the precision with which the quantity is known: The more significant figures, the more precise is the value. Generally, if you report the value of a measured quantity with three significant figures, you indicate that the value of the third of these figures may be off by as much as one half. Thus, if you report a mass as 8.3 g (two significant figures), you indicate that the mass lies somewhere between 8.25 and 8.35 g, whereas if you give the value as 8.300 g (four significant figures) you indicate that the mass lies between 8.2995 and 8.3005 g.

Note, however, that this rule applies only to measured quantities or numbers calculated from measured quantities. If a quantity is known precisely—like a pure integer (2) or a counted rather than measured quantity (16 oranges)—its value implicitly contains an infinite number of significant figures (5 cows really means 5.0000 . . . cows).

A rule of thumb is: *When two or more quantities are combined by multiplication and/or division, the number of significant figures in the result should equal the lowest number of significant figures of any of the multiplicands or divisors.* If the initial result of a calculation violates this rule, you must round off the result to reduce the number of significant figures to its maximum allowed value, although if several calculations are to be performed in sequence it is advisable to keep extra significant figures of intermediate quantities and to round off only the final result. Examples:

$$
\begin{array}{cccc}
(3) & (4) & (7) & (3) \\
\end{array}
$$
$$(3.57)(4.286) = 15.30102 \Rightarrow 15.3$$

$$
\begin{array}{cccccc}
(2) & (4) & (3) & (9) & (2) & (2) \\
\end{array}
$$
$$(5.2 \times 10^{-4})(0.1635 \times 10^7)/(2.67) = 318.42966 \Rightarrow 3.2 \times 10^2 = 320$$

(The raised quantities in parentheses denote the number of significant figures in the given numbers.)

The rule for addition and subtraction concerns the position of the last significant figure in the sum—that is, the location of this figure relative to the decimal point. The rule is: *When two or more numbers are added or subtracted, the positions of the last significant figures of each number should be compared. Of these positions, the one farthest to the left is the position of the last permissible significant figure of the sum or difference.*

Several examples of this rule follow, in which arrows (↓) denotes the last significant figure of each number.

$$\downarrow$$
$$1530 \ \downarrow$$
$$-2.56$$
$$\overline{1527.44} \Rightarrow 1530$$

$$\downarrow \quad \uparrow\downarrow \quad \downarrow \quad \downarrow$$
$$1.0000 + 0.036 + 0.22 = 1.2560 \Rightarrow 1.26$$

$$\downarrow \qquad \downarrow$$
$$2.75 \times 10^6 + 3.400 \times 10^4 = (2.75 + 0.03400) \times 10^6$$

$$\downarrow$$
$$= 2.78400 \times 10^6 \Rightarrow 2.78 \times 10^6$$

Finally, a rule of thumb for rounding off numbers in which the digit to be dropped is a 5 is always to make the last digit of the rounded-off number even:

$$1.35 \Rightarrow 1.4$$
$$1.25 \Rightarrow 1.2$$

TEST YOURSELF

1. Express the following quantities in scientific notation, and indicate how many significant figures each has.
 (a) 12 200
 (b) 12 200.0
 (c) 0.003040

2. Express the following quantities in standard decimal form, and indicate how many significant figures each has.
 (a) 1.34×10^5
 (b) 1.340×10^{-2}
 (c) 0.00420×10^6

3. How many significant figures would the solution of each of the following problems have? What are the solutions of (c) and (d)?
 (a) $(5.74)(38.27)/(0.001250)$
 (b) $(1.76 \times 10^4)(0.12 \times 10^{-6})$
 (c) $1.000 + 10.2$
 (d) $18.76 - 7$

4. Round off each of the following numbers to three significant figures.
 (a) 1465
 (b) 13.35
 (c) 1.765×10^{-7}

5. When the value of a number is given, the significant figures provide an indication of the uncertainty in the value; for example, a value of 2.7 indicates that the

number lies between 2.65 and 2.75. Give ranges within which each of the following values lie.

(a) 4.3

(b) 4.30

(c) 2.778×10^{-3}

(d) 2500

(e) 2.500×10^{3}

2.7 PROCESS DATA REPRESENTATION AND ANALYSIS

The operation of any chemical process is ultimately based on the measurement of process variables—temperatures, pressures, flow rates, concentrations, and so on. It is sometimes possible to measure these variables directly, but as a rule, indirect techniques must be used.

Suppose, for example, that you wish to measure the concentration, C, of a solute in a solution. To do so, you normally measure a quantity, X—such as a thermal or electrical conductivity, or a light absorbance, or the volume of a titer—that varies in a known manner with C, and then you calculate C from the measured value of X. The relationship between C and X is determined in a separate **calibration** experiment, in which solutions of known concentration are prepared and X is measured for each solution.

Consider a calibration experiment in which a variable, y, is measured for several values of another variable, x.

x	1.0	2.0	3.0	4.0
y	0.3	0.7	1.2	1.8

In the terms of the first paragraph, y might be a reactant concentration or some other process variable and x would be a readily measured quantity (such as a conductivity) whose value correlates with the value of y. Our object is to use the calibration data to estimate the value of y for a value of x between tabulated points (**interpolation**) or beyond the range of the data table (**extrapolation**).

A number of interpolation and extrapolation methods are commonly used, including two-point linear interpolation (which you learned in high school algebra in conjunction with reading tables of logarithms), graphical interpolation, and curve fitting. Which one is most appropriate depends on the nature of the relationship between x and y.

Figure 2.7-1 shows several illustrative (x, y) plots. If the plot of a given data set looks like that shown in plot (a) or (b) of this figure, a straight line would probably be fitted to the data and used as the basis for subsequent interpolation or extrapolation. On the other hand, if the plot is distinctly curved as in plot (c), a curve could be drawn in by inspection and used as the basis of interpolation, or straight-line segments could be fitted to successive pairs of points, or a nonlinear function $y(x)$ that fits the data could be sought.

FIGURE 2.7-1

Representative plots of experimental data

The technique of drawing a line or curve through the data by inspection is self-explanatory. The other listed methods are reviewed in the sections that follow.

2.7a Two-Point Linear Interpolation

The equation of the line through (x_1, y_1) and (x_2, y_2) on a plot of y versus x is

$$y = y_1 + \frac{x - x_1}{x_2 - x_1}(y_2 - y_1) \qquad \text{(2.7-1)}$$

(Can you prove it?) You may use this equation to estimate y for an x between x_1 and x_2; you may also use it to estimate y for an x outside this range (i.e., to *extrapolate* the data), but with a much greater risk of inaccuracy.

If the points in a table are relatively close together, linear interpolation should provide an accurate estimate of y for any x and vice versa; on the other hand, if the points are widely separated or if the data are to be extrapolated, one of the curve-fitting techniques to be outlined in the next section should be used.

TEST YOURSELF

1. Values of a variable (f) are measured at several times (t):

f	1	4	8
t	1	2	3

Show that if two-point linear interpolation is used (a) $f(t = 1.3) \approx 1.9$; (b) $t(f = 5) \approx 2.25$.

2. If a function $y(x)$ appears as shown in each of the diagrams shown below, would two-point linear interpolation yield estimates of y that are too high, too low, or correct? If the two-point linear interpolation formula (Equation 2.7-1)

were used to estimate $y(x_3)$ from the tabulated values of (x_1, y_1) and (x_2, y_2) in plot (b), would the estimated value be too high or too low?

(a) (b) (c)

2.7b Fitting a Straight Line

A convenient way to indicate how one variable depends on another is with an equation:

$$y = 3x + 4$$
$$y = 4.24(x - 3)^2 - 23$$
$$y = 1.3 \times 10^7 \sin(2x)/(x^{1/2} + 58.4)$$

If you have an analytical expression for $y(x)$ like those shown above, you can calculate y for any given x or (with a somewhat greater effort) determine x for any given y, or you can program a computer to perform these calculations.

Suppose the values of a dependent variable y have been measured for several values of an independent variable x, and a plot of y versus x on rectangular coordinate paper yields what appears to be a straight line. The equation you would use to represent the relationship between x and y is then

$$y = ax + b \tag{2.7-2}$$

If the points show relatively little scatter, like those of Figure 2.7-1a, a line may then be drawn through them by inspection, and if (x_1, y_1) and (x_2, y_2) are two points—which may or may not be data points—on the line, then

Slope:
$$a = \frac{y_2 - y_1}{x_2 - x_1} \tag{2.7-3}$$

Intercept:
$$b \begin{cases} = y_1 - ax_1 \\ = y_2 - ax_2 \end{cases} \tag{2.7-4}$$

Once a has been calculated from Equation 2.7-3 and b has been determined from either of Equations 2.7-4, it is good practice to check the result by verifying that Equation 2.7-2 is satisfied at the point—(x_1, y_1) or (x_2, y_2)—*not* used in the calculation of b.

EXAMPLE 2.7-1 Fitting a Straight Line to Flowmeter Calibration Data

Rotameter calibration data (flow rate versus rotameter reading) are as follows:

Flow Rate V(liters/min)	Rotameter Reading R
20.0	10
52.1	30
84.6	50
118.3	70
151.0	90

1. Draw a calibration curve and determine an equation for $V(R)$.
2. Calculate the flow rate that corresponds to a rotameter reading of 36.

Solution

1. The calibration curve appears as follows:

A line drawn through the data by visual inspection passes through the points $(R_1 = 10, V_1 = 20)$ and $(R_2 = 60, V_2 = 101)$. Therefore,

$$V = aR + b \quad \text{(Since the data fall on a line)}$$

$$a = \frac{V_2 - V_1}{R_2 - R_1} = \frac{101 - 20}{60 - 10} = 1.62 \quad \text{(From Equation 2.7-3)}$$

$$b = V_1 - aR_1 = 20 - (1.62)(10) = 3.8 \quad \text{(From Equation 2.7-4)}$$

The result is, therefore,

$$\boxed{V = 1.62R + 3.8}$$

Check: At point (2),

$$aR_2 + b = (1.62)(60) + 3.8 = 101 = V_2$$

2. At $R = 36$, $V = (1.62)(36) + 3.8 = \boxed{62.1 \text{ liters/min}}$

2.7c Fitting Nonlinear Data

During a recent week at a major university, 423 experimenters separately measured and plotted data, and found that their data points did not lie on straight lines; 416 of these people shrugged their shoulders, said "Looks like a line to me," and drew one; the other 7 went about finding an equation other than $y = ax + b$ to relate the variables.

Fitting a nonlinear equation (anything but $y = ax + b$) to data is usually much harder than fitting a line; however, with some nonlinear equations you can still use straight-line fitting if you plot the data in a suitable manner. Suppose, for example, that x and y are related by the equation $y^2 = ax^3 + b$. A plot of measured y versus x data would clearly be curved; however, a plot of y^2 versus x^3 would be a straight line with slope a and intercept b. More generally, if any two quantities are related by an equation of the form

$$\text{(Quantity 1)} = a\,\text{(Quantity 2)} + b$$

then a plot of the first quantity (y^2 in the above example) versus the second (x^3) on rectangular coordinate paper yields a straight line with slope a and intercept b.

Here are several additional examples of plots that yield straight lines:

1. $y = ax^2 + b$. Plot y versus x^2.

2. $y^2 = \dfrac{a}{x} + b$. Plot y^2 versus $\dfrac{1}{x}$.

3. $\dfrac{1}{y} = a(x - 3) + b$. Plot $\dfrac{1}{y}$ versus $(x - 3)$.

4. $\mathrm{Sin}\ y = a(x^2 - 4)$. Plot $\sin y$ versus $(x^2 - 4)$. The line through the data must be drawn through the origin. (Why?)

Even if the original equation is not in a proper form to generate a linear plot, you can sometimes rearrange it to get it into such a form:

5. $y = \dfrac{1}{C_1 x - C_2} \Rightarrow \dfrac{1}{y} = C_1 x - C_2$

 Plot $\dfrac{1}{y}$ versus x. Slope $= C_1$, intercept $= -C_2$.

6. $y = 1 + (mx^2 + n)^{1/2} \Rightarrow (y - 1)^2 = mx^2 + n$
 Plot $(y - 1)^2$ versus x^2. Slope $= m$, intercept $= n$.

Let us summarize the procedure. If you have (x, y) data that you wish to fit with an equation that can be written in the form $f(y) = ag(x) + b$,

1. Calculate $f(y)$ and $g(x)$ for each tabulated (x, y) point, and plot $f(y)$ versus $g(x)$.

2. If the plotted points fall on a straight line, the equation fits the data. Choose two points on the line—(g_1, f_1) and (g_2, f_2)—and calculate a and b as outlined in the previous section.

$$a = \frac{f_2 - f_1}{g_2 - g_1} \qquad b = f_1 - ag_1 \qquad \text{or} \qquad f_2 - ag_2$$

EXAMPLE 2.7-2 Linear Curve-Fitting of Nonlinear Data

A mass flow rate $F(g/s)$ is measured as a function of temperature $T(K)$.

T	10	20	40	80
F	14.76	20.14	27.73	38.47

There is reason to believe that F varies linearly with the square root of T.

$$F = aT^{1/2} + b$$

Use a straight-line plot to verify this formula, and determine a and b.

Solution

If the formula is correct, a plot of F versus $T^{1/2}$ should be linear, with slope $= a$ and intercept $= b$. The data table is augmented by adding a $T^{1/2}$ row:

T	10	20	40	80
$T^{1/2}$	3.162	4.472	6.325	8.944
F	14.76	20.14	27.73	38.47

and F is plotted versus $T^{1/2}$.

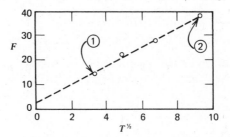

Since the plot is linear, the proposed formula is verified. A line drawn through the data points passes through the first and last points, so these points may be used to calculate the slope and intercept.

$$F = aT^{1/2} + b$$
$$\left\Vert \begin{array}{l} (T_1^{1/2} = 3.162, F_1 = 14.76) \\ (T_2^{1/2} = 8.944, F_2 = 38.47) \end{array} \right.$$

Slope: $a = \dfrac{F_2 - F_1}{T_2^{1/2} - T_1^{1/2}} = \dfrac{38.47 - 14.76}{8.944 - 3.162} = 4.10 \text{ g·s}^{-1}\cdot\text{K}^{-1/2}$

Intercept: $b = F_1 - aT_1^{1/2} = 14.76 - (4.10)(3.162) = 1.80 \text{ g·s}^{-1}$

(verify the units), so that

$$F = 4.10T^{1/2} + 1.80$$

Check: At point (2), $4.10T_2^{1/2} + 1.80 = (4.10)(8.944) + 1.80 = 38.47 = F_2$

Two nonlinear functions that often occur in process analysis are the **exponential function,** $y = ae^{bx}$ [or $y = a \exp(bx)$], where $e \approx 2.7182818$, and the **power law,** $y = ax^b$. Before we describe how the parameters of these functions may be determined by linear curve fitting, let us review some algebra.

The natural logarithm (ln) is the inverse of the exponential function.

$$P = \exp(Q) \Leftrightarrow \ln P = Q \tag{2.7-5}$$

It follows that

$$\ln[\exp(Q)] = Q \quad \text{and} \quad \exp(\ln P) = P \tag{2.7-6}$$

The natural logarithm of a number may be calculated from the common logarithm (\log_{10} or just log) using the relation

$$\ln x = 2.302585 \log_{10} x \tag{2.7-7}$$

The familar rules for taking logarithms of products and powers are applicable to natural logarithms: if $y = ax$ then $\ln y = \ln a + \ln x$, and if $y = x^b$ then $\ln y = b \ln x$. These properties suggest ways to fit exponential and power law functions to (x, y) data:

$$\begin{cases} y = a \exp(bx) \Rightarrow \ln y = \ln a + bx \\ \text{Plot } \ln y \text{ versus } x. \text{ Slope} = b, \text{ intercept} = \ln a. \end{cases} \tag{2.7-8}$$

$$\begin{cases} y = ax^b \Rightarrow \ln y = \ln a + b \ln x \\ \text{Plot } \ln y \text{ versus } \ln x. \text{ Slope} = b, \text{ intercept} = \ln a. \end{cases} \tag{2.7-9}$$

Once you have determined $\ln a$ as the intercept of either of these plots, you can calculate a from Equation 2.7-6 as $\exp(\ln a)$; for example, if $\ln a = 3$, then $a = \exp(3) = 20.1$.

TEST YOURSELF

1. The following plot is made from experimental (x, y) data.

$$x^2 - 2$$

What equation would you use to relate x and y?

2. How would you plot (x, y) data to get a straight line, and how would you determine a and b for each of the following functions?

 (a) $y = a\sqrt{x} + b$. (*Solution:* Plot y versus \sqrt{x}; let $(\sqrt{x_1}, y_1)$ and $(\sqrt{x_2}, y_1)$ be two points on the line; calculate $a = (y_2 - y_1)/(\sqrt{x_2} - \sqrt{x_1})$, $b = y_1 - a\sqrt{x_1}$.)

 (b) $1/y = a(x - 3)^2 + b$.

 (c) $y = (ax^2 - b)^{1/3}$.

 (d) $\sin(y) = (ax + b)^{-2}$.

 (e) $y = ae^{bx}$.

 (f) $y = ax^b$.

2.7d Logarithmic Coordinates

Suppose you wish to fit an exponential function $y = a\exp(bx)$ to measured (x, y) data. If there are many data points, calculating the logarithm of each y value (needed to plot $\ln y$ versus x) could take more time than the fitting procedure itself. However, suppose that an additional scale were drawn parallel to the $\ln y$ axis, on which values of y were shown adjacent to the corresponding values of $\ln y$ on the first scale. (See Figure 2.7-2.) Now, instead of having to calculate $\ln y$ for each tabulated y to locate the data points on the graph, you can find the y values on the second scale and locate the points directly. If the same type of scale (called a **logarithmic scale**) were drawn parallel to the other axis, you could also plot $\ln x$ values without having to calculate them from tabulated x values. Graph paper with logarithmic scales on both axes is called **log paper,** and paper with one logarithmic and one rectangular axis is called **semilog paper.**

FIGURE 2.7-2

Construction of a logarithmic scale

When you plot values of a variable y on a logarithmic scale you are in effect plotting the logarithm of y on a rectangular scale. Suppose, for example, that y and x are related by the equation $y = a \exp(bx)$ ($\ln y = \ln a + bx$). To determine a and b you may plot y versus x on semilog paper, choosing two points (x_1, y_1) and (x_2, y_2) on the resulting line, or you may plot $\ln y$ versus x on rectangular coordinate paper, drawing the line through the corresponding two points $(x_1, \ln y_1)$ and $(x_2, \ln y_2)$. In either case, b and a are obtained as

$$b = \frac{\ln y_2 - \ln y_1}{x_2 - x_1} = \frac{\ln(y_2/y_1)}{x_2 - x_1}$$

$$\ln a = \ln y_1 - bx_1$$

$$\text{or} \qquad \Rightarrow [a = \exp(\ln a)]$$

$$\ln a = \ln y_2 - bx_2$$

In summary

1. If y versus x data appear linear when plotted on semilog paper, then $\ln y$ versus x would be linear on rectangular paper, and the data can therefore be correlated by an exponential function $y = a \exp(bx)$. (See Equation 2.7-8.)

2. If y versus x data appear linear when plotted on log paper, then $\ln y$ versus $\ln x$ would be linear on rectangular paper, and the data can therefore be correlated by a power law $y = ax^b$. (See Equation 2.7-9.)

3. When you plot values of a variable z on a logarithmic axis and your plot yields a straight line through two points with coordinate values z_1 and z_2, replace $z_2 - z_1$ with $\ln(z_2/z_1) = \ln z_2 - \ln z_1$ in the formula for the slope.

4. *Do not plot values of* $\ln z$ *on a logarithmic scale and expect anything useful to result.*

EXAMPLE 2.7-3 Curve Fitting on Semilog and Log Paper

A plot of F versus t yields a line that passes through the points $(t_1 = 15, F_1 = 0.298)$ and $(t_2 = 30, F_2 = 0.0527)$ on (1) semilog paper, (2) log paper. In each case, calculate the equation that relates F and t.

Solution

1. *Semilog plot*

$$\ln F = bt + \ln a \qquad \text{(since the plot appears linear)}$$

$$\Downarrow$$

$$F = ae^{bt}$$

$$b = \frac{\ln(F_2/F_1)}{t_2 - t_1} = \frac{\ln(0.0527/0.298)}{(30 - 15)} = -0.1155$$

$$\ln a = \ln F_1 - bt_1 = \ln (0.298) + (0.1155)(15) = 0.5218$$
$$\Downarrow$$
$$a = \exp (0.5218) = 1.685$$

or

$$\boxed{F = 1.685 \exp (-0.1155t)}$$

Check: $F(t_2) = 1.685 \exp (-0.1155 \times 30) = 0.0527$

2. *Log plot*

$$\ln F = b \ln t + \ln a \qquad \text{(since the plot appears linear)}$$
$$\Downarrow$$
$$F = at^b$$

$$b = \frac{\ln (F_2/F_1)}{\ln (t_2/t_1)} = \frac{\ln (0.0527/0.298)}{\ln (30/15)} = -2.50$$

$$\ln a = \ln F_1 - b \ln t_1 = \ln (0.298) + 2.5 \ln (15) = 5.559$$
$$\Downarrow$$
$$a = \exp (5.559) = 260$$

or

$$\boxed{F = 260t^{-2.5}}$$

Check: $F(t_2) = 260(30)^{-2.5} = 0.0527$

TEST YOURSELF

1. The following plots yield straight lines. What are the equations that relate the variables?
 (a) P versus t on rectangular coordinate paper.
 (b) P versus t on semilog paper.
 (c) P versus t on log paper.
 (d) $y^2 - 3$ versus $1/x^2$ on semilog paper. (Express the answer as an exponential function.)
 (e) $1/F$ versus $t^2 - 4$ on log paper. (Express the answer as a power law.)

2. What would you plot against what on what kind of paper to get a straight line for the following relationships (a and b are constants)?
 (a) $P = a \exp (bt)$
 (b) $P = at^b$
 (c) $P^2 = \exp (at^3 + b)$
 (d) $1/P = a(t - 4)^{-b}$

2.7e Fitting a Line to Scattered Data

There is little problem fitting a line to data that look like this:

Life being the way it is, however, you are much more likely to come up with something more like this:

When data points are as scattered as these are, you can draw any number of lines that appear to fit the points equally well (or equally poorly, depending on your point of view). The question is which line to use.

A number of statistical techniques exist for fitting a function to a set of scattered data. The application of the most common of these techniques—the method of least squares—to the fitting of a straight line to a series of y versus x data points is outlined and illustrated in Appendix A.1 and the use of this technique is required for the solution of Problems 35 through 40 at the end of this chapter.

PROBLEMS

1. Using dimensional equations, convert
 (a) 3 wk to milliseconds
 (b) 26 miles/h to ft/s
 (c) 26.7 $cm^4/min^2 \cdot g$ to $m^4/day^2 \cdot kg$

2. Using the table of conversion factors on the inside front cover, convert
 (a) 340 m/s to miles/h
 (b) 57.5 lb_m/ft^3 to kg/m^3
 (c) 120 hp to kJ/min

3. Using a single dimensional equation, estimate the number of steps it would take you, walking at your normal stride, to walk from the Earth to Alpha Centauri, a distance of 4.3 light-years. The speed of light is 186 000 miles/s.

4. A frustrated chemistry professor once claimed that if all the reports he had graded in his career were stacked on top of one another, they would reach from the Earth to the moon. Assume that an average report is the thickness of about 10 sheets of lined loose-leaf paper, and use a single dimensional equation to estimate the number of reports the professor would have had to grade for his claim to be valid.

5. A quantity is given as 175 N·m/kg·s. Express this quantity in
(a) CGS units and (b) American engineering units.

6. You are trying to decide which of two automobiles to buy. The first costs $9,500 and has a rated gasoline mileage of 28 miles/gal. The second car is of European manufacture, costs $12,700, and has a rated mileage of 19 km/liter. If the cost of gasoline is $1.10/gal and if the cars actually deliver their rated mileage, how many miles would you have to drive to recover the cost difference between the two cars. (Use a single dimensional equation for the calculation, if possible.)

7. A supersonic transport (SST) consumes 5320 imperial gallons of kerosene per hour of flight and flies an average of 14 h/day. It takes roughly 7 tons of crude oil to produce 1 ton of kerosene. Kerosene has a density of 0.965 g/cm^3. How may SST's would it take to consume the entire world production of kerosene in a year when the production of crude oil is 4.02×10^9 metric tons? (One metric ton = 1000 kg.)

8. Calculate
(a) The weight in lb_f of a 10.0-lb_m object
(b) The weight in N of a 10.0-kg object
(c) The weight in lb_f of a 4539-g object.

9. A column of mercury (density = 13.6 g/cm^3) is 3 mm in diameter and 72 cm high. Calculate the column weight in both newtons and lb_f and its mass in lb_m.

10. A horizontal cylindrical drum is 2.00 m in diameter and 4.00 m long. The drum is slowly filled with benzene (density = 0.879 g/cm^3). Derive a formula for W, the weight in newtons of the benzene in the tank, as a function of h, the depth of the liquid in centimeters.

11. According to Archimedes' principle, the mass of a floating object equals the mass of the fluid displaced by the object. Use this principle to solve the following problems.
(a) A wooden cylinder 30.0 cm high floats vertically in a tub of water (density = 1.00 g/cm^3). The top of the cylinder is 14.1 cm above the surface of the liquid. What is the density of the wood?
(b) The same cylinder floats vertically in a liquid of unknown density. The top of the cylinder is 20.7 cm above the surface of the liquid. What is the liquid density?

12. A right circular cone of base radius R, height H, and known density ρ_s floats base down in a liquid of unknown density ρ_f. A height h of the cone is above the liquid surface. Derive a formula for ρ_f in terms of ρ_s, R, and (h/H), simplifying it algebraically to the greatest possible extent. [Recall Archimedes' principle, stated in the preceding problem, and note that the volume of a cone equals (base area)(height)/3.]

13. The fern is defined as the unit of force required to accelerate a unit of mass, called the bung, with the gravitational acceleration on the surface of the moon (in m/s²), which is one-sixth of the normal gravitational acceleration on earth.
 (a) What is g_c in this system? (Give both its numerical value and its units.)
 (b) What is the weight in ferns of a 3-bung object on the moon? What does the same object weigh in Pittsburgh?

14. Is the equation

$$\Delta P = 14L\bar{v}\mu/D^2$$

dimensionally homogeneous? (ΔP is pressure drop, lb_f/ft^2; 14 is a dimensionless constant; L is pipe length, ft; \bar{v} is fluid velocity, ft/s; μ is fluid viscosity, $lb_m/ft\cdot s$; D is pipe diameter, ft.) If so, are its units consistent? If not, what factor must be added to the right side of the equation to provide consistency?

15. The **Prandtl number** is a dimensionless group important in heat transfer calculations. It is defined as $C_p\mu/k$, where C_p is the heat capacity of a fluid, μ is the fluid viscosity, and k is the thermal conductivity. For a particular fluid, $C_p = 0.583$ J/g·°C, $k = 0.286$ W/m·°C, and $\mu = 1936$ $lb_m/ft\cdot h$. Calculate the Prandtl number. (Remember, it is dimensionless.)

16. The **Reynolds number** is a dimensionless group defined for a fluid flowing in a pipe as

$$Re = Dv\rho/\mu$$

where D = pipe diameter, v = fluid velocity, ρ = fluid density, and μ = fluid viscosity. When the value of the Reynolds number is less than about 2100, the flow is **laminar**—the fluid flows in smooth streamlines. For Reynolds numbers above 2100, the flow is **turbulent**—a great deal of internal agitation takes place.

 Methyl ethyl ketone (MEK) flows through a 2.067-in. ID pipe at 20°C, at which temperature its density is 0.805 g/cm³ and its viscosity is 0.43 centipoise (1 cP = 1.00 × 10^{-3} kg/m·s). The average fluid velocity is 0.048 ft/s. Determine whether the flow is laminar or turbulent.

17. The following dimensionless equation correlates the values of variables in a system in which solid particles are suspended in a flowing gas.

$$\frac{k_g d_p y}{D} = 2.00 + 0.600(\mu/\rho D)^{1/3}(d_p v\rho/\mu)^{1/2}$$

Both $(\mu/\rho D)$ and $(d_p v\rho/\mu)$ are dimensionless groups. k_g is a coefficient that expresses the rate at which a particular species transfers from the gas to the solid particles.

The value of k_g is needed to design a catalytic reactor. Since this coefficient is difficult to determine directly, values of the other variables are measured or estimated, and k_g is calculated from the given correlation. The variable values are as follows.

$$d_p = 5.00 \text{ mm}$$
$$y = 0.100 \quad \text{(dimensionless)}$$
$$D = 0.100 \text{ cm}^2/\text{s}$$
$$\mu = 1.00 \times 10^{-5} \text{ N} \cdot \text{s/m}^2$$
$$\rho = 1.00 \times 10^{-3} \text{ g/cm}^3$$
$$v = 10.0 \text{ m/s}$$

What is the estimated value of k_g? (Show units.)

18. After considerable experimentation and data analysis, the following empirical equation for the pressure drop through a particular type of column packing was obtained.

$$\Delta P = 3.61 \frac{\mu^{0.15} N \rho^{0.85} v^{1.85}}{D^{1.15}}$$

where ΔP = pressure drop, N/m^2
 μ = fluid viscosity, kg/m·s
 N = depth of packing, m
 ρ = fluid density, kg/m^3
 v = fluid velocity, m/s
 D = packing diameter, m

(a) What are the units of 3.61?

(b) Suppose you are given data for μ, N, ρ, v, and D in American engineering units (mass in lb$_m$, length in ft), which you prefer to use directly rather than converting them to SI units for substitution in the formula. Modify the formula so that the pressure drop would still be calculated in N/m^2 but the other terms would be in American engineering units. (See Example 2.5-1b, and the three-step procedure that follows that example.)

19. A seed crystal of diameter D (mm) is placed in a solution of dissolved salt, and new crystals are observed to nucleate (form) at a constant rate r (crystals/min). Experiments run with seed crystals of different sizes show that the rate of nucleation varies with the seed crystal diameter as

$$r \text{ (crystals/min)} = 200D - 10D^2 \quad (D \text{ in mm})$$

(a) What are the units of the constants 200 and 10? (Assume this is a valid and therefore dimensionally homogeneous equation.)

(b) Derive the formula for r (crystals/s) in terms of D (inches).

20. The density of a fluid is given by the empirical equation

$$\rho = 1.13 \exp (1.2 \times 10^{-10} P)$$

where ρ = density (g/cm^3), and P = pressure (N/m^2).

 (a) What are the units of 1.13 and 1.2×10^{-10}?
 (b) Derive the formula for $\rho(\text{lb}_m/\text{ft}^3)$ as a function of $P(\text{lb}_f/\text{in}^2)$.

21. The volume of a microbial culture is observed to increase according to the formula

$$V(\text{cm}^3) = e^t$$

where t is time in seconds.
 (a) Calculate the expression for $V(\text{in}^3)$ in terms of $t(h)$.
 (b) It was observed in Section 2.5 that both the exponential function and its argument must be dimensionless. The given equation seems to violate both of these rules, and yet the equation is valid. Explain this paradox. (*Hint:* Observe the result of part a.)

22. The following expression has occurred in a unit conversion problem.

$$Q = \frac{(34\ 040)(0.8300)(5)(9.230 \times 10^7)}{(3.14159)(0.620 \times 10^{15})}$$

The factor 5 is a pure integer. Calculate Q, expressing your answer in both scientific and decimal notation, making sure it has the correct number of significant figures.

23. Perform the following calculations, paying attention to significant figures.
 (a) $(3.4)(2.75)$
 (b) $(5.400 \times 10^3)/27$
 (c) $1.487 + 23.8$
 (d) $(4.750 \times 10^4) - (8.60 \times 10^2)$

24. A concentration C (mol/liter) varies with time (min) according to the equation

$$C = 3.00 \exp(-2.00t)$$

 (a) What are the units of 3.00 and 2.00?
 (b) Suppose the concentration is measured at $t = 0$ and $t = 1$ min. Use two-point linear interpolation or extrapolation to estimate $C(t = 0.6 \text{ min})$ and $t(C = 0.10$ mol/liter) from the measured values, and compare these results with the true values of these quantities.
 (c) Sketch a curve of C versus t, and show graphically the points you determined in part b.

***25.** The vapor pressures of 1-chlorotetradecane at several temperatures are tabulated here.

$T(°C)$	98.5	131.8	148.2	166.2	199.8	215.5
p (mm Hg)	1	5	10	20	60	100

 (a) Use two-point linear interpolation to estimate the value of p at $T = 185°C$.
 (b) Write a computer subroutine to estimate the vapor pressure of 1-chlorotetradecane for any temperature between 98.5°C and 215.5°C using two-point linear interpolation. The subroutine must determine which two tabulated temperatures bracket the given temperature, and apply the interpolation to estimate $p(T)$. Then write a main program to read and store the values of p and T given in the table and to

* Computer problem.

generate a table of vapor pressures at temperatures $T = 100°C, 105°C, 110°C, \ldots,$ 215°C, calling your subroutine to estimate p at each temperature. Check your program using the result of part a.

26. Calculate the equation for $y(x)$ from the given information, noting that when a coordinate point (a, b) is specified, a refers to the abscissa value and b to the ordinate value.

(a) A plot of y versus x on rectangular coordinate paper yields a line that passes through $(1, 2)$ and $(3, 1)$.

(b) A plot of $1/(y + 1)$ versus x^2 on rectangular coordinate paper yields a line that passes through $(5, 6)$ and $(1, 3)$. (*Note:* These are abscissa and ordinate values, not values of x and y.)

27. A hygrometer, which measures the amount of moisture in a gas stream, is to be calibrated using the following diagrammed apparatus.

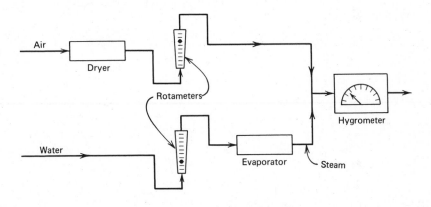

Steam and dry air are fed at known flow rates and mixed to form a gas stream with a known water content, and the hygrometer reading is recorded; the flow rate of either the water or the air is changed to produce a stream with a different water content and the new reading is recorded, and so on. The following data are taken.

Mass Fraction of Water, y	Hygrometer Reading, R
0.011	5
0.044	20
0.083	40
0.126	60
0.170	80

(a) Draw a calibration curve, and determine an equation for $y(R)$.

(b) Suppose a sample of a stack gas is inserted in the sample chamber of the hygrometer, and a reading $R = 43$ is obtained. If the mass flow rate of the stack gas is 1200 kg/h, what is the molar flow rate of water vapor in the gas?

28. Calculate the equations for $y(x)$ from the given information. The plots are all straight lines.

 (a) A plot of ln y versus x on rectangular coordinate paper passes through (1.0, 0.693) and (2.0, 0.0). (The first coordinate in each pair is the abscissa, the second is the ordinate.)

 (b) A plot of y versus x on semilog paper passes through (1, 2) and (2, 1).

 (c) A plot of y versus x on log paper passes through (1, 2) and (2, 1).

 (d) A plot of $(y^2 - 1)$ versus $(x - 2)$ on semilog paper passes through (1.0, 40.2) and (2.0, 807.0).

 (e) A plot of $(y^2 - 1)$ versus $(x - 2)$ on log paper passes through (1.0, 40.2) and (2.0, 807.0)

29. What would you plot on what kind of paper to get a straight line if experimental (x, y) data are to be correlated by the following functions, and what would the slopes and intercepts be in terms of the function parameters? (The solution to part a is given as an example.)

 (a) $y^2 = mx^3 - n$. *Solution:* Plot y^2 versus x^3 on rectangular coordinate paper. Slope = m, intercept = $-n$.

 (b) $1/y^2 = a(x + 5) + b$

 (c) $y = r \exp(-b/x^2)$

 (d) $\sin y = a(x^2 - 2)^b$

 (e) $y^3 = [ax^b + 3]^{-1}$

30. A chemical reaction $A \rightarrow B$ is carried out in a closed vessel. The following data are taken for the concentration of A, C_A(mol/liter), as a function of time, t(min), from the start of the reaction.

t(min)	0	36	65	100	160	∞
C_A(mol/liter)	0.1823	0.1453	0.1216	0.1025	0.0795	0.0495

\uparrow C_{A0} \uparrow C_{Ae}

A proposed reaction mechanism predicts that C_A and t should be related by the expression

$$\ln \frac{C_A - C_{Ae}}{C_{A0} - C_{Ae}} = -kt$$

where k is the **reaction rate constant.** Do the data support this prediction? If so, determine the value of k. (Use semilog paper in your solution of this problem.)

31. The temperature in a process unit is controlled by passing cooling water at a measured rate through a jacket that encloses the unit.

The exact relationship between the unit temperature $T(°C)$ and the cooling water flow rate ϕ (liters/s) is extremely complex, and it is desired to derive a simple empirical formula to approximate this relationship over a limited range of flow rates and temperatures. Data are taken for T versus ϕ. Plots of T versus ϕ on rectangular coordinate and semilog paper are distinctly curved (ruling out $T = a\phi + b$ and $T = ae^{b\phi}$ as possible empirical functions), but a plot on log paper appears as follows.

A line drawn through the data goes through the points $(\phi_1 = 25, T_1 = 210)$ and $(\phi_2 = 40, T_2 = 120)$.

(a) What is the empirical relationship between ϕ and T?

(b) Using your derived equation, calculate the flow rate needed to maintain the temperature at 175°C.

32. The relationship between the pressure (P) and volume (V) of the air in a cylinder during the upstroke of a piston in an air compressor can be expressed as

$$PV^k = C$$

where k and C are constants. During a compression test, the following data are taken.

P(mm Hg)	760	1140	1520	2280	3040	3800
V(cm^3)	48.3	37.4	31.3	24.1	20.0	17.4

Determine the values of k and C that best fit the data. (Give both numerical values and units.)

33. In modeling the effect of an impurity on crystal growth, the following equation was derived.

$$\frac{G - G_L}{G_0 - G} = \frac{1}{K_L C^m}$$

where C is impurity concentration, G_L is a limiting growth rate, G_0 is the growth rate of the crystal with no impurity present, and K_L and m are model parameters.

In a particular experiment, $G_0 = 3.0 \times 10^{-3}$ mm/min, and $G_L = 1.8 \times 10^{-3}$ mm/min. Growth rates are measured for several impurity concentrations C (parts per million, or ppm), with the following results.

C(ppm)	50.0	75.0	100.0	125.0	150.0
G(mm/min) $\times 10^3$	2.50	2.20	2.04	1.95	1.90

(Thus, for example, when $C = 50.0$ ppm, $G = 2.50 \times 10^{-3}$ mm/min). Determine K_L and m, giving both numerical values and units.

34. A process instrument reading, Z(volts), is thought to be related to a process stream flow rate V(liters/s) and pressure P(kPa) by the following expression.

$$Z = aV^b P^c$$

Process data have been obtained in two sets of runs—one with V held constant, the other with P held constant. The data are as follows:

Point	1	2	3	4	5	6	7
V(liters/s)	0.65	1.02	1.75	3.43	1.02	1.02	1.02
P(kPa)	11.2	11.2	11.2	11.2	9.1	7.6	5.4
Z(volts)	2.27	2.58	3.72	5.21	3.50	4.19	5.89

(a) Suppose you had only performed runs 2, 3, and 5. Calculate a, b, and c algebraically from the data for these three runs.

(b) Now use a graphical method and all the data to calculate a, b, and c. Comment on why you would have more confidence in this result than in that of part a. (*Hint:* You will need at least two plots.)

35. Fit (a) a line, (b) a line through the origin to the following data using the method of least squares (Appendix A.1).

x	0.3	1.9	3.2
y	0.4	2.1	3.1

On a single plot, show both fitted lines and the data points.

36. A quantity p varies with another quantity q according to the formula

$$p^2 = 1/(aq + b)$$

The following data are taken for p and q.

q	1.00	2.00	3.00	5.00
p	0.95	0.59	0.43	0.35

(a) What plot would yield a straight line if the given equation is correct?

(b) Estimate a and b by the method of least squares. (See Example A.1 of Appendix A.1.)

37. *x-y* data appear as follows.

x	0.5	14	84
y	2.20	4.30	6.15

(a) Plot the data on log paper.

(b) Determine the coefficients of a power law expression $y = ax^b$ using the method of least squares. (Remember what you are really plotting—there is no way to avoid taking logarithms of the data point coordinates in this case.)

(c) Draw your calculated line on the same plot as the data.

38. Suppose you have *n* data points $(x_1, y_1), (x_2, y_2), \ldots, (x_n, y_n)$, and you wish to fit a line through the origin $(y = ax)$ to these data using the method of least squares. Derive Equation A.1-6 (Appendix A.1) for the slope of the line by writing the expression for the vertical distance d_i from the *i*th data point (x_i, y_i) to the line, then writing the expression for $\phi = \sum d_i^2$, and finding by differentiation the value of *a* that minimizes this function.

*39. Write a computer program to fit a straight line $y = ax + b$ to tabulated (x, y) data, assuming that no more than 100 data points will be taken in any one run. Your program should read in and store the data, evaluate the slope *a* and intercept *b* of the best line through the data using Equations A.1-3 through A.1-5 in Appendix A, then print out the values of these quantities. Finally, the program should print out the measured values of $x(x_i)$ and $y(y_i)$ and calculated values of $y(ax_i + b)$ for each tabulated value of *x*.

Test your program by fitting a line to the data in the following table.

x	1.0	1.5	2.0	2.5	3.0
y	2.35	5.53	8.92	12.15	15.38

40. The rate at which a substance passes through a semipermeable membrane is determined by the **diffusivity** $D(\text{cm}^2/\text{s})$ of the gas. *D* varies with the membrane temperature $T(K)$ according to the **Arrhenius law.**

$$D = D_0 \exp(-E/RT)$$

where D_0 = the **preexponential factor**
E = the **activation energy** for diffusion
R = 1.987 cal/mol·K

Diffusivities of SO_2 in a fluorosilicone rubber tube are measured at several temperatures, with the following results:

T(K)	D(cm²/s) × 10⁶
347.0	1.34 ⟵ (so that $D = 1.34 \times 10^{-6}$ cm²/s)
374.2	2.50
396.2	4.55
420.7	8.52
447.7	14.07
471.2	19.99

* Computer problem.

(a) What are the units of D_0 and E?

(b) How should the data be plotted to obtain a straight line on rectangular coordinate paper?

(c) Plot the data in the manner indicated in part b, and determine D_0 and E from the resulting line.

*(d) Write a computer program to read in the (T, D) data, and to calculate D_0 and E using the method of least squares. Then run the program and print out the results.

* Computer problem.

Chapter Three

Processes and Process Variables

A **process** is any operation or series of operations that causes a physical or chemical change in a substance or a mixture of substances. The material that enters a process is referred to as the **input** or **feed** to the process, and that which leaves is called the **output** or **product**. A **process unit** is an apparatus in which one of the operations that constitute a process is carried out. Each process unit has associated with it a set of input and output **process streams,** which consist of the materials that enter and leave the unit.

As a chemical engineer, you might be called on to design individual process units (such as reactors, distillation columns, heat exchangers), supervise the operation of a process, or modify a process design to accommodate a change in the feed or in the desired product characteristics. As a rule, to do any of these things you must know the amounts, compositions, and conditions of the materials that enter and leave each process unit, and if you are working with an existing unit, you must be able to measure enough of these quantities to verify that the process is doing what it was designed to do.

In this chapter we present definitions, illustrative measurement techniques, and methods of calculating variables that characterize the operation of processes and individual process units. In later chapters, we discuss how you can use measured values of some of these variables to calculate process-related quantities that cannot be measured directly but that must be known before the process can be fully designed or evaluated.

3.1 MASS AND VOLUME

The **density** of a substance is the mass per unit volume of the substance (kg/m^3, g/cm^3, lb$_m$/ft^3, etc.). The **specific volume** of a substance is the volume per unit mass of the substance (m^3/kg, ft^3/lb$_m$, etc.) and is, therefore, the inverse of the density. Densities of pure solids and liquids are relatively independent of temperature and pressure and may be found in standard references (such as *Perry's Chemical Engineers' Handbook*,[1] pp. 3-6 through 3-44). Methods for calculating densities of gases and mixtures of liquids are given in Chapter 5 of this book.

The density of a substance can be used as a conversion factor to relate the mass and the volume of a quantity of the substance. For example, the density of carbon

[1] R. H. Perry and D. W. Green, Eds., *Perry's Chemical Engineers' Handbook*, 6th Edition, McGraw-Hill, New York, 1984.

tetrachloride is 1.595 g/cm^3: the mass of 20.0 cm^3 of CCl$_4$ is, therefore,

$$\frac{20.0 \text{ cm}^3 \left| 1.595 \text{ g}\right.}{\left| \text{cm}^3\right.} = 31.9 \text{ g}$$

and the volume of 6.20 lb$_m$ of CCl$_4$ is

$$\frac{6.20 \text{ lb}_m \left| 454 \text{ g} \right| 1 \text{ cm}^3}{\left| 1 \text{ lb}_m \right| 1.595 \text{ g}} = 1760 \text{ cm}^3$$

The **specific gravity** of a substance is the ratio of the density ρ of the substance to the density ρ_{ref} of a reference substance at a specific condition

$$SG = \rho/\rho_{\text{ref}} \tag{3.1-1}$$

The reference most commonly used for solids and liquids is water at 4.0°C, which has the following density:

$$
\begin{aligned}
\rho_{\text{ref}}(H_2O, 4°C) &= 1.000 \text{ g/cm}^3 \\
&= 1000 \text{ kg/m}^3 \\
&= 62.43 \text{ lb}_m/\text{ft}^3
\end{aligned}
\tag{3.1-2}
$$

Note that the density of a liquid or solid in g/cm^3 is numerically equal to the specific gravity of that substance. The notation

$$SG = 0.6 \frac{20°}{4°}$$

signifies that the specific gravity of a substance at 20°C with reference to water at 4°C is 0.6.

If you are given the specific gravity of a substance, multiply it by the reference density in any units to get the density of the substance in the same units. For example, if the specific gravity of a liquid is 2.00, its density is 2000 kg/m^3 or 2.00 g/cm^3 or 125 lb$_m$/ft^3. Specific gravities of liquids and solids are given in Table B.1.

Note: Special density units called degrees Baumé (°Bé), degrees API (°API), and degrees Twaddell (°Tw) are occasionally used, particularly in the petroleum industry. Definitions of and conversion factors for these units are given on p. 1-19 of *Perry's Chemical Engineers' Handbook.*

TEST YOURSELF

1. What are the units of specific gravity?
2. A liquid has a specific gravity of 0.50. What is its density in g/cm^3? What is its specific volume in cm^3/g? What is its density in lb$_m$/ft^3? What is the mass of 3.0 cm^3 of this liquid? What volume is occupied by 18 g?

3. If substance A and substance B each have a density of 1.34 g/cm³, must 3 cm³ of A have the same mass as 3 cm³ of B?

4. If substance A and substance B each have a specific gravity of 1.34, must 3 cm³ of A have the same mass as 3 cm³ of B? Why not?

5. You are given a small rock, a graduated cylinder (with markings in milliliters) wide enough for the rock to fit in, a balance calibrated in grams, and water, and are given the task of determining the specific gravity of the rock. Outline the procedure and calculations.

EXAMPLE 3.1-1 Mass, Volume, and Density

Calculate the density of mercury in lb_m/ft^3 from a tabulated specific gravity, and calculate the volume in ft^3 occupied by 215 kg of mercury.

Solution

Table B.1 lists the specific gravity of mercury at 20°C as 13.546. Therefore,

$$\rho_{Hg} = (13.546)\left(62.43 \, \frac{lb_m}{ft^3}\right) = \boxed{845.7 \, \frac{lb_m}{ft^3}}$$

$$V = \frac{215 \text{ kg}}{} \left| \frac{1 \text{ lb}_m}{0.454 \text{ kg}} \right| \frac{1 \text{ ft}^3}{845.7 \text{ lb}_m} = \boxed{0.560 \text{ ft}^3}$$

3.2 FLOW RATE

3.2a Mass and Volumetric Flow Rate

Continuous processes involve the movement of material from one point to another—sometimes between process units, sometimes from a production facility to a transportation depot, or vice versa. The rate at which a material is transported through a process line is the **flow rate** of that material.

The flow rate of a process stream may be expressed as a **mass flow rate** (mass/time) or as a **volumetric flow rate** (volume/time). Suppose a fluid (gas or liquid) flows in the cylindrical pipe shown below, where the shaded area represents a section perpendicular to the direction of flow. If the mass flow rate of the fluid is $m(kg/s)$, then every second m kilograms of the fluid pass through the cross section. If the volumetric flow rate of the fluid at the given cross section is $V(m^3/s)$, then every second V cubic meters

of the fluid pass through the cross section. However, the mass m and the volume V of a fluid—in this case, the fluid that passes through the cross section each second—are not independent quantities but are related through the fluid density ρ:

$$\rho = m/V \qquad (3.2\text{-}1)$$

Thus: *The density of a fluid can be used to convert a known volumetric flow rate of a process stream to the mass flow rate of that stream or vice versa.*

The mass flow rates of process streams must be known for many process calculations, but it is frequently more convenient to measure volumetric flow rates. A common procedure is, therefore, to measure V and calculate m from V and the density of the stream fluid.

TEST YOURSELF

1. The mass flow rate of *n*-hexane ($\rho = 0.659$ g/cm^3) in a pipe is 6.59 g/s. What is the volumetric flow rate of the hexane?
2. The volumetric flow rate of CCl$_4$ ($\rho = 1.595$ g/cm^3) in a pipe is 100.0 cm^3/min. What is the mass flow rate of the CCl$_4$?
3. Suppose a gas is flowing through a cone-shaped pipe.

 How do the mass flow rates of the gas at the inlet and outlet compare? (Remember the law of conservation of mass). If the density of the gas is constant, how do the volumetric flow rates at these two points compare? What if the density decreases from inlet to outlet?

3.2b Flow Rate Measurement

A **flowmeter** is a device mounted in a process line that provides a continuous reading of the flow rate in the line. Two commonly used flowmeters—the **rotameter** and the **orifice meter**—are shown in Figure 3.2-1. *Perry's Chemical Engineers' Handbook*, pp. 5-7 through 5-17, describes many others.

The rotameter is a tapered vertical tube containing a float; the larger the flow rate, the higher the float rises in the tube. The orifice meter is an obstruction in the flow channel with a narrow opening through which the fluid passes. The fluid pressure drops from the upstream side of the orifice to the downstream side; the pressure drop (measured with a differential manometer, which is discussed in the next section) varies with the flow rate—the greater the flow rate, the larger the pressure drop.

(a) *(b)*

FIGURE 3.2-1

Flowmeters: (*a*) rotameter and (*b*) orifice meter

Problems at the end of this chapter illustrate the calibration and use of both types of flowmeters.

TEST YOURSELF

1. A steadily flowing stream of water is funneled into a graduated cylinder for exactly 30 s, during which time 50 ml are collected. What is the volumetric flow rate of the stream? The mass flow rate?

2. What is a rotameter? An orifice meter?

3. A rotameter calibration curve (flow rate versus float position) obtained using a liquid is mistakenly used to measure a gas flow rate. Would you expect the gas flow rate determined in this manner to be too high or too low?

CREATIVITY EXERCISE

Given below is the first of a series of exercises contained in this book that we call Creativity Exercises. These exercises differ considerably from the kinds of problems you are accustomed to seeing in homework assignments and quizzes. In the latter, you are generally given information and asked to find the single correct solution to a problem. In the creativity exercises, you are asked to think of many possible responses without spending too much time doing so. There is no such thing as the "correct" response or even a "good" or "bad" response. The idea is to go for quantity rather than quality, imagination (even humor) rather than "correctness." Try to suspend your critical judgment completely and come up with as many ideas as you can, regardless of how likely they are to be efficient, cost-effective, or even to work at all.

In a real industrial situation, this "brainstorming" approach to creative problem solving is frequently used as the first step in solving the most difficult problems facing the company. Doing these exercises will help sharpen the skills you will need for successful brainstorming and at the same time will give you a broader understanding of the concepts contained in the text.

Here, then, is the first exercise. Invent as many devices as you can that might function as flowmeters for gases and/or liquids. In each case, describe the device and state what would be measured. (*Example:* Put a propeller in a flowing stream and measure its rotational speed.)

3.3 CHEMICAL COMPOSITION

3.3a Moles and Molecular Weight

The **atomic weight** of an element is the mass of an atom on a scale that assigns ^{12}C (the isotope of carbon whose nucleus contains six protons and six neutrons) a mass of exactly 12. The atomic weights of all of the elements in their naturally occurring isotopic proportions are listed in the table at the back of the book. The **molecular weight** of a compound is the sum of the atomic weights of the atoms that constitute a molecule of the compound: atomic oxygen (O), for example, has an atomic weight of approximately 16, and therefore molecular oxygen (O_2) has a molecular weight of approximately 32. Molecular weights of a number of compounds are given in Table B.1.

A **gram-mole** (g-mole, or mol in the SI system) of a species is the amount of that species whose mass in grams is numerically equal to its molecular weight. (If the species is an element, it is technically correct to refer to a gram-atom rather than a gram-mole. We will not observe this distinction, but will use moles for both elements and compounds.) Other types of moles (e.g., kg-moles or kmol, lb-moles, ton-moles) are similarly defined. Carbon monoxide (CO), for example, has a molecular weight of 28; 1 mol of CO therefore contains 28 g, 1 lb-mole contains 28 lb_m, 1 ton-mole contains 28 tons, and so on. In short: *If the molecular weight of a substance is M, then there are M kg/kmol, M g/mol, and M lb_m/lb-mole of this substance.* The molecular weight may thus be used as a conversion factor that relates the mass and the number of moles of a quantity of the substance. For example, 34 kg of ammonia (NH_3: $M = 17$) are equivalent to

$$\frac{34 \text{ kg NH}_3 \mid 1 \text{ kmol NH}_3}{\mid 17 \text{ kg NH}_3} = 2 \text{ kmol NH}_3 \qquad (3.3\text{-}1)$$

and 4 lb-moles of ammonia are equivalent to

$$\frac{4 \text{ lb-moles NH}_3 \mid 17 \text{ lb}_m \text{ NH}_3}{\mid 1 \text{ lb-mole NH}_3} = 68 \text{ lb}_m \text{ NH}_3 \qquad (3.3\text{-}2)$$

(It is often helpful in mass-mole conversions to include the chemical formula in the dimensional equation, as illustrated above.)

The same factors used to convert masses from one unit to another may be used to convert the equivalent molar units: there are 454 g/lb_m, for example, and therefore there are 454 mol/lb-mole, regardless of the substance involved. (Prove it—convert 1 lb-mole of a substance with molecular weight M to gram-moles.)

One gram-mole of any species contains 6.02×10^{23} (Avogadro's number) molecules of that species.

EXAMPLE 3.3-1 Conversion Between Mass and Moles

How many of each of the following are contained in 100.0 g of CO_2 ($M = 44.01$)? (1) mol CO_2; (2) lb-moles CO_2; (3) mol (or g-atoms) C; (4) mol O; (5) mol O_2; (6) g O_2; (7) molecules of CO_2.

Solution

1. $$\frac{100.0 \text{ g } CO_2}{44.01 \text{ g } CO_2} \left| \frac{1 \text{ mol } CO_2}{} \right. = \boxed{2.273 \text{ mol } CO_2}$$

2. $$\frac{2.273 \text{ mol } CO_2}{} \left| \frac{1 \text{ lb-mole}}{453.6 \text{ mol}} \right. = \boxed{5.011 \times 10^{-3} \text{ lb-mole } CO_2}$$

Each molecule of CO_2 contains one atom of C, one molecule of O_2, and two atoms of O. Therefore, each 6.02×10^{23} molecules of CO_2 (1 mol) contains 1 mol C, 1 mol O_2, and 2 mol O. Thus

3. $$\frac{2.273 \text{ mol } CO_2}{} \left| \frac{1 \text{ mol C}}{1 \text{ mol } CO_2} \right. = \boxed{2.273 \text{ mol C}}$$

4. $$\frac{2.273 \text{ mol } CO_2}{} \left| \frac{2 \text{ mol O}}{1 \text{ mol } CO_2} \right. = \boxed{4.546 \text{ mol O}}$$

5. $$\frac{2.273 \text{ mol } CO_2}{} \left| \frac{1 \text{ mol } O_2}{1 \text{ mol } CO_2} \right. = \boxed{2.273 \text{ mol } O_2}$$

6. $$\frac{4.546 \text{ mol O}}{} \left| \frac{16.0 \text{ g O}}{1 \text{ mol O}} \right. = \boxed{72.7 \text{ g O}}$$

7. $$\frac{2.273 \text{ mol } O_2}{} \left| \frac{32.0 \text{ g } O_2}{1 \text{ mol } O_2} \right. = \boxed{72.7 \text{ g } O_2}$$

8. $$\frac{2.273 \text{ mol } CO_2}{} \left| \frac{6.02 \times 10^{23} \text{ molecules}}{1 \text{ mol}} \right. = \boxed{1.37 \times 10^{24} \text{ molecules}}$$

Note: Parts 6 and 7 could also be done by observing from the molecular formula that each 44.0 g of CO_2 contains 32.0 g of O_2 or O, so that

$$\frac{100.0 \text{ g } CO_2}{} \left| \frac{32.0 \text{ g } O_2}{44.0 \text{ g } CO_2} \right. = 72.7 \text{ g } O_2$$

The molecular weight of a species can be used to relate the mass flow rate of a continuous stream of this species to the corresponding molar flow rate. For example, if carbon dioxide (CO_2: $M = 44.0$) flows through a pipeline at a rate of 100 kg/h, the

molar flow rate of the CO_2 is

$$\frac{100 \text{ kg } CO_2}{h} \left| \frac{1 \text{ kmol } CO_2}{44.0 \text{ kg } CO_2} \right. = 2.27 \frac{\text{kmol } CO_2}{h} \qquad (3.3\text{-}3)$$

If the output stream from a chemical reactor contains CO_2 flowing at a rate of 850 lb-moles/min, the corresponding mass flow rate is

$$\frac{850 \text{ lb-moles } CO_2}{} \left| \frac{44.0 \text{ lb}_m \text{ } CO_2}{\text{lb-mole}} \right. = 37\,400 \frac{\text{lb}_m \text{ } CO_2}{\text{min}} \qquad (3.3\text{-}4)$$

TEST YOURSELF

1. What is a mol of a species of molecular weight M, in terms of **(a)** a number of molecules? **(b)** a mass?
2. What is a ton-mole of a species?
3. How many lb-moles and lb_m of (a) H_2, (b) H are contained in 1 lb-mole of H_2O?
4. How many gram-moles of C_3H_8 are contained in 2 kmol of this substance?
5. One hundred kilograms of molecular hydrogen (H_2) are fed into a reactor each hour. What is the molar flow rate of this stream in gram-moles/hour?

3.3b Mass and Mole Fractions and Average Molecular Weight

Process streams occasionally contain one substance, but more often they consist of mixtures of liquids or gases, or solutions of one or more solutes in a liquid solvent.

The following terms may be used to define the composition of a mixture of substances including a species A.

$$\textit{Mass fraction:} \quad x_A = \frac{\text{mass of A}}{\text{total mass}} \left(\frac{\text{kg A}}{\text{kg total}} \text{ or } \frac{\text{g A}}{\text{g total}} \text{ or } \frac{\text{lb}_m \text{ A}}{\text{lb}_m \text{ total}} \right) \qquad (3.3\text{-}5)$$

$$\textit{Mole fraction:} \quad y_A = \frac{\text{moles of A}}{\text{total moles}} \left(\frac{\text{kmol A}}{\text{kmol}} \text{ or } \frac{\text{mol A}}{\text{mol}} \text{ or } \frac{\text{lb-moles A}}{\text{lb-mole}} \right) \qquad (3.3\text{-}6)$$

The **percent by mass** of A is $100x_A$, and the *mole percent* of A is $100y_A$.

EXAMPLE 3.3-2 Conversions Using Mass and Mole Fractions

A solution contains 15% A by mass ($x_A = 0.15$) and 20 mole% B ($y_B = 0.20$).

1. Calculate the mass of A in 175 kg of the solution.

$$\frac{175 \text{ kg solution}}{} \left| \frac{0.15 \text{ kg A}}{\text{kg solution}} \right. = \boxed{26 \text{ kg A}}$$

2. Calculate the mass flow rate of A in a stream of the solution flowing at a rate of 53 lb_m/h.

$$\frac{53 \text{ lb}_m}{h} \, \bigg| \, \frac{0.15 \text{ lb}_m \text{ A}}{\text{lb}_m} = \boxed{8.0 \frac{\text{lb}_m \text{ A}}{h}}$$

(If a mass or molar unit—such as lb_m in 53 lb_m/h—is not followed by the name of a species, the unit should be understood to refer to the total mixture or solution rather than to a specific component.)

3. Calculate the molar flow rate of B in a stream flowing at a rate of 1000 mol/min.

$$\frac{1000 \text{ mol}}{\text{min}} \, \bigg| \, \frac{0.20 \text{ mol B}}{\text{mol}} = \boxed{200 \frac{\text{mol B}}{\text{min}}}$$

4. Calculate the total solution flow rate that corresponds to a molar flow rate of 28 kmol B/s.

$$\frac{28 \text{ kmol B}}{s} \, \bigg| \, \frac{1 \text{ kmol solution}}{0.20 \text{ kmol B}} = \boxed{140 \frac{\text{kmol solution}}{s}}$$

5. Calculate the mass of the solution that contains 300 lb_m of A.

$$\frac{300 \text{ lb}_m \text{ A}}{} \, \bigg| \, \frac{1 \text{ lb}_m \text{ solution}}{0.15 \text{ lb}_m \text{ A}} = \boxed{2000 \text{ lb}_m \text{ solution}}$$

Note that *the numerical value of a mass or a mole fraction does not depend on the mass units in the numerator and denominator as long as these units are the same.* If the mass fraction of benzene in a mixture is 0.25, then $x_{C_6H_6}$ equals 0.25 kg C_6H_6/kg total, 0.25 g C_6H_6/g total, 0.25 lb_m C_6H_6/lb_m total, and so on.

A set of mass fractions may be converted to an equivalent set of mole fractions by (a) assuming as a **basis of calculation** a mass of the mixture (e.g., 100 kg or 100 lb_m); (b) using the known mass fractions to calculate the mass of each component in the basis quantity, and converting these masses to moles, and (c) taking the ratio of the moles of each component to the total number of moles. The same procedure is followed to convert mole fractions to mass fractions, except that a total number of moles (e.g., 100 mol or 100 lb-moles) is taken as a basis of calculation.

EXAMPLE 3.3-3 **Conversion from a Composition by Mass to a Molar Composition**

A mixture of gases has the following composition by mass:

O_2	16%	$(x_{O_2} = 0.16 \text{ g } O_2/\text{g total})$
CO	4.0%	
CO_2	17%	
N_2	63%	

What is the molar composition?

Solution

Basis: 100 g of the mixture.

Calculate how many moles of each substance are present.[2]

$$n_{O_2} = \frac{100 \text{ g total}}{} \left| \frac{0.16 \text{ g } O_2}{\text{g total}} \right| \frac{1 \text{ mol } O_2}{32.0 \text{ g } O_2} = 0.500 \text{ mol}$$

$$n_{CO} = (100)(0.040)/28.0 = 0.143 \text{ mol}$$

$$n_{CO_2} = (100)(0.17)/44.0 = 0.386 \text{ mol}$$

$$n_{N_2} = (100)(0.63)/28.0 = 2.250 \text{ mol}$$

$$n_t = n_{O_2} + n_{CO} + n_{CO_2} + n_{N_2} = 3.279 \text{ mol}$$

\Downarrow

$$y_{O_2} = \frac{n_{O_2}}{n_t} = \frac{0.500}{3.279} = 0.15 \; \frac{\text{mol } O_2}{\text{mol total}}$$

$$y_{CO} = 0.143/3.279 = 0.044 \; \frac{\text{mol CO}}{\text{mol total}}$$

$$y_{CO_2} = 0.386/3.279 = 0.12 \; \frac{\text{mol } CO_2}{\text{mol total}}$$

$$y_{N_2} = 2.25/3.279 = 0.69 \; \frac{\text{mol } N_2}{\text{mol total}}$$

Check: $\sum y_i = 1.000$

The **average molecular weight** (or mean molecular weight) of a mixture, \bar{M} (kg/kmol, lb_m/lb-mole, etc.), is the ratio of the mass of a sample of the mixture (m_t) to the number of moles of all species (n_t) in the sample. If y_i is the mole fraction of the ith component of the mixture and M_i is the molecular weight of this component, then

$$\bar{M} = y_1 M_1 + y_2 M_2 + \cdots = \sum_{\substack{\text{all} \\ \text{components}}} y_i M_i \qquad (3.3\text{-}7)$$

Exercise: Derive this formula by taking a basis of 1 mol of the mixture and calculating m_t following the procedure of Example 3.3-3. If x_i is the mass fraction of the ith component, then

$$\frac{1}{\bar{M}} = \frac{x_1}{M_1} + \frac{x_2}{M_2} + \cdots = \sum_{\substack{\text{all} \\ \text{components}}} \frac{x_i}{M_i} \qquad (3.3\text{-}8)$$

(Prove it!)

EXAMPLE 3.3-4 Calculation of an Average Molecular Weight

Calculate the average molecular weight of air (1) from its approximate molar composition of 79% N_2, 21% O_2 and (2) from its approximate composition by mass of 76.7% N_2, 23.3% O_2.

[2] We will retain more than the allowed number of two significant figures in the intermediate calculations, but will only show two in the final answers.

Solution

1. From Equation 3.3-7, with $y_{N_2} = 0.79$, $y_{O_2} = 0.21$,

$$\overline{M} = y_{N_2}M_{N_2} + y_{O_2}M_{O_2}$$

$$= \frac{0.79 \text{ kmol N}_2}{\text{kmol}} \left| \frac{28 \text{ kg N}_2}{\text{kmol}} + \frac{0.21 \text{ kmol O}_2}{\text{kmol}} \right| \frac{32 \text{ kg O}_2}{\text{kmol}}$$

$$= \boxed{29 \frac{\text{kg}}{\text{kmol}}} \left(= 29 \frac{\text{lb}_\text{m}}{\text{lb-mole}} = 29 \frac{\text{g}}{\text{mol}} \right)$$

2. From Equation 3.3-8,

$$\frac{1}{\overline{M}} = \frac{0.767 \text{ g N}_2/\text{g}}{28 \text{ g N}_2/\text{mol}} + \frac{0.233 \text{ g O}_2/\text{g}}{32 \text{ g O}_2/\text{mol}} = 0.035 \frac{\text{mol}}{\text{g}}$$

$$\Downarrow$$

$$\boxed{\overline{M} = 29 \text{ g/mol}}$$

Note: Air contains small amounts of carbon dioxide, argon, and other gases that have been neglected in this calculation, but whose presence does not significantly affect the calculated value of \overline{M}.

TEST YOURSELF

1. The molecular weight of atomic hydrogen is approximately 1, and that of atomic bromine is 80. What is (a) the mass fraction, (b) the mole fraction of bromine in pure HBr?

2. 100 lb_m/min of A($M = 2$) and 300 lb_m/min of B($M = 3$) flow through a pipe. What are the mass fractions and mole fractions of A and B, the mass flow rate of A, the molar flow rate of B, the total mass flow rate and the total molar flow rate of the mixture?

3.3c Concentration

The **mass concentration** of a component of a mixture or solution is the mass of this component per unit volume of the mixture (g/cm^3, $\text{lb}_\text{m}/\text{ft}^3$, kg/in.^3, ...). The **molar concentration** of a component is the number of moles of the component per unit volume of the mixture (kmol/m^3, lb-moles/ft^3, ...). The **molarity** of a solution is the value of the molar concentration of the solute expressed in gram-moles solute/liter solution (e.g., a 2-molar solution of A contains 2 mol A/liter of solution).

The concentration of a substance in a mixture or solution can be used as a conversion factor to relate the mass (or moles) of a component in a sample of the mixture to the sample volume, or to relate the mass (or molar) flow rate of a component of a continuous stream to the total volumetric flow rate of the stream. Consider, for

example, a 0.02-molar solution of NaOH (i.e., a solution containing 0.02 mol NaOH/liter): 5 liters of this solution contains

$$\frac{5 \text{ liters}}{} \left| \frac{0.02 \text{ mol NaOH}}{\text{liter}} \right. = 0.1 \text{ mol NaOH}$$

and if a stream of this solution flows at a rate of 2 liters/min, the molar flow rate of NaOH is

$$\frac{2 \text{ liters}}{\text{min}} \left| \frac{0.02 \text{ mol NaOH}}{\text{liter}} \right. = 0.04 \frac{\text{mol NaOH}}{\text{min}}$$

TEST YOURSELF

A solution whose volume is V (liters) contains n (mol) of a solute A whose molecular weight is M(g A/mol). In terms of V, n, and M:

1. What is the molar concentration of A?
2. What is the mass concentration of A?

In terms of C_A(mol A/liter) and c_A(g A/liter),

3. What volume of the solution contains 20 mol of A?
4. What is the mass flow rate of A in a stream whose volumetric flow rate is 120 liters/h?

EXAMPLE 3.3-5 Conversion Between Mass, Molar, and Volumetric Flow Rates of a Solution

A 0.50-molar aqueous solution of sulfuric acid flows into a process unit at a rate of 1.25 m³/min. The specific gravity of the solution is 1.03. Calculate (1) the mass concentration of H_2SO_4 in kg/m³, (2) the mass flow rate of H_2SO_4 in kg/s, and (3) the mass fraction of H_2SO_4.

Solution

1. $c \left(\dfrac{\text{kg } H_2SO_4}{\text{m}^3} \right) = \dfrac{0.50 \text{ mol } H_2SO_4}{\text{liter}} \left| \dfrac{98 \text{ g}}{\text{mol}} \right| \dfrac{1 \text{ kg}}{10^3 \text{ g}} \left| \dfrac{10^3 \text{ liter}}{1 \text{ m}^3} \right.$

 $= \boxed{49 \dfrac{\text{kg } H_2SO_4}{\text{m}^3}}$

2. $q \left(\dfrac{\text{kg } H_2SO_4}{\text{s}} \right) = \dfrac{1.25 \text{ m}^3}{\text{min}} \left| \dfrac{49 \text{ kg } H_2SO_4}{\text{m}^3} \right| \dfrac{1 \text{ min}}{60 \text{ s}} = \boxed{1.0 \dfrac{\text{kg } H_2SO_4}{\text{s}}}$

3. The mass fraction of H_2SO_4 equals the ratio of the mass flow rate of H_2SO_4—which we know—to the total mass flow rate, which can be calculated from the total volumetric flow rate and the solution density.

$$\rho_{\text{solution}} = (1.03)\left(1000 \frac{\text{kg}}{\text{m}^3}\right) = 1030 \frac{\text{kg}}{\text{m}^3}$$

$$\Downarrow$$

$$Q_{\text{solution}}\left(\frac{\text{kg}}{\text{s}}\right) = \frac{1.25 \text{ m}^3 \text{ solution}}{\text{min}} \left| \frac{1030 \text{ kg}}{\text{m}^3 \text{ solution}} \right| \frac{1 \text{ min}}{60 \text{ s}} = 21.46 \frac{\text{kg}}{\text{s}}$$

$$\Downarrow$$

$$x_{H_2SO_4} = \frac{q_{H_2SO_4}}{Q_{\text{total}}} = \frac{1.02 \text{ kg } H_2SO_4/\text{s}}{21.46 \text{ kg solution}/\text{s}} = \boxed{0.048 \frac{\text{kg } H_2SO_4}{\text{kg solution}}}$$

CREATIVITY EXERCISE

Itemize as many ways as you can think of to measure the concentration of a solute in a solution. (*Example:* If the solute absorbs light of a specific wavelength, pass a beam of light of this wavelength through the solution and measure the fractional light absorption.)

3.4 PRESSURE

3.4a Fluid Pressure and Hydrostatic Head

A **pressure** is the ratio of a force to the area on which the force acts. Accordingly, pressure units are force units divided by area units (e.g., N/m^2, dynes/cm^2, and $lb_f/in.^2$ or psi). The SI pressure unit N/m^2 is called a **pascal** (Pa).

Consider a fluid (gas or liquid) contained in a closed vessel or flowing through a pipe, and suppose that a hole of area A is made in the wall of the containing vessel, as in Figure 3.4-1. The **fluid pressure** may be defined as the ratio F/A, where F is the

FIGURE 3.4-1
Fluid pressure in a tank and a pipe

FIGURE 3.4-2
Pressure at the base of a fluid column

minimum force that would have to be exerted on a plug in the hole to keep the fluid from emerging.

We must introduce an additional definition of fluid pressure to explain the concept of atmospheric pressure and to discuss common methods for measuring pressures of fluids in tanks and pipes. Suppose a vertical column of fluid is h(m) high and has a uniform cross-sectional area A(m^2). Further suppose that the fluid has a density of ρ(kg/m^3), and that a pressure P_0(N/m^2) is exerted on the upper surface of the column. (See Figure 3.4-2.) The pressure P of the fluid at the base of the column—called the **hydrostatic pressure** of the fluid—is, by definition, the force F exerted on the base divided by the base area A. F thus equals the force on the top surface plus the weight of the fluid in the column. It is not difficult to show that

$$P = P_0 + \rho \frac{g}{g_c} h \qquad (3.4\text{-}1)$$

(See if you can derive this equation.) Since A does not appear in this equation, the formula is applicable to a column of fluid as thin as a test tube or as wide as the ocean.

In addition to being expressible as a force per unit area, a pressure may be expressed as a **head** of a particular fluid—that is, as the height of a hypothetical column of this fluid that would exert the given pressure at its base if the pressure at the top were zero. You can thus speak of a pressure of 14.7 psi or equivalently of a pressure (or head) of 33.9 ft of water (33.9 ft H$_2$O) or 76 cm of mercury (76 cm Hg). *The equivalence between a pressure P(force/area) and the corresponding head P_h (height of a fluid) is given by Equation 3.4-1 with $P_0 = 0$:*

$$P\left(\frac{\text{force}}{\text{area}}\right) = \rho_{\text{fluid}} \frac{g}{g_c} P_h(\text{head of fluid}) \qquad (3.4\text{-}2)$$

EXAMPLE 3.4-1 Calculation of a Pressure as a Head of Fluid

Express a pressure of 2.00×10^5 Pa in terms of mm Hg.

Solution

Solve Equation 3.4-2 for P_h(mm Hg), assuming that $g/g_c = 9.807$ N/kg and noting that the density of mercury is 13.6×1000 kg/m$^3 = 13\,600$ kg/m^3.

$$P_h = P(g_c/g)/\rho_{Hg}$$

$$= \frac{2.00 \times 10^5 \text{ N}}{\text{m}^2} \left| \frac{\text{m}^3}{13\,600 \text{ kg}} \right| \frac{\text{kg}}{9.807 \text{ N}} \left| \frac{10^3 \text{ mm}}{1 \text{ m}} \right. = \boxed{1.50 \times 10^3 \text{ mm Hg}}$$

Note: We will henceforth use an unsubscripted P to denote pressure expressed as either (force/area) or as the head of a fluid.

The relationship between the pressure at the base of a column of fluid of height h and the pressure at the top is particularly simple if these pressures are expressed as heads of the given fluid: If the column is mercury, for example, then

$$\boxed{P(\text{mm Hg}) = P_0(\text{mm Hg}) + h(\text{mm Hg})} \qquad (3.4\text{-}3)$$

Any other length unit and chemical species may be substituted for mm Hg in this equation.

The conversion table on the inside front cover of this book lists values of a pressure expressed in several common force/area units and as heads of mercury and water. The use of this table for pressure unit conversion is illustrated by the conversion of 20.0 psi to cm Hg.

$$\frac{20.0 \text{ psi}}{} \left| \frac{76.0 \text{ cm Hg}}{14.696 \text{ psi}} \right. = 103 \text{ cm Hg}$$

EXAMPLE 3.4-2 Pressure Below the Surface of a Fluid

What is the pressure 30.0 m below the surface of a lake? Atmospheric pressure (the pressure at the surface) is 10.4 m H_2O, and the density of water is 1000.0 kg/m^3. Assume that g/g_c is 9.807 N/kg.

Solution

First, the hard way, using Equation 3.4-1.

$$P = P_0 + \rho(g/g_c)h$$
$$\Downarrow$$

$$P(\text{N/m}^2) = \frac{10.4 \text{ m } H_2O}{} \left| \frac{1.013 \times 10^5 \text{ N/m}^2}{10.33 \text{ m } H_2O} \right.$$

$$+ \frac{1000.0 \text{ kg/m}^3}{} \left| \frac{9.807 \text{ N}}{\text{kg}} \right| \frac{30.0 \text{ m}}{} = \boxed{3.96 \times 10^5 \text{ N/m}^2 \text{ (Pa)}}$$

Next, the easy way, using Equation 3.4-3.

$$P(\text{m } H_2O) = 10.4 \text{ m } H_2O + 30.0 \text{ m } H_2O = \boxed{40.4 \text{ m } H_2O}$$

(Verify that the two calculated pressures are equivalent.)

TEST YOURSELF

1. Define (a) the pressure of a fluid flowing in a pipe, (b) hydrostatic pressure, and (c) a head of a fluid corresponding to a given pressure.

2. Consider the tank in Figure 3.4-1. Does the pressure depend on the height of the opening in the tank? (*Hint:* Yes.) Why? Would you expect the difference between the pressure at the top and that at the bottom to be very large if the fluid were air? How about water? Mercury?

3. Suppose the pressure in the tank of Figure 3.4-1 is given as 1300 mm of mercury. Does this tell you anything about the height of the tank? If you were given the area of the hole (say 4 cm^2), how would you calculate the force needed to hold the plug in?

4. Suppose the pressure at a point within a column of mercury in a tube is 74 mm Hg. What is the pressure 5 mm below this point? (If this takes you more than one second, you are probably doing it wrong.)

3.4b Atmospheric Pressure, Absolute Pressure, and Gauge Pressure

The pressure of the atmosphere can be thought of as the pressure at the base of a column of fluid (air) located at the point of measurement (at sea level, for example). The pressure P_0 at the top of the column equals 0, and ρ and g are average values of the density of air and the acceleration of gravity between the top of the atmosphere and the measurement point.

A typical value of the atmospheric pressure at sea level, 760.0 mm Hg, has been designated as a standard pressure of 1 atmosphere. The conversion table on the inside front cover lists equivalent values of this pressure in various units.

The fluid pressures referred to so far are all **absolute pressures,** in that a pressure of zero corresponds to a perfect vacuum. Many pressure measuring devices give the **gauge pressure** of a fluid, or the pressure relative to atmospheric pressure. A gauge pressure of zero indicates that the absolute pressure of the fluid is equal to atmospheric pressure. The relationship for converting between absolute and gauge pressure is

$$\boxed{P_{\text{absolute}} = P_{\text{gauge}} + P_{\text{atmospheric}}} \qquad (3.4\text{-}4)$$

The abbreviations psia and psig are commonly used to denote absolute and gauge pressure in $lb_f/in.^2$. Also, it is common to refer to negative gauge pressures (absolute pressures less than atmospheric) as positive amounts of vacuum: For example,

a gauge pressure of −1 cm Hg (75.0 cm Hg absolute if atmospheric pressure is 76.0 cm Hg) may also be called 1 cm of vacuum.

TEST YOURSELF

1. Is atmospheric pressure always equal to 1 atm?
2. What is absolute pressure? Gauge pressure?
3. The gauge pressure of a gas is −20 mm Hg at a point where atmospheric pressure is 755 mm Hg. How else can the gas pressure be expressed in terms of mm Hg? (Give two values.)
4. A mercury column is open to the atmosphere on a day when atmospheric pressure is 29.9 in. Hg. What is the gauge pressure 4 in. below the surface? The absolute pressure? (Give the answers in in. Hg.)

3.4c Fluid Pressure Measurement

Several mechanical devices are used for the measurement of fluid pressures. The most common of these devices is a **Bourdon gauge,** which is a hollow tube closed at one end and bent into a C configuration. The open end of the tube is exposed to the fluid whose pressure is to be measured. As the pressure increases, the tube tends to straighten, causing a pointer attached to the tube to rotate. The position of the pointer on a calibrated dial gives the *gauge* pressure of the fluid. A schematic diagram of a Bourdon gauge is shown in Figure 3.4-3.

Bourdon gauges are used to measure fluid pressures from nearly perfect vacuums to about 7000 atm. More accurate measurements of pressures below about 3 atm are provided by **manometers.**

A manometer is a U-shaped tube partially filled with a liquid of known density (the **manometer fluid**). When the ends of the tube are exposed to different pressures

(a) (b)

FIGURE 3.4-3
Bourdon gauge

(a) Open–end (b) Differential (c) Sealed–end

FIGURE 3.4-4

Manometers

the fluid level drops in the high pressure arm and rises in the low pressure arm. The difference between the pressures can be calculated from the measured difference between the liquid levels in each arm.

Manometers are used in several different ways, as shown in Figure 3.4-4. In each diagram, pressure P_1 is greater than pressure P_2.

Figure 3.4-4a shows an **open-end manometer:** One end is exposed to a fluid whose pressure is to be measured, and the other is open to the atmosphere. Figure 3.4-4b shows a **differential manometer,** used to measure the pressure difference between two points in a process line. Figure 3.4-4c shows a **sealed-end manometer,** which has a vacuum enclosed at one end. If the open end of a sealed-end manometer is exposed to the atmosphere ($P_1 = P_{atm}$) the device functions as a **barometer.**

The formula that relates the pressure difference $P_1 - P_2$ to the difference in manometer fluid levels is based on the principle that the fluid pressure must be the same at any two points at the same height in a continuous fluid. In particular, the pressure at the height of the lower surface of a manometer fluid is the same in both arms of the manometer. (See Figure 3.4-5.) Writing and equating expressions for the pressures at points (a) and (b) in Figure 3.4-5 yields the general manometer equation

General manometer equation:
$$P_1 + \rho_1 \frac{g}{g_c} d_1 = P_2 + \rho_2 \frac{g}{g_c} d_2 + \rho_f \frac{g}{g_c} h \qquad \text{(3.4-5)}$$

In a differential manometer, fluids 1 and 2 are the same, and consequently $\rho_1 = \rho_2 = \rho$. The general manometer equation then reduces to

Differential manometer equation:
$$P_1 - P_2 = (\rho_f - \rho) \frac{g}{g_c} h \qquad \text{(3.4-6)}$$

If either fluid 1 or 2 is a gas at a moderate pressure (e.g., if one arm is open to the atmosphere), the density of this fluid is 100 to 1000 times lower than the density of the manometer fluid, so that the corresponding $\rho g d / g_c$ term in Equation 3.4-5 may

$$\frac{1}{P} = 3.94t^{\frac{1}{2}} - 0.517$$

FIGURE 3.4-5

Manometer variables

be neglected. If *both* fluids are gases, then the equation becomes

$$P_1 - P_2 = \rho_f \frac{g}{g_c} h$$

and if both P_1 and P_2 are expressed as heads of the manometer fluid, then

Manometer formula for gases: $\boxed{P_1 - P_2 = h}$ (3.4-7)

If P_2 is atmospheric pressure, then the gauge pressure at point 1 is simply the difference in the levels of the manometer fluid.

EXAMPLE 3.4-3 Pressure Measurement with Manometers

1. A differential manometer is used to measure the drop in pressure between two points in a process line containing water. The specific gravity of the manometer fluid is 1.05. The measured levels in each arm are shown below. Calculate the pressure drop between points 1 and 2 in dynes/cm^2.

2. The pressure of gas being pulled through a line by a vacuum pump is measured with an open-end mercury manometer. A reading of -2 in. is obtained. What is the gas gauge pressure in inches of mercury? What is the absolute pressure if $P_{atm} = 30$ in. Hg?

Solution

1. $h = (382 - 374)$ mm $= 8$ mm. From Equation 3.4-6,

$$P_1 - P_2 = (\rho_f - \rho)\frac{g}{g_c}h$$

$$= \frac{(1.05 - 1.00)\,g}{cm^3}\left|\frac{980.7\,cm}{s^2}\right|\frac{1\,dyne}{1\,g\cdot cm/s^2}\left|\frac{8\,mm}{}\right|\frac{1\,cm}{10\,mm}$$

$$= \boxed{40\,\frac{dynes}{cm^2}}$$

2. From Equation 3.4-7 and the definition of a gauge pressure

$$P_1 - P_{atm} = P_{gauge} = \boxed{-2\ \text{in. Hg}}$$

$$\Downarrow$$

$$P_1 = P_{atm} + P_{gauge} = 30 - 2 = \boxed{28\ \text{in. Hg}}$$

TEST YOURSELF

1. What is a Bourdon gauge? What range of pressures can it be used to measure? As it is normally calibrated, does it measure gauge or absolute pressure?

2. What is an open-end manometer? A differential manometer? A sealed-end manometer?

3. True or false?
 (a) An open-end manometer provides a direct reading of the gauge pressure of a gas.
 (b) A closed-end manometer provides a direct reading of the absolute pressure of a gas.

(c) The reading of a differential manometer does not depend on the density of the fluid in the pipeline but only on that of the manometer fluid.

4. The pressure of a gas in a pipeline is measured with an open-end mercury manometer. The mercury level in the arm connected to the line is 14 mm *higher* than the level in the open arm. What is the gauge pressure of the gas in the line?

CREATIVITY EXERCISE

Think of several devices that might be used to measure fluid pressures; be as imaginative as you can. (*Example:* Allow a gas at the pressure to be measured to fill a balloon, and measure the final balloon diameter.)

3.5 TEMPERATURE

The temperature of a substance in a particular state of aggregation (solid, liquid, or gas) is a measure of the average kinetic energy possessed by the substance molecules. Since this energy cannot be measured directly, the temperature must be determined indirectly by measuring some physical property of the substance whose value depends on temperature in a known manner. Such properties and the temperature measuring devices based on them include electrical resistance of a conductor (**resistance thermometer**), voltage at the junction of two dissimilar metals (**thermocouple**), spectra of emitted radiation (**pyrometer**), and volume of a fixed mass of a fluid (**thermometer**).

Temperature scales can be defined in terms of any of these properties, or in terms of physical phenomena, such as freezing and boiling, that take place at fixed temperatures and pressures. You might refer, for example, to "the temperature at which the resistivity of a copper wire is 1.92×10^{-6} ohms/cm^3" or to "the temperature two-thirds of the way from the boiling point of water at 1 atm to the melting point of NaCl."

It is convenient to have, in addition to these physical scales, a simple numerical temperature scale—among other reasons, so that you do not have to use 25 words to give the value of a single temperature. A defined temperature scale is obtained by arbitrarily assigning numerical values to two reproducibly measurable temperatures; for example, assign a value of 0 to the freezing point of water and a value of 100 to the boiling point of water at a pressure of 1 atm. This completely specifies the scale, since in addition to locating the two points it specifies that the length of a unit temperature interval (called a **degree**) is $\frac{1}{100}$ of the distance between the two reference points on the scale.

The two most common temperature scales are defined using the freezing point (T_f) and boiling point (T_b) of water at a pressure of 1 atm.

Celsius (or centigrade) scale: T_f is assigned a value of 0°C, and T_b is assigned a value of 100°C. *Absolute zero* (theoretically the lowest temperature attainable in nature) on this scale falls at -273.15°C.

Fahrenheit scale: T_f is assigned a value of 32°F, and T_b is assigned a value of 212°F. Absolute zero falls at -459.67°F.

The **Kelvin** and **Rankine** scales are defined such that absolute zero has a value of 0 and the size of a degree is the same as a Celsius degree (Kelvin scale) or a Fahrenheit degree (Rankine scale).

The following relationships may be used to convert a temperature expressed in one defined scale unit to its equivalent in another.

$$T(K) = T(°C) + 273.15 \qquad \text{(3.5-1)}$$
$$T(°R) = T(°F) + 459.67 \qquad \text{(3.5-2)}$$
$$T(°R) = 1.8T(K) \qquad \text{(3.5-3)}$$
$$T(°F) = 1.8T(°C) + 32 \qquad \text{(3.5-4)}$$

Equations like those given here are always in the form of the equation of a line $(y = ax + b)$. If $(°A)$ and $(°B)$ stand for any two temperature units, to derive the equation for $T(°B)$ in terms of $T(°A)$ you must know equivalent values on each scale of two temperatures—say T_1 and T_2. Then

1. Write $T(°B) = aT(°A) + b$.
2. Substitute $T_1(°B)$ and $T_1(°A)$ in the equation—you then have one equation in two unknowns (a and b). Substitute $T_2(°B)$ and $T_2(°A)$ to get the second equation in the two unknowns, and solve for a and b.

EXAMPLE 3.5-1 Derivation of a Temperature Conversion Formula

Derive Equation 3.5-4 for $T(°F)$ in terms of $T(°C)$. Use $T_1 = 0°C$ (32°F) and $T_2 = 100°C$ (212°F).

Solution

$$T(°F) = aT(°C) + b$$

Substitute T_1: $32 = (a)(0) + b \Rightarrow b = 32$

Substitute T_2: $212 = (a)(100) + 32 \Rightarrow a = 1.8$

$$\Downarrow$$
$$T(°F) = 1.8T(°C) + 32$$

A degree is both a temperature and a temperature interval, a fact that sometimes leads to confusion. Consider the temperature interval from 0°C to 5°C. There are nine Fahrenheit and Rankine degrees in this interval, and only five Celsius and Kelvin degrees. An interval of 1 Celsius or Kelvin degree therefore contains 1.8

Fahrenheit or Rankine degrees, leading to the conversion factors

$$\frac{1.8°F}{1°C}, \frac{1.8°R}{1 K}, \frac{1°F}{1°R}, \frac{1°C}{1 K}$$ (3.5-5)

Note: These conversion factors refer to temperature intervals, not temperature.[3] For example, to find the number of Celsius degrees between 32°F and 212°F you can say that

$$\Delta T(°C) = \frac{(212 - 32)°F}{} \left| \frac{1°C}{1.8°F} \right. = 100°C$$

but to find the Celsius temperature corresponding to 32°F you must use Equation 3.5-4; you cannot say

$$T(°C) = \frac{32°F}{} \left| \frac{1°C}{1.8°F} \right.$$

A temperature A temperature
interval

TEST YOURSELF

1. Suppose you were given a glass tube with mercury in it, but no scale markings, and you were given only a beaker of water, a freezer, and a bunsen burner to work with. How would you calibrate the thermometer to provide readings in °C?

2. Which is higher, a temperature of 1°C or 1°F?

3. Which is greater, a temperature interval of 1°C or 1°F?

[3] Some authors have proposed varying the position of the degree symbol to indicate whether a temperature or temperature interval is denoted; that is, 5°C refers to a temperature of five degrees Celsius, and 5C° means an interval of five Celsius degrees. This idea, while an excellent one in principle, has not caught on, so you will have to get used to making the distinction yourself from the context in which the unit appears.

EXAMPLE 3.5-2 Temperature Conversion

Consider the interval from 20°F to 80°F.

1. Calculate the equivalent temperatures in °C and the interval between them.
2. Calculate directly the interval in °C between the temperatures.

Solution

1. From Equation 3.5-4

$$T(°C) = \frac{T(°F) - 32}{1.8}$$

so that

$$T_1 = 20°F = \frac{20 - 32}{1.8} = -6.7°C$$

$$T_2 = 80°F = \frac{80 - 32}{1.8} = 26.6°C$$

and

$$T_2 - T_1 = 26.6 - (-6.7) = 33.3°C$$

2. From Equation 3.5-5

$$\Delta T(°C) = \frac{\Delta T(°F)}{\left|\,\dfrac{1°C}{1.8°F}\right.} = \frac{(80 - 20)°F}{\left|\,\dfrac{1°C}{1.8°F}\right.} = 33.3°C$$

which checks with part 1, as we would hope.

EXAMPLE 3.5-3 Temperature Conversion and Dimensional Homogeneity

The heat capacity of ammonia, defined as the amount of heat required to raise the temperature of a unit mass of ammonia by 1° at constant pressure, is, over a limited range of temperatures, given by the expression

$$C_p\left(\frac{Btu}{lb_m \cdot °F}\right) = 0.487 + 2.29 \times 10^{-4} T(°F)$$

Determine the expression for C_p in (J/g·°C) in terms of $T(°C)$.

Solution

The °F in the units of C_p refers to a temperature interval, while the unit of T is a temperature. The calculation is best done in two steps.

1. Substitute for $T(°F)$ and simplify the resulting equation.

$$C_p\left(\frac{Btu}{lb_m \cdot °F}\right) = 0.487 + 2.29 \times 10^{-4}[1.8T(°C) + 32]$$

$$= 0.494 + 4.12 \times 10^{-4} T(°C)$$

2. Convert to the desired temperature interval unit using Equation 3.5-5.

$$C_p\left(\frac{J}{g \cdot {}^\circ C}\right) =$$

$$[0.494 + 4.12 \times 10^{-4}T({}^\circ C)]\frac{Btu}{lb_m \cdot {}^\circ F}\left|\frac{1.8 {}^\circ F}{1.0 {}^\circ C}\right|\frac{1\,J}{9.486 \times 10^{-4}\,Btu}\left|\frac{1\,lb_m}{454\,g}\right.$$

$$\Downarrow$$

$$\boxed{C_p\left(\frac{J}{g \cdot {}^\circ C}\right) = 2.06 + 1.72 \times 10^{-3}T({}^\circ C)}$$

CREATIVITY EXERCISES

1. Invent several temperature-measuring devices. For each, describe the device, and state what you would measure. (*Example:* Put a guinea pig on a treadmill in a room and measure the rate at which he runs to keep warm.) (Well, it *could* work.)

2. Think of as many ways as you can to use a solid block as a temperature-measuring device. (*Example:* Put it in a furnace equipped with a window and observe the color with which it glows.)

PROBLEMS

1. Calculate the density in kg/m^3 of:
 (a) A liquid with a density $\rho = 68.5 \, lb_m/ft^3$ (i) using a conversion factor table; (ii) using Equation 3.1-2.
 (b) A solid with a specific gravity of 7.8.

2. The specific gravity of nitrobenzene is 1.20.
 (a) Calculate the mass in kg of 250 liters of nitrobenzene.
 (b) Calculate the volumetric flow rate in ml/min corresponding to a mass flow rate of 30.0 lb_m nitrobenzene/h.
 (c) Calculate the density of a mixture formed by mixing 20.0 liters of nitrobenzene with 3.00 ft^3 of benzene (S.G. = 0.879), assuming that the volume of the mixture is the sum of the two feed component volumes.

3. At 10°C, pure nitric acid has a specific gravity of 1.531, and an aqueous solution containing 40.0 wt% HNO_3 has a specific gravity of 1.256. You need a quantity of the 40% solution that contains exactly 31.50 g of HNO_3.
 (a) Calculate the required volume (liters) of the 40% solution using the given specific gravity.

(b) Suppose you did not have the specific gravity data for aqueous nitric acid. Calculate the required volume from the pure component specific gravities, assuming volume additivity. What percentage error is introduced by this assumption?

4. A rectangular block of solid carbon (graphite) floats at the interface of two immiscible liquids. The bottom liquid is a relatively heavy lubricating oil, and the top liquid is water. Of the total block volume, 54.3% is immersed in the oil and the balance is in the water. What is the specific gravity of the oil? (*Suggestion:* Recall Archimedes' principle, and do a force balance on the block.)

5. A friend of yours bought several dozen pure silver rings on a recent trip abroad, but now suspects that they might not be all that pure. He brings them to you to check. You weigh 10 of the rings and find that they have a combined mass of 7.38 g; you then place them in a graduated cylinder 2.5 cm in diameter and observe that the water level rises by 8.45 mm. Are the rings pure silver?

6. An object of density ρ_a, volume V_a, and weight W_a is thrown from a rowboat floating on the surface of a small pond, and sinks to the bottom. The weight of the rowboat without the jettisoned object is W_b. Before the object was thrown out, the depth of the pond was h_{p1}, and the bottom of the boat was a distance h_{b1} above the pond bottom. After the object sinks, the values of these quantities are h_{p2} and h_{b2}. The area of the pond is A_p; that of the boat is A_b. A_b may be assumed constant, so that the volume of water displaced by the boat is $A_b(h_p - h_b)$.

 (a) Derive an expression for the change in the pond depth ($h_{p2} - h_{p1}$). Does the liquid level of the pond rise or fall, or is it indeterminate?

 (b) Derive an expression for the change in the height of the bottom of the boat above the bottom of the pond ($h_{b2} - h_{b1}$). Does the boat rise or fall relative to the pond bottom, or is it indeterminate?

7. Before a rotameter can be used to determine an unknown flow rate, a **calibration curve** of flow rate versus rotameter reading must be prepared. A calibration technique for liquids is illustrated below. A flow rate is set by adjusting the pump speed; the rotameter reading is recorded, and the effluent from the rotameter is collected in a graduated cylinder for a timed interval. The procedure is repeated for several pump settings to generate a smooth curve of flow rate versus rotameter reading.

The following data are taken in a rotameter calibration run with water as the fluid.

Rotameter Reading	Collection Time (min)	Volume Collected (cm^3)
2	1	297
2	1	301
4	1	454
4	1	448
6	0.5	300
6	0.5	298
8	0.5	371
8	0.5	377
10	0.5	440
10	0.5	453

(a) Prepare a calibration curve.

(b) What is the mass flow rate in kg/s of a water stream that gives a reading of 5.3 on this rotameter?

8. How many of each of the following are contained in 10.0 kmol of propane? **(a)** g C_3H_8; **(b)** mol C_3H_8; **(c)** lb-moles C_3H_8; **(d)** mol (g-atoms) C; **(e)** mol H; **(f)** mol H_2; **(g)** g C; **(h)** g H; **(i)** g H_2; **(j)** molecules of C_3H_8.

9. Sixty-five cubic meters per hour of benzene are fed into a reactor.
 (a) What is the mass flow rate of this stream in kg/h?
 (b) What is the molar flow rate in mol/s?
 (c) Suppose you learn that your answers to parts a and b are too large by a factor of 1000. What did you probably assume incorrectly?

10. A mixture of methane and ethane contains 30 wt% ethane.
 (a) Calculate the gram-moles of methane in 200 kg of this mixture. (Use a single dimensional equation)
 (b) What must the mass flow rate of this mixture be for the molar flow rate of methane to be 25.0 lb-moles/s?

11. The feed to an ammonia synthesis reactor contains 25 mole% nitrogen and the balance hydrogen. The flow rate of the stream is 3000 kg/h. Calculate the rate of flow of nitrogen into the reactor in kg/h. (*Suggestion:* First calculate the average molecular weight of the mixture.)

12. A suspension of calcium carbonate particles in water flows through a pipe. Your assignment is to determine both the flow rate and composition by weight of this slurry. You proceed to collect the stream in a graduated cylinder for 1.00 min; you then weigh the cylinder, evaporate the collected water, and reweigh the cylinder. The following results are obtained.

Mass of empty cylinder: 65.0 g
Mass of cylinder + collected slurry: 565 g
Volume collected: 455 ml
Mass of cylinder after evaporation: 215 g

Calculate:
(a) The volumetric flow rate and mass flow rate of the suspension
(b) The density of the suspension
(c) The mass fraction of $CaCO_3$ in the suspension

13. Things were going smoothly at the Philpott Pharmaceuticals pilot plant facility during the midnight to 8 A.M. shift until Henry Anthropoid, the reactor operator, let the run instruction sheet get too close to the Bunsen burner that was keeping the coffee hot. What followed ended in a total loss of the run sheet, the coffee, and a substantial portion of Anthropoid's left shirtsleeve.

Remembering with a grimace the last time he telephoned his supervisor in the middle of the night, Anthropoid decided to rely on his memory of the new run settings. He adjusted the flow rate of cyclohexane ($M = 84$, SG = 0.779) to 11.0 liters/min and that of tetrahydrofreefurol ($M = 216$, SG = 1.30) to 5.0 liters/min.

Anthropoid should have known two things. (1) The mass fraction of THF in the combined reactant stream was supposed to be 0.25; (2) If the mole fraction of THF exceeds 0.3 the reaction mixture explodes.
(a) Was his memory correct?
(b) If not, did he learn about his mistake immediately?

14. A mixture of gases contains equimolar amounts of ethylene, ethane, and propane. Calculate the mass fractions of the components in the mixture and the average molecular weight.

15. A mixture of ethanol (C_2H_5OH) and water is analyzed and found to contain 60% water by mass. Assuming volume additivity of the components, calculate the liters of this mixture required to provide 150 mol of ethanol.

16. A mixture of methane and air is capable of being ignited only if the mole% of methane is between 5% and 15%. A mixture containing 9.0 mole% methane in air flowing at a rate of 700 kg/h is to be diluted with pure air to reduce the methane concentration to the lower flammability limit. Calculate the required flow rate of air in mol/h and the percent *by mass* of oxygen in the product gas. (*Note:* Air may be taken to consist of 21 mole% O_2 and 79% N_2 and to have an average molecular weight of 29.0)

17. A gas containing CO, CO_2, and CH_4 is analyzed with a gas chromatograph. The output of the gas chromatograph appears on a strip-chart recorder as shown here.

For each of the three species, the area under the peak is roughly proportional to the number of moles of the indicated substance in the sample. Use the information shown to determine the mole fractions and mass fractions of CO, CO_2, and CH_4 in the gas.

***18.** A **gas chromatograph** (GC) is a device to separate the components of a sample of a gas or liquid mixture and to provide a measure of the amount of each component in the sample. The output from a chromatographic analysis typically takes the form of a series of peaks on a strip-chart recorder. (For example, see the preceding problem.) Each peak corresponds to a specific component, and the area under the peak is proportional to the amount of that component in the sample $[n_i(\text{mol}) = k_i A_i$, where A_i is the area of the peak corresponding to the ith species]. The proportionality constants (k_i) are determined in separate calibration experiments in which known amounts of the components are injected into the GC sample port and the corresponding peak areas are measured.

Write a computer program to calculate the composition of a mixture from the set of peak areas obtained with a chromatograph. The program should perform the following tasks:

(a) Read in N, the number of species to be analyzed

(b) Read in $M_1, M_2, M_3, \ldots, M_N$, the molecular weights of the N species

(c) Read in $k_1, k_2, k_3, \ldots, k_N$, the calibration constants for the N species

(d) Read in N_a, the number of chromatographic analyses performed

(e) For the first analysis, read in the measured peak areas A_1, A_2, \ldots, A_N

(f) Calculate and print out (i) the sample number, (ii) the mole fractions of each species in the sample, (iii) the mass fractions of each species in the sample

(g) Repeat steps e and f for each of the remaining analyses

Test your program on a set of analyses of mixtures of (1) methane, (2) ethane, (3) propane, and (4) n-butane, with $k_1 = 0.150$ (mmol/area), $k_2 = 0.287$ (mmol/area), $k_3 = 0.467$ (mmol/area), and $k_4 = 0.583$ (mmol/area). The peak areas measured in five analyses are given here.

Sample	A_1	A_2	A_3	A_4
1	3.6	2.8	2.4	1.7
2	7.8	2.4	5.6	0.4
3	3.4	4.5	2.6	0.8
4	4.8	2.5	1.3	0.2
5	6.4	7.9	4.8	2.3

19. A 0.01-molar solution of benzoin ($M = 212$) in methyl methacrylate ($\rho = 950 \text{ kg/m}^3$) flows through a pipe at a mass flow rate of 100 kg/min.

(a) Calculate the volumetric flow rate of the solution and the mass flow rate of benzoin. Assume $\rho_{\text{soln}} = \rho_{\text{MMA}}$.

(b) How long would it take to fill a 2-m ID × 4-m-high tank at this flow rate?

20. A gas stream containing 20 mole% benzene and the balance nitrogen flows into a condenser. The condensate, which is pure benzene, is collected in a graduated cylinder. In 30.0 s, exactly 755 cm^3 is collected. A rotameter in the gas pipeline *leaving* the condenser

* Computer problem.

indicates that the total molar flow rate of this stream is 83.0 mol/min. Calculate
- (a) The mass flow rate (g/min) of the feed to the condenser
- (b) The mass fraction of benzene in the exiting gas stream, and the average molecular weight of this stream.

21. The little-known rare earth element Nauseum (atomic weight = 172) has the interesting property of being completely insoluble in everything but 12-year-old bourbon. This curious fact was discovered in the laboratory of Professor Ludwig von Schlimazel, the eminent German chemist whose invention of the bathtub ring won him the Nobel Prize. Having unsuccessfully tried to dissolve Nauseum in 7642 different solvents over a 10-year period, Schlimazel finally came to the 30-ml of Old Aardvark Bottled-in-Bond that was the only remaining liquid in his laboratory. Always willing to suffer personal loss in the name of science, Schlimazel calculated the amount of Nauseum needed to make up a 0.03-molar solution, put the Aardvark bottle on the desk of his faithful technician Edgar P. Settera, weighed out the calculated amount of Nauseum and put it next to the bottle, and then wrote the message that has become part of history:

<p style="text-align:center">"Ed Settera. Add Nauseum!"</p>

How many grams of Nauseum did he weigh out?

22. The reaction $A \rightarrow B$ is carried out in a laboratory reactor. According to a proposed reaction rate law the concentration of A should vary with time as follows.

$$C_A = C_{A0} \exp(-kt)$$

where C_{A0} is the initial concentration of A in the reactor and k is a constant.
- (a) If C_A and C_{A0} are in lb-moles/ft^3 and t is in minutes, what are the units of k?
- (b) The following data are taken for $C_A(t)$.

t(min)	C_A(lb-moles/ft^3)
0.5	1.02
1.0	0.84
1.5	0.69
2.0	0.56
3.0	0.38
5.0	0.17
10.0	0.02

Verify the proposed rate law graphically, and calculate C_{A0} and k.
- (c) Convert the formula with the calculated constants included to an expression for the molarity of A in the reaction mixture in terms of t(seconds). Calculate the molarity at $t = 200$ s.

23. Perform the following pressure conversions, assuming when necessary that atmospheric pressure is 1 atm. Unless otherwise stated, the given pressures are absolute.
- (a) 1250 mm Hg to psi
- (b) 25.0 ft H$_2$O to kPa

(c) 3.00 atm to N/cm^2
(d) 100.0 cm Hg to $dynes/m^2$
(e) 30.0 cm Hg of vacuum to atm
(f) 10.0 psig to mm Hg (gauge)
(g) 10.0 psig to mm Hg (absolute)
(h) 650 mm Hg to mm Hg (gauge)
(i) 10 in. Hg of vacuum to in. Hg (gauge)
(j) 20.0 psi to cm of carbon tetrachloride (CCl_4)

24. A vat of acetone is filled to a depth of 5.0 m on a day when atmospheric pressure is 752 mm Hg.
 (a) Calculate the gauge pressure at the bottom of the vat in meters of acetone.
 (b) Calculate the absolute pressure at the bottom of the vat in kilopascals.

25. The viewing window in a diving suit has an area of roughly 65 cm^2. If an attempt were made to maintain the pressure on the inside of the suit at 1 atm, what force (N, and lb_f) would the window have to withstand if the diver descended to a depth of 150 m. Take the specific gravity of the water to be 1.05.

26. A gravity-feed system is to be used to supply water to a heat exchanger.

 A water pressure of at least 30 psig is needed at the inlet of the heat exchanger.
 (a) What is the minimum elevation of the water level above the exchanger?
 (b) What elevation would be required if the fluid were *n*-heptane?

27. A closed-end manometer and an open-end manometer (both mercury) are each used to measure the pressure in a gas pipe. The reading on the first manometer is 1570 mm, and that on the second is 814 mm. Calculate P_{gauge}, $P_{absolute}$, and $P_{atmosphere}$.

28. A Bourdon gauge calibrated to read in psi is mounted on a gas tank. The gauge reads 0 when the tank is open to the atmosphere, and 75 after the tank is filled.
 (a) Does the reading represent the absolute pressure or the gauge pressure of the gas in the tank?
 (b) If the gauge were replaced by an open-end mercury manometer, what would be the difference (in inches) between the two mercury column levels?

29. An open-end manometer containing water as the manometer fluid is connected to an open tank of liquid benzene. The manometer reading is 50.0 cm when the water surface

in the open leg of the manometer is level with the manometer tap in the tank. (See diagram.)

How many centimeters below the benzene surface is the tap?

30. A fluid of unknown density is used in two manometers—one closed-end, the other across an orifice in a water pipeline. The readings shown here are obtained on a day when barometric pressure is 763 mm Hg.

What is the pressure drop (mm Hg) from point (*a*) to point (*b*)?

31. An open-end water manometer is connected to a stoppered vessel containing a gas. The gauge pressure of the gas is 50.0 cm H_2O. The manometer has been in service for many years, and the arm connected to the gas vessel has become so encrusted that it is impossible to see the manometer fluid level in it.

(a) If you could see the manometer fluid level in the left arm, how much higher or lower (state which) would it be than the level in the right arm?

(b) Suppose you notice that after several days, the level in the right arm has dropped. Give five possible explanations. For each one, state whether the level in the left arm would rise, drop, or remain the same.

32. An **inclined manometer** is a useful device for measuring small pressure differences.

The formulas for the pressure difference in terms of the liquid level difference h given in Section 3.4 remain valid, but while h would be small and difficult to read for a small pressure drop if the manometer were vertical, L can be made quite large for the same pressure drop by making the angle of inclination, θ, small.

(a) Derive a formula for h in terms of L and θ.

(b) Suppose the manometer fluid is mercury, the process fluid is a gas, the inclination of the manometer is $\theta = 15°$, and a reading $L = 8.7$ cm is obtained. What is the pressure difference between points 1 and 2?

33. Three distinct liquids are used in the manometer shown here.

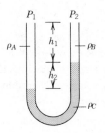

(a) Derive an expression for P_1-P_2 in terms of ρ_A, ρ_B, ρ_C, h_1, and h_2.

(b) Suppose fluid A is methanol, B is water, and C is a manometer fluid with a specific gravity of 1.37; pressure $P_2 = 121.0$ kPa; $h_1 = 30.0$ cm, and $h_2 = 24.0$ cm. Calculate P_1(kPa).

34. An orifice meter (see Figure 3.2-1) is to be calibrated for the measurement of the flow rate of a stream of liquid acetone. The differential manometer fluid has a specific gravity of 1.10.

The calibration is accomplished by connecting the orifice meter in series with a rotameter that has previously been calibrated for acetone, adjusting a valve to set the flow rate, and recording the flow rate (determined from the rotameter reading and the rotameter calibration curve) and the differential manometer reading, h. The procedure is repeated for several valve settings to generate an orifice meter calibration curve of flow rate versus h. The following data are taken.

Flow rate Q(ml/s)	Manometer reading h(mm)
0	0
245	5
346	10
424	15
490	20
548	25
600	30

(a) For each of the given readings, calculate the pressure drop across the orifice, ΔP(mm Hg).

(b) The flow rate through an orifice should be related to the pressure drop across the orifice by the formula

$$Q = K(\Delta P)^n$$

Verify graphically that the given orifice calibration data are correlated by this relationship, and determine the values of K and n that best fit the data.

(c) Suppose the orifice meter is mounted in a process line containing acetone and a reading $h = 17$ mm is obtained. Determine the volumetric, mass, and molar flow rates of acetone in the line.

35. Convert the following temperatures (parts a and b) and temperature intervals (parts c and d)

(a) $T = 20°C$ to °F, K, °R

(b) $T = -15°F$ to °C, °R, K

(c) $\Delta T = 20°C$ to °F, K, °R

(d) $\Delta T = 100°R$ to °F, °C, K

36. The volume of an iron bar varies with temperature according to the formula:

$$V(cm^3) = 100 + 3.55 \times 10^{-3}\ T(°C)$$

(a) What are the units of the constants 100 and 3.55×10^{-3}?

(b) Calculate the volume of the bar in cubic inches at a temperature $T = 212°F$.

(c) Derive an equation for $V(\text{in.}^3)$ as a function of $T(°F)$. Use it to check your answer to part b.

37. A temperature scale that never quite caught on was formulated by the Austrian chemist Johann Sebastian Farblunget. The reference points on this scale were $0°FB$ ($4°C$), the temperature below which Farblunget's postnasal drip began to bother him, and $1000°FB$ ($98°C$), the boiling point of beer.

(a) Derive an equation for $T(°C)$ in terms of $T(°FB)$. (See Example 3.5-1.)

(b) Derive an equation for $T(°FB)$ in terms of $T(°R)$.

(c) What is a temperature interval of 10.0 Farblunget degrees equivalent to in °C, K, °F, and °R?

38. A **thermocouple** is a temperature measurement device that consists of two dissimilar metal wires joined at one end.

A voltage generated at the metal junction is read on a potentiometer or millivoltmeter. When certain metals are used, the voltage varies linearly with the temperature at the junction of the two metals.

$$V(\text{mV}) = aT(°C) + b$$

An iron–constantan thermocouple (constantan is an alloy of copper and nickel) is calibrated by inserting its junction in boiling water and measuring a voltage $V = 5.27$ mV, and then inserting the junction in silver chloride at its melting point and measuring $V = 24.88$ mV.

(a) Derive the linear equation for $V(\text{mV})$ in terms of $T(°C)$. Then convert it to an equation for T in terms of V.

(b) If the thermocouple is mounted in a chemical reactor and the voltage is observed to go from 10.0 mV to 13.6 mV in 20 s, what is the average value of the rate of change of temperature, dT/dt, during the measurement period?

39. A thermostat control with dial markings from 0 to 100 is used to regulate a process temperature. A calibration plot of the temperature $T(°C)$ versus the dial setting, R, is a straight line on log paper that passes through the points ($R_1 = 20.0$, $T_1 = 220.0$) and ($R_2 = 40.0$, $T_2 = 150.0$).

(a) Derive a formula for $T(°C)$ in terms of R.

(b) Estimate the thermostat setting needed to obtain a temperature of $346°F$.

40. A gas contains 40 wt% oxygen, with the balance CO_2. The Bourdon gauge on a 400-ft³ tank of this gas reads 784 psi on a day when the barometric pressure is 760 mm Hg and the temperature is 37°C. The number of moles of gas in the tank can be calculated from the ideal gas law,

$$n(\text{kmol}) = PV/0.08206T$$

where P = absolute pressure in atm, V = volume in m³, and T = temperature in degrees kelvin.

(a) Convert the given formula to an equation for n(lb-moles) in terms of P(psig), V(ft³), and T(°F).

(b) How many pounds of oxygen are in the tank?

41. Streams of methane and air (79 mole% N_2, the balance O_2) are combined at the inlet of a combustion furnace preheater. The pressures of each stream are measured with open-end mercury manometers, the temperatures are measured with resistance thermometers, and the volumetric flow rates are measured with orifice meters.

Data:
Flowmeter 1: $V_1 = 947$ m³/h
Flowmeter 2: $V_2 = 195$ m³/min
Manometer 1: $h_1 = 232$ mm
Manometer 2: $h_2 = 156$ mm
Manometer 3: $h_3 = 74$ mm
Resistance thermometer 1: $r_1 = 26.159$ ohm
Resistance thermometer 2: $r_2 = 26.157$ ohm
Resistance thermometer 3: $r_3 = 44.789$ ohm
Atmospheric pressure: A closed-end mercury manometer reads $h = 29.76$ in.

The resistance thermometers were calibrated by measuring their resistances at the freezing and boiling points of water, with the following results:

$$T = 0°C: \quad r = 23.624 \text{ ohm}$$
$$T = 100°C: \quad r = 33.028 \text{ ohm}$$

A straight-line relationship between T and r may be assumed.

The relationship between the total molar flow rate of a gas and its volumetric flow rate is, to a good approximation, given by a form of the ideal gas law.

$$Q\left(\frac{\text{kmol}}{s}\right) = \frac{12.186P(\text{atm})V(\text{m}^3/\text{s})}{T(\text{K})}$$

where P is the *absolute* pressure of the gas.

(a) Derive the resistance thermometer calibration formula for $T(°C)$ in terms of $r(\text{ohm})$.

(b) Convert the given gas law expression to an expression for $Q(\text{kmol/min})$ in terms of $P(\text{mm Hg})$, $T(°C)$, and $V(\text{m}^3/\text{min})$.

(c) Calculate the temperatures and pressures at points 1, 2, and 3.

(d) Calculate the molar flow rate of the combined gas stream.

(e) Calculate the reading of flowmeter 3 in m^3/min.

(f) Calculate the total mass flow rate and the mass fraction of methane at point 3.

***42.** You are performing an experiment in which the concentration, C_A, of a reactive species is measured as a function of time, t, at several temperatures, T. At a fixed temperature, C_A varies with t according to the relation

$$1/C_A = 1/C_{A0} + kt \qquad (1)$$

where $C_A(\text{mol/liter})$ is the concentration of A at time $t(\text{min})$, $C_{A0}(\text{mol/liter})$ is the initial concentration of A, and $k(\text{liters/mol min})$ is the reaction **rate constant.** The rate constant in turn depends on temperature, according to the formula

$$k = k_0 \exp\left(-E/8.314T\right) \qquad (2)$$

where k_0 is a constant, $T(\text{K})$ is the reactor temperature, and $E(\text{J/mol})$ is the reaction **activation energy.**

Write a computer program that will carry out the following tasks:

(a) Read in M_A, the molecular weight of A, and N_T, the number of temperatures at which measurements were made.

(b) For the first temperature, read in the value of T in °C; the number of data points, N; and the concentrations and times (t_1, C_{A1}), (t_2, C_{A2}), ..., (t_N, C_{AN}), where the times are in minutes and the concentrations are in grams of A/liter.

(c) Convert the temperatures to degrees kelvin and the concentrations to mol A/liter.

(d) Use the method of least squares (Appendix A.1) in conjunction with Equation 1 to find the value of k that best fits the data. (*Hint:* First cast the equation in the form $y = kx + b$.) Store the values of k and T in arrays.

(e) Print out in a neat format the values of $T(\text{K})$, the converted concentrations (mol/liter) and times, and k.

(f) Repeat steps b through d for the other temperatures.

[*For extra credit:* Use the method of least squares again in conjunction with Equation 2 to determine the value of E that best fits the calculated (T, k) values. Again, start by casting Equation 2 in the form $y = ax + b$.]

* Computer problem.

It will be convenient to perform the least-squares slope calculation in a subroutine, since it must be done repeatedly.

Test your program on the following data.

$$M_A = 65.0 \text{ g/mol}$$

t(min)	$T = 94°C$ C_A(g/l)	$T = 110°C$ C_A(g/l)	$T = 127°C$ C_A(g/l)	$T = 142°C$ C_A(g/l)
10	8.1	3.5	1.5	0.72
20	4.3	1.8	0.76	0.36
30	3.0	1.2	0.50	0.24
40	2.2	0.92	0.38	0.18
50	1.8	0.73	0.30	0.15
60	1.5	0.61	0.25	0.12

PART TWO
MATERIAL BALANCES

Chapter Four

Fundamentals of Material Balances

Certain restrictions imposed by nature must be taken into account when designing a new process or analyzing an existing one. You cannot, for example, specify an input to a reactor of 1000 g of lead and an output of 2000 g of lead or gold or anything else. Similarly, if you know that 1500 lb_m of sulfur are contained in the coal burned each day in a power plant boiler, you do not have to analyze the ash and stack gases to know that 1500 lb_m of sulfur per day leave the furnace in one form or another.

The basis for both of these observations is the law of conservation of mass, which states that mass can neither be created nor destroyed. (We will not be concerned in this book with the almost infinitesimal conversions between mass and energy associated with chemical reactions.) Statements based on the law of conservation of mass such as "total mass of input = total mass of output," or "(lb_m sulfur/day)$_{in}$ = (lb_m sulfur/day)$_{out}$" are examples of **mass balances** or **material balances.** The design of a new process or analysis of an existing one is not complete until it is established that the inputs and outputs of the entire process and of each individual unit satisfy balance equations applied to each process material.

Part 2 of this book, which begins with this chapter, outlines procedures for writing material balances on individual process units and multiple-unit processes. In this chapter we present methods for organizing known information about process variables, setting up material balance equations, and solving these equations for unknown variables. In Chapters 5 and 6 we introduce various physical properties and laws that govern the behavior of process materials and indicate how these properties and laws are taken into account (as they must be) in formulating material balances.

4.1 PROCESS CLASSIFICATION

Chemical processes may be classified as **batch, continuous,** or **semibatch,** and as either **steady state** or **transient.** Before writing a material balance on a process system, you must know which of these categories the process falls into.

1. *Batch process.* The feed is charged into the system at the beginning of the process, and the products are removed all at once sometime later. No mass crosses the system boundaries between the time the feed is charged and the time the product is removed.

Example: Rapidly add reactants to a tank, and remove the products and un-consumed reactants sometime later when the system has come to equilibrium.

2. **Continuous process.** The inputs and outputs flow continuously throughout the duration of the process.
 Example: Pump a mixture of liquids into a distillation column at a constant rate, and steadily withdraw vapor and liquid streams from the top and bottom of the column.

3. **Semibatch process** (also referred to as **semicontinuous process**). Any process that is neither batch nor continuous.
 Examples: Allow the contents of a pressurized gas container to escape to the atmosphere; slowly blend several liquids in a tank from which nothing is being withdrawn.

If the values of all the variables in a process (i.e., all temperatures, pressures, volumes, flow rates, etc.) do not change with time, except possibly for minor fluctuations about constant mean values, the process is said to be operating at **steady state.** If any of the process variables change with time, **transient** or **unsteady-state** operation is said to exist. By their nature, batch and semibatch processes are unsteady-state operations (why?), whereas continuous processes may be either steady state or transient.

Batch processing is commonly used when relatively small quantities of a product are to be produced on any single occasion, while continuous processing is better suited to large production rates. Continuous processes are usually run as close to steady state as possible; unsteady-state (transient) conditions exist during the start-up of a process and following changes—intentional or otherwise—in process operating conditions.

TEST YOURSELF

Classify the following processes as batch, continuous, or semibatch, and transient or steady state.

1. A balloon is filled with air at a steady rate of 2 g/min.
2. A bottle of milk is taken from the refrigerator and left on the kitchen table.
3. Water is boiled in an open flask.
4. Carbon monoxide and steam are fed into a tubular reactor at a steady rate and react to form carbon dioxide and hydrogen. Products and unused reactants are withdrawn at the other end. The reactor contains air when the process is started up. The temperature of the reactor is constant, and the composition and flow rate of the entering reactant stream are also independent of time. Classify the process (a) initially, (b) after a long period of time has elapsed.

4.2 BALANCES

4.2a The General Balance Equation

Suppose methane is a component of both the input and output streams of a continuous process unit, and that in an effort to determine whether the unit is performing as designed, the mass flow rates of methane in both streams are measured and found to be different.

There are only four possible explanations for the observed difference between the measured flow rates.

1. Methane is leaking from the unit.
2. Methane is either being consumed as a reactant or being generated as a product within the unit.
3. Methane is accumulating in the unit—possibly adsorbing on the walls.
4. The measurements are wrong.

If the measurements are correct and there are no leaks, the other possibilities—generation or consumption in a reaction and accumulation within the process unit—are all that can account for a difference between the input and output flow rates.

A **balance** (or inventory) on a material in a system (a single process unit, a collection of units, or an entire process) may be written in the following general way:

Input	+	*generation*	−	*output*	−	*consumption*	=	*accumulation*
(enters		(produced		(leaves		(consumed		(buildup
through		within		through		within		within
system		system)		system		system)		system)
boundaries)				boundaries)				

$$(4.2\text{-}1)$$

This **general balance equation** may be written for any material that enters or leaves any process system; it can be applied to the total mass of this material or to any molecular or atomic species involved in the process. The meaning of each term of the equation is illustrated in the following example.

EXAMPLE 4.2-1 The General Balance Equation

Each year 50 000 people move into a city, 75 000 people move out, 22 000 are born and 19 000 die. Write a balance on the population of the city.

Solution

Let P denote people.

$$\text{Input} + \text{generation} - \text{output} - \text{consumption} = \text{accumulation}$$

$$50\,000\,\frac{P}{\text{yr}} + 22\,000\,\frac{P}{\text{yr}} - 75\,000\,\frac{P}{\text{yr}} - 19\,000\,\frac{P}{\text{yr}} = A\left(\frac{P}{\text{yr}}\right)$$

$$\Downarrow$$

$$A = -22\,000\,\frac{P}{\text{yr}}$$

Each year the city's population decreases by 22 000 people.

Two types of balances may be written:

1. *Differential balances,* or balances that indicate what is happening in a system at an instant of time. Each term of the balance equation is then a **rate** (rate of input, rate of generation, etc.) and has units of the balanced quantity unit divided by a time unit (people/yr, g SO_2/s, barrels/day). This is the type of balance usually applied to a continuous process. (See Example 4.2-1.)

2. *Integral balances,* or balances that describe what happens between two instants of time. Each term of the equation is then an **amount** of the balanced quantity and has the corresponding unit (people, g SO_2, barrels). This type of balance is usually applied to a batch process, with the two instants of time being the moment after the input takes place and the moment before the product is withdrawn.

We are concerned in this text primarily with differential balances applied to continuous steady-state systems and integral balances applied to batch systems between their initial and final states. In Chapter 11, we consider general balances on unsteady-state systems and show how integral and differential balances are related—in fact, how each can be derived from the other.

The generation and consumption terms in a balance equation equal zero if the balanced quantity is neither a reactant nor a reaction product. These terms are always zero if the balanced quantity is total mass, since (neglecting nuclear reactions) mass can neither be created nor destroyed.

4.2b Balances on Continuous Steady-State Processes

Suppose hydrogen appears as a component in the input and output streams of a continuous process at steady state. The accumulation term in a balance on hydrogen must equal zero; if it did not, the amount of hydrogen in the process system would necessarily change with time, and the process would then by definition not be at steady state. The same argument holds for every substance involved in the process, so that for continuous steady-state processes the general balance equation, Equation 4.2-1,

simplifies to

$$\boxed{\text{Input} + \text{generation} = \text{output} + \text{consumption}} \qquad (4.2\text{-}2)$$

EXAMPLE 4.2-2 Material Balances on a Continuous Distillation Process

One thousand kilograms per hour of a mixture of benzene (B) and toluene (T) that contains 50% benzene by mass are separated by distillation into two fractions. The mass flow rate of benzene in the top stream is 450 kg B/h, and that of toluene in the bottom stream is 475 kg T/h. The operation is at steady state. Write balances on benzene and toluene to calculate the unknown component flow rates in the output streams.

Solution

The process can be depicted schematically as shown here.

Since the process is at steady state there can be no buildup of anything in the system, so the accumulation term equals zero in all material balances. In addition, since no chemical reactions occur, there can be no nonzero generation or consumption terms. For all balances, Equation 4.2-2 therefore takes the simple form input = output.

Benzene Balance

$$500 \text{ kg B/h} = 450 \text{ kg B/h} + q_2$$
$$\Downarrow$$
$$\boxed{q_2 = 50 \text{ kg B/h}}$$

Toluene Balance

$$500 \text{ kg T/h} = q_1 + 475 \text{ kg T/h}$$
$$\Downarrow$$
$$\boxed{q_1 = 25 \text{ kg T/h}}$$

Check the calculation:

Total Mass Balance

$$1000 \text{ kg/h} = 450 + q_1 + q_2 + 475 \text{ (all kg/h)}$$
$$\Downarrow q_1 = 25, q_2 = 50$$
$$1000 \text{ kg/h} = 1000 \text{ kg/h} \quad \checkmark$$

4.2c Integral Balances on Batch Processes

Ammonia is produced from nitrogen and hydrogen in a batch reactor. At time $t = 0$ there are m_0 mol of NH_3 in the reactor, and at a later time t_f the reaction terminates and the contents of the reactor, which include m_f mol of ammonia, are withdrawn. Between t_0 and t_f no ammonia enters or leaves through the reactor boundaries, so the general balance equation (Equation 4.2-1) is simply generation = accumulation. Moreover, the quantity of ammonia that builds up (accumulates) in the reactor between t_0 and t_f is simply $m_f - m_0$, the final amount minus the initial amount.

The same reasoning may be applied to any substance participating in a batch process to obtain

Accumulation = final output − initial input (by definition)

= generation − consumption (from Equation 4.2-1)

Equating these two expressions for the accumulation yields

$$\boxed{\text{Initial input} + \text{generation} = \text{final output} + \text{consumption}} \qquad \textbf{(4.2-3)}$$

This equation is identical to Equation 4.2-2 for continuous steady-state processes, except that in this case the input and output terms denote the initial and final amounts of the balanced substance rather than flow rates of the balanced substance in continuous feed and product streams.

EXAMPLE 4.2-3 Balances on a Batch Mixing Process

Two methanol–water mixtures are contained in separate flasks. The first mixture contains 40 wt% methanol, and the second contains 70 wt% methanol. If 200 g of the first mixture are combined with 150 g of the second, what are the mass and composition of the product?

Solution

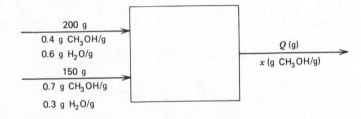

Observe that the input and output streams shown on the chart denote the initial and final states for this batch process. Since no reactions are involved, the generation and consumption terms of Equation 4.2-3 may be omitted so that all balances have the simple form "input = output."

Total Mass Balance

$$200 \text{ g} + 150 \text{ g} = Q(\text{g})$$
$$\Downarrow$$
$$\boxed{Q = 350 \text{ g}}$$

Methanol Balance

$$\frac{200 \text{ g}}{} \left| \frac{0.400 \text{ g CH}_3\text{OH}}{\text{g}} \right. + \frac{150 \text{ g}}{} \left| \frac{0.700 \text{ g CH}_3\text{OH}}{\text{g}} \right. = \frac{Q(\text{g})}{} \left| \frac{x(\text{g CH}_3\text{OH})}{\text{g}} \right.$$
$$\Downarrow Q = 350$$
$$\boxed{x = 0.529 \text{ g CH}_3\text{OH/g}}$$

We now know everything about the product, including the mass fraction of water (what is it?) A water balance serves only to check the calculation.

Water Balance

$$\text{Input} = \text{output}$$
$$(200)(0.600) + (150)(0.300) = (350)(1 - 0.529) \quad \text{(Verify!)}$$
$$\Downarrow$$
$$165 \text{ g H}_2\text{O} = 165 \text{ g H}_2\text{O} \quad \checkmark$$

4.2d Integral Balances on Semibatch and Continuous Processes

Integral balances can also be written for semibatch and continuous processes. The procedure is to write a differential balance on the system and then to integrate it between two instants of time. (A general discussion of the procedure is presented in Chapter 11.) In most cases the required calculations are more complex than those we have seen so far; however, some problems of this type are relatively straightforward, such as the one in the next example.

EXAMPLE 4.2-4 Integral Balance on a Semibatch Process

Air is bubbled through a drum of liquid hexane at a rate of 0.100 kmol/min. The gas stream leaving the drum contains 10.0 mole% hexane vapor. Air may be considered insoluble in liquid hexane. Use an integral balance to estimate the time required to vaporize 10.0 m^3 of hexane.

Solution

We begin with a differential balance on air (input = output).

$$0.100 \text{ kmol air/min} = 0.900Q \Rightarrow Q = 0.111 \text{ kmol/min}$$

A hexane balance takes the form accumulation = − output (verify). The accumulation term in the integral balance is the total amount of hexane lost from the system, which is

$$\Delta n = \frac{-10.0 \text{ m}^3}{} \frac{0.659 \times 1000 \text{ kg}}{\text{m}^3} \frac{1}{86.2 \text{ kg}} \frac{\text{kmol}}{} = -76.45 \text{ kmol}$$

The output term in the balance is obtained by integrating the differential output rate $(0.100Q \text{ kmol C}_6\text{H}_6/\text{min})$ from the initial process time $(t = 0)$ to the final time (t_f), which is the quantity we seek to calculate. The balance (accumulation = − output) is

$$-76.45 \text{ kmol} = -\int_0^{t_f} 0.100Q \, dt = -0.100Qt_f$$

$$\Downarrow Q = 0.111$$

$$t_f = \boxed{6880 \text{ min}}$$

TEST YOURSELF

Balances are to be written for each of the quantities listed below for a continuous process. For each case, state the conditions under which the balance equation takes the simple form "input = output." (The solutions to the first two parts are given as illustrations.)

1. Total mass. (Steady state)
2. Mass of species A. (Steady-state, A is nonreactive.)
3. Total moles.
4. Moles of species A.
5. Volume. (The answer provides an indication of why volumes are almost always converted to masses or moles before balances are written.)

CREATIVITY EXERCISE

Sulfur dioxide is contained in the feed and effluent streams of a chemical reactor, but it is neither a reactant nor a product. The volumetric flow rates of both streams (liters/min) are measured with rotameters, and the concentrations of SO_2 in both streams (mol/liter) are determined with a gas chromatograph. The molar flow rate of SO_2 in the reactor effluent (determined as the product of volumetric flow rate and concentration) is 20% lower than the molar flow rate of SO_2 in the feed. Think of as many possible explanations for the discrepancy as you can.

4.3 MATERIAL BALANCE CALCULATIONS

All material balance problems are variations on a single theme: Given values of some input and output stream variables, calculate values of others. Solving such problems requires the derivation and solution of equations for unknown stream variables. Solving the derived equations is usually a matter of simple algebra, but deriving them from a description of a process and a collection of process data may present considerable difficulties. It may not be at all obvious from the problem statement just what is known and what is required, for example, and it is not uncommon to find students (particularly on quizzes) scratching their heads and staring vacantly for an hour at a problem that should take five minutes to solve completely.

In this section we outline a procedure for reducing a description of a process to a set of equations that can be solved for unknown process variables. The approach to be outlined is not the only conceivable way to attack material balance problems, but it always works and it is effective keeping the time spent on vacant staring and head scratching to a minimum.

4.3a Flowcharts

In this book and in years to come, you will be confronted with prose of the following type.

The catalytic dehydrogenation of propane is carried out in a continuous packed bed reactor. One thousand pounds per hour of pure propane are fed to a preheater where they are heated to a temperature of 670°C before they pass into the reactor. The reactor effluent gas, which includes propane, propylene, methane, and hydrogen, is cooled from 800°C to 110°C and fed to an absorption tower where the propane and propylene are dissolved in oil. The oil then goes to a stripping tower in which it is heated, releasing the dissolved gases; these gases are recompressed and sent to a high-pressure distillation column in which the propane and propylene are separated. The propane stream is recycled back to join the feed to the reactor preheater. The product stream from the distillation column contains 98% propylene, and the recycle stream is 97% propane. The stripped oil is recycled to the absorption tower.

When you are presented with a description of a process like this one and are then asked to determine something about the process, it is essential to organize the given information in a way that is convenient for subsequent calculations. The best way to do this is to draw a **flowchart** of the process, using boxes or other symbols to represent process units (reactors, mixers, separation units, etc.) and lines with arrows to represent inputs and outputs.

For example, suppose a gas containing N_2 and O_2 is combined with propane in a batch combustion chamber in which some (but not all) of the O_2 and C_3H_8 react to form CO_2 and H_2O, and the product is then cooled, condensing the water. The flowchart of this two-unit process might appear as shown in Figure 4.3-1.

FIGURE 4.3-1

Flowchart of a combustion-condensation process

When used properly, the flowchart of a process can help immeasurably to get material balance calculations started and to keep them moving. To accomplish this, the chart must be fully *labeled* when it is first drawn, with values of known process variables and symbols for unknown variables being written for each input and output stream. Thereafter, the chart functions as a scoreboard for the problem solution: As each unknown variable is determined its value is filled in, so that the chart provides a continuous record of where the solution stands and what must still be done.

Several suggestions follow for labeling a flowchart to get the greatest possible benefit from it in material balance calculations.

1. *Write the values and units of all known stream variables at the locations of the streams on the chart.* For example, a stream containing 21 mole% O_2 and 79% N_2 at 320°C and 1.4 atm flowing at a rate of 400 mol/h might be labeled

$$400 \text{ mol/h}$$

0.21 mol O_2/mol
0.79 mol N_2/mol
$T = 320°C, P = 1.4$ atm

When you have done this for every stream on the chart, you have a summary of the known information about the process, each item being conveniently associated with the part of the process to which it relates.

The stream variables of primary interest in material balance problems are those that indicate how much of each component is present in the stream (for a

batch process) or the flow rate of each component (for a continuous process). This information can be given in two ways: as the total amount or flow rate of the stream and the fractions of each component, or directly as the amount or flow rate of each component.

100 kmol/min

——————————————→ ⇔ ——————————————→

60 kmol N_2/min 0.6 kmol N_2/kmol
40 kmol O_2/min 0.4 kmol O_2/kmol

10 lb_m mixture

——————————————→ ⇔ ——————————————→

3.0 lb_m CH_4 0.3 lb_m CH_4/lb_m
4.0 lb_m C_2H_4 0.4 lb_m C_2H_4/lb_m
3.0 lb_m C_2H_6 0.3 lb_m C_2H_6/lb_m

Once you have labeled a stream one way, it is easy to calculate the quantities that correspond to the alternative way of labeling. (Verify this for the two examples just given.)

2. *Assign algebraic symbols to unknown stream variables* (such as Q kg solution/min, x lb_m N_2/lb_m, n kmol C_3H_8, etc.) *and write these variable names and their associated units on the chart.* For example, if you did not know the flow rate of the stream described in the illustration of step 1, you might label the stream

Q(mol/h)

————————————————————→

0.21 mol O_2/mol
0.79 mol N_2/mol
$T = 320°C$, $P = 1.4$ atm

while if the flow rate were known and the mole fractions were not, the stream might be labeled

400 mol/h

————————————————————→

x (mol O_2/mol)
$1 - x$ (mol N_2/mol)
$T = 320°C$, $P = 1.4$ atm

You may ultimately have to derive and solve an equation for each unknown that appears on the chart, and it is therefore to your advantage to keep the number of labeled unknowns to a minimum. When labeling component mass or mole fractions of a stream, for example, variable names need only be assigned to all but one fraction, since the last one must be 1 minus the sum of the others. If you are given that the mass of stream 1 is half that of stream 2, label the masses of these streams Q and $2Q$ rather than Q_1 and Q_2; if you know that there is three times as much nitrogen (by mass) in a stream as oxygen, label the mass fractions of O_2 and N_2 x(g O_2/g) and $3x$(g N_2/g) rather than x and y.

If a volumetric flow rate of a stream is given, it is generally useful to label the mass or molar flow rate of this stream or to calculate it directly, since balances are not normally written on volumetric quantities.

EXAMPLE 4.3-1 Flowchart of an Air Humidification and Oxygenation Process

An experiment on the growth rate of certain organisms requires the establishment of an environment of humid air enriched in oxygen. Three input streams are fed into an evaporation chamber to produce an output stream with the desired composition.

A: Liquid water, fed at a rate of 20.0 cm³/min.
B: Air (21 mole% O_2, the balance N_2).
C: Pure oxygen, with a molar flow rate one fifth of the molar flow rate of stream B.

The output gas is analyzed and is found to contain 1.5 mole% water. Draw and label a flowchart of the process, and calculate all unknown stream variables.

Solution

Notes on the labeling:

1. Since the one known flow rate (20 cm³ H_2O/min) is given on a per minute basis, it is most convenient to label all stream flow rates on this basis.
2. Once the variable name Q is chosen for the air flow rate, the given information about the ratio of the air and O_2 flow rates may be used to label the O_2 flow rate $0.2Q$.
3. The mole fractions of the components of any stream must add up to 1. Since the mole fraction of H_2O in the outlet stream is known to be 0.015, once the mole fraction of O_2 is labeled x, that of N_2 must be $1 - (x + 0.015) = (0.985 - x)$ mol N_2/mol.

The quantity W may be calculated from the given volumetric flow rate and the density of liquid water:

$$W = \frac{200 \text{ cm}^3 \text{ H}_2\text{O}}{\text{min}} \left| \frac{1.00 \text{ g H}_2\text{O}}{\text{cm}^3} \right| \frac{1 \text{ mol}}{18.02 \text{ g}} \Rightarrow \boxed{W = 1.11 \frac{\text{mol H}_2\text{O}}{\text{min}}}$$

The three remaining unknowns (Q, F, and x) may be determined from balances, all of which have the simple form input = output for this nonreactive steady-state process. The balances are easily written by referring to the flowchart.

H_2O Balance

$$W\left(\frac{\text{mol } H_2O}{\text{min}}\right) = F\frac{\text{mol}}{\text{min}}\left|\frac{0.015 \text{ mol } H_2O}{\text{mol}}\right.$$

$$\Downarrow W = 1.11$$

$$\boxed{F = 74.1 \frac{\text{mol}}{\text{min}}}$$

Total Mole Balance

$$(0.2Q + Q + W)\frac{\text{mol}}{\text{min}} = F\frac{\text{mol}}{\text{min}}$$

$$\Downarrow \begin{array}{l} W = 1.11 \\ F = 74.1 \end{array}$$

$$\boxed{Q = 60.8 \frac{\text{mol}}{\text{min}}}$$

N_2 Balance

$$\frac{Q \text{ mol}}{\text{min}}\left|\frac{0.79 \text{ mol } N_2}{\text{mol}}\right. = \frac{F \text{ mol}}{\text{min}}\left|\frac{(0.985 - x) \text{ mol } N_2}{\text{mol}}\right.$$

$$\Downarrow$$

$$0.79Q = F(0.985 - x)$$

$$\Downarrow \begin{array}{l} Q = 60.8 \\ F = 74.1 \end{array}$$

$$\boxed{x = 0.337 \text{ mol } O_2/\text{mol}}$$

TEST YOURSELF

Several labeled process streams are shown below. Calculate the indicated quantities in terms of the labeled stream variables. The solution of the first problem is given as an illustration.

1. 100 lb-moles

 \longrightarrow Calculate q (lb-moles CH_4)

 0.300 lb-mole CH_4/lb-mole m (lb$_m$ C_2H_4)

 0.400 lb-mole C_2H_4/lb-mole

 0.300 lb-mole C_2H_6/lb-mole

Solution

$$q = (0.300)(100) = 30.0 \text{ lb-moles CH}_4$$

$$m = \frac{(0.400)(100)\text{lb-moles C}_2\text{H}_4}{} \frac{28.0 \text{ lb}_m}{\text{lb-mole}} = 1120 \text{ lb}_m \text{ C}_2\text{H}_4$$

2. 250 kg/h

\longrightarrow Calculate W (kg C_7H_8/min)
in terms of x

x kg C_6H_6/kg
$1 - x$ kg C_7H_8/kg

3. 75 ml CCl_4 (liquid)

\longrightarrow Calculate R (mol CCl_4)
(specific gravity = 1.595)

4. \longrightarrow Calculate B (kg total/s), q (kg CO/s),
and x (kg CO_2/kg total) in terms of Q

50 kg H_2O/s
$\begin{cases} Q \text{ kg dry gas/s} \\ 0.25 \text{ kg CO/kg dry gas} \\ 0.75 \text{ kg CO}_2\text{/kg dry gas} \end{cases}$

4.3b Flowchart Scaling and Basis of Calculation

Suppose a kilogram of benzene is mixed with a kilogram of toluene. The output from this simple process is obviously 2 kg of a mixture that is 50% benzene by mass.

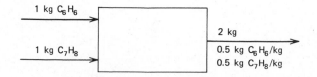

The process depicted by this flowchart is said to be **balanced,** since material balances on both system components—C_6H_6 and C_7H_8—are satisfied. [1 kg in $= (2 \times 0.5)$ kg out in both cases.]

Observe now that the masses (but not the mass fractions) of all streams could be multiplied by a common factor and the process would remain balanced; moreover, the stream masses could be changed to mass flow rates, and the mass units of all stream variables (including the mass fractions) could be changed from kg to g or lb$_m$ or any other mass unit, and still the process would be balanced.

The procedure of changing the values of all stream amounts or flow rates by a proportional amount while leaving the stream compositions unchanged is referred to as **scaling** the flowchart—**scaling up** if the final stream quantities are larger than the original quantities, **scaling down** if they are smaller.

Suppose you have balanced a process and the amount or flow rate of one of the process streams is n_1. You can scale the flowchart to make the amount or flow rate of this stream n_2 by multiplying all stream amounts or flow rates by the ratio n_2/n_1. You cannot, however, scale masses or mass flow rates to molar quantities or vice versa by simple multiplication; conversions of this type must be carried out using the methods of Section 3.3b.

EXAMPLE 4.3-2 Scale-up of a Separation Process Flowchart

A 60–40 mixture (by moles) of A and B is separated into two fractions. A flowchart of the process is shown here.

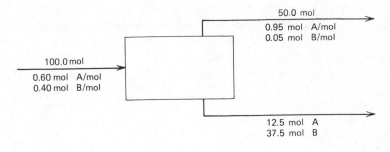

It is desired to achieve the same separation with a continuous feed of 1250 lb-moles/h. Scale the flowchart accordingly.

Solution

The scale factor is

$$\frac{1250 \text{ lb-moles/h}}{100 \text{ mol}} = 12.5 \frac{\text{lb-moles/h}}{\text{mol}}$$

The masses of all streams in the batch process are converted to flow rates as follows:

Feed: $\dfrac{100 \text{ mol}}{} \left| \dfrac{12.5 \text{ lb-moles/h}}{\text{mol}} = 1250 \dfrac{\text{lb-moles}}{\text{h}} \right.$ (as specified)

Top product stream: $(50.0)(12.5) = 625$ lb-moles/h

Bottom product stream: $(12.5)(12.5) = 156$ lb-moles A/h

$(37.5)(12.5) = 469$ lb-moles B/h

The units of the mole fractions in the top product stream may be changed from mol/mol to lb-mole/lb-mole, but their values remain the same. The flowchart for the scaled-up process is shown here.

Since a balanced process can always be scaled, material balance calculations can be performed on the basis of any convenient set of stream amounts or flow rates, and the results can afterward be scaled to any desired extent. The first step in balancing a process is to choose an amount (mass or moles) or a mass or molar flow rate of a stream or a stream component as a **basis of calculation;** all unknown variables are then determined to be consistent with this basis.

If a stream amount or flow rate is given in a problem statement, it is usually most convenient to use this quantity as a basis of calculation; all subsequently calculated stream quantities are then automatically scaled to the correct values for the process. *If no stream amounts or flow rates are known, assume one.* In this case, choose an amount of a stream with a known composition: If mass fractions are known, choose a total mass as a basis; if mole fractions are known, choose a total number of moles.

TEST YOURSELF

1. What is a balanced process? How do you scale a flowchart? What is a basis of calculation?

2. The processes shown below were balanced using the indicated bases of calculation. Scale as directed, and draw flowcharts for the scaled processes.
 (a) Mix C_2H_6 with air. *Basis of calculation: 100 mol C_2H_6.*

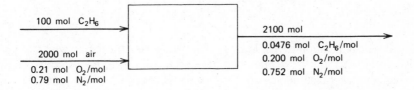

Scale up to a feed of 1000 kmol C_2H_6/h.

 (b) Distill a benzene–toluene mixture. *Basis: 1 g feed.*

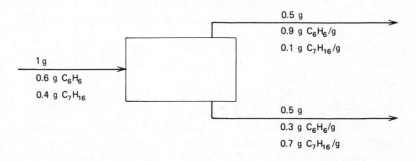

Scale up to a flow rate of 100 lb_m/min of the upper product stream.

4.3c Balancing a Process

Suppose that 3 kg/min of benzene and 1 kg/min of toluene are mixed. The process flowchart might be drawn and labeled as follows.

There are two unknown quantities—Q and x—associated with the process, so two equations are needed to calculate them.

Material balance equations for this nonreactive process all have the simple form input = output. Three possible balances can be written—on total mass, benzene, and

toluene—any two of which provide the equations needed to determine Q and x. For example,

Total Mass Balance: $3 \dfrac{\text{kg}}{\text{min}} + 1 \dfrac{\text{kg}}{\text{min}} = Q \Rightarrow \boxed{Q = 4 \text{ kg/min}}$

Benzene Balance: $3 \dfrac{\text{kg C}_6\text{H}_6}{\text{min}} = \dfrac{Q \text{ kg}}{\text{min}} \left| \dfrac{x \text{ kg C}_6\text{H}_6}{\text{kg}} \right.$

$$\Downarrow Q = 4$$

$$\boxed{x = 0.75 \text{ kg C}_6\text{H}_6/\text{kg}}$$

A logical question to ask at this point is how far you can go with this procedure—if one of the input flow rates had also been unknown, for example, could another balance (on toluene) have been written to solve for it? Other points to consider are which balances to use when a choice exists and the order in which these balances should be taken.

The answers to these questions are not at all obvious when chemical reactions are involved in the process, and we will temporarily postpone consideration of this subject. The following rules apply to nonreactive processes.

1. *The maximum number of independent equations that can be derived by writing balances on a nonreactive system equals the number of chemical species in the input and output streams.*

 In the given example two substances—benzene and toluene—make up the input and output streams of the process; you can write mass or mole balances on benzene and toluene and a total mass or mole balance, but only two of these three equations are independent—writing the third accomplishes nothing. (If you wrote all three equations in an effort to determine three unknown variables, you would probably go through an elaborate algebraic exercise to prove that $1 = 1$ or something equally uninformative.)

2. *Write balances first that involve the fewest unknown variables.*

 In the example, a total mass balance involves only one unknown, Q, while benzene and toluene balances each involve both Q and x. By writing first a total balance and then a benzene balance, we were able to solve first one equation in one unknown, then a second equation, also in one unknown. If we had instead written benzene and toluene balances, we would have had to solve two simultaneous equations in two unknowns; the same answer would have been obtained, but with greater effort.

EXAMPLE 4.3-3 Balances on a Mixing Unit

An aqueous solution of sodium hydroxide contains 20% NaOH by mass. It is desired to produce an 8% NaOH solution by diluting a stream of the 20% solution with a stream of pure water.

1. Calculate the ratios (g H_2O/g feed solution) and (g product solution/g feed solution).

2. Determine the feed rates of 20% solution and diluting water needed to produce 2310 lb_m/min of the 8% solution.

Solution

We could take a basis of 2310 lb_m product/min, but for illustrative purposes and to have neater numbers to work with let us choose a different basis and scale the final results.

1. **Basis: 100 g Feed Solution**

 Draw and label the flowchart, remembering that the amount of the product stream is now unknown.

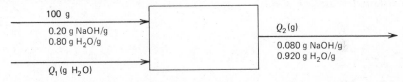

100 g
0.20 g NaOH/g
0.80 g H_2O/g

Q_1 (g H_2O)

Q_2(g)
0.080 g NaOH/g
0.920 g H_2O/g

(Since the known stream amount is given in grams, it is convenient to label all unknown amounts with this unit.)

There are two unknowns—Q_1 and Q_2—and since there are two substances—NaOH and H_2O—in the input and output streams, two balances may be written to solve for them. A total mass balance and a water balance each involve both unknowns, but an NaOH balance involves only one.

NaOH Balance

$$(\text{g NaOH})_{in} = (\text{g NaOH})_{out}$$
$$\Downarrow$$
$$(0.20)(100 \text{ g}) = 0.080Q_2 \Rightarrow \boxed{Q_2 = 250 \text{ g}}$$

It is a good practice to write calculated variable values on the flowchart as soon as they are known for ease of use in later calculations; at this point, 250 would therefore be written in place of Q_2 on the chart.

Total Mass Balance

$$100 \text{ g} + Q_1 = Q_2$$
$$\Downarrow Q_2 = 250 \text{ g}$$
$$\boxed{Q_1 = 150 \text{ g } H_2O}$$

The desired ratios may now be calculated:

$$\frac{Q_1(\text{g } H_2O)}{100 \text{ g feed solution}} \xrightarrow{\;Q_1 = 150\;} \boxed{1.5 \; \frac{\text{g } H_2O}{\text{g feed solution}}}$$

$$\frac{Q_2(\text{g product})}{100 \text{ g feed solution}} \xrightarrow{\;Q_2 = 250\;} \boxed{2.5 \; \frac{\text{g product}}{\text{g feed solution}}}$$

2. The scale factor is obtained as the true flow rate of the product stream divided by the rate calculated on the assumed basis.

$$\frac{2310 \text{ lb}_m \text{ product/min}}{250 \text{ g product}} = 9.24 \frac{\text{lb}_m/\text{min}}{\text{g}}$$

Feed Solution Flow Rate

$$\frac{100 \text{ g}}{} \left| \frac{9.24 \text{ lb}_m/\text{min}}{\text{g}} \right. = \boxed{924 \frac{\text{lb}_m \text{ feed solution}}{\text{min}}}$$

Dilution Water Flow Rate

$$\frac{150 \text{ g}}{} \left| \frac{9.24 \text{ lb}_m/\text{min}}{\text{g}} \right. = \boxed{1386 \frac{\text{lb}_m \text{ H}_2\text{O}}{\text{min}}}$$

Check: $(924 + 1386) \text{ lb}_m/\text{min} = 2310 \text{ lb}_m/\text{min}$

TEST YOURSELF

1. Prove that the following flowchart depicts a balanced process by writing three balances.

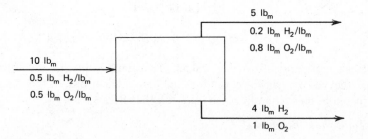

2. Indicate the balances you would write and the order in which you would write them to solve for the unknown stream variables in the following process.

4.3d Problem Bookkeeping

Everyone who has done material balance calculations has had the frustrating experi-
ence of spending a long time deriving and solving equations for unknown process
variables, only to discover that not enough information is available to solve the
problem. A properly drawn and labeled flowchart can be used to determine—*before
any calculations are performed*—whether a given problem can be solved from the given
information.

We refer to the procedure as **problem bookkeeping.** It consists of counting the
number of unknown variables on the chart, and then counting the number of inde-
pendent equations relating them. If the numbers are equal, the problem can in prin-
ciple be solved. If there are more unknowns than equations, then either the problem
is underspecified or relations have been forgotten. Trying to solve the problem would
then be a waste of time: You must either come up with the needed number of addi-
tional relations or better define the problem. If there are more equations than un-
knowns, the flowchart is probably incompletely labeled; otherwise, the problem is
overspecified, and again there is little point wasting time trying to solve it.

The relations that can be used to relate process stream variables include the
following:

1. *Material balances.* For a nonreactive process, up to N material balances may
 be written, where N is the number of independent species involved in the process.
 For a reactive process, the maximum number of material balances is reduced by
 the number of independent chemical reactions among the process species.

2. *An energy balance* (Chapters 7 through 9). If the amount of energy exchanged
 between the system and its surroundings is specified or if it is one of the unknown
 process variables, an energy balance provides a relationship.

3. *Process specifications.* The problem statement may tell you how several pro-
 cess variables are related. For example, you may be told the fraction of some
 entering stream component that appears in one of the product streams. This in-
 formation can be translated into an equation involving the amounts of the com-
 ponent in both streams.

4. *Physical properties and laws.* Two of the unknown variables may be the mass
 and volume of a stream material, in which case a tabulated specific gravity (for
 liquids and solids) or a gas law (Chapter 5) may provide a relationship between
 the variables. In other instances, saturation or equilibrium conditions for one or
 more of the streams (Chapter 6) may provide a needed relation.

5. *Physical constraints.* For example, if the mole fractions of the component
 streams are separately labeled (e.g., x, y, and z), then a relation among them is
 that their sum equals 1. (If instead of z you label the last fraction $1 - x - y$, then
 you have one less variable and one less relation to worry about, which is why we
 have suggested this procedure.)

A more formal procedure for problem bookkeeping termed **degree-of-freedom analysis**
is discussed in Chapter 10.

EXAMPLE 4.3-4 Problem Bookkeeping

A stream of humid air enters a condenser, in which 95% of the water vapor in the air is condensed. The flow rate of the condensate is measured and is found to be 225 liters/h. Dry air may be taken to contain 21 mole% oxygen, with the balance nitrogen. Calculate the flow rate of the gas stream leaving the condenser and the mole fractions of oxygen, nitrogen, and water in this stream.

Solution

Basis: 225 liters/h Condensate

Both oxygen and nitrogen enter the process in the feed stream and exit in the same proportion in the gas stream leaving the condenser. They may therefore be considered a single species (dry air) and could have been shown as such on the flowchart.

We may now do the problem bookkeeping. There are six unknowns on the chart—n_1 through n_6. We are allowed up to three material balances—one for each species. We must therefore find three additional relations to solve for all unknowns. One is the relationship between the volumetric and molar flow rates of the condensate: We can determine n_3 from the given volumetric flow rate and the known specific gravity and molecular weight of liquid water. A second is the given fact that 95% of the water is condensed. This provides a relationship between n_3 and n_2 ($n_3 = 0.95n_2$).

However, no information in the problem statement provides a basis for obtaining a sixth relation. The problem is therefore underspecified, and there is no point in attempting to solve it. Without the flowchart, it would have been extremely difficult to see this, and a great deal of time could have been wasted in a futile effort.

Suppose now that we had been given an additional piece of information—for example, that the entering air contains 10 mole% water. The flowchart would then appear as follows:

The bookkeeping tells us that there are five unknowns and that we have five relations to solve for them (three balances, the relationship between V_2 and n_2, and the fractional condensation). The problem is therefore solvable in principle. We may then lay out the solution, still before proceeding to any numerical calculations:

1. Solve for n_2 from V_2 and the specific gravity and molecular weight of liquid water.
2. Write the equation that expresses the given fractional condensation $[n_2 = 0.95(0.1n_1)]$. Solve for n_1.
3. Write balances on oxygen, nitrogen, and water. Solve for n_3, n_4, and n_5, respectively.
4. Calculate the flow rate of the exit gas as the sum of the component flow rates and the component mole fractions as n_i/n_{tot}.

The details of the calculations are left as an exercise. You might also try to treat air as a single species on the flowchart and see how the problem bookkeeping would be affected.

4.3e Outline of a Procedure for Material Balance Calculations

The techniques introduced in the previous sections and several additional procedural suggestions are summarized in the outline below. Once again, we emphasize that this is not the only possible method to use for solving material balance problems; your course instructor may suggest another approach, and you will develop your own methods as you gain experience. The important thing is to adopt and become accustomed to a systematic procedure so that you do not have to reinvent the wheel on every new problem you encounter.

Given a description of a process, the values of several process variables, and a list of quantities to be determined:

1. *Draw a flowchart, and fill in all given variable values.*
2. *Choose as a basis of calculation an amount or flow rate of one of the process streams.* If one such quantity is given, it is usually most convenient to use it as the basis of calculation; all subsequently calculated quantities will then be correctly scaled. If several amounts or flow rates are known, always use them as a basis. If no amount or flow rate is specified, assume an amount of a stream with a known composition and write the assumed quantity on the flowchart.
3. *Label unknown stream variables on the chart,* particularly mass or molar flow rates and mass or mole fractions of stream components. Try to incorporate given relationships between unknown quantities in the labeling.
4. *Do the problem bookkeeping.* Count unknowns and relations among them. If the numbers are equal, lay out the problem solution (see Example 4.3-4); if not, figure out what's wrong (the chart is incompletely labeled, an additional relation exists that was not counted,)
5. *Convert known stream volumes or volumetric flow rates to mass or molar quantities, using tabulated densities* (Section 3.2) *or gas laws* (Chapter 5).

6. *If you are given mixed mass and mole units for a stream* (e.g., a total mass flow rate and component mole fractions or conversely) *convert all quantities to one basis or the other using the methods of Section 3.3b.*

7. *If any information is given in the problem statement that has not been used in labeling the flowchart, translate it into equations in the variables defined in step 3.*

8. *Write material balance equations.* If no reactions occur and there are N species present, you may write at most N balances, either on each species or on total mass or moles and all but one species. *Write balances in an order such that those that involve the fewest unknowns are written first.* At this point you should have as many equations as unknowns; if not, either the problem is not well defined or a relationship between the variables has been forgotten.

9. *Solve the equations derived in steps 7 and 8 for the unknown quantities to be determined. When the value of an unknown has been calculated, write it on the flowchart immediately* and then substitute this value in any equations in which the unknown appears.

10. *If a stream quantity or flow rate Q_g was given in the problem statement and another value Q_c was either taken as a basis or calculated for this stream, scale the balanced process by the ratio Q_g/Q_c to obtain the final result.*

The following example illustrates this procedure.

EXAMPLE 4.3-3 Material Balances on a Distillation Column

A mixture containing 45% benzene (B) and 55% toluene (T) by mass is fed to a distillation column. An overhead stream of 95 wt% B is produced, and 8% of the benzene fed to the column leaves in the bottom stream. The feed rate is 2000 kg/h. Determine the overhead flow rate and the mass flow rates of benzene and toluene in the bottom stream.

Solution

Basis: Given Feed Rate

The labeled flowchart is as follows.

D kg/h
0.95 kg B/kg
0.05 kg T/kg

2000 kg/h
0.45 kg B/kg
0.55 kg T/kg

Contains 8% of the B in the feed

w_B kg B/h
w_T kg T/h

There are three unknowns on the chart—D, w_B, and w_T—and therefore three equations are needed. We are entitled to write only two material balances since two species are involved in the process; the third equation must therefore come from additional given information (specifically, the relative amount of benzene in the bottom stream.) The latter relation is

B in bottom stream = 0.080 (B in feed)

$$\Downarrow$$

$$w_B(\text{kg B/h}) = 0.080[(0.45)(2000) \text{ kg B/h}]$$

$$\Downarrow$$

$$\boxed{w_B = 72 \text{ kg B/h}} \qquad \text{(Write 72 in place of } w_B \text{ on the chart)}$$

Toluene and total mass balances each involve two unknowns, D and w_1, but a benzene balance involves only one, D.

Benzene Balance

$$(0.45)(2000) \frac{\text{kg B}}{\text{h}} = 0.95D + w_B$$

$$\Downarrow w_B = 72 \text{ kg B/h}$$

$$\boxed{D = 870 \text{ kg/h}} \qquad \text{(Write it on the chart)}$$

Total Mass Balance (A toluene balance could be used equally well)

$$2000 \frac{\text{kg}}{\text{h}} = D + w_B + w_T$$

$$\Downarrow \begin{array}{l} D = 870 \text{ kg/h} \\ w_B = 72 \text{ kg/h} \end{array}$$

$$\boxed{w_T = 1060 \text{ kg } T/\text{h}}$$

If you wanted to, you could now easily calculate the total flow rate of the bottom stream and the mass fractions of benzene and toluene in this stream. (What are they?)

4.4 BALANCES ON MULTIPLE-UNIT PROCESSES

Industrial chemical processes rarely involve just one unit. One or more chemical reactors are often present, as are units for mixing reactants, blending products, heating and cooling process streams, and separating products from each other and from unconsumed reactants. The process engineer's job involves keeping track of materials that flow to and from the overall process and also of materials that flow from one unit to another within the process.

In the preceding sections, we referred rather loosely to "the system", as in the statement "At steady state the rate at which benzene enters the system equals the rate at which it leaves." Not much was said about what "the system" was. Up to now, however, not much needed to be said, since we have considered only processes involving a single unit—a mixer, or a reactor, or a distillation column—and that unit necessarily constituted the system.

When several units are involved in a process, the definition of the system is not so obvious. A system may be defined as any portion of the process one chooses to consider: the entire process, an interconnected combination of some of the process units, a single unit, or a point at which two or more process streams come together or one stream splits into branches. Determining all unknown stream amounts (or flow rates) and compositions for a process usually requires choosing several such systems and successively writing balances on each of them.

A flowchart for a two-unit process might appear as shown in Figure 4.4-1. Five boundaries drawn about portions of the process shown in this diagram define systems on which balances may be written. Boundary A contains the entire process; the system defined by this boundary has all feeds to the process as inputs and all product streams as outputs. Boundary B contains a feed stream mixing point, boundary C contains the first process unit, boundary D contains a stream splitting point, and boundary E contains the second unit. Balances taken on system A (the entire process) are referred to as **overall balances.**

FIGURE 4.4-1

Flowchart of a two-unit process. Dashed lines denote boundaries of systems about which balances can be written

The procedure for solving material balance problems is essentially that outlined in Section 4.3, whether single-unit or multiple-unit processes are involved, except that with multiple-unit processes you may have to isolate and write balances on several subsystems to obtain enough equations to determine all unknown stream variables. It is generally advisable to take overall balances first to determine unknown process feed and product stream variables; thereafter, balances on various process subsystems may be used to determine interior stream variables.

When choosing subsystems on which to write balances, you can simplify the calcula-tions by choosing boundaries that intersect streams containing as few unknown variables as possible.

EXAMPLE 4.4-1 Two-Unit Distillation Process

A labeled flowchart of a continuous steady-state two-unit distillation process is shown below. Each stream contains two components, A and B, in different proportions. Three streams whose flow rates and/or compositions are not known are labeled 1, 2, and 3.

Calculate the unknown flow rates and compositions of streams 1, 2, and 3.

Solution

The systems about which balances might be taken are shown on the following repre-sentation of the flowchart.

The outer boundary encompasses the entire process and has as input and output streams all of the streams that enter and leave the process. Two of the interior boundaries surround the individual process units, and the third encloses a stream junction point.

Basis: Given flow rates

There are two unknowns in the streams that enter and leave the total process, Q_3 and x_3, and since there are two independent components in these streams (A and B) we may write two balances.

Overall Mass Balance

$$(100 + 30)\frac{\text{kg}}{\text{h}} = (40 + 30)\frac{\text{kg}}{\text{h}} + Q_3$$

$$\Downarrow$$

$$\boxed{Q_3 = 60 \text{ kg/h}}$$

Overall Balance On A

$$(0.500)(100) + (0.300)(30) = (0.900)(40) + (0.600)(30) + (60)(x_3)$$

$$\Downarrow$$

$$\boxed{x_3 = 0.0833 \text{ kg A/kg}}$$

To determine the flow rate and composition of a connecting stream, we must write balances on a subsystem whose boundary intersects this stream. Of the three such boundaries shown in the flowchart, the middle one (about the stream junction) would not be a good one to use at this point since its input and output streams contain four unknown quantities (Q_1, x_1, Q_2, x_2), while the boundaries about the process units each intersect streams that contain a total of two unknowns.

Let us choose the boundary about unit 1 for the next set of balances. There are two unknowns, Q_1 and x_1, in the streams that intersect this boundary, and up to two balances may be written.

Total Mass Balance on Unit 1

$$100 \text{ kg/h} = 40\frac{\text{kg}}{\text{h}} + Q_1$$

$$\Downarrow$$

$$\boxed{Q_1 = 60 \text{ kg/h}}$$

Balance on A in Unit 1

$$(0.500)(100) = (0.900)(40) + 60x_1$$

$$\boxed{x_1 = 0.233 \text{ kg A/kg}}$$

To find Q_2 and x_2, we could write balances about either the stream mixing point or Unit 2. Let us choose the first alternative.

Mass Balance About Mixing Point

$$Q_1 + 30 \text{ kg/h} = Q_2$$
$$\Downarrow Q_1 = 60 \text{ kg/h}$$

$$\boxed{Q_2 = 90 \text{ kg/h}}$$

Balance on A About Mixing Point

$$Q_1 x_1 + (0.300)(30) \text{ kg A/h} = Q_2 x_2$$
$$\begin{Vmatrix} Q_1 = 60 \text{ kg/h} \\ x_1 = 0.233 \text{ kg A/kg} \\ \Downarrow Q_2 = 90 \text{ kg/h} \end{Vmatrix}$$

$$\boxed{x_2 = 0.255 \text{ kg A/h}}$$

The situation becomes still more complicated when three or more process units are involved. In such cases, balances may be written not only for the overall process and individual process units, but also for combinations of units. Finding the right combinations can lead to a considerable gain in computational efficiency.

The next example illustrates such a case.

EXAMPLE 4.4-2 An Extraction–Distillation Process

A mixture containing 50% acetone and 50% water by weight is to be separated into its components. The separation process consists of extraction of the acetone from the water into a solvent that dissolves acetone but is nearly immiscible with water, followed by distillation to separate the acetone from the solvent. The solvent to be used is methyl isobutyl ketone (MIBK). The description that follows introduces some of the terms commonly used in reference to liquid extraction processes.

The acetone (**solute**)–water (**feed solvent**) mixture is first contacted with the MIBK (**wash solvent**) in a mixer that provides good contact between the two liquid phases. Most of the acetone in the feed transfers from the aqueous (water) phase to the organic (MIBK) phase in this step. The mixture passes into a settling tank, where the phases separate and are separately withdrawn. The phase rich in the feed solvent (water, in this process) is referred to as the **raffinate,** and the phase rich in the wash solvent (MIBK) is the **extract.**

The raffinate passes to a second extraction stage, where it is contacted with a second stream of pure MIBK, leading to the transfer of more acetone. The second-stage raffinate is discarded. The extracts from the two mixer-settler stages are combined and fed to a distillation column. The vapor product (overhead) from the tower is rich in acetone and is the product of the process. The bottoms stream is rich in MIBK, and in a real process it would probably be treated further and recycled back to the extractors (see Section 4.5), but we will not consider recycle in this example.

In a pilot plant study, for every 100 kg of acetone–water fed to the first extractor, 100 kg of MIBK are fed to the first extractor and 75 kg are fed to the second extractor.

The extract from the first stage is found to contain 27.5% acetone. The second-stage raffinate has a mass of 43.1 kg, and consists of 5.3% acetone, 1.6% MIBK, and 93.1% water, and the second-stage extract contains 9.0% acetone, 88% MIBK, and 3% water. The overhead product from the distillation column contains 2.0% MIBK, 1.0% water, and the balance acetone.

For an assumed basis of 100 kg acetone–water feed, calculate the masses and compositions (component percentages by mass) of the stage 1 raffinate and extract, the stage 2 extract, the combined extract, and the distillation overhead and bottoms products.

Solution

As always, we begin by drawing and labeling the flowchart. (Verify that the chart reflects the information given in the process description.)

Balances might be written for a number of systems, including the entire process, the individual extraction units, the distillation column, and the extract mixing point. You may easily verify, however, that at least four unknown stream variables are involved in each of these systems, while the maximum number of independent balances permitted for any system is three. (Why?)

However, consider the system consisting of the two extraction units, defined by the dashed boundary on the flowchart. For this system, only three unknowns are involved—E_{CM}, E_{CA}, and E_{CW}. Since three independent balances may be written, this system provides a convenient starting point for the process analysis.

The solution plan is as follows:

1. Balances around E_1-E_2: Solve for E_{CM}, E_{CA}, and E_{CW}.
2. Balances around extract mixing point: Solve for E_1, x_{M1}, and E_2.
3. Balances around E_1 (or E_2): Solve for R_{1M}, R_{1A}, and R_{1W}.

At this point, we can see that we are stuck. There are four remaining unknowns—V, B_A, B_M, and B_W. Whether we choose the overall process or the distillation column as our system, we will only have three equations (the three independent material balances), and so we will be unable to solve the problem. Moreover, since A, M, and W all appear in both outlet streams, there is no way to solve for any one of the individual unknowns. (If there were no water in the overhead, for example, we would be able to deduce that $B_W = E_{CW}$.)

The problem is thus underdefined; unless another piece of information is furnished, the flow rates and compositions of the still products are indeterminate. The world—unlike most textbooks—often presents you with problems that don't come neatly packaged with all the information you need to solve them.

The calculations that can be performed are show below. All balances have the form "input = output" (Why?), and every additive term of every balance has units of kilograms of the balanced species.

Balances Around E_1–E_2

M: $100 + 75 = E_{CM} + (0.016)(43.1) \Rightarrow E_{CM} = 174.3$ kg
A: $(0.50)(100.0) = E_{CA} + (0.053)(43.1) \Rightarrow E_{CA} = 47.7$ kg
W: $(0.50)(100.0) = E_{CW} + (0.931)(43.1) \Rightarrow E_{CW} = 9.87$ kg

Balances Around Mixing Point

Mass: $E_1 + E_2 = 174.3 + 47.7 + 9.87 = 232$ $\left.\right\}$ \Rightarrow $E_1 = 145$ kg
A: $\quad\; 0.275E_1 + 0.0900E_2 = 47.7$ $\qquad\quad E_2 = 87.0$ kg
M: $\quad\; x_{M1}(145) + (0.88)(87.0) = 174.3 \Rightarrow x_{M1} = 0.674$

Balances Around E_1

M: $100 = (0.674)(145) + R_{1M} \Rightarrow R_{1M} = 2.27$ kg
A: $50 = (0.275)(145) + R_{1A} \Rightarrow R_{1A} = 10.1$ kg
W: $50 = (0.725 - 0.674)(145) + R_{1W} \Rightarrow R_{1W} = 42.6$ kg

If we knew any of the variables V, B_A, B_M, or B_W, we could calculate the remaining three. Since we do not, we must terminate the calculations at this point.

4.5 RECYCLE AND BYPASS

It is rare that a chemical reaction A \rightarrow B proceeds to completion in a continuous reactor; no matter how much A is present to begin with or how long the reaction mixture remains in the reactor, some A is normally found in the product.

Unfortunately, you have to pay for all the reactant fed into a process, not just the amount that reacts, and any A that leaves with the product therefore represents wasted resources. Suppose, however, you could find a way to separate most or all of the unconsumed reactant from the product stream; you could then sell the resulting relatively pure product and **recycle** the unconsumed reactant back to the reactor. You would, of course, have to pay for the separation and recycle equipment, but you

would compensate for this cost by having to purchase less fresh reactant and being able to sell the purified product at a higher price.

A labeled flowchart of a chemical process involving reaction, product separation, and recycle is shown in Figure 4.5-1. Note the distinction between the **fresh feed** to the process and the feed to the reactor, which is the sum of the fresh feed and recycle stream. If some of the stream variables shown in Figure 4.5-1 were unknown, you could determine them by writing balances on the overall process and about the reactor, separator, and mixing point.

FIGURE 4.5-1

Flowchart of a reactor with separation and recycle of unconsumed reactant

Students frequently have difficulty when they first encounter the concept of recycle, because they find it hard to understand that material can circulate in a system without an accumulation of mass. If you have this difficulty, you might find it helpful to refer back to the flowchart of Figure 4.5-1. Observe that even though there is material circulating within the system, there is no net accumulation: 110 kg of material enters the system each minute, and the same mass leaves each minute. Within the system there is a net circulation rate of 120 kg/min, but this has no effect on the overall process material balance.

There are several reasons for using recycle in a chemical process, besides the one given previously (recovering and reusing unconsumed reactants). These reasons include the following:

1. *Recovery of catalyst.* Many reactors use catalysts to increase the rate of the reaction. Catalysts are usually expensive, and the processes generally include provisions for recovering them from the product stream and recycling them. They may be recovered with the unconsumed reactants or recovered separately in special facilities designed for this purpose.

2. *Dilution of a process stream.* Suppose a slurry (a suspension of solids in a liquid) is fed to a filter. If the concentration of solids in the slurry is too high,

the slurry is difficult to handle and the filter will not operate properly. Rather than diluting the feed with fresh liquid, a portion of the filtrate can be recycled to dilute the feed to the desired solids concentration.

3. ***Control of a process variable.*** Suppose a reaction releases an extremely large amount of heat, making the reactor difficult and expensive to control. The rate of heat generation can be reduced by lowering the reactant concentration, which can in turn be accomplished by recycling a portion of the reactor effluent to the inlet. In addition to acting as a diluent for the reactants, the recycled material also serves as a capacitance for the heat released: The greater the mass of the reaction mixture, the lower the temperature to which that mass will be raised by a fixed amount of heat.

4. ***Circulation of a working fluid.*** The most common example of this application is the refrigeration cycle used in household refrigerators and air conditioners. In these devices, a single material is reused indefinitely, with only small makeup quantities being added to the system to replenish working fluid that may be lost through leaks or spills.

The following example illustrates a material balance calculation for a nonreactive process that includes recycle.

EXAMPLE 4.5-1 An Evaporative Crystallization Process

The flowchart of a steady-state process to recover crystalline potassium chromate (K_2CrO_4) from an aqueous solution of this salt is shown below.

Forty-five hundred kilograms per hour of a solution that is $\frac{1}{3}K_2CrO_4$ by mass is joined by a recycle stream containing 36.36% K_2CrO_4, and the combined stream is fed into an evaporator. The concentrated stream that leaves the evaporator contains 49.4% K_2CrO_4; this stream is fed into a crystallizer in which it is cooled (causing crystals of K_2CrO_4 to come out of solution) and then filtered. The filter cake consists of K_2CrO_4 crystals and a solution that contains 36.36% K_2CrO_4 by mass; the crystals account for 95% of the total mass of the filter cake. The solution that passes through the filter, also 36.36% K_2CrO_4, is the recycle stream.

Calculate the weight of water removed in the evaporator, the rate of production of crystalline K_2CrO_4, the ratio (kg recycle)/(kg fresh feed), and the feed rates that the evaporator and crystallizer must be designed to handle.

Solution

It is almost mandatory to draw and label carefully the flowchart for problems of this sort to avoid drowning in a sea of data and relationships among process variables.

Basis: 4500 kg/h Fresh Feed

Let K denote K_2CrO_4 and W denote water.

It is generally desirable to begin by analyzing the overall process. Let us therefore do the bookkeeping on this system, to determine whether enough information is available to solve for the unknown variables. Referring to the flowchart, we see that three unknowns are involved (Q_w, P_c, and P_s). We are allowed to write two balances, and the information that the filter cake contains 95% solid crystals provides the third relation. We may then proceed with the solution.

We first write all given relationships among stream variables as equations in the labeled unknown quantities. The only such relationship in this problem is that between the mass of the solid crystals and that of the solution in the filter cake.

$$P_s(\text{kg/h}) = 0.05\underbrace{(P_c + P_s)}(\text{kg/h})$$

$$\text{Mass of} \quad\quad \text{Total}$$
$$\text{solution} \quad\Downarrow\quad \text{mass}$$

$$P_s = 0.05263P_c \qquad\qquad (1)$$

Now we write material balances, beginning with overall process balances.

Overall K Balance

$$(0.3333)(4500) \text{ kg K/h} = P_c + 0.3636P_s$$

$$\Downarrow P_s = 0.05263P_c$$

$$P_c = \frac{1500}{1 + (0.01914)} = \boxed{1472\ \frac{\text{kg crystals}}{\text{h}}}$$

and from Equation (1)

$$P_s = (0.05263)(1472) = 77.5 \text{ kg/h}$$

Overall Total Mass Balance

$$4500 \text{ kg/h} = Q_w + P_c + P_s$$
$$\Downarrow P_c = 1472, P_s = 77.5$$

$$\boxed{Q_w = 2950 \text{ kg/h}}$$

We have gone as far as we can go with overall balances. (Two components means no more than two independent balances.) Of the subsystems that can be chosen next, the mixing point where the fresh feed and the recycle stream combine involves three unknowns (R, F_e, and x_e), the evaporator also involves three unknowns (F_e, x_e, and F_c), but the crystallizer-filter involves only two—F_c and R. Therefore, we choose this subsystem for the next set of balances.

Mass Balance on Crystallizer

$$F_c(\text{kg/h}) = (R + P_c + P_s)(\text{kg/h})$$
$$\Downarrow \begin{array}{l} P_c = 1472 \\ P_s = 77.5 \end{array}$$
$$F_c = R + 1549.9 \text{ kg/h} \tag{2}$$

Water Balance on Crystallizer

$$0.506F_c(\text{kg W/h}) = 0.6364P_s + 0.6364R$$
$$\Downarrow P_s = 77.5$$
$$F_c = 97.35 \text{ kg W/h} + 1.258R \tag{3}$$

Solving equations (2) and (3) simultaneously yields

$$\boxed{F_c = 7180 \text{ kg/h}}$$
$$R = 5630 \text{ kg/h}$$

and hence

$$\frac{R \text{ (kg recycle/h)}}{4500 \text{ (kg fresh feed/h)}} = \frac{5630}{4500} = \boxed{1.25 \; \frac{\text{kg recycle}}{\text{kg fresh feed}}}$$

Finally, we can determine the feed to the evaporator by writing a mass balance about either the stream mixing point or the evaporator. The former balance is

$$4500 \text{ kg/h} + R = F_e(\text{kg/h})$$
$$\Downarrow R = 5630$$
$$\boxed{F_e = 10\,130 \text{ kg/h}}$$

Check: A mass balance about the evaporator yields

$$F_e = Q_w + F_c$$
$$\Downarrow$$
$$10\,130 = 2950 + 7180 \quad \checkmark$$

A procedure that has several features in common with recycle is **bypass,** in which a fraction of the feed to a process unit is diverted around the unit and combined with the output stream from the unit. (For an example of a process that uses bypass, see Problem 34 at the end of this chapter.) By varying the fraction of the feed that is bypassed, we can vary the composition and properties of the product.

Bypass and recycle calculations are approached in exactly the same manner: The flowchart is drawn and labeled, and overall balances and balances around the process unit or the stream mixing point following the process unit are used to determine unknown variables.

4.6 BALANCES ON REACTIVE SYSTEMS

The occurrence of a chemical reaction in a process brings several complications into the material balance procedures described in the previous sections. The stoichiometric equation of the reaction imposes constraints on the relative amounts of reactants and products in the input and output streams (if A → B, for example, you cannot start with 1 mol of pure A and end with 2 mol of B). In addition, a material balance on a reactive substance does not have the simple form input = output, but must include a generation or consumption term.

In this section we review reaction terminology and outline procedures for carrying out material balance calculations on reactive systems.

4.6a Stoichiometry

Stoichiometry is the theory of the proportions in which chemical species combine with one another. The **stoichiometric equation** of a chemical reaction is a statement of the relative number of molecules or moles of reactants and products that participate in the reaction. For example, the stoichiometric equation

$$2\,SO_2 + O_2 \rightarrow 2\,SO_3$$

indicates that two molecules (g-moles, lb-moles) of SO_2 react with one molecule (g-mole, lb-mole) of O_2 to produce two molecules (g-moles, lb-moles) of SO_3. The numbers that precede the formulas for each species are the **stoichiometric coefficients** of the reaction components.

A valid stoichiometric equation must be balanced; that is, the number of atoms of each atomic species must be the same on both sides of the equation, since atoms cannot be created or destroyed in chemical reactions. The equation

$$SO_2 + O_2 \rightarrow SO_3$$

cannot be valid, for example, since it indicates that three atoms of atomic oxygen (O) are produced for every four atoms that enter into the reaction, for a net loss of one atom, but

$$SO_2 + \tfrac{1}{2}O_2 \rightarrow SO_3 \qquad \begin{pmatrix} 1\,S \rightarrow 1\,S \\ 3\,O \rightarrow 3\,O \end{pmatrix}$$

and

$$2\,SO_2 + O_2 \rightarrow 2\,SO_3 \qquad \begin{pmatrix} 2\,S \rightarrow 2\,S \\ 6\,O \rightarrow 6\,O \end{pmatrix}$$

are balanced.

The **stoichiometric ratio** of two molecular species participating in a reaction is the ratio of their stoichiometric coefficients in the balanced reaction equation. This ratio can be used as a conversion factor to calculate the amount of a particular reactant (or product) that was consumed (or produced), given a quantity of another reactant or product that participated in the reaction. For the reaction

$$2\,SO_2 + O_2 \rightarrow 2\,SO_3$$

you can write the stoichiometric ratios

$$\frac{2 \text{ mol } SO_3 \text{ produced}}{1 \text{ mol } O_2 \text{ reacted}}, \quad \frac{2 \text{ lb-moles } SO_2 \text{ reacted}}{2 \text{ lb-moles } SO_3 \text{ produced}}$$

and so on. If you know, for example, that 1600 kg/h of SO_3 are to be produced, you can calculate the amount of oxygen required as

$$\frac{1600 \text{ kg } SO_3 \text{ produced}}{h} \left| \frac{1 \text{ kmol } SO_3}{80 \text{ kg } SO_3} \right| \frac{1 \text{ kmol } O_2 \text{ reacted}}{2 \text{ kmol } SO_3 \text{ produced}}$$

$$= 10 \frac{\text{kmol } O_2}{h}$$

It is good practice to include the terms "reacted" and "produced" when performing conversions of this sort: Simply writing 1 mol O_2/2 mol SO_3 could be taken to mean that 2 mol of SO_3 contain 1 mol of O_2, which is not true.

TEST YOURSELF

Consider the reaction

$$C_4H_8 + 6\,O_2 \rightarrow 4\,CO_2 + 4\,H_2O$$

1. Is the stoichiometric equation balanced?
2. What is the stoichiometric coefficient of CO_2?
3. What is the stoichiometric ratio of H_2O to O_2? (Include units.)
4. How many lb-moles of O_2 react to form 400 lb-moles of CO_2? (Use a dimensional equation.)
5. One hundred g-moles/min of C_4H_8 are fed into a reactor, and 50% reacts. At what rate is water formed?

4.6b Limiting and Excess Reactants, Fractional Conversion, and Extent of Reaction

Two reactants, A and B, are said to be present in **stoichiometric proportion** if the ratio (moles A present)/(moles B present) equals the stoichiometric ratio obtained from the balanced reaction equation. For the reactants in the reaction

$$2\,SO_2 + O_2 \rightarrow 2\,SO_3$$

to be present in stoichiometric proportion, there must be 2 moles of SO_2 for every mole of O_2 (so that $n_{SO_2}/n_{O_2} = 2/1$) present in the feed to the reactor.

If reactants are fed to a chemical reactor in stoichiometric proportion and the reaction proceeds to completion, all of the reactants are consumed. In the above reaction, for example, if 200 mol of SO_2 and 100 mol of O_2 are initially present, the SO_2 and O_2 would disappear at the same instant. It follows that if you start with 100 mol of O_2 and less than 200 mol of SO_2 (i.e., if the SO_2 is present in less than its stoichiometric proportion) the SO_2 disappears before the O_2, while if there are more than 200 mol of SO_2 initially present the O_2 disappears first.

The reactant that would disappear first if a reaction proceeded to completion is called the **limiting reactant;** the other reactants are termed **excess reactants.** *A reactant is limiting if it is present in less than its stoichiometric proportion relative to every other reactant.* If all reactants are present in stoichiometric proportion, then no reactant is limiting (or they all are, depending on how you choose to look at it).

Suppose n moles of an excess reactant are present, and that n_s moles would correspond to stoichiometric proportion (so that n must be greater than n_s). The **fractional excess** of this reactant is defined as $(n - n_s)/n_s$, and the **percentage excess** is 100 times the fractional excess. Consider, for example, the reaction

$$H_2 + Br_2 \rightarrow 2\,HBr$$

and suppose that 25 mol H_2/h and 20 mol Br_2/h are fed to a reactor. Br_2 is clearly the limiting reactant, and to be in stoichiometric proportion n_{H_2} would have to be

20 mol/h. The fractional excess of H_2 is therefore

$$\frac{25 - 20}{20} = 0.25$$

and H_2 is said to be in excess by 25%.

Chemical reactions do not take place instantaneously, and indeed often proceed rather slowly. In such cases, it is not practical to design the reactor for complete conversion of the limiting reactant; instead, the reactor effluent emerges with some of the reactant still present, and is then usually subjected to a separation process to remove the unconverted reactant from the product. The separated reactant is then recycled back to the reactor inlet. The **fractional conversion** of a reactant is the ratio

$$f = \frac{\text{moles reacted}}{\text{moles fed}} \qquad \text{(4.6-1)}$$

The fraction unreacted is accordingly $1 - f$. If 100 moles of a reactant are fed and 90 of them react, the fractional conversion is 0.9 (the **percentage conversion** is 90%) and the fraction unreacted is 0.1. If 20 mol/min of a reactant are fed and the percentage conversion is 80%, then $(20)(0.8) = 16$ mol/min have reacted and $(20)(0.2) = 4$ mol/min remain unreacted.

Considering the reaction discussed above ($H_2 + Br_2 \rightarrow 2HBr$), suppose we start with 100 mol of hydrogen, 50 of bromine, and 50 of hydrogen bromide, and that 30 mol of hydrogen react. How much of each species will be present in the product?

Clearly, if you start with 100 mol of H_2 and 30 of them react, you will be left with 70 mol H_2. Also, if 30 mol of H_2 react, 30 mol of Br_2 also react (Why?), leaving you with 20 mol Br_2. Finally, the 30 mol of H_2 that react form 60 mol of HBr, which when added to the 50 you started with gives you 110 mol HBr.

More generally, if ξ mol of H_2 react, we may follow the same reasoning and write

$$n_{H_2} = (n_{H_2})_0 - \xi$$
$$n_{Br_2} = (n_{Br_2})_0 - \xi$$
$$n_{HBr} = (n_{HBr})_0 + 2\xi$$

Once you know how much hydrogen (or bromine) reacts or how much hydrogen bromide is formed, you can determine ξ from one of these equations and then determine the amounts of each component present from the other equations.

Next, we may generalize this result to any chemical reaction. Suppose you start out with n_{i0} mols of a particular reactant or product A_i (batch process), or a feed rate of n_{i0} mol/s (continuous steady-state process); v_i is the stoichiometric coefficient of this species (always positive); and n_i is the amount or flow rate of A_i at the reactor outlet. For the types of reactors we are considering here—batch and continuous steady-state—we can write

$$n_i = n_{i0} + \beta_i \xi \qquad \text{(4.6-2)}$$

$$\begin{aligned} \beta_i &= v_i && \text{if } A_i \text{ is a product} \\ &= -v_i && \text{if } A_i \text{ is a reactant} \\ &= 0 && \text{if } A_i \text{ is inert} \end{aligned}$$

(Verify that this expression yields the previously given formulas for the HBr reaction.) The quantity ξ, which has the same units as n, is called the **extent of reaction**. If you know the feed quantities (n_{i0} for all i), and you also know any one of the n_i's, you can calculate ξ by applying Equation 4.6-2 to the component for which n_i is known. You can then calculate the remaining n_i's by applying Equation 4.6-2 to the other species, substituting the known value of ξ.

For example, consider the ammonia formation reaction.

$$N_2 + 3H_2 = 2NH_3$$

Suppose the feed to a continuous reactor consists of 100 mol/s of nitrogen, 300 mol/s of hydrogen, and 1 mol/s of argon (an inert gas). From Equation 4.6-2, we may write for the reactor outlet flow rates

$$n_{N_2} = 100 \text{ mol/s} - \xi$$
$$n_{H_2} = 300 \text{ mol/s} - 3\xi$$
$$n_{NH_3} = 2\xi$$
$$n_A = 1 \text{ mol/s}$$

If you know the flow rate of any component or the fractional conversion of nitrogen or hydrogen, you can calculate ξ, and then the other two unknown outlet flow rates. *Try it:* For a fractional hydrogen conversion of 60%, calculate the outlet flow rate of hydrogen, the extent of reaction, and the outlet flow rates of nitrogen and ammonia. The last value should be 120 mol/s.

TEST YOURSELF

The oxidation of ethylene to produce ethylene oxide proceeds according to the equation

$$2\,C_2H_4 + O_2 \rightarrow 2\,C_2H_4O$$

The feed to a reactor contains 100 kmol C_2H_4 and 100 kmol O_2.

1. Which reactant is limiting?
2. What is the percentage excess of the excess reactant?
3. If the reaction proceeds to completion, how much of the excess reactant will be left; how much C_2H_4O will be formed; and what is the extent of reaction?
4. If the reaction proceeds to a point where the fractional conversion of the limiting reactant is 50%, how much of each reactant and product is present at the end, and what is the extent of reaction?
5. If the reaction proceeds to a point where 60 mol of O_2 are left, what is the fractional conversion of C_2H_4? The fractional conversion of O_2? The extent of reaction?

CREATIVITY EXERCISE

A single chemical reaction, A → 2B, takes place in a batch reactor, with known initial quantities of A and B. Think of as many process variables as you can that might vary with the extent of reaction; then suggest means by which each of these variables might be measured, thereby providing a measure of the extent of reaction versus time. For example, the refractive index of the reaction mixture might vary with the mixture composition; the experiment would then be to pass a beam of light through the reaction vessel and measure the angle of refraction as a function of time.

EXAMPLE 4.6-1 Reaction Stoichiometry

Acetonitrile is produced by the reaction of propylene, ammonia, and oxygen.

$$C_3H_6 + NH_3 + \tfrac{3}{2}O_2 \rightarrow C_3H_3N + 3\,H_2O$$

The feed contains 10 mole% propylene, 12% ammonia, and 78% air. A fractional conversion of 30% of the limiting reactant is achieved. Determine which reactant is limiting, the percentage by which each of the other reactants is in excess, and the molar flow rates of all product gas constituents for a 30% conversion of the limiting reactant, taking 100 mol of feed as a basis.

Solution

Basis: 100 mol Feed

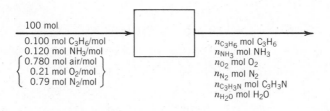

The feed to the reactor contains

$$(C_3H_6)_0 = 10.0 \text{ mol}$$

$$(NH_3)_0 = 12.0 \text{ mol}$$

$$(O_2)_0 = \frac{78.0 \text{ mol air}}{} \left| \frac{0.210 \text{ mol } O_2}{\text{mol air}} \right. = 16.4 \text{ mol}$$

$$\Downarrow$$

$$\left. \begin{array}{l} (NH_3/C_3H_6)_0 = 12.0/10.0 = 1.20 \\ (NH_3/C_3H_6)_{\text{stoich}} = 1/1 = 1 \end{array} \right\} \Rightarrow NH_3 \text{ is in excess}$$

$$\left. \begin{array}{l} (O_2/C_3H_6)_0 = 16.4/10.0 = 1.64 \\ (O_2/C_3H_6)_{\text{stoich}} = 1.5/1 = 1.5 \end{array} \right\} \Rightarrow O_2 \text{ is in excess}$$

Since propylene is fed in less than stoichiometric proportion relative to the two other reactants, *propylene is the limiting reactant.*

To determine the percentages by which ammonia and oxygen are in excess, we must first determine the stoichiometric amounts of these reactants corresponding to the amount of propylene in the feed (10 mol).

$$(NH_3)_{stoich} = \frac{10.0 \text{ mol } C_3H_6 \,\Big|\, 1 \text{ mol } NH_3}{\Big|\, 1 \text{ mol } C_3H_6} = 10.0 \text{ mol } NH_3$$

$$(O_2)_{stoich} = \frac{10.0 \text{ mol } C_3H_6 \,\Big|\, 1.5 \text{ mol } O_2}{\Big|\, 1 \text{ mol } C_3H_6} = 15.0 \text{ mol } O_2$$

$$(\% \text{ excess})_{NH_3} = \frac{(NH_3)_0 - (NH_3)_{stoich}}{(NH_3)_{stoich}} \times 100\%$$

$$= (12 - 10)/10 \times 100\% = \boxed{20\% \text{ excess } NH_3}$$

$$(\% \text{ excess})_{O_2} = (16.4 - 15.0)/16.4 \times 100\% = \boxed{9.3\% \text{ excess } O_2}$$

If the fractional conversion of C_3H_6 is 30%, then

$$(C_3H_6)_{out} = 0.700(C_3H_6)_0 = \boxed{7.0 \text{ mol } C_3H_6}$$

But from Equation 4.6-2, $n_{C_3H_6} = 10.0 - \xi$. The extent of reaction is therefore $\xi = 3.0$. Then, also from Equation 4.6-2,

$$
\begin{array}{rcl|l}
n_{NH_3} = 12.0 - \xi & = & 9.0 \text{ mol } NH_3 \\
n_{O_2} = 16.4 - 1.5\xi & = & 11.9 \text{ mol } O_2 \\
n_{C_3H_3N} = \xi & = & 3.00 \text{ mol } C_3H_3N \\
n_{N_2} = (N_2)_0 & = & 61.6 \text{ mol } N_2 \\
n_{H_2O} = 3\xi & = & 9.0 \text{ mol } H_2O
\end{array}
$$

4.6c Chemical Equilibrium

Two of the fundamental questions of chemical process engineering are, given a set of reactive species and reaction conditions, (1) what will be the final (equilibrium) composition of the reaction mixture, and (2) how long will the system take to reach a specified state short of equilibrium? The discipline of **chemical equilibrium thermodynamics** concerns itself with the first question, and **chemical kinetics** deals with the second.

Some reactions are essentially **irreversible;** that is, the reaction proceeds only in a single direction (from reactants to products) and the concentration of the limiting reactant eventually approaches zero (although "eventually" could mean seconds for some reactions and years for others). The equilibrium composition for such a reaction

is therefore the composition corresponding to complete consumption of the limiting reactant.

Other reactions (or the same reactions at different conditions) are **reversible;** reactants form products and products undergo the reverse reactions to reform the reactants. For example, consider the reaction in which ethylene is hydrolyzed to form ethanol:

$$C_2H_4 + H_2O \rightleftharpoons C_2H_5OH$$

If you start with ethylene and water, the forward reaction occurs; then once ethanol is present, the reverse reaction begins to take place. As the concentrations of C_2H_4 and H_2O decrease, the rate of the forward reaction decreases, and as the C_2H_5OH concentration increases, the rate of the reverse reaction increases. Eventually a point is reached at which the rates of the forward and reverse reactions are equal. At this point no further composition change takes place and the reaction mixture is in chemical equilibrium.

A full discussion of the relations that can be used to determine equilibrium compositions of reactive mixtures is beyond the scope of this text; however, at this point you have enough knowledge to be able to calculate equilibrium compositions if the relations are given to you. The next example illustrates such a calculation.

EXAMPLE 4.6-2 Calculation of an Equilibrium Composition

If the water–gas shift reaction,

$$CO(g) + H_2O(g) \rightleftharpoons CO_2(g) + H_2(g)$$

proceeds to equilibrium at a temperature $T(K)$, the mole fractions of the four reactive species satisfy the relation

$$\frac{y_{CO_2}y_{H_2}}{y_{CO}y_{H_2O}} = K(T)$$

where $K(T)$ is the reaction **equilibrium constant.** At $T = 1105$ K, $K = 1.00$.

Suppose the feed to a reactor contains 1.00 mol of CO, 2.00 mol of H_2O, and no CO_2 or H_2, and the reaction mixture comes to equilibrium at 1105 K. Calculate the equilibrium composition and the fractional conversion of the limiting reactant.

Solution

The strategy is to express all mole fractions in terms of a single variable (ξ_e, the extent of reaction at equilibrium), substitute in the equilibrium relation, solve for ξ_e, and back-substitute to calculate the mole fractions and any other desired quantity. From Equation 4.6-2,

$$n_{CO} = 1.00 - \xi_e \quad \text{(number of gram-moles of CO present at equilibrium)}$$
$$n_{H_2O} = 2.00 - \xi_e$$
$$n_{CO_2} = \xi_e$$
$$\frac{n_{H_2} = \xi_e}{n_{total} = 3.00}$$

from which

$$y_{CO} = (1.00 - \xi_e)/3.00$$

$$y_{H_2O} = (2.00 - \xi_e)/3.00$$

$$y_{CO_2} = \xi_e/3.00$$

$$y_{H_2} = \xi_e/3.00$$

Substitution of these expressions into the equilibrium relation (with K = 1.00) yields

$$\frac{\xi_e^2}{(1.00 - \xi_e)(2.00 - \xi_e)} = 1.00$$

This may be rewritten as a standard quadratic equation and solved to yield $\xi_e = 0.667$. This quantity may in turn be substituted back into the expressions for $y_i(\xi_e)$ to yield

$$y_{CO} = 0.111, \quad y_{H_2O} = 0.444, \quad y_{CO_2} = 0.222, \quad y_{H_2} = 0.222$$

The limiting reactant in this case is CO (Verify). At equilibrium,

$$n_{CO} = 1.00 - 0.667 = 0.333$$

The fractional conversion of CO at equilibrium is therefore

$$f_{CO} = (1.00 - 0.333)\text{mol reacted}/(1.00 \text{ mol fed}) = \boxed{0.667}$$

4.6d Multiple Reactions, Yield, and Selectivity

In most chemical processes, reactants are brought together with the object of producing a desired product in a single reaction. Unfortunately, reactants can usually combine in more than one way, and the product once formed may itself react to yield something less desirable. The result of these side reactions is an economic loss: Less of the desired product is obtained for a given quantity of raw materials, or a greater quantity of raw materials must be fed to the reactor to obtain a specified product yield.

For example, ethylene can be produced by the dehydrogenation of ethane:

$$C_2H_6 \rightarrow C_2H_4 + H_2$$

Once some hydrogen is produced, it can react with ethane to produce methane:

$$C_2H_6 + H_2 \rightarrow 2\,CH_4$$

Moreover, ethylene can react with ethane to form propylene and methane:

$$C_2H_4 + C_2H_6 \rightarrow C_3H_6 + CH_4$$

Since the object of the process is to produce ethylene, only the first of these reactions can be regarded as desirable; the second one consumes the reactant without yielding the desired product, and the third consumes both the reactant and the desired product. The engineer designing the reactor and specifying operating conditions for it must consider not only how to maximize the production of the desired product (C_2H_4), but also how to minimize the production of undesired byproducts (CH_4, C_3H_6).

The terms **yield** and **selectivity** are used to describe the degree to which a desired reaction predominates over competing side reactions.

$$\textit{Yield:} \quad \frac{\text{Moles of desired product formed}}{\substack{\text{Moles that would have been formed if there were} \\ \text{no side reactions and the limiting reactant had} \\ \text{reacted completely}}} \qquad (4.6\text{-}3)$$

$$\textit{Selectivity:} \quad \frac{\text{Moles of desired product formed}}{\text{Moles of undesired product formed}} \qquad (4.6\text{-}4)$$

The yield defined by Equation 4.6-3 is always a fraction; it may also be expressed as a percentage by multiplying by 100%. If A is the desired product and B is an undesired product, one then refers to the *selectivity of A relative to B*. High values of the yield and selectivity signify that the undesired side reactions have been successfully suppressed relative to the desired reaction.

Besides being defined by Equation 4.6-3, yield is also sometimes defined as moles of desired product divided by either moles of reactant fed or moles of reactant consumed in the reactor. For a given feed and product composition, the yields defined in these three manners may have completely different values, so that when you see a reference to a yield, it is important to be aware of the working definition.

The concept of extent of reaction can be extended to multiple reactions, only now each independent reaction has its own extent. If a set of reactions takes place in a batch or continuous steady-state reactor, and v_{ij} is the stoichiometric coefficient of substance A_i in reaction j, we may then write

$$n_i = n_{i0} + \sum_j \beta_{ij}\xi_j \qquad (4.6\text{-}5)$$

$$\begin{aligned} \beta_{ij} &= v_{ij} \quad && \text{if } A_i \text{ is a product in reaction } j \\ &= -v_{ij} \quad && \text{if } A_i \text{ is a reactant in reaction } j \\ &= 0 \quad && \text{if } A_i \text{ does not appear in reaction } j \end{aligned}$$

For a single reaction, this equation reduces to Equation 4.6-2.

For example, consider the pair of reactions in which ethylene is oxidized either to ethylene oxide or to carbon dioxide.

$$C_2H_4 + \tfrac{1}{2}O_2 \rightarrow C_2H_4O \qquad (1)$$
$$C_2H_4 + 3O_2 \rightarrow 2CO_2 + 2H_2O \qquad (2)$$

The moles (or molar flow rates) of each of the five species involved in these reactions can be expressed in terms of the feed values and extents of reaction using Equation 4.6-5:

$$(C_2H_4)_{out} = (C_2H_4)_0 - \xi_1 - \xi_2$$
$$(O_2)_{out} = (O_2)_0 - 0.5\xi_1 - 3\xi_2$$
$$(C_2H_4O)_{out} = (C_2H_4O)_0 + \xi_1$$
$$(CO_2)_{out} = (CO_2)_0 + 2\xi_2$$
$$(H_2O)_{out} = (H_2O)_{out} + 2\xi_2$$

If values of any two outlet amounts are given, the values of ξ_1 and ξ_2 may be determined from the corresponding two equations, and the remaining amounts may in turn be calculated from the remaining three equations. The next example illustrates this procedure.

EXAMPLE 4.6-3 Yield and Selectivity in a Dehydrogenation Reactor

The reactions

$$C_2H_6 \rightarrow C_2H_4 + H_2$$
$$C_2H_6 + H_2 \rightarrow 2\,CH_4$$

take place in a continuous reactor at steady state. The feed contains 85.0 mole% ethane (C_2H_6) and the balance inerts (I). The fractional conversion of ethane is 0.501, and the fractional yield of ethylene is 0.471. Calculate the molar composition of the product gas and the selectivity of ethylene to methane production.

Solution

Basis: 100 mol Feed

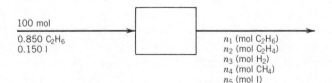

From Equation 4.6-5, the outlet component amounts in terms of extents of reaction are as follows:

$$n_1(\text{mol } C_2H_6) = 85.0 - \xi_1 - \xi_2$$
$$n_2(\text{mol } C_2H_4) = \xi_1$$
$$n_3(\text{mol } H_2) = \xi_1 - \xi_2$$
$$n_4(\text{mol } CH_4) = 2\xi_2$$
$$n_5(\text{mol } I) = 15.0$$

Ethane Conversion

$$n_1 = \frac{(1.000 - 0.501)\ \text{mol } C_2H_6 \text{ unreacted}}{\text{mol } C_2H_6 \text{ fed}} \left| \frac{85.0\ \text{mol } C_2H_6 \text{ fed}}{} \right.$$

$$= 42.4\ \text{mol } C_2H_6 = 85.0 - \xi_1 - \xi_2 \tag{1}$$

Ethylene Yield

$$\text{Maximum possible ethylene} = \frac{85.0\ \text{mol } C_2H_6 \text{ fed}}{} \left| \frac{1\ \text{mol } C_2H_4}{1\ \text{mol } C_2H_6} \right. = 85.0\ \text{mol}$$

$$\Downarrow$$

$$n_2 = 0.471(85.0\ \text{mol } C_2H_4) = 40.0\ \text{mol } C_2H_4 = \xi_1$$

Substituting 40.0 for ξ_1 in (1) yields $\xi_2 = 2.6$ mol. Then

$$n_3 = \xi_1 - \xi_2 = 37.4 \text{ mol } H_2$$
$$n_4 = 2\xi_2 = 5.2 \text{ mol } CH_4$$
$$n_5 = 15.0 \text{ mol } I$$
$$n_{tot} = (42.4 + 40.0 + 37.4 + 5.2 + 15.0) = 140.0 \text{ mol}$$

$$\Downarrow$$

$$\boxed{30.3\% \; C_2H_6, \quad 28.6\% \; C_2H_4, \quad 26.7\% \; H_2, \quad 3.7\% \; CH_4, \quad 10.7\% \; I}$$

$$\text{Selectivity} = (40.0 \text{ mol } C_2H_4)/(5.2 \text{ mol } CH_4)$$

$$= \boxed{7.7 \; \frac{\text{mol } C_2H_4}{\text{mol } CH_4}}$$

TEST YOURSELF

Consider the following pair of reactions.

$$A \to 2B \text{ (desired)}$$
$$A \to C \text{ (undesired)}$$

Suppose 100 mol of A are fed to a batch reactor and the final product contains 10 mol of A, 160 mol of B, and 10 mol of C. Calculate

1. The fractional conversion of A.
2. The percentage yield of B.
3. The selectivity of B relative to C.
4. The extents of the first and second reactions.

CREATIVITY EXERCISES

The reactions

$$A \to B$$
$$B \to C$$

take place in a continuous reactor, with pure A as the feed. B is the desired product, and C is an undesired byproduct. The fractional conversion of A can be varied by changing the reactor volume: the larger the volume, the greater the conversion.

1. Think of as many reasons as you can why C might be undesirable.
2. Think of as many possible ways as you can to run the reactor so as to increase the yield of B. (For example, if A is a gas, run the reactor at a temperature and

pressure such that B condenses and is removed before it has a chance to react further to C.)

3. Suggest as many possible ways as you can to deal with C, once it has been recovered from the reactor effluent stream.

4.6e Balances on Atomic and Molecular Species

Figure 4.6-1 shows a flowchart for the dehydrogenation of ethane in a steady-state continuous reactor. The reaction is

$$C_2H_6 \rightarrow C_2H_4 + H_2$$

One hundred gram-moles of ethane are fed to the reactor. The product stream is analyzed for hydrogen, and it is found that the molar flow rate of H_2 is 40 mol/min.

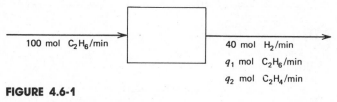

100 mol C_2H_6/min

40 mol H_2/min

q_1 mol C_2H_6/min

q_2 mol C_2H_4/min

FIGURE 4.6-1

Dehydrogenation of ethane

A number of different balances could be written on this process, including balances on total mass, C_2H_6, C_2H_4, and H_2. Of these, only the first has the simple form input = output; since the three given species all participate in the reaction, the balance equation for each of them must include a generation term (for C_2H_4 and H_2) or a consumption term (for C_2H_6).

Notice, however, that balance equations may also be written for atomic carbon and atomic hydrogen, regardless of the molecular species in which the carbon and hydrogen atoms happen to be found. Balances on atomic species *can* be written input = output, since atoms can neither be created (generation = 0) nor destroyed (consumption = 0) in a chemical reaction.

Before illustrating these balances, let us resolve an ambiguity. When we speak of a hydrogen balance, it could mean two completely different things: a balance on molecular hydrogen, existing as an independent species, or a balance on total hydrogen— bound and unbound. We will henceforth use the terms **molecular hydrogen balance** and **atomic hydrogen balance** to represent these two types of balances.

Some of the balances that can be written for the process shown in Figure 4.6-1 are as follows. (Recall that the general balance equation for a steady-state process is input + generation = output + consumption.)

Molecular H_2 Balance: generation = output

$$\text{Gen}_{H_2}\left(\frac{\text{mol } H_2 \text{ produced}}{\text{min}}\right) = 40 \text{ mol } H_2/\text{min}$$

C_2H_6 Balance: input = output + consumption

$$\frac{100 \text{ mol } C_2H_6}{\text{min}} = q_1\left(\frac{\text{mol } C_2H_6}{\text{min}}\right) + \text{Cons}_{C_2H_6}\left(\frac{\text{mol } C_2H_6 \text{ reacted}}{\text{min}}\right)$$

C_2H_4 Balance: generation = output

$$\text{Gen}_{C_2H_4}\left(\frac{\text{mol } C_2H_4 \text{ produced}}{\text{min}}\right) = q_2\left(\frac{\text{mol } C_2H_4}{\text{min}}\right)$$

Atomic C Balance: input = output

$$\frac{100 \text{ mol } C_2H_6}{\text{min}} \left|\frac{2 \text{ mol C}}{1 \text{ mol } C_2H_6}\right. = q_1 \frac{\text{mol } C_2H_6}{\text{min}} \left|\frac{2 \text{ mol C}}{1 \text{ mol } C_2H_6}\right.$$

$$+ q_2 \frac{\text{mol } C_2H_4}{\text{min}} \left|\frac{2 \text{ mol C}}{1 \text{ mol } C_2H_4}\right.$$

$$\Downarrow$$

$$100 = q_1 + q_2$$

Atomic H Balance: input = output

$$\frac{100 \text{ mol } C_2H_6}{\text{min}} \left|\frac{6 \text{ mol H}}{1 \text{ mol } C_2H_6}\right. = \frac{40 \text{ mol } H_2}{\text{min}} \left|\frac{2 \text{ mol H}}{1 \text{ mol } H_2}\right.$$

$$+ q_1 \frac{\text{mol } C_2H_6}{\text{min}} \left|\frac{6 \text{ mol H}}{1 \text{ mol } C_2H_6}\right. + q_2 \frac{\text{mol } C_2H_4}{\text{min}} \left|\frac{4 \text{ mol H}}{1 \text{ mol } C_2H_4}\right.$$

$$\Downarrow$$

$$600 = 80 + 6q_1 + 4q_{2,}$$

There are two ways to solve for q_1 and q_2: directly from the two atomic balances, or by using the three molecular species balances in conjunction with the stoichiometric equation for the reaction.

In general, material balance calculations on reactive systems may be set up and solved using balances on molecular species or atomic species or by using extents of reaction, as in the preceding examples. Each method must of course lead to the same result; however, one is frequently more convenient than the others, so it is a good idea to become comfortable with all three. The example that follows illustrates all three approaches.

EXAMPLE 4.6-4 Combustion of Methane

Methane is burned with oxygen to yield carbon dioxide and water. The feed contains 20 mole% CH_4, 60% O_2, and 20% CO_2, and a 90% conversion of the limiting reactant is achieved. Calculate the molar composition of the product stream using (1) balances on molecular species, (2) atomic balances, (3) the extent of reaction.

Solution

Basis: 100 mol Feed

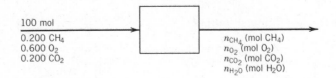

$$CH_4 + 2\,O_2 \rightarrow CO_2 + 2\,H_2O$$

Since a 2:1 ratio of O_2 to CH_4 would be stoichiometric and the actual feed ratio is 3:1, CH_4 is the limiting reactant and O_2 is in excess.

Before balances are written the given process information should be used to determine unknown variables or relations between them. In this case, the fact that the methane conversion is 90% tells us that 10% of the methane fed to the reactor emerges in the product, or

$$n_{CH_4} = 0.100(20.0\ \text{mol CH}_4\ \text{fed}) = 2.0\ \text{mol CH}_4$$

Now all that remains are the balances. We will proceed by each of the indicated methods.

1. ***Molecular Balances***

$$CH_4\ \text{reacted} = \frac{20.0\ \text{mol CH}_4\ \text{fed}}{}\ \bigg|\ \frac{0.900\ \text{mol react}}{\text{mol fed}} = 18\ \text{mol CH}_4\ \text{react}$$

CO_2 Balance (input + generation = output)

$$\frac{100\ \text{mol}}{}\ \bigg|\ \frac{0.200\ \text{mol CO}_2}{\text{mol fed}}\ +\ \frac{18\ \text{mol CH}_4\ \text{react}}{}\ \bigg|\ \frac{1\ \text{mol CO}_2\ \text{produced}}{1\ \text{mol CH}_4\ \text{react}} = n_{CO_2}$$

$$\Downarrow$$

$$n_{CO_2} = 38\ \text{mol CO}_2$$

H_2O Balance (generation = output)

$$\frac{18\ \text{mol CH}_4\ \text{react}}{}\ \bigg|\ \frac{2\ \text{mol H}_2O\ \text{produced}}{1\ \text{mol CH}_4\ \text{react}} = n_{H_2O}$$

$$\Downarrow$$

$$n_{H_2O} = 36\ \text{mol H}_2O$$

Molecular O_2 Balance (input = output + consumption)

$$\frac{100\ \text{mol}}{}\ \bigg|\ \frac{0.600\ \text{mol O}_2}{\text{mol}} = n_{O_2}\ +\ \frac{18\ \text{mol CH}_4\ \text{react}}{}\ \bigg|\ \frac{2\ \text{mol O}_2\ \text{react}}{1\ \text{mol CH}_4\ \text{react}}$$

$$\Downarrow$$

$$n_{O_2} = (60 - 36)\ \text{mol O}_2 = 24\ \text{mol O}_2$$

In summary, the output quantities are 2 mol CH_4, 24 mol O_2, 38 mol CO_2, and 36 mol H_2O, for a total of 100 mol. (Since 3 moles of products are produced for every 3 moles of reactants consumed, it should come as no surprise that total moles in = total moles out.) The mole fractions of the product gas components are thus:

$$2 \text{ mol } CH_4/100 \text{ mol} = \boxed{\begin{array}{l} 0.02 \text{ mol } CH_4/\text{mol}, \quad 0.24 \text{ mol } O_2/\text{mol}, \\ 0.38 \text{ mol } CO_2/\text{mol}, \quad 0.36 \text{ mol } H_2O/\text{mol} \end{array}}$$

2. *Atomic Balances*

Referring to the flowchart, we see that a balance on atomic carbon involves only one unknown (n_{CO_2}), and a balance on atomic hydrogen also involves one unknown (n_{H_2O}), but a balance on atomic oxygen involves three unknowns. We should therefore write the C and H balances first, and then the O balance to determine the remaining unknown variable. All atomic balances have the form input = output. (We will just determine the component amounts; calculation of the mole fractions then follows as in part 1.)

C Balance

$$\frac{20.0 \text{ mol } CH_4}{} \left| \frac{1 \text{ mol C}}{1 \text{ mol } CO_2} \right. + \frac{20 \text{ mol } CO_2}{} \left| \frac{1 \text{ mol C}}{1 \text{ mol } CO_2} \right.$$

$$= \frac{2.0 \text{ mol } CH_4}{} \left| \frac{1 \text{ mol C}}{1 \text{ mol } CH_4} \right. + \frac{n_{CO_2} \text{ mol } CO_2}{} \left| \frac{1 \text{ mol C}}{1 \text{ mol } CO_2} \right.$$

$$\Downarrow$$

$$n_{CO_2} = 38 \text{ mol } CO_2$$

H Balance

$$\frac{20 \text{ mol } CH_4}{} \left| \frac{4 \text{ mol H}}{1 \text{ mol } CH_4} \right. = \frac{2 \text{ mol } CH_4}{} \left| \frac{4 \text{ mol H}}{1 \text{ mol } CH_4} \right.$$

$$+ \frac{n_{H_2O} \text{ mol } H_2O}{} \left| \frac{2 \text{ mol H}}{1 \text{ mol } H_2O} \right.$$

$$\Downarrow$$

$$n_{H_2O} = 36 \text{ mol } H_2O$$

O Balance

$$\frac{60 \text{ mol } O_2}{} \left| \frac{2 \text{ mol O}}{1 \text{ mol } O_2} \right. + \frac{20 \text{ mol } CO_2}{} \left| \frac{2 \text{ mol O}}{1 \text{ mol } CO_2} \right.$$

$$= (n_{O_2} \text{ mol } O_2)(2) + (38 \text{ mol } CO_2)(2) + (36 \text{ mol } H_2O)(1)$$

$$\Downarrow$$

$$n_{O_2} = 24 \text{ mol } O_2$$

confirming the results obtained using molecular balances.

3. ***Extent of Reaction***

According to Equation 4.6-2, for all reactive species

$$n_{out} = n_{in} + \beta_i \xi$$

For the species in this problem, we may write

$$n_{CH_4} = 20.0 \text{ mol} - \xi \xrightarrow[\qquad]{n_{CH_4} = 2 \text{ mol } CH_4} \xi = 18.0 \text{ mol}$$

$$n_{O_2} = 60.0 \text{ mol} - 2\xi = 60.0 - (2)(18.0) = 24.0 \text{ mol } O_2$$

$$n_{CO_2} = 20.0 \text{ mol} + \xi = 20.0 + 18.0 = 38.0 \text{ mol } CO_2$$

$$n_{H_2O} = 0 + 2\xi = (2)(18.0) = 36.0 \text{ mol } H_2O$$

Again, the previous solutions are obtained.

As a rule, balances on atomic and molecular species are equally convenient to use when only one chemical reaction takes place, while balances on atomic species are preferable if several simultaneous reactions take place.

TEST YOURSELF

Methane is burned to form carbon dioxide and water in a batch reactor:

$$CH_4 + 2\,O_2 \rightarrow CO_2 + 2\,H_2O$$

The feed to the reactor and the products obtained are shown in the following flow chart.

100 mol CH$_4$	40 mol CH$_4$
250 mol O$_2$	130 mol O$_2$
	60 mol CO$_2$
	120 mol H$_2$O

1. How much CH$_4$ reacted? What is the fractional conversion of this substance?
2. Write the following balances, and verify that they are satisfied. Your solutions should resemble those given for parts (a) and (b).
 (a) CH$_4$. {I = O + C. 100 mol CH$_4$ = (40 + 60) mols CH$_4$ ⇒ 100 = 100}
 (b) Atomic O. {I = O. (250)(2) mol O = [(130 × 2) + (60)(2) + 120] mol O ⇒ 500 = 500}
 (c) Molecular O$_2$.
 (d) H$_2$O.
 (e) Atomic hydrogen.

4.6f Product Separation and Recycle

Two definitions of reactant conversion are used in the analysis of chemical reactors with product separation and recycle of unconsumed reactants:

Overall Conversion: $\dfrac{\text{reactant input to process} - \text{reactant output from process}}{\text{reactant input to process}}$

$$(4.6\text{-}6)$$

Single-Pass Conversion: $\dfrac{\text{reactant input to reactor} - \text{reactant output from reactor}}{\text{reactant input to reactor}}$

$$(4.6\text{-}7)$$

As usual, the corresponding percentage conversions are obtained by multiplying these quantities by 100%.

For example, consider the following labeled flowchart for a simple chemical process based on the reaction A → B.

The overall conversion of A is from Equation 4.6-6.

$$\frac{(75 \text{ mol A/min})_{\text{in}} - (0 \text{ mol/min})_{\text{out}}}{75 \text{ mol A/min}} \times 100\% = 100\%$$

The single-pass conversion is from Equation 4.6-7

$$\frac{(100 \text{ mol A/min})_{\text{in}} - (25 \text{ mol A/min})_{\text{out}}}{100 \text{ mol A/min}} \times 100\% = 75\%$$

This provides another illustration of the object of recycle. We have achieved complete use of the reactant for which we are paying—the fresh feed—even though only 75% of the reactant entering the reactor is consumed before emerging. The reason the overall conversion is 100% is that perfect separation was assumed: any A that does not react gets sent back to the reactor. If a less-than-perfect separation were achieved and some A left with the product stream, the overall conversion would be less than 100%, although it would always be greater than the single-pass conversion.

TEST YOURSELF

What are the overall and single-pass conversions for the process shown in Figure 4.5-1?

EXAMPLE 4.6-5 Dehydrogenation of Propane

Propane is dehydrogenated to form propylene in a catalytic reactor

$$C_3H_8 \rightarrow C_3H_6 + H_2$$

The process is to be designed for a 95% overall conversion of propane. The reaction products are separated into two streams: the first, which contains H_2, C_3H_6, and 0.555% of the propane that leaves the reactor, is taken off as product; the second

stream, which contains the balance of the unreacted propane and 5% of the propylene in the product stream, is recycled to the reactor. Calculate the composition of the product, the ratio (moles recycled)/(moles fresh feed), and the single-pass conversion.

Solution

Basis: 100 mol Fresh Feed

Note: In labeling the feed stream to the reactor, we have implicitly used balances on propane and propylene about the stream junction.

In terms of the labeled variables, the quantities to be calculated are

Mole Fractions of Product Stream Components: $\dfrac{Q_1}{Q_1 + Q_2 + Q_3}, \ldots$

Recycle Ratio: $\dfrac{Q_{r_1} + Q_{r_2}}{100}$

Single-Pass Conversion: $\dfrac{(100 + Q_{r_1}) - P_1}{(100 + Q_{r_1})} \times 100\%$

We must therefore calculate the values of $Q_1, Q_2, Q_3, Q_{r_1}, Q_{r_2}$ and P_1.
Let us first examine the overall process.

$$\text{Overall conversion} = 95\% \Rightarrow \frac{100 \text{ mol } C_3H_8 - Q_1}{100 \text{ mol } C_3H_8} = 0.95$$

$$\Downarrow$$

$$\boxed{Q_1 = 5 \text{ mol } C_3H_8}$$

Overall Propane Balance: input = output + consumption:

$$100 \text{ mol } C_3H_8 = 5 \text{ mol } C_3H_8 + C \text{ (mol } C_3H_8 \text{ reacted)}$$

$$\Downarrow$$

$$C = 95 \text{ mol } C_3H_8 \text{ reacted}$$

Overall Propylene Balance: output = generation:

$$Q_2 = \frac{95 \text{ mol } C_3H_8 \text{ reacted}}{} \frac{1 \text{ mol } C_3H_6 \text{ formed}}{1 \text{ mol } C_3H_8 \text{ reacted}}$$

$$\Downarrow$$

$$\boxed{Q_2 = 95 \text{ mol } C_3H_6}$$

Similarly, an *overall H_2 balance* yields

$$Q_3 = 95 \text{ mol } H_2$$

The analysis of the product is, therefore,

$$
\begin{array}{l|l}
5 \text{ mol } C_3H_8 & 2.6\% \ C_3H_8 \\
95 \text{ mol } C_3H_6 \Rightarrow & 48.7\% \ C_3H_6 \\
95 \text{ mol } H_2 & 48.7\% \ H_2 \\
\hline
195 \text{ mol total}
\end{array}
$$

We may now proceed to the analysis of the interior streams including the recycle stream. First, we summarize the information given in the problem statement.

$$Q_1 = 0.00555P_1 \xrightarrow{\quad Q_1 = 5 \text{ mol } C_3H_8 \quad} \boxed{P_1 = 900 \text{ mol } C_3H_8}$$

$$Q_{r_2} = 0.05Q_2 \xrightarrow{\quad Q_2 = 95 \text{ mol } C_3H_6 \quad} \boxed{Q_{r_2} = 4.75 \text{ mol } C_3H_6}$$

Next, write balances around the separation unit (which is nonreactive, so that input = output for all species.)

C_3H_8 Balance About Separation Unit

$$P_1 = Q_1 + Q_{r_1}$$
$$\Big\Downarrow \begin{array}{l} P_1 = 900 \text{ mol } C_3H_8 \\ Q_1 = 5 \text{ mol } C_3H_8 \end{array}$$
$$\boxed{Q_{r_1} = 895 \text{ mol } C_3H_8}$$

We now have all the variable values we need. The desired quantities are

$$\text{Recycle ratio} = \frac{(Q_{r_1} + Q_{r_2}) \text{ mol recycle}}{100 \text{ mol fresh feed}}$$
$$\Big\Downarrow \begin{array}{l} Q_{r_1} = 895 \text{ mol} \\ Q_{r_2} = 4.75 \text{ mol} \end{array}$$
$$\boxed{9.00 \ \frac{\text{mol recycle}}{\text{mol fresh feed}}}$$

$$\text{Single-pass conversion} = \frac{(100 + Q_{r_1}) - P_1}{(100 + Q_{r_1})} \times 100\%$$
$$\Big\Downarrow \begin{array}{l} Q_{r_1} = 895 \text{ mol } C_3H_8 \\ P_1 = 900 \text{ mol } C_3H_8 \end{array}$$
$$\boxed{9.55\%}$$

4.6g Purging

A problem may arise in processes that involve recycling. Suppose a material that enters with the fresh feed or is produced in a reaction remains entirely in a recycle stream, rather than being carried out in a process product. If nothing were done to prevent this occurrence, the substance would continuously enter the process and would have no way of leaving; it would therefore steadily accumulate, making the attainment of steady state impossible. To prevent this buildup, a portion of the recycle stream must be withdrawn as a **purge stream** to rid the process of the undesired material.

The flowchart given in Figure 4.6-2 for the production of ethylene oxide from ethylene illustrates this situation. The reaction is $2\,C_2H_4 + O_2 \rightarrow 2\,C_2H_4O$.

FIGURE 4.6-2

Process with recycle and purge

A mixture of ethylene and air constitutes the fresh feed to the process. The effluent from the reactor passes to an absorber, in which the ethylene oxide passes into a liquid solution. The gas stream leaving the absorber, which contains nitrogen and unreacted ethylene and oxygen, is recycled to the reactor.

If there were no nitrogen (or any other inert substance) in the feed, there would be no need for a purge stream. The recycle would contain only ethylene and oxygen; the fresh feed would contain just enough of these substances to make up for the amount lost in the reaction, and the system would be at steady state. However, there is nitrogen. It enters the system at a rate of 113 mol/s and leaves the system at the same rate in the purge stream. If the system were not purged, nitrogen would accumulate at this rate until something—probably unpleasant—occurred to shut the process down.

Material balance calculations on systems involving recycle and purge follow the procedures given in previous sections. When labeling the flowchart, note that the purge stream and the recycle stream before and after the purge takeoff all have the same composition.

EXAMPLE 4.6-6 Recycle and Purge in the Synthesis of Methanol

Methanol may be produced by the reaction of carbon dioxide and hydrogen.

$$CO_2 + 3\,H_2 \rightarrow CH_3OH + H_2O$$

The fresh feed to the process contains hydrogen and carbon dioxide in stoichiometric proportion, and 0.5 mole% inerts (I). The reactor effluent passes to a condenser, which removes essentially all of the methanol and water formed, and none of the reactants or inerts. The latter substances are recycled to the reactor. To avoid buildup of the inerts in the system, a purge stream is withdrawn from the recycle. The feed to the reactor contains 2% inerts, and the single-pass conversion is 60%. Calculate the molar flow rates of the fresh feed, the total feed to the reactor, and the purge stream for a methanol production rate of 1000 mol/h.

Solution

Basis: 100 mol Combined Feed to the Reactor

As a general rule, the combined feed to the reactor is a convenient stream to use as a basis of calculation for recycle problems, provided that its composition is known. Since in this process the reactants are fed in stoichiometric proportion and they are never separated from each other, they must be present in stoichiometric proportion throughout the process; that is, $(CO_2/H_2) = 1/3$. The feed thus contains 2 mol I (2% of 100 mol), and 98 mol $CO_2 + H_2$, of which 24.5 mol are CO_2 (1/4 of 98) and 73.5 mol are H_2.

Take a moment to examine the chart labeling. In particular, notice that we have built in the facts that CO_2 and H_2 are always present in a 1:3 ratio and that the compositions of the gaseous effluent from the condenser, the purge stream, and the recycle stream are all identical. The more of this sort of information you can incorporate in the chart labeling, the easier the subsequent calculations become.

Let us outline the solution, referring to the flowchart.

1. Calculate n_1, n_2, and n_3 from the feed to the reactor, the single-pass conversion, and reactor balances.

2. Calculate n_4 from a total mole balance about the condenser, and then x_4 from a CO_2 balance about the condenser.

3. Calculate n_0 and n_6 from balances on total moles and I about the recycle-fresh-feed mixing point. (Two equations in two unknowns.) Then calculate n_5 from a mole balance about the recycle–purge split point.

4. Scale up the calculated flows of fresh feed, combined reactor feed, and purge streams by the factor $(1000/n_2)$. The results will be the flow rates in mol/h corresponding to a methanol production of 1000 mol/h.

60% Single-Pass Conversion

$$n_1 = 0.400(24.5) = 9.80 \text{ mol } CO_2$$
$$3n_1 = 29.4 \text{ mol } H_2$$
$$0.600(24.5) = 14.7 \text{ mol } CO_2 \text{ react}$$
$$\Downarrow \text{ Balances on } CH_3OH, H_2O: \text{ output} = \text{generation}$$
$$n_2 = 14.7 \text{ mol } CH_3OH$$
$$n_3 = 14.7 \text{ mol } H_2O$$

Mole Balance About Condenser

$$2 + 4n_1 + n_2 + n_3 = n_2 + n_3 + n_4$$
$$\Downarrow \text{ substitute for } n_1$$
$$n_4 = 41.2 \text{ mol}$$

CO_2 Balance About Condenser

$$n_1 = n_4 x_4$$
$$\Downarrow$$
$$x_4 = (9.80/41.2) = 0.2379 \text{ mol } CO_2/\text{mol}$$

Mole Balance About Mixing Point

$$n_0 + n_6 = 100 \tag{1}$$

I Balance About Mixing Point

$$0.00500n_0 + n_6(1 - 4x_4) = 2$$
$$\Downarrow x_4 = 0.2379$$
$$0.00500n_0 + 0.0484n_6 = 2 \tag{2}$$

Solving (1) and (2) simultaneously yields

$$n_0 = 65.4 \text{ mol fresh feed}$$
$$n_6 = 34.6 \text{ mol recycle}$$

Mole Balance About Purge Takeoff

$$n_4 = n_5 + n_6$$
$$\Downarrow$$
$$n_5 = 41.2 - 34.6 = 6.6 \text{ mol purge}$$

The required scale factor is

$$[(1000 \text{ mol/h})/n_2 \text{ mol}] \xrightarrow{n_2 = 14.7} 68.03 \text{ h}^{-1}$$

The desired flow rates are therefore

$65.4 \times 68.03 =$	4450 mol fresh feed/h
$100.0 \times 68.03 =$	6803 mol/h fed to reactor
$6.6 \times 68.03 =$	449 mol/h purge

4.7 COMBUSTION REACTIONS

Combustion—the rapid reaction of a fuel with oxygen—is perhaps more important than any other class of industrial chemical reactions, despite the fact that combustion products (CO_2, H_2O, and frequently CO and SO_2) are relatively worthless compared with the fuels burned to obtain them. The significance of these reactions lies in the tremendous quantities of heat they release—heat that is used to produce steam, which is then used to drive the turbines that generate most of the world's electrical power. The job of designing power generation equipment usually falls to mechanical engineers, but the analysis of combustion reactions and reactors and the abatement and control of environmental pollution caused by combustion products are problems with which chemical engineers are heavily involved.

In the sections that follow, we introduce terminology commonly used in the analysis of combustion reactors and discuss material balance calculations for such reactors. Methods of determining the heat that can be obtained from combustion reactions are given in Chapter 9.

4.7a Combustion Chemistry

Most of the fuel used in power plant combustion furnaces is either coal (carbon, some hydrogen and sulfur, and various noncombustible materials), fuel oil (mostly heavy hydrocarbons, some sulfur), gaseous fuel (such as *natural gas*, which is primarily methane), or *liquefied petroleum gas*, which is usually propane and/or butane.

When a fuel is burned, carbon in the fuel reacts to form either CO_2 or CO, hydrogen forms H_2O, and sulfur forms SO_2. At temperatures greater than approximately

1800°C, some of the nitrogen in the air reacts to form nitric oxide (NO). A combustion reaction in which CO is formed from a hydrocarbon is referred to as **partial combustion** or **incomplete combustion** of the hydrocarbon.

Examples:

$$C + O_2 \rightarrow CO_2 \qquad \text{Complete combustion of carbon}$$

$$C_3H_8 + 5\,O_2 \rightarrow 3\,CO_2 + 4\,H_2O \qquad \text{Complete combustion of propane}$$

$$C_3H_8 + \tfrac{7}{2}\,O_2 \rightarrow 3\,CO + 4\,H_2O \qquad \text{Partial combustion of propane}$$

$$CS_2 + 3\,O_2 \rightarrow CO_2 + 2\,SO_2 \qquad \text{Complete combustion of carbon disulfide}$$

Perry's Chemical Engineers' Handbook, pp. 9-3 through 9-54,[2] provides a thorough discussion of fuels and combustion.

For obvious economic reasons, air is the source of oxygen in most combustion reactors. Air has the following molar composition.

N_2	78.03%	
O_2	20.99%	
Ar	0.94%	Average molecular weight = 29.0
CO_2	0.03%	
H_2, He, Ne, Kr, Xe	0.01%	
	100.00%	

In most combustion calculations, it is acceptable to simplify this composition to 79% N_2, 21% O_2 ⇒ 79 moles N_2/21 moles O_2 = 3.76 moles N_2/mole O_2.

The term **composition on a wet basis** is commonly used to denote the component mole fractions of a gas that contains water; **composition on a dry basis** signifies the component mole fractions of the same gas without the water. For example, a gas that contains 33.3 mole% CO_2, 33.3% N_2 and 33.3% H_2O (wet basis) contains 50% CO_2 and 50% N_2 on a dry basis.

The product gas that leaves a combustion chamber is referred to as the **stack gas** or **flue gas.** When the flow rate of a gas in a stack is measured, it is the total flow rate of the gas including water; on the other hand, common techniques for analyzing stack gases provide compositions on a dry basis. You must, therefore, be able to convert a composition on a dry basis to its corresponding composition on a wet basis before writing material balances on the combustion reactor. The procedure to convert a composition from one basis to another is similar to the one used to convert mass fractions to mole fractions and vice versa: Assume an amount of the stack gas (e.g., 100 mol of wet gas if composition on a wet basis is known, 100 mol of dry gas if a dry basis composition is given), calculate how much of each component is present, and use this information to calculate mole fractions on the basis desired.

[2] R. H. Perry and D. W. Green, Eds., *Perry's Chemical Engineers' Handbook*, 6th Edition, McGraw-Hill, New York, 1984.

EXAMPLE 4.7-1 Composition on Wet and Dry Bases

1. *Wet Basis ⇒ Dry Basis*

A stack gas contains 60 mole% N_2, 15% CO_2, 10% O_2, and the balance H_2O. Calculate the molar composition of the gas on a dry basis.

Solution

Basis: 100 mol Wet Gas

$$\begin{array}{l} 60 \text{ mol } N_2 \\ 15 \text{ mol } CO_2 \\ \underline{10 \text{ mol } O_2} \\ 85 \text{ mol dry gas} \end{array}$$

$$\frac{60}{85} = 0.706 \frac{\text{mol } N_2}{\text{mol dry gas}}$$

$$\frac{15}{85} = 0.176 \frac{\text{mol } CO_2}{\text{mol dry gas}}$$

$$\frac{10}{85} = 0.118 \frac{\text{mol } O_2}{\text{mol dry gas}}$$

2. *Dry Basis ⇒ Wet Basis*

An **Orsat analysis** (a technique for stack gas analysis) yields the following dry basis composition:

N_2	65%
CO_2	14%
CO	11%
O_2	10%

A humidity measurement shows that the mole fraction of H_2O in the stack gas is 0.07. Calculate the stack gas composition on a wet basis.

Solution

Basis: 100 lb-moles Dry Gas

$$0.07 \frac{\text{lb-mole } H_2O}{\text{lb-mole wet gas}} \Leftrightarrow 0.93 \frac{\text{lb-mole dry gas}}{\text{lb-mole wet gas}}$$

$$\frac{0.07 \text{ lb-mole } H_2O/\text{lb-mole wet gas}}{0.93 \text{ lb-mole dry gas}/\text{lb-mole wet gas}} = 0.0753 \frac{\text{lb-mole } H_2O}{\text{lb-mole dry gas}}$$

Hence the gas in the assumed basis contains

$$\frac{100 \text{ lb-moles dry gas} \mid 0.0753 \text{ lb-mole H}_2\text{O}}{\mid \text{lb-mole dry gas}} = 7.53 \text{ lb-moles H}_2\text{O}$$

$$\frac{100 \text{ lb-moles dry gas} \mid 0.650 \text{ lb-mole N}_2}{\mid \text{lb-mole dry gas}} = 65.0 \text{ lb-moles N}_2$$

$$(100)(0.140) = 14.0 \text{ lb-moles CO}_2$$
$$(100)(0.110) = 11.0 \text{ lb-moles CO}$$
$$(100)(0.100) = \underline{10.0 \text{ lb-moles O}_2}$$
$$107.5 \text{ lb-moles wet gas}$$

The mole fractions of each stack gas component may now easily be calculated:

$$y_{\text{H}_2\text{O}} = \frac{7.53}{107.5} \frac{\text{lb-moles H}_2\text{O}}{\text{lb-moles wet gas}} = 0.070 \frac{\text{lb-mole H}_2\text{O}}{\text{lb-mole wet gas}}, \ldots$$

TEST YOURSELF

1. What is the approximate molar composition of air? What is the mole ratio of N_2 to O_2 in air?
2. A gas contains 1 mol H_2, 1 mol O_2, and 2 mol H_2O. What is the molar composition of this gas on a wet basis? On a dry basis?
3. A flue gas contains 5 mole% H_2O. Calculate:
 (a) kmol flue gas/kmol H_2O
 (b) kmol dry flue gas/kmol flue gas
 (c) kmol H_2O/kmol dry flue gas

4.7b Theoretical and Excess Air

If two reactants participate in a reaction and one is considerably more expensive than the other, the usual practice is to feed the less expensive reactant in excess of the valuable one. This has the effect of increasing the conversion of the valuable reactant at the expense of the cost of the excess reactant and additional pumping costs.

The extreme case of an inexpensive reactant is air, which is free. Combustion reactions are therefore invariably run with more air than is needed to supply oxygen in stoichiometric proportion to the fuel. The following terms are commonly used to describe the proportions of fuel and air fed to a reactor.

Theoretical Oxygen: The moles (batch) or molar flow rate (continuous) of O_2 needed for complete combustion of all of the fuel fed to the reactor, assuming that all carbon in the fuel is oxidized to CO_2 and all hydrogen is oxidized to H_2O.

Theoretical Air: The quantity of air that contains the theoretical oxygen.

Excess Air: The amount by which the air fed to the reactor exceeds the theoretical air.

Percent Excess Air: $\dfrac{(\text{moles air})_{\text{fed}} - (\text{moles air})_{\text{theoretical}}}{(\text{moles air})_{\text{theoretical}}} \times 100\%$ **(4.7-1)**

If you know the fuel feed rate and the stoichiometric equation for complete combustion of the fuel, you can calculate the theoretical O_2 and air feed rates. If in addition you know the actual feed rate of air, you can calculate the percent excess air from Equation 4.7-1. It is also easy to calculate the air feed rate from the theoretical air and a given value of the percentage excess: If 50% excess air is supplied, for example, then

$$(\text{moles air})_{\text{fed}} = 1.5 \, (\text{moles air})_{\text{theoretical}}$$

EXAMPLE 4.7-2 Theoretical and Excess Air

One hundred mol per hour of butane (C_4H_{10}) and 5000 mol per hour of air are fed into a combustion reactor. Calculate the percent excess air.

Solution

First, calculate the theoretical air from the feed rate of fuel and the stoichiometric equation for complete combustion of butane.

$$C_4H_{10} + \tfrac{13}{2} O_2 \rightarrow 4\,CO_2 + 5\,H_2O$$

$$(O_2)_{\text{theoretical}} = \frac{100 \text{ mol } C_4H_{10}}{h} \,\bigg|\, \frac{6.5 \text{ mol } O_2 \text{ required}}{\text{mol } C_4H_{10}}$$

$$= 650 \frac{\text{mol } O_2}{h}$$

$$(\text{air})_{\text{th}} = \frac{650 \text{ mol } O_2}{h} \,\bigg|\, \frac{4.76 \text{ mol air}}{\text{mol } O_2} = 3094 \frac{\text{mol air}}{h}$$

Hence

$$\% \text{ excess air} = \frac{(\text{air})_{\text{fed}} - (\text{air})_{\text{th}}}{(\text{air})_{\text{th}}} \times 100\% = \frac{5000 - 3094}{3094} \times 100\% = \boxed{61.6\%}$$

If instead you had been *given* 61.6% excess air, you could have calculated $(\text{air})_{\text{fed}} = 1.616\,(\text{air})_{\text{th}} = 1.616\,(3094) = 5000$ mol/h.

Two points of confusion often arise in the calculation of theoretical and excess air, both of which are caused by ignoring the definitions of these terms.

1. ***The theoretical air required to burn a given quantity of fuel does not depend on how much is actually burned.*** The fuel may not react completely, and it may be partially burned to form both CO and CO_2, but the theoretical air is still that which would be required to react with *all* of the fuel to form CO_2 only.

2. *The value of the percent excess air depends only on the theoretical air and the air feed rate, and not on how much O_2 is consumed in the reactor or whether combustion is complete or partial.*

TEST YOURSELF

Methane burns in the reactions:

$$CH_4 + 2\,O_2 \rightarrow CO_2 + 2\,H_2O$$
$$CH_4 + \tfrac{3}{2}\,O_2 \rightarrow CO + 2\,H_2O$$

One hundred mol of methane per hour are fed to a reactor.

1. What is the theoretical O_2 flow rate if complete combustion occurs in the reactor?
2. What is the theoretical O_2 flow rate assuming that 70% of the methane reacts to form CO? (Careful!)
3. What is the theoretical air flow rate?
4. If 100% excess air is supplied, what is the flow rate of air entering the reactor?
5. If the actual flow rate of air is such that 300 mol O_2/h enter the reactor, what is the percent excess air?

CREATIVITY EXERCISES

1. Years ago it was common to operate boiler furnaces with air fed in 20% excess or more, while today improved boiler designs enable the use of 5–10% excess air. Cite as many possible negative consequences as you can think of for the air-to-fuel feed ratio being (a) too low; (b) too high.
2. The costs of petroleum and natural gas have increased dramatically in the last decade, and there is some question about their continued long-term availability. List as many alternative energy sources as you can think of, being as creative as you can, and then go back and suggest possible drawbacks to each one.

4.7c Material Balances on Combustion Reactors

The procedure for writing and solving material balances for a combustion reactor is essentially the same as that for any other reactive system. Bear in mind these additional points, however.

1. When you draw and label the flowchart, remember to include the N_2 at both the inlet and outlet of the reactor, and unreacted fuel and O_2 as well as combustion products (CO_2, H_2O, CO) at the outlet.
2. If you are given the percent excess air, the O_2 actually fed may be calculated by multiplying the theoretical O_2—determined from the fuel feed rate and composi-

tion and the reaction stoichiometry for complete combustion—by (1 + fractional excess air). The N_2 fed can then be calculated as $3.76 \times O_2$ fed, and total air fed is $4.76 \times O_2$ fed.

3. If only one reaction is involved, it is no more or less difficult to write balances on molecular species than on atomic species to calculate unknown stream variables; if several reactions occur simultaneously, however—such as combustion to form both CO and CO_2—atomic balances are usually more convenient.

EXAMPLE 4.7-3 Combustion of Ethane

Ethane (C_2H_6) is burned with 50% excess air. The percentage conversion of the ethane is 90%; of the ethane burned, 25% reacts to form CO and the balance to form CO_2. Calculate the composition of the flue gas and the ratio of water to dry flue gas.

Solution

Basis: 100 mol Ethane Fed

$$C_2H_6 + \frac{7}{2} O_2 \rightarrow 2\ CO_2 + 3\ H_2O$$

$$C_2H_6 + \frac{5}{2} O_2 \rightarrow 2\ CO + 3\ H_2O$$

Notes:

1. Since no product stream mole fractions are known, subsequent calculations are easier if individual component amounts rather than a total amount and mole fractions are labeled.

2. Known information about the composition of air has been used to label the incoming N_2 stream $(3.76 = \frac{79}{21})$.

3. If the ethane reacted completely, q_1 would be omitted.

4. Since excess air is supplied, O_2 *must* appear in the product stream.

Theoretical O_2

$$\frac{100 \text{ mol } C_2H_6}{} \left| \frac{3.50 \text{ mol } O_2}{1 \text{ mol } C_2H_6} \right. = 350 \text{ mol}$$

O_2 Fed

$$Q = (1.5)(350) = \boxed{525 \text{ mol } O_2 \text{ fed}}$$

N_2 Fed

$$(3.76)(525) = \boxed{1974 \text{ mol } N_2 \text{ fed}}$$

C_2H_6 Reacted

$$(0.90)(100) = 90 \text{ mol } C_2H_6 \text{ reacted}$$

C_2H_6 Balance

$$\text{Output} = \text{input} - \text{consumption} \quad \boxed{q_1 = 100 - 90 = 10 \text{ mol}}$$

N_2 Balance

$$\text{Input} = \text{output} \quad \boxed{1974 \text{ mol } N_2 = q_3}$$

CO Balance

$$\text{Output} = \text{generation}$$

$$q_4 = \frac{0.25 \times 90 \text{ mol } C_2H_6 \text{ react to form CO}}{} \left| \frac{2 \text{ mol CO formed}}{1 \text{ mol } C_2H_6 \text{ reacted}} \right.$$

$$\Downarrow$$

$$\boxed{q_4 = 45 \text{ mol CO}}$$

Atomic C Balance

$$\text{Input} = \text{output}$$

$$ C_2H_6 \quad CO \quad CO_2$$

$$\frac{100 \text{ mol } C_2H_6}{} \left| \frac{2 \text{ mol C}}{1 \text{ mol } C_2H_6} \right. = \overbrace{(q_1)(2)} + \overbrace{(q_4)(1)} + \overbrace{(q_5)(1)}$$

$$\Downarrow \begin{array}{l} q_1 = 10 \\ q_4 = 45 \end{array}$$

$$\boxed{q_5 = 135 \text{ mol } CO_2}$$

Atomic H Balance

$$\text{Input} = \text{output}$$

$$\frac{100 \text{ mol } C_2H_6}{} \left| \frac{6 \text{ mol H}}{1 \text{ mol } C_2H_6} \right. = \frac{10 \text{ mol } C_2H_6}{} \left| \frac{6 \text{ mol H}}{1 \text{ mol } C_2H_6} \right.$$

$$+ \frac{q_6 \text{ (mol } H_2O)}{} \left| \frac{2 \text{ mol H}}{1 \text{ mol } H_2O} \right.$$

$$\Downarrow$$

$$\boxed{q_6 = 270 \text{ mol } H_2O}$$

Atomic O Balance

$$Input = output$$

$$\frac{525 \text{ mol } O_2}{} \left| \frac{2 \text{ mol } O}{1 \text{ mol } O_2} = \frac{q_2(\text{mol } O_2)}{} \right| \frac{2 \text{ mol } O}{1 \text{ mol } O_2} + \frac{45 \text{ mol } CO}{} \left| \frac{1 \text{ mol } O}{1 \text{ mol } CO} \right.$$

$$+ \frac{135 \text{ mol } CO_2}{} \left| \frac{2 \text{ mol } O}{1 \text{ mol } CO_2} \right.$$

$$+ \frac{270 \text{ mol } H_2O}{} \left| \frac{1 \text{ mol } O}{1 \text{ mol } H_2O} \right.$$

$$\Downarrow$$

$$\boxed{q_2 = 232 \text{ mol } O_2}$$

The analysis of the stack gas is now complete. Summarizing:

$$q_1 = 10 \text{ mol } C_2H_6$$
$$q_2 = 232 \text{ mol } O_2$$
$$q_3 = 1974 \text{ mol } N_2$$
$$q_4 = 45 \text{ mol } CO$$
$$q_5 = \underline{135 \text{ mol } CO_2}$$
$$2396 \text{ mol dry gas}$$
$$+ q_6 = \underline{270 \text{ mol } H_2O}$$
$$2666 \text{ mol total}$$

Hence the flue gas composition on a dry basis is

$$y_1 = \frac{10 \text{ mol } C_2H_6}{2396 \text{ mol dry gas}} = 0.00417 \frac{\text{mol } C_2H_6}{\text{mol}}$$

$$y_2 = \frac{232 \text{ mol } O_2}{2396 \text{ mol dry gas}} = 0.0970 \frac{\text{mol } O_2}{\text{mol}}$$

$$y_3 = \frac{1974 \text{ mol } N_2}{2396 \text{ mol dry gas}} = 0.824 \frac{\text{mol } N_2}{\text{mol}}$$

$$y_4 = \frac{45 \text{ mol } CO}{2396 \text{ mol dry gas}} = 0.019 \frac{\text{mol } CO}{\text{mol}}$$

$$y_5 = \frac{135 \text{ mol } CO_2}{2396 \text{ mol dry gas}} = 0.0563 \frac{\text{mol } CO_2}{\text{mol}}$$

and the mole ratio of water to dry flue gas is

$$\frac{270 \text{ mol } H_2O}{2396 \text{ mol dry flue gas}} = \boxed{0.113 \frac{\text{mol } H_2O}{\text{mol dry flue gas}}}$$

Frequently a fuel with an unknown molecular composition is burned. By analyzing the products of the combustion reaction and writing material balances, you can determine the ratio of hydrogen to carbon in the fuel; this information may enable you to deduce the approximate fuel composition. A calculation of this sort is illustrated in Example 4.7-4.

EXAMPLE 4.7-4 Combustion of a Natural Gas

A natural gas of an unknown composition is burned with air. An analysis of the product gas yields the following results:

$$0.130 \text{ mole } H_2O/\text{mole wet gas}$$

Orsat analysis of flue gas: 1.5% CO
6.0% CO_2
8.2% O_2
84.3% N_2

Calculate the ratio of hydrogen to carbon in the gas, and speculate on what the gas might be.

Solution

Basis: 100 mol Dry Flue Gas

Since the composition of the fuel is unknown, label it as though carbon and hydrogen were entering separately.

From the given mole fraction of water,

$$Q_w = \frac{100 \text{ mol dry gas}}{} \left| \frac{0.130 \text{ mol } H_2O}{0.870 \text{ mol dry gas}} = 14.9 \text{ mol } H_2O \right.$$

Atomic Carbon Balance

$$q_C = \frac{100 \text{ mol flue gas}}{} \left| \frac{0.015 \text{ mol CO}}{\text{mol flue gas}} \right| \frac{1 \text{ mol C}}{1 \text{ mol CO}}$$

$$+ \overbrace{(100)(0.06)(1)}^{\text{C in } CO_2} = 7.5 \text{ mol C}$$

Atomic H Balance

$$q_H = \frac{Q_w \,(\text{mol H}_2\text{O})}{} \left| \frac{2 \text{ mol H}}{\text{mol H}_2\text{O}} = 2 \times 14.9 = 29.8 \text{ mol H} \right.$$

from which

$$\frac{q_H}{q_C} = \frac{29.8 \text{ mol H}}{7.5 \text{ mol C}} = 3.97 \frac{\text{mol H}}{\text{mol C}}$$

The natural gas may therefore be written with the formula $(\text{CH}_{3.97})_n$. Since there is only one hydrocarbon for which the ratio of H to C is close to 3.97—that is, CH_4—we may conclude in this case that the natural gas is essentially pure methane, with perhaps trace amounts of other hydrocarbons. [If we had obtained, say, $q_H/q_C \approx 2$, we could have gone no further than to write the fuel as $(\text{CH}_2)_n$; from the information given, there would have been no way to distinguish between C_2H_4, C_3H_6, a mixture of CH_4 and C_2H_2, etc.]

PROBLEMS

1. Water enters a two-liter tank at a rate of 3.00 g/s and is withdrawn at a rate of 5.00 g/s. The tank is initially half full.
 (a) Is this process continuous, batch, or semibatch? Is it transient or steady state?
 (b) Write a mass balance for this process, letting $n(t)$ be the mass of water in the tank at time t. (See Example 4.2-1.) Identify the terms of the general balance equation that are present in your equation, and state the reason for omitting any terms not present.
 (c) Solve the balance equation to determine how long it will take the tank to drain completely.

2. A liquid-phase chemical reaction $A \rightarrow B$ takes place in a well-stirred tank. The concentration of A in the feed is $C_{A0}(\text{mol/cm}^3)$, and that in the tank and outlet stream is $C_A(\text{mol/cm}^3)$. Neither concentration varies with time. The volume of the tank contents is $V(\text{cm}^3)$, and the volumetric flow rate of the inlet and outlet streams is $Q(\text{cm}^3/\text{s})$.

The **reaction rate** (the rate at which A is consumed by reaction in the tank) is given by the expression

$$r(\text{mol/s}) = kVC_A$$

where k is a constant.

 (a) Is this process continuous, batch, or semibatch? Transient or steady state?

 (b) Write a differential balance on A to derive the following relation between the inlet and outlet reactant concentrations.

$$C_A = \frac{C_{A0}}{1 + kV/Q}$$

3. A liquid mixture of benzene and toluene contains 50% benzene by mass. A portion of the mixture is vaporized to yield a vapor containing 85.0% benzene and a residual liquid containing 10.6% benzene by mass.

 (a) Suppose the process is carried out continuously and at steady state, with a mixture feed rate of 100.0 kg/h. Let Q_v(kg/h) and Q_l(kg/h) be the flow rates of the vapor and liquid product streams. Draw and label a process flowchart, then write and solve differential balances on total mass and on benzene to determine Q_v and Q_l. (See Example 4.2-2.)

 (b) Next, suppose the process is carried out in a closed container, which initially contains 100.0 kg of the liquid mixture. Let Q_v and Q_l be the masses of the final vapor and liquid phases. Draw and label a flowchart, then write and solve integral balances on total mass on benzene to determine Q_v and Q_l.

4. Draw and label the given streams, and derive expressions for the quantities indicated in terms of labeled variables. The solution of part a is given as an illustration.

 (a) A continuous stream contains 40 mole% benzene and the balance toluene. Derive expressions for the molar and mass flow rates of benzene in terms of the total molar flow rate Q(mol/s) of the stream.

 Solution

 Q (mol/s)

 \longrightarrow

 0.4 mol C_6H_6/mol
 0.6 mol C_7H_8/mol

$$q_{C_6H_6} = \boxed{0.4Q \text{ (mol } C_6H_6/s)}$$

$$\frac{0.4Q \text{ (mol } C_6H_6)}{} \left| \frac{78.1 \text{ g } C_6H_6}{\text{mol}} \right. = \boxed{31.2Q \text{ (g } C_6H_6/s)}$$

 (b) The feed to a batch process contains equimolar amounts of hydrogen and oxygen. Derive an expression for the kilograms of oxygen in terms of the total moles Q(mol) of this stream.

 (c) A stream containing hydrogen, ethylene, and nitrogen has a mass flow rate of 100 kg/min. Derive an expression for the mass flow rate of H_2 (g/h) in terms of the mass fraction of this species, x_H.

 (d) A continuous stream of humid air contains water vapor and dry air, the latter to a good approximation containing 21 mole% O_2 and 79% N_2. Derive expressions

for the molar flow rate of O_2, and for the mole fractions of H_2O and O_2 in the gas in terms of Q_1 (lb-moles H_2O/s) and Q_2 (lb-moles dry air/s).

(e) The product stream from a batch reactor contains NO, NO_2, and N_2O_4. The mole fraction of NO_2 is 0.300. Derive an expression for the gram-moles of N_2O_4 in terms of Q(mol mixture) and x(mol NO/mol).

5. Draw a flowchart for the catalytic dehydrogenation of propane from the description of this process that begins Section 4.3a. Label all feed, product, and connecting streams between units.

6. Suppose a distillation column operates as shown in the following flowchart.

(a) How many independent material balances may be written for this system?

(b) How many unknown flow rates and/or mole fractions must be specified before the others may be calculated? (See Example 4.3-4. Also, remember what you know about the component mole fractions of a mixture—e.g., the relationship between x_1 and y_1.) Briefly explain your answer.

(c) Suppose values are given for Q_1 and x_1. Give a series of equations, each involving only a single unknown, for the remaining variables. (Once a variable has been calculated in one of these equations, it may of course appear in subsequent equations without being counted as an unknown.)

7. Following is a labeled flowchart of an extraction process in which a solute (A) is transferred from one solvent (S) to another (T) in which it is more soluble.

(a) What is the maximum number of independent material balances that can be written for this process?

(b) Calculate W, Q, and R, using the given feed as a basis and writing balances in an order such that you never have an equation that involves more than one unknown quantity.

(c) Calculate the difference between the amount of A in the feed solution and that in the 2% A solution, and show that it equals the amount that leaves in the 20% A solution.

(d) Calculate the ratio (g A in final S solution/g A in feed solution).

8. Strawberries contain about 15% solids and 85% water. To make strawberry jam, crushed strawberries and sugar are mixed in a 45:55 ratio, and the mixture is heated to evaporate water until the residue contains one-third water by mass. Draw and label a flowchart of this process, and use it to calculate how many pounds of strawberries are needed to make a pound of jam.

9. One thousand kilograms per hour of a mixture containing equal parts by mass of benzene and toluene are distilled. The flow rate of the overhead product stream is 488 kg/h, and the bottom stream contains 7.11 wt% benzene. Draw and label a flowchart of the process. Then calculate the mass and mole fractions of benzene and the molar flow rates of benzene and toluene (mol/h) in the overhead product stream.

10. A stream containing 25 wt% methanol in water is to be diluted with a second stream containing 10% methanol to form a product containing 17% methanol.

(a) Choose a convenient basis of calculation, draw and label a flowchart of this process, and calculate the ratio (kg 17% solution/kg 25% solution).

(b) What feed rate of the 10% solution would be required to produce 1250 kg/h of the product?

11. Eggs are sorted into two sizes (large and extra large) at the Cheerful Chicken Dairy. Unfortunately, business has not been good lately, and since the Cheerful Chicken's 40-year-old egg-sorting machine finally gave up the ghost there have been no funds available to replace it. Instead, Old Fred, one of the firm's sharper-eyed employees, has been equipped with a "Large" rubber stamp in his right hand and an "X-large" stamp in his left, and is assigned to stamp each egg with the appropriate label as it goes by on the conveyor belt. Down the line, another employee puts the eggs into one of two hoppers, each egg according to its stamp. The system works reasonably well, all things considered, except that Old Fred has a heavy hand, and on the average breaks 30% of the 120 eggs that pass by him each minute. At the same time, a check of the "X-large" stream reveals a flow rate of 70 eggs/min, of which 25 eggs/min are broken.

(a) Draw and label a flowchart for this process.

(b) Write and solve balances about the egg sorter on total eggs and broken eggs.

(c) How many "large" eggs leave the plant each minute?

(d) What fraction of the "large" eggs are broken?

(e) Is Old Fred right- or left-handed.

12. A paint mixture containing 25% of a pigment and the balance water sells for $12.00/kg, and a mixture containing 10% pigment sells for $7.00/kg. If a paint retailer produces a

blend containing 15% pigment, for how much should he sell it ($/kg) to make a 10% profit?

13. Liquid water and air flow into a humidification chamber in which the water evaporates completely. The entering air contains 1.00 mole% $H_2O(v)$, 20.8% O_2, and the balance N_2, and the humidified air contains 10.0 mole% H_2O. Calculate the volumetric flow rate (ft^3/min) of liquid required to humidify 200 lb-moles/min of the entering air.

14. The house special at Oswald's Oasis is a mixture containing 75% C_2H_5OH by mass, and the balance water. The cost of alcohol has been increasing, however, and Oswald has decided that perhaps a 60% blend would be just as effective. He has on hand a vat containing 300 gal of the 75% mixture (SG = 0.877) and can purchase any desired amount of a 40% mixture (SG = 0.952). How many gallons of the latter mixture must he buy?*

15. In a tannery, mangrove bark is extracted by treating the finely ground wood with hot water. The original bark contains 4% moisture, 37% tannin, and 23% soluble nontannin material. The residue (spent bark) removed from the extraction tanks contains 62% moisture, 2.8% tannin, and 0.9% soluble nontannin material. What percentage of the tannin in the original bark remains unextracted in the residue?

16. A sedimentation process is to be used to separate pulverized coal from slate. A suspension of finely divided particles of galena (lead sulfide, SG = 7.44) in water is prepared, with an overall SG of 1.48. A mixture of coal and slate is placed in the suspension; the coal rises to the top, and the slate sinks.
 (a) What can you conclude about the specific gravities of coal and slate?
 (b) How much galena must be added to 2500 kg of water to produce a suspension of the desired specific gravity?

17. A gas containing equal parts (on a molar basis) of H_2, N_2, and H_2O is passed through a column of calcium chloride pellets, which absorb 97% of the water and none of the other gases. The column packing was initially dry and had a mass of 2.00 kg. Following 6 h of continuous operation, the pellets are reweighed and are found to have a mass of 2.21 kg. Calculate the molar flow rate (mol/h) of the feed gas and the mole fraction of water vapor in the product gas.

18. Wet sugar that contains 20% water is sent through a dryer in which 75% of the water is removed.
 (a) Taking a basis of 100 kg feed, calculate the mass fraction of dry sugar in the wet sugar that leaves the dryer.
 (b) Calculate the ratio (kg H_2O removed/kg wet sugar leaving the dryer).
 (c) If 1000 ton/day of wet sugar are fed to the dryer, how much additional water must be removed from the outlet sugar to dry it completely, and how much revenue can be expected if dry sugar sells for $0.25/lb?

* You may wonder why Oswald does not simply add water to dilute the mixture. Here are three possible reasons—guess which one is correct: (1) He did not think of it; (2) he could, but the problem would then be less interesting; (3) we only thought about this after the manuscript was finished, and we were not about to change anything at that point.

19. One hundred milliliters per minute of a 5-molar aqueous solution of NaOH (SG = 1.18) are mixed with a 10-molar aqueous solution of NaOH (SG = 1.37). It is desired to produce a solution containing 11.7 mole% NaOH. Calculate the required volumetric flow rate (ml/min) of the 10-molar solution.

20. A dilute aqueous solution of H_2SO_4 (Solution A) is to be concentrated with a solution containing 90% H_2SO_4 by mass (Solution B) to produce a 75% solution (Solution C).

The flow rate and concentration of Solution A change periodically, so that it is necessary to adjust the flow rate of Solution B to keep the product H_2SO_4 concentration constant.

Flowmeters A and B have linear calibration plots of mass flow rate (q) versus meter reading (R), which pass through the following points:

$$\text{Flowmeter A:} \quad q_A = 150 \text{ lb}_m/\text{h}, \quad R_A = 20$$
$$q_A = 500 \text{ lb}_m/\text{h}, \quad R_A = 60$$
$$\text{Flowmeter B:} \quad q_B = 200 \text{ lb}_m/\text{h}, \quad R_B = 20$$
$$q_B = 800 \text{ lb}_m/\text{h}, \quad R_B = 60$$

The analyzer calibration is a straight line on a plot of $\%H_2SO_4(x)$ on a logarithmic scale versus meter reading (R_x) on a linear scale. The line passes through the points ($x = 20\%$, $R_x = 3.0$) and ($x = 100\%$, $R_x = 10.0$).

(a) Calculate the flow rate of Solution B needed to concentrate 300 lb_m/h of 55% H_2SO_4 (Solution A), and the resulting flow rate of Solution C. (The calibration data are not needed for this part.)

(b) Derive the calibration equations for $q_A(R_A)$, $q_B(R_B)$, and $x(R_x)$. Calculate the values of R_A, R_B, and R_x corresponding to the flow rates and concentrations of part a.

(c) The process technician's job is to read Flowmeter A and the analyzer periodically, and then to adjust the flow rate of Solution B to its required value. Derive a formula that the technician can use for R_B in terms of R_A and R_x, and then check it by substituting the values of part a.

21. A feed stream to a process is to consist of 1000 kg/h of a gas containing 25.0 mole% N_2 and the balance H_2. The stream is obtained by blending gases from two tanks, Tank A and Tank B. The gases in both tanks are mixtures of nitrogen and hydrogen, with nitrogen mole fractions x_A and x_B, respectively.

(a) Suppose $x_A = 0.10$ and $x_B = 0.50$. Calculate the required molar flow rates (mol/h) of the two gas mixtures.

(b) What mathematical condition or conditions involving x_A and x_B must be satisfied for a 25 mole% blend to be obtainable from the two tank gases? (*For example:* If $x_A = 0.10$ and $x_B = 0.20$, there is no way to produce a 25% mixture. Convince yourself that this statement is true, then come up with the requested condition.)

***(c)** Write a computer program to do the following things.

 1. Read (input) values of x_A and x_B.

 2. Quit if $x_A = -1.0$.

 3. See whether it is possible to make up a 25 mole% mixture from the two given feed gases. If it is not possible, print out an error message and go back to step 1.

 4. Calculate and print the molar flow rates of gases A and B needed to make 1000 kg/h of the 25% blend.

 5. Go back to step 1.

Run your program with the following input data.

x_A	0.05	0.10	0.10	0.10	0.20	0.20	0.20	0.25
x_B	0.50	0.50	0.35	0.20	0.50	0.35	0.20	0.50

22. An artificial kidney is a device that removes water and waste metabolites from blood. (See Example 3 of Chapter 1.) In one such device, the **hollow fiber hemodialyzer,** blood flows from an artery through the insides of a bundle of hollow cellulose acetate fibers, and dialyzing fluid, which consists of water and various dissolved salts, flows on the outside of the fibers. Water and waste metabolites—principally urea, creatinine, uric acid, and phosphate ions—pass through the fiber walls into the dialyzing fluid, and the purified blood is returned to a vein. Suppose that at some time during a dialysis the arterial and venous blood conditions are as follows.

	Arterial (entering) Blood	Venous (exiting) Blood
Flow Rate	200.0 ml/min	195.0 ml/min
Urea (H_2NCONH_2) Concentration	1.90 mg/ml	1.75 mg/ml

* Computer problem.

(a) Calculate the rates at which urea and water are being removed from the blood.

(b) If the dialyzing fluid enters at a rate of 1500 ml/min, calculate the concentration of urea in the exiting dialysate. (Neglect the urea volume.)

(c) Suppose we want to reduce the patient's urea level from an initial value of 2.7 mg/ml to a final value of 1.1 mg/ml. If the total blood volume is 5.0 liters and the average rate of urea removal is that calculated in part a, how long must the patient be dialyzed? (Neglect the loss in total blood volume due to the removal of water in the dialyzer.)

23. Carbon disulfide is to be recovered from a gas containing 15.0 mole% CS_2, 17.8% O_2, and 67.2% N_2. The gas is fed to a continuous absorption tower, where it contacts liquid benzene, which absorbs CS_2 but not O_2 or N_2. Benzene is fed to the column in a 2:1 mole ratio to the feed gas. Some of the benzene entering as liquid evaporates and leaves the top of the tower as vapor. The gas leaving the absorber contains 2% CS_2 and 2% benzene.

(a) Calculate the fraction of the CS_2 fed to the column recovered in the liquid outlet stream, the mole fraction of CS_2 in this stream, and the fraction of benzene fed to the column lost in the vapor product.

(b) What processing steps are the liquid and vapor effluent streams likely to be subjected to after they leave the absorption tower?

24. Alunite is an aluminum-containing compound with the formula $3Al_2O_3 \cdot K_2O \cdot 4SO_3 \cdot 6H_2O$. An ore containing alunite is analyzed and found to contain 23 wt% alumina (Al_2O_3). The ore is subjected to a series of chemical treatments that precipitate most of the aluminum in the ore as $Al(OH)_3$. The precipitate is separated and roasted to produce alumina, which is then dissolved in a molten salt bath and electrolyzed to form pure aluminum. About 88% of the aluminum in the ore is eventually recovered as the metal. Calculate the metric tons of ore required to produce a metric ton of aluminum.

25. Pipes through which process materials are transported in industrial plants are often too large in diameter for flow-metering devices such as rotameters and orifice meters to be used. A technique to measure flow rates in such pipes is the **indicator dilution method,** in which a continuous stream of an easily measured substance (the **tracer**) is injected into the process stream at a known rate, and the concentration of the tracer is measured at a point far enough downstream of the injection point for the tracer to be completely mixed with the process stream. The larger the flow rate of the process stream, the lower the concentration of tracer at the measurement point.

A gas stream that contains 1.00 mole% CO_2 and the balance methane flows through a pipeline. Fifty kilograms of CO_2 per minute are injected into the line, and a sample of the gas downstream of this point is found to contain 1.7 mole% CO_2.

(a) Calculate the molar flow rate of the gas prior to injection of the tracer.

(b) Suppose samples are withdrawn at several locations downstream of the injection point, and the measured CO_2 concentration at the first three of these points varies, but from the fourth point on the concentration is relatively steady at 1.7%. Explain this behavior.

26. A variation of the indicator dilution technique (see preceding problem) is used to measure

total blood volume. A known amount of a tracer is injected into the bloodstream and disperses uniformly throughout the circulatory system. A blood sample is then withdrawn, the tracer concentration in the sample is measured, and the measured concentration [which equals (tracer injected)/(total blood volume) if no tracer is lost through blood vessel walls] is used to determine the total blood volume.

In one such experiment, 0.60 cm³ of a solution containing 5.00 mg/liter of a dye is injected into an artery of a grown man. About 10 min later, after the tracer has had time to distribute itself uniformly throughout the bloodstream, a blood sample is withdrawn and placed in the sample chamber of a spectrophotometer. A beam of light passes through the chamber, and the spectrophotometer measures the intensity of the transmitted beam and displays the value of the solution absorbance (a quantity that increases with the amount of light absorbed by the sample). The value displayed is 0.18. A calibration curve of absorbance A versus tracer concentration C (micrograms dye/liter blood) is a straight line through the origin and the point ($A = 0.9$, $C = 3$ μg/liter). Estimate the patient's total blood volume from these data.

27. A gas consisting of NO_2 in air is bubbled through a scrubbing tower into which water is fed at a rate of 1 m³/h. The water absorbs most of the NO_2 and none of the air. The

volumetric flow rate of the feed gas is determined with an orifice meter, with a differential mercury manometer being used to measure the pressure drop across the orifice. Calibration data for this meter are tabulated here.

h(mm Hg)	ϕ(m³/h)
100	142
200	204
300	247
400	290

The molar density of the feed gas may be determined from the formula

$$\rho\left(\frac{mol}{liter}\right) = \frac{12.2P(atm)}{T(K)}$$

where P and T are the absolute pressure and temperatuure of the gas. An electrochemical detector is used to measure the NO_2 concentration in the inlet and outlet gas streams: NO_2 in the sampled gas is absorbed in a solution across which a fixed voltage is applied and the mole fraction of NO_2 in the gas is determined from the resulting current. The calibration curve for the analyzer is a straight line on a semilog plot of y (mol NO_2/mol total) versus R (analyzer reading), which passes through the points

y(log scale)	R(rectangular scale)
0.00166	20
0.1107	90

The following data are taken:

$$T = 75°F$$
$$P = 150 \text{ psig}$$
$$h \text{ (orifice meter)} = 210 \text{ mm}$$ (Feed gas)
$$R \text{ (NO}_2 \text{ analyzer)} = 82.4$$

$$R \text{ (NO}_2 \text{ analyzer)} = 11.6 \quad \text{(Outlet gas)}$$

(a) Determine the orifice meter calibration formula by plotting ϕ versus h on log paper.

(b) Determine the NO_2 analyzer calibration formula.

(c) Draw and label a process flowchart.

(d) Calculate the mole fraction of NO_2 in the liquid effluent stream. (Neglect the small quantity of water that leaves with the exit gas.)

(e) How much NO_2 is removed each hour from the feed gas?

***28.** Consider the NO_2 absorption tower described in Problem 27.

(a) Derive an algorithm (a series of equations) to determine the inlet water flow rate $Q(m^3/h)$ required to yield a specified mole fraction of dissolved NO_2 in the product solution [x(mol NO_2/mol)] for given values of h(mm Hg), P(psig), T(°F), R_{in} and R_{out}.

(b) Write a computer program to read in values of h, P, T, R_{in}, and R_{out}, print out these values, then read in a value of x, calculate Q, print out x and Q, go back and read in a new value of x, and so forth. Have the program terminate if a value of $x = 0.0$ is read in.

* Computer problem

(c) Test your program by reading in the values of h, P, T, R_{in}, and R_{out} and the value of x calculated in Problem 27. Then calculate the inlet water flow rates required to yield $x = 0.01$, 0.05, 0.1, and 0.2 for the same inlet and outlet gas parameters.

29. Following is a labeled flowchart for a steady-state two-unit process, on which boundaries are shown to denote subsystems about which balances can be taken. State the maximum number of balances that can be written for each subsystem and the order in which you would write balances to determine the unknown process variables. (See Example 4.4-1.)

30. Seawater containing 3.50 wt% salt passes through a series of 10 evaporators. Roughly equal quantities of water are vaporized in each of the 10 units and are then condensed and combined to obtain a product stream of fresh water. The brine at the outlet of the last evaporator contains 5.00 wt% salt. If 30 000 kg/h of seawater are fed to the process, calculate the fractional yield of fresh water and the weight percent of salt in the solution leaving the fourth evaporator.

31. A liquid mixture containing 30 mole% benzene (B), 25% toluene (T), and 45% xylene (X) is fed at a rate of 1275 kmol/h to a distillation unit consisting of two columns. The bottoms product from the first column is to contain 99 mole% X and no B, and 98% of the X in the feed is to be recovered in this stream. The overhead product from the first column is fed to a second column. The overhead product from the second column contains 99 mole% B and no X. The benzene recovered in this stream represents 96% of the B in the feed to this column. Calculate the molar flow rates (kmol/h) and component mole fractions in each product stream from both columns.

32. Coffee beans contain components that are soluble in water and others that are not. Instant coffee is produced by dissolving the soluble portion in boiling water (i.e., by making coffee) in large percolators, then feeding the coffee to a spray dryer in which the water is evaporated, leaving the soluble coffee as a dry powder. The insoluble portion of the coffee beans (the spent grounds) passes through several drying operations, and the dried grounds are either burned or used as landfill. The solution removed from the grounds in the first stage of the drying operation is fed to the spray dryer to join the effluent from the percolators. A flowchart of this process is shown here.* The symbols S and I

* D. F. Rudd, G. J. Powers, and J. J. Siirola, *Process Synthesis.* © 1973, pp. 69–70. Adapted by permission of Prentice-Hall, Inc., Englewood Cliffs, N.J.

denote the soluble and insoluble components of the coffee beans, W is water, and C is a solution containing 35% S and 65% W by mass.

(a) Calculate the flow rates (kg/h) of each of streams 1 to 8.

(b) If the liquid effluent from the press could be fed to the spray dryer without affecting the taste of the product, by what percentage would the production rate of instant coffee be increased?

33. A liquid mixture containing 40 mole% benzene and 60% toluene is separated in a distillation column. The vapor leaving the top of the column, which contains 95% benzene, is condensed and split into two equal fractions: One is taken off as the overhead product stream, and the other (the **reflux**) is recycled to the top of the column. The overhead product stream contains 90% of the benzene fed to the column. The liquid leaving the bottom of the column is fed to a partial reboiler, in which 45% of the liquid is vaporized. The vapor generated in the reboiler is returned to the bottom of the column, and the residual liquid is taken off as the bottoms product stream. The compositions of the streams leaving the reboiler are governed by the relation

$$\frac{y_B/(1 - y_B)}{x_B/(1 - x_B)} = 2.25$$

where y_B and x_B are the mole fractions of benzene in the vapor and liquid streams, respectively. For a basis of 100 mol fed to the column, calculate the molar flow rate of the overhead product stream, and the molar flow rates and mole fractions of benzene in the bottoms product stream, the vapor stream recycled to the column from the reboiler, and the liquid stream leaving the bottom of the column.

34. Fresh orange juice contains 12.0 wt% solids and the balance water, and concentrated orange juice contains 42% solids. Initially a single evaporation process was used for the concentration, but volatile constituents of the juice escaped with the water, leaving the concentrate with a flat taste. The present process overcomes this problem by bypassing the evaporator with a fraction of the fresh juice; the juice that enters the evaporator is concentrated to 58% solids, and the product is mixed with the bypassed fresh juice to achieve the desired final concentration of solids.

 (a) Calculate the amount of concentrated juice produced per 100-kg fresh juice fed to the process, and the fraction of the feed that bypasses the evaporator.

 (b) The volatile ingredients that provide the taste are contained in the fresh juice that bypasses the evaporator. You could get more of these ingredients in the final product by evaporating to (say) 90% solids instead of 58%; you could then bypass a greater fraction of the fresh juice, and you would thereby obtain an even better tasting product. Suggest possible drawbacks to this proposal.

35. An evaporation–crystallization process of the type described in Example 4.5-1 is used to obtain solid potassium sulfate from an aqueous solution of this salt. The fresh feed to the process contains 18.6 wt% K_2SO_4. The wet filter cake consists of solid K_2SO_4 crystals and a 40.0 wt% K_2SO_4 solution, in a ratio 10 kg crystals/kg solution. The filtrate, also a 40% solution, is recycled to join the fresh feed. Of the water fed to the evaporator, 42.66% is evaporated. The evaporator has a maximum capacity of 155 kg water evaporated/min.

 (a) Calculate the maximum production rate of solid K_2SO_4, the rate at which fresh feed must be supplied to achieve this production rate, and the ratio kg recycle/kg fresh feed.

 (b) Calculate the composition and feed rate of the stream entering the crystallizer if the process is scaled to 75% of its maximum capacity.

 (c) The wet filter cake is subjected to another operation after leaving the filter. Suggest what it might be. Also, what do you think the principal operating costs for this process might be?

36. A gas stream containing 25 mole% CO_2 and 75% CH_4 is treated in a gas conditioning plant. The stream is fed to an absorber at a rate of 50 kmol/h and is contacted in the absorber with a liquid solvent containing 0.500 mole% dissolved CO_2 and the balance methanol. The gas leaving the absorber contains 1.00 mole% CO_2 and essentially all of the methane fed to the unit. The CO_2-rich solvent leaving the absorber goes to a stripping column; in which a stream of nitrogen gas contacts the solvent, removing 90% of the dissolved CO_2. The regenerated solvent is then recycled to the absorption column. Methanol can be assumed to be nonvolatile—that is, none enters the vapor phase in either process unit.

 (a) Calculate the fractional CO_2 removal (moles absorbed/mole fed), and the molar flow rate and composition of the liquid feed to the stripping column.

 (b) Calculate the molar feed rate to the absorber required to produce an absorber product gas flow rate of 1000 kg/h.

 (c) How would the process have to be modified if the volatility of methanol could not be neglected?

37. In the production of a bean oil, beans containing 10 wt% oil and 90% solids are ground and fed to a stirred tank, where they are suspended in liquid n-hexane. The feed ratio is (3 kg hexane/kg beans). Essentially all of the oil in the beans is extracted into the hexane. The mixer effluent passes to a filter. The filter cake contains 75 wt% bean solids and the balance bean oil and hexane, the latter in the same ratio in which they emerge from the mixer. The filter cake is discarded, and the liquid filtrate is fed to a vacuum evaporator, in which the hexane is vaporized and thereby separated from the oil. The hexane vapor is subsequently condensed and recycled to the stirred tank.

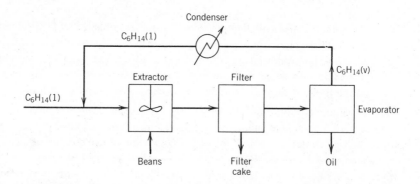

 (a) Calculate the yield of bean oil product (kg oil/kg beans fed), the required fresh hexane feed (kg C_6H_{14}/kg beans fed), the recycle to fresh feed ratio (kg hexane recycled/kg fresh feed), and the compositions (component mass fractions) of the slurry leaving the extractor and the filtrate.

 (b) Suggest possible changes in process conditions or additional processing steps that might improve the economics of this process.

38. Following is a diagram of the shirt-cleaning process used by the Floods of Suds One-Day Laundry Service, Inc. The shirts are soaked in an agitated tub containing Whizzo, the Wonder Detergent, and are then wrung out and sent to a rinse stage. The dirty Whizzo is sent to a filter in which most of the dirt is removed; the cleaned detergent is recycled

back to join a stream of pure Whizzo, with the combined stream serving as the feed to the washtub.

Data:

1. Each 100 kg of dirty shirts contains 2 kg of dirt.
2. The washing removes 95% of the dirt.
3. For each 100 kg of dirty shirts, 25 kg of Whizzo leave with the clean shirts, of which 22 kg are wrung back into the tub.
4. The detergent that enters the tub contains 97% Whizzo, and that which enters the filter contains 90%. The wet dirt that leaves the filter contains 10% Whizzo.

(a) How much pure Whizzo must be supplied per 100 kg of dirty shirts?

(b) What is the composition of the recycled stream?

39. A drug (D) is obtained in a three-stage extraction from the leaves of a tropical plant. About 1000 kg of leaf is required to produce 1 kg of the drug. The extraction solvent (S) is a mixture containing 16.5 wt% ethanol (E) and the balance water (W). The following process is carried out to extract the drug and recover the solvent.

1. 3300 kg of S and 620 kg of leaf are charged to a mixer and stirred for several hours, during which a portion of the drug contained in the leaf goes into solution. The contents of the mixer are then discharged through a filter. The liquid filtrate, which carries over roughly 1% of the leaf fed to the mixer, is pumped to a holding tank, and the solid cake (spent leaf and entrained liquid) is sent to a second mixer. The entrained liquid has the same composition as the filtrate and a mass equal to 15% of the mass of liquid charged to the mixer. The extracted drug has a negligible effect on the total mass and volume of the spent leaf and the filtrate.

2. The second mixer is charged with the spent leaf from the first mixer, and with the filtrate from the previous batch in the third mixer. The leaf is extracted for several more hours, and the contents of the mixer are then discharged to a second filter. The filtrate, which contains 1% of the leaf fed to the second mixer, is pumped to the same holding tank that received the filtrate from the first mixer, and the solid cake—spent leaf and entrained liquid—is sent to the third mixer. The entrained liquid mass is 15% of the mass of liquid charged to the second mixer.

3. The third mixer is charged with the spent leaf from the second mixer and with 2720 kg of solvent S. The mixer contents are filtered; the filtrate, which contains 1% of the leaf fed to the third mixer, is recycled to the second mixer; and the solid cake is discarded. As before, the mass of the entrained liquid in the solid cake is 15% of the mass of liquid charged to the mixer.

4. The contents of the filtrate holding tank are filtered to remove the carried-over spent leaf, and the wet cake is pressed to recover entrained liquid. A negligible amount of liquid is lost in this step. The filtrate, which contains D, E, and W, is pumped to an extraction unit (another mixer).

5. In the extraction unit, the alcohol–water–drug solution is contacted with another solvent (F), which is almost but not completely immiscible with ethanol and water. Essentially all of the drug (D) is extracted into the second solvent, from which it is eventually separated by a process of no concern in this problem. Some ethanol but no water is also contained in the extract. The solution from which the drug has

been extracted (the **raffinate**) contains 13.0 wt% E, 1.5% F, and 85.5% W. It is fed to a steam stripper for recovery of the ethanol.

6. The feeds to the stripping column are the solution just described and steam. The two streams are fed in a ratio such that the overhead product steam from the column contains 20.0 wt% E and 2.6% F, and the bottom product steam contains 1.3 wt% E and the balance W.

Draw and label a flowchart of the process, taking as a basis one batch of leaf processed. Then calculate

(a) The masses of the components of the filtrate holding tank
(b) The masses of the components D and E of the extract stream from the extraction unit
(c) The mass of steam fed to the stripper, and the masses of the stripper overhead and bottoms products

40. Calculate the following quantities for the reaction in which NO is oxidized to form N_2O_4.
(a) The stoichiometric ratio of O_2 to NO
(b) Mol N_2O_4 produced/mol NO reacted
(c) The mass ratio of NO to O_2 if the two gases are fed to a reactor in stoichiometric proportion
(d) The pounds of NO required to produce 1 ton of N_2O_4

41. Ammonia is burned to form nitric oxide in the following reaction:

$$4\,NH_3 + 5\,O_2 \rightarrow 4\,NO + 6\,H_2O$$

Calculate the following quantities.
(a) Mol NO produced/mol O_2 reacted
(b) The molar flow rate of oxygen corresponding to a feed rate of 50 lb-moles NH_3/h if the reactants are fed in stoichiometric proportion
(c) The mass of NO produced if 200 kg of ammonia and 200 kg of oxygen react and the reaction proceeds to completion.

42. Ethane is burned with oxygen to form carbon dioxide and water. One hundred mol/h of each reactant are fed to the furnace. Determine
(a) The stoichiometric ratio of O_2 to C_2H_6
(b) The limiting reactant
(c) The percentage by which the other reactant is in excess

*43. A stream containing H_2S and inert gases and a second stream of pure SO_2 are fed to a sulfur recovery reactor, where the reaction

$$2\,H_2S + SO_2 \rightarrow 3\,S + 2\,H_2O$$

takes place. The feed rates are adjusted so that the ratio of H_2S to SO_2 in the combined feed is always stoichiometric.
(a) If the first stream contains 85 mol% H_2S and enters the unit at a rate of 400 kmol/h, what must the molar flow rate of the SO_2 be?

* Computer problem.

(b) It is desired to automate the SO_2 flow rate adjustment using a computer. The flow rate of the H_2S stream will be measured with an electronic flowmeter that transmits a signal R_f directly proportional to the molar flow rate q. A flow rate of 100 kmol/h results in a signal $R_f = 20$ mV. The mole fraction of H_2S in this stream is measured with a thermal conductivity detector, which transmits a signal R_a. Calibration data for this analyzer are as follows:

R_a(mV)	0	20	40	60	80	100
x(mol H_2S/mol total)	0.00	0.14	0.32	0.63	0.89	1.00

The computer transmits a voltage signal R_c to a flow control valve in the SO_2 line. The relationship between R_c and the SO_2 flow rate is a direct proportionality, with a signal $R_c = 50$ mV corresponding to a molar flow rate of 100 kmol/h. Write a computer program to calculate the value of R_c from specified (input) values of R_f and R_a. Use two-point linear interpolation to determine from the calibration data the mole fraction of H_2S corresponding to the specified value of R_a. Test the program using the flow rate and composition data of part a.

44. Ethane is fed to a dehydrogenation reactor at a rate of 150 kmol/h. The reaction products are acetylene and hydrogen. A fractional conversion of 80% is achieved. Calculate the following properties of the final product gas.
(a) Total molar flow rate
(b) Mole ratio of hydrogen to acetylene
(c) Average molecular weight
(d) Mass flow rate of acetylene

45. The reaction between ethylene and hydrogen bromide to form ethyl bromide is carried out in a continuous reactor. The product stream is analyzed and is found to contain 50 mole% C_2H_5Br and 33.3% HBr. The feed to the reactor contains only ethylene and hydrogen bromide. Calculate the fractional conversion of the limiting reactant and the percentage by which the other reactant is in excess.

46. In the Deacon process for the manufacture of chlorine, HCl and O_2 react to form Cl_2 and H_2O. Sufficient air (21 mole% O_2, 79% N_2) is fed to provide 25% excess oxygen,

and the fractional conversion of HCl is 70%. Calculate the mole fractions of the product stream components.

47. Titanium dioxide (TiO_2) is used extensively as a white pigment. It is produced from an ore that contains ilmenite ($FeTiO_3$) and ferric oxide (Fe_2O_3). The ore is digested with an aqueous sulfuric acid solution to produce an aqueous solution of titanyl sulfate [$(TiO)SO_4$] and ferrous sulfate ($FeSO_4$). Water is added to hydrolyze the titanyl sulfate to H_2TiO_3, which precipitates, and H_2SO_4. The precipitate is then roasted, driving off water and leaving a residue of pure titanium dioxide. (Several steps to remove iron from the intermediate solutions as iron sulfate have been omitted from this description.)

Suppose an ore containing 24.3% Ti by mass is digested with an 80% H_2SO_4 solution, supplied in 50% excess of the amount needed to convert all the ilmenite to titanyl sulfate and all the ferric oxide to ferric sulfate [$Fe_2(SO_4)_3$]. Further suppose that 89% of the ilmenite actually decomposes. Calculate the masses (kg) of ore and 80% sulfuric acid solution that must be fed to produce 1000 kg of pure TiO_2.

48. Coal containing 5.0 wt% S is burned at a rate of 1250 lb_m/min in a boiler furnace. All of the sulfur in the coal is oxidized to SO_2. The product gas is sent to a scrubber in which most of the SO_2 is removed, and the scrubbed gas then passes out of a stack. An Environmental Protection Agency regulation requires that the gas in the stack must contain no more than 0.018 lb_m SO_2/lb_m coal burned. To test compliance with this regulation a flowmeter and an SO_2 analyzer are mounted in the stack. The volumetric flow rate of the scrubbed gas is found to be 2867 ft^3/s, and the SO_2 analyzer reading is 37. Calibration data for the analyzer are given in the table below.

SO_2 Analyzer Calibration Data

C(g SO_2/m^3 gas)	Reading (0–100 scale)
0.30	10
0.85	28
2.67	48
7.31	65
18.2	81
30.0	90

(a) What is the equation that relates SO_2 concentration in lb_m/ft^3 to the analyzer reading?

(b) Is the EPA regulation being complied with?

(c) What percentage of the SO_2 produced in the furnace is removed in the scrubber?

49. The gas-phase reaction between methanol and acetic acid to form methyl acetate and water

$$CH_3OH + CH_3COOH \rightleftharpoons CH_3COOCH_3 + H_2O$$
$$\quad\;(A) \qquad\quad (B) \qquad\qquad (C) \qquad\quad (D)$$

takes place in a batch reactor and proceeds to equilibrium. At the reaction temperature, the mole fractions of the four reactive species at equilibrium satisfy the relation

$$\frac{y_C y_D}{y_A y_B} = 3.87$$

(a) Suppose the feed to the reactor consists of n_{A0}, n_{B0}, n_{C0}, n_{D0}, and n_{I0} mol of A, B, C, D, and an inert gas, I, respectively. Let ξ = extent of reaction. Write expressions for the gram-moles of each reactive species in the final product, $n_A(\xi)$, $n_B(\xi)$, $n_C(\xi)$, and $n_D(\xi)$. Then use these expressions and the given equilibrium relation to derive an equation for ξ_e, the equilibrium extent of reaction, in terms of n_{A0}, ..., n_{I0}. (See Example 4.6-2.)

(b) If the feed to the reactor contains equimolar quantities of methanol and acetic acid and no other species, calculate the equilibrium fractional conversion.

(c) It is desired to produce 70 mol of methyl acetate starting with 80 mol of acetic acid. If the reaction proceeds to equilibrium, how much methanol must be fed? What is the final product gas composition?

50. At low to moderate pressures, the equilibrium state of the water gas shift reaction

$$CO + H_2O \rightleftharpoons CO_2 + H_2$$

is approximately described by the relation

$$\frac{y_{CO_2} y_{H_2}}{y_{CO} y_{H_2O}} = 0.0247 \exp\left[4020/T(\mathrm{K})\right]$$

The feed to a continuous shift reactor contains 30 mole% CO, 10% CO_2, 40% water, and the balance an inert gas. The reaction proceeds to equilibrium.

(a) Calculate the gram-moles of hydrogen produced per mol of water fed if the reaction is carried out at 850°C.

(b) Calculate the temperature at which the reaction should be run to achieve a fractional CO conversion of 75%.

51. Methanol can be formed from carbon monoxide and hydrogen in the gas-phase reaction

$$CO + 2\,H_2 \rightleftharpoons CH_3OH$$
$$(A) \quad (B) \qquad (C)$$

The mole fractions of the reactive species at equilibrium satisfy the relation

$$\frac{y_C}{y_A y_B^2} \frac{1}{P^2} = K_e(T)$$

where y = mole fraction, P = total pressure (atm), K_e = the reaction equilibrium constant (atm^{-2}), and T = temperature (K). The equilibrium constant K_e equals 10.5 at 373 K, and 2.316×10^{-4} at 573 K. A semilog plot of K_e (logarithmic scale) versus $1/T(\mathrm{K})$ (rectangular scale) is approximately linear between T = 300 K and T = 600 K.

(a) Derive a formula for $K_e(T)$, and use it to show that $K_e(450\ \mathrm{K}) = 0.0548$ atm^{-2}.

(b) Write expressions for n_A, n_B, and n_C (mol of each species), and then y_A, y_B, and y_C, in terms of n_{A0}, n_{B0}, n_{C0}, and ξ, the molar extent of reaction. Then derive an equation involving only n_{A0}, n_{B0}, n_{C0}, P, T, and ξ_e, where ξ_e is the value of ξ at equilibrium.

(c) Suppose you begin with equimolar quantities of CO and H_2, and no CH_3OH, and the reaction proceeds to equilibrium at 423 K and 2.00 atm. Calculate the molar composition of the product (y_A, y_B, and y_C), and the fractional conversion of CO.

*(d) Write a computer program to take as input an initial molar composition (y_{A0}, y_{B0}, and y_{CO}), and a final temperature and pressure, and to calculate and print out the equilibrium molar composition of the product gas and the fractional conversion of CO. Test your program with the data of part c. (*Suggestion:* Write the equilibrium relation derived in part b as a cubic equation in ξ_e, and use Newton's rule—Appendix A.2—to obtain the solution.)

52. Methane and oxygen react in the presence of a catalyst to form formaldehyde. In a parallel side reaction, some of the methane is instead oxidized to carbon dioxide and water.

$$CH_4 + O_2 \rightarrow HCHO + H_2O$$
$$CH_4 + 2\,O_2 \rightarrow CO_2 + 2\,H_2O$$

The feed to the reactor contains equimolar amounts of methane and oxygen.

(a) The fractional conversion of methane is 95%, and the fractional yield of formaldehyde is 90%. Calculate the molar composition of the reactor output stream and the selectivity of formaldehyde production relative to carbon dioxide production.

(b) The reactor output stream contains 45 mole% formaldehyde, 1% carbon dioxide, 4% methane, and the balance oxygen and water. Calculate the fractional conversion of methane, the fractional yield of formaldehyde, and the selectivity of formaldehyde production relative to carbon dioxide production.

53. Ethanol is produced commercially by the hydration of ethylene:

$$C_2H_4 + H_2O \rightarrow C_2H_5OH$$

Some of the product is converted to diethyl ether in the side reaction

$$2\,C_2H_5OH \rightarrow (C_2H_5)O + H_2O$$

The feed to a reactor contains ethylene and steam in a 3:2 mole ratio, and 10 mole% inerts. The fractional conversion of ethylene is 5%, and the selectivity of ethanol production relative to ether production is 18.0 mol/mol. Calculate the molar composition of the reactor output stream.

54. Ethane is chlorinated in a continuous reactor:

$$C_2H_6 + Cl_2 \rightarrow C_2H_5Cl + HCl$$

Some of the product monochloroethane is further chlorinated in an undesired side reaction:

$$C_2H_5Cl + Cl_2 \rightarrow C_2H_4Cl_2 + HCl$$

(a) Suppose your principal objective is to maximize the selectivity of monochloroethane production relative to dichloroethane production. Would you design the reactor for a high or low conversion of ethane? Which of the two reactants would you feed in excess? Briefly explain.

(b) Suppose the conversion of ethane is 13%; selectivity is (13.3 moles C_2H_5Cl)/ (mole $C_2H_4Cl_2$); and the product contains a negligible amount of chlorine. Calcu-

* Computer problem.

late the moles of all species present in the product stream per 100 moles of monochloroethane produced.

55. Solid calcium fluoride (CaF_2) reacts with sulfuric acid to form solid calcium sulfate and gaseous hydrogen fluoride. The HF is then dissolved in water to form hydrofluoric acid. A source of calcium fluoride is fluorite ore containing 96 wt% CaF_2 and 4% SiO_2.

In a typical hydrofluoric acid manufacturing process, fluorite ore is reacted with 93 wt% aqueous sulfuric acid, supplied in 15% excess of the stoichiometric amount. Ninety-five percent of the ore dissolves in the acid. Some of the HF formed reacts with the dissolved silica in the reaction.

$$6 \text{ HF} + SiO_2(aq) \rightarrow H_4SiF_6(s) + 2 \text{ H}_2O(l)$$

The hydrogen fluoride exiting from the reactor is subsequently dissolved in enough water to produce 60 wt% hydrofluoric acid. Calculate the quantity of fluorite ore needed to produce a metric ton of acid.

56. Chlorobenzene (C_6H_5Cl), an important solvent and intermediate in the production of many other chemicals, is produced by bubbling chlorine gas through liquid benzene in the presence of ferric chloride catalyst. In an undesired side reaction, the product is further chlorinated to dichlorobenzene, and in a third reaction the dichlorobenzene is chlorinated to trichlorobenzene.

The feed to a chlorination reactor consists of essentially pure benzene and a technical grade of chlorine (98 wt% Cl_2, the balance impurities with an average molecular weight of 25.0). The liquid output from the reactor contains 65.0 $wt\%$ C_6H_6, 32.0% C_6H_5Cl, 2.5% $C_6H_4Cl_2$, and 0.5% $C_6H_3Cl_3$. The gaseous output contains only HCl and the impurities that entered with the chlorine.

(a) Determine the percentage by which benzene is fed in excess, the fractional conversion of benzene, the fractional yield of monochlorobenzene (gram-moles formed per mol that could have been formed theoretically), and the mass ratio of the gas feed to the liquid feed.

(b) Why would benzene be fed in excess and the fractional conversion be kept as low as it is?

(c) What might be done with the gaseous effluent?

57. The following two reactions take place in a gas-phase batch reactor.

$$2 \text{ CO}_2 \rightleftharpoons 2 \text{ CO} + O_2$$
$$\text{(A)} \qquad \text{(B)} \qquad \text{(C)}$$

$$O_2 + N_2 \rightleftharpoons 2 \text{ NO}$$
$$\text{(C)} \quad \text{(D)} \qquad \text{(E)}$$

If the system comes to equilibrium at 3000 K and 1 atm, the product gas mole fractions satisfy the relations

$$\frac{y_B^2 y_C}{y_A^2} = 0.1071 \qquad \frac{y_E^2}{y_C y_D} = 0.01493$$

(a) Let n_{A0}, \ldots, n_{E0} be the initial number of gram-moles of each species. Write expressions for the number of gram-moles of each species at equilibrium and then

for the mole fractions in terms of the initial moles of each species and the extents of the two reactions, ξ_{e1} and ξ_{e2} (see Equation 4.6-5). Then substitute in the equilibrium relations to derive two simultaneous equations for ξ_{e1} and ξ_{e2}.

(b) One third of a mol each of CO_2, O_2, and N_2 are charged into a batch reactor, and the reactor contents equilibrate at 3000 K and 1 atm. Calculate the component mole fractions of the reactor contents at equilibrium. (*Suggestion:* Use the Newton-Raphson method, as outlined in Section A.2.h of Appendix A. Guess initial values of 0.1 for both ξ_{e1} and ξ_{e2}, and iterate until successive estimates of these variables differ by less than 0.1%.)

***(c)** Write a computer program to implement the procedure of part b for an arbitrary starting composition. The program should (i) read values of n_{A0}, n_{B0}, n_{C0}, n_{D0}, and n_{E0}; (ii) terminate if the value of n_{A0} is negative; (iii) print out the initial moles of each species, and calculate and print out the total moles and mole fractions of each species at equilibrium; (iv) go back to step i. Use a relative convergence criterion $\varepsilon = 1.0 \times 10^{-5}$. Run the program for the following feed mixtures.

n_{A0}	n_{B0}	n_{C0}	n_{D0}	n_{E0}
1/3	0	1/3	1/3	0
1/2	0	0	1/2	0
1/5	1/5	1/5	1/5	1/5

58. A catalytic reactor is used to produce formaldehyde from methanol by the reaction

$$CH_3OH \rightarrow HCHO + H_2$$

A single-pass conversion of 70% is achieved in the reactor. The methanol in the reactor product is separated from the formaldehyde and hydrogen in a multiple-unit process. The production rate of formaldehyde is 600.0 kg/h. Calculate the required feed rate of methanol (mol/h) to the process for the following two cases.

(a) No recycle

(b) The recovered methanol is recycled to the reactor

59. Methanol is produced by reacting carbon monoxide and hydrogen. A fresh feed stream containing CO and H_2 joins a recycle stream, and the combined stream is fed to a reactor. A portion of the methanol leaving the reactor is condensed, and the uncon-sumed CO and H_2 and uncondensed CH_3OH are recycled. The streaming going from the reactor to the condenser flows at a rate of 275 mol/min, and contains 10.6 wt% H_2, 64.0 wt% CO, and 25.4 wt% CH_3OH. The mole fraction of methanol in the recycle stream is 0.004. Calculate the molar flow rates of CO and H_2 in the fresh feed, and the production rate of liquid methanol.

60. Methane reacts with chlorine to produce methyl chloride and hydrogen chloride.

$$CH_4 + Cl_2 \rightarrow CH_3Cl + HCl$$

Once formed, the methyl chloride may undergo further chlorination to form methylene chloride (CH_2Cl_2), chloroform ($CHCl_3$), and carbon tetrachloride (CCl_4).

* Computer problem.

In a methyl chloride production process, methane and chlorine are fed to a reactor in a mole ratio of 5:1. (The ratio is kept high to minimize polysubstitution.) A single-pass chlorine conversion of 100% may be assumed. The mole ratio of CH_3Cl to CH_2Cl_2 in the product is 4:1, and negligible amounts of chloroform and carbon tetrachloride are produced. The product gases are cooled, condensing the CH_3Cl and CH_2Cl_2, which are then separated in a distillation column. The gas leaving the condenser goes to a scrubber, in which HCl is absorbed. The gas leaving the scrubber, which may be considered pure methane, is recycled back to the reactor.

For a production rate of 1000 kg CH_3Cl/h, calculate

(a) The flow rate and molar composition of the fresh feed.

(b) The rate at which HCl must be removed in the scrubber.

(c) The molar flow rate of the recycle stream.

61. Ethylene oxide is produced by the catalytic oxidation of ethylene

$$2\,C_2H_4 + O_2 \rightarrow 2\,C_2H_4O$$

An undesired competing reaction is the combustion of ethylene

$$C_2H_4 + 3\,O_2 \rightarrow 2\,CO_2 + 2\,H_2O$$

The feed to the reactor (*not* the fresh feed to the process) contains 3 moles of ethylene per mole of oxygen. The single-pass conversion of ethylene is 20%, and for every 100 moles of ethylene consumed in the reactor, 80 moles of ethylene oxide emerge in the reactor products. A multiple-unit process is used to separate the products: Ethylene and oxygen are recycled back to the reactor, ethylene oxide is sold as a product, and carbon dioxide and water are discarded. Calculate the molar flow rates of ethylene and oxygen in the fresh feed needed to produce 1500 kg C_2H_4O/h, and the overall conversion of ethylene.

62. The fresh feed to an ammonia production process contains 24.75 mole% nitrogen, 74.25 mole% hydrogen, and the balance inerts (I). The feed is combined with a recycle stream containing the same species, and the combined stream is fed to a reactor in which a 25% single-pass conversion of nitrogen is achieved. The products pass through a condenser in which essentially all of the ammonia is removed, and the remaining gases are recycled. However, to prevent buildup of the inerts in the system, a purge stream must be taken off. (See Section 4.6g.) The recycle stream contains 12.5 mole% inerts. Calculate the overall conversion of nitrogen, the ratio (moles purge gas/mole of gas leaving the condenser), and the ratio (moles fresh feed/mole fed to the reactor).

63. The fresh feed to a methanol synthesis unit of the type described in Example 4.6-6 contains 32.0 mole% CO, 64.0% H_2, and 4.0% N_2, and flows at a rate of 100 mol/h. The fresh feed is mixed with a recycle stream flowing at a rate of 400 mol/h to produce a reactor feed containing 13.0 mole% N_2. The product stream leaving the condenser contains only liquid methanol. To prevent a buildup of nitrogen in the system, a purge stream is withdrawn from the gas stream leaving the condenser. The gases not purged constitute the stream recycled to the reactor. Calculate the production rate of methanol (mol/h), the molar flow rate and composition of the purge gas, and the overall and single-pass conversions.

64. Isooctane is produced by the reaction of butylene and isobutane in an emulsion with concentrated sulfuric acid.

$$C_4H_{10} + C_4H_8 \rightarrow C_8H_{18}$$

The fresh feed to the process flows at a rate of 40 000 kg/h, and contains 25.00 mole% isobutane, 25.00% butylene, and 50.00% n-butane, which is chemically inert in this process. Additional isobutane is recycled to the reactor, and essentially all of the butylene fed to the reactor is consumed. A portion of the reactor effluent is recycled to the reactor inlet, and the remainder passes to a settler, in which the aqueous (sulfuric acid) and hydrocarbon phases are allowed to separate. The acid is recycled to the reactor, and the hydrocarbons pass to a distillation column. The overhead from the column contains isooctane and n-butane, and the bottoms product, which is recycled to the reactor, contains only isobutane. The stream entering the reactor contains 200 moles of isobutane per mole of butylene, and 2 kg of 91 wt% H_2SO_4(aq) per kg of hydrocarbon. The stream obtained by combining the fresh feed and isobutane recycle contains 5.0 moles of isobutane per mole of butylene.

Determine the flow rates (kmol/h) of each component of the product stream and the three recycle streams.

***65.** Ethyl acetate (A) undergoes a saponification reaction with sodium hydroxide (B)

$$CH_3COOC_2H_5 + NaOH \rightarrow CH_3COONa + C_2H_5OH$$
$$\text{(A)} \qquad\qquad \text{(B)}$$

* Computer problem.

The reaction is carried out at a steady state in a series of stirred-tank reactors. The output from the ith reactor is the input to the $(i + 1)$st reactor. The volumetric flow rate between the reactors is constant at ϕ(liters/min), and the volume of each tank is V(liters).

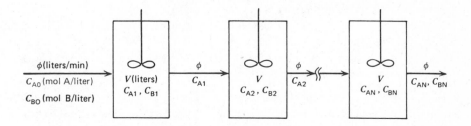

The concentrations of A and B in the feed to the first tank are C_{A0} and C_{B0} (mol/liter). The tanks are stirred sufficiently for their contents to be uniform throughout, so that C_A and C_B in a tank equal C_A and C_B in the stream leaving that tank. The rate of reaction is given by the expression

$$r\left(\frac{\text{mol A or B reacting}}{\text{min} \cdot \text{liter}}\right) = kC_A C_B$$

where k(liter/mol·min) is the **reaction rate constant.**

(a) Write a material balance on A in the ith tank, and show that it yields

$$C_{A,i-1} = C_{Ai} + k\tau C_{Ai} C_{Bi}$$

where $\tau = V/\phi$ is the **mean residence time** in each tank. Then write a balance on B in the ith tank and subtract the two balances, using the result to prove that

$$C_{Bi} - C_{Ai} = C_{B0} - C_{A0}, \quad \text{all } i$$

(b) Use the equations derived in part a to prove that

$$C_{A,i-1} = C_{Ai} + k\tau C_{Ai}(C_{Ai} + C_{B0} - C_{A0})$$

and from this relation derive an equation of the form

$$aC_{Ai}^2 + \beta C_{Ai} + \gamma = 0$$

where α, β, and γ are functions of k, C_{A0}, C_{B0}, $C_{A,i-1}$, and τ. Then write the solution of this equation for C_{Ai}.

(c) Write a computer program to calculate the number of tanks needed to achieve a fractional conversion of not less than x_A at the outlet of the final reactor. Your program should execute the following steps.
1. Read in values of k, ϕ, V, C_{A0}, C_{B0}, and x_A.
2. Calculate C_{AF}, the maximum permissible concentration of A at the outlet of the last tank.
3. Use the equation for C_{Ai} derived in part b to calculate C_{A1}, C_{A2}, etc. After each calculation, print out the value of $C_{Ai,}$ and compare this value with

C_{AF}. If $C_{Ai} > C_{AF}$, continue to the next tank: if $C_{Ai} \leq C_{AF}$, print out the number of tanks and stop.

(d) Suppose the reaction is to be carried out at 20°C, at which temperature $k = 92.4$, and that the other process variables have the following values:

Feed concentrations: $C_4H_8O_2(A) = 5$ g/liter
NaOH(B) = 0.1 molar

Throughput: $\phi = 430$ liters/min

Tank volume: $V = 200$ liters

Minimum final fractional conversion: $x_A = 0.95$

Use your program to calculate the required number of tanks and the final fractional conversion of ethyl acetate. (*Remember*, C_{A0} and C_{B0} must be expressed in mol/liter.)

66. A gas contains 5 wt% C_3H_8, 5 wt% C_4H_{10}, 16 wt% O_2, 38 wt% N_2, and the balance H_2O. Calculate the molar composition of this gas on both a wet and a dry basis, and the ratio (moles H_2O/mole dry gas).

67. A natural gas contains 92 mole% methane, 5% ethane, and 3% propane. If 100 kmol/h of this fuel are to be burned completely with 125% excess air, what is the required air feed rate? How would the answer change if the combustion were only 75% complete?

68. One hundred twenty kmol of propane are fed to a combustion furnace along with 4284 kmol of air. Not all of the propane is burned in the furnace, and both CO and CO_2 are formed. If it is possible to do so without additional information, calculate the percent excess air supplied to the furnace; if more information is needed, state what it is.

69. Three liters of liquid *n*-hexane and four liters of liquid *n*-heptane are mixed and burned with 60% excess air. Calculate the gram-moles of air fed to the burner.

70. A fuel gas produced by gasifying coal is to be burned with 15% excess air. The gas contains 50 mole% nitrogen and the balance carbon monoxide and hydrogen. A sample of the gas is passed through an infrared spectrometer, which registers a signal R proportional (but not directly proportional) to the mole fraction of carbon monoxide in the sample, and a reading $R = 44.6$ is recorded.

Analyzer calibration data are as follows:

x(mol CO/mol)	0.05	0.10	0.40	0.80	1.00
R	10.0	17.0	49.4	73.6	99.7

It is believed that a power law $(x = aR^b)$ should be suitable for fitting the calibration data. Derive the equation relating x and R, and then calculate the molar flow rate of air required for a fuel feed rate of 130 kmol/h, assuming that CO and H_2 are oxidized but N_2 is not.

71. Natural gas containing a mixture of methane, ethane, propane, and butane is burned in a furnace with excess air.

(a) One hundred mol/h of a gas containing 94.4 mole% methane, 3.40% ethane, 0.60% propane, and 0.50% butane are to be burned with 17% excess air. Calculate the required molar flow rate of the air.

(b) Let

$$n_f(\text{mol/h}) = \text{molar flow rate of the fuel gas}$$

$$x_1, x_2, x_3, x_4 = \text{mole fractions of methane, ethane, propane,}$$
$$\text{and butane, respectively, in the fuel}$$

$$P_{xs} = \text{percent excess air}$$

$$n_a(\text{mol/h}) = \text{molar flow rate of the air fed to the furnace}$$

Derive an expression for n_a in terms of the other variables defined above. Check your formula with the results of part a.

*(c) Suppose the feed rate and composition of the fuel gas are subject to periodic variations, and a process control computer is to be used to adjust the flow rate of air to maintain a constant percentage excess. A calibrated electronic flowmeter in the fuel gas line transmits a signal R_f that is directly proportional to the flow rate, with a flow rate of 75.0 mol/h yielding a signal $R_f = 60$. The fuel gas composition is obtained with an on-line gas chromatograph. A sample of the gas is injected into the GC, and signals A_1, A_2, A_3, and A_4, which are directly proportional to the moles of methane, ethane, propane, and butane, respectively, in the sample, are transmitted. (Assume the same proportionality constant for all species.) The control computer processes these data to determine the required air flow rate, and then sends a signal R_a to a control valve in the air line. The relationship between R_a and the resulting air flow rate n_a is another direct proportionality, with a signal $R_a = 25$ leading to an air flow rate of 550 mol/h.

Write a computer program to perform the following tasks: (i) Read the desired percentage excess, P_{xs}; (ii) read values of R_f, A_1, A_2, A_3, and A_4; (iii) quit if R_f is negative; (iv) calculate R_a; (v) Print n_f, x_1, x_2, x_3, x_4, and n_a; (vi) go back to step ii.

Test your program on the following data, assuming that 17% excess air is required in all cases.

R_f	A_1	A_2	A_3	A_4
80	302.1	10.88	1.92	1.60
74	276.2	23.20	9.90	1.54
113	284.8	16.61	6.05	2.11

72. Propane is burned with air. Determine the molar composition of the product gas if no CO is formed, assuming:

(a) Theoretical air supplied, 100% conversion of fuel.

(b) 20% excess air, 100% conversion.

(c) 20% excess air, 90% conversion. Calculate the molar composition of the stack gas on a dry basis.

73. A mixture of 70 mole% butane and 30 mole% hydrogen is burned with 25% excess air. Fractional conversions of 85% of the butane and 90% of the hydrogen are achieved;

* Computer problem.

of the butane that reacts, 95% goes to form CO_2 and the balance goes to form CO. Calculate the mole fraction of water in the product gas.

74. *n*-Hexane is burned with excess air. On a dry basis, the product gas contains 0.268 mole% hexane, 7.60% oxygen, 9.10% carbon dioxide, and the balance nitrogen. Calculate the percent excess air fed to the reactor and the fractional conversion of hexane.

75. Liquid methanol is fed to a space heater at a rate of 10.0 liters/h and is burned with excess air. The product gas is analyzed, and the following dry-basis mole percentages are determined: $CH_3OH = 0.84\%$, $CO_2 = 7.10\%$, $CO = 2.40\%$.

 (a) Calculate the fractional conversion of methanol, the percentage excess air fed, and the flow rate of water (mol/h) in the product gas.

 (b) Suppose the combustion products are released directly into a room. What problems do you see with the process, and what remedies can you suggest?

76. A gas containing methane, ethane, and carbon dioxide is analyzed with a gas chromatograph (GC) and flame ionization detector (FID): the GC separates the components of the gas, and the FID registers a signal proportional to the amount of hydrocarbon (but not CO_2) in its sample chamber. The output of the FID is as follows:

The area under each peak is proportional to the number of carbon atoms in the sample, so that 1 mol of ethane would yield a peak with twice the area of a peak corresponding to 1 mol of methane.

One hundred mol/min of the gas are burned with 921 mol/min of air. A 0.50 mol sample of the product gas is passed through a condenser, in which essentially all of the water is removed. The condensed water is weighed and is found to have a mass of 1.134 g. The dry gas leaving the condenser is analyzed and found to contain 11.9 mole% CO_2 and no CO or hydrocarbons.

 (a) Calculate the mole fractions of CH_4, C_2H_6, and CO_2 in the product gas.

 (b) Calculate the percent excess air fed to the reactor.

77. You have been given the job of analyzing a fuel gas mixture containing propane and butane. On discovering that the gas chromatograph is broken again, you set up a combustion experiment to make the measurement. You charge a quantity of the gas into an evacuated chamber and then add air to the chamber. You ignite the mixture, allow the combustion to proceed to completion, and perform an Orsat analysis on a sample of the product gas. The dry basis composition you determine is 84.8 mole% N_2, 11.4% CO_2, and 3.8% O_2.

(a) What is the mole fraction of propane in the fuel gas?

(b) Suppose you had been told that the fuel gas contains a mixture of hydrocarbons, but you had not been given the component species. What could you have determined about the fuel gas from the given data?

78. A fuel oil is analyzed and found to contain 87 wt% carbon, 11% hydrogen, 1.4% sulfur, and the remainder noncombustible matter. The oil is burned with 20% excess air, based on complete combustion of the carbon to CO_2, the hydrogen to H_2O, and the sulfur to SO_2. The oil is burned completely, but 5% of the carbon forms CO. Calculate the molar composition of the stack gas.

79. The analysis of a coal indicates 70 wt% C, 20% H, 2% S, and the balance noncombustible ash. The coal is burned at a rate of 5000 kg/h, and the feed rate of air to the furnace is 50 kmol/min. All of the ash and 6% of the carbon in the fuel leave the furnace as a molten slag; the remainder of the carbon leaves in the stack gas as CO and CO_2; the hydrogen in the coal is oxidized to water, and the sulfur emerges as SO_2. The selectivity of CO_2 to CO production is 10:1.

(a) Calculate the percent excess air fed to the reactor.

(b) Calculate the mole fractions of the gaseous pollutants—CO and SO_2—in the stack gas.

(c) Emitted sulfur dioxide by itself is a health hazard, but it is a more serious threat to the environment as a precursor to **acid rain.** Under the catalytic action of sunlight, the sulfur dioxide is oxidized to sulfur trioxide, which in turn combines with water vapor to form sulfuric acid, which eventually returns to earth as rainfall. Acid rain formed in this manner has caused extensive damage to forests, fields, and lakes in many parts of the world. For the furnace described above, calculate the rate of formation of sulfuric acid (kg/h) if all the emitted SO_2 is converted in the indicated manner.

80. The composition of a coal is determined either by a **proximate analysis**—a procedure described below—or a more thorough **ultimate analysis,** described in Problem 48 of Chapter 9. In a proximate analysis, the coal is finely ground and air-dried. Samples of the dried coal are then subjected to several operations, with the sample weights being recorded before and after each operation. **Moisture content** is determined as the weight loss when a sample is held at 105°C in an oxygen-free atmosphere for roughly 2 h, added to that lost in the initial drying step. **Volatile matter** (primarily organic tars) is determined by holding a sample at 925°C in an oxygen-free atmosphere for 7 min and subtracting the moisture loss from the total weight loss. **Ash** (or **mineral matter**—oxides and sulfates of silicon, aluminum, iron, calcium, sulfur, and trace minerals) is the residue that remains after a sample has been heated to 800°C in an oxygen-containing atmosphere until all the organic matter has been burned away. **Fixed carbon** is what is present in coal besides moisture, volatile matter, and ash.

(a) Use the following proximate analysis data to determine the percentages by mass of moisture, fixed carbon, volatile matter, and ash in a coal.

$$1.207 \text{ g} \xrightarrow[\substack{25°C, \ 12 \text{ h}}]{\text{air-dry}} 1.147 \text{ g}$$

The remaining tests are performed on air-dried samples.

$$1.234 \text{ g} \xrightarrow[\text{2 h}]{\text{105°C, N}_2} 1.204 \text{ g}$$

$$1.347 \text{ g} \xrightarrow[\text{7 min}]{\text{925°C, N}_2} 0.811 \text{ g}$$

$$1.175 \text{ g} \xrightarrow[\text{1 h}]{\text{800°C, air}} 0.111 \text{ g}$$

(b) If the mass ratio of C to H in the volatile matter is 6:1, calculate the gram-moles of air theoretically required to burn 1 metric ton of this coal.

81. The product gas from a combustion reaction has the following dry-basis molar composition: 80.35 mole% CO_2, 4.73% CO, 0.033% SO_2, 14.89% O_2. Pure oxygen is fed to the furnace in 10% excess of that required to burn the fuel completely. There is no oxygen in the fuel. Calculate the elemental composition (mole% of the various elements) of the fuel.

82. A fuel oil is fed to a furnace and burned with 25% excess air. The oil contains 87.0 wt% C, 10.0% H, and 3.0% S. Analysis of the furnace exhaust gas shows only N_2, O_2, CO_2, SO_2, and H_2O. The sulfur dioxide emission rate is to be controlled by passing the exhaust gas through a scrubber, in which most of the SO_2 is absorbed in an alkaline solution. The gases leaving the scrubber (all of the N_2, O_2, and CO_2, and some of the H_2O and SO_2 entering the unit) pass out to a stack. The scrubber has a limited capacity, however, so that a fraction of the exhaust gas must be bypassed directly to the stack.

At one point during the operation of the process, the scrubber removes 90% of the SO_2 in the gas fed to it, and the combined stack gas contains 612.5 ppm (parts per million) SO_2 on a dry basis; that is, every million moles of dry stack gas contains 612.5 moles of SO_2. Calculate the fraction of the exhaust bypassing the scrubber at this moment.

83. You have been sent by the Environmental Protection Agency to measure SO_2 emissions from a small industrial power plant. You withdraw and analyze a gas sample from the boiler stack and obtain the following composition: 75.66% N_2, 10.24% CO_2, 8.27% H_2O, 5.75% O_2, and 0.0825% SO_2. You show these figures to the plant superintendent next day, and she insists they must be wrong, since the fuel was a natural gas containing methane and ethane, and no sulfur. You ask if they ever burn another fuel, and the superintendent replies that they sometimes use a fuel oil but the plant log shows that they were not doing so when the measurements were made. You do some calculations and prove that the oil and not the gas must have been the fuel; the superintendent checks further and discovers that the plant log is in error and you are right.

(a) Calculate the mole ratio of carbon to hydrogen in the fuel, and use the result to prove that the fuel could not have been the natural gas.

(b) Calculate the *mass* ratio of carbon to hydrogen and the weight% of sulfur in the fuel, assuming that C, H, and S are the only elements present. Then use the results in conjunction with the ultimate fuel oil analyses in Table 9.10 on p. 9-10 of *Perry's Handbook*, 6th Edition, to deduce the most probable classification of the fuel oil.

84. Fuel oils contain primarily organic compounds and sulfur. Suppose the molar composition of the organic fraction of a fuel oil may be represented by the formula $C_pH_qO_r$; the

mass fraction of sulfur in the fuel is x_s (kg S/kg fuel); and the percentage excess air P is defined in terms of the theoretical air required to burn only the organic materials in the fuel. Complete combustion of the fuel to form CO_2, SO_2, and H_2O may be assumed.

(a) For a high-sulfur No. 6 fuel oil, to a good approximation $p = 0.71$, $q = 1.1$, $r = 0.003$, and $x_s = 0.04$. Calculate the composition of the flue gas on a dry basis if 21% excess air is supplied.

***(b)** Write a program to calculate the mole fractions of the flue gas components on a dry basis for specified values of p, q, r, x_s, and P. The program should read in a run number and the values of p, q, r, x_s, and P, calculate the mole fractions, print out the data and the calculated quantities, and return to read in the data for the next sample. Termination should occur if a run number of 0 is read in. Test your program using the following data.

Run	p	q	r	x_s	P
1	0.71	1.10	0.003	0.04	21%
2	0.71	1.10	0.003	0.04	39%
3	0.73	1.26	0.001	0.002	21%

* Computer problem.

Chapter Five

Single-Phase Systems

Most of the material balance problems in Chapter 4 could be solved entirely from information given in the problem statements. As you will come to discover, real problems in process analysis are rarely so conveniently self-contained; before you can carry out a complete material balance on a process, you must usually determine various physical properties of the process materials and use these properties to derive additional relations among the system variables.

Suppose, for instance, you know the composition, temperature, pressure, and mass flow rate of a process stream, and you wish to calculate the volumetric flow rate in order to size the equipment (pumps, valves and fittings, etc.) that will be used to transport the stream. Clearly you need the process fluid density to perform this calculation. The following methods can be used to determine this quantity:

(a) *Look It Up.* When you need a value for a physical property of a substance—whether it be a density, vapor pressure, solubility, or heat capacity—there is a good chance that someone somewhere has measured this property and published the result. Since experiments are usually costly and time consuming, a reliable source of physical property data is an invaluable asset in process analysis.

Four excellent sources of data are the following:

1. *Perry's Chemical Engineers' Handbook*, 6th Edition, R. H. Perry and D. W. Green Eds., McGraw-Hill, New York, 1984.
2. *Handbook of Chemistry and Physics*, 66th Edition, R. C. Weast, Ed., Chemical Rubber Company, Boca Raton, FL, 1985.
3. *Selected Values of Chemical Thermodynamic Properties*, National Bureau of Standards, NBS Circular 500 (1952), currently being issued in revised form as NBS Technical Note 270, Washington, D.C.
4. *Journal of Physical and Chemical Reference Data*, Supplement 2, Volume 11. A joint publication of the American Chemical Society, American Institute of Physics, and the National Bureau of Standards, American Chemical Society, Washington, D.C., 1982.

If the desired information cannot be found in these references, the species in question can be looked up in the index of *Chemical Abstracts* in an effort to locate data in the open literature.

(b) *Estimate It.* There are a relatively small number of atomic elements and a much larger but still countable number of molecular species of interest to chemical

engineers. Chemical species can be combined into mixtures and solutions in an infinite number of ways, however, and it is clearly impossible to tabulate physical property data for even a small fraction of the possible combinations. Moreover, even when data are found they are likely to have been determined for conditions other than the ones for which you need them. Reid, Prausnitz, and Sherwood[1] summarize a large number of empirical correlations that express physical properties of a mixture in terms of the pure component properties and the mixture composition. These correlations can be used to estimate physical properties when no data are available and to extrapolate available data to conditions other than the ones at which they were obtained.

(c) *Measure It.* When no information on a particular physical property of a substance can be found in the literature or when the property must be known with a precision greater than that provided by general estimation formulas, the only recourse is to determine the property experimentally. *Perry's Chemical Engineers' Handbook*[2] contains a great deal of information on experimental techniques for measuring physical properties; details can also be found in any of numerous texts on experimental physical, organic, and analytical chemistry.

All three sources of information on physical properties—published data, predictive correlations, and experimental measurements—are illustrated in the sections that follow. Sections 5.1 through 5.4 deal with the relationships between the density or specific volume of a substance and the composition, temperature, and pressure of the substance: Section 5.1 concerns solids and liquids; Section 5.2 deals with ideal gases; and Sections 5.3 and 5.4 extend the discussion to real (nonideal) gases.

5.1 LIQUID AND SOLID DENSITIES

Specific gravities of many solids and liquids at a single temperature are listed in Table B.1 in Appendix B. More extensive tabulations can be found in *Perry's Chemical Engineers' Handbook* on pp. 3-6 through 3-44 and p. 3-95.

When you heat a liquid or a solid it normally expands (i.e., its density decreases). In most process applications, however, it can usually be assumed with little error that solid and liquid densities are independent of temperature—provided, of course, that no phase changes occur. Similarly, changes in pressure do not cause significant changes in liquid or solid densities; these substances are therefore termed **incompressible.**

Perry's Chemical Engineer's Handbook (pp. 3-75 and 3-77) lists densities of liquid water and mercury at different temperatures and gives expressions on pp. 3-104

[1] R. C. Reid, J. M. Prausnitz, and T. K. Sherwood, *The Properties of Gases and Liquids*, 3rd Edition, McGraw-Hill, New York, 1977.

[2] R. H. Perry and D. W. Green, Eds., *Perry's Chemical Engineers' Handbook*, 6th Ed., McGraw-Hill, New York, 1984.

through 3-106 that may be used to calculate densities for many other substances at different temperatures. Reid, Prausnitz, and Sherwood present a number of methods to estimate the density of a liquid for which tabulated densities are not available. Some of these estimation formulas are also given in *Perry's Chemical Engineers' Handbook* on pp. 3-273 and 3-274.

The density of a mixture of liquids can be estimated by assuming that the component volumes are additive: for example, if 2 ml of a liquid A and 3 ml of liquid B are mixed, the resulting volume would be assumed to be 5 ml. This leads to a simple averaging formula for the mixture density:

$$1/\bar{\rho} = \sum_{i=1}^{n} x_i/\rho_i \qquad \qquad (5.1\text{-}1)$$

where n is the number of components in the mixture, ρ_i is the density of the ith component, and x_i is the mass fraction of this component. This formula works best for mixtures of components with similar molecular structures (e.g., all straight-chain hydrocarbons of nearly equal molecular weight). For other mixtures or for solutions of solids in liquids, data must either be found in the literature or measured experimentally. If you need the density of a mixture of two substances of which one is water, try *Perry's Chemical Engineers' Handbook*, pp. 3-78 through 3-94, and note the additional references listed on p. 3-78 of the *Handbook*.

TEST YOURSELF

1. The specific gravity of water is 1.0000 at 4°C. A stream of water at 4°C has a mass flow rate of 200 g/s. What is its volumetric flow rate? If the stream temperature is raised to 75°C, does the mass flow rate change? How would you expect the volumetric flow rate to change? How would you calculate the volumetric flow rate at 75°C without actually measuring it?

2. The pressure equivalence $14.696 \text{ lb}_f/\text{in.}^2 \Leftrightarrow 760 \text{ mm Hg}$ is not complete—it should be stated as

$$14.696 \text{ lb}_f/\text{in.}^2 \Leftrightarrow 760 \text{ mm Hg at } 0°C$$

Why it is necessary to state a temperature?

3. Suppose you mix m_1 g of liquid A_1 (density ρ_1 g/cm^3), m_2 g of liquid A_2 (density ρ_2), . . . , and m_n g of liquid A_n (density ρ_n). Assume that the volumes are additive. Show that the density of the mixture is given by Equation 5.1-1.

EXAMPLE 5.1-1 Determination of a Solution Density

Determine the density in g/cm^3 of a 50 weight% aqueous solution of H_2SO_4 at 20°C, both by looking up a tabulated value and by assuming volume additivity of the solution components.

Solution

1. ***Look It Up.*** *Perry's Chemical Engineers' Handbook*, pp. 3-85 and 3-86, tabulates specific gravity data for sulfuric acid. From this table,

$$\rho(50\% \ H_2SO_4, 20°C) = 1.3951 \ g/cm^3$$

2. ***Estimate It.*** The pure-component densities are

$$\rho(H_2O, 20°C) = 0.998 \ g/cm^3 \ (Handbook, \ p. \ 3\text{-}75)$$
$$\rho(H_2SO_4, 18°C) = 1.834 \ g/cm^3 \ (Handbook, \ p. \ 3\text{-}23)$$

We will neglect the density change for H_2SO_4 between 18°C and 20°C, although *Perry's Chemical Engineers' Handbook*, p. 3-106, gives thermal expansion data for H_2SO_4 that could be used to make this minor correction. We then estimate from Equation 5.1-1:

$$1/\bar{\rho} = 0.5/0.998 + 0.5/1.834 = 0.7736 \ cm^3/g$$
$$\Downarrow$$
$$\bar{\rho} = 1.29 \ g/cm^3$$

The density estimated assuming volume additivity thus differs from the true density [given in part (a)] by 7.3%. The acceptability of this level of accuracy depends on the intended use of the estimated quantity.

5.2 IDEAL GASES

Looking up a density or specific volume at one temperature and pressure and using it at another temperature and pressure usually works well for a solid or a liquid, but not at all for a gas. To solve most problems that involve gases, an expression is needed that relates specific volume to temperature and pressure, so that if any two of these quantities are known the third can be calculated.

Typical problems that require a *PVT* relationship for their solution include the following.

1. Propane at 120°C and 2.3 bars passes through a flow meter that reads 250 liters/min. What is the mass flow rate of the gas?

2. A pure hydrocarbon gas fills a 2-liter vessel at 30°C with an absolute pressure of 25 atm. How many gram-moles of gas are contained in the vessel? If the mass of the gas is 60 g, what might the gas be?

3. A 20-ft^3 gas cylinder in your laboratory can hold pressures up to 400 atm. The Bourdon gauge on the cylinder reads 380 atm one morning when the temperature is 55°F. How high can the temperature go before it becomes advisable to move to a different laboratory?

An equation that relates the quantity (mass or moles) and volume of a gas to the gas temperature and pressure is called an **equation of state.** The simplest and most widely used equation of state is the **ideal gas law,** which is adequate for many engineering calculations over a wide range of conditions. At high pressures and low temperatures, however, all gases deviate from ideal behavior and more complex equations of state are needed to describe their *PVT* relations.

In this section we discuss the ideal gas law and show how it is applied to systems containing single gaseous substances and mixtures of gases. Section 5.3 outlines methods used for a single real gas (by definition, a gas for which the ideal gas law does not work) and for mixtures of real gases.

5.2a The Ideal Gas Law

The ideal gas law can be derived from the kinetic theory of gases by assuming that gas molecules have a negligible volume, exert no forces on one another, and collide elastically with the walls of their container. The law usually appears in the form

$$\boxed{PV = nRT} \tag{5.2-1}$$

where

$P = $ *absolute* pressure of a gas
$V = $ volume or volumetric flow rate of the gas
$n = $ number of moles or molar flow rate of the gas
$R = $ the *gas constant*, whose value depends on the units of P, V, n, and T
$T = $ *absolute* temperature of the gas

The equation may also be written as

$$P\hat{V} = RT \tag{5.2-2}$$

where $\hat{V} = V/n$ is the molar volume of the gas.

A gas whose *PVT* behavior is well represented by Equation 5.2-1 is said to behave as an **ideal gas** or **perfect gas.** The use of this equation does not require a knowledge of the species of the gas: 1 mol of an ideal gas at 0°C and 1 atm occupies 22.415 liters, whether the gas is argon, nitrogen, a mixture of propane and air, or any other single species or mixture of gases.

The gas constant R has units of (pressure)(volume)/(mole)(temperature); moreover, since pressure times volume has units of energy (prove it), R may also be expressed in units of (energy)/(mole)(temperature). Values of the gas constant expressed in various units are listed in the table on the inside back cover of this book.

EXAMPLE 5.2-1 The Ideal Gas Law

Calculate the volume in liters occupied by 100 g of nitrogen at 23°C and 3 psig, assuming ideal gas behavior.

Solution

The ideal gas law relates absolute temperature, absolute pressure, and the quantity of a gas in moles. We therefore first calculate

$$n = \frac{100 \text{ g N}_2}{28.0 \text{ g/mol}} = 3.57 \text{ mol N}_2$$

$$T = 296 \text{ K}$$

and (assuming $P_{atm} = 14.7$ psia) $P = 17.7$ psia. Then from the ideal gas law

$$V(\text{liters}) = \frac{nRT}{P}$$

$$= \frac{(3.57 \text{ mol})(296 \text{ K})}{(17.7 \text{ psia})} \left| \frac{R(\text{liter} \cdot \text{psia})}{(\text{mol} \cdot \text{K})} \right.$$

Unfortunately, the table of gas constants at the back of this book does not list the value of R with this particular set of units. In its absence, we use a value that comes close and carry out the necessary additional unit conversions.

$$V = \frac{(3.57 \text{ mol})(296 \text{ K})}{(17.7 \text{ psi})} \left| \frac{0.08206 \text{ liter} \cdot \text{atm}}{\text{mol} \cdot \text{K}} \right| \frac{14.7 \text{ psi}}{\text{atm}} = 72.0 \text{ liters}$$

TEST YOURSELF

1. What is an equation of state? What is the ideal gas law? What is a law? (You won't find the answer to the last one in the text, but it's something you should know anyway.)

2. Two gas cylinders have identical volumes and contain gases at identical temperatures and pressures. Cylinder A contains hydrogen, and cylinder B contains carbon dioxide. If both gases behave ideally, how does the number of moles of H_2 compare with the number of moles of CO_2? How do the masses of the two substances compare?

3. One hundred grams per hour of C_2H_4 flow through a pipe at 1.2 atm and 70°C, and 100 g/h of C_4H_8 flow through a second pipe at 1.2 atm and 70°C. Which stream has the higher volumetric flow rate? How much higher is it?

4. Derive a formula for the density of an ideal gas of molecular weight M at temperature T and pressure P.

CREATIVITY EXERCISE

An unlabeled gas cylinder is equipped with a sensitive pressure gauge. Devise several experiments that could be used to estimate the molecular weight of the gas, using only equipment and materials likely to be found in the average home or purchasable at a neighborhood hardware store. (You may have a sensitive balance if you need one, but you may not assume that the average home has its own chemistry laboratory.)

5.2b Standard Conditions

Doing PVT calculations by substituting into the ideal gas law is straightforward, but to use this method you must have on hand either a table of values of R with different units or an extremely good memory. A way to avoid these requirements is to write the gas law for the process conditions

$$PV = nRT \qquad (5.2\text{-}3)$$

and for a set of arbitrarily chosen reference conditions

$$P_s V_s = n_s R T_s \qquad (5.2\text{-}4)$$

and then to divide the first equation by the second:

$$\frac{PV}{P_s V_s} = \frac{nT}{n_s T_s} \qquad (5.2\text{-}5)$$

Since the reference conditions are known, Equation 5.2-5 may then be used to solve for any one of the variables P, V, n, and T from given values of the other three. (Note that you do not need a value for R, which cancels out when the equations are divided.) Reference conditions normally used are shown in Table 5.2-1. The given

TABLE 5.2-1

Standard Conditions for Gases

System	T_s	P_s	V_s	n_s
SI	273 K	1 atm	0.022415 m^3	1 mol
CGS	273 K	1 atm	22.415 liters	1 mol
American Engineering	492°R	1 atm	359.05 ft^3	1 lb-mole

reference temperature (0°C, 273 K, 492°R) and pressure (1 atm) are commonly referred to as **standard temperature and pressure** (STP). These are easy values to remember; you should also commit to memory the quantities

$$\frac{V_s}{n_s} = 0.0224 \; \frac{\text{m}^3(\text{STP})}{\text{mol}} \Leftrightarrow 22.4 \; \frac{\text{liters}(\text{STP})}{\text{mol}} \Leftrightarrow 359 \; \frac{\text{ft}^3(\text{STP})}{\text{lb-mole}}$$

EXAMPLE 5.2-2 Conversion from Standard Conditions

Butane (C_4H_{10}) at 360°C and 3.00 atm (absolute) flows into a reactor at a rate of 1100 kg/h. Calculate the volumetric flow rate of this stream using conversion from standard conditions.

Solution

As always, molar quantities and absolute temperature and pressure must be used.

$$n = \frac{1100 \text{ kg/h}}{58.1 \text{ kg/kmol}} = 19.0 \text{ kmol/h}$$

$$T = 633 \text{ K}, \qquad P = 3.00 \text{ atm}$$

From Equation 5.2-5

$$\frac{PV}{P_s V_s} = \frac{nT}{n_s T_s} \Rightarrow V = n\frac{V_s}{n_s}\frac{T}{T_s}\frac{P_s}{P}$$

$$\Downarrow$$

$$V = \frac{19.0\text{ kmol}}{\text{h}} \left| \frac{22.4\text{ m}^3\text{(STP)}}{\text{kmol}} \right| \frac{633\text{ K}}{273\text{ K}} \left| \frac{1.00\text{ atm}}{3.00\text{ atm}} \right| = \boxed{329\,\frac{\text{m}^3}{\text{h}}}$$

You will frequently encounter problems that involve gases at two different states (sets of conditions)—for example, at the inlet and outlet of a process unit. A convenient way to determine an unknown variable (P, V, n, or T) of the gas at one point or another is to write the gas law for both states and divide one equation by the other. Any variables that are the same at both points will cancel, leaving an equation containing only the variable you wish to determine and known quantities.

EXAMPLE 5.2-3

Ten cubic feet of air at 70°F and 1.00 atm are heated to 610°F and compressed to 2.50 atm. What volume does the gas occupy in its final state?

Solution

Let 1 denote the initial state of the gas and 2 the final state. Note that $n_1 = n_2$ (the number of moles of the gas does not change). Assume ideal gas behavior.

n lb–moles air
$V_1 = 10\text{ ft}^3$
$T_1 = 70°F = 530°R$
$P_1 = 1\text{ atm}$

Process

n lb–moles air
$V_2\text{ ft}^3$
$T_2 = 610°F = 1070°R$
$P_2 = 2.5\text{ atm}$

$$\begin{aligned} P_2 V_2 &= nRT_2 \\ P_1 V_1 &= nRT_1 \end{aligned} \Rightarrow \frac{P_2 V_2}{P_1 V_1} = \frac{T_2}{T_1}$$

$$V_2 = V_1\left(\frac{P_1}{P_2}\right)\left(\frac{T_2}{T_1}\right) = \frac{10.0\text{ ft}^3}{} \left| \frac{1.00\text{ atm}}{2.50\text{ atm}} \right| \frac{1070°R}{530°R} = \boxed{8.08\text{ ft}^3}$$

TEST YOURSELF

1. What are standard temperature and pressure? What are the values of V_s/n_s in the SI, CGS, and American Engineering systems?

2. What happens to the volume of an ideal gas when you double the pressure at a fixed temperature? When you double the temperature at a fixed pressure?

3. What happens to the density of a fixed quantity of an ideal gas when you raise the temperature at a fixed pressure? When you raise the temperature at a fixed volume?

4. Consider two different ideal gases in equal volume containers at the same temperature and pressure. Which of the following quantities are necessarily the same for each gas? (a) Number of moles. (b) Mass. (c) Molar density. (d) Mass density. (e) Number of molecules.

5.2c Ideal Gas Mixtures

Suppose n_A moles of substance A, n_B moles of B, n_C moles of C, and so on, are contained in a volume V at a temperature T and total pressure P. The **partial pressure** p_A and **pure component volume** v_A of A in the mixture are defined as follows:

p_A: The pressure that would be exerted by n_A moles of A alone in the same total volume V at the same temperature T.

v_A: The volume that would be occupied by n_A moles of A alone at the total pressure P and temperature T of the mixture.

Suppose next that each of the individual mixture components and the mixture as a whole behave in an ideal manner. (This is the definition of an **ideal gas mixture**.) If there are n moles of all species in the volume V at a pressure P and temperature T, then

$$PV = nRT \qquad (5.2\text{-}6)$$

In addition, from the definition of partial pressure,

$$p_A V = n_A RT \qquad (5.2\text{-}7)$$

Dividing the second equation by the first yields

$$\frac{p_A}{P} = \frac{n_A}{n} = y_A \quad \text{(the mole fraction of A in the gas)} \qquad (5.2\text{-}8)$$

or

$$\boxed{p_A = y_A P} \qquad (5.2\text{-}9)$$

That is, *the partial pressure of a component in an ideal gas mixture is the mole fraction of that component times the total pressure.*[3] Moreover, since $y_A + y_B + \cdots = 1$,

$$p_A + p_B + \cdots = (y_A + y_B + \cdots)P = P \qquad (5.2\text{-}10)$$

or, *the partial pressures of the components of an ideal gas mixture add up to the total pressure.*

[3] Equation 5.2-9 is often used as the definition of the partial pressure. For an ideal gas mixture, the definition given and Equation 5.2-9 are equivalent; for a nonideal gas the concept of partial pressure has little utility.

A similar series of calculations can be performed for pure component volumes.

$$Pv_A = n_A RT \tag{5.2-11}$$

$$\Downarrow \text{ Divide by } PV = nRT$$

$$\frac{v_A}{V} = \frac{n_A}{n} = y_A \tag{5.2-12}$$

or

$$\boxed{v_A = y_A V} \tag{5.2-13}$$

and

$$v_A + v_B + \cdots = V \tag{5.2-14}$$

The quantity v_A/V is the **volume fraction** of A in the mixture, and 100 times this quantity is the **percentage by volume** of this component. As Equation 5.2-12 shows, *the volume fraction of a substance in an ideal gas mixture equals the mole fraction of this substance.* Stating, for example, that an ideal gas mixture contains 30% CH_4 and 70% C_2H_6 by volume, is equivalent to specifying 30 mole% CH_4 and 70 mole% C_2H_6.

TEST YOURSELF

1. A cylinder containing a mixture of N_2 and O_2 is taken from the basement to the top of a 50-story building. Assuming that the temperature remains constant, which of the following properties of the gas change during the ascent? **(a)** mass, **(b)** weight, **(c)** absolute pressure, **(d)** gauge pressure, **(e)** partial pressure of N_2, **(f)** pure component volume of O_2, **(g)** density, **(h)** specific molar volume.

2. An ideal gas mixture at $P = 10$ bar and $T = 200°C$ in a 100-m^3 tank contains 50 mole% H_2 and 50 mole% N_2. What is the partial pressure of H_2? What is the pure component volume of H_2? What would happen to p_{H_2} and v_{H_2} if the temperature were raised?

3. An ideal gas mixture at $P = 10$ atm contains 50 weight% H_2 and 50 weight% N_2. Is the partial pressure of H_2 less than, equal to, or greater than 5 atm?

EXAMPLE 5.2-4 Material Balances on an Evaporator-Compressor

Liquid acetone (C_3H_6O) is fed at a rate of 400 liters/min into a heated chamber, where it evaporates into a nitrogen stream. The gas leaving the heater is diluted by another nitrogen stream flowing at a measured rate of 419 m^3 (STP)/min. The combined gases are then compressed to a total pressure $P_{gauge} = 6.3$ atm at a temperature of 325°C. The partial pressure of acetone in this stream is $p_a = 501$ mm Hg. Atmospheric pressure is 763 mm Hg.

1. What is the molar composition of the stream leaving the compressor?
2. What is the volumetric flow rate of the nitrogen entering the evaporator if the temperature and pressure of this stream are $T = 27°C$ and $P_{gauge} = 475$ mm Hg?

Solution

Basis: Given Feed Rates

Assume ideal gas behavior. Let q_1, q_2, \ldots (mol/min) be the molar flow rates of each stream.

You should be able to examine the flowchart and see exactly how the solution will proceed.

1. Calculate q_2 (From the given volumetric flow rate and a tabulated density of liquid acetone)
 q_3 (Ideal gas law)
 y_a (p_a/P)
2. Calculate q_4 (Overall acetone balance)
 then q_1 (Overall mole balance)
 then V_1 (Ideal gas law)

Calculate Molar Flow Rate of Acetone

From Table B.1 in Appendix B, the density of liquid acetone is 0.792 g/cm^3 = 792 g/liter, so that

$$q_2 = \frac{400 \text{ liters}}{\text{min}} \left| \frac{792 \text{ g}}{\text{liter}} \right| \frac{1 \text{ mol}}{58.1 \text{ g}} = 5470 \frac{\text{mol C}_3\text{H}_6\text{O}}{\text{min}}$$

Determine Mole Fractions from Partial Pressures

In the stream leaving the compressor,

$$\frac{p_a}{P} = y_a \left(\frac{\text{mol C}_3\text{H}_6\text{O}}{\text{mol}} \right)$$

$$P = P_{gauge} + P_{atm} = \frac{6.3 \text{ atm}}{1} \left| \frac{760 \text{ mm Hg}}{1 \text{ atm}} \right. + 763 \text{ mm Hg} = 5550 \text{ mm Hg}$$

so that

$$y_a = \frac{501 \text{ mm Hg}}{5550 \text{ mm Hg}} = 0.0903 \frac{\text{mol C}_3\text{H}_6\text{O}}{\text{mol}}$$

$$1 - y_a = 0.9097 \frac{\text{mol N}_2}{\text{mol}}$$

Calculate q_3 from PVT Information

$$q_3 = \frac{419 \text{ m}^3(\text{STP})}{\text{min}} \left| \frac{1 \text{ mol}}{0.0224 \text{ m}^3(\text{STP})} = 18\ 700 \frac{\text{mol}}{\text{min}} \right.$$

Overall Mole Balance on Acetone

$$q_2 = q_4 y_a$$
$$\left\Vert \begin{array}{l} q_2 = 5470 \\ y_a = 0.0903 \end{array} \right.$$
$$q_4 = 60\ 600 \text{ mol/min}$$

Overall Mole Balance

$$q_1 + q_2 + q_3 = q_4$$
$$\left\Vert \begin{array}{l} q_2 = 5470 \\ q_3 = 18\ 700 \\ q_4 = 60\ 600 \end{array} \right.$$
$$q_1 = 36\ 400 \text{ mol/min}$$

$$\left\Vert \begin{array}{l} \text{Ideal gas law} \\ T_1 = 27°\text{C} = 300 \text{ K} \\ P_1 = 475 \text{ mm Hg gauge} = 1238 \text{ mm Hg} \end{array} \right.$$

$$V_1 = q_1 \frac{V_s}{n_s} \frac{T_1}{T_s} \frac{P_s}{P_1}$$

$$= \frac{36\ 400 \text{ mol/min}}{} \left| \frac{0.0224 \text{ m}^3}{1 \text{ mol}} \right| \frac{300 \text{ K}}{273 \text{ K}} \left| \frac{760 \text{ mm Hg}}{1238 \text{ mm Hg}} \right.$$

$$\Downarrow$$

$$\boxed{V_1 = 550 \text{ m}^3 \text{ N}_2/\text{min}}$$

5.3 REAL GASES

As the temperature of a gas decreases and the pressure increases, the ideal gas law provides an increasingly poor description of the gas behavior, and it becomes necessary to turn to one of dozens of more complex equations of state. The ideal gas law should yield an error of 1% or less if the molar volume $\hat{V}(= RT/P)$ is greater than

5 liters/mol (80 ft³/lb-mole) for diatomic gases and 20 liters/mol (320 ft³/lb-mole) for other gases.

One approach to representing the PVT behavior of real gases is to use equations of state that contain from 2 to 17 constants whose values differ for different gases. A summary of these equations and suggestions about when to use them are given by Reid, Prausnitz, and Sherwood; one such equation is presented in Section 5.3a, and others are contained in problems and examples in the rest of this chapter.

Another approach to real gas PVT calculations is to correct for deviations from ideal behavior by including a term called the **compressibility factor** in the ideal gas law. The resulting equation is called the compressibility factor equation of state and is discussed in Sections 5.3b and c. This approach can also be easily adapted to PVT calculations for real gas mixtures, as outlined in Section 5.3d.

5.3a Virial Equations of State

A **virial equation of state** is an equation of the form

$$\frac{P\hat{V}}{RT} = 1 + \frac{B(T)}{\hat{V}} + \frac{C(T)}{\hat{V}^2} + \cdots \tag{5.3-1}$$

The coefficient $B(T)$ is called the **second virial coefficient,** $C(T)$ is the **third virial coefficient,** and so on. If $B = C = D = \cdots = 0$, the equation reduces to the ideal gas law.

Methods exist for estimating at least the second and third virial coefficients B and C for simple gases, but they are difficult to use and often require information not known for the gas in question. Another approach is to measure P for a number of values of \hat{V} and T and then to determine empirical expressions for B, C, and so on, that fit the data. A typical result of the latter approach is the **Benedict–Webb–Rubin** (BWR) equation, which contains eight constants.

$$\frac{P\hat{V}}{RT} = 1 + \frac{B}{\hat{V}} + \frac{C}{\hat{V}^2} + \frac{D}{\hat{V}^4} + \frac{E}{\hat{V}^5} \tag{5.3-2}$$

where

$$B = B_0 - \frac{A_0}{RT} - \frac{C_0}{RT^3} \tag{5.3-3}$$

$$C = b - \frac{a}{RT} + \frac{ce^{-\gamma/V^2}}{RT^3} \tag{5.3-4}$$

$$D = \frac{c\gamma e^{-\gamma/\hat{V}^2}}{RT^3} \tag{5.3-5}$$

$$E = \frac{a\alpha}{RT} \tag{5.3-6}$$

Constants: A_0, B_0, C_0, a, b, c, α, γ

Reid, Prausnitz, and Sherwood tabulate values of the constants for CO_2, N_2, SO_2, H_2S, and several light hydrocarbons. Some of these values are given in Table 5.3-1.

TABLE 5.3-1

Constants for Benedict–Webb–Rubin Equation[a]

For P(atm), T(K), V(liters/mol), and $R = 0.08206$(liter·atm/mol·K)

Gas	A_0	B_0	$C_0 \times 10^{-6}$	a	b	$c \times 10^{-6}$	$\alpha \times 10^3$	$\gamma \times 10^2$
Methane	1.85500	0.042600	0.02257	0.049400	0.00338004	0.002545	0.124359	0.6000
Ethylene	3.33958	0.0556833	0.131140	0.259000	0.0086000	0.021120	0.178000	0.923000
Ethane	4.15556	0.0627724	0.179592	0.345160	0.0111220	0.0327670	0.243389	1.18000
Propane	6.87225	0.0973130	0.508256	0.947700	0.0225000	0.129000	0.607175	2.20000
Propylene	6.11220	0.0850647	0.439182	0.774056	0.0187059	0.102611	0.455696	1.82900
n-Butane	10.0847	0.124361	0.992830	1.88231	0.0399983	0.316400	1.10132	3.40000
n-Pentane	12.1794	0.156751	2.12121	4.07480	0.0668120	0.824170	1.81000	4.75000
Benzene	6.50997	0.069464	3.42997	5.570	0.07663	1.17642	0.7001	2.930
CO_2	2.51606	0.044885	0.14744	0.136814	0.0041239	0.014918	0.08466	0.52533
N_2	1.05364	0.047426	0.008059	0.035102	0.0023277	0.0007284	0.1272	0.5300
SO_2	2.12054	0.0261827	0.793879	0.844395	0.0146542	0.113362	0.0719604	0.592390

[a] Reprinted from E. J. Henley and E. M. Rosen, *Material and Energy Balance Computations*, Wiley, New York, 1969, with constants taken from M. Benedict, et al., *Chem. Eng. Progr.* 478, 419, 1951. Constants for additional species are given by H. W. Cooper and J. C. Goldfrank, *Hydrocarbon Processing*, 4612, 141, 1967.

EXAMPLE 5.3-1 The Benedict–Webb–Rubin Equation

Two gram-moles of nitrogen are placed in a 3-liter tank at $-150°C$. Estimate the tank pressure using the ideal gas law and the BWR equation.

Solution

$\hat{V} = 1.50$ liters/mol and $T = 123$ K. From the ideal gas law

$$P_{\text{ideal}} = \frac{RT}{\hat{V}} = \frac{0.08206 \text{ liter·atm}}{\text{mol·K}} \left| \frac{123 \text{ K}}{} \right| \frac{1 \text{ mol}}{1.50 \text{ liter}} = \boxed{6.73 \text{ atm}}$$

Substituting the constants for N_2 of Table 5.3-1 into Equations 5.3-3 through 5.3-6 yields

$$B = -0.11092, \qquad C = -0.01278, \qquad D = 2.692 \times 10^{-5}, \qquad E = 4.3 \times 10^{-6}$$

and

$$1 + \frac{B}{\hat{V}} + \frac{C}{\hat{V}^2} + \frac{D}{\hat{V}^4} + \frac{E}{\hat{V}^5} = 0.920$$

$$\Downarrow \text{ Equation 5.3-2}$$

$$P = \frac{RT}{\hat{V}} \left(1 + \frac{B}{\hat{V}} + \cdots \right) = (6.73 \text{ atm})(0.920) = \boxed{6.19 \text{ atm}}$$

Equations of state of the type just given are relatively accurate for gases containing a single species and are also relatively cumbersome. They should be used when considerable precision is needed in a PVT calculation; in addition, since computers can easily be programmed to solve them, they are useful when many PVT calculations are to be performed for a single species.

5.3b Cubic Equations of State

A number of equations of state are linear in pressure and cubic in volume. Truncation of the virial equation of state after the third term (C/\hat{V}^2), for example, results in such an equation. Cubic equations of state are simple, and they have been remarkably useful in describing PVT properties of a wide variety of gases, and vapor-liquid equilibrium conditions under limited circumstances.

One of the most widely used cubic equations of state is the Soave modification of the Redlich-Kwong (SRK) equation.

$$P = RT/(\hat{V} - b) - \alpha a/\hat{V}(\hat{V} + b) \tag{5.3-6}$$

where α, a, and b are system-dependent parameters. (You can show that this equation is indeed cubic in \hat{V} by finding a common denominator for the right side and rearranging the resulting expression.) The parameters a and b can be evaluated by the

TABLE 5.3-2

Pitzer Acentric Factors[a]

Compound	Acentric Factor
Ammonia	0.250
Argon	−0.004
Carbon dioxide	0.225
Carbon monoxide	0.049
Chlorine	0.073
Ethane	0.098
Hydrogen sulfide	0.100
Methane	0.008
Methanol	0.559
Nitrogen	0.040
Oxygen	0.021
Propane	0.152
Sulfur dioxide	0.251
Water	0.344

[a] R. C. Reid, J. M. Prausnitz, and T. K. Sherwood, *The Properties of Gases and Liquids*, 3rd Edition, McGraw-Hill, New York, 1977.

relationships

$$a = 0.42747 R^2 T_c^2 / P_c \qquad \text{(5.3-7)}$$

$$b = 0.08664 R T_c / P_c \qquad \text{(5.3-8)}$$

where T_c and P_c are, respectively, the **critical temperature** and **critical pressure** of the gas. We will discuss the physical significance of these constants later in this section; for now, we simply note that values for them are listed in Table B.1 of Appendix B and on pp. 630 to 665 of Reid, Prausnitz, and Sherwood, and methods for estimating them are given in Chapter 2 of Reid, Prausnitz, and Sherwood.

The parameter α of the SRK equation is obtained from a correlation based on experimental data. We first look up the value of a constant called the Pitzer acentric factor, ω, which reflects the geometric complexity and polarity of the gas molecule. Table 5.3-2 has selected values of ω; a more complete list may be found in Reid, Prausnitz, and Sherwood. We then calculate

$$m = 0.48508 + 1.55171\omega - 0.15613\omega^2 \qquad \text{(5.3-9)}$$

$$\alpha = [1 + m(1 - \sqrt{T/T_c})]^2 \qquad \text{(5.3-10)}$$

Once α, a, and b have been estimated, Equation 5.3-6 may be solved for whichever of the variables P, T, or \hat{V} is unknown.

EXAMPLE 5.3-2 Use of the SRK Equation of State

A gas cylinder having a volume of 2.50 m^3 contains 1.00 kmol of carbon dioxide at 200 K. Use the SRK equation of state to estimate the gas pressure in atm.

Solution

The system temperature is given, and the specific volume is easily calculated to be

$$\hat{V} = V/n = \frac{2.5 \text{ m}^3}{1 \text{ kmol}} \left| \frac{1 \text{ kmol}}{1000 \text{ mol}} \right. = 2.5 \times 10^{-3} \text{ m}^3/\text{mol}$$

From Table B.1, $P_c = 72.9$ atm (7.38×10^6 Pa), $T_c = 304.2$ K, and from Table 5.3-2, $\omega = 0.225$. Parameters in the SRK equation of state are evaluated using Equations 5.3-7 through 5.3-10.

$$a = \frac{0.42747}{} \left| \frac{(8.314 \text{ m}^3 \text{ Pa/mol K})^2}{7.38 \times 10^6 \text{ Pa}} \right| \frac{(304.2 \text{ K})^2}{} = 0.370 \text{ m}^6 \text{ Pa/mol}^2$$

$$b = \frac{0.08664}{} \left| \frac{(8.314 \text{ m}^3 \text{ Pa/mol K})}{7.38 \times 10^6 \text{ Pa}} \right| \frac{(304.2 \text{ K})}{} = 2.97 \times 10^{-5} \text{ m}^3/\text{mol}$$

$$m = 0.826 \quad \text{(from Equation 5.3-9)}$$
$$\alpha = 1.34 \quad \text{(from Equation 5.3-10)}$$

The SRK equation can now be solved for the tank pressure.

$$P = RT/(\hat{V} - b) - \alpha a/\hat{V}(\hat{V} + b)$$

$$= \frac{(8.314 \text{ m}^3 \text{ Pa/mol K})(200 \text{ K})}{(2.5 \times 10^{-3} \text{ m}^3/\text{mol} - 2.97 \times 10^{-5} \text{ m}^3/\text{mol})}$$

$$- \frac{1.34(0.370 \text{ m}^6 \text{ Pa/mol}^2)}{2.5 \times 10^{-3} \text{ m}^3/\text{mol} (2.5 \times 10^{-3} + 2.97 \times 10^{-5}) \text{ m}^3/\text{mol}}$$

$$P = 5.95 \times 10^5 \text{ Pa} \Rightarrow \boxed{P = 5.87 \text{ atm}}$$

Use of the ideal gas law results in an estimate of 6.57 atm for the pressure, a deviation of 12% from the more accurate SRK-determined value.

Evaluation of a system volume from a given temperature and pressure using a cubic equation of state requires a trial-and-error procedure. Several such procedures are described in Appendix A.2; one of them, Newton's method, is illustrated in the following example.

EXAMPLE 5.3-3 Use of the SRK Equation of State to Estimate Volume

A stream of carbon dioxide at 200 K and 6.8 atm is fed to a process at a rate of 100 kmol/h. Use the SRK equation of state to estimate the volumetric flow rate of gas in this stream.

Solution

The parameters in the SRK equation of state have the same values as in the previous example. Rearranging Equation 5.3-6, the SRK equation of state can be written in

the form

$$f(\hat{V}) = P - \frac{RT}{\hat{V} - b} + \frac{\alpha a}{\hat{V}(\hat{V} + b)} = 0$$

Substituting known quantities into this equation gives

$$f(\hat{V}) = 6.89 \times 10^5 - \frac{1.66 \times 10^3}{\hat{V} - 2.97 \times 10^{-5}} + \frac{0.4958}{\hat{V}(\hat{V} + 2.97 \times 10^{-5})} = 0$$

where each term in $f(\hat{V})$ has units of Pa and \hat{V} is in m³/mol. An initial estimate of \hat{V} can be obtained from the ideal gas law.

$$\hat{V} = \frac{RT}{P} = \frac{8.314 \text{ m}^3 \text{ Pa}}{\text{mol K}} \left| \frac{200 \text{ K}}{6.8 \text{ atm}} \right| \frac{1 \text{ atm}}{1.013 \times 10^5 \text{ Pa}} = 2.413 \times 10^{-3} \frac{\text{m}^3}{\text{mol}}$$

This gives $f(\hat{V}) = 7.660 \times 10^4$. Using Newton's method of approximation, a new estimate of \hat{V} can be obtained from the derivative $f'(\hat{V})$

$$f'(\hat{V}) = RT/(\hat{V} - b)^2 - \alpha a(2\hat{V} + b)/[\hat{V}(\hat{V} + b)]^2$$

and the expression (Equation A.2-2 of Appendix A)

$$\hat{V}_{new} = \hat{V} - f(\hat{V})/f'(\hat{V})$$

The iterative calculation proceeds as shown in the following table.

\hat{V}	$f(\hat{V})$	$f'(\hat{V})$	$\Delta\hat{V} = \hat{V}_{new} - \hat{V}$
2.413×10^{-3}	7.660×10^4	2.230×10^8	-3.435×10^{-4}
2.069×10^{-3}	-1.070×10^4	2.895×10^8	3.696×10^{-5}
2.106×10^{-3}	-1.574×10^2	2.810×10^8	5.601×10^{-7}

Additional iterations would refine the solution to the problem, but for this example, we may take \hat{V} to equal 2.11×10^{-3} m³/mol. Since the molar flow rate is 100 kmol/h, the volumetric flow rate of the gas is 211 m³/h. Use of the ideal gas law would have led to a volumetric flow rate of approximately 241 m³/h, a value 14% higher than that obtained with the SRK equation.

The physical significance of the critical constants is as follows. If the pressure of a single-component gas at a relatively low temperature is increased, a point is reached at which the gas begins to condense. If the gas is brought to a higher temperature and the process is repeated, a higher pressure is required to start condensation; moreover, the densities of the gas and liquid phases are closer together than they were at the lower temperature. Eventually a temperature T_c and pressure P_c are reached at which the two phases have exactly the same density. (At this point there is technically only one phase.) If T is greater than T_c, no phase change occurs, no matter how high the pressure is raised. T_c is defined as the critical temperature of the substance, and P_c is the critical pressure; a substance at $T = T_c$ and $P = P_c$ is said to be at the **critical state.**

TEST YOURSELF

1. What are the dimensions of a and b in the SRK equation of state?
2. What are the physical properties of a particular gas that you must know in order to use the SRK equation of state?
3. Verify that the derivative of $f(\hat{V})$ in Example 5.3-3 is as shown.

5.3c Compressibility Factor Equation of State

An equation of state that retains the simplicity of the ideal gas law but describes the PVT behavior of real gases over a much wider range of conditions is the following:

$$\boxed{PV = znRT} \quad \text{or} \quad \boxed{P\hat{V} = zRT} \tag{5.3-11}$$

The coefficient z is called the **compressibility factor,** and the equation is known as the **compressibility factor equation of state.** A value of $z = 1$ corresponds to ideal gas behavior.

The compressibility factor of a gas depends on the gas temperature and pressure and differs for different gases at a given T and P. Values of $z(T, P)$ for air, argon, CO_2, CO, H_2, CH_4, N_2, O_2 and steam are given in *Perry's Chemical Engineers' Handbook*, pp. 3-112 through 3-118. A method of estimating z when a table is not available is given in Section 5.3d.

PVT calculations based on the compressibility factor equation of state are performed exactly like calculations based on the ideal gas law, except that z must be determined (measured, looked up, or estimated) for each set of conditions. The equation may be applied directly with R expressed in the proper units, or conversion from standard conditions may be used. The following example illustrates the latter procedure.

EXAMPLE 5.3-4 The Compressibility Factor Equation of State

Two hundred kilograms of nitrogen are contained in a closed tank at $-100°C$. A pressure gauge on the tank reads 79 atm. Estimate the tank volume using the compressibility factor equation of state.

Solution

Use conversion from standard conditions.

$$\begin{aligned} PV &= znRT \\ P_sV_s &= n_sRT_s \end{aligned} \Rightarrow \frac{PV}{P_sV_s} = z\frac{n}{n_s}\frac{T}{T_s}$$

or

$$V = zn \frac{V_s}{n_s} \frac{P_s}{P} \frac{T}{T_s}$$

From *Perry's Chemical Engineers' Handbook*, p. 3-113, z for N_2 at $-100°C$ (173 K) and 80 atm (≈ 81 bar) is approximately 0.73, and

$$n = \frac{200 \text{ kg} \ \left| \ 1000 \text{ mol}}{\left| \ 28 \text{ kg}} = 7140 \text{ mol}$$

so that

$$V = \frac{0.73 \ \left| \ 7140 \text{ mol} \ \right| \ 0.0224 \text{ m}^3 \ \left| \ 1 \text{ atm} \ \right| \ 173 \text{ K}}{\left| \ 1 \text{ mol} \ \right| \ 80 \text{ atm} \ \left| \ 273 \text{ K}} = \boxed{0.925 \text{ m}^3}$$

TEST YOURSELF

1. What are the advantages and disadvantages of multiple-constant equations of state such as the BWR and SRK equations?

2. What is the compressibility factor equation of state? What are the units of the compressibility factor?

3. Suppose the compressibility factor of a gas is 0.9 at a given temperature and pressure. If the volumetric flow rate of a stream of this gas is calculated to be 90 liters/min using the ideal gas law, what is the correct volumetric flow rate?

5.3d Estimation of Compressibility Factors: The Law of Corresponding States

It would be convenient if the compressibility factor at a specific temperature and pressure were the same for all gases, so that a single chart or table of $z(T, P)$ could be used for all *PVT* calculations. Nature is not that accommodating, however: z for N_2 at 0°C and 100 atm is 0.9848, for example, while z for CO_2 at the same temperature and pressure is 0.2020.

It is possible, however, to construct a single universal chart for the estimation of compressibility factors for all gases if independent variables other than T and P are used. The chart—portions of which are shown in Figures 5.3-1 through 5.3-3—is called the **generalized compressibility chart.**

Suppose you have values for two of the variables P, V, and T for a gas and wish to calculate the third; suppose further that conditions are such that the ideal gas law is not likely to hold and that tabulated compressibility factors are not available. The procedure for solving this problem using the generalized compressibility chart is as follows.

FIGURE 5.3-1

Generalized compressibility chart, low pressures. (From D. M. Himmelblau, *Basic Principles and Calculations in Chemical Engineering*, 3rd edition, copyright © 1974, p. 175. Reprinted by permission of Prentice-Hall, Inc., Englewood Cliffs, N.J.)

FIGURE 5.3-2

Generalized compressibility chart, medium pressures. (From D. M. Himmelblau, Basic Principles and Calculations in Chemical Engineering, 3rd edition, copyright © 1974, p. 176. Reprinted by permission of Prentice-Hall, Inc., Englewood Cliffs, N.J.)

FIGURE 5.3-3

Generalized compressibility chart, high pressures. (From D. M. Himmelblau, *Basic Principles and Calculations in Chemical Engineering*, 3rd edition, copyright © 1974, p. 177. Reprinted by permission of Prentice-Hall, Inc., Englewood Cliffs, N.J.)

1. Look up or estimate values of the critical temperature T_c and the critical pressure P_c (Table B.1).

2. If the gas is either hydrogen or helium, determine **pseudocritical constants** from the empirical formulas

$$(T_c)_{\text{adjusted}} = T_c + 8K \tag{5.3-12}$$

$$(P_c)_{\text{adjusted}} = P_c + 8 \text{ atm} \tag{5.3-13}$$

Equations 5.3-12 and 5.3-13 are known as **Newton's corrections.**

3. Calculate **reduced values** of the two known variables $(T$ and P, T and \hat{V}, or P and $\hat{V})$ using the formulas

$$\textit{Reduced Temperature:} \quad T_r = T/T_c \tag{5.3-14}$$

$$\textit{Reduced Pressure:} \quad P_r = P/P_c \tag{5.3-15}$$

$$\textit{Ideal Reduced Volume}^4: \quad V_r = \frac{\hat{V}}{RT_c/P_c} = \frac{\hat{V}P_c}{RT_c} \tag{5.3-16}$$

Perform whatever unit conversions may be necessary to make the calculated reduced variables dimensionless. If the gas is H_2 or He, substitute for T_c and P_c the adjusted values of these quantities calculated in Step 2. *All temperatures and pressures substituted into these formulas must be absolute.*

4. Use the compressibility charts to determine the compressibility factor, and then solve for the unknown gas variable from the compressibility factor equation of state.

The compressibility factor equation of state used in conjunction with the generalized compressibility chart is not generally as accurate as a multiple-constant equation of state for PVT calculations under highly nonideal conditions, and it cannot be readily adapted to computer calculations. Its advantages include relative computational simplicity and (as will be shown) ease of adaptability to multicomponent gas mixtures.

EXAMPLE 5.3-5 The Generalized Compressibility Chart

One hundred mol of nitrogen are contained in a 5-liter vessel at $-20.6°C$. Estimate the pressure in the cylinder.

Solution

From Table B.1 in Appendix B, the critical temperature and pressure of nitrogen are

$$T_c = 126.2 \text{ K}, \qquad P_c = 33.5 \text{ atm}$$

[4] RT_c/P_c is the specific molar volume that would be calculated by the ideal gas law for a temperature T_c and a pressure P_c. It has no physical significance for the gas in question, unlike T_c, P_c, and the *critical volume* V_c, another property of the gas.

The reduced temperature and volume are calculated from Equations 5.3-14 and 5.3-16 as

$$T_r = \frac{T}{T_c} = \frac{(-20.6 + 273.2)(\text{K})}{126.2 \text{ K}} = 2.00$$

$$V_r = \frac{\hat{V}P_c}{RT_c} = \frac{5 \text{ liters}}{100 \text{ mol}} \left| \frac{33.5 \text{ atm}}{126.2 \text{ K}} \right| \frac{\text{mol} \cdot \text{K}}{0.08206 \text{ liter} \cdot \text{atm}} = 0.161$$

From Figure 5.3-3, the intersection of $T_r = 2$ and $V_r = 0.161$ occurs at approximately $z = 1.77$. From the compressibility factor equation of state, we can then calculate

$$P = \frac{zRT}{\hat{V}} = \frac{1.77}{} \left| \frac{0.08206 \text{ liter} \cdot \text{atm}}{\text{mol} \cdot \text{K}} \right| \frac{252.4 \text{ K}}{0.05 \text{ liter/mol}} = \boxed{733 \text{ atm}}$$

Note: You could also read the value of P_r at the intersection and calculate $P = P_r P_c$; however, it is generally more accurate to calculate unknown variables by first determining z and then using the equation of state, as was done above.

The basis for the generalized compressibility chart is the experimentally observed fact that *the values of certain physical properties of gases—such as the compressiblity factor—depend to a great extent on how near the gas is to its critical state.* The reduced temperature and pressure provide a measure of this proximity; the closer T_r and P_r are to 1, the closer the gas is to its critical state. This suggests that a plot of z versus T_r and P_r should be approximately the same for all substances, which is the case.[5]

The assertion that different substances behave in the same manner at the same reduced conditions is known as the **law of corresponding states.** Table 5.3-3 illustrates this law by comparing the compressibility factor for N_2 and CO_2 (1) at the same T and P, (2) at the same T_r and P_r. As the table shows, the values of z at the same temperature and pressure are widely separated, while at the same reduced conditions they are almost equal.

The application of this and other three-parameter correlations is discussed in Section 3.3 of Reid, Prausnitz, and Sherwood.

TABLE 5.3-3

Illustration of the Law of Corresponding States

Substance	T	P	$T_r = T/T_c$	$P_r = P/P_c$	z
N_2	223 K	400 atm	1.76	11.94	1.265
(1) CO_2	223 K	400 atm	(0.733)	(5.49)	~0.72
(2) CO_2	(535 K)	(870 atm)	1.76	11.96	~1.3

Data: N_2-$T_c = 126.2$ K, $P_c = 33.5$ atm
CO_2-$T_c = 304.2$ K, $P_c = 72.9$ atm

[5] An extension of the generalized charts that provides somewhat greater accuracy also allows for a dependence of $z(T, P)$ on z_c, the compressibility factor at the critical point, which generally varies between 0.25 and 0.29.

TEST YOURSELF

1. Suppose you need to know the volume that would be occupied by 10 kmol of H_2 at $-190°C$ and 300 atm. Would you consider it reasonable to use a value calculated from the ideal gas law? How would you use the generalized compressibility chart for this calculation?

2. Why would a chart such as that in Figure 5.3-1 be useless if T and P were the chart parameters?

3. What is the law of corresponding states, and how does it provide a basis for the generalized compressibility chart?

5.3e Real Gas Mixtures: Kay's Rule

Techniques exist for adapting both multiparameter equations of state such as the BWR and SRK equations and the compressibility factor equation of state for use in PVT calculations for real gas mixtures. A number of these techniques are summarized in Chapter 2 of Reid, Prausnitz, and Sherwood.

A technique that is both simple and reasonably accurate is *Kay's rule*, which uses the compressibility factor equation of state

$$PV = z_m nRT \quad \text{or} \quad P\hat{V} = z_m RT \tag{5.3-17}$$

with a **mean compressibility factor** z_m. To determine z_m for a mixture of gases A, B, C, ... with mole fractions y_a, y_b, y_c, \ldots, calculate the **pseudocritical constants of the mixture** as average values of the critical constants of the mixture components.

Pseudocritical Temperature: $\quad T_c' = y_a T_{ca} + y_b T_{cb} + y_c T_{cc} + \cdots$ (5.3-18)

Pseudocritical Pressure: $\quad P_c' = y_a P_{ca} + y_b P_{cb} + y_c P_{cc} + \cdots$ (5.3-19)

If the temperature T and total pressure P of the mixture are known, calculate the **pseudoreduced temperature** $T_r' = T/T_c'$ and the **pseudoreduced pressure** $P_r' = P/P_c'$, and use the generalized compressibility chart to obtain z_m, which may then be substituted in Equation 5.3-17 to determine \hat{V} for the mixture. If you know T and \hat{V} or P and \hat{V}, you may determine the third variable by trial and error using the same method, iterating until the calculated value of \hat{V} equals the given value.

Note: The pseudocritical constants of a mixture are simply empirical parameters that have been found useful for correlating physical properties of a mixture. Unlike T_c and P_c for a single component, T_c' and P_c' do not have any physical significance.

Like the theory of corresponding states on which it is based, Kay's rule provides only approximate values of the quantities it is used to calculate. It works best when used for mixtures of nonpolar compounds whose critical temperatures and pressures are within a factor of two of one another. Reid, Prausnitz, and Sherwood provide more complex but more accurate mixing rules for systems that do not fall into this category.

TEST YOURSELF

What is Kay's rule? How would you use it to calculate the specific molar volume of an equimolar mixture of gases at a given temperature and pressure? For what types of gases would you have the greatest confidence in the answer?

EXAMPLE 5.3-6 Kay's Rule

A mixture of 75% H_2 and 25% N_2 (molar basis) is contained in a tank at $P = 800$ atm and $T = -70°C$. Estimate the specific volume of the mixture in liters/mol using Kay's rule.

Solution

Critical Constants: From Table B.1.

H_2: $T_c = 33$ K
 Adjusted $T_c = 33 + 8 = 41$ K (Newton's correction: Equation 5.3-12)
 $P_c = 12.8$ atm
 Adjusted $P_c = 12.8 + 8 = 20.8$ atm (Newton's correction: Equation 5.3-13)
N_2: $T_c = 126.2$ K
 $P_c = 33.5$ atm

Pseudocritical Constants: From Equations 5.3-18 and 5.3-19

$$T'_c = y_{H_2}(T_c)_{H_2} + y_{N_2}(T_c)_{N_2} = (0.75 \times 41 \text{ K}) + (0.25 \times 126.2 \text{ K}) = 62.3 \text{ K}$$
$$P'_c = y_{H_2}(P_c)_{H_2} + y_{N_2}(P_c)_{N_2} = (0.75 \times 20.8 \text{ atm}) + (0.25 \times 33.5 \text{ atm}) = 24.0 \text{ atm}$$

Reduced Conditions: $T = (-70 + 273) = 203$ K, $P = 800$ atm

$$T_r = \frac{T}{T'_c} = \frac{203 \text{ K}}{62.3 \text{ K}} = 3.26$$

$$P_r = \frac{P}{P'_c} = \frac{800 \text{ atm}}{24.0 \text{ atm}} = 33.3$$

Mean Compressibility: From Figure 5.3-3, z_m ($T_r = 3.26$, $P_r = 33.3$) = 1.86

Calculation of Specific Volume

$$P\hat{V} = z_m RT$$
$$\Downarrow$$
$$\hat{V}\left(\frac{\text{liters}}{\text{mol}}\right) = \frac{z_m T(\text{K})}{P(\text{atm})} \times R \frac{\text{liter·atm}}{\text{mol·K}} = \frac{(1.86)(203)(0.08206)}{800}$$

$$= \boxed{0.0387 \frac{\text{liter}}{\text{mol}}}$$

PROBLEMS

1. A liquid mixture containing 40 mole% benzene and the balance toluene flows into a tank mounted on a balance. The mass in kilograms indicated by the scale is plotted against time, and a straight line results that passes through the points ($t = 3$ min, $m = 200$ kg) and ($t = 5$ min, $m = 280$ kg).
 (a) Estimate the volumetric flow rate of the liquid mixture.
 (b) What does the empty tank weigh?

2. When a liquid or a gas occupies a volume, it may be assumed to fill the volume completely. On the other hand, when solid particles occupy a volume, there are always spaces (voids) among the particles. The **porosity** or **void fraction** of a bed of particles is the ratio (void volume)/(total bed volume). The **bulk density** of the solids is the ratio (mass of solids)/ (total bed volume), and the **absolute density** of the solids has the usual definition, (mass of solids)/(volume of solids).

 Suppose 600 g of a crushed ore is placed in a graduated cylinder, filling it to the 184 cm^3 level. One hundred cm^3 of water is then added to the cylinder, whereupon the water level is observed to be at the 233.5 cm^3 mark. Calculate the porosity of the dry particle bed, the bulk density of the ore in this bed, and the absolute density of the ore.

3. Using the ideal gas law, calculate the specific volume in m^3/mol and the density in kg/m^3 of hydrogen at 50°C and 200 mm Hg (gauge).

4. The pressure gauge on a 20-ft^3 tank of nitrogen at 75°F reads 82 psi. Estimate the mass of nitrogen in the tank by
 (a) Direct solution of the ideal gas law.
 (b) Conversion from standard conditions. (See Example 5.2-2.)

5. Steam flows through a pipe at a rate of 400 liters/min, a pressure of 5 atm (gauge), and a temperature of 400°C. Farther downstream, the steam pressure has dropped to atmospheric pressure, and the temperature is 350°C. A barometer in the plant reads 757 mm Hg. What is the volumetric flow rate at the second point?

6. A stream of air enters a 7.50-cm ID pipe at a velocity of 60.0 m/s at 27°C and 1.80 bars (gauge). At a point downstream, the air flows through a 5.00-cm ID pipe at 30°C and 1.53 bars (gauge). What is the velocity of the gas at this point?

7. The label has come off a cylinder of gas in your laboratory. You know only one species of gas is contained in the cylinder, but you do not know whether it is hydrogen, oxygen, or nitrogen. To find out, you evaluate a 5-liter flask, seal it and weigh it, then let gas from the cylinder flow into it until the gauge pressure equals 1.00 atm. The flask is reweighed, and the mass of the added gas is found to be 13.0 g. Room temperature is 27°C, and barometric pressure is 1.00 atm. What is the gas?

8. A gas container filled with nitrogen at standard temperature and pressure has a mass of 37.289 g. The same container filled with carbon dioxide at STP has a mass of 37.440 g. When filled with an unknown gas at STP, the container mass is 37.062 g. Calculate the molecular weight of the unknown gas, and then state its most likely identity.

9. A gas chromatograph is to be calibrated for the analysis of carbon tetrachloride. Pure nitrogen enters one side of a mixing tee, and nitrogen containing 280 mg CCl$_4$/liter gas

enters the other at a rate of 0.200 liter/min. An open-end water manometer at the tee reads 8.0 cm. Farther downstream the gas mixture is sent through the chromatograph at a flow rate of 10.0 liters/min at atmospheric pressure. The temperature of the entire system is constant at 24°C. What is the concentration of CCl_4 (mg/liter) in the gas entering the chromatograph?

10. A nitrogen rotameter is calibrated by feeding N_2 from a compressor through a pressure regulator, a needle valve, the rotameter, and a **dry test meter**, a device that measures the total volume of gas that passes through it. A water manometer is used to measure the gas pressure at the rotameter outlet. A flow rate is set using the needle valve, the rotameter reading, Q, is noted, and the change in the dry gas meter reading (ΔV) for a measured running time (Δt) is recorded.

The following calibration data are taken on a day when the temperature is 23°C and the barometric pressure is 763 mm Hg.

Q	Δt (min)	ΔV (liters)
5.0	10	1.50
9.0	10	2.90
12.0	5	2.00

(a) Prepare a calibration chart of Q versus ϕ [cm^3(STP)/min]. (The abscissa is the flow rate at standard conditions equivalent to the actual flow rate at the measurement conditions.)

(b) Suppose the rotameter is used to measure the flow rate of a nitrogen stream at the calibration conditions and a reading of 7.5 is obtained. What is the mass flow rate of the stream (g/s)?

11. The flow rate required to yield a specified reading on an orifice meter varies inversely as the square root of the fluid density; that is, if a fluid with density ρ_1 (g/cm^3) flowing at a rate v_1 (cm^3/s) yields a meter reading R, then the flow rate of a fluid with density ρ_2 required to yield the same reading is

$$v_2 = v_1(\rho_1/\rho_2)^{1/2}$$

(a) Suppose an orifice meter calibrated with nitrogen at 25°C and 758 mm Hg is used to measure the flow rate of a process stream of hydrogen at 50°C and 1800 mm Hg, and the value read from the calibration chart is 350 cm^3/s. Calculate the true volumetric flow rate of the gas.

(b) Repeat part (a), only suppose the process fluid contains 25 mole% SO_2 and the balance SO_3.

12. A venturi meter is calibrated with a gas of molecular weight M_c at absolute temperature and pressure T_c and P_c. The volumetric flow rate, V_c, of the gas at each calibration point is converted to V_{cs}, the equivalent flow rate at standard temperature and pressure (T_s, P_s). The calibration chart is a plot of V_{cs} versus manometer reading h. Ideal gas behavior may be assumed.

(a) Suppose one point on the calibration plot was obtained with the calibration gas at 27°C and 1.1 atm flowing at a rate of 2.75 m³/h. Calculate V_{cs} for this point.

(b) The volumetric flow rate corresponding to a specified meter reading varies inversely with the square root of the fluid density. (See the preceding problem.) The meter is used to measure the flow rate of a stream of gas of molecular weight M at temperature and pressure T and P, and a value V_{cs} is read from the calibration chart. Derive a formula for the volumetric flow rate of the gas corrected to STP, V_s, in terms of V_{cs} and the other process variables defined above. (*Suggestion:* First determine V_c, then V, the true volumetric flow rate of the process gas, and finally V_s.)

13. A gas rotameter is calibrated with air ($M = 29.0$) at 20.0 cm Hg (gauge) and 25°C. The calibration data are well correlated by the linear relation

$$v(\text{SCMH}) = 1.25R - 0.70 \tag{1}$$

for $3 \le R \le 15$. (R is the rotameter reading and v is the volumetric flow rate in cubic meters per hour at standard conditions equivalent to the actual flow rate at the calibration conditions.)

This formula is only applicable to a gas with a density equal to that of the calibration gas. Fortunately, it is not necessary to have separate calibration curves for each new set of conditions, since the volumetric flow rate corresponding to a fixed rotameter reading is inversely proportional to the square root of the gas density. Thus, a flow rate v_m of a gas whose density is ρ_m and a flow rate v_c of a gas at the calibration conditions (density ρ_c) yield the same meter reading if

$$v_m/v_c = (\rho_c/\rho_m)^{1/2} \tag{2}$$

The measurement procedure is therefore to read the rotameter, use the calibration formula (Equation 1) to obtain a value of v_c, and use the square root correction formula (Equation 2) to obtain the desired value, v_m.

(a) Use the ideal gas law to convert Equation 2 into a formula for v_m in terms of the process and calibration gas absolute temperatures (T_m, T_c), absolute pressures (P_m, P_c), and molecular weights (M_m, M_c).

(b) Suppose H_2 at 29°C and 460 mm Hg (absolute) passes through the rotameter and a meter reading of 10.0 is obtained. Use the given calibration formula and the formula of part (a) to calculate the standard and actual volumetric flow rates (m³/h) and the mass flow rate (kg/h) of this stream.

***(c)** Write a program to read in values of T_m(°C), P_m(mm Hg), M_m, and R for an arbitrary number of measurement conditions, and to calculate and print out the standard and actual volumetric flow rates (m³/h) and mass flow rate (kg/h). Run your program for (i) H_2 at 29°C and 460 mm Hg, $R = 10.0$; (ii) air ($M = 29.0$) at 55°C and 1520 mm Hg, $R = 5.0$; (iii) methane at 15°C and 925 mm Hg, $R = 12.3$.

14. One hundred g/min of air (21 mole% O_2, the rest N_2) is mixed with a stream of CO_2. The product stream is analyzed and found to contain 56.7 wt% N_2. Calculate the volumetric flow rate of the CO_2 feed stream (liters/min at STP) and the mole fractions of the product stream components.

* Computer problem.

15. **Spray drying** is a process in which a liquid containing dissolved or suspended solids is injected into a chamber through a spray nozzle or centrifugal disk atomizer. The resulting mist is contacted with hot air, which evaporates most or all of the liquid, leaving the dried solids to fall to a conveyor belt at the bottom of the chamber.

 Powdered milk is produced in a spray dryer 6 m in diameter by 6 m high. Air enters at 167°C and -40 cm H_2O. The milk fed to the atomizer contains 70% water by mass, all of which evaporates. The outlet gas contains 12 mole% water and leaves the chamber at 83°C and 1 atm (absolute) at a rate of 311 m³/min. Calculate the production rate of dried milk and the volumetric flow rate of the inlet air.

16. A tank in a room at 19°C is initially open to the atmosphere on a day when the barometric pressure is 102 kPa. A block of dry ice (solid CO_2) with a mass of 15.7 kg is dropped into the tank, which is then sealed. The reading on the tank pressure gauge initially rises very quickly, then much more slowly, eventually reaching a value of 3.27 MPa.

 (a) How many mols of air were in the tank initially? Neglect the volume occupied by CO_2 in the solid state, and assume that a negligible amount of CO_2 escapes prior to the sealing of the tank.

 (b) What is the final density (g/liter) of the gas in the tank?

 (c) Explain the observed variation of pressure with time.

17. In a **froth flotation tank,** air is bubbled through wastewater to which a foaming agent has been added. The air bubbles carry grease, oil, and light solids in the wastewater up to the surface, where they are trapped in the foam. The foam and collected wastes are skimmed off the surface and disposed of by burning or burying.

 Ten thousand gallons per hour of sewage are treated in an open froth flotation tank. About 0.03 ft³ (STP) of air are required per gallon of sewage. The air is bubbled into the tank at a depth of 10 ft; the tank temperature is constant at 70°F; and the barometric pressure is 29.7 in. Hg. The bubbles enter the water with an average diameter of 1 mm at a pressure equal to the hydrostatic pressure of the water at the point of entry.

 (a) What is the volumetric flow rate of the air at its entering conditions? (*Note:* The air does not enter at standard temperature and pressure.)

 (b) What is the average diameter of the bubbles as they reach the surface?

18. A balloon 20 m in diameter is filled with helium at a gauge pressure of 2.0 atm. A man is standing in a basket suspended from the bottom of the balloon. A restraining cable attached to the basket keeps the balloon from rising. The balloon (not including the gas

it contains), the basket, and the man have a combined mass of 150 kg. The temperature is 24°C that day, and the barometer reads 760 mm Hg.

(a) Calculate the mass (kg) and weight (N) of the helium in the balloon.

(b) How much force is exerted on the balloon by the restraining cable? (*Recall:* The buoyant force on a submerged object equals the weight of the fluid—in this case, the air—displaced by the object. Neglect the volume of the basket and its contents.)

(c) Calculate the initial acceleration of the balloon when the restraining cable is released.

(d) Why does the balloon eventually stop rising? What would you need to know to calculate the altitude at which it stops?

(e) Suppose at its point of suspension in midair the balloon is heated, raising the temperature of the helium. What happens and why?

19. The Cookenwythe Gas Company pumps propane gas to the nearby Noxious Chemicals, Inc., polypropylene production plant. The gas is metered at the Noxious plant at 12 500 ft^3/h @ 4.7 atm (gauge) and 70°F. The pressure at the Cookenwythe plant is 8.5 atm (gauge), and the temperature is also 70°F. Noxious pays Cookenwythe at the rate of $0.26/lb$_m$ C$_3$H$_8$.

One dark night Sebastian Goniff, a Noxious engineer who is really a spy for the Rancid Plastics Corporation—Noxious' principal competitor and a bunch of really rotten guys—puts into effect his plan to divert propane from the Cookenwythe–Noxious line into an underground pipe that leads to a secret Rancid tank-truck loading station in the middle of a nearby abandoned garbage dump. To cover the operation, Goniff obtains a broken pressure gauge that is stuck at 4.7 atm and substitutes it for the gauge at the Noxious plant. He adjusts the gas pressure regulator so that the real gauge pressure is 1.8 atm, instructs his associate in the field via walkie-talkie to open the Rancid line gradually, and tells him to hold it when the flowmeter at Noxious reads 12 500 ft^3/h. To the unsuspecting Noxious meter reader, the flow rate and pressure would therefore appear to be normal.

The plan goes according to schedule, until the associate smells gas, suspects a leak near the valve, and lights a match to see if he can pinpoint the location.

(a) What should the flowmeter read at the Cookenwythe end of the pipeline?

(b) How much does Noxious pay Cookenwythe each month?

(c) What flow rate of propane (lb$_m$/h) should the Rancid trucks be prepared to handle?

(d) What happened?

20. An adult takes about 12 breaths per minute, inhaling roughly 500 ml of air with each breath. The molar compositions of the inspired and expired gases are as follows.

Species	Inspired Gas (%)	Expired Gas (%)
O_2	20.6	15.1
CO_2	0.0	3.7
N_2	77.4	75.0
H_2O	2.0	6.2

The inspired gas is at 24°C and 1 atm, and the expired gas is at body temperature and pressure—37°C and 1 atm. Nitrogen is not transported into or out of the blood in the lungs, so that $(N_2)_{in} = (N_2)_{out}$.

(a) Calculate the masses of O_2, CO_2, and H_2O transferred from the pulmonary gases to the blood or vice versa (specify which) per minute.

(b) Calculate the volume of air exhaled per milliliter inhaled.

(c) At what rate (g/min) is this individual losing weight by merely breathing?

21. As everyone who has used a fireplace knows, when a fire burns in a furnace a **draft**, or slight vacuum, is induced that causes the hot combustion gases and entrained particulate matter to flow up and out of the stack. The reason is that the hot gas in the stack is less dense than air at ambient temperature, leading to a lower hydrostatic head inside the furnace than outside it. The **theoretical draft** $D(N/m^2)$ is the difference in these hydrostatic heads; the **actual draft** takes into account pressure losses undergone by the gases flowing in the stack.

Let $T_s(K)$ be the average temperature in a stack of height $L(m)$, and T_a the outside temperature, and let M_s and M_a be the average molecular weights of the gases inside and outside the stack. Assume that the pressures inside and outside the stack are both equal to atmospheric pressure, $P_a(N/m^2)$ (in fact, the pressure inside the stack is normally a little lower).

(a) Use the ideal gas law to prove that the theoretical draft is given by the expression

$$D(N/m^2) = \frac{P_a L}{R} \frac{g}{g_c} \left(\frac{M_a}{T_a} - \frac{M_s}{T_s} \right)$$

(b) Suppose the gas in a 53-m stack has an average temperature of 655 K and contains 18 mole% CO_2, 2% O_2, and 80% N_2 on a day when barometric pressure is 755 mm Hg and the outside temperature is 294 K. Calculate the theoretical draft (cm H_2O) induced in the furnace.

22. A fuel gas containing 81% methane, 8% ethane, 5% propane, 4% n-butane, and 2% isobutane by volume flows to a furnace at a rate of 1450 m^3/h @ 32°C and 25 kPa (gauge), where it is burned with 15% excess air. Calculate the required flow rate of air in SCMH (standard cubic meters per hour).

*23. The flow of air to a gas-fired boiler furnace is regulated by a minicomputer controller. The fuel gases used in the furnace are mixtures of methane (A), ethane (B), propane (C), n-butane (D), and isobutane (E). At periodic intervals the temperature, pressure, and volumetric flow rate of the fuel gas are measured, and voltage signals proportional to their values are transmitted to the computer. In addition, when a new feed gas is used, a sample of the gas is analyzed, and the mole fractions of each of the five components are determined and read into the computer. The desired percent excess air is then specified, and the computer calculates the required volumetric flow rate of air and transmits the appropriate signal to a flow control valve in the air line.

The linear proportionalities between the input and output signals and the corresponding process variables may be determined from the following calibration data.

* Computer problem.

Fuel Temperature: $\quad T = 25.0°C, \qquad R_T = 14$
$\qquad\qquad\qquad\qquad\quad T = 35.0°C, \qquad R_T = 27$

Fuel Pressure: $\quad P_{gauge} = 0\ kPa, \qquad R_P = 0$
$\qquad\qquad\qquad\quad P_{gauge} = 20.0\ kPa, \qquad R_P = 6$

Fuel Flow Rate: $\quad V_F = 0\ m^3/h, \qquad\qquad R_F = 0$
$\qquad\qquad\qquad\qquad V_F = 2.00 \times 10^3\ m^3/h, \qquad R_F = 10$

Air Flow Rate: $\quad V_A = 0\ m^3(STP)/h, \qquad\qquad R_A = 0$
$\qquad\qquad\qquad\quad V_A = 1.0 \times 10^5\ m^3(STP)/h, \qquad R_A = 25$

(a) Write a program to read in values of R_S, R_T, R_P, the fuel gas component mole fractions x_A, x_B, x_C, x_D, and x_E, and the percent excess air PX, and to calculate and print out the required value of R_A.

(b) Run your program for the following data.

R_F	R_T	R_P	x_A	x_B	x_C	x_D	x_E	PX
7.25	23.1	7.5	0.81	0.08	0.05	0.04	0.02	15%
5.80	7.5	19.3	0.58	0.31	0.06	0.05	0.00	23%
2.45	46.5	15.8	0.00	0.00	0.65	0.25	0.10	33%

24. The oxidation of sulfur dioxide

$$2\ SO_2 + O_2 \rightleftharpoons 2\ SO_3$$

is carried out in an isothermal constant-volume batch reactor at 610°C. At time $t = 0$ the reactor contains only SO_2 and O_2 in stoichiometric proportion, at an absolute pressure of 1.21 atm.

(a) If the reaction were irreversible and went to completion, what would the final pressure be?

(b) Suppose the reactor pressure drops and eventually levels out at 0.91 atm. Calculate the fractional conversion of SO_2 at equilibrium.

25. Liquid ammonia (SG = 0.817) is vaporized and decomposed to form nitrogen and hydrogen. How many liters of product gas at 20°C and 1.8 bars (absolute) would be formed from 50.0 liters of liquid ammonia?

26. Propylene is hydrogenated in a batch reactor:

$$C_3H_6(g) + H_2(g) \rightarrow C_3H_8(g)$$

Equimolar amounts of propylene and hydrogen are fed into the reactor at 25°C and a total absolute pressure of 32.0 atm, and some time later the temperature is 250°C. You may assume ideal gas behavior for this problem, although at the high pressures involved this assumption constitutes a crude approximation at best.

(a) If the reaction goes to completion at 250°C, what would the final pressure be?

(b) If the pressure is 35.1 atm when the temperature is 250°C, what percentage of the propylene fed has reacted?

27. A natural gas contains 92 wt% CH_4 and the balance C_2H_6. Five hundred liters per minute of this gas at 40°C and 1.1 bars are to be burned with 25% excess air. The air flowmeter is calibrated to read the volumetric flow rate at standard temperature and pressure. What should the meter read when the flow rate is set to the desired value?

28. An ideal gas mixture contains 40% N_2, 30% CO, and 30% H_2 by volume at $P = 2$ atm (absolute) and $T = 65°C$. Calculate
 (a) The partial pressure of each component.
 (b) The mass fraction of nitrogen.
 (c) The average molecular weight of the gas.
 (d) The density of the gas (g/liter).

29. A process gas flowing at a rate of 20 000 ft^3/h at 1.00 atm and 200°F contains H_2S with a partial pressure of 0.0294 psi. The gas passes to a scrubber in which 92% of the H_2S is removed. Calculate the H_2S removal rate in lb_m/h.

30. A stream of hot dry nitrogen flows through a process unit containing liquid acetone. A substantial portion of the acetone vaporizes and is carried off by the nitrogen. The combined gases leave the recovery unit at 220°C and 1.1 bars and enter a cooler-compressor in which a portion of the acetone is condensed. The remaining gas leaves the condenser at 10°C and 40 bars. The partial pressure of acetone in the feed to the condenser is 0.110 bar, and that in the effluent gas is 0.133 bar. Assume ideal gas behavior.
 (a) Calculate for a basis of 1 m^3 of gas fed to the condenser the mass of acetone condensed (kg) and the volume of gas leaving the condenser (m^3).
 (b) Suppose the volumetric flow rate of the gas leaving the condenser is 20 m^3/h. Calculate the rate (kg/h) at which acetone is vaporized in the solvent recovery unit.

31. Ammonia is one of the chemical constituents of industrial waste that must be removed in a treatment plant before the wastewater can be properly discharged into a river or estuary. A two-part process is frequently carried out to accomplish this removal. Lime [$Ca(OH)_2$] is first added to the wastewater to drive the equilibrium

$$NH_3 + H_2O \rightleftharpoons NH_4^+ + OH^-$$

to the left. Air is then bubbled through the alkaline wastewater, stripping out the ammonia.

One million gallons per day of alkaline wastewater containing 0.02 mole NH_3/mole H_2O (assumes all free NH_3 as opposed to NH_4^+) are fed to a stripping tower that operates at 68°F. Air at 68°F and 21.3 psia is bubbled through the tower. The air-to-wastewater feed ratio is 300 ft^3/gal, and 90% of the ammonia in the wastewater is stripped out. Calculate the volumetric flow rate of the gas leaving the tower and the partial pressure of ammonia in this gas.

32. You have purchased a gas cylinder that is supposed to contain 5.0 mole% HCl ($\pm 0.1\%$) and 95% N_2. The experiments you have been running are not yielding reasonable results, however, and you suspect that the percentage of HCl in the cylinder gas is, in fact, much higher than the specified value.

To check this hypothesis you bubble the gas through 20 liters of water for exactly 1 h. The inlet gas is metered at a gauge pressure of 510 mm Hg and a temperature of

23°C. Prior to entering the absorber, the gas passes through a flowmeter that indicates a flow rate of 2.00 liters/min. At the conclusion of the run a sample of the absorber solution is titrated, and it is found that the HCl concentration is 0.0274 mol/liter. What is the concentration of HCl in the cylinder gas? (Assume that all of the HCl is absorbed in the water.)

33. Two humid gas streams are combined in a heated mixing chamber. The first stream contains 23.5 mole% ethane and 76.5% ethylene on a dry basis and enters the chamber at 25°C and 105 kPa at a rate of 125 liters/min. The second stream is humid air, which enters at 75°C and 115 kPa at a rate of 355 liters/min. The product stream emerges at 65°C and 1.00 atm. A *hygrometer* is used to measure the water contents of the two feed streams and the combined product stream. The calibration curve for the hygrometer is a straight line on a semilog plot of y (mole fraction of water) versus R (hygrometer reading), which passes through the points $(y = 10^{-4}, R = 5)$ and $(y = 0.2, R = 90)$. The following readings are recorded.

Hydrocarbon Feed Stream: $R = 86.0$

Air Feed Stream: $R = 12.8$

(a) What is the equation that relates y to R?
(b) Calculate the volumetric flow rate of the product stream and the molar composition of the product gas on a dry basis.
(c) Calculate the partial pressure of water in the product gas and the hygrometer reading for this stream.

34. A solid carbohydrate $(C_aH_bO_c)$ with a specific gravity of 1.59 is placed in a 1.000-liter combustion chamber. The chamber is evacuated and then charged with pure oxygen. Complete combustion of the carbohydrate takes place. A sample of the product gas is cooled to condense all the water formed in the combustion, and the remaining gas is analyzed by gas chromatography. The following data are obtained:

Mass of Carbohydrate Charged: 3.42 g

Chamber Conditions Prior to Combustion: $T = 26.8°C, P = 499.9$ kPa

Chamber Conditions Following Combustion: $T = 483.4°C, P = 1950.0$ kPa

Product Gas Analysis: 38.7 mole% CO_2, 25.8% O_2, 35.5% H_2O

Assume (i) none of the carbohydrate is lost when the chamber is evacuated; (ii) the ideal gas law is applicable; (iii) the vapor pressure of the carbohydrate at 27°C is negligible. Do *not* neglect the volume of the carbohydrate.

(a) Determine at least two possible molecular formulas for the carbohydrate (i.e., sets of integer values of a, b, and c) consistent with the given data.
(b) If the molecular weight of the carbohydrate is independently determined to be in the range 300 to 350, what is its molecular formula?

35. The **adiabatic flame temperature** of a fuel is the temperature achieved if the fuel is burned completely in a perfectly insulated container.

You are performing an experiment to measure the adiabatic flame temperature of cyclopentane. You put 10.0 ml of liquid cyclopentane into a well-insulated steel vessel with a volume of 11.2 liters, and pressurize the vessel with air to achieve a stoichiometric

ratio of oxygen to cyclopentane. You then ignite the fuel, planning to record the final temperature. The vessel is equipped with a thermocouple and a pressure gauge.

(a) If room temperature is 27°C and barometric pressure is 1.00 bar, what should the pressure gauge read prior to ignition?

(b) Suppose you discover after the combustion has been carried out that the thermo-couple is not functioning properly. Use the final pressure gauge reading of 75.3 bars to estimate the adiabatic flame temperature of cyclopentane, assuming ideal gas behavior (a poor assumption in this case).

36. In a sulfuric acid production process, an ore containing 87 wt% iron pyrites, FeS_2, and 13% inert matter is burned in a furnace with dry air to form $Fe_2O_3(s)$, $SO_2(g)$, and $SO_3(g)$. The gas leaving the furnace passes to a catalytic converter, in which most of the SO_2 is oxidized to SO_3. The effluent from the converter emerges at 450°C and 1 atm.

Data: 85% conversion of pyrites

40% excess air fed based on complete oxidation of sulfur to SO_3

40% of the FeS_2 burned goes to form SO_2, and the remainder goes to form SO_3.

96% of the SO_2 that enters the converter is oxidized to SO_3

Calculate the volume of product gas in m^3(STP) produced per 100 kg of ore burned and the molar composition of this gas.

37. Most of the concrete used in the construction of buildings, roads, dams, and bridges is made from **portland cement,** a substance obtained by pulverizing the hard, granular residue (**clinker**) from the roasting of a mixture of clay and limestone, and adding other materials to modify the setting properties of the cement and the mechanical properties of the concrete.

The charge to a portland cement rotary kiln contains 20% of a dried building clay (72 wt% SiO_2, 16% Al_2O_3, 7% Fe_2O_3, 1.7% K_2O, 3.3% Na_2O) and 80% limestone (95 wt% $CaCO_3$, 5% impurities). When the solid temperature reaches about 900°C, *calcination* of the limestone to lime (CaO) and carbon dioxide occurs. As the temperature continues to rise to about 1450°C, the lime reacts with the minerals in the clay to form such compounds as $3\,CaO\cdot SiO_2$, $3\,CaO\cdot Al_2O_3$, and $4\,CaO\cdot Al_2O_3\cdot Fe_2O_3$. The flow rate of CO_2 from the kiln is 1250 m^3/h at 1000°C and 1 atm. Calculate the feed rates of clay and limestone (kg/h) and the weight percent of Fe_2O_3 in the final cement.

38. The ultimate analysis of a No. 2 fuel oil is 87.3 wt% carbon, 12.6% hydrogen, 0.09% sulfur, and the balance inerts. This oil is burned in a steam generating furnace with 20% excess air. The air is preheated to 175°C and enters the furnace at a gauge pressure of 180 mm Hg. The sulfur and hydrogen in the fuel are completely oxidized, 5% of the carbon is oxidized to CO, and the balance forms CO_2.

(a) Calculate the feed ratio (m^3 air)/(kg oil).

(b) Calculate the mole fractions (dry basis) and the ppm (parts per million on a wet basis, or moles contained in 10^6 moles of the wet stack gas) of the stack gas species that might be considered environmental hazards.

39. A stream of liquid *n*-pentane flows at a rate of 50.4 liters/min into a heating chamber, where it evaporates into a stream of air 25% in excess of the amount needed to burn the

pentane completely. The temperature and pressure of the entering air are 336 K and 413.6 kPa (gauge). The heated gas leaves the chamber at 518 K and 310.2 kPa (gauge) and passes into a combustion chamber, in which 90% of the pentane is burned. The product gas goes to a condenser, in which essentially all of the water formed in the combustion and the unreacted pentane are liquefied. The gas leaving the condenser is at 275.0 K and 1 atm (absolute).

(a) Estimate the volumetric flow rate of the gas leaving the heater.

(b) Estimate the volumetric flow rate of the liquid leaving the condenser.

40. A natural gas containing 97% methane by volume enters an electric arc pyrolysis furnace at 400°C and 130 kPa. Methane is converted to acetylene and hydrogen in the furnace. The conversion is held to 15% to avoid excessive decomposition of methane into its elements. The acetylene in the product gas is separated from the residual methane and hydrogen in a series of absorption and fractionation steps and is bubbled through an aqueous solution of ferric and mercurous sulfates, in which hydration of the acetylene to acetaldehyde occurs. The yield of acetaldehyde is 0.60 mol formed per mol of acetylene fed. The gas that emerges may be considered to contain only acetylene and acetaldehyde and flows at a rate of 450 m^3/h at 85°C and 120 kPa. The acetaldehyde is separated and fed to a catalytic oxidation reactor, entering at 60°C and 500 kPa. The acetaldehyde reacts with oxygen to form acetic acid, with a yield of 0.90 mol acetic acid formed per mol of acetaldehyde fed. Calculate the volumetric flow rates (m^3/h) of the natural gas feed to the pyrolysis reactor and the acetaldehyde feed to the oxidation reactor, and the production rate (kg/h) of acetic acid.

41. Sulfur dioxide is catalytically oxidized to sulfur trioxide in the production of sulfuric acid. (See Problem 36.) If conditions are such that ideal gas behavior is followed, the reaction proceeds to an equilibrium point at which the component partial pressures satisfy the relation

$$\frac{p_{SO_3}}{p_{SO_2}p_{O_2}^{1/2}} = K_p(T)$$

At 883 K, $K_p = 8.50$ atm$^{-1/2}$. Calculate the fractional conversion of sulfur dioxide if SO_2 and air are fed to a reactor in stoichiometric proportion, the reactor operates at 883 K and 4.0 atm, and the reaction proceeds to equilibrium.

42. You have been assigned the task of measuring the equilibrium constant for the reaction $N_2O_4 \rightleftharpoons 2NO_2$ as a function of temperature. To do so, you obtain a rigid 2-liter vessel equipped with a pressure gauge, evacuate and then fill the container with a mixture of NO_2 and N_2O_4, and heat the vessel to $T_0 = 473$ K, a temperature at which you know the gas is essentially pure NO_2. The gauge pressure at this point is noted to be 1.00 atm. You then decrease the temperature in stages, recording the equilibrium gauge pressure at each temperature. The data are as follows:

T(K)	473	350	335	315	300
P_{gauge}(atm)	1.00	0.314	0.111	−0.097	−0.224

Pure NO_2

(a) How many mol of NO_2 are in the vessel at 473 K?

(b) The reaction equilibrium constant is

$$K_p = p_{NO_2}^2/p_{N_2O_4}$$

where p_{NO_2} and $p_{N_2O_4}$ are the equilibrium partial pressures of NO_2 and N_2O_4. Derive an equation or a series of equations for calculating K_p (atm) from specified values of T and P_{gauge}. (*Suggestion:* Begin by defining n_1 and n_2 as the moles of NO_2 and N_2O_4 present at equilibrium.) Then calculate K_p for $T = 350$ K, 335 K, 315 K, and 300 K.

(c) The equilibrium constant should vary with temperature according to the relation

$$K_p = ae^{-b/T}$$

Use the results of part (b) to determine the values of a and b by a graphical curve-fitting procedure.

*43. A reaction with stoichiometry $A \rightleftharpoons 2B$ is carried out in a batch reactor of volume V(liters), beginning with pure B at a temperature T_0(K) and gauge pressure P_0(atm). The temperature is changed in steps, and the equilibrium gauge pressure at each temperature is noted. Write a computer program to perform the following steps.

1. Read in values of V, T_0, P_0, N (the number of temperature–pressure pairs) and the equilibrium values of T(K) and P_{gauge}(atm).

2. Calculate and print out the values of the equilibrium constant

$$K_p(\text{atm}) = p_B^2/p_A$$

at each temperature. (See Section 4.6c.)

3. Calculate and print out the values of a and b in the expression

$$K_p = ae^{-b/T}$$

by linearizing this expression (take logarithms of both sides) and using the method of least squares (Appendix A.1).

Test your program using the data of Problem 42 for the NO_2–N_2O_4 reaction.

44. One mol each of CO_2, O_2, and N_2 are fed to a batch reactor and heated to 3000 K and 5.0 atm. The two reactions given here proceed to equilibrium (also shown are the equilibrium constants at 3000 K).

$$CO_2 \rightleftharpoons CO + \tfrac{1}{2}O_2$$

$$K_1 = (p_{CO}p_{O_2}^{1/2})/p_{CO_2} = 0.3272 \text{ atm}^{1/2}$$

$$\tfrac{1}{2}O_2 + \tfrac{1}{2}N_2 \rightleftharpoons NO$$

$$K_2 = p_{NO}/(p_{O_2}p_{N_2})^{1/2} = 0.1222$$

Calculate the equilibrium composition (component mole fractions) of the reactor contents. [*Suggestion:* Express K_1 and K_2 in terms of the extents of the two reactions, ξ_1 and ξ_2. (See Section 4.6b.) Then use a trial-and-error procedure, such as Newton's rule (Appendix A.2) to solve for ξ_1 and ξ_2, and use the results to determine the equilibrium mole fractions.]

* Computer problem.

45. The van der Waals equation of state is

$$\left(P + \frac{a}{\hat{V}^2}\right)(\hat{V} - b) = RT$$

The values of the constants a and b for air are

$$a = 1.33 \text{ atm}(\text{liter/mol})^2$$
$$b = 0.0366 \text{ liter/mol}$$

Estimate using the van der Waals equation and any trial-and-error method of your choice (Appendix A.2) the volume occupied by 2 kg of air at 223 K and 50 atm. Obtain an initial estimate from the ideal gas law.

***46.** Write a computer program to solve van der Waals equation of state (see Problem 45) for \hat{V} from given values of P and T using Newton's method (Appendix A.2), stopping when the fractional change in \hat{V} from one iteration to the next is 1.0×10^{-5}. Have the program (i) read in a and b; (ii) read in P and T; (iii) terminate if P is negative; (iv) calculate \hat{V}; (v) print out P, T, \hat{V}, and the number of iterations it took to achieve convergence; (vi) return to step (ii). Make provision for an upper limit of 10 Newton iterations, so that the program cannot get into an infinite loop. Test your program using the values of a and b given for air in Problem 45, a temperature of $-50°C$, and $P(\text{atm}) = 1, 10, 20, 50,$ and 100.

47. A 10.0 m^3 tank is charged with 350 kg of propane gas at 25°C. Use the SRK equation of state to estimate the pressure in the tank; then calculate the percentage error that would result from the use of the ideal gas law for the calculation.

48. The absolute pressure within a 140.0-liter gas cylinder should not exceed 51.0 atm. Suppose the cylinder contains 200 mol of a gas. Use the SRK equation of state to calculate the maximum permissible cylinder temperature if the gas is (i) carbon dioxide; (ii) argon. Finally, calculate the values that would be predicted by the ideal gas law.

49. A stream of oxygen at $-65°C$ and 8.3 atm flows at a rate of 185 kg/h. Use the SRK equation of state to estimate the volumetric flow rate of this stream. (See Example 5.3-3.)

***50.** Write a computer program to calculate the specific volume of a gas at a given pressure and temperature using the SRK equation of state. Your program should perform the following tasks.

 (a) Read in the name of the gas (you may omit this if you are not familiar with character-string manipulations) and the values of the critical temperature (K), the critical pressure (atm), and the Pitzer acentric factor for the gas.

 (b) Read in the temperature $T(K)$ and pressure $P(\text{atm})$ for which the specific volume \hat{V} is to be calculated. Terminate if T is negative.

 (c) Calculate the SRK parameters a, b, m, and α.

 (d) Use the ideal gas law to provide an initial estimate of \hat{V}.

 (e) Use Newton's rule to refine the estimate of \hat{V}, stopping when successive estimates are within 0.01% of each other (see Appendix A.2 and Example 5.3-3). Have the program print out the successive estimates and values of $f(\hat{V})$ at each of them.

* Computer problem.

Build in an upper limit of 15 iterations; if convergence is not achieved within this limit, print out an error message and quit. When convergence is achieved, print out the values of T, P, and the final value of \hat{V}.

(f) Go back to step (b).

Test your program by finding \hat{V} for CO_2 at (i) 200 K and 6.8 atm, (ii) 250 K and 12.3 atm; (iii) 300 K and 21.5 atm.

51. Use the generalized compressibility chart to estimate z for
(a) Nitrogen at 40°C and 40 MPa.
(b) Helium at -200°C and 350 atm. (Don't forget Newton's corrections.)

52. A certain gas has a molecular weight of 30, a critical temperature of 310 K, and a critical pressure of 4.5 MPa. Calculate the density in kg/m^3 of this gas at 465 K and 9.0 MPa
(a) If the gas is ideal.
(b) If the gas obeys the law of corresponding states.

53. Fifty pounds of CO_2 are contained in a 5.0-ft^3 tank. The safety limit of the tank is 1600 psig. Use the compressibility chart to estimate the maximum permissible gas temperature.

54. A stream of oxygen enters a compressor at 298 K and 1.00 atm at a rate of 87 m^3/h and is compressed to 358 K and 1000 atm. Calculate the flow rate of compressed O_2, using the compressibility factor equation of state.

55. A 20-liter cylinder containing oxygen at 175 atm (absolute) is used to supply O_2 to an oxygen tent. The cylinder can be used until its absolute pressure drops to 1.1 atm. Assuming a constant temperature of 27°C, calculate the moles of O_2 that can be obtained from the cylinder, using the compressibility factor equation of state when appropriate.

56. A gas consists of 20 mole% CH_4, 30% C_2H_6, and 50% C_2H_4. Twenty kilograms of this gas are to be compressed to a pressure of 200 bars at 90°C. Using Kay's rule, calculate the final volume of the gas.

57. A 120-liter cylinder is evacuated and filled with 20.0 kg of a gas containing 10.0 mole% N_2O and the balance N_2. The gas temperature is 24°C. Use the compressibility chart to solve the following problems.
(a) What is the gauge pressure (atm) of the cylinder gas after the tank is filled?
(b) A fire breaks out in the plant where the cylinder is kept, and the cylinder valve ruptures when the gas gauge pressure reaches 273 atm. What was the gas temperature (°C) at the moment before the rupture occurred?

58. The product gas from a coal gasification plant consists of 60 mole% CO and the balance H_2; it leaves the plant at 150°C and 2000 psia. The gas is expanded through a turbine, and the outlet gas is fed to a boiler furnace at 100°C and 1 atm at a rate of 20 000 ft^3/min. Estimate the inlet flow rate to the turbine in ft^3/min, using Kay's rule. What percentage error would result from the use of the ideal gas law at the turbine inlet?

59. A 100.0-liter cylinder of a gas containing 97 mole% CO and 3% CO_2 is delivered to your plant. You sign the receipt for it, noting that the pressure gauge on the tank reads 2000 psi. Several days later you notice that the gauge reads 1875 psi, indicating a serious leak. The storage room in which the cylinder is kept has a volume of 21.4 m^3 and is very poorly ventilated. Calculate the maximum mole% of CO in the room at the time the leak is discovered, assuming that the leaking gas spreads uniformly throughout the room and that the room temperature is a constant 30°C, using Kay's rule when appropriate.

60. Methanol is produced by reacting carbon monoxide and hydrogen at 644 K over a $ZnO-Cr_2O_3$ catalyst. A mixture of CO and H_2 in a ratio (2 mol H_2/mol CO) is compressed and fed to the catalyst bed at 644 K and 34.5 MPa (absolute). The product gases are passed through a condenser, in which the methanol is liquefied. A single-pass conversion of 25% is obtained. The **space velocity,** or ratio of the volumetric flow rate of the feed gas to the volume of the catalyst bed, is (25 000 m^3/h)/(1 m^3 catalyst bed).

(a) You are designing a reactor to produce 545 kmol CH_3OH/h. Estimate the volumetric flow rate that the compressor must be capable of delivering if no gases are recycled, and the required volume of the catalyst bed. (Use Kay's rule for pressure-volume calculations.)

(b) If (as is done in practice) the gases from the condenser are recycled to the reactor, the compressor is then required to deliver only the fresh feed. What volumetric flow rate must it deliver assuming that the methanol produced is completely recovered in the condenser? (In practice it is not; moreover, a purge stream must be taken off to prevent the buildup of impurities in the system.)

***61.** You need to evaluate a series of volumes corresponding to specified pressures and temperatures for pure ethylene. A great deal of precision is required for these calculations, and you decide to use the Benedict-Webb-Rubin equation. Since Equation 5.3-2 cannot be solved explicitly for \hat{V}, you must use a trial-and-error technique. The complexity of the functional dependences on \hat{V} within the equation lead you to rule out Newton's method. Instead, you adopt the following marching technique:

1. Write the BWR equation in the form $f(\hat{V}) = 0$ by transposing all terms to one side.
2. Choose an initial estimate \hat{V}_0 (a reasonable choice should occur to you quickly), and evaluate $f_0 = f(\hat{V}_0)$.
3. Choose a step size $\Delta\hat{V} = 0.02\hat{V}_0$, let $\hat{V}_1 = \hat{V}_0 + \Delta\hat{V}$, and evaluate $f_1 = f(\hat{V}_1)$. If f_1 is farther away from 0 than f_0 was, change the sign of $\Delta\hat{V}$ and repeat this step.
4. Let $\hat{V}_n = \hat{V}_{n-1} + \Delta\hat{V}$, and continue to evaluate $f(\hat{V}_n)$ until the sign of f changes from one iteration to the next. The desired value of \hat{V} must lie between \hat{V}_n and \hat{V}_{n-1}.
5. Let $(\Delta\hat{V})_{new} = -0.1(\Delta\hat{V})_{old}$, and repeat the stepping procedure from \hat{V}_n until f again changes sign.
6. Estimate the value of \hat{V} for which $f = 0$ by interpolating between the last two (f, \hat{V}) pairs. It is not difficult to show that this leads to the formula

$$\hat{V} = \hat{V}_{n-1} - \frac{f_{n-1}}{f_n - f_{n-1}}(\hat{V}_n - \hat{V}_{n-1})$$

(a) Derive the above formula.

(b) Write a computer program to implement this procedure. Read in the number of (P, T) pairs for which \hat{V} is to be calculated, and then for each pair calculate \hat{V}. Write a function subroutine to perform the evaluation of $f(\hat{V})$. Have the program print out each calculated value of \hat{V} and f so you can see how the search proceeds.

(c) Test your program by evaluating \hat{V} (liters/mol) for $T = 340$ K and $P = 126$ atm. Compare the result with the results you would obtain using the ideal gas law and the generalized compressibility chart.

* Computer problem.

Chapter Six
Multiphase Systems

Although chemical reactions are at the heart of many processes of interest to chemical engineers, most of the individual operations in industrial processes do not involve reactions. Some process units are simply mixers that blend streams of different compositions—possibly reactants prior to entry into a reactor, possibly components of a product blend. Other units accomplish physical separations, in which components of a mixture are separated into two or more fractions.

As a rule, separation processes involve two phases: either a gas and a liquid, a solid and a liquid, a gas and a solid, or two immiscible liquids. It is relatively easy to separate phases from each other—in fact, they tend to separate themselves due to large density differences—but it is difficult to separate different components of a single phase. The idea behind most processes designed to separate mixture components is therefore to get the components into different phases, and then to separate the phases.

Suppose, for example, you wish to split a mixture containing equal amounts of benzene and methanol into its components. As long as the mixture is entirely a liquid or a gas there is no efficient way to remove individual molecules of one species or the other. Several possibilities exist for accomplishing the separation in a two-phase system, however, including the following:

1. Heat a liquid mixture of the two substances, and remove and condense the vapor that comes off (**distillation**). Both benzene and methanol evaporate to some extent, but methanol is more volatile; the vapor that comes off is consequently rich in methanol, and the residual liquid is rich in benzene. Under the proper conditions, repeated applications of this process can yield close to a complete separation of the two components.

2. Cool a liquid mixture, and filter out the crystals that form (**crystallization**). Benzene freezes at 5.5°C and methanol freezes at −97°C, so that under the right conditions nearly pure benzene crystals may be recoverable.

3. Contact a liquid mixture with a solvent that dissolves one component and is immiscible with the other, then separate the two phases (**liquid extraction**). For example, water is miscible with methanol, but not with benzene. If water is mixed with the original solution, it will therefore pick up a great deal of the methanol and almost no benzene; if the mixture is then allowed to settle, the water (+methanol) and benzene (+methanol) form two phases, which may easily be separated.

To design a separation process unit, you must know certain physical properties of the system with which you will be working. Before you can design a distillation column, for example, you must know how volatile each feed component is, which requires knowing the **vapor pressure** of each substance as a function of temperature. To design a crystallizer to precipitate and remove dissolved salts from a solution, you must know the **solubilities** of the salts as functions of temperature. To design an aqueous extraction column for the benzene-methanol separation process, you must know the **distribution coefficient** of methanol, or the equilibrium ratio of the mole fraction of methanol in the water phase to that in the benzene phase.

In this chapter we discuss the physical properties and laws that govern the behavior of two-phase systems, showing how these laws must be included in the formulation of material balances on separation process units. Section 6.1 discusses properties of multiphase systems that contain only one component; the remaining sections extend the discussion to multicomponent systems. Section 6.2 introduces the **phase rule,** a general relationship between the number of components and phases in a system and the number of system variables whose values can be specified arbitrarily. Sections 6.3 through 6.6 cover multicomponent gas–liquid systems, solutions of solids in liquids, and systems containing two immiscible or partially miscible liquids.

TEST YOURSELF

Suggest a method that might be suitable to achieve each of the following separations.

1. Separate crude oil into volatile low molecular weight compounds (naphthas, used to make gasoline and light chemicals), intermediate molecular weight compounds (used for heating oils), and nonvolatile high molecular weight compounds (used for lubricating oils).

2. Remove water from a bleached wood pulp slurry.

3. Obtain fresh water from a salt solution (such as seawater).

4. Separate NH_3 from a mixture of N_2, H_2, and NH_3. Ammonia is highly soluble in water; also, it condenses at $-33.4°C$.

CREATIVITY EXERCISE

A gas contains two species, A and B. Suggest as many methods as you can think of, both conventional and unconventional, for separating the two species. Briefly indicate the conditions required for each method to work. (For example, find a third substance, C, that reacts with A to form a solid, and introduce C into the mixture. The A will react and deposit out, leaving B in the gas phase.)

6.1 SINGLE-COMPONENT PHASE EQUILIBRIUM

6.1a Phase Diagrams

At most specific temperatures and pressures, a single pure substance at equilibrium normally exists entirely as a solid, liquid, or gas, but at certain temperatures and pressures two phases and even all three phases may coexist. Pure water is a gas at 130°C and 100 mm Hg, for example, and a solid at −40°C and 10 atm, but at 100°C and 1 atm it may be a gas, a liquid, or a mixture of both, and at approximately 0.0098°C and 4.58 mm Hg it may be a solid, a liquid, a gas, or any combination of the three.

A **phase diagram** of a pure substance is a plot of pressure versus temperature that shows the conditions at which the substance exists as a solid, a liquid, and a gas. The boundaries between the single-phase regions represent the PT values at which two phases may coexist. The phase diagrams of water and carbon dioxide are shown in Figure 6.1-1.

What the phase diagram means and what can be done with it are illustrated by a hypothetical experiment in which pure water is placed in a leakproof evacuated cylinder fitted with a movable piston, as shown in the diagram below. Heat can be added to or withdrawn from the cylinder, so that the temperature in the chamber can be adjusted to any desired value, and the absolute pressure of the cylinder contents (which equals $(F + W)/A$, where W is the weight of the piston) can similarly be adjusted by varying the force F on the piston.

Suppose the system starts out at $T = 20°C$, and the force is set at a value such that the system pressure is 3 mm Hg. As the phase diagram shows, water can only exist as a vapor at these conditions, so that any liquid that may initially have been in the

FIGURE 6.1-1

Phase diagrams of H_2O and CO_2 (not drawn to scale)

chamber evaporates (boils), until finally the chamber contains only water vapor at 20°C and 3 mm Hg (point A on Figure 6.1-1a).

Now suppose the force on the piston is slowly increased while the system temperature is held constant at 20°C until the pressure in the cylinder reaches 760 mm Hg, and thereafter heat is added to the system with the pressure remaining constant until the temperature reaches 130°C. The state of the water throughout this process can be

determined by following path A → B → C → D → E on Figure 6.1-1a. The conditions of the system at various stages of the process are shown in the diagram below.

Notice that all phase transitions—condensation at point B, and evaporation at point D—take place at boundaries on the phase diagram; the system cannot move off these boundaries until the transitions are complete.

Several familiar terms may be defined with reference to the phase diagram.

1. If T and P correspond to a point on the vapor–liquid equilibrium curve for a substance, P is the **vapor pressure** of the substance at temperature T, and T is the **boiling point** (more precisely, the **boiling point temperature**) of the substance at pressure P.

2. The boiling point of a substance at $P = 1$ atm is the **normal boiling point** of that substance.

3. If (T, P) falls on the solid–vapor equilibrium curve, then P is the vapor pressure of the solid at temperature T, and T is the **sublimation point** at pressure P.

4. If (T, P) falls on the solid–liquid equilibrium curve, then T is the **melting point** or **freezing point** at pressure P.

5. The point (T, P) at which solid, liquid, and vapor phases can all coexist is called the **triple point** of the substance.

As you can see from Figure 6.1-1, the freezing point of water decreases with increasing pressure. (Verify.) This behavior is extremely unusual; most substances, including carbon dioxide, exhibit the opposite behavior. Note also, however, that the changes are very slight; the solid–liquid equilibrium curves are in fact nearly vertical. Their slopes have been exaggerated in Figure 6.1-1 for illustrative purposes.

Normal boiling and melting points for many substances are given in Table B.1 of Appendix B and for many more substances on pp. 3-6 through 3-44 of *Perry's Chemical Engineers' Handbook*,[1] and vapor pressures are tabulated on pp. 3-45 through 3-73 of the *Handbook*. Techniques for estimating vapor pressures are presented in Section 6.1b.

TEST YOURSELF

(Referring to Figure 6.1-1)

1. What is the sublimation point of H_2O at 3 mm Hg? What is the vapor pressure of ice at $-5°C$?
2. What is the triple point of CO_2?
3. Describe what happens when the pressure of pure CO_2 is raised from 1 atm to 9.9 atm at $-78.5°C$, and then the temperature is raised from $-78.5°C$ to $0°C$ at 9.9 atm.
4. What is the vapor pressure of CO_2 at $-78.5°C$? At $-40°C$?
5. What is the sublimation point of CO_2 at 1 atm? The melting point at 9.9 atm? The boiling point at the latter pressure?

6.1b Estimation of Vapor Pressures: The Clausius–Clapeyron Equation, Cox Charts, and the Antoine Equation

In process analysis, one frequently needs to know the vapor pressure, p^*, of a process stream component. Occasionally, however, a search of the literature turns up vapor pressure data only at temperatures other than the one of interest, or possibly no data at all.

One solution to this problem is to measure p^* at the desired temperature; to do so is not always convenient, however, and is clearly a great deal of trouble if only an

[1] R. H. Perry and D. W. Green, Eds., *Perry's Chemical Engineers' Handbook*, 6th Edition, McGraw-Hill, New York, 1984.

approximate value is sought. An alternative is to estimate the vapor pressure using an empirical equation for $p^*(T)$. Reid, Prausnitz, and Sherwood[2] summarize and compare vapor pressure estimation formulas, several of which are given in the paragraphs that follow.

A theoretical relationship between p^*, the vapor pressure of a pure substance, and T, the absolute temperature, is the **Clapeyron equation**

$$\frac{dp^*}{dT} = \frac{\Delta \hat{H}_v}{T(\hat{V}_g - \hat{V}_l)} \tag{6.1-1}$$

where

T is *absolute* temperature

\hat{V}_g and \hat{V}_l are the specific molar volumes of the vapor and liquid, respectively

$\Delta \hat{H}_v$ is the **latent heat of vaporization,** or the energy required to vaporize one mole of the liquid (to be defined more precisely in Chapter 8).

Unless the pressure is extremely high, the specific volume of the liquid is negligible relative to that of the vapor (i.e., $\hat{V}_g - \hat{V}_l \approx \hat{V}_g$). If we assume that this is the case, apply the ideal gas law to the vapor (so that \hat{V}_g is replaced with RT/p^* in Equation 6.1-1) and rearrange the resulting equation with the aid of elementary calculus, we obtain

$$\frac{d(\ln p^*)}{d(1/T)} = -\frac{\Delta \hat{H}_v}{R} \tag{6.1-2}$$

(Convince yourself that Equation 6.1-2 is correct by working backward to derive Equation 6.1-1.)

If the vapor pressure of a substance is measured at several temperatures, and $\ln p^*$ is plotted versus $1/T$ (or p^* is plotted versus $1/T$ on semilog paper), then from Equation 6.1-2, the slope of the resulting curve at a given temperature equals $-\Delta \hat{H}_v/R$. This is the method most commonly used to determine heats of vaporization experimentally.

Suppose now that the heat of vaporization of a substance is independent of temperature, or nearly so in the temperature range over which vapor pressures are available. Equation 6.1-2 may then be integrated to yield the **Clausius–Clapeyron equation**

$$\ln p^* = -\frac{\Delta \hat{H}_v}{RT} + B \tag{6.1-3}$$

where B is a constant that varies from one substance to another. According to this equation, a plot of $\ln p^*$ versus $1/T$ (or a semilog plot of p^* versus $1/T$) should be a straight line, with slope $-\Delta \hat{H}_v/R$ and intercept B.

If you know $\Delta \hat{H}_v$ and p^* at a single temperature T_0, you can solve the Clausius–Clapeyron equation for B, and thereafter use this equation to estimate p^* at any

[2] R. C. Reid, J. H. Prausnitz, and T. K. Sherwood, *The Properties of Gases and Liquids*, 3rd Edition, McGraw-Hill, New York, 1977.

temperature close to T_0. If you have p^* versus T data, you can plot $\ln p^*$ versus $1/T$ and determine $\Delta \hat{H}_v/R$ and B graphically or by the method of least squares (see Appendix A.1).

**EXAMPLE 6.1-1 Vapor Pressure Estimation
Using the Clausius–Clapeyron Equation**

The vapor pressure of benzene is measured at two temperatures, with the following results:

$$T_1 = 7.6°C, \qquad p_1^* = 40 \text{ mm Hg}$$
$$T_2 = 15.4°C, \qquad p_2^* = 60 \text{ mm Hg}$$

Calculate the parameters of the Clausius–Clapeyron equation, and then estimate p^* at 42.2°C using this equation.

Solution

p^*	$T(°C)$	$T(K)$
40 mm Hg	7.6°C	280.8
60 mm Hg	15.4°C	288.6

The slope of the line through the two data points on a plot of $\ln p^*$ versus $1/T$ is

$$-\frac{\Delta \hat{H}_v}{R} = \frac{\ln (p_2^*/p_1^*)}{[(1/T_2) - (1/T_1)]} = \frac{T_1 T_2 \ln (p_2^*/p_1^*)}{(T_1 - T_2)}$$

$$= \frac{(280.8 \text{ K})(288.6 \text{ K}) \ln (60 \text{ mm Hg}/40 \text{ mm Hg})}{(280.8 - 288.6)(\text{K})} = -4213 \text{ K}$$

The intercept B is obtained from Equation 6.1-3 as

$$B = \ln p_1^* + \frac{\Delta \hat{H}_v}{RT_1}$$

$$= \ln 40 + (4213/280.8) = 18.69$$

The Clausius–Clapeyron equation is, therefore,

$$\boxed{\ln p^* = -\frac{4213}{T(\text{K})} + 18.69}, \quad p^* \text{ in mm Hg}$$

Check: $T = 15.4°C = 288.6$ K

$$\ln p^* = -\frac{4213}{288.6} + 18.69 = 4.093$$

$$\Downarrow$$

$$p^* = \exp (4.093) = 60 \text{ mm Hg} \quad \checkmark$$

Finally, at $T = 42.2°C = 315.4$ K

$$\ln p^* = -\frac{4213}{315.4} + 18.69 = 5.334$$

$$\Downarrow$$

$$p^* = \exp(5.334) = \boxed{207 \text{ mm Hg}}$$

Perry's Chemical Engineers' Handbook, p. 3-50, lists the vapor pressure of benzene at 42.2°C as 200 mm Hg, so that the use of the Clausius–Clapeyron equation results in an estimation error of approximately 3.5%.

Note: If desired, the heat of vaporization of benzene $\Delta\hat{H}_v$ may be estimated from the slope of the Clausius–Clapeyron plot $(-\Delta\hat{H}_v/R)$ as

$$\Delta\hat{H}_v = (\Delta\hat{H}_v/R)(R)$$

$$= \frac{4213 \text{ K}}{} \left| \frac{8.314 \text{ J}}{\text{mol} \cdot \text{K}} \right. = 35\,030 \text{ J/mol}$$

In 1923, E. R. Cox proposed a method of plotting vapor pressure data to obtain straight lines over wide temperature ranges. The method is to choose a reference substance (water is most commonly used) for which the vapor pressure is known accurately as a function of temperature, and to plot log p^* versus log p_r^*, where p^* and p_r^* are the vapor pressures at the same temperature of the substance of interest and the reference substance, respectively (see Figure 6.1-2).

To construct such a plot—called an **equal-temperature reference substance plot**—from two or more known values of $p^*(T)$, you must look up the vapor pressure p_r^* of the reference substance at a series of temperatures $\{T_i\}$, and plot $p^*(T_i)$ versus $p_r^*(T_i)$ on log paper. You can usually draw a straight line through the points plotted in this manner, which may then be used to determine p^* for any T from the known value of $p_r^*(T)$.

FIGURE 6.1-2

Reference substance plot for vapor pressure correlation

FIGURE 6.1-3

Reference substance plot: temperature scale for water on abscissa

An additional step is normally taken in the construction of plots of this type. The normal use of the plot is to determine p^* for a given T; p_r^* is only an intermediate quantity used to relate these variables. The necessity of looking up p_r^* for each given temperature can be avoided if values of $T(p_r^*)$ are shown on a second abscissa scale (see Figure 6.1-3). Now to find $p^*(T)$ you need only find T on the new abscissa scale; the value of $p_r^*(t)$ will be located at the same abscissa value, and the curve of p^* versus p_r^* may then be used to determine $p^*(T)$.

Notice, however, that there is no longer a need for the p_r^* scale, since once you find T on the abscissa you can proceed directly to the curve. The p_r^* scale can therefore be omitted, leaving what has come to be called a **Cox chart**—log p^* on the ordinate, and the temperature scale shown in Figure 6.1-3 on the abscissa. Special graph paper—sometimes called **vapor pressure paper**—which has these coordinate axes is available commercially. Cox chart plots for various substances are shown in Figure 6.1-4.

It is also possible to get straight line correlations with **Duhring plots,** which are plots of the temperature at which a substance has a certain vapor pressure versus the temperature at which a reference substance has the same vapor pressure. The principles of the preparation and use of these charts are identical to those of the Cox chart.

A relatively simple empirical equation that correlates vapor pressure-temperature data extremely well is the **Antoine equation**

$$\log_{10} p^* = A - \frac{B}{T + C} \tag{6.1-4}$$

When you have calculated $q = \log_{10} p^*$, you may recover p^* as the antilogarithm of q, or 10^q. Values of A, B, and C for several compounds are listed in Table 6.1-1.

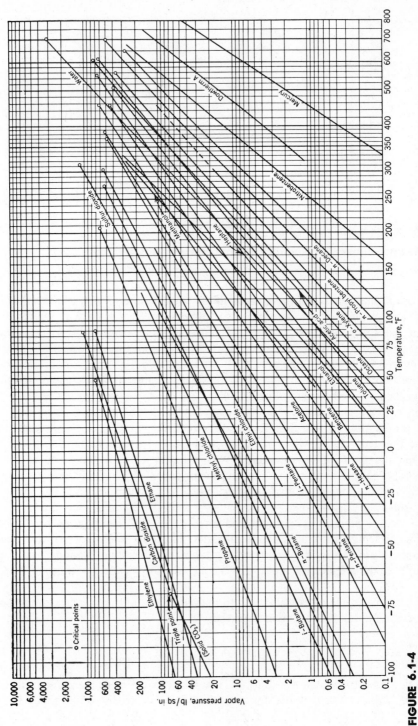

FIGURE 6.1-4

Cox chart vapor pressure plots. (From A. S. Foust et al., *Principles of Unit Operations*., Wiley, New York, 1960, p. 550.)

TABLE 6.1-1

Antoine Equation Constantsa

$$\log_{10} p^* = A - \frac{B}{T + C} \qquad p^* \text{ in mm Hg} \qquad T \text{ in } °C$$

Substance	Formula	Range, °C	A	B	C
Acetaldehyde	C_2H_4O	-45 to $+70$	6.81089	992.0	230
Acetic acid	$C_2H_4O_2$	0 to $+36$	7.80307	1651.2	225
		$+36$ to $+170$	7.18807	1416.7	211
Acetone	C_3H_6O	—	7.02447	1161.0	224
Ammonia	NH_3	-83 to $+60$	7.55466	1002.711	247.885
Benzene	C_6H_6	—	6.90565	1211.033	220.790
Carbon tetrachloride	CCl_4	—	6.93390	1242.43	230.0
Chlorobenzene	C_6H_5Cl	0 to $+42$	7.10690	1500.0	224.0
		$+42$ to $+230$	6.94504	1413.12	216.0
Chloroform	$CHCl_3$	-30 to $+150$	6.90328	1163.03	227.4
Cyclohexane	C_6H_{12}	-50 to $+200$	6.84498	1203.526	222.863
Ethyl acetate	$C_4H_8O_2$	-20 to $+150$	7.09808	1238.71	217.0
Ethyl alcohol	C_2H_6O	—	8.04494	1554.3	222.65
Ethylbenzene	C_8H_{10}	—	6.95719	1424.255	213.206
n-Heptane	C_7H_{16}	—	6.90240	1268.115	216.900
n-Hexane	C_6H_{14}	—	6.87776	1171.530	224.366
Methyl alcohol	CH_4O	-20 to $+140$	7.87863	1473.11	230.0
Methyl ethyl ketone	C_4H_8O	—	6.97421	1209.6	216
n-Pentane	C_5H_{12}	—	6.85221	1064.63	232.000
Isopentane	C_5H_{12}	—	6.78967	1020.012	233.097
Styrene	C_8H_8	—	6.92409	1420.0	206
Toluene	C_7H_8	—	6.95334	1343.943	219.377
Water	H_2O	0 to 60	8.10765	1750.286	235.0
		60 to 150	7.96681	1668.21	228.0

a Reprinted from *Lange's Handbook of Chemistry*, 9th Edition, Handbook Publishers, Inc., Sandusky, Ohio, 1956. For an expanded list of Antoine constants see J. A. Dean, *Lange's Handbook of Chemistry*, 12th Edition, McGraw-Hill, New York, 1979, pp. 10–29 through 10–54.

TEST YOURSELF

1. You need to know the vapor pressure of n-hexane at 87°C. Name two ways you could obtain or estimate this quantity using material readily available to you.

2. Suppose you are given the vapor pressure p^* of a substance at three closely spaced temperatures T_1, T_2, and T_3, and you wish to determine p^* at a fourth temperature T_4 far removed from the other three. If you use the Clausius–Clapeyron equation to correlate p^* and T, how would you plot the data and extrapolate it to T_4?

3. Why would it be preferable to use a Cox chart to plot and extrapolate the data of question 2 rather than the Clausius–Clapeyron equation?

CREATIVITY EXERCISE

Give as many reasons as you can think of, however farfetched, for wanting to know the vapor pressure of a substance at a given temperature. (*Example:* You want to know whether you can leave it overnight in an unstoppered flask without most of it evaporating.)

6.2 THE GIBBS PHASE RULE

When two phases are brought into contact with each other, a redistribution of the components of each phase normally takes place—species evaporate, condense, dissolve, or precipitate until a state of equilibrium is reached in which the temperatures and pressures of both phases are the same and the compositions of each phase no longer change with time.

Suppose you have a closed vessel containing three components A, B, and C distributed between gas and liquid phases, and you wish to describe this system to someone else in sufficient detail so that person could duplicate it exactly. Specifying the system temperature and pressure, the masses of each phase, and two mass or mole fractions for each phase would certainly be sufficient; however, these variables are not all independent—once some of them are specified, others may be calculated from physical properties of the system components.

The variables that describe the condition of a process system fall into two categories: **extensive variables,** which depend on the size of the system, and **intensive variables,** which do not. Mass and volume are examples of extensive variables; intensive variables include temperature, pressure, density and specific volume, and mass and mole fractions of individual system components.

The number of intensive variables that can be specified independently for a system at equilibrium is called the **number of degrees of freedom** of the system. Let

$$\Pi = \text{the number of phases in a system}$$

$$m = \text{the number of chemical species}$$

$$F = \text{the number of degrees of freedom}$$

The relationship among F, Π, and m is given by the **Gibbs phase rule.** If no reactions occur among the system components, the phase rule is

$$\boxed{F = 2 + m - \Pi} \tag{6.2-1}$$

If r independent reactions occur among the system components and the reactions proceed to equilibrium, then the value of m in this equation must be reduced by r. (*Note: Perry's Chemical Engineers' Handbook*, p. 4-78, presents a proof of the phase rule and outlines a method for determining how many independent reactions may occur among the components of a system.)

The example that follows illustrates the application of the phase rule to several systems. The remainder of the chapter presents the equilibrium relationships that are used to determine the remaining intensive system variables once the allowed number of these variables has been specified.

EXAMPLE 6.2-1 The Phase Rule

Specify the number of degrees of freedom of each of the following equilibrium systems:

1. Pure liquid water

$$\text{One phase } (\Pi = 1), \quad \text{one component } (m = 1)$$
$$\Downarrow$$
$$F = 2 + 1 - 1 = 2$$

Two intensive variables (for example, T and P) can therefore be specified arbitrarily (recognizing, however, that only within certain ranges of T and P can water exist as a liquid).

2. A mixture of ice and liquid water

$$\Pi = 2, \quad m = 1$$
$$\Downarrow$$
$$F = 2 + 1 - 2 = 1$$

Thus only one variable (T or P) can be specified. If (say) a value of T is chosen, the value of P may be uniquely determined. (It can be read from the solid-liquid equilibrium curve on the H_2O phase diagram.)

3. Pure liquid acetone in equilibrium with acetone vapor and nitrogen

$$\Pi = 2, \quad m = 2$$
$$\Downarrow$$
$$F = 2 + 2 - 2 = 2$$

Two variables can therefore be specified: T and P, or a gas-phase mass or mole fraction (of either acetone or N_2) and either T or P, and the third may then be determined. (How to determine it is the subject of Section 6.3.)

TEST YOURSELF

1. Define and give examples of extensive and intensive variables. Define "number of degrees of freedom of a system." What is the phase rule?

2. Use the phase rule to determine the number of degrees of freedom of each of the following equilibrium systems.
 (a) A mixture of liquid, ice, and water vapor.
 (b) NaCl crystals in an aqueous NaCl solution.
 (c) Humid air in equilibrium with condensed water (dry air may be considered a single species).

(d) A mixture of four liquid hydrocarbons in equilibrium with solid crystals of one of the hydrocarbons and a gas phase containing the vapors of all four species.

(e) A gaseous mixture of H_2, Br_2, and HBr, given that the single reaction

$$H_2 + Br_2 \rightleftarrows 2\,HBr$$

has proceeded to equilibrium. (See sentence following Equation 6.2-1.)

6.3 GAS–LIQUID SYSTEMS: ONE CONDENSABLE COMPONENT

Systems containing several components, of which only one is capable of existing as a liquid at the process conditions, are common in industrial processes. Separation processes that involve such systems include **evaporation, drying,** and **humidification—** all of which involve transfer of liquid into the gas phase—and **condensation** and **dehumidification,** which involve transfer of the condensable species from the gas to the liquid phase.

Suppose liquid water is introduced into a chamber that initially contains dry air and that the temperature and pressure in the system are kept constant at 75°C and 760 mm Hg. Initially the gas phase contains no water ($p_{H_2O} = 0$), and water molecules consequently begin to evaporate. The mole fraction of water in the gas phase, y_{H_2O}, increases, and hence so does $p_{H_2O} = y_{H_2O}P$. (In writing the latter expression we have assumed that the gas phase behaves ideally.) Eventually, however, the amount of water in the gas phase is such that the rate at which water molecules enter the gas phase by evaporation equals the rate at which they leave by condensation, and thereafter no change occurs in the amount or composition of either phase. The gas phase is then said to be **saturated** with water—it contains all the water it can hold at the system temperature and pressure—and the water in the gas phase is referred to as a **saturated vapor.**

Let us apply the phase rule to this equilibrium system. Since there are two phases and two components,

$$F = 2 + m - \Pi = 2$$

It follows that only two of the three intensive variables T, P, and y_{H_2O} can be specified arbitrarily and that some relationship must exist that uniquely determines the value of the third variable once the first two have been fixed.

A law that describes the behavior of gas–liquid systems over a wide range of conditions provides the desired relationship. *If a gas at temperature T and pressure P contains a saturated vapor whose mole fraction is y_v (moles vapor/mole total gas), and if this vapor is the only species that would condense if the temperature were slightly lowered, then the partial pressure of the vapor in the gas equals the pure component vapor pressure $p_v^*(T)$ at the system temperature.*

Saturation Condition, Single Condensable Species: $\quad p_v = \boxed{\; y_v P = p_v^*(T) \;}$ \qquad **(6.3-1)**

Equation 6.3-1 is a limiting case of **Raoult's law,** which will be introduced in a more general context in Section 6.4. It is the fundamental relation used in the analysis of gas–liquid systems containing one condensable component. A wide variety of problems occur in connection with such systems, but they all eventually involve knowing two of the variables y_v, P, and T and having to determine the third using Equation 6.3-1.

EXAMPLE 6.3-1 Composition of a Saturated Gas–Vapor System

Air and liquid water are contained at equilibrium in a closed chamber at $T = 75°C$ and $P = 760$ mm Hg. Calculate the molar composition of the gas phase.

Solution

Since the gas and liquid are in equilibrium, the air must be saturated with water vapor (if it were not, more water would evaporate), so that Equation 6.3-1 may be applied.

$$y_{H_2O} = p^*_{H_2O}(75°C)/P$$

From Table B.3 in Appendix B, $p^*_{H_2O}(75°C) = 289$ mm Hg. Consequently,

$$y_{H_2O} = \frac{289 \text{ mm Hg}}{760 \text{ mm Hg}} = \boxed{0.380 \frac{\text{mol } H_2O}{\text{mol}}}$$

$$y_{\text{dry air}} = 1 - y_{H_2O} = \boxed{0.620 \frac{\text{mol dry air}}{\text{mol}}}$$

Several important points concerning the behavior of gas–liquid systems and several terms used to describe the state of such systems are summarized here.

1. A gas in equilibrium with a liquid must be saturated with that liquid.
2. The partial pressure of a vapor at equilibrium in a gas mixture containing a single condensable component cannot exceed the vapor pressure of the pure component at the system temperature. If $p_v = p^*_v$, the vapor is saturated; any attempt to increase p_v—either by adding more vapor to the gas phase or by increasing the total pressure—must instead lead to condensation.
3. A vapor present in a gas in less than its saturation amount is referred to as a **superheated vapor.** For such a vapor,

Superheated Condition, Single Condensable Species:

$$p_v = \boxed{y_v P < p^*_v(T)} \tag{6.3-2}$$

Since only a saturated vapor can condense (why?), to achieve condensation in a system containing a superheated vapor one or more of the variables of (6.3-2)

must be changed so that the inequality becomes an equality. This can be done in several ways, such as by increasing the pressure at constant temperature (the left side increases, while the right side stays constant) or by decreasing the temperature at constant pressure (the left side remains constant, and the right side decreases).

4. If a gas containing a single superheated vapor is cooled at constant pressure, the temperature at which the vapor becomes saturated is referred to as the **dew point** of the gas. From the saturation condition of Equation 6.3-1, it follows that

$$p_v = \boxed{y_v P = p_v^*(T_{dp})} \tag{6.3-3}$$

The difference between the temperature of a gas and its dew point is called the **degrees of superheat** of the gas. If any two of the quantities y_v, P, and T_{dp} (or, equivalently, the temperature of the gas and the degrees of superheat) are known, the third quantity may be determined from Equation 6.3-3 and a table or graph of p_v^* versus T.

EXAMPLE 6.3-2 Material Balances on a Condenser

A stream of air at 100°C and 5260 mm Hg contains 10% water by volume.

1. Calculate the dew point of the air and degrees of superheat of the water vapor.
2. Calculate the percentage of the vapor that condenses and the final composition of the gas phase if the gas is cooled to 80°C at constant pressure.
3. Calculate the percentage condensation and the final gas phase composition if instead of being cooled the gas is compressed isothermally to 8500 mm Hg.

Solution

1. $p_{H_2O} = y_{H_2O}P = (0.1)(5260 \text{ mm Hg}) = 526 \text{ mm Hg}$

 $p_{H_2O}^*(100°C) = 760 \text{ mm Hg} > p_{H_2O} \Rightarrow$ the vapor is superheated

 From Equation 6.3-3

$$p_{H_2O} = p_{H_2O}^*(T_{dp}) = 526 \text{ mm Hg}$$
$$\Downarrow \text{ Table B.3}$$
$$\boxed{T_{dp} = 90°C}$$

 and the vapor has $100°C - 90°C = \boxed{10°C \text{ of superheat}}$

2. Since the air becomes saturated at 90°C, further cooling must lead to condensation. Since the products are liquid water in equilibrium with a gas phase, the water vapor in the gas must remain saturated.

 On the following flowchart, the symbol BDA stands for **bone-dry air,** a term used to signify the water-free component of an air–water vapor mixture.

Basis: 100 mol Feed Gas

Let us first do the problem bookkeeping. Three unknown variables appear on the chart—Q_1, Q_2, and y. Since only two species are involved in the process, we can only write two independent material balances, leaving us one equation short. If we had neglected to observe that the gas at the condenser outlet is saturated with water, solution of the problem would be impossible; however, the saturation condition supplies us with the needed third equation.

The solution outline is as follows: Apply the saturation condition at the outlet to determine y, the mole fraction of water in the outlet gas; then use a dry air balance to determine Q_2 and a total mole balance or a water balance to determine the final unknown, Q_1.

Saturation Condition at Outlet

$$yP = p_{H_2O}^*(T)$$
$$\Downarrow$$
$$y = \frac{p_{H_2O}^*(80°C)}{P} = \frac{355 \text{ mm Hg}}{5260 \text{ mm Hg}} = \boxed{0.0675 \frac{\text{mol } H_2O}{\text{mol}}}$$

Balance on Dry Air

$$\frac{100 \text{ mol} \mid 0.9 \text{ mol BDA}}{\mid \text{mol}} = Q_2(1 - y)$$
$$\Downarrow y = 0.0675$$
$$Q_2 = 96.5 \text{ mol}$$

Total Mole Balance

$$100 \text{ mol} = Q_1 + Q_2$$
$$\Downarrow Q_2 = 96.5 \text{ mol}$$
$$Q_1 = 3.5 \text{ mol } H_2O \text{ condensed}$$

Percentage Condensation

$$\frac{3.5 \text{ mol } H_2O \text{ condensed}}{(0.1 \times 100) \text{ mol } H_2O \text{ fed}} \times 100\% = \boxed{35\%}$$

3. Initially $y_{H_2O}P < p^*_{H_2O}(100°C)$. Saturation occurs when P is high enough for the inequality to become an equality, or

$$P_{saturation} = \frac{p^*_{H_2O}(100°C)}{y_{H_2O}} = \frac{760 \text{ mm}}{0.1} = 7600 \text{ mm Hg}$$

Any increase in P above 7600 mm Hg must cause condensation, so that the products from the compression to 8500 mm Hg must include a liquid stream.

Basis: 100 mol Feed Gas

Before going through the solution, try to outline it in the way done following the flowchart of part (2).

Saturation Condition $y = \dfrac{p^*_{H_2O}(100°C)}{P} = \dfrac{760 \text{ mm Hg}}{8500 \text{ mm Hg}} = \boxed{0.0894 \dfrac{\text{mol } H_2O}{\text{mol}}}$

Balance on Bone-Dry Air $(100)(0.900) = Q_2(1 - y)$
$$\Downarrow y = 0.0894$$
$$Q_2 = 98.8 \text{ mol}$$

Total Mole Balance $100 = Q_1 + Q_2$
$$\Downarrow Q_2 = 98.8 \text{ mol}$$
$$Q_1 = 1.2 \text{ mol } H_2O \text{ condensed}$$

Percentage Condensation $\dfrac{1.2 \text{ mol } H_2O \text{ condensed}}{(0.1 \times 100) \text{ mol } H_2O \text{ fed}} \times 100\% = \boxed{12\%}$

The mechanism of evaporation of a liquid depends on the relative values of the vapor pressure of the liquid and the total system pressure. If evaporation takes place at a temperature such that $p^* < P$, the process involves transfer of molecules from the surface of the liquid to the gas above the surface, while if $p^* = P$ vapor bubbles form rapidly throughout the entire liquid: that is, the liquid *boils*. The temperature at which $p^* = P$ is the **boiling point** of the liquid at the given pressure.

TEST YOURSELF

1. If water vapor is in equilibrium with liquid water, must the vapor be saturated? Can a vapor be saturated if there is no liquid present in the system?

2. The vapor pressure of acetone is 200 mm Hg at 22.7°C. Liquid acetone is kept in a sealed flask at 22.7°C; the gas above the liquid contains air and acetone vapor at a pressure of 960 mm Hg. What is (a) the partial pressure of acetone in the gas, (b) the partial pressure of N_2, (c) the mole fraction of acetone in the gas? How would you determine the boiling point of the acetone, assuming a constant total pressure of 960 mm Hg?

 Suppose you have a curve of $p_{H_2O}^*$ versus T, and you are given the temperature and pressure (T_0 and P_0) of a mixture of water and noncondensable gases.

3. Define the dew point of the gas. Would the vapor be saturated or superheated if $T_0 > T_{dp}$? If $T_0 = T_{dp}$?

4. If you are told that the gas is saturated, how would you calculate the mole fraction of water in it? What would happen to the vapor if you (a) heated the gas isobarically (at constant pressure), (b) cooled the gas isobarically, (c) compressed the gas isothermally (at constant temperature), (d) expanded the gas isothermally?

5. If you are given the mole fraction of water in the gas, how would you calculate the dew point of the gas?

6. If you are given the degrees of superheat of the gas, how would you calculate the mole fraction of water in the gas?

Several quantities besides those introduced in the previous section are commonly used to describe the state and composition of a gas containing a single condensable vapor. In the definitions to be given, the term "saturation" refers to any gas–vapor combination, while "humidity" refers specifically to an air–water system.

Suppose a gas at temperature T and pressure P contains a vapor whose partial pressure is p_v, and whose vapor pressure is $p_v^*(T)$.

Relative Saturation (Relative Humidity):

$$s_r(h_r) = \frac{p_v}{p_v^*(T)} \times 100\% \qquad (6.3\text{-}4)$$

A relative humidity of 40%, for example, signifies that the partial pressure of water vapor equals $\frac{4}{10}$ of the vapor pressure of water at the system temperature.

Molal Saturation (Molal Humidity)·

$$s_m(h_m) = \frac{p_v}{P - p_v} = \frac{\text{moles of vapor}}{\text{moles of vapor-free (dry) gas}} \qquad (6.3\text{-}5)$$

(Can you prove the second equality?)

Absolute Saturation (Absolute Humidity):

$$s_a(h_a) = \frac{p_v M_v}{(P - p_v)M_{dry}} = \frac{\text{mass of vapor}}{\text{mass of dry gas}} \qquad (6.3\text{-}6)$$

where M_v is the molecular weight of the vapor and M_{dry} is the average molecular weight of the dry (vapor-free) gas.

Percentage saturation (percentage humidity):

$$s_p(h_p) = 100\,\frac{s_m}{(s_m)^*} = 100\,\frac{p_v/(P - p_v)}{p_v^*/(P - p_v^*)} \qquad (6.3\text{-}7)$$

If you are given any of these quantities for a gas at a given temperature and pressure, you can solve the defining equation to calculate the partial pressure or mole fraction of the vapor in the gas; thereafter, you can proceed by the method given previously to calculate the dew point, degrees of superheat, and so on.

TEST YOURSELF

The vapor pressure of styrene is 100 mm Hg at 82°C and 200 mm Hg at 100°C. A gas that consists of 10% styrene and 90% noncondensables is contained in a tank at 100°C and 1000 mm Hg. Calculate:

1. The dew point of the gas.
2. The relative saturation.
3. The molal saturation and percentage saturation.

CREATIVITY EXERCISES

1. Suppose you know the temperature and barometric pressure on a given day. List as many ways as you can think of to determine—exactly or approximately—the mole fraction of water vapor in the air.
2. Repeat question 1, only this time limit yourself to methods in which a stuffed bear plays a part in the determination. (*Example:* Saturate the bear with water, and measure the rate at which it loses weight due to evaporation.)

EXAMPLE 6.3-3

Humid air at 75°C, 1.1 bar, and 30% relative humidity is fed into a process unit at a rate of 1000 m^3/h. Determine (1) the molar flow rates of water, dry air, and oxygen entering the process unit, (2) the molal humidity, absolute humidity, and percentage humidity of the air, (3) the dew point.

Solution

1. $h_r = 100 p_{H_2O}/p^*_{H_2O}(75°C)$

 $\quad\Big\| \quad h_r = 30\%$

 $\quad\Downarrow \quad p^*_{H_2O}(75°C) = 289$ mm Hg (from Table B.3)

 $p_{H_2O} = (0.3)(289$ mm Hg$) = 86.7$ mm Hg

 $\quad\Big\| \quad y_{H_2O} = p_{H_2O}/P$

 $\quad\Downarrow \quad P = 1.1$ bar $= 825$ mm Hg

 $y_{H_2O} = (86.7$ mm Hg$)/(825$ mm Hg$) = 0.105$ mol H_2O/mol

 The molar flow rate of wet air is given by the ideal gas law as

 $$n = PV/RT = \frac{1000 \text{ m}^3}{\text{h}} \left| \frac{1.1 \text{ bar}}{348 \text{ K}} \right| \frac{\text{kmol} \cdot \text{K}}{0.0831 \text{ m}^3 \cdot \text{bar}} = 38.0 \frac{\text{kmol}}{\text{h}}$$

 Consequently,

 $$n_{H_2O} = \frac{38.0 \text{ kmol}}{\text{h}} \left| \frac{0.105 \text{ kmol } H_2O}{\text{kmol}} \right. = \boxed{3.99 \frac{\text{kmol } H_2O}{\text{h}}}$$

 $$n_{BDA} = \frac{38.0 \text{ kmol}}{\text{h}} \left| \frac{(1 - 0.105) \text{ kmol BDA}}{\text{kmol}} \right. = \boxed{34.0 \frac{\text{kmol BDA}}{\text{h}}}$$

 $$n_{O_2} = \frac{34.0 \text{ kmol BDA}}{\text{h}} \left| \frac{0.21 \text{ kmol } O_2}{\text{kmol BDA}} \right. = \boxed{7.14 \frac{\text{kmol } O_2}{\text{h}}}$$

2. $h_m = \dfrac{p_{H_2O}}{P - p_{H_2O}} = \dfrac{86.7 \text{ mm Hg}}{(825 - 86.7) \text{ mm Hg}} = \boxed{0.117 \dfrac{\text{mol } H_2O}{\text{mol BDA}}}$

 The same result could have been obtained from the results of part (a) as $(3.99$ kmol H_2O/h$)/(34.0$ kmol BDA/h$)$.

 $$h_a = \frac{0.117 \text{ kmol } H_2O}{\text{kmol BDA}} \left| \frac{18.0 \text{ kg } H_2O}{\text{kmol}} \right| \frac{1 \text{ kmol}}{29.0 \text{ kg BDA}} = \boxed{0.0726 \frac{\text{kg } H_2O}{\text{kg BDA}}}$$

 $(h_m)^* = \dfrac{p^*_{H_2O}}{P - p^*_{H_2O}} = \dfrac{289 \text{ mm Hg}}{(825 - 289) \text{ mm Hg}} = 0.539 \dfrac{\text{kmol } H_2O}{\text{kmol BDA}}$

 $h_p = 100(h_m)/(h_m)^* = (100)(0.117)/(0.539) = \boxed{21.7\%}$

3. $p_{H_2O} = 86.7$ mm Hg $= p^*_{H_2O}(T_{dp})$

 $\quad\Downarrow$ Table B.3

 $\boxed{T_{dp} = 48.7°C}$

6.4 MULTICOMPONENT GAS–LIQUID SYSTEMS

Gas–liquid processes that involve several components in each phase include many chemical reactions, distillation, and transfer of one or more species from a gas to a liquid (**absorption** or **scrubbing**) or vice versa (**stripping**).

When multicomponent gas and liquid phases are in equilibrium, a limited number of intensive system variables may be specified arbitrarily (the number is given by the phase rule), and the remaining variables must then be determined using equilibrium relationships for the distribution of components between the two phases. In this section we define several such relationships and illustrate how they are used in the solution of material balance problems.

6.4a Tabulated Vapor–Liquid Equilibrium Data

The best way to evaluate equilibrium compositions is from tabulated data. *Perry's Chemical Engineers' Handbook*, pp. 3-64 through 3-70, gives partial pressures of vapors over various liquid solutions; Example 6.4-1 illustrates the use of such data.

EXAMPLE 6.4-1 Absorption of SO$_2$

One hundred pound-moles per hour of an SO$_2$–air mixture containing 45 mole% SO$_2$ is contacted with liquid water in a continuous absorber at 30°C. The liquid is analyzed and found to contain 2 g of SO$_2$ per 100 g of H$_2$O. Assuming that the gas and liquid streams leaving the contactor are in equilibrium at 30°C and 1 atm, calculate the fraction of the entering SO$_2$ absorbed in the water and the required water feed rate.

Solution

Basis: Given Feed Rate of Gas

From Table 3-12 on p. 3-65 of *Perry's Chemical Engineers' Handbook*, the equilibrium partial pressures of H$_2$O and SO$_2$ over a solution of the indicated composition are

$$p_{H_2O} = 31.6 \text{ mm Hg}$$

$$p_{SO_2} = 176 \text{ mm Hg}$$

so that the composition of the exit gas stream is

$$y_{H_2O} = \frac{31.6 \text{ mm Hg}}{760 \text{ mm Hg}} = 0.0416 \frac{\text{lb-mole } H_2O}{\text{lb-mole}}$$

$$y_{SO_2} = \frac{176 \text{ mm Hg}}{760 \text{ mm Hg}} = 0.232 \frac{\text{lb-mole } SO_2}{\text{lb-mole}}$$

$$y_{air} = 1 - y_{H_2O} - y_{SO_2} = 0.727 \frac{\text{lb-mole air}}{\text{lb-mole}}$$

Three unknown process variables remain—L_1, G_2, and L_2—and since a total of three independent balances can be written, the system is determinate.

Air Balance

$$(0.55 \times 100) \frac{\text{lb-moles air}}{h} = \frac{G_2 \text{ lb-moles}}{h} \bigg| y_{air}$$

$$\Downarrow y_{air} = 0.727$$
$$G_2 = 75.7 \text{ lb-moles/h}$$

To write the remaining two balances, it is necessary to determine the mass fractions of SO_2 and H_2O in the liquid effluent.

$$\frac{2 \text{ lb}_m \, SO_2}{100 \text{ lb}_m \, H_2O} \Rightarrow \frac{2 \text{ lb}_m \, SO_2}{102 \text{ lb}_m \text{ total}} \Rightarrow x_{SO_2} = 0.0196 \text{ lb}_m \, SO_2/\text{lb}_m$$

$$\Downarrow$$
$$x_{H_2O} = 0.9804 \text{ lb}_m \, H_2O/\text{lb}_m$$

SO₂ Balance

$$\frac{100 \text{ lb-moles}}{h} \bigg| \frac{0.45 \text{ lb-mole } SO_2}{\text{lb-mole}} = G_2 y_{SO_2} + \frac{L_2 \text{ lb}_m}{h} \bigg| \frac{x_{SO_2} \text{ lb}_m \, SO_2}{\text{lb}_m} \bigg| \frac{\text{lb-mole}}{64 \text{ lb}_m \, SO_2}$$

$$\bigg\Vert \begin{array}{l} G_2 = 75.7 \\ y_{SO_2} = 0.232 \\ x_{SO_2} = 0.0196 \end{array}$$
$$L_2 = 89\,600 \text{ lb}_m/h$$

H₂O Balance

$$L_1(\text{lb}_m \, H_2O/h) = \frac{G_2 \text{ lb-moles}}{h} \bigg| \frac{y_{H_2O} \text{ lb-mole } H_2O}{\text{lb-mole}} \bigg| \frac{18 \text{ lb}_m \, H_2O}{\text{lb-mole}} + L_2 x_{H_2O}$$

$$\bigg\Vert \begin{array}{l} G_2 = 75.7 \\ y_{H_2O} = 0.0416 \\ L_2 = 89\,600 \\ x_{H_2O} = 0.9804 \end{array}$$

$$\boxed{L_1 = 87\,900 \text{ lb}_m \, H_2O/h} \quad \text{(Feed to absorber)}$$

Fraction SO$_2$ Absorbed

$$SO_2 \text{ absorbed} = \frac{89\,600 \text{ lb}_m \text{ liquid effluent}}{h} \left| \frac{0.0196 \text{ lb}_m \text{ SO}_2}{\text{lb}_m} \right.$$

$$= 1756 \frac{\text{lb}_m \text{ SO}_2 \text{ absorbed}}{h}$$

$$SO_2 \text{ fed} = \frac{100 \text{ lb-moles}}{h} \left| \frac{0.45 \text{ lb-mole SO}_2}{\text{lb-mole}} \right| \frac{64 \text{ lb}_m}{\text{lb-mole}} = 2880 \frac{\text{lb}_m \text{ SO}_2 \text{ fed}}{h}$$

$$\Downarrow$$

$$\frac{1756 \text{ lb}_m \text{ SO}_2 \text{ absorbed}/h}{2880 \text{ lb}_m \text{ SO}_2 \text{ fed}/h} = \boxed{0.610 \frac{\text{lb}_m \text{ SO}_2 \text{ absorbed}}{\text{lb}_m \text{ SO}_2 \text{ fed}}}$$

6.4b Raoult's and Henry's laws

If you apply the phase rule to a multicomponent gas–liquid system at equilibrium, you will discover that the compositions of the two phases at a given temperature and pressure are not independent. Once a set of mole fractions or (for the gas phase) partial pressures is specified for either phase, the composition of the other phase is uniquely determined by physical properties of the system components.

The general relationships that govern the distribution of a substance between a gas and a liquid phase are the subject matter of **phase equilibrium thermodynamics,** and, for the most part, fall beyond the scope of this text. Several simple approximate relationships are frequently used, however, which in many cases yield reasonably accurate results.

Suppose A is a substance contained in a gas–liquid system in equilibrium at temperature T and pressure P. Two simple expressions—**Raoult's law** and **Henry's law**—describe under different circumstances the relationship between p_A, the partial pressure of A in the gas phase, and x_A, the mole fraction of A in the liquid phase.

Raoult's law: $$p_A \equiv \boxed{y_A P = x_A p_A^*(T)} \tag{6.4-1}$$

where p_A^* is the vapor pressure of pure liquid A at temperature T and y_A is the mole fraction of A in the gas phase.

Henry's law: $$p_A \equiv \boxed{y_A P = x_A H_A(T)} \tag{6.4-2}$$

where $H_A(T)$ is the **Henry's law constant** for A in a specific solvent.

Raoult's law is an approximation that is generally valid when x_A is close to 1—that is, when the liquid phase is almost pure A. It is sometimes valid over the entire range of compositions for mixtures of similar substances, such as straight-chain hydrocarbons of similar molecular weights.

Note: When $x_A = 1$—that is, when the liquid is pure A—Raoult's law reduces to the expression $p_A = p_A^*(T)$ given previously for systems with only one condensable component.

Henry's law is valid for solutions in which x_A is close to 0 (dilute solutions of A) provided that A does not dissociate, ionize, or react in the liquid phase. The law is often applied to solutions of noncondensable gases. Values of Henry's law constants (or closely related quantities) are given for several gases in water on pp. 3-101 through 3-103 of *Perry's Chemical Engineers' Handbook,* and references to other data sources are given on p. 14-6 of the *Handbook.*

Another way in which equilibrium data may be presented is in the form of **distribution coefficients,** defined for a distributed species A as $K_A = y_A/x_A$, where the value of K_A usually depends on the system temperature and pressure but not on the system composition. Values of distribution coefficients are given on pp. 13-16 through 13-22 of *Perry's Chemical Engineers' Handbook.*

TEST YOURSELF

1. What is Raoult's law, and when is it most likely to be valid?
2. What is Henry's law, and when is it most likely to be valid?
3. What is a distribution coefficient?
4. A gas containing CO_2 is in equilibrium with liquid water containing a small amount of dissolved CO_2 at 30°C and 3 atm. Would you use Raoult's law or Henry's law to estimate the relationship between (a) x_{CO_2} and p_{CO_2}, (b) x_{H_2O} and p_{H_2O}, where x denotes mole fraction in the liquid and p denotes partial pressure in the gas? In each case, what would you look up and where would you look it up?

EXAMPLE 6.4-2 Raoult's and Henry's Laws

Use either Raoult's law or Henry's law to solve the following problems.

1. A gas containing 1 mole% ethane is in contact with water at 20°C and 20 atm. Estimate the mole fraction of dissolved ethane.
2. An equimolar liquid mixture of benzene (B) and toluene (T) is in equilibrium with its vapor at 30°C. What is the system pressure and the composition of the vapor?

Solution

1. Hydrocarbons normally are relatively insoluble in water, so that the solution of ethane is probably extremely dilute. Let us therefore apply Henry's law. Page 3-102 of *Perry's Chemical Engineers' Handbook* gives the Henry's law constant for ethane as 2.63×10^4 atm/mole fraction. From Equation 6.4-2

$$x_{C_2H_6} = \frac{y_{C_2H_6}P}{H_{C_2H_6}} = \frac{(0.01)(20 \text{ atm})}{2.63 \times 10^4 \text{ atm/mole fraction}} = \boxed{7.6 \times 10^{-6} \frac{\text{mol } C_2H_6}{\text{mol}}}$$

2. Since benzene and toluene are structurally similar compounds, we may apply Raoult's law. From Table 6.1-1:

$$\log_{10} p_B^* = 6.906 - \frac{1211}{T + 220.8} \xrightarrow{T = 30°C} p_B^* = 119 \text{ mm Hg}$$

$$\log_{10} p_T^* = 6.9533 - \frac{1343.9}{T + 219.38} \xrightarrow{T = 30°C} p_T^* = 36.7 \text{ mm Hg}$$

Using Equation 6.4-1

$$p_B = x_B p_B^* = (0.5)(119) = 59.5 \text{ mm Hg}$$
$$p_T = x_T p_T^* = (0.5)(36.7) = 18.35 \text{ mm Hg}$$

$$P = p_B + p_T = \boxed{77.9 \text{ mm Hg}}$$

$$y_B = p_B/P = \boxed{0.764 \text{ mole benzene/mole}}$$

$$y_T = p_T/P = \boxed{0.236 \text{ mole toluene/mole}}$$

CREATIVITY EXERCISE

Give as many cases as you can think of where it would be useful or necessary to know the Henry's law solubility of a gas in a liquid. (*Example:* You wish to calculate the pressure needed to achieve a specified level of carbonation in bottled soda.) In your list, include several examples with environmental relevance.

6.4c Vapor–Liquid Equilibrium Calculations for Ideal Solutions

Suppose heat is added slowly to a closed vessel that contains a liquid, and that the pressure in the vessel is held constant. We have already examined what happens in a situation of this sort if the liquid is a single species: The temperature increases until the boiling point of the liquid is reached, and thereafter the liquid vaporizes at a constant temperature. Once the vaporization is complete, further addition of heat raises the temperature of the vapor.

Consider now what happens if the liquid is a mixture of several components. As heat is added, the liquid temperature rises until a temperature is reached at which the first bubble of vapor forms. Up to this point, the process looks like that for a single component. However, if the liquid is a mixture, the vapor generated will generally have a composition different from that of the liquid. As vaporization proceeds, therefore, the composition of the remaining liquid continuously changes, and hence so does its vaporization temperature. A similar phenomenon occurs if a mixture of vapors is subjected to a condensation process at constant pressure: At some temperature the first droplet of liquid forms, and thereafter the composition of the vapor and the condensation temperature both change.

It is obvious that to design an evaporation or condensation process you must know the conditions at which the transition from vapor to liquid or liquid to vapor takes place. In addition, design of other separation processes such as distillation, absorption, and stripping also require information on the conditions at which phase transitions occur, and on the compositions of the resulting phases. This section outlines the required calculations for a relatively simple class of mixtures.

The temperature at which the first bubble of vapor forms when a liquid is heated slowly at constant pressure is the **bubble point** of the liquid at the given pressure. Similarly, if a gas–vapor mixture is slowly cooled at constant pressure, the temperature at which the first droplet of liquid forms is the **dew point** of the gas at the given pressure. Calculating bubble point and dew point temperatures can be a relatively complex task for an arbitrary mixture of components. However, if the liquid phase behaves as an **ideal solution** (one for which Raoult's or Henry's law is obeyed for all compositions), the calculations are relatively straightforward.

Suppose we have an ideal liquid solution (one that follows Raoult's law) containing several species—A, B, C, . . .—with known component mole fractions x_A, x_B, x_C, If the mixture is heated at a constant pressure, P, to its bubble point, T_{bp}, the further addition of a slight amount of heat will lead to the formation of a vapor bubble. Since this bubble is in equilibrium with the liquid and the solution has been presumed to be ideal, the partial pressures of the components are given by Raoult's law, Equation 6.4-1.

$$p_i = x_i p_i^*(T_{bp}), \qquad i = A, B, \ldots \tag{6.4-3}$$

where p_i^* is the vapor pressure of component i. Moreover, since we have assumed that only A, B, C, . . . are present in the system, the sum of the partial pressures must be the total system pressure, P; hence,

$$P = x_A p_A^*(T_{bp}) + x_B p_B^*(T_{bp}) + \cdots \tag{6.4-4}$$

The bubble point temperature may be calculated by trial-and-error as the value of T_{bp} that satisfies this equation; all that is needed is a set of relationships for $p_i^*(T)$, such as the Antoine equation or vapor pressure charts or tables. Once T_{bp} is known, the composition of the vapor phase can easily be determined by evaluating the partial pressures of each component from Equation 6.4-3 and determining each vapor–phase mole fraction as $y_i = p_i/P$.

The dew point of a mixture of gases and vapors may be found by a similar method. Again, suppose that the gas phase contains A, B, C, . . .—all of which may condense— and G, which cannot, at a fixed pressure, P, and let y_i be the mole fraction of component i in the gas phase. If the gas mixture is cooled to its dew point, T_{dp}, it will be in equilibrium with the first droplet of liquid that forms. If the liquid is an ideal solution, then the compositions of the droplet and the gas phase are related through Raoult's law, and the liquid-phase mole fractions may be calculated as

$$x_i = y_i P/p_i^*(T_{dp}), \qquad i = A, B, \ldots \tag{6.4-5}$$

The dew point of the vapor mixture may be determined by noting that the mole fractions of the liquid components must sum to 1.

$$x_A + x_B + x_C + \cdots = 1$$

$$\Downarrow \text{Equation 6.4-5}$$

$$\frac{y_A P}{p_A^*(T_{dp})} + \frac{y_B P}{p_B^*(T_{dp})} + \cdots = 1 \tag{6.4-6}$$

The value of T_{dp} can be found by trial-and-error once expressions for $p_i^*(T)$ have been substituted. The composition of the liquid phase may then be determined from Equation 6.4-5.

EXAMPLE 6.4-3 Bubble and Dew Point Calculations

1. Calculate the bubble point temperature and the equilibrium vapor composition associated with a 40 mole% benzene–60 mole% toluene liquid mixture at 1 atm.
2. Calculate the dew point temperature and the equilibrium liquid composition associated with a gas mixture containing 10 mole% benzene, 10 mole% toluene, and the balance nitrogen (noncondensable) at 1 atm.

Solution

1. Let A = benzene and B = toluene in Equations 6.4-3 and 6.4-4. The latter equation may be written in the form:

$$f(T_{bp}) = 0.40 p_A^*(T_{bp}) + 0.60 p_B^*(T_{bp}) - 760 \text{ mm Hg} = 0$$

The solution procedure is to choose a temperature; evaluate p_A^* and p_B^* (mm Hg) for that temperature from the Antoine equation, Table 6.1-1; evaluate $f(T_{bp})$ from the above equation; and repeat the calculation until a temperature is found for which $f(T_{bp})$ is sufficiently close to 0.

Several techniques are discussed in Appendix A.2 for accelerating trial-and-error searches of this sort. One of them, the *regula-falsi* method, has been used to generate the trial temperatures shown here for the third and subsequent trials.

$T(°C)$	$f(T)$
80	-282.2
100	114.0
94.2	-20.7
95.1	-0.93
95.14	-0.05

The solution is taken to be $\boxed{T_{bp} = 95.14°C.}$ At this temperature, Equation 6.4-3 yields

$$p_A = (0.40)(1181 \text{ mm Hg}) = 472.5 \text{ mm Hg}$$

$$p_B = (0.60)(479 \text{ mm Hg}) = 287.5 \text{ mm Hg}$$

$$\Downarrow$$

$$P = 472.5 + 287.5 = 760.0 \text{ mm Hg} \quad (\textit{Check})$$

$$y_A = 472.5/760.0 = 0.622 \text{ mol benzene/mol}$$
$$y_B = 1 - y_A = 0.378 \text{ mol toluene/mol}$$

2. With A again denoting benzene and B toluene, Equation 6.4-6 may be written

$$f(T_{dp}) = \frac{(0.10)(760 \text{ mm Hg})}{p_A^*(T_{dp})} + \frac{(0.10)(760 \text{ mm Hg})}{p_B^*(T_{dp})} - 1.00 = 0$$

A trial-and-error procedure similar to that used in part (a) leads to the result $\boxed{T_{dp} = 52.44°\text{C}}$ (for which the function $f = 0.00013$). Then, from Equation 6.4-5,

$$x_A = (0.10)(760)/p_A^*(52.44°\text{C}) = 0.256 \text{ mol benzene/mol}$$
$$x_B = 1 - x_A = 0.744 \text{ mol toluene/mol}$$

Note: Bubble and dew point calculations can also be performed for a specified system temperature (rather than pressure). For example, suppose you are given the composition and temperature of an ideal liquid and asked to calculate the equilibrium vapor–phase composition and system pressure. Equations 6.4-3 and 6.4-4 could be used as in the preceding example, but the calculations would not be trial-and-error.

Vapor–liquid equilibrium calculations for binary (two-component) systems can be simplified considerably using a graphic device known as a *Txy* **diagram.** Suppose the bubble point of an ideal solution of two components—A and B—is determined as illustrated above for a fixed pressure, P, and several values of x_A, and the equilibrium vapor composition y_A is determined for each liquid composition. The *Txy* diagram is a plot of the equilibrium temperature versus the mole fraction of one of the components—usually the more volatile one—with curves being drawn for both the liquid phase (T versus x_A) and the vapor phase (T versus y_A). A plot of this type is shown in Figure 6.4-1a for the benzene-toluene system at $P = 1$ atm. Alternatively, equilibrium pressure can be plotted against mole fraction for a fixed temperature to generate a *Pxy* **diagram** (Figure 6.4-1b). (The term **boiling-point diagram** is also used for plots of this type.)

Once you have a *Txy* diagram like that of Figure 6.4-1, bubble and dew point calculations become trivial. To determine a bubble point temperature for a given liquid-phase composition, you need only go to the liquid curve on the *Txy* diagram for the system pressure and read the desired temperature from the ordinate scale. (If you are not sure why this works, go back and consider again how the curve was generated.) You can then immediately determine from the vapor curve the composition of the vapor in equilibrium with the given liquid at that temperature.

The dew point of a vapor mixture of A and B at P can be determined from the *Txy* diagram if there are no species other than A and B in the gas phase. If this is the case, you need only look up the specified mole fraction of A in the vapor phase,

(a) Txy diagram

(b) Pxy diagram

FIGURE 6.4-1

***Txy* and *Pxy* diagrams for benzene–toluene**

read the dew point temperature from the corresponding ordinate value of the vapor curve, and move over to the liquid curve and down to read the composition of the liquid in equilibrium with this vapor. If a noncondensable species is present in the gas phase, however, you must use Equation 6.4-4 to find the dew point, as in the previous example.

What happens as a liquid mixture is continuously vaporized is easily seen in reference to the Txy diagram. Consider the benzene-toluene system again, and suppose heat is added to a 60 mol% benzene and 40 mol% toluene liquid mixture at a fixed pressure of 1 atm. As Figure 6.4-1 shows, the mixture will begin to boil at 90°C, and the vapor generated will contain 78% benzene. However, once a small amount of liquid has been vaporized, the remainder no longer contains 60% benzene; it contains less, since the vapor produced is relatively rich in this component. Consequently, the temperature of the liquid steadily rises as more and more liquid is vaporized, and the compositions of both phases change continuously during the process.

EXAMPLE 6.4-4 Bubble and Dew Point Calculations using *Txy* Diagrams

1. Using the Txy diagram, determine the bubble point temperature and the equilibrium vapor composition associated with a 40 mole% benzene–60 mole% toluene liquid mixture at 1 atm. If the mixture is steadily vaporized until the remaining liquid contains 25% benzene, what is the final temperature?

2. Using the Txy diagram, determine the dew point temperature and the equilibrium liquid composition associated with a mixture of benzene and toluene vapors containing 40 mole% benzene at 1 atm. If condensation proceeds until the remaining vapor contains 60% benzene, what is the final temperature?

Solution

1. From Figure 6.4-1, for $x_B = 0.40$, $\boxed{T_{bp} = 95°C,}$ and $\boxed{y_B = 0.62.}$ (This matches the result of the much lengthier solution of the previous example.) When $x_B = 0.25$, $\boxed{T_{bp} = 101°C.}$ The temperature thus rises by 6°C as the vaporization proceeds.

2. From Figure 6.4-1, for $y_B = 0.40$, $\boxed{T_{dp} = 102°C,}$ and $\boxed{x_B = 0.20.}$ When $y_B = 0.60$, $\boxed{T_{bp} = 96°C.}$

The term "boiling" is sometimes incorrectly used to describe any process involving the transition from liquid to vapor. In fact, boiling refers to a specific type of vaporization process wherein vapor bubbles form at a heated container surface and evolve from the liquid; it does *not* refer to molecular evaporation of liquid from a gas–liquid interface, which may occur at temperatures below the boiling point. (Recall your experience with leaving a container of liquid, coffee perhaps, in a room, and finding the liquid partially or completely evaporated on your return.)

In our discussion of single-component systems, we considered the case of a liquid being heated in a container exposed to the atmosphere and observed that the liquid boils when its temperature is such that the vapor pressure of the liquid equals the total pressure of the atmosphere above it. A similar phenomenon occurs for liquid mixtures. If a mixture is heated slowly in an open container, vapor bubbles will form at the heated surface and emerge into the gas phase when the vapor pressure of the liquid equals the pressure above the liquid.[3] Some reflection should convince you that the temperature at which this occurs is the bubble point of the liquid at that pressure. For an ideal liquid solution, the boiling point may therefore be determined *approximately* from Equation 6.4-7.

$$x_A p_A^*(T_{bp}) + x_B p_B^*(T_{bp}) + \ldots = P \qquad (6.4\text{-}7)$$

EXAMPLE 6.4-5 Boiling Point of a Mixture

A mixture that is 75 mole% benzene and 25 mole% toluene is to be distilled in a batch distillation column. The column startup procedure calls for charging the still and slowly adding heat until boiling begins. Estimate the temperature at which boiling begins and the initial composition of the vapor generated, assuming the system pressure is 760 mm Hg.

Solution

From the Txy diagram, the mixture will boil at approximately $\boxed{87°C}$. The initial vapor composition is approximately $\boxed{88 \text{ mole% benzene}}$ and 12 mole% toluene.

[3] This is only an approximation, albeit generally a good one. In fact, the vapor pressure must be slightly greater than the gas-phase pressure to overcome the effects of liquid surface tension and the hydrostatic head of liquid at the heated surface.

We conclude this discussion with one final reminder. The vapor–liquid equilibrium calculations we have shown in this section, and the *Txy* diagram we illustrated in Figure 6.4-1, are all based on the ideal solution assumption and the corresponding use of Raoult's law. Many commercially important systems involve nonideal solutions, or systems of immiscible or partially miscible liquids, for which Raoult's law is inapplicable and the *Txy* diagram looks nothing like the one shown for benzene and toluene. Calculations for systems of this sort are considered in texts on phase equilibrium thermodynamics.

TEST YOURSELF

1. What is the bubble point of a liquid mixture at a given pressure? The dew point of a vapor mixture at a given pressure?

2. At what temperature will an equimolar liquid mixture of benzene and toluene begin to boil at 1 atm? What is the mole fraction of benzene in the vapor phase?

3. At what temperature will an equimolar vapor mixture of benzene and toluene at 1 atm begin to condense? What is the mole fraction of benzene in the condensate? What happens to the system temperature as the condensation proceeds?

4. Would you expect the bubble point of a liquid mixture to increase, decrease, or remain the same as the pressure increases? What about the dew point of a vapor mixture?

5. When you raise a liquid to its boiling point the pressure beneath the liquid surface where the bubbles form is different from the gas phase pressure (why?), so that the boiling does not take place at precisely the calculated temperature. Explain. If the bottom of a vat of water is heated on a day when atmospheric pressure is 1 atm, and the height of the water is 5 ft, how would you estimate the boiling temperature?

6.5 SOLUTIONS OF SOLIDS IN LIQUIDS

6.5a Solubility and Saturation

There is an upper limit on the amount of a solid that can be dissolved in a liquid. This limit varies considerably from one solute–solvent pair to another: For example, 100 g of water at 20°C can dissolve 222 g of $AgNO_3$, 0.003 g of $AgCO_3$, and 0.00002 g of AgBr. The limit also depends strongly on temperature: The solubility of $AgNO_3$ in 100 g of water increases from 222 g at 20°C to 952 g at 100°C. *Perry's Chemical Engineers' Handbook* on pp. 3-6 through 3-44 and 3-97 through 3-100 gives the solubilities of many substances in water, ethyl alcohol, and diethyl ether at specified temperatures.

A solution that contains as much of a dissolved species as it can hold is said to be **saturated** with that species. If a saturated solution is cooled, the solubility of the

solute generally decreases, and some solute precipitates out in the form of crystals. *A solution in equilibrium with crystals of a solute must be saturated with that solute.* (What would happen if it were not?)

EXAMPLE 6.5-1 Crystallization and Filtration

One hundred fifty kilograms of a saturated solution of $AgNO_3$ at 100°C are cooled to 20°C, and the crystals are filtered out. The wet filter cake, which contains 80% solid crystals and 20% saturated solution by mass, passes to a dryer in which the remaining water is removed. Calculate the fraction of the $AgNO_3$ in the feed stream eventually recovered as dry crystals and the amount of water that must be removed in the drying stage.

Solution

Basis: 150 kg Feed

Both the filtrate and the liquid retained in the filter cake are in equilibrium with solid $AgNO_3$ crystals and must therefore be saturated with $AgNO_3$ at 20°C. The compositions of saturated solutions of silver nitrate at 100°C and 20°C are given at the beginning of this section.

$$100°C: \quad \frac{952 \text{ g AgNO}_3}{100 \text{ g H}_2\text{O}} \Rightarrow \frac{952 \text{ g AgNO}_3}{(100 + 952) \text{ g}} = 0.905 \text{ g AgNO}_3/\text{g}$$
$$\Downarrow$$
$$0.095 \text{ g H}_2\text{O}/\text{g}$$

$$20°C: \quad \frac{22 \text{ g AgNO}_3}{100 \text{ g H}_2\text{O}} \Rightarrow \frac{222 \text{ g AgNO}_3}{(100 + 222) \text{ g}} = 0.689 \text{ g AgNO}_3/\text{g}$$
$$\Downarrow$$
$$0.311 \text{ g H}_2\text{O}/\text{g}$$

Composition of the Filter Cake

$$Q_2 = 0.8(Q_2 + Q_3) \Rightarrow Q_2 = 4Q_3$$

H_2O Balance About the Cooler and Filter

$$(0.095 \times 150) \text{ kg } H_2O = 0.311Q_1 + 0.311Q_3$$

Mass Balance About the Cooler and Filter

$$150 \text{ kg} = Q_1 + Q_2 + Q_3$$

The simultaneous solution of these three equations yields

$$Q_1 = 20 \text{ kg}$$
$$Q_2 = 104 \text{ kg}$$
$$Q_3 = 26 \text{ kg}$$

Overall $AgNO_3$ Balance

$$(0.905 \times 150) \text{ kg } AgNO_3 = 0.689Q_1 + Q_5$$
$$\Downarrow Q_1 = 20 \text{ kg}$$
$$Q_5 = 122 \text{ kg } AgNO_3 \text{ crystals recovered}$$

Percentage Recovery

$$\frac{122 \text{ kg } AgNO_3 \text{ recovered}}{(0.905 \times 150) \text{ kg } AgNO_3 \text{ fed}} \times 100\% = \boxed{89.9\%}$$

Overall Mass Balance

$$150 \text{ kg} = Q_1 + Q_4 + Q_5$$
$$\Downarrow \begin{array}{l} Q_1 = 20 \text{ kg} \\ Q_5 = 122 \text{ kg} \end{array}$$
$$Q_4 = \boxed{8 \text{ kg } H_2O \text{ removed in the dryer}}$$

6.5b Phase Diagrams

Solubility data can be presented efficiently and clearly in **phase diagrams,** which show equilibrium compositions in a system as functions of one or more system variables. The usual phase diagram for a solid–liquid system is a plot of temperature versus mass fraction of the solid.

A phase diagram for the $MgSO_4$–H_2O system is shown in Figure 6.5-1. The lines *ab, bc, cd, de,* and *ef* on this figure represent saturated solutions of $MgSO_4$ in equilibrium with solid crystals whose compositions are given by the vertical lines *ag, hi, jk, lm,* and *no.*

When certain salts are formed by crystallization from an aqueous solution, the salt molecules may have bonded to them one or more molecules of water. For example,

FIGURE 6.5-1

Phase diagram for $MgSO_4$–H_2O. (Adapted from R. H. Perry and D. W. Green,
Eds., *Perry's Chemical Engineers' Handbook*, 6th Edition, p. 19–26. Copyright ©
1984 by McGraw-Hill, Inc. Used with permission of McGraw-Hill Book Co.)

solid magnesium sulfate exists in several forms:

$MgSO_4$ anhydrous magnesium sulfate
$MgSO_4 \cdot H_2O$ magnesium sulfate monohydrate
$MgSO_4 \cdot 6\,H_2O$ magnesium sulfate hexahydrate
$MgSO_4 \cdot 7\,H_2O$ magnesium sulfate heptahydrate
$MgSO_4 \cdot 12\,H_2O$ magnesium sulfate dodecahydrate

Solid magnesium sulfate heptahydrate, for example, contains 7 moles of water per mole of $MgSO_4$. The water contained in such substances is called **water of hydration,** and the substances themselves are called **hydrated salts.**

As an illustration of the use of the phase diagram, let us consider a 27.5 wt% solution of $MgSO_4$ at 100°F (point A). If the solution is cooled, the system follows a vertical path to point B (80°F), at which $MgSO_4 \cdot 7\,H_2O$ begins to precipitate. (The mass fraction of $MgSO_4$ in the heptahydrate salt is 0.488, the number written on the vertical line to the right of point B.) If the system is further cooled to 60°F (point D), it will consist of a saturated solution of $MgSO_4$ with a mass fraction of 0.245 (point C) in equilibrium with solid crystals of $MgSO_4 \cdot 7\,H_2O$ (point E).

Material balances on solid–liquid systems for which phase diagrams are available can be simplified by a device known as the **lever rule.** Suppose a system with an overall mass fraction x_F (g MgSO/g) (e.g., point D on Figure 6.5-1) separates into a saturated solution with mass fraction x_s (point C) and solid crystals with mass fraction x_c (point E). (See Figure 6.5-2.) An overall balance on this process yields

$$F = S + C \tag{6.5-1}$$

and an $MgSO_4$ balance yields

$$Fx_F = Sx_s + Cx_c \tag{6.5-2}$$

$$\Downarrow \text{ Equation 6.5-1}$$

$$(S + C)x_F = Sx_s + Cx_c \tag{6.5-3}$$

$$\Downarrow \text{ rearranging}$$

$$\frac{C}{S} = \frac{x_F - x_s}{x_c - x_F} \tag{6.5-4}$$

But the right side is the ratio of distances $\overline{CD}/\overline{DE}$ on Figure 6.5-1, so that

$$\frac{C\ (\text{kg crystals})}{S\ (\text{kg solution})} = \overline{CD}/\overline{DE} \tag{6.5-5}$$

This is the lever rule. It is also not difficult to show that

$$\frac{\text{Mass of crystals}}{\text{Total of mass}} = \overline{CD}/\overline{CE} \tag{6.5-6}$$

(Show it!)

FIGURE 6.5-2

Schematic of MgSO₄ crystallization process

TEST YOURSELF

1. What does the line jk on Figure 6.5-1 represent?

2. What is a hydrated salt?

3. At what temperature would a solution containing 35 wt% $MgSO_4$ be saturated? What would be the composition of crystals in equilibrium with this solution?

4. Use the lever rule to determine the mass ratio of crystals to saturated solution for the system at point D on Figure 6.5-1. Then determine the mass fraction of the crystals.

EXAMPLE 6.5-2 Application of the Lever Rule

A solution of $MgSO_4$ at 220°F containing 43 g of $MgSO_4$ per 100 g of water is fed to a cooling crystallizer operating at 50°F. If the solution leaving the crystallizer is saturated, determine the rate at which solution must be fed to the crystallizer to produce 1 ton of $MgSO_4 \cdot 7\,H_2O$ per hour.

Solution

Let M denote $MgSO_4$.

Basis: 1 ton crystals/h

The mass fraction of $MgSO_4$ in the product solution ($x_s = 0.232$) was determined from Figure 6.5-1. From the lever rule, Equation 6.5-4,

$$\frac{C}{S} = \frac{0.301 - 0.232}{0.488 - 0.301} = 0.369 \Rightarrow S = \frac{1 \text{ ton/h}}{0.369} = 2.71 \frac{\text{ton}}{\text{h}}$$

and from an overall mass balance

$$\boxed{F = C + S = 3.71 \frac{\text{tons}}{\text{h}}}$$

6.5c Colligative Solution Properties

Physical properties of a solution generally differ from the same properties of the pure solvent. Under certain conditions, the changes in the values of several properties—such as vapor pressure, boiling point, and freezing point—depend only on the concentration of solute in the solution, and not on what the solute and solvent are. These three properties are referred to as **colligative solution properties.** (A fourth colligative property—osmotic pressure—will not concern us in this text.)

An understanding of colligative properties is important in the determination of process operating conditions. For example, a process can be designed to recover pure water by evaporation or freezing of seawater. In the first instance, pure water is recovered by condensation of vapor from the evaporator, while in the second it is recovered by separating and melting ice. An engineer who intends to design an evaporator or an ice crystallizer clearly must know the temperature at which the phase transition takes place—the boiling point in the first case, the freezing point in the second. In addition, measured values of colligative solution properties are often used to deduce properties of either the solvent or the solute, such as molecular weight, which cannot easily be determined by more direct means.

This section presents an introductory view of this topic, considering only the simple case of a solution in which the solute is nonvolatile (i.e., has a negligible vapor pressure at the solution temperature), and the dissolved solute neither dissociates (which rules out ionizing acids, bases, and salts) nor reacts with the solvent. Discussions of more complex systems can be found in most physical chemistry texts.

Consider a solution in which the solute mole fraction is x and the vapor pressure of the pure solvent at the solution temperature is p_s^*. Applying Raoult's law (Equation 6.4-1) to the solution, we obtain for the partial pressure of the solvent

$$p_s(T) = (1 - x)p_s^*(T) \tag{6.5-7}$$

If the liquid is pure solvent ($x = 0$), this equation predicts that the partial pressure of the solvent vapor equals the vapor pressure of the solvent, as we would expect. Since the solute is nonvolatile, the solvent is the only component in the vapor in equilibrium with the solution. The pressure exerted by this vapor is referred to as the *effective solvent vapor pressure*:

$$(p_s^*)_e = p_s = (1 - x)p_s^* \tag{6.5-8}$$

Since x—and hence $(1 - x)$—is less than one, we find that the effect of the solute is to lower the effective solvent vapor pressure. This effect is often defined in terms of **vapor pressure lowering,** which is defined as the difference between the vapor pressure of the pure component and the effective vapor pressure of the solvent.

$$\Delta p_s^* = p_s^* - (p_s^*)_e = x p_s^* \tag{6.5-9}$$

The simplicity and generality of Equation 6.5-9 are surprising. According to the equation, if a solution contains 20 mole% solute, then the solvent partial pressure is 80% of the vapor pressure of pure solvent at the system temperature, regardless of the temperature, the pressure, *and what the solute and solvent are.* (Hence, vapor pressure lowering is a colligative property, by definition.) The only stipulations are that Raoult's law holds, and the solute is nonvolatile, nonreactive, and nondissociative.

The lowering of the solvent vapor pressure has two important consequences. *The solvent in a solution at a given pressure boils at a higher temperature and freezes at a lower temperature than does the pure solvent at the same pressure.* The validity of these statements can be seen by referring to Figure 6.5-3, a phase diagram for an arbitrary solute–solvent system. Shown on this figure are the vapor–liquid equilibrium curves for pure solvent (solid curve) and for a solution of a fixed solute concentration (dashed curve). The latter curve lies below the former, reflecting the fact that the effective vapor pressure at a given temperature is lower than that of the pure solvent. The higher the solute concentration, the farther is the solution curve below that for the pure solvent.

The effect of the solute on the solution boiling point is easy to see from the diagram. Recall that the boiling point of a liquid at a given pressure is the intersection of a horizontal line at that pressure with the vapor–liquid equilibrium curve. At pressure P_o, the pure solvent boils at temperature T_{bo}, while the solution boils at a higher temperature, T_{bs}.

The change in the solvent freezing point is a little less obvious. First, consider the triple point—the intersection of the solid–vapor and the liquid–vapor equilibrium curves. It is clear from Figure 6.5-3 that the effect of the vapor pressure lowering is to lower the triple point of the solution relative to pure solvent. If in addition the solid–liquid equilibrium curve for the solution is (like that for the pure solvent) almost ver-

FIGURE 6.5-3

Phase equilibrium curves for pure solvent and solution

tical, then the freezing point at an arbitrary pressure P_o also drops—on the diagram, from T_{mo} for the pure solvent to T_{ms} for the solution.

Since we know how the solution vapor pressure varies with concentration (the relationship being given by Equation 6.5-8) and temperature (through the Clausius–Clapeyron equation, Equation 6.1-3), we can determine the relationships between concentration and both boiling point elevation and freezing point depression. The relationships are particularly simple for dilute solutions ($x \sim 0$, where x is solute mole fraction).

$$\Delta T_b = T_{bs} - T_{bo} = \frac{RT_{bo}^2}{\Delta \hat{H}_v} x \qquad (6.5\text{-}10)$$

$$\Delta T_m = T_{mo} - T_{ms} = \frac{RT_{mo}^2}{\Delta \hat{H}_m} x \qquad (6.5\text{-}11)$$

In these equations, $\Delta \hat{H}_v$ refers to the heat of vaporization of the pure solvent at its boiling point T_{bo}, and $\Delta \hat{H}_m$ denotes the heat of fusion of the pure solvent at its melting point T_{mo}. These properties of the solvent can be looked up in data tables, such as Table B.1 of this text. Their physical significance is discussed further in Chapter 8. The derivation of Equation 6.5-10 constitutes the subject matter of Problem 71 at the end of this chapter.

Since the coefficients of x in these two equations are constant, it follows that for dilute solutions of nonvolatile, nonreactive, nondissociative solutes, both boiling point elevation and freezing point depression vary linearly with solute mole fraction.

The next example concerns applications of Equations 6.5-8 through 6.5-11 to the determination of a vapor pressure and phase transition temperatures for a known solution concentration, and to the inverse calculation of a solution composition and solute molecular weight from a measured colligative property.

TEST YOURSELF

1. What is a colligative solution property? Name three of them.

2. The vapor pressure of a solvent at 120°C is 1000 mm Hg. A solution contains 15% of a solute in this solvent at 120°C. If the behavior described in this section is followed, what is the effective vapor pressure of the solvent? What conditions must be met for your answer to be valid?

3. The solution of question 2 is heated to a temperature at which boiling occurs at a total pressure of 1000 mm Hg. Is the boiling temperature greater than, less than, or equal to 120°C. What is the vapor pressure of the pure solvent at the boiling point?

4. Explain why salt is scattered on roads and sidewalks during a snowstorm.

5. One gram-mole of a nondissociating solute dissolved in 10 mol of a solvent lowers the solvent vapor pressure by 20 mm Hg. Suppose now that you dissolve 1 mol of a solute that dissociates (ionizes) in 10 mol of the same solvent. What would you guess would happen to the solvent vapor pressure?

EXAMPLE 6.5-3 Colligative Property Calculations

A solution of 5.000 g of a solute and 100.0 g of water is heated slowly at a constant pressure of 1.00 atm and is observed to boil at 100.421°C. Estimate the molecular weight of the solute, the effective solvent vapor pressure at 25°C, and the solution freezing point at 1 atm. The necessary properties of water can be found in Table B.1.

Solution

If the values of the normal boiling point and heat of vaporization of pure water (from Table B.1) and the gas constant are substituted into Equation 6.5-10, the result is

$$\Delta T_b = \frac{(8.314 \text{ J/mol·K})(373.16 \text{ K})^2 x}{40\ 656 \text{ J/mol}} = 28.5x$$

Thus, from the measured boiling point elevation, $\Delta T_b = 0.421$ K, we may deduce that the mole fraction of the solute in the solution is $x = 0.421/28.5 = 0.0148$. But since the solution is known to contain $(5.000/M_s)$ mol of solute, where M_s is the solute molecular weight, and $(100.0/18.016) = 5.551$ mol of water, we may write

$$0.0148 = (5.000/M_s)/[(5.000/M_s + 5.551)$$
$$\Downarrow$$

$$\boxed{M_s = 60.1 \text{ g/mol}}$$

From Equation 6.5-8 the effective solvent vapor pressure at 25°C is determined from the vapor pressure of pure water at this temperature (obtained in Table B.3) as

$$(p_s^*)_e = (1.000 - 0.0148)(23.756 \text{ mm Hg}) = \boxed{23.40 \text{ mm Hg}}$$

Finally, substituting values of the melting point and heat of fusion of water (from Table B.1) and the gas constant into Equation 6.5-11, we obtain

$$\Delta T_m = \frac{(8.314 \text{ J/mol K})(273.16 \text{ K})^2(0.0148)}{(6009.5 \text{ J/mol})} = 1.53 \text{ K} = 1.53°C$$
$$\Downarrow$$

$$T_{ms} = (0.000 - 1.53)°C = \boxed{-1.53°C}$$

CREATIVITY EXERCISE

A solution contains an unknown amount of table salt dissolved in water. List as many ways as you can think of to measure or estimate the concentration of salt in the solution without leaving the kitchen of your home. The only instruments you are allowed to bring home from work are a thermometer that covers the range $-10°C$ to $120°C$ and a small laboratory balance. (*Example:* Make up several solutions with known salt concentrations, and compare their tastes with that of the unknown solution.)

6.6 IMMISCIBLE AND PARTIALLY MISCIBLE LIQUIDS

6.6a Miscibility and Distribution Coefficients

If water and methyl isobutyl ketone (MIBK) are mixed at 25°C, a single phase results if the mixture contains more than 98% water or 97.7% MIBK by mass; otherwise, the mixture separates into two liquid phases, one of which contains 98% H_2O and 2% MIBK and the other 97.7% MIBK and 2.3% H_2O. Water and MIBK are examples of **partially miscible** liquids; they would be termed **immiscible** if one phase contained a negligible amount of water and the other a negligible amount of MIBK.

If a third substance is added to a two-phase liquid mixture, it distributes itself according to its relative solubility in each phase. For example, acetone is soluble in both water and chloroform—two nearly immiscible liquids—but much more so in chloroform. If a mixture of acetone and water is contacted with chloroform, a substantial portion of the acetone enters the chloroform-rich phase. Separation of the acetone and water may then be easily accomplished by allowing the mixture to settle and separating the two phases. This is an illustration of the separation process of **liquid extraction.**

Suppose A and S are two nearly immiscible liquids, and B is a solute distributed between the phases of an A–S mixture. The **distribution coefficient** for component B is the ratio of the mass fraction of B in the S phase to that in the A phase. *Perry's Chemical Engineers' Handbook* on pp. 15-9 through 15-13 lists distribution coefficients for a number of ternary liquid systems. Example 6.6-1 illustrates the use of this physical property in a material balance calculation.

EXAMPLE 6.6-1 Extraction of Acetone from Water

Two hundred cubic centimeters of an acetone–water mixture that contains 10 wt% acetone are mixed with 400 cm^3 of chloroform at 25°C, and the phases are then allowed to settle. What percentage of the acetone is transferred from the water to the chloroform?

Solution

Basis: Given Quantities

The densities of the pure substances are given in Table B.1

$$
\begin{array}{ll}
\text{Acetone (A):} & 0.792 \text{ g/cm}^3 \\
\text{Chloroform (C):} & 1.489 \text{ g/cm}^3 \\
\text{Water (W):} & 1.000 \text{ g/cm}^3
\end{array}
$$

Since density data for the acetone–water system are not readily available, let us use Equation 5.1-1 to estimate the density of the feed solution:

$$
\frac{1}{\bar{\rho}} = \frac{x_A}{\rho_A} + \frac{x_W}{\rho_W} = \frac{0.100}{0.792} + \frac{0.900}{1.000} = 1.026 \frac{\text{cm}^3}{\text{g}}
$$

$$
\Downarrow
$$

$$
\bar{\rho} = 0.974 \text{ g/cm}^3
$$

The mass of the solution fed is, therefore,

$$\frac{200 \text{ cm}^3 \mid 0.974 \text{ g}}{\mid \text{cm}^3} = 195 \text{ g}$$

and that of the chloroform is

$$\frac{400 \text{ cm}^3 \mid 1.489 \text{ g}}{\mid \text{cm}^3} = 596 \text{ g}$$

Let us next assume that chloroform and water are immiscible. (How to take into account the partial miscibility of the two solvents is discussed in Section 6.6b.) In drawing the flowchart for this problem, it is advisable to label the quantities of each component in each of the two output streams, rather than the total stream masses and component mass fractions. (Labeling in the latter manner results in having to solve four simultaneous equations in four unknowns.)

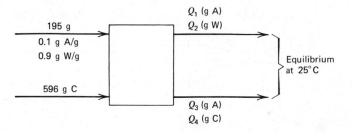

The distribution coefficient for the ACW system is given on p. 15-11 of *Perry's Chemical Engineers' Handbook*. If x denotes mass fraction of acetone,

$$K = \frac{(x)_{C \text{ phase}}}{(x)_{W \text{ phase}}} = \frac{Q_3/(Q_3 + Q_4)}{Q_1/(Q_1 + Q_2)} = 1.72 \qquad \textbf{(6.6-1)}$$

This provides one equation in the four unknowns $Q_1, Q_2, Q_3,$ and Q_4. The others are provided by material balances.

C Balance

$$596 \text{ g} = Q_4$$

W Balance

$$(0.900)(195 \text{ g}) = Q_2 \Rightarrow Q_2 = 175.5 \text{ g}$$

A Balance

$$(0.100)(195 \text{ g}) = Q_1 + Q_3$$

Substituting the known values of Q_2 and Q_4 into the first equation yields (with the acetone balance) two equations in two unknowns, which may be solved to obtain

$$Q_1 = 2.7 \text{ g A in water phase}$$
$$Q_3 = 16.8 \text{ g A in chloroform phase}$$

The percentage of acetone transferred is, therefore,

$$\frac{16.8 \text{ g acetone in chloroform phase}}{(0.100 \times 195) \text{ g acetone fed}} \times 100\% = \boxed{86.1\%}$$

In practice, extraction is often carried out in several consecutive stages, with the solution leaving each unit being contacted with additional solvent in the next unit. If enough stages are used, almost complete transfer of the solute can be achieved. Problem 74 at the end of this chapter illustrates this method of operation.

TEST YOURSELF

1. What is a distribution coefficient? What is liquid extraction?
2. The distribution coefficient for the system water–acetic acid–vinyl acetate is

$$\frac{\text{Mass fraction of acetic acid in vinyl acetate}}{\text{Mass fraction of acetic acid in water}} = 0.294$$

Is acetic acid more or less soluble in vinyl acetate than in water? If you use vinyl acetate to extract acetic acid from water, how would the relative masses of the two phases compare? ($m_{VA} \ll m_W$, $m_{VA} \approx m_W$ or $m_{VA} \gg m_A$?)

6.6b Phase Diagrams for Ternary Systems

The behavior of partially miscible ternary (three-component) systems is frequently represented on a **triangular phase diagram,** such as the one shown in Figure 6.6-1 for H_2O–MIBK–acetone at 25°C. Each apex of the triangle represents a single component and edges represent binary solutions. Edge b, for example, represents solutions of H_2O and acetone. Point K represents a mixture that is 20% MIBK, 65% acetone, and 15% H_2O. Any mixture falling in region A, such as that at point K, is a single-phase liquid, whereas any mixture whose overall composition falls in region B separates into two phases.

The lines shown within region B—called **tie lines**—connect phases in equilibrium with each other. For example, if MIBK, H_2O, and acetone are blended so that a mixture with overall composition at point M (55% H_2O, 15% acetone, 30% MIBK) results, the mixture separates into phases having compositions given by points L (85.3% H_2O, 11.9% acetone, 2.8% MIBK) and N (3.7% H_2O, 20.0% acetone, 76.3% MIBK). When a mixture does not fall on a tie line, interpolation is necessary to determine the composition of each phase.

The lever rule—derived in Section 6.5b for solid–liquid systems—also works for liquid–liquid systems. The ratio of the mass of the phase at point L to that at point N is thus $\overline{MN}/\overline{LM}$.

FIGURE 6.6-1

Triangular phase diagram for water–acetone–methyl isobutyl ketone at 25°C. (From D. F. Othmer, R. E. White, and E. Trueger, *Ind. Eng. Chem.* 33, 1240, 1941.) (Composition in wt%).

TEST YOURSELF

1. What is a tie line on a triangular phase diagram?
2. Show that a mixture with a composition of 4% acetone, 51% MIBK, and 45% H_2O separates into two phases. What is the composition of each phase? Use the lever rule to calculate the ratio of the mass of the MIBK-rich phase to the mass of the H_2O-rich phase.

EXAMPLE 6.6-2 Extraction of Acetone from Water: Use of the Phase Diagram

One thousand kilograms of a 30 wt% solution of acetone in water and a second stream of pure methyl isobutyl ketone are fed to a mixer. The mixture is then fed to a settler where two phases form and are withdrawn separately at 25°C. What

quantity of MIBK must be fed to the process to reduce the acetone concentration in the water-rich phase to 5 wt%, assuming that the fluids remain in the settler long enough for equilibrium to be achieved?

Solution

Since the two product streams are in equilibrium, their compositions must lie on the phase envelope and must be connected by a tie line. Accordingly, the composition of R is 5% acetone, 92.5% H_2O, and 2.5% MIBK and that of Q is 10% acetone, 87% MIBK, and 3% H_2O. Overall balances can now be used to determine Q, R, and S.

Mass Balance

$$S + 1000 \text{ kg} = Q + R$$

Acetone Balance

$$(0.30)(1000 \text{ kg}) = 0.10Q + 0.05R$$

H_2O Balance

$$(0.70)(1000 \text{ kg}) = 0.03Q + 0.925R$$

Solving these three equations yields

$$Q = 2665 \text{ kg}$$
$$R = 670 \text{ kg}$$

and

$$\boxed{S = 2335 \text{ kg MIBK}}$$

PROBLEMS

1. The normal boiling point of water is 100°C. Four liters of pure water vapor at 100°C and 0.80 atm are compressed isothermally. The pressure of the system is monitored, and the following behavior is observed:

(a) What is happening during each of steps AB, BC, and CD? What is the pressure corresponding to the horizontal portion of the curve?

(b) Estimate the volume occupied by the water at points B, C, and D. (Assume the vapor behaves like an ideal gas.)

2. One cubic centimeter of liquid chloroform ($CHCl_3$) is introduced into a 2-liter flask that had previously been evacuated. The flask and its contents are allowed to equilibrate at 25.9°C, at which temperature the vapor pressure of chloroform is 200 mm Hg.

(a) How many gram-moles of chloroform does the flask contain?

(b) Is there any liquid left in the flask after equilibrium is achieved? What is the final pressure in the flask?

3. The vapor pressure of methanol is 100 mm Hg at 21.2°C, and 760 mm Hg at 64.7°C. Estimate the vapor pressure of methanol at 40°C using (i) the Antoine equation; (ii) the Clausius-Clapeyron equation and the two given data points; (iii) linear interpolation between the two given points. Taking the first estimate to be correct, calculate the percentage error associated with the second and third estimates.

4. The vapor pressures of *n*-heptane at several temperatures are given here.

T(°C)	30.6	41.8	58.7	78.0	98.4	124.8
p^*(mm Hg)	60	100	200	400	760	1520

Use a semilog plot based on the Clausius-Clapeyron equation to derive a formula for $p^*(T)$, where p^* is in mm Hg and T is in °C, and to estimate the heat of vaporization (kJ/mol) of *n*-heptane. (*Remember:* Absolute temperatures must be used in the Clausius-Clapeyron equation.)

***5.** You are given vapor pressure data in the form of (T, p^*) pairs, where T is in °C and p^* is in mm Hg. Write a computer program to perform the following tasks.

(a) Read in N, the number of (T, p^*) pairs in the data set.

(b) Read in $T_1, p_1^*, T_2, p_2^*, \ldots, T_N, p_N^*$, and echo (print out) the data in a neat tabular format.

(c) Fit the Clausius-Clapeyron equation to the data using the method of least squares (Appendix A.1). That is, find a and b in the formula $y = ax + b$, where $y = \ln p^*$, and $x = 1/(T + 273.2)$. Print out the values of a and b.

* Computer problem.

(d) Read in an arbitrary number of temperatures (°C) and another arbitrary number of vapor pressures (mm Hg). Use the formula derived in part (c) to estimate the vapor pressure corresponding to each given temperature, and then to estimate the boiling point corresponding to each given pressure. Print out each calculated (T, p^*) pair.

Test your program by fitting the data for *n*-heptane given in Problem 4 and estimating the vapor pressures of this substance at 50°C, 80°C, and 110°C, and the boiling points at 760 mm Hg and 2000 mm Hg. Then indicate which of the last two values you would have the least confidence in, and why.

6. The apparatus shown here is used to measure the vapor pressure of chlorobenzene.

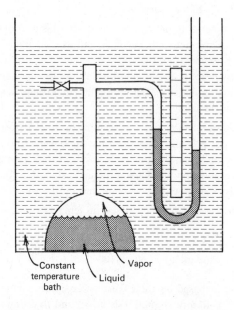

Constant temperature bath

Vapor

Liquid

A flask containing pure chlorobenzene is placed in a constant temperature bath, and the pressure of the vapor is measured with a mercury manometer. This process is repeated at several bath temperatures. The following readings are taken on a day when the atmospheric pressure is 755 mm Hg.

$T(°C)$	Mercury Level, Left Arm (mm)	Mercury Level Right Arm (mm)
35.3	767	32
58.3	747	52
89.4	677	122
110.0	577	222
160.2	17	782

(a) Calculate p^* for chlorobenzene at each temperature.

(b) Construct a semilog plot of p^* versus $1/T$, and use it to estimate the normal boiling point and the heat of vaporization of chlorobenzene.

(c) Does the Clausius-Clapeyron equation appear to be justified for chlorobenzene in the temperature range 35.3–160.2°C? Explain.

7. Estimate the vapor pressure of acetone in mm Hg at $T = 25°C$

(a) From data in *Perry's Chemical Engineers' Handbook*.

(b) Using the Cox chart (Figure 6.1-4).

(c) Using the Antoine equation.

8. Liquid methyl ethyl ketone (MEK) is introduced into a chamber containing nitrogen at 45°C. The pressure rises, and then levels off at 1060 mm Hg with liquid still present. For the system at equilibrium, calculate:

(a) The number of degrees of freedom according to the phase rule.

(b) The partial pressures and mole fractions of each component in the gas (see Example 6.3-1).

(c) (g MEK vapor/liter gas).

9. The vapor pressure of an organic solvent is 50 mm Hg at 20°C and 200 mm Hg at 45°C. Some of this solvent is contained in a closed flask at 60°C. Liquid is present in the flask. The volume of the gas above the liquid is 120 cm^3, and the pressure of the gas is 2.1 atm. How many moles of the solvent are contained in the gas phase? (Use what you know about how vapor pressure depends on temperature.)

10. Pure diethylamine (DEA) is contained in a flask attached to an open-ended mercury manometer. When the flask contents are at 21.0°C, the height of mercury in the arm of the manometer connected to the flask is 697 mm and that in the arm open to the atmosphere is 130 mm. At 38.0°C, the mercury level in the flask arm is 597 mm Hg and that in the other arm is 230 mm Hg. Atmospheric pressure is 767 mm Hg.

(a) Extrapolate the data to determine the vapor pressure of DEA at 45°C, using the Clausius-Clapeyron equation as the basis of the extrapolation.

(b) Air saturated with DEA at 45°C and 1 atm is cooled to 21°C at constant pressure. Calculate the percentage of the DEA originally in the vapor phase that condenses. (See Example 6.3-2.)

11. Calculate the density (g/liter) of humid air at 75°C, 1.00 atm, and a relative humidity of 60%.

12. The latest weather report includes the following data. "The temperature is 31°C. Barometric pressure is 75.4 cm Hg. The relative humidity is 98%." Calculate the mole fraction of water, dew point, molal humidity, absolute humidity, and percentage humidity of the air.

13. Air containing 20 mole% water vapor is cooled isobarically at 1 atm from 200°C to 15°C.

(a) Calculate the dew point and degrees of superheat of the feed air.

(b) How much water condenses per cubic meter of feed gas? (See Example 6.3-2.)

(c) Suppose a sample of the air at 200°C is put in a closed chamber containing a mirror and the pressure is raised at constant temperature until a mist forms on the mirror. At what pressure (atm) would the mist form? (Assume ideal gas behavior.)

14. Air at 90°C and 1.00 atm (absolute) contains 10.0 mole% water. A continuous stream of this air enters a compressor-condenser, in which the temperature is lowered to 15.6°C and the pressure is raised to 3.00 atm. The air leaving the condenser is then heated isobarically to 100°C. Calculate the fraction of water in the air that is condensed, the relative humidity of the air at 100°C, and the ratio m³ outlet air @ 100°C/m³ feed air @ 90°C.

15. Dry air is bubbled through 40 liters of water at a rate of 25 liters (STP)/min. The exit gas is saturated with water at 25°C. The total pressure in the gas phase is 1.00 atm. How long will it take for all of the water to be carried off?

16. Humid air at 323 K and 102 kPa with 5 K of superheat is fed to a condenser at a rate of 150 kmol/h. Gas and liquid streams leave the condenser in equilibrium at 283 K. Calculate how long it would take for the condensate to fill a 300-liter drum.

17. An organic vapor is to be recovered from a gas by condensation. The partial pressure of this vapor in the gas mixture entering the condenser is 1900 mm Hg, and its partial pressure in the outlet stream is 1000 mm Hg. The total pressure in the system is 2000 mm Hg, and the temperature at the outlet is 200°C. Calculate the m³ of gas leaving the condenser per 100 kmol of condensate.

18. A process stream containing acetone vapor in nitrogen enters a cooler-compressor at 65°C and 1.00 atm, with the relative saturation of acetone equal to 50%. The gas is compressed and cooled to 20°C to recover 95% of the acetone as a liquid. Calculate the outlet pressure and the ratio (liters condensate/m³ gas fed).

19. An adult takes roughly 12 breaths each minute, inhaling approximately 500 cm³ with each breath. The exhaled air is saturated with water vapor at body temperature, 37°C. Oxygen and carbon dioxide are exchanged in the lungs. The amount of nitrogen exhaled equals the amount inhaled, and the mole fraction of nitrogen in the expired air is 0.75. Estimate the mass of water that a body loses by breathing during a 24-h period if the temperature of the inhaled air is 23°C and the relative humidity of this air is 30%.

20. A gas stream containing equimolar quantities of a solvent vapor (S) and helium leaves a chemical plant at 90°C and 1.0 atm. The solvent is expensive, and you have been asked to develop a process for its recovery. You propose cooling the gas stream, condensing a portion of the solvent, and discharging the remaining gas. Two cooling fluids are available: cooling tower water at 20°C, and a refrigerant at −35°C. The vapor pressure of the solvent is approximately equal to that of benzene at the same temperature.

 (a) Calculate the percentage recoveries of solvent obtainable when each of the two coolants is used, assuming the final condenser temperature equals the coolant temperature.

 (b) What more would you have to know to be able to decide which coolant to use?

21. On a hot summer day the temperature is 36°C, barometric pressure is 763 mm Hg, and the relative humidity is 96%. An air conditioner draws in outside air, cools it to 18°C, and delivers it at a rate of 12 500 liters/h. Calculate the rate of moisture condensation (kg/h), and the volumetric flow rate of the air drawn from the outside.

22. An air conditioner is designed to bring 10 000 ft³/min of outside air at 90°F, 29.8 in. Hg, and 98% relative humidity to 40°F, thereby condensing a portion of the water vapor,

and then to reheat the air, releasing it into a room at 60°F. Calculate the rate of condensation (gallons H_2O/min), and the volumetric flow rate of the air delivered to the room. (*Suggestion:* On the flowchart, treat the cooling-condensation and the reheating as separate process stages.)

23. The air in a building is to be maintained at 25°C and a relative humidity of 45%. This is accomplished by passing the air through a water spray. The air enters the spray chamber at 32°C and 77% relative humidity, leaves the chamber cooled and saturated with water vapor, and is then reheated to 25°C. Calculate the temperature of the air leaving the spray chamber and the kilograms of water added to or removed from (specify which) each kilogram of dry air processed.

24. A hygrometer is to be used to measure the amount of water in humid air. The instrument calibration curve should be a straight line on a plot of y—the mole fraction of water in the air—on a logarithmic scale versus H—the hygrometer reading—on a linear scale.

 A sample of room air is charged into the hygrometer sample chamber on a day when the temperature is 22°C, the barometric pressure is 1.00 atm, and the relative humidity is 40%. The resulting meter reading is $H = 5.0$. A second measurement is then made by heating liquid water to 50°C in a sealed flask containing air, allowing the system to equilibrate at a pressure of 839 mm Hg with liquid still present, drawing a sample of the air above the liquid, and injecting it into the hygrometer chamber (which is heated to prevent condensation). The meter reading in this case is $H = 48$.

 (a) Determine the equation for y as a function of H.

 (b) Suppose you wish to condition air at 35°C and 1 atm to produce air at 22°C, 1 atm, and 40% relative humidity. The air conditioner first cools the air, condensing the necessary amount of water, and then reheats the remaining air to 22°C. A sample of the outside air is injected into the hygrometer chamber, and the resulting reading is $H = 30$. Calculate the temperature to which the air must be cooled before it is reheated and the kilograms of water condensed per cubic meter of conditioned air delivered.

25. A liquid stream consisting of 20 mole% butane and the balance a heavy, nonvolatile hydrocarbon is fed to the top of a stripping column, where it is contacted with an upward-flowing air stream. Butane comes out of solution and is carried along with the air stream. The combined air–butane stream leaves the column at 80°F and 1 atm. The liquid stream emerging from the bottom of the column contains all of the heavy hydrocarbon and 2% of the butane in the entering liquid stream. Air can be assumed insoluble in the hydrocarbon.

 For the process to remove the specified amount of butane from the liquid, a minimum air-to-liquid feed ratio is required. When the minimum air flow is fed, the gas stream leaving the top of the stripper is in equilibrium with the entering liquid stream. (This condition can be approached but never exactly achieved, since it would require an infinitely long column.) Use Raoult's law and assume that the hydrocarbon vapor pressure is effectively zero to perform the following calculations.

 (a) Determine the butane mole fraction in the exit gas.

 (b) Calculate the minimum air-to-liquid molar feed ratio.

(c) Calculate the composition of the exiting gas stream if the feed ratio is 1.7 times the minimum, assuming that the same percentage of the entering butane is stripped from the liquid.

(d) Increasing the air flow rate for a given butane recovery decreases the cost of the process in one way and increases it in another. Explain.

26. A gas stream containing 35 mole% hydrogen, 30% carbon monoxide, 30% carbon dioxide, and 5% methane flowing at a rate of 133 m³/min is cooled from 1273 K to 275 K at a constant absolute pressure of 35.0 atm. The gas leaving the cooler is fed to an absorber where it is contacted with refrigerated liquid methyl alcohol, which absorbs essentially all of the CO_2, 98% of the methane, and none of the other feed gas components. The methanol is fed to the absorber at a molar flow rate 1.2 times that of the inlet gas. The gas leaving the absorber leaves saturated with methanol vapor at 261 K and passes into a cross-country pipeline.

(a) Calculate the volumetric flow rate of methanol entering the absorber (m³/min) and the molar flow rate of methanol in the gas leaving the absorber. Do not assume ideal gas behavior when doing PVT calculations.

(b) What is the likely intended use of the product gas? Why is it desirable to remove the CO_2 from the gas prior to feeding it into the pipeline?

27. Recovery of a solvent vapor from a gas stream by condensation can be achieved by cooling the gas, by compressing it, or by a combination of these operations. The greater the compression, the less cooling is needed.

(a) Suppose a gas at pressure P_0 and temperature T_0 containing a single condensable vapor is the feed to a recovery process, that the dew point of this gas is T_{do}, that a fraction f of the vapor is to be condensed, and that the vapor pressure $p^*(T)$ may be expressed as a function of temperature by the Antoine equation. Assume a basis of 1 mol of the feed gas, draw and label a flowchart, and derive the following relation for the final condenser pressure in terms of the final temperature T_f and the specified feed conditions and fractional solvent recovery:

$$P_f = \frac{p^*(T_f)[1 - fp^*(T_{do})/P_0]}{(1 - f)p^*(T_{do})/P_0}$$

*(b) The cost of refrigeration equipment can be estimated using the empirical formula

$$C_t(\$) = 2000 + 27(\Delta T)^2$$

where $\Delta T = T_0 - T_f$, and the cost of the compressor can be estimated as

$$C_p(\$) = 4500 + 5.58(\Delta P)$$

where $\Delta P = P_f - P_0$.** Write a program to determine the optimal final temperature T_f and final pressure P_f; that is, the values of T_f and P_f that correspond to a minimum total equipment cost. Your program should carry out the following steps.

* Computer problem.

** These formulas are fictitious. Real cost estimation formulas can be found in the references on p. 25-5 of *Perry's Chemical Engineers' Handbook.*

(i) Read in values of feed parameters T_0, P_0, and T_{d0}, the desired fractional recovery f, and the Antoine equation constants A, B, and C.

(ii) As an initial guess, set the value of T_f equal to T_0, calculate P_f from the equation of part (a), and determine the total cost corresponding to these conditions.

(iii) Repeat step (ii) at 2°-decrements in temperature, continuing until the cost rises from one step to the next. Print out the temperature and total cost at each step.

(iv) Print out the final temperature, final pressure, and total equipment cost at the point where the cost is at its minimum value.

(c) Test your program by calculating the optimal final temperature and pressure for a 95% recovery of ethylbenzene from a stream containing this vapor in nitrogen at 50°C and 765 mm Hg, with a dew point of 40°C. The Antoine equation constants for ethylbenzene are listed in Table 6.1-1.

(d) One way to refine the estimate of the optimum would be to use a smaller temperature interval in the vicinity of the point obtained using 2°-decrements. Suggest another method that would use only costs already evaluated. (*Hint:* Use the costs evaluated at three temperatures.)

28. Twenty-four kilograms per minute of wet wood pulp containing 0.5 kg H_2O/kg dry pulp pass through a hot air dryer. Air containing 0.1% water by volume enters the dryer at 95°C and 52 cm Hg (gauge) at a metered flow rate of 162 m^3/min and leaves the dryer at 83°C and 5.2 cm Hg (gauge). A sample of the exiting air is drawn into a chamber containing a mirror, and the chamber is cooled slowly, keeping the gauge pressure at 5.2 cm Hg. A mist is observed to form on the mirror at a temperature of 30.6°C. Calculate the mass (kg) of water removed from the pulp each minute, the volumetric flow rate of the exiting air, and the weight percent of water in the exiting pulp.

29. Wet leather enters a continuous tunnel dryer with a moisture content of 61% and leaves with a moisture content of 6% at a rate of 485 lb_m/h. Dry air at 140°F and 1 atm enters the dryer, and air and evaporated water leave the dryer at 130°F and 1 atm. Calculate the rate at which wet leather enters the dryer and the volumetric flow rate of the inlet air if the outlet air has a relative humidity of 50%.

30. Soybeans from which oil has been extracted into a hexane solvent leave the extraction unit containing 0.80 kg hexane/kg soybean solids. The solids are to be dried by being contacted with nitrogen, which enters the dryer at 90°C. The solids leaving the dryer should contain 0.05 kg hexane/kg solids. The gas leaving the dryer exits at 80°C and 1.0 atm with a relative saturation of 60%, and is fed to a condenser in which it is cooled to 25°C and compressed to 4.0 atm, recovering part of the vaporized hexane as a condensate. Calculate the fractional recovery of hexane (kg condensed/kg entering with the soybeans).

31. *n*-Hexane is burned with excess air. An analysis of the product gas yields the following *dry-basis* molar composition: N_2—82.1%, CO_2—6.9%, CO—2.1%, O_2—8.6%, C_6H_{14}—0.265%. The stack gas emerges at 500°C and 1.00 atm. Calculate the percentage conversion of hexane, the percentage excess air fed to the reactor, and the dew point of the stack gas.

32. A fuel gas containing methane and ethane flows to a furnace at a rate of 250 liters/min at 303 K and 1.04 bars (absolute) and is burned with excess air. The stack gas flows at a rate of 9250 liters/min at 573 K and 1.01 bars (absolute) and contains CO_2 at a partial pressure of 48.3 mm Hg and no CO or unburned hydrocarbons. Calculate the mole fraction of methane in the fuel and the dew point of the stack gas.

33. A mixture of propane and butane is burned with air. A partial analysis of the stack gas yields the following dry basis volume percentages.

$$C_3H_8 = 0.0527\%$$
$$C_4H_{10} = 0.0527\%$$
$$CO = 1.48\%$$
$$CO_2 = 7.12\%$$

The absolute pressure of the stack gas is 776 mm Hg, and the dew point of the gas is 46.5°C. Calculate the molar composition of the fuel.

34. An important parameter in the design of gas absorbers is the ratio of the flow rate of the feed liquid to that of the feed gas. The lower the value of this ratio, the lower the cost of solvent required to process a given quantity of gas.

 In the absorber shown here, propane is recovered from a 5 mole% propane–95% air mixture by contacting the mixture with liquid n-decane. A negligible amount of decane is vaporized in the process, and 98.1% of the propane entering the unit is absorbed.

(a) The minimum value of the liquid-to-gas feed ratio L_1/G_2 corresponds to the two bottom streams being in equilibrium. (This condition requires an infinitely large absorber, so it cannot be achieved in practice.) Calculate this minimum ratio, using Raoult's law to relate the compositions of the entering gas and the exiting liquid.

(b) Calculate the mole fraction of propane in the exiting liquid if $L_1/G_2 = 1.5(L_1/G_2)_{min}$.

35. A stream of outside air at temperature $T_0(°C)$, P_0(mm Hg), and relative humidity h_{r0} is to be conditioned at constant temperature and pressure to a relative humidity of h_{rf}. A portion of the air is passed through a bed of silica gel particles, and a fraction z_r of

the moisture in the air is removed by adsorption onto the particles. The remaining air bypasses the bed. The bypass stream and the stream leaving the bed are combined to obtain the conditioned air stream.

(a) Suppose 45 m^3/min of outside air at $T_0 = 300$ K, $P_0 = 760$ mm Hg, and $h_{r0} = 85\%$ are to be conditioned to a relative humidity $h_{rf} = 30\%$, and the silica gel removes 95% of the moisture in the air fed to it. Calculate z_b, the fraction of the feed air that bypasses the silica gel bed.

(b) For the general case, derive a formula for the bypass fraction z_b (moles air bypassed/ mol air fed) in terms of some or all of the variables T_0(K), P_0(mm Hg), h_{r0}(%), h_{rf}(%), z_r (mol water adsorbed/mol water entering bed), and $p^*(T_0)$ (mm Hg), the vapor pressure of water at T_0. Verify your solution to part (a) using this formula.

(c) In practice, even if the temperature and humidity of the outside air remain constant, the fraction of air bypassed would have to decrease with time to keep the relative humidity of the conditioned air constant, and after a long enough period of operation, the process would simply not work. Explain.

36. Nitric acid is used extensively for the production of inorganic and organic nitrates, for metal treatments of various kinds, and for photoengraving. It is produced by oxidizing ammonia to nitric oxide over a platinum-rhodium catalyst, oxidizing the nitric oxide to nitrogen dioxide, and dissolving the NO_2 in water.

$$4 NH_3(g) + 5 O_2(g) \rightarrow 4 NO(g) + 6 H_2O(g)$$
$$2 NO(g) + O_2(g) \rightarrow 2 NO_2(g)$$
$$3 NO_2(g) + H_2O(1) \rightarrow 2 HNO_3(aq) + NO(g)$$

A side reaction that lowers the product yield is the oxidation of ammonia to nitrogen and water vapor.

$$4 NH_3(g) + 3 O_2(g) \rightarrow 2 N_2(g) + 6 H_2O(g)$$

Saturated ammonia vapor, produced by vaporizing pure liquid ammonia at 780 kPa (absolute), is mixed with air, and the combined stream enters a converter. Prior to being mixed with the ammonia, the air is compressed and passed through a preheater. It enters the compressor at 30°C and 1 atm with a dew point of 25°C, and it exchanges heat in the preheater with the gases emerging from the converter. The quantity of oxygen in the feed is the amount theoretically required to convert all of the ammonia to HNO_3.

In the converter, the ammonia reacts completely, with 97% forming NO and the balance forming N_2. In the short time in which the reaction mixture is in the presence of the catalyst (less than 0.001 s), a negligible amount of NO_2 is formed. The product gas is subjected to a series of cooling and hydration steps, in which the NO is completely oxidized to NO_2, which in turn combines with water (some of which is present in the product gas, the rest of which is added) to form a 62 wt% aqueous nitric acid solution. The NO formed in the latter reaction is reoxidized and the added NO_2 is hydrated to form still more HNO_3. The product gas from the process may be taken to contain only N_2 and O_2. A simplified flowchart of the process follows.

(a) Taking a basis of 100 mol of ammonia fed to the process, calculate (i) the volumetric flows (m^3) of the ammonia vapor and of the air fed to the compressor intake, using the compressibility factor equation of state for the ammonia calculation; (ii) the moles and molar composition of the gas leaving the converter; (iii) the required feed of liquid water (m^3) to the absorber.

(b) Scale up the results calculated in part (a) to a new basis of 1000 metric tons of 62% nitric acid solution produced.

37. A dry gas containing 14% NH_3 by volume is contacted with water at 32°F in a continuous mixer. The effluent liquid and gas streams are in equilibrium with each other. The effluent liquid, whose specific gravity is 0.960, is titrated with acid and is found to contain 5.6 mol dissolved ammonia per liter of solution. The outlet gas pressure is 1 atm. Using tabulated data from *Perry's Chemical Engineers' Handbook* (pp. 3-72 and 3-73) for partial pressures of water and ammonia over aqueous ammonia solutions, calculate the percentage of the ammonia in the feed that is removed in the scrubber.

38. Sulfur trioxide (SO_3) dissolves in and reacts with water to form an aqueous solution of sulfuric acid (H_2SO_4). The vapor in equilibrium with such a solution contains both SO_3 and H_2O. If enough SO_3 is added, all of the water reacts and the solution becomes pure H_2SO_4. If still more SO_3 is added, it dissolves to form a solution of SO_3 in H_2SO_4, called **oleum** or **fuming sulfuric acid.** The vapor in equilibrium with oleum is pure SO_3. A 20% oleum by definition contains 20 kg of dissolved SO_3 and 80 kg H_2SO_4 per hundred kilograms of solution. Alternatively, the oleum composition can be expressed as %SO_3, with the constituents of the oleum considered to be SO_3 and H_2O.

(a) Prove that a 25% oleum contains 86.2% SO_3.

(b) Suppose a gas stream at 40°C and 1.2 atm containing 90% SO_3 by volume contacts a liquid stream of 98 wt% H_2SO_4(aq), producing 25% oleum at the tower outlet. Tabulated equilibrium data indicate that the partial pressure of SO_3 in equilibrium with this oleum is 4.95 mm Hg. Calculate (i) the mole fraction of SO_3 in the outlet gas if this gas is in equilibrium with the liquid product at 40°C and 1 atm, and (ii) the ratio (m^3 gas feed)/(kg liquid feed).

39. A gas containing nitrogen, benzene, and toluene is in equilibrium with a 30 mole% benzene, 70 mole% toluene liquid mixture at 75°C and 10 atm. Calculate the gas phase composition (mole fractions) using Raoult's law.

40. Using Raoult's law or Henry's law for each substance (whichever one you think appropriate), calculate the gas phase composition (mole fractions) and pressure of a solution containing 0.1 mole% N_2 and 99.9 mole% water in equilibrium with N_2 gas and water vapor at 90°C.

41. The **solubility coefficient** of a gas is defined as the number of cubic centimeters (STP) of the gas that dissolves in 1 cm^3 of a solvent under a partial pressure of 1 atm. The solubility coefficient of CO_2 in water at 20°C is 0.0901.

 (a) Calculate the Henry's law constant in atm/mole fraction for CO_2 in H_2O at 20°C from the given solubility coefficient.

 (b) How many grams of CO_2 can be dissolved in a 12-oz bottle of soda at 20°C if the gas above the soda is pure CO_2 at a gauge pressure of 2.5 atm (1 liter = 33.8 fluid ounces). Assume the liquid properties are those of water.

 (c) What volume would the dissolved CO_2 occupy if it were released from solution at body temperature and pressure—37°C and 1 atm.

42. A gas containing 1000 ppm (parts per million by volume) SO_2 in dry air is being used to calibrate a stack gas SO_2 analyzer. The flow rate of the gas is 2.00 liters/min at 20°C and 10.0 mm Hg (gauge). Before the gas is vented to the atmosphere, it is passed through a scrubber containing 120 liters of water, which absorbs most of the SO_2. The quantities of SO_2 in the water and the gas leaving the scrubber may be assumed to be related by Henry's law, and the exiting gas may be taken to be saturated with water at 20°C and 1 atm. The fractional amount of SO_2 in the exhaust gas should not exceed 100 ppm. The scrubbing solution must therefore be changed whenever the concentration of dissolved SO_2 reaches the value that corresponds to 100 ppm SO_2 in the gas.

 (a) *Perry's Chemical Engineers' Handbook*, p. 3-65, gives the following data for aqueous solutions of SO_2 at 20°C.

r (g SO_2/100 g H_2O)	p_{SO_2} (mm Hg)
0.0	0
0.5	29
1.0	59
1.5	90

 Use these data to estimate the Henry's law constant for SO_2 in water at 20°C, in units of (mm Hg)/(mole fraction).

 (b) Use the result of part (a) to calculate the moles of SO_2 dissolved when the solution must be changed, and the molarity of the solution at this point (mol SO_2/liter).

 (c) After how many hours of operation must the solution be changed? (For simplicity, neglect the change in liquid volume due to evaporation of water.)

 (d) Suggest changes (in both scrubbing conditions and the scrubbing solution) that might lead to an increased removal of SO_2 from the feed gas.

43. Nitrogen is bubbled through a container holding a liquid mixture of acetone and ethanol at 10°C and 1 atm. The dry air flow rate is 10 liters (STP)/min and the exiting gas is saturated with both acetone and ethanol vapors.

 (a) Suppose the initial liquid composition is 50 mole% acetone and 50 mole% ethanol. What is the composition of the first gas to emerge from the liquid?

 (b) What will happen to the exiting gas composition as the process continues?

44. A liquid mixture contains 50% benzene and 50% toluene by mass.

(a) Using Raoult's law, calculate the total pressure and mole fractions of each substance in a gas phase containing only benzene and toluene vapor in equilibrium with the given liquid mixture at 60°C.

(b) Prove that the bubble point of the liquid mixture at $P = 0.715$ atm is 80°C.

(c) Briefly describe what would happen if the mixture were heated slowly, eventually reaching a temperature of 80°C.

(d) Suppose you had not been given the bubble point in part (b), but had been asked to determine it. How would you have done so?

45. Calculate

(a) The bubble point temperature and equilibrium vapor composition associated with an equimolar mixture of liquid n-hexane and n-heptane at 1.0 atm.

(b) The dew point temperature and equilibrium liquid composition associated with a gas mixture containing 30% n-hexane, 30% n-heptane, and 40% air at 1.0 atm.

46. A liquid mixture containing 30.0 mole% benzene and 70.0 mole% styrene is to be separated in a distillation column. The column produces an overhead product (distillate) and a bottoms product. The bottoms product is 99 mole% styrene and contains 2.0% of the benzene fed to the column.

The liquid stream leaving the bottom of the column (not the bottoms product) goes to a partial reboiler, in which a portion of it is vaporized at 150°C and returned to the bottom of the column. The residual liquid from the reboiler is the bottoms product. The vapor and liquid streams exiting the reboiler are in equilibrium. The **boilup ratio,** or mole ratio of the vapor and liquid streams leaving the reboiler, is 2.5:1. Using Raoult's law where appropriate, calculate the compositions (component mole fractions) of the distillate product, the vapor returned to the column from the reboiler, and the liquid feed to the reboiler, and estimate the required operating pressure of the reboiler.

47. In distillation, the vapor leaving the upper portion of the distillation column goes to a condenser, where some or all of it is condensed. If some is condensed, the condensate is returned to the top of the column as **reflux,** and the uncondensed vapor, which is in equilibrium with the reflux, is taken off as the overhead product. If all of the vapor is condensed, a fraction of the condensate is returned to the column and the remainder constitutes the overhead product.

Suppose the overhead product from a methanol–water distillation column (but not necessarily the vapor leaving the column) is 90 mole% methanol, and that the lowest attainable temperature in the condenser is 40°C.

(a) Using Raoult's law, estimate the minimum pressure at which the condenser can operate as a partial condenser, and the minimum pressure at which it can operate as a total condenser. In terms of dew points and bubble points, what does 40°C represent for each of these pressures?

(b) Suppose the condenser operates as a total condenser at 40°C, the production rate of overhead product is 100 mol/s, and the mole ratio of overhead product to reflux is 1.5:1. Calculate the molar flow rates and compositions of the reflux stream and the vapor feed to the condenser.

(c) Now suppose that the condenser operates as a partial condenser at 40°C and that the overhead product rate and reflux ratio have the values given in part (b). Calcu-

late the operating pressure of the condenser and the molar flow rates and composi-
tions of the reflux stream and the vapor feed to the condenser.

*48. Write a computer program to estimate the dew point of a mixture of up to nine com-
ponents. Have the program carry out the following steps.

(a) Read in the total pressure $P(mm\ Hg)$, the number of condensable substances N;
and for each such substance, the gas phase mole fraction y and the Antoine equa-
tion constants A_i, B_i, and C_i (Table 6.1-1).

(b) Read in an initial guess for the dew point, T.

(c) Calculate

$$S = 1 - P \sum_i y_i/p_i^*(T)$$

using the Antoine equation to evaluate p_i^*. The goal is to find T such that $S = 0$
(see Equation 6.4-6). If S is positive, decrease T by 2°C and repeat the calculation,
and if S is negative increase T by 2°C and repeat the calculation. Keep changing
T in 2°C steps until the sign of S changes. The dew point must lie between the
last two temperatures. Build a provision into your program that will print an error
message and stop execution if the temperature falls below 0°C or above 160°C
without the sign of S changing.

(d) Estimate the dew point by linear interpolation between the last two points. (*Sugges-
tion:* Let $S_a(T_a)$ and $S_b(T_b)$ be these two points. Derive a formula for the T-intercept
of a line through $S_a(T_a)$ and $S_b(T_b)$ on a plot of S versus T, and use this formula
in your program.)

(e) Calculate the liquid-phase mole fraction x_i of each condensable component.

Test your program by calculating the dew point and condensate composition of a gas
mixture at 760 mm Hg containing 11.64 mole% benzene, 3.40% ethylbenzene, 8.39%
toluene, and the balance noncondensable gases.

49. The **relative volatility** of two components is a measure of the ease with which these
components can be separated by distillation. It is defined by the equation

$$\alpha_{i,j} = \frac{y_i/x_i}{y_j/x_j}$$

where $\alpha_{i,j}$ is the volatility of component i with reference to component j, and x and y are
liquid and vapor mole fractions, respectively.

(a) Show that the relative volatility of two components in a system that follows
Raoult's law is equal to the ratio of vapor pressures of the two components.

(b) The vapor pressures of benzene (B) and toluene (T) at 80°C are 757.7 and 291.2
mm Hg, respectively. Assuming that these compounds form ideal liquid and vapor
mixtures, and that their relative volatility is constant, evaluate equilibrium vapor
mole fractions at 80°C corresponding to $x_B = 0$, 0.2, 0.4, 0.6, 0.8 and 1.0.

50. Use Raoult's law to estimate the bubble point temperature (°F) of a liquid at 45 psia con-
taining 20 mole% *n*-butane, 50% *n*-pentane, and 30% *n*-hexane. Then obtain a better

* Computer problem.

estimate using the nomograph of distribution coefficients in Figure 13-14b on p. 13-18 of *Perry's Chemical Engineers' Handbook.* Use the latter result to estimate the composition of the vapor phase in equilibrium with the given liquid.

51. A **stage** of a separation process is defined as an operation in which components of one or more feed streams divide themselves between two phases, and the phases are taken off separately. In an **ideal stage** or **equilibrium stage,** the two effluent streams are in equilibrium with each other. Distillation columns often consist of a series of vertically distributed stages. Vapor flows upward and liquid flows downward between adjacent stages; some of the liquid fed to each stage vaporizes, and some of the vapor fed to each stage condenses. A portion of the top section of a distillation column is shown here.

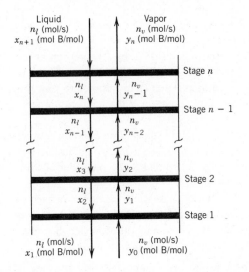

Consider a distillation column operating at 1 atm in which benzene and styrene are being separated. A vapor stream containing 65 mole% benzene and 35 mole% styrene enters the first stage shown at a rate of 200 mol/h, and liquid containing 55 mole% benzene and 45 mole% styrene leaves this stage at a rate of 150 mol/h. You may assume (1) the stages are ideal; (2) Raoult's law can be used to relate the compositions of the streams leaving each stage; (3) the vapor and liquid molar flow rates do not change by a significant amount from one stage to the next.

(a) How would you expect the mole fraction of benzene in the liquid at each stage to vary from one stage to another, beginning with the bottom stage and moving up the column? In light of your answer and considering that the pressure remains constant from one stage to another, how would you then expect the temperature to vary at progressively higher stages? Briefly explain.

(b) Estimate the temperature at the bottom stage and the compositions of the vapor stream leaving this stage and the liquid stream entering it. Then repeat these calculations for the second stage from the bottom.

(c) Describe how you would calculate the number of ideal stages required to reduce the styrene content of the vapor to less than 5 mole%.

52. Absorption of n-hexane by a heavy oil is an important step in the recovery of hexane from systems in which it is used as a solvent. In one such process, a gas stream containing 5 mole% hexane and 95 mole% nitrogen enters at the bottom of an absorber in which it is contacted with a nonvolatile oil. The ratio of oil fed to the system at the top of the column to gas fed at the bottom is 2 mol oil/1 mol gas. The absorber operates at an approximately constant temperature and pressure of 49°C and 1 atm. Of the hexane entering the absorber, 99.9% is absorbed by the oil.

The absorber consists of a series of ideal stages (see Problem 51) arranged so that gas flows upward from one stage to the next and liquid flows downward from one stage to the next. The liquid and gas streams leaving each stage are in equilibrium with each other (by the definition of an ideal stage), with compositions related by Raoult's law. Since absorption of all of the hexane would only change the gas molar flow rate by 5% and the liquid flow rate by 2.5%, the molar flow rates of both the gas and liquid streams entering and leaving each stage of the column may for simplicity be considered constant and equal to their average values in the column.

(a) Taking as a basis a gas feed rate to the column of 100 mol/s, determine the mole fractions of hexane in the gas and liquid streams leaving the absorber, and calculate the average value of the liquid molar flow rate in the column, n_l, and that of the gas, n_v.

(b) Estimate the mole fractions of hexane in the gas leaving the bottom stage of the column and in the liquid entering this stage.

(c) Suppose that x_i and y_i are the mole fractions of hexane in the liquid and gas streams leaving stage i. Derive the following formulas.

$$y_i = x_i p_i^*(T)/P \tag{1}$$

$$x_{i+1} = x_i + \frac{n_v}{n_l}(y_i - y_{i-1}) \tag{2}$$

Verify that these formulas yield the answers you calculated in part (b).

*(d) Write a computer program to determine the number of stages required to reduce the mole fraction of hexane to its required final value or less. The program should perform the following tasks.

 (i) Read in values of T, P, y_0, n_v and n_l, and R, the latter being the desired percentage recovery of hexane. ($R = 99.9\%$ in the problem statement.) Terminate if $T \le 0$; otherwise print the values read.

 (ii) Calculate $p_H^*(T)$, the vapor pressure of n-hexane at the absorber temperature; y_f, the mole fraction of hexane in the vapor leaving the absorber; and x_1, the mole fraction of hexane in the liquid leaving the absorber.

 (iii) Evaluate y_i and x_{i+1} for successive stages of the column from equations 1 and 2. For each stage, print the stage number, x_i, and y_i. Stop the calculations when y_i is less than or equal to y_f, and return to step i.

Run your program using the values of P, y_0, n_v, n_l, and R given in the problem statement and three temperatures: 50°C, 60°C, and 70°C.

* Computer problem.

53. A vapor mixture containing 50 mole% benzene and 50% toluene at 1 atm is cooled iso-barically in a closed container from an initial temperature of 150°C. Use the Txy diagram of Figure 6.4-1 to answer the following questions.

 (a) At what temperature does the first drop of condensate form, and what is its composition?

 (b) At one point during the process the system temperature is 95°C. Determine the mole fractions of benzene in the vapor and liquid phases at this point, and calculate the ratio (moles vapor/mole liquid).

 (c) At what temperature does the last drop of condensate form, and what is its composition?

54. Four gram-moles of benzene and six gram-moles of toluene are placed in a closed cylinder equipped with a piston. The cylinder is immersed in a boiling water bath, so that its temperature is maintained at 100°C. The force exerted on the piston can be varied to adjust the cylinder pressure to any desired value. The pressure is initially 1000 mm Hg and is gradually lowered to 600 mm Hg. Use the Pxy diagram of Figure 6.4-1 to convince yourself that the cylinder initially contains only liquid benzene and toluene, and to answer the following questions.

 (a) At what pressure does the first vapor bubble form, and what is its composition?

 (b) At what pressure does the last droplet of liquid evaporate, and what is its composition?

 (c) What are the liquid and vapor compositions in equilibrium with each other when the pressure is 800 mm Hg, and what is the ratio (moles vapor/mole liquid) at this point?

 (d) Estimate the volume of the cylinder contents when the pressure is (i) 1000 mm Hg; (ii) 800 mm Hg; (iii) 600 mm Hg.

55. Consider a binary mixture of two substances, A and B, with x_{Am} equalling the mass fraction of A. Suppose the mixture is placed in a closed container and is brought to a temperature T_0 and pressure P_0 such that vapor and liquid mixtures of A and B coexist at equilibrium, with x_{Av} and x_{Al} equalling the mass fractions of A in the vapor and liquid phases, respectively. The Txy diagram for the AB system is shown here.

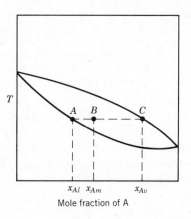

Mole fraction of A

Prove that the vapor-to-liquid mass ratio can be obtained graphically using the **lever rule:**

$$\frac{\text{Mass of vapor}}{\text{Mass of liquid}} = \frac{|AB|}{|BC|}$$

where $|AB|$ and $|BC|$ denote the lengths of the line segments on the Txy plot. (*Suggestion:* Begin by drawing and labeling a flowchart for a process in which the original mixture is split up into two product streams.)

56. Vapor–liquid equilibrium data for methanol–water mixtures at a pressure of 1 atm are given here.

$T(°C)$	x_M	y_M	$T(°C)$	x_M	y_M
100.0	0.000	0.000	75.3	0.400	0.729
96.4	0.020	0.134	73.1	0.500	0.779
93.5	0.040	0.230	71.2	0.600	0.825
91.2	0.060	0.304	69.3	0.700	0.870
89.3	0.080	0.365	67.6	0.800	0.915
87.7	0.100	0.418	66.0	0.900	0.958
84.4	0.150	0.517	65.0	0.950	0.979
81.7	0.200	0.579	64.5	1.000	1.000
78.0	0.300	0.665			

(a) Construct a Txy diagram for this system.

(b) A thermocouple inserted into a two-phase methanol–water mixture at equilibrium reads 70°C. The system pressure is 1 atm. What are the liquid and vapor compositions?

(c) An equimolar mixture of methanol and water is fed to an evacuated vaporizer and allowed to come to equilibrium at 80°C and 1 atm. What are the molar compositions of each phase of the system? What percentage of the final mixture is vapor?

(d) A liquid mixture containing 65 mole% methanol and 35% water is fed to an equilibrium flash tank operating at 1 atm. The system conditions are such that 20% of the entering feed is vaporized. Determine the compositions of the liquid and vapor streams leaving the tank, and the temperature at which the tank operates.

***57.** Your job entails preparing and using Txy diagrams for a variety of systems and conditions.

(a) Write a computer program to generate a Txy diagram of the type shown in Figure 6.4-1 for an ideal binary solution. Have the program read in the values of the Antoine equation constants A, B, and C for each substance and the pressure (atm) for which the diagram is to be generated. Then have the program evaluate the bubble point temperature at the given pressure for $x = 0, 0.05, 0.10, 0.15, \ldots, 0.95$, 1.0, where x is the mole fraction of the more volatile component in a binary liquid mixture. Each such evaluation will yield one point (T, x) on the liquid equilibrium

* Computer problem.

curve. At the same time, once the bubble point temperature has been determined, the mole fraction, y, of the more volatile component in the equilibrium vapor can also be calculated, yielding a point (T, y) on the vapor equilibrium curve. Use any trial-and-error method you choose for the bubble point calculation, such as the regula-falsi method outlined in Appendix A.2.

(b) Test your program by generating and drawing the Txy diagram for chloroform and benzene at $P = 1.00$ atm. If possible, have the computer draw the graph.

(c) The following vapor–liquid equilibrium data have been obtained for mixtures of chloroform (C) and benzene (B) at 1 atm.

$T(°C)$	80.6	79.8	79.0	77.3	75.3	71.9	68.9	61.4
x_C	0	0.08	0.15	0.29	0.44	0.66	0.79	1.00
y_C	0	0.10	0.20	0.40	0.60	0.80	0.90	1.00

Plot these data on the graph generated in part (b). Estimate the percentage errors in the Raoult's law values of the bubble point and y_C corresponding to $x_C = 0.44$, taking the tabulated values to be correct. (*Caution:* Percentage errors should be based on absolute temperatures.)

58. A liquid mixture containing 40 mole% methanol and 60% ethanol is placed in an open vessel and heated slowly. Estimate the temperature at which the mixture begins to boil. If heat is continuously supplied, how will the liquid temperature and composition change with time?

59. Acetaldehyde is produced by the catalytic dehydrogenation of ethanol.

$$C_2H_5OH(g) \rightarrow CH_3CHO(g) + H_2(g)$$

A fresh feed (pure ethanol) and a recycle stream (95 mole% ethanol and 5 mole% acetaldehyde) combine and enter a reactor at 280°C. The exit gases are cooled to $-40°C$ to condense the acetaldehyde and unreacted ethanol. The gas leaving the condenser is scrubbed to remove the remaining organics, and hydrogen is sold as a valuable reaction byproduct. The condensate, which is 50 mole% ethanol, passes to a distillation column. The distillate, which contains 96 mole% acetaldehyde, is sold for use as is or for additional refining, and the bottom stream is recycled to the reactor. The pressure of each gas stream may be taken to be 1 atm. The production rate of the 96% acetaldehyde stream is 1500 kg/h. Calculate the molar flow rates of the fresh feed, the recycle stream, and the hydrogen; the volumetric flow rate of the feed to the reactor; the overall and single-pass conversions; and the rates at which ethanol and acetaldehyde are removed in the hydrogen scrubber. (*Suggestion:* Use Raoult's law and the Antoine equation in your analysis of the condenser.)

60. A two-stage process is used to separate H_2S from a gas containing 96% H_2 and 4% H_2S by volume. The H_2S is absorbed in a solvent, which passes to a stripping column. Air is bubbled through the stripping column, and the effluent solvent is recycled to the absorber. The Henry's law constant for the absorption of H_2S in the solvent at 0°C is 27 atm/mole fraction. Calculate the molar flow rate of pure solvent and the volumetric flow rate of the gas at G_4, neglecting evaporation of solvent in both columns. (See flowchart.)

Data on Process Streams

G_1: 100 mol/h
 96% H_2, 4% H_2S, 1.8 atm

G_2: 99.9% H_2, 0.1% H_2S

G_3: 200 mol air/h

G_4: 40°C, 1 atm

L_1: 0.200 mole% H_2S

L_2: Assume at equilibrium with
 G_1 at 0°C

61. The solubility of sodium bicarbonate in water is 9.6 g/100 g H_2O at 20°C, and 16.4 g/100 g H_2O at 60°C. If a saturated solution of $NaHCO_3$ at 60°C is cooled to 20°C, what percentage of the dissolved salt crystallizes out?

62. One hundred liters per minute of an aqueous solution containing 50 g $Ba(NO_3)_2$/liter flows into a vacuum evaporator operating at 40°C. At what rate must water be evaporated to crystallize out 80% of the salt in the feed solution. (The solubility of barium nitrate in water at 40°C may be found in *Perry's Chemical Engineers' Handbook*. Assume that the density of the feed solution is approximately that of pure water.)

63. A stream of 1275 kg/h of an aqueous solution of potassium hydroxide is to be cooled to 0°C and evaporated under vacuum to produce crystals of $KOH \cdot 2H_2O$. An aliquot of 5 g of this solution is titrated to neutrality with 22.4 ml of 1 M H_2SO_4. The solubility of KOH at 0°C is 97 g KOH/100 g H_2O. At what rate must water be evaporated to precipitate out 70% of the KOH in the feed?

64. A salt A is soluble in a solvent S. A conductivity meter used to measure the solute concentration in A-S solutions is calibrated by dissolving a known quantity of A in S, adding more S to bring the solution volume to a fixed value, and noting the conductivity meter reading. The data given below are taken at 30°C.

Solute Dissolved g	Solution Volume ml	Meter Reading R
0	100.0	0
20.0	100.0	30
30.0	100.0	45

The following experiment is performed. One hundred sixty grams of A are dissolved in S at 30°C. S is added until a final solution volume of 500 ml is obtained. The solution

is cooled slowly to 0°C while being stirred, and is maintained at this temperature long enough for precipitation of crystals to be complete. The concentration of A in the supernatant liquid is then measured with the conductivity meter, yielding $R = 17.5$. The solution is next reheated in small temperature increments. The last crystal is observed to dissolve at 10.2°C. A specific gravity of 1.10 may be assumed for all A-S solutions.

(a) Derive an expression for C(g A/ml solution) in terms of R.

(b) Calculate the solubilities (g A/100 g S) at 10.2°C and 0°C and the mass of solid crystals in the beaker at 0°C.

(c) If half the solvent in the flask were to evaporate at 0°C, how much more A would come out of solution?

65. Potassium dichromate ($K_2Cr_2O_7$) is to be recovered from a 13 wt% aqueous solution in a continuous crystallization operation. The solution is joined by a recycle stream and is fed to a vacuum evaporator, where water is removed, and the remaining solution is cooled to 20°C, at which temperature the solubility of the salt is 0.390 mol $K_2Cr_2O_7$/kg H_2O. The solution and suspended potassium dichromate crystals flow to a centrifuge. The crystals and 5% of the solution constitute the solid effluent from the centrifuge, and the remaining solution constitutes the liquid discharge. The liquid discharge is recycled to the evaporator. The solid effluent, which contains 90 wt% crystals and 10% entrained solution, is fed to a dryer, where it is contacted with hot air. The remaining water in the effluent is evaporated, leaving pure potassium dichromate crystals. The air leaves the dryer at 90°C and 1 atm with a dew point of 39.2°C. For a production rate of 1000 kg solid $K_2Cr_2O_7$/h, calculate the required feed rate (kg/h) of 13% solution, the rate of evaporation of water in the evaporator (kg/h), the flow rate (kg/h) of the recycle stream, and the feed rate of air (standard liters/h).

66. An aqueous solution containing 30% $MgSO_4$ is cooled to 50°F. Use Figure 6.5-1 to calculate the percentage of the dissolved $MgSO_4$ that precipitates as the heptahydrate salt and the composition of the remaining solution.

67. Sixty grams of $MgSO_4$ are dissolved in 140 g of water at 200°F. Use Figure 6.5-1 to perform the following calculations.

(a) Determine the amount and composition of each phase if the solution is cooled to equilibrium at 130°F.

(b) Determine the amount and composition of each phase if the solution is cooled to equilibrium at 80°F.

(c) Determine the amount and composition of each phase if the solution is brought to equilibrium at 30°F in a process during which 5% of the total water in the system evaporates.

68. A solution is made by dissolving 60.0 g of diphenyl (MW = 154.2) in 570 ml of liquid benzene. Estimate the effective vapor pressure of the solution at 25°C, and the normal melting point and boiling point of the solution.

69. An aqueous solution of urea (MW = 60.06) freezes at -5°C and 1 atm. Estimate the normal boiling point of this solution and the grams of urea that would have to be added to 1000 g of the solution to raise the normal boiling point by an additional 7°C.

70. An organic solvent has a molecular weight of 94.10. A solution is prepared by dissolving 0.5550 g of a solute with a molecular weight of 110.1 in 100.0 g of this solvent. The freezing

point of this solution is observed to be 0.382°C lower than that of the pure solvent. A second solution is prepared by dissolving 0.4372 g of a solute with an unknown molecular weight in 96.50 g of the same solvent. A freezing point depression of 0.467°C is obtained in this case. Determine the molecular weight of the second solute and the heat of fusion (kJ/mol) of the solvent. The melting point of the pure solvent is -5.000°C.

71. Derive Equation 6.5-10 for the boiling point elevation of a dilute solution of concentration x(mol solute/mol solution) $\ll 1$. To do so, suppose that when the pressure is P_0, the pure solvent boils at temperature T_0, and the solution boils at $T_s > T_0$. The object is to obtain an expression for $\Delta T_b = T_s - T_0$. Further suppose that at temperature T_0 the vapor pressure of the solution is P_s, which must be less than P_0. (See diagram.)

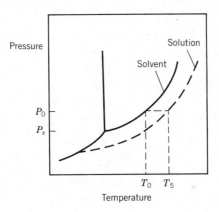

The procedure is as follows.

(a) Write the Clausius-Clapeyron equation (6.1-3) for $P_s(T_0)$, the solution vapor pressure at T_0, and then for $P_0(T_s)$, the solution vapor pressure at T_s. Subtract the two equations. Simplify the resulting equation algebraically, assuming that at the low solute concentrations in question the heats of vaporization are the same for the pure solvent and the solution, and that $T_s T_0 \approx T_0^2$.

(b) Substitute the Raoult's law expression (Equation 6.5-8) for P_s in terms of P_0. Observe that if $x \ll 1$, then $\ln(1 - x) \approx -x$. The desired result follows.

72. The distribution coefficient for n-heptane distributed between benzene and ethylene glycol at 25°C is 0.30 (mass fraction in glycol phase/mass fraction in benzene phase). Fifty grams of n-C_7H_{16} are added to a flask containing 400 g C_6H_6 and 550 g $C_2H_4(OH)_2$. How much n-C_7H_{16} is contained in each phase at equilibrium, assuming that the two solvents are immiscible?

73. A stream of 15 wt% acetone in water enters a single equilibrium stage extraction unit at a rate of 200 liters/min. It is desired to extract 90% of the acetone with chloroform at 25°C. At what rate must chloroform be fed to the unit? Assume water and chloroform are immiscible, and use phase equilibrium data given in Example 6.6-1.

74. Acetone is to be extracted with water from a 50 wt% acetone–50 wt% n-C_6H_{14} mixture at 25°C. A two-stage batch process is to be used. A quantity of the feed mixture is combined with an equal mass of water and the combined fluids are shaken and allowed to

settle. The aqueous phase is withdrawn, the same mass of water used in the first stage of the process is added to the hexane phase, and the process is repeated. What percentage of the acetone in the feed solution remains in the hexane following this process? Assume that hexane and water are immiscible, and use phase equilibrium data given in *Perry's Chemical Engineers' Handbook*.

75. A mixture of water, acetone, and methyl isobutyl ketone contains equal parts by mass of each substance. Use Figure 6.6-1 to estimate the composition of each phase of this mixture at equilibrium at 25°C, and the mass of each phase if the total mass of the mixture is 250 g.

76. Five hundred grams of a 40% acetone–60% H_2O mixture are added to 300 g of a 20% acetone–80% MIBK mixture at 25°C. Use Figure 6.6-1 to calculate the mass and composition of each phase of the resulting mixture.

77. A stream containing acetone and water flowing at a rate of 350 kg/min and a second stream of pure methyl isobutyl ketone are blended in a stirred tank, and the product stream passes to a settler at 25°C. Two phases form and are separated in the settler. One of them, whose flow rate is determined to be 310 kg/min, is analyzed and found to contain 64.5% MIBK and 30% acetone. Use Figure 6.6-1 to determine the flow rate and composition of the second product stream and the rate at which MIBK is fed to the unit.

78. Water is used to extract acetone from a 35% acetone–65% MIBK mixture flowing at a rate of 200 kg/min. A two-equilibrium-stage operation at 25°C is used. If 300 kg H_2O/min are fed to each extraction unit, what fraction of the acetone in the feed solution would be removed, and what would be the composition of the combined extract?

PART THREE
ENERGY BALANCES

Chapter Seven

Energy and Energy Balances

Energy is expensive. We have not yet learned how to use efficiently the endless supply of "free" energy provided by such sources as the sun, winds, and tides; nuclear power generation is possible, but the need to dispose of radioactive wastes from nuclear reactors safely is a serious unresolved problem; and there are not nearly enough waterfalls and dams to provide sufficient hydroelectric power to meet the world's energy needs. This leaves us with fuel combustion—burning a combustible gas, oil, or solid fuel, and using the heat released as a source of thermal or (indirectly) electrical energy.

Process industries have always recognized that wasting energy leads to reduced profits, but in the past the cost of energy was often an insignificant part of the overall process cost, and gross operational inefficiencies were tolerated. In recent years, however, a dramatic decrease in the availability of natural gas and petroleum has raised the cost of energy severalfold and has intensified the need to eliminate unnecessary energy consumption. If a plant uses more energy than its competitors, its product could now be priced out of the marketplace.

As an engineer designing a process, one of your principal jobs would therefore be to account carefully for the energy that flows into and out of each process unit and to determine the overall energy requirement for the process. You would do this by writing **energy balances** on the process, in much the same way that you write material balances to account for the mass flows to and from the process and its units. Typical problems that may be solved using energy balances include the following:

1. How much power (energy/time) is required to pump 1250 m^3/h of water from a storage vessel to a process unit? (The answer determines the size of the required pump motor.)

2. How much heat is required to convert 2000 kg of water at 30°C to steam at 180°C?

3. A hydrocarbon mixture is distilled, producing a liquid and a vapor stream, each with a known or calculable flow rate and composition. The heat input to the still is provided by condensing saturated steam at a pressure of 15 bars. At what rate must steam be supplied to process 2000 mol/h of the feed mixture?

4. A highly exothermic chemical reaction A → B takes place in a continuous reactor. If a 75% conversion of A is to be achieved, at what rate must heat be removed from the reactor to keep the contents at a constant temperature?

5. How much coal must be burned each day to produce enough heat to generate the steam to run the turbines to produce enough electricity to meet the daily power requirements of a city of 500 000 people?

295

6. A chemical process consists of four reactors, 25 pumps, and a number of compressors, distillation columns, mixing tanks, evaporators, filter presses, and other materials handling and separation units. Each individual unit either requires or releases energy in the form of work or heat.

 (a) How can the process operation be designed to minimize the total energy requirement? (For example, can the energy released by an energy-emitting process unit be used as input to an energy-absorbing process unit?)

 (b) What is the total energy requirement for the process as it is finally designed, and how much will it cost to provide this energy? (The answer could determine whether or not the process is economically feasible.)

In this chapter we show how energy balances are formulated and applied. Section 7.1 defines the types of energy a process system can possess and the ways in which energy can be transferred to and from a system. Section 7.2 reviews the procedure for calculating the kinetic energy and gravitational potential energy of a process stream. Sections 7.3 and 7.4 derive the general energy balance equation for closed (batch) systems and open (continuous) systems, and various applications of these equations are illustrated in Sections 7.5 through 7.7.

7.1 FORMS OF ENERGY: THE FIRST LAW OF THERMODYNAMICS

The total energy of a system has three components:

1. *Kinetic energy:* Energy due to the motion of the system as a whole relative to some frame of reference (usually the Earth's surface).

2. *Potential energy:* Energy due to the position of the system in a potential field (such as a gravitational or electromagnetic field) or due to the conformation of the system relative to some equilibrium conformation (e.g., due to the extension of a spring).

3. *Internal energy:* Energy due to the motion of molecules relative to the center of mass of the system, to the rotational and vibrational motion and the electromagnetic interactions of the molecules, and to the motion and interactions of the atomic and subatomic constituents of the molecules.

Suppose a process system is **closed,** meaning that no mass is transferred across its boundaries while the process is taking place. Energy may be transferred between such a system and its surroundings in two ways:

1. As **heat,** or energy that flows as a result of a temperature difference between a system and its surroundings. The direction of flow is always from a higher temperature to a lower one. *Heat is defined as positive when it is transferred to the system from the surroundings.*

2. As **work,** or energy that flows in response to any driving force other than a temperature difference, such as a force, a torque, or a voltage. For example, if a gas in a cylinder expands and moves a piston against a restraining force, the gas does work on the piston (energy is transferred as work from the gas to its surroundings, which include the piston). *In this text, work—like heat—is defined as positive when it is done by the surroundings on the system.* (*Note:* The opposite sign convention is also frequently used. The choice is arbitrary, as long as it is used consistently; however, to avoid confusion when reading thermodynamics references, you should be sure which convention has been adopted.)

The terms "work" and "heat" refer only to energy that is being transferred: You can speak of the heat or work added to a system or given off by a system, but it is meaningless to speak of the heat or work possessed by or contained within a system.

Energy, like work, has units of force times distance: for example, joules (N·m), ergs (dyne·cm), and ft·lb$_f$. It is also common to use energy units defined in terms of the amount of heat that must be transferred to a specified mass of water to raise the temperature of the water by a specified temperature interval at a constant pressure of 1 atm. The most common of these units are tabulated here.

Unit	Symbol	Mass of Water	Temperature Interval
Kilogram-calorie or kilocalorie	kcal	1 kg	15°C to 16°C
Gram-calorie or calorie	cal	1 g	15°C to 16°C
British thermal unit	Btu	1 lb$_m$	60°F to 61°F

Conversion between these and other energy units may be performed using the conversion factors in the table on the inside front cover.

The principle that underlies all energy balance calculations is the law of conservation of energy, which states that energy can neither be created nor destroyed. This law is also called the **first law of thermodynamics.** In its most general form, the first law states that the rate at which energy (kinetic + potential + internal) is carried into a system by the input streams, plus the rates at which it enters as heat and work, minus the rate at which it is transported out of the system by the output streams, equals the rate of accumulation of energy in the system. (That is, accumulation = input − output, as would be expected.)

Instead of presenting the equation in its most general form at this point, we will build up to it in stages. The next section reviews how to evaluate the kinetic and potential energy of an object and shows how the calculation can readily be extended to determine the rates at which kinetic and potential energy are transported by a flowing stream. Section 7.3 presents an integrated form of the transient balance equation that describes the behavior of a system between an initial state and a final state. This form of the equation is particularly useful for analyzing batch process systems. In Section 7.4 the first law is developed for a continuous steady-state process.

Most process systems are conveniently analyzed using one of the two forms of the energy balance equation presented in Sections 7.3 and 7.4. To perform energy

balance calculations on other types of processes, such as semibatch processes or continuous processes that are being started up or shut down, the full transient energy balance equation is required. This equation is discussed in an introductory fashion in Chapter 11, where it is shown how the simplified equations of Sections 7.3 and 7.4 are derived. A more thorough treatment of the full equation is deferred to thermodynamics courses and texts.

TEST YOURSELF

1. What forms of energy may a system possess? How may energy be transferred to and from a closed system?
2. Why is it meaningless to speak of the heat possessed by a system?
3. Suppose the initial energy of a system (internal + kinetic + potential) is E_i, the final energy is E_f, an amount of energy Q is transferred from the environment to the system as heat and an amount W is transferred from the environment to the system as work. According to the first law of thermodynamics, how must E_i, E_f, Q, and W be related?

7.2 KINETIC AND POTENTIAL ENERGY

The kinetic energy of an object of mass m moving with velocity v relative to the surface of the Earth is

$$E_k = \frac{mv^2}{2g_c} \tag{7.2-1}$$

If a fluid enters a system at a mass flow rate m and uniform velocity v, then E_k calculated from Equation 7.2-1 may be thought of as the rate at which kinetic energy is transported into the system by the fluid.

EXAMPLE 7.2-1 Kinetic Energy Transported by a Flowing Stream

Water flows into a process unit through a 2-cm ID pipe at a rate of 2 m³/h. Calculate E_k for this stream in joules/second.

Solution

First calculate the linear velocity (which equals the volumetric flow rate divided by the cross-sectional area of the pipe) and the mass flow rate of the fluid.

$$v = \frac{2 \text{ m}^3}{\text{h}} \left| \frac{(100)^2 \text{ cm}^2}{\text{m}^2} \right| \frac{1}{\pi(1)^2 \text{ cm}^2} \left| \frac{1 \text{ h}}{3600 \text{ s}} \right. = 1.77 \text{ m/s}$$

$$m = \frac{2 \text{ m}^3}{\text{h}} \left| \frac{1000 \text{ kg}}{\text{m}^3} \right| \frac{1 \text{ h}}{3600 \text{ s}} = 0.556 \text{ kg/s}$$

Then, from Equation 7.2-1,

$$E_k = \frac{0.556 \text{ kg/s}}{2} \left| \frac{(1.77)^2 \text{ m}^2}{\text{s}^2} \right| \frac{1 \text{ N}}{\text{kg-m/s}^2} = 0.870 \text{ N} \cdot \text{m/s} = \boxed{0.870 \text{ J/s}}$$

The gravitational potential energy of an object of mass m is

$$E_p = m \frac{g}{g_c} z \tag{7.2-2}$$

where z is the height of the object above a reference plane at which E_p is arbitrarily defined to be zero. If m is the mass flow rate of a flowing stream, then E_p may be thought of as the rate at which the stream transports potential energy.

EXAMPLE 7.2-2 Potential Energy Increase of a Flowing Fluid

Crude oil is pumped at a rate of 15.0 kg/s from a well 220 m deep to a storage tank 20 m above ground level. Calculate the attendant rate of increase of potential energy.

Solution

$$\Delta z = z_{\text{final}} - z_{\text{initial}} = +240 \text{ m}$$

$$\Delta E_p = m \frac{g}{g_c} \Delta z$$

$$= \frac{15.0 \text{ kg}}{\text{s}} \left| \frac{9.81 \text{ N}}{\text{kg}} \right| \frac{240 \text{ m}}{} = 35\ 300 \text{ N} \cdot \text{m/s} = \boxed{35\ 300 \text{ J/s}}$$

A pump would have to deliver at least this much power to raise the oil at the given rate.

TEST YOURSELF

A gas flows through a long pipe of constant diameter. The outlet of the pipe is higher than the inlet, and the pressure of the gas at the outlet is less than the inlet pressure. The gas temperature is constant throughout the pipe and the system is at steady state.

1. How do the mass flow rates at the inlet and outlet compare? The densities? The volumetric flow rates?

2. Is the change in potential energy of the gas from inlet to outlet positive, negative, or zero? What about the change in kinetic energy?

7.3 ENERGY BALANCES ON CLOSED SYSTEMS

A system is referred to as **open** or **closed** according to whether or not mass crosses the system boundary during the period of time covered by the energy balance. A batch process system is, by definition, closed, and semibatch and continuous systems are open.

An integral energy balance equation may be derived for a closed system between two instants of time. Since energy can neither be created nor destroyed, the generation and consumption terms of the general balance equation (4.2-1) drop out, leaving

$$\text{Accumulation} = \text{input} - \text{output} \tag{7.3-1}$$

In deriving the integral mass balance equation for a closed system in Section 4.2c we eliminated the input and output terms, since by definition no mass crosses the boundaries of a closed system. It is possible, however, for energy to be transferred across the boundaries as heat or work, so that the right side of Equation 7.3-1 may not be eliminated automatically. As with mass balances, however, the accumulation term equals the final value of the balanced quantity (in this case, the system energy) minus the initial value of this quantity. Equation 7.3-1 may therefore be written

$$\begin{array}{ll} \text{final} & \text{initial} \quad \text{net energy transferred} \\ \text{system} - \text{system} = \text{to the system} \\ \text{energy} & \text{energy} \quad (\text{in–out}) \end{array} \tag{7.3-2}$$

Now

$$\text{Initial system energy} = U_i + E_{ki} + E_{pi}$$
$$\text{Final system energy} = U_f + E_{kf} + E_{pf}$$
$$\text{Energy transferred} = Q + W$$

where the subscripts i and f refer to the initial and final states of the system and U, E_k, E_p, W, and Q represent internal energy, kinetic energy, potential energy, work done on the system by its surroundings, and heat transferred to the system from its surroundings. Equation 7.3-2 then becomes

$$(U_f - U_i) + (E_{kf} - E_{ki}) + (E_{pf} - E_{pi}) = Q + W \tag{7.3-3}$$

or, if the symbol Δ is used to signify (final–initial),

$$\boxed{\Delta U + \Delta E_k + \Delta E_p = Q + W} \tag{7.3-4}$$

Equation 7.3-4 is the basic form of the first law of thermodynamics for a closed system. When applying this equation to a given process, you should be aware of the following points:

1. The internal energy of a system depends almost entirely on the chemical composition, state of aggregation, and temperature of the system materials: It is independent of pressure for ideal gases and nearly independent of pressure for

liquids and solids. Therefore, if no changes in temperature, phase, or chemical composition occur in a process, and if the process materials are all either solids, liquids, or ideal gases, then $\Delta U \approx 0$.

2. If a system and its surroundings are at the same temperature (or if the system is perfectly insulated), then $Q = 0$. The system is then termed **adiabatic.**

3. Work done on or by a closed system is accomplished by movement of the system boundary against a resisting force, or by the generation of an electrical or radiation current that passes beyond the system boundary. (Examples of the first type of work are the motion of a piston or the rotation of a shaft that projects through the system boundary.) If there are no moving parts or generated currents in a closed system, then $W = 0$.

4. If potential energy changes other than those due to change in height are involved in a process (e.g., movement against an elastic restraining force or an electric or magnetic field), terms to account for them must be included in the ΔE_p term of Equation 7.3-4.

TEST YOURSELF

1. What do the terms closed system and open system mean? What is an adiabatic system?

2. If 250 J are added to a system as heat, what is the value of Q in the energy balance equation? If 250 J of work are done by the system on its surroundings, what is the value of W?

3. If a closed system has an internal energy of 100 kcal at the beginning of a process and 50 kcal at the end, what is ΔU?

4. Under what circumstances might U be considered independent of pressure for a pure substance?

EXAMPLE 7.3-1 Energy Balance on a Closed System

A gas is contained in a cylinder fitted with a movable piston.

The initial gas temperature is 25°C.

The cylinder is placed in boiling water with the piston held in a fixed position. Heat in the amount of 2 kcal is absorbed by the gas, which equilibrates at 100°C (and a higher pressure). The piston is then released, and the gas does 100 J of work

in moving the piston to its new equilibrium position. The final gas temperature is 100°C.

Write the energy balance equation for each of the two stages of this process, and in each case solve for the unknown energy term in the equation. In solving this problem, consider the gas in the cylinder to be the system, neglect the change in potential energy of the gas as the piston moves vertically, and assume the gas behaves ideally. Express all energies in joules.

Solution

1.

<div align="center">

25°C 100°C

Initial state Final state

</div>

$$\Delta U + \Delta E_k + \Delta E_p = Q + W \text{ (Equation 7.3-4)}$$

$$\Delta E_k = 0 \text{ (The system is stationary)}$$
$$\Delta E_p = 0 \text{ (No vertical displacement)}$$
$$W = 0 \text{ (No moving boundaries)}$$

$$\Delta U = Q$$

$$Q = 2 \text{ kcal}$$

$$\Delta U = \frac{2 \text{ kcal}}{} \left| \frac{10^3 \text{ cal}}{\text{kcal}} \right| \frac{1 \text{ J}}{0.23901 \text{ cal}} = \boxed{8368 \text{ J} = \Delta U}$$

The gas thus gains 8368 J of internal energy in going from 25 to 100°C.

2.

<div align="center">

100°C 100°C

</div>

$$\Delta U + \Delta E_k + \Delta E_p = Q + W$$

$$\Delta E_k = 0 \text{ (The system is stationary at the initial and final states)}$$
$$\Delta E_p = 0 \text{ (Assumed negligible by hypothesis)}$$
$$\Delta U = 0 \text{ (U depends only on T for an ideal gas, and T does not change)}$$

$$0 = Q + W$$

$$W = -100 \text{ J (Why is it negative?)}$$

$$\boxed{Q = 100 \text{ J}}$$

Thus an additional 100 J of heat are absorbed by the gas as it expands and reequilibrates at 100°C.

7.4 ENERGY BALANCES ON OPEN SYSTEMS AT STEADY STATE

An open process system by definition has mass crossing its boundaries as the process occurs. Work must be done on such a system to force mass in, and work is done on the surroundings by mass that emerges; both work terms must be included in the energy balance equation.

In Section 7.4a we outline the calculation of the work required to move fluid through a continuous process system, and in Section 7.4b we review the concepts of intensive and extensive variables introduced in Chapter 6 and introduce the concept of specific properties of a substance. Section 7.4c uses the results of the two preceding sections to derive the energy balance equation for an open system at steady state.

7.4a Flow Work and Shaft Work

The net work W done on an open system by its surroundings may be written as

$$W = W_s + W_f \tag{7.4-1}$$

where

$W_s =$ **shaft work,** or work done on the process fluid by a moving part within the system (e.g., a pump rotor)

$W_f =$ **flow work,** or work done on the fluid at the system inlet minus work done by the fluid at the system outlet

To derive an expression for W_f, we initially consider the single inlet–single outlet system shown here.

Fluid at a pressure $P_{in}(N/m^2)$ enters a pipe at a volumetric flow rate $V_{in}(m^3/s)$ and exits at a pressure $P_{out}(N/m^2)$ and volumetric flow rate $V_{out}(m^3/s)$. The fluid that enters the system has work done on it by the fluid just behind it at a rate

$$W_{in}(N \cdot m/s) = P_{in}(N/m^2)V_{in}(m^3/s) \tag{7.4-2}$$

while the fluid leaving the system performs work on the surroundings at a rate

$$W_{out} = P_{out}V_{out} \tag{7.4-3}$$

The net rate at which work is done on the system at the inlet and outlet is, therefore,

$$W_f = P_{in}V_{in} - P_{out}V_{out} \tag{7.4-4}$$

If several input and output streams enter and leave the system, the PV products for each must be summed to determine W_f.

TEST YOURSELF

An incompressible liquid flows through a straight horizontal pipe. Friction of the fluid within the pipe causes a small amount of heat to be generated; to compensate for this, flow work must be done on the fluid to move it through the system (so that W_f is greater than zero).

1. How are V_{in} and V_{out} related, where V is the volumetric flow rate of the liquid? (Remember, the fluid is incompressible.)
2. How must the pressures P_{in} and P_{out} be related? ($P_{in} > P_{out}$, $P_{in} = P_{out}$, or $P_{in} < P_{out}$?)

7.4b Specific Properties: Enthalpy

As we observed in Section 6.2, the properties of a process material are either extensive (proportional to the quantity of the material) or intensive (independent of the quantity). Mass, number of moles, and volume (or mass flow rate, molar flow rate, and volumetric flow rate for a continuous stream), kinetic energy, potential energy, and internal energy (or the rates of transport of these quantities by a continuous stream) are extensive properties, while temperature, pressure, and density are intensive.

A **specific property** is an intensive quantity obtained by dividing an extensive stream property by the total amount (or flow rate) of the stream material. Thus, if the volume of a fluid is 200 cm³ and the mass of the fluid is 200 g, the **specific volume** of the fluid is 1 cm³/g. Similarly, if the mass flow rate of a stream is 100 kg/min and the volumetric flow rate is 150 liters/min, the specific volume of the stream material is $(150/100) = 1.5$ liters/kg; if the rate at which kinetic energy is transported by this stream is 300 J/min, then the **specific kinetic energy** of the stream material is $(300 \text{ J/min})/(100 \text{ kg/min}) = 3$ J/kg. *We will use the symbol ^ to denote a specific property:* \hat{V} will denote specific volume, \hat{U} specific internal energy, and so on.

If the temperature and pressure of a process material are such that the specific internal energy of the material is \hat{U}(J/kg), then a mass m(kg) of this material has a total internal energy

$$U(\text{J}) = m(\text{kg})\hat{U}(\text{J/kg}) \tag{7.4-5}$$

Similarly, a continuous stream of this material with a mass flow rate m(kg/s) transports internal energy at a rate

$$U(\text{J/s}) = m(\text{kg/s})\hat{U}(\text{J/kg}) \tag{7.4-6}$$

A property that occurs in the energy balance equation for open systems (Section 7.4c) is the **specific enthalpy**, defined as

$$\hat{H} = \hat{U} + P\hat{V} \tag{7.4-7}$$

where P is total pressure and \hat{U} and \hat{V} are specific internal energy and volume. The gas constants tabulated on the inside back cover provide a convenient source for the conversion factors needed to evaluate \hat{H} from Equation 7.4-7, as the following example shows.

EXAMPLE 7.4-1 Calculation of Enthalpy

The specific internal energy of helium at 300 K and 1 atm is 3800 J/mol, and the specific molar volume at the same temperature and pressure is 24.63 liters/mol. Calculate the specific enthalpy of helium at this temperature and pressure, and the rate at which enthalpy is transported by a stream of helium at 300 K and 1 atm whose molar flow rate is 250 kmol/h.

Solution

$$\hat{H} = \hat{U} + P\hat{V} = 3800 \text{ J/mol} + (1 \text{ atm})(24.63 \text{ liters/mol})$$

To convert the second term to joules we need the factor (J/liter·atm). From the gas constant table on the inside back cover,

$$0.08206 \text{ liter·atm/mol·K} = 8.314 \text{ J/mol·K}$$

Dividing the right side by the left side yields the desired factor

$$\frac{8.314 \text{ J/mol·K}}{0.08206 \text{ liter·atm/mol·K}} = 101.3 \text{ J/liter·atm}$$

Therefore,

$$\hat{H} = 3800 \text{ J/mol} + \frac{24.63 \text{ liter·atm}}{\text{mol}} \left| \frac{101.3 \text{ J}}{\text{liter·atm}} \right. = \boxed{6295 \text{ J/mol}}$$

If $n = 250$ kmol/h

$$H = n\hat{H} = \frac{250 \text{ kmol}}{\text{h}} \left| \frac{10^3 \text{ mol}}{\text{kmol}} \right| \frac{6295 \text{ J}}{\text{mol}} = \boxed{1.57 \times 10^9 \text{ J/h}}$$

TEST YOURSELF

The specific internal energy of a fluid is 200 cal/g.

1. What is the internal energy of 30 g of this fluid?
2. If the fluid leaves a system at a flow rate of 5 g/min, at what rate does it transport internal energy out of the system?
3. What would you need to know to calculate the specific enthalpy of this fluid?

7.4c The Steady-State Open System Energy Balance

The first law of thermodynamics for an open system at steady state has the form

$$\text{Input} = \text{output} \tag{7.4-8}$$

(Why do the accumulation, generation, and consumption terms of the general balance equation vanish?) "Input" here signifies the total rate of transport of kinetic,

potential, and internal energy by all process input streams plus the rates at which energy is transferred as heat and work, and "output" is the total rate of energy transport by the output streams.

If E_j denotes the total rate of energy transport by the jth input or output stream of a process, and Q and W are again defined as the rates of flow of heat and work into the process, then Equation 7.4-8 may be written

$$Q + W + \sum_{\substack{\text{input} \\ \text{streams}}} E_j = \sum_{\substack{\text{output} \\ \text{streams}}} E_j$$

$$\Downarrow$$

$$\sum_{\substack{\text{output} \\ \text{streams}}} E_j - \sum_{\substack{\text{input} \\ \text{streams}}} E_j = Q + W \qquad (7.4\text{-}9)$$

If m_j, E_{kj}, E_{pj}, and U_j are the mass flow rate, kinetic energy, potential energy, and internal energy of the jth process stream, then the rate at which energy is transported into or out of the system by this stream is

$$E_j = U_j + E_{kj} + E_{pj}$$

$$\left\|\begin{array}{l} U_j = m_j \hat{U}_j \\ E_{kj} = m_j v_j^2 / 2g_c \\ E_{pj} = m_j(g/g_c)z_j \end{array}\right.$$

$$E_j = m_j\left(\hat{U}_j + \frac{v_j^2}{2g_c} + \frac{g}{g_c}z_j\right) \qquad (7.4\text{-}10)$$

where v_j is the velocity of the jth stream and z_j is the height of this stream relative to a reference plane at which $E_p = 0$.

The total work W done on the system by its surroundings equals the shaft work W_s plus the flow work W_f (Equation 7.4-1). If V_j is the volumetric flow rate of the jth stream, and P_j is the pressure of this stream, then, as was shown in Section 7.4a,

$$W_f = \sum_{\substack{\text{input} \\ \text{streams}}} P_j V_j - \sum_{\substack{\text{output} \\ \text{streams}}} P_j V_j$$

$$\Downarrow V_j = m_j \hat{V}_j$$

$$W = W_s + \sum_{\substack{\text{input} \\ \text{streams}}} m_j P_j \hat{V}_j - \sum_{\substack{\text{output} \\ \text{streams}}} m_j P_j \hat{V}_j \qquad (7.4\text{-}11)$$

Substituting the expression for E_j of Equation 7.4-10 and that for W of Equation 7.4-11 into Equation 7.4-9 and bringing the $P\hat{V}$ terms to the left side yields

$$\sum_{\substack{\text{output} \\ \text{streams}}} m_j\left[\hat{U}_j + (P_j\hat{V}_j) + \frac{v_j^2}{2g_c} + \frac{g}{g_c}z_j\right]$$

$$- \sum_{\substack{\text{input} \\ \text{streams}}} m_j\left[\hat{U}_j + (P_j\hat{V}_j) + \frac{v_j^2}{2g_c} + \frac{g}{g_c}z_j\right] = Q + W_s \qquad (7.4\text{-}12)$$

Equation 7.4-12 could be used for all steady-state open system energy balance problems. As a rule, however, the term $\hat{U}_j + P_j\hat{V}_j$ is combined and written as \hat{H}_j, the variable previously defined as the specific enthalpy. In terms of this variable, Equation 7.4-12 becomes

$$\sum_{\substack{\text{output} \\ \text{streams}}} m_j\left(\hat{H}_j + \frac{v_j^2}{2g_c} + \frac{g}{g_c}z_j\right) - \sum_{\substack{\text{input} \\ \text{streams}}} m_j\left(\hat{H}_j + \frac{v_j^2}{2g_c} + \frac{g}{g_c}z_j\right) = Q + W_s \quad (7.4\text{-}13)$$

Finally, let us use the symbol Δ to denote total output minus total input, so that

$$\sum_{\substack{\text{output} \\ \text{streams}}} m_j\hat{H}_j - \sum_{\substack{\text{input} \\ \text{streams}}} m_j\hat{H}_j = \Delta H \qquad (7.4\text{-}14a)$$

$$\sum_{\substack{\text{output} \\ \text{streams}}} m_j v_j^2/2g_c - \sum_{\substack{\text{input} \\ \text{streams}}} m_j v_j^2/2g_c = \Delta E_k \qquad (7.4\text{-}14b)$$

$$\sum_{\substack{\text{output} \\ \text{streams}}} m_j z_j(g/g_c) - \sum_{\substack{\text{input} \\ \text{streams}}} m_j z_j(g/g_c) = \Delta E_p \qquad (7.4\text{-}14c)$$

In terms of these quantities, Equation 7.4-13 becomes

$$\boxed{\Delta H + \Delta E_k + \Delta E_p = Q + W_s} \qquad (7.4\text{-}15)$$

We will use Equation 7.4-15 as the starting point for most energy balance calculations on open systems at steady state.

Notice that if a process has a single input stream and a single output stream, and there is no accumulation of total mass in the system (so that $n_{\text{in}} = n_{\text{out}} = n$), the expression for ΔH of Equation 7.4-14a simplifies to

$$\Delta H = n(\hat{H}_{\text{out}} - \hat{H}_{\text{in}}) = n\,\Delta\hat{H} \qquad (7.4\text{-}16)$$

Also notice that if a specific variable has the same value for all input and output streams, the corresponding term of Equation 7.4-15 drops out. For example, if \hat{H}_j is the same for all streams, then from (7.4-14a)

$$\Delta H = \hat{H}\left[\sum_{\substack{\text{output} \\ \text{streams}}} m_j - \sum_{\substack{\text{input} \\ \text{streams}}} m_j\right] \qquad (7.4\text{-}17)$$

But from a total mass balance the quantity in brackets (which is simply total mass in minus total mass out) equals zero, and hence $\Delta H = 0$, as claimed.

TEST YOURSELF

How would you simplify Equation 7.4-15 in each of the following cases?

1. There are no moving parts in the system.
2. The system and its surroundings are at the same temperature.

3. The linear velocities of all streams are the same.

4. All streams enter and leave the process at a single height.

EXAMPLE 7.4-2 Energy Balance on a Turbine

Five hundred kilograms per hour of steam drive a turbine. The steam enters the turbine at 44 atm and 450°C at a linear velocity of 60 m/s and leaves at a point 5 m below the turbine inlet at atmospheric pressure and a velocity of 360 m/s. The turbine delivers shaft work at a rate of 700 kW, and the heat loss from the turbine is estimated to be 10^4 kcal/h. Calculate the specific enthalpy change associated with the process.

Solution

From Equation 7.4-15

$$\Delta H = Q + W_s - \Delta E_k - \Delta E_p$$

Normally heat, work, and kinetic and potential energy terms are determined in different units. To evaluate ΔH, we will convert each term to kW(kJ/s), using conversion factors given on the inside front cover, first noting that $m = (500/3600) = 0.139$ kg/s.

$$\Delta E_k = \frac{m}{2g_c}(v_2^2 - v_1^2)$$

$$= \frac{0.139 \text{ kg/s}}{2} \left| \frac{1 \text{ N}}{1 \text{ kg·m/s}^2} \right| \frac{(360^2 - 60^2) \text{ m}^2}{\text{s}^2} \left| \frac{1 \text{ W}}{1 \text{ N·m/s}} \right| \frac{1 \text{ kW}}{10^3 \text{ W}}$$

$$= 8.75 \text{ kW}$$

$$\Delta E_p = \frac{mg}{g_c}(z_2 - z_1)$$

$$= \frac{0.139 \text{ kg/s}}{} \left| \frac{9.81 \text{ N}}{\text{kg}} \right| \frac{(-5) \text{ m}}{} \left| \frac{1 \text{ kW}}{10^3 \text{ N·m/s}} \right. = -6.81 \times 10^{-3} \text{ kW}$$

$$Q = \frac{-10^4 \text{ kcal}}{\text{h}} \left| \frac{1 \text{ J}}{0.239 \times 10^{-3} \text{ kcal}} \right| \frac{1 \text{ h}}{3600 \text{ s}} \left| \frac{1 \text{ kW}}{10^3 \text{ J/s}} \right. = -11.6 \text{ kW}$$

$$W_s = -700 \text{ kW}$$

$$\Downarrow$$

$$\Delta H = Q + W_s - \Delta E_k - \Delta E_p = -720 \text{ kW}$$

But

$$\Delta H = m(\hat{H}_2 - \hat{H}_1) \quad \text{(from Equation 7.4-16)}$$
$$\Downarrow$$
$$\hat{H}_2 - \hat{H}_1 = \Delta H/m$$

$$= \frac{-720 \text{ kJ/s}}{0.139 \text{ kg/s}} = \boxed{-5180 \text{ kJ/kg}}$$

7.5 TABLES OF THERMODYNAMIC DATA

7.5a Reference States and State Properties

It is not possible to know the absolute value of \hat{U} for a process material, but you can determine the *change* in \hat{U} corresponding to a specified change of state (temperature, pressure, and phase). This may be done, for example, by bringing a known mass m of a substance through the specified change of state in such a way that all terms of the energy balance except ΔU (i.e., heat, work, and changes in potential and kinetic energy) are known. Once $\Delta\hat{U} = (\Delta U/m)$ has been determined, $\Delta\hat{H}$ for the same change in state can be calculated as $\Delta\hat{U} + \Delta P\hat{V}$.

A convenient way to tabulate measured changes in \hat{U} or \hat{H} is to choose a temperature, pressure, and state of aggregation as a **reference state,** and to list $\Delta\hat{U}$ or $\Delta\hat{H}$ for changes from this state to a series of other states. Suppose, for example, that the enthalpy changes for carbon monoxide going from a reference state of 0°C and 1 atm to two other states are measured, with the following results:

$$CO(g, 0°C, 1 \text{ atm}) \rightarrow CO(g, 100°C, 1 \text{ atm}): \Delta\hat{H}_1 = 2\,919 \text{ J/mol}$$
$$CO(g, 0°C, 1 \text{ atm}) \rightarrow CO(g, 500°C, 1 \text{ atm}): \Delta\hat{H}_2 = 15\,060 \text{ J/mol}$$

Since \hat{H} cannot be known absolutely, for convenience we may assign a value $\hat{H}_0 = 0$ to the reference state; then $\Delta\hat{H}_1 = \hat{H}_1 - 0 = \hat{H}_1$, $\Delta\hat{H}_2 = \hat{H}_2$, and so on. A table may then be constructed for CO at 1 atm:

T(°C)	\hat{H}(J/mol)
0	0
100	2 919
500	15 060

Note that the value 2919 J/mol for \hat{H} at 100°C does *not* mean that the absolute value of the specific enthalpy of CO at 100°C and 1 atm is 2919 J/mol—we cannot know the absolute value of \hat{H}—but rather means that the change in \hat{H} when CO goes from the reference state to 100°C and 1 atm is 2919 J/mol. We then say *the enthalpy of CO at 100°C and 1 atm relative to 0°C and 1 atm is 2919 J/mol.*

Some enthalpy tables give the reference states on which the listed values of \hat{H} are based and others do not; however, you do not have to know the reference state to calculate $\Delta\hat{H}$ for the transition from one tabulated state to another. If \hat{H}_1 is the

enthalpy at state 1 and \hat{H}_2 is that at state 2, then $\Delta \hat{H}$ for the transition from state 1 to state 2 equals $\hat{H}_2 - \hat{H}_1$, regardless of the reference state on which \hat{H}_1 and \hat{H}_2 are based. (*Caution.* If two different tables are used, be sure \hat{H}_1 and \hat{H}_2 are based on the same reference state.) For example, $\Delta \hat{H}$ for CO going from 100 to 500°C at 1 atm is $(15\,060 - 2919)$ J/mol $= 12\,141$ J/mol. If another reference state had been used to generate the specific enthalpies of CO at 100°C and 500°C, \hat{H}_1 and \hat{H}_2 would have different values but $\hat{H}_2 - \hat{H}_1$ would still be 12 141 J/mol.

This convenient result is a consequence of the fact that \hat{H}, like \hat{U}, is a **state property,** or a property of a system component whose value depends only on the state of the system (temperature, pressure, phase, and composition) and not on how the system reached that state.[1] We will have more to say about this concept in Chapter Eight.

EXAMPLE 7.5-1 Use of Tabulated Enthalpy Data

The following entries are taken from a data table for saturated methyl chloride.

State	$T(°F)$	$P(\text{psia})$	$\hat{V}(\text{ft}^3/\text{lb}_m)$	$\hat{H}(\text{Btu}/\text{lb}_m)$
Liquid	-40	6.878	0.01553	0.000
Vapor	0	18.90	4.969	196.23
Vapor	50	51.99	1.920	202.28

1. What reference state was used to generate the given enthalpies?
2. Calculate $\Delta \hat{H}$ and $\Delta \hat{U}$ for the transition of saturated methyl chloride vapor from 50°F to 0°F.

Solution

1. Liquid at $-40°$F and 6.878 psia (the state at which $\hat{H} = 0$). You do not need this information to solve part 2.
2. $\Delta \hat{H} = \hat{H}(0°F) - \hat{H}(50°F) = (196.23 - 202.28) = -6.05$ Btu/lb$_m$

 $\Delta \hat{U} = \Delta \hat{H} - \Delta P \hat{V} = \Delta \hat{H} - (P_{\text{final}}\hat{V}_{\text{final}} - P_{\text{initial}}\hat{V}_{\text{initial}})$

 $= -6.05$ Btu/lb$_m$

 $$-\frac{[(18.90)(4.969) - (51.99)(1.920)] \text{ ft}^3 \cdot \text{psia}/\text{lb}_m}{} \left| \frac{1.987 \text{ Btu}}{10.73 \text{ ft}^3 \cdot \text{psia}} \right.$$

 $= \boxed{-4.96 \text{ Btu}/\text{lb}_m}$

The value of the conversion factor (Btu/ft^3·psia) was obtained from the table of gas constants in back of the book. (Verify it!)

[1] We will not prove our claim that \hat{U} and \hat{H} satisfy this condition. All references on thermodynamics discuss this point in detail.

Tables of enthalpies and other state properties of many substances may be found in *Perry's Chemical Engineers' Handbook*[2] on pp. 3-160 through 3-246 and in other references listed at the end of this chapter.

TEST YOURSELF

1. What is a state property?
2. The enthalpy of a vapor A relative to liquid A at 0°C and 1 atm is 5000 J/kg at 0°C and 1 atm, and 7500 J/kg at 30°C and 1 atm.
 (a) What is \hat{H} of A(l) at 0°C and 1 atm?
 (b) What is the approximate value of \hat{H} for A(v) at 0°C and 5 atm?
 (c) What is $\Delta\hat{H}$ for the process

$$A(v, 30°C, 1 \text{ atm}) \rightarrow A(v, 0°C, 1 \text{ atm})$$

 Does the answer depend on the reference state used to generate the table of enthalpies? Why not?

7.5b Steam Tables

For many years, compilations of physical properties of water issued in **steam tables** have been standard references for mechanical and chemical engineers involved with steam cycles for electrical power generation.

Steam tables are contained in Appendix B in this book. The reference state for these tables is liquid water at the triple point—0.01°C and 0.00611 bar—at which state \hat{U} is defined to be zero. Tables B.4 and B.5 list specific volumes, internal energies, and enthalpies of saturated steam—Table B.4 by temperature, and Table B.5 by pressure. If you specify a temperature, for example, you can look up in Table B.4 \hat{V}, \hat{U}, \hat{H}, and the pressure of saturated steam corresponding to that temperature. (Recall that from the phase rule, specifying one intensive variable of a saturated one-component two-phase system completely fixes all other intensive quantities.) Table B.6 gives data for the same properties of superheated steam as functions of both temperature and pressure.

The use of the steam tables is illustrated in Example 7.5-2.

EXAMPLE 7.5-2 Steam Tables

Steam at 10 bars (absolute) with 190°C of superheat is fed to a turbine at a rate $m = 2000$ kg/h. The turbine operation is adiabatic, and the effluent is saturated steam at 1 bar. Calculate the work output of the turbine in kilowatts, neglecting kinetic and potential energy changes.

[2] R. H. Perry and D. W. Green, Eds., *Perry's Chemical Engineers' Handbook*, 6th Edition, McGraw-Hill, New York, 1984.

Solution

The energy balance for this steady-state open system is

$$\dot{W}_s = \Delta \dot{H} = \dot{m}(\hat{H}_{out} - \hat{H}_{in})$$

(Why was the heat term dropped?)

Inlet Steam

Table B.6 indicates that steam at 10 bars is saturated at 180°C (verify this), so that the inlet steam temperature is 180°C + 190°C = 370°C. Interpolating in the same table,

$$\hat{H}_{in}(10 \text{ bars, } 370°C) = 3201 \text{ kJ/kg}$$

Outlet Steam

From either Table B.5 or B.6, you can find that the enthalpy of saturated steam at 1 bar is

$$\hat{H}_{out}(1 \text{ bar, saturated}) = 2675 \text{ kJ/kg}$$

Energy Balance

$$\dot{W}_s = \Delta \dot{H} = \frac{2000 \text{ kg}}{\text{h}} \left| \frac{(2675 - 3201) \text{ kJ}}{\text{kg}} \right| \frac{1 \text{ h}}{3600 \text{ s}}$$

$$= -292 \text{ kJ/s} = \boxed{-292 \text{ kW}}$$

The turbine thus delivers 292 kW of work to its surroundings.

The superheated steam table, Table B.6, lists values for both liquid water and steam. If you wish to determine \hat{H} for liquid water at a temperature T and pressure P that cannot easily be found in this table, you may calculate it in the following manner: (1) Look up \hat{U} and \hat{V} for *saturated* liquid at the specified temperature in Table B.4; (2) Assume these values are independent of pressure, and calculate $\hat{H}(P, T) = \hat{U} + P\hat{V}$. Furthermore, if the pressure is not excessive (say, less than 50 bars) *or if it is unknown*, neglect the $P\hat{V}$ correction and use the saturated liquid enthalpy $\hat{H}(T)$ given in Table B.4.

7.6 ENERGY BALANCE PROCEDURES

A properly drawn and labeled flowchart is essential for the efficient solution of energy balance problems. When labeling the flowchart, be sure to include all of the information you will need to determine the specific enthalpy of each stream component, including known temperatures and pressures. In addition, show states of aggregation of process materials when they are not obvious: Do not simply write H_2O, for example, but rather $H_2O(s)$, $H_2O(l)$ or $H_2O(v)$, according to whether water is present as a solid, a liquid, or a vapor.

EXAMPLE 7.6-1 Energy Balance on a One-Component Process

Two streams of water are mixed to form the feed to a boiler. Process data are given here.

> Feed stream 1: 120 kg/min @ 30°C
> Feed stream 2: 175 kg/min @ 65°C
> Boiler pressure: 17 bars (absolute)

The exiting steam emerges from the boiler through a 6-cm ID pipe. Calculate the required heat input to the boiler in kilojoules per minute if the emerging steam is saturated at the boiler pressure. Neglect the kinetic energies of the liquid inlet streams.

Solution

1. *A first step in solving problems of this sort is to determine (if possible) the flow rates of all stream components by means of material balances.* In this case, it is a trivial matter to write and solve a mass balance on water to determine that the flow rate of the emerging steam is 295 kg/min.

2. *Next, determine the specific enthalpies of each stream component.* Steam Tables B.4 and B.5 were used to determine \hat{H} for liquid water at 30°C and 65°C and for saturated steam at a pressure of 17 bars. The latter entry in the table also furnished the temperature of the saturated steam corresponding to this pressure (204°C). Note that the entries for liquid water correspond to pressures that may or may not equal the actual pressures of the inlet streams (which we do not know); we assume, however, that the enthalpy of liquid water is approximately independent of pressure, and use the tabulated values.

3. *The final step is to write the appropriate form of the energy balance equation and solve it for the desired quantity.* For this open process system,

$$Q + W_s = \Delta H + \Delta E_k + \Delta E_p$$

$\quad\Vert\quad W_s = 0$: no moving parts

$\quad\Vert\quad \Delta E_p = 0$: generally assumed unless displacements through large heights are involved

$$Q = \Delta H + \Delta E_k$$

Evaluate ΔH

From Equation 7.4-14a

$$\Delta H = \sum_{\text{outlet}} n_i \hat{H}_i - \sum_{\text{inlet}} n_i \hat{H}_i$$

$$= \frac{295 \text{ kg}}{\text{min}} \left| \frac{2793 \text{ kJ}}{\text{kg}} - \frac{120 \text{ kg}}{\text{min}} \right| \frac{125.7 \text{ kJ}}{\text{kg}} - \frac{175 \text{ kg}}{\text{min}} \left| \frac{271.9 \text{ kJ}}{\text{kg}} \right.$$

$$= 7.61 \times 10^5 \text{ kJ/min}$$

Evaluate ΔE_k

From Table B.5, the specific volume of saturated steam at 17 bars is 0.1166 m^3/kg, and the cross-sectional area of the 6-cm ID pipe is

$$A = \pi R^2 = \frac{3.1416}{} \left| \frac{(3.00)^2 \text{ cm}^2}{} \right| \frac{1 \text{ m}^2}{10^4 \text{ cm}^2} = 2.83 \times 10^{-3} \text{ m}^2$$

The steam velocity is

$$v(\text{m/s}) = V(\text{m}^3/\text{s})/A(\text{m}^2)$$

$$= \frac{295 \text{ kg}}{\text{min}} \left| \frac{1 \text{ min}}{60 \text{ s}} \right| \frac{0.1166 \text{ m}^3}{\text{kg}} \left| \frac{}{2.83 \times 10^{-3} \text{ m}^2} \right.$$

$$= 202 \text{ m/s}$$

Then, since the kinetic energies of the inlet streams are assumed negligible,

$$\Delta E_k \approx (E_k)_{\text{steam}} = mv^2/2g_c$$

$$\Delta E_k = \frac{295 \text{ kg/min}}{2} \left| \frac{(202)^2 \text{ m}^2}{\text{s}^2} \right| \frac{\text{N} \cdot \text{s}^2}{1 \text{ kg} \cdot \text{m}} \left| \frac{1 \text{ kJ}}{10^3 \text{ N} \cdot \text{m}} \right. = 6.02 \times 10^3 \text{ kJ/min}$$

Finally

$$Q = \Delta H + \Delta E_k$$

$$= [(7.61 \times 10^5) + (6.02 \times 10^3)] \text{ kJ/min}$$

$$= \boxed{7.67 \times 10^5 \text{ kJ/min}}$$

Observe that the kinetic energy change is only a small fraction—roughly 0.8%—of the total heat requirement for the process. This is a typical result, and it is not uncommon to neglect kinetic and potential energy changes (at least as a first approximation) relative to enthalpy changes for processes that involve phase changes, chemical reactions, or large temperature changes.

When process streams contain several components, the specific enthalpies of each component must be determined separately and substituted in the energy balance

equation when ΔH is evaluated. *For mixtures of gases or of liquids with similar molecular structures* (e.g., mixtures of straight chain hydrocarbons), *you may assume that \hat{H} for a mixture component is the same as \hat{H} for the pure substance at the same temperature and pressure*. Procedures to follow for solutions of gases or solids in liquids and for mixtures of dissimilar liquids are outlined in Chapter 8.

EXAMPLE 7.6-2 Energy Balance on a Two-Component Process

A gas stream containing 60 wt% ethane and 40% butane is to be heated from 150 to 250 K at a pressure of 5 bars. Calculate the required heat input per kg of the mixture, neglecting potential and kinetic energy changes and using tabulated enthalpy data for C_2H_6 and C_4H_{10}.

Solution

Basis: 1 kg Mixture

The enthalpies of butane at 150 K and 5 bars, and 250 K and 5 bars are given on p. 3-178 of *Perry's Chemical Engineers' Handbook*, and those of ethane at the same conditions are given on p. 3-188 of the *Handbook*. The tabulated enthalpy values are shown below on the flowchart.

No material balances are necessary since there is only one input stream and one output stream, so we may proceed directly to the energy balance.

$$Q + W_s = \Delta H + \Delta E_k + \Delta E_p$$

$$\Big\Downarrow \begin{array}{l} W_s = 0 \text{ (no moving parts)} \\ \Delta E_k = 0, \Delta E_p = 0 \text{ (by hypothesis)} \end{array}$$

$$Q = \Delta H$$

Since the process materials are all gases, we may set the enthalpies of each stream equal to the sums of the individual component enthalpies and write

$$Q = \Delta H = \underset{\substack{\text{outlet} \\ \text{components}}}{\sum} n_i \hat{H}_i - \underset{\substack{\text{inlet} \\ \text{components}}}{\sum} n_i \hat{H}_i$$

$$= \frac{0.6 \text{ kg } C_2H_6}{} \left| \frac{973.3 \text{ kJ}}{\text{kg}} + \frac{0.4 \text{ kg } C_4H_{10}}{} \right| \frac{237.0 \text{ kJ}}{\text{kg}}$$

$$- (0.6)(314.3) - (0.4)(30.0) = \boxed{478.2 \text{ kJ}}$$

In the two previous examples, it was possible to perform complete material balances before undertaking the energy balance. In another class of problems one more stream amount or flow rate is unknown than can be determined by material balances alone. To solve problems of this type, you must write and solve material and energy balances simultaneously.

EXAMPLE 7.6-3 Simultaneous Material and Energy Balances

Saturated steam at 1 atm is discharged from a turbine at a rate of 1000 kg/h. Superheated steam at 300°C and 1 atm is needed as a feed to a heat exchanger; to produce it, the turbine discharge stream is mixed with superheated steam available from a second source at 400°C and 1 atm. The mixing unit operates adiabatically. Calculate the amount of steam at 300°C produced, and the required volumetric flow rate of the 400°C steam.

Solution

Specific enthalpies of the two feed streams and the product stream are obtained from the steam tables, and are shown below on the flowchart.

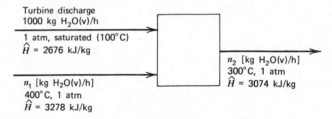

There are two unknown quantities in this process—n_1 and n_2—and only one permissible material balance. (Why?) The material and energy balances must therefore be solved simultaneously to determine the two flow rates.

Mass Balance on Water

$$1000 + n_1 = n_2 (\text{kg/h}) \tag{1}$$

Energy Balance

$$Q + W_s = \Delta H + \Delta E_k + \Delta E_p$$

$$Q = 0 \text{ (process is adiabatic)}$$
$$W_s = 0 \text{ (no moving parts)}$$
$$\Delta E_k \approx 0, \Delta E_p \approx 0 \text{ (hypothesis)}$$

$$\Delta H = \sum_{\text{outlet}} n_i \hat{H}_i - \sum_{\text{inlet}} n_i \hat{H}_i = 0$$

$$\frac{1000 \text{ kg}}{h} \left| \frac{2676 \text{ kJ}}{kg} \right. + (n_1)(3278) = (n_2)(3074) \tag{2}$$

Solving (1) and (2) simultaneously yields

$$n_1 = 1951 \text{ kg/h}$$

$$\boxed{n_2 = 2951 \text{ kg/h}} \quad \text{(Product flow rate)}$$

From Table B.6, the specific volume of steam at $400°C$ and 1 atm (≈ 1 bar) is $3.11 \text{ m}^3/\text{kg}$. The volumetric flow rate of this stream is therefore

$$\frac{1951 \text{ kg}}{\text{h}} \left| \frac{3.11 \text{ m}^3}{\text{kg}} \right. = \boxed{6070 \text{ m}^3/\text{h}}$$

If specific volume data were not available, the ideal gas law could be used as an approximation for the last calculation.

7.7 MECHANICAL ENERGY BALANCES

In chemical process units such as reactors, distillation columns, evaporators, and heat exchangers, shaft work and kinetic and potential energy changes tend to be negligible compared with heat flows and internal energy and enthalpy changes. Energy balances on such units are therefore usually written ignoring the former terms, and so take the simple form $Q = \Delta U$ (closed system) or $Q = \Delta H$ (open system).

Another important class of operations is one for which the opposite is true—heat flows and internal energy changes are secondary in importance to kinetic and potential energy changes and shaft work. Most of these operations involve the flow of fluids to, from, and between tanks, reservoirs, wells, process units, transportation depots, and waste discharges.

In this section we consider energy balances on continuous processes of this type. We will restrict our consideration to operations involving a single incompressible fluid; if an operation involves gases, the full energy balance equation we have used previously should again be applied.

Consider an open system with one input and one output stream, each with mass flow rate m, and suppose that the process fluid is a liquid, so that $\hat{V}_{\text{in}} = \hat{V}_{\text{out}} = \hat{V}$. If \hat{V} is replaced by $1/\rho$, where ρ is the liquid density, then the open system energy balance equation (Equation 7.4-12) may be written as

$$\frac{\Delta P}{\rho} + \frac{\Delta v^2}{2g_c} + \frac{g}{g_c} \Delta z + (\Delta \hat{U} - Q/m) = W_s/m \tag{7.7-1}$$

The shaft work W_s is the work done on the fluid by turbines or pumps in the process line.

In many cases only slight amounts of heat are transferred to or from the surroundings, there is little change in temperature from inlet to outlet, and no phase changes or reactions occur. Even under these circumstances, some kinetic or potential energy is always converted to thermal energy as a result of friction due to the movement of

the fluid through the system. In consequence, the quantity $(\Delta \hat{U} - Q/m)$ always has a positive component, termed the **friction loss** and given the symbol F. Equation 7.7-1 may therefore be written

$$\frac{\Delta P}{\rho} + \frac{\Delta v^2}{2g_c} + \frac{g}{g_c}\Delta z + F = \frac{W_s}{m} \qquad \text{(7.7-2)}$$

Equation 7.7-2 is referred to as the **mechanical energy balance equation.** It is valid for flow of an incompressible fluid through an open process system, under conditions where heat losses and internal energy changes due to anything but friction are negligible.

Methods of estimating friction losses for flow through straight pipes, orifices, nozzles, elbows, and so on, are given in Section 5 of *Perry's Chemical Engineers' Handbook*, but will not be discussed in this text. In the balance of this book we consider only processes in which friction losses are either specified or neglected.

A simplied form of the mechanical energy balance equation is obtained for frictionless processes $(F \approx 0)$ in which no shaft work is performed $(W_s = 0)$:

$$\frac{\Delta P}{\rho} + \frac{\Delta v^2}{2g_c} + \frac{g}{g_c}\Delta z = 0 \qquad \text{(7.7-3)}$$

Equation 7.7-3 is called the **Bernoulli equation.**

EXAMPLE 7.7.1 The Bernoulli Equation

Water flows through the system shown here at a rate of 20 liters/min. Estimate the pressure required at point 1 if friction losses are negligible.

Solution

50 m

② 1 cm ID pipe P_2 = 1 atm

① 0.5 cm ID tube 20 liters H_2O/min P_1 = ?

All of the terms of the Bernoulli equation, Equation 7.7-3, are known except ΔP, the variable to be determined, and Δv^2, which must be calculated from the known liquid flow rate and the diameters of the inlet and outlet pipes.

Velocities

$$v(\text{m/s}) = V(\text{m}^3/\text{s})/A(\text{m}^2)$$

The volumetric flow rate must be the same at points 1 and 2. (Why?)

$$v_1 = \frac{20 \text{ liters}}{\text{min}} \left| \frac{1 \text{ m}^3}{10^3 \text{ liters}} \right| \frac{1}{\pi(0.25)^2 \text{ cm}^2} \left| \frac{10^4 \text{ cm}^2}{\text{m}^2} \right| \frac{1 \text{ min}}{60 \text{ s}} = 17.0 \text{ m/s}$$

$$v_2 = \frac{20 \text{ liters}}{\text{min}} \left| \frac{1 \text{ m}^3}{10^3 \text{ liters}} \right| \frac{1}{\pi(0.5)^2 \text{ cm}^2} \left| \frac{10^4 \text{ cm}^2}{1 \text{ m}^2} \right| \frac{1 \text{ min}}{60 \text{ s}} = 4.24 \text{ m/s}$$

$$\Downarrow$$

$$\Delta v^2 = (v_2^2 - v_1^2) = (4.24^2 - 17.0^2) \text{ m}^2/\text{s}^2 = -271.0 \text{ m}^2/\text{s}^2$$

Bernoulli Equation (Equation 7.7-3)

$$\frac{\Delta P(\text{N/m}^2)}{\rho(\text{kg/m}^3)} + \frac{\Delta v^2(\text{m}^2/\text{s}^2)}{2 \cdot g_c(\text{kg·m/s}^2 \cdot \text{J})} + \frac{g(\text{m/s}^2)\Delta z(\text{m})}{g_c(\text{kg·m/s}^2 \cdot \text{J})} = 0$$

$$\begin{Vmatrix} \Delta P = P_2 - P_1 \\ \rho = 1000 \text{ kg/m}^3 \\ \Delta v^2 = -271.0 \text{ m}^2/\text{s}^2 \\ g_c = 1 \text{ kg·m/s}^2 \cdot \text{J} \\ g = 9.81 \text{ m/s}^2 \\ \Delta z = z_2 - z_1 = 50 \text{ m} \end{Vmatrix}$$

$$\frac{P_2 - P_1}{1000} - 135.5 + 490 = 0 \text{ (J/kg)}$$

$$\Downarrow P_2 = 1 \text{ atm} = 1.01325 \times 10^5 \text{ N/m}^2$$

$$P_1 = 4.56 \times 10^5 \text{ N/m}^2 = 4.56 \times 10^5 \text{ Pa}$$

$$= \boxed{4.56 \text{ bars}}$$

A common type of problem to which the mechanical energy balance equation is applicable is one that involves drainage or siphoning of a liquid from a container. The correct choice of the locations of points 1 and 2 greatly simplifies such problems; it is convenient to choose as point 1 a location on the liquid surface in the tank being drained and to take point 2 at the discharge stream outlet. If the container is being drained relatively slowly, the kinetic energy at point 1 may be neglected. Example 7.7-2 illustrates the computational procedure for such problems.

EXAMPLE 7.7-2　Siphoning

Gasoline ($\rho = 50 \text{ lb}_m/\text{ft}^3$) is to be siphoned from a tank. The friction loss in the line is $F = 0.8 \text{ ft·lb}_f/\text{lb}_m$. Estimate how long it will take to siphon 5 gal, neglecting the change in liquid level in the gasoline tank during this process and assuming that both point 1 (at the liquid surface in the gas tank) and point 2 (in the tube just prior to the exit) are at 1 atm.

Solution

$$\text{Point 1: } P_1 = 1 \text{ atm}, v_1 \approx 0 \text{ ft/s}, z_1 = 2.5 \text{ ft}$$
$$\text{Point 2: } P_2 = 1 \text{ atm}, v_2 = ?, z_2 = 0 \text{ ft}$$

Mechanical Energy Balance (Equation 7.7-2)

$$(\Delta P/\rho) + (\Delta v^2/2g_c) + [(g/g_c)(\Delta z)] + F = W_s/m$$

$$
\begin{Vmatrix}
\begin{aligned}
&\Delta P = 0 \\
&\Delta v^2 \approx v_2^2 (\text{ft}^2/\text{s}^2) \\
&g_c = 32.174 \text{ lb}_m \cdot \text{ft/lb}_f \cdot \text{s}^2 \\
&g/g_c = 1 \text{ lb}_f/\text{lb}_m \\
&\Delta z = -2.5 \text{ ft} \\
&F = 0.8 \text{ ft} \cdot \text{lb}_f/\text{lb}_m \\
&W_s = 0
\end{aligned}
\end{Vmatrix}
$$

$$\frac{v_2^2}{(2)(32.174)} - 2.5 + 0.8 = 0 \ (\text{ft} \cdot \text{lb}_f/\text{lb}_m)$$

$$\Downarrow$$

$$v_2 = 10.5 \text{ ft/s}$$

The volumetric flow rate of the fluid in the tube is

$$V(\text{ft}^3/\text{s}) = v_2(\text{ft/s}) \cdot A(\text{ft}^2)$$

$$= \frac{10.5 \text{ ft}}{\text{s}} \left| \frac{\pi(0.125)^2 \text{ in.}^2}{} \right| \frac{1 \text{ ft}^2}{144 \text{ in.}^2} = 3.58 \times 10^{-3} \text{ ft}^3/\text{s}$$

$$t(s) = \frac{\text{volume to be drained (ft}^3)}{\text{volumetric flow rate (ft}^3/\text{s})}$$

$$= \frac{(5 \text{ gal})(0.1337 \text{ ft}^3/\text{gal})}{3.58 \times 10^{-3} \text{ ft}^3/\text{s}} \left| \frac{187 \text{ s}}{60 \text{ s/min}} \right| = \boxed{3.11 \text{ min}}$$

The last illustration we will consider is one in which the potential energy lost by falling water is converted to electrical energy by means of a turbine and generator. The work done by the water in rotating the turbine must be included as shaft work in the energy balance equation. (Its sign must be negative.)

EXAMPLE 7.7-3 Hydraulic Power Generation

Water flows from an elevated reservoir through a conduit to a turbine at a lower level and out of the turbine through a similar conduit. At a point 100 m above the turbine the pressure is 207 kPa, and at a point 3 m below the turbine the pressure is 124 kPa. What must the water flow rate be if the turbine output is 1.00 MW?

Solution

No data are given for the friction loss, so we will set $F = 0$, recognizing that doing so introduces an error into the calculation. Since the diameters of the conduits at points 1 and 2 are the same and water may be considered incompressible, $\Delta v^2 = 0$. Equation 7.7-2 then becomes

$$\frac{\Delta P}{\rho} + \frac{g}{g_c} \Delta z = \frac{W_s}{m}$$

$$\Downarrow$$

$$m = W_s \bigg/ \left(\frac{\Delta P}{\rho} + \frac{g}{g_c} \Delta z \right)$$

$$W_s = \frac{-1.00 \text{ MW}}{} \left| \frac{10^6 \text{ N·m/s}}{1 \text{ MW}} \right. = -1.00 \times 10^6 \text{ N·m,}$$

$$\Delta P = \frac{(124 - 207) \text{ kPa}}{} \left| \frac{1000 \text{ N/m}^2}{1 \text{ kPa}} \right. = -8.30 \times 10^4 \text{ N/m}$$

$$\rho = 1.00 \times 10^3 \text{ kg/m}^3$$

$$g/g_c = 9.807 \text{ N/kg}$$

$$\Delta z = -103 \text{ m}$$

$$\boxed{m = 915 \text{ kg/s}}$$

TEST YOURSELF

1. Under what conditions is the mechanical energy balance equation, Equation 7.7-2, applicable. What is the physical significance of the friction loss term, F, in this equation?

2. Under what conditions is the Bernoulli equation, Equation 7.7-3, applicable?

REFERENCES: SOURCES OF THERMODYNAMIC DATA

The references given below either contain thermodynamic data or list data sources. The open literature includes numerous journals in which you may find data on materials not listed in these references; *Chemical Abstracts* is a good place to begin a literature search. If the search is unsuccessful, you must either estimate the desired property by using techniques to be described in Chapter 8, or measure it.

Haywood, R. W., *Thermodynamic Tables in SI (Metric) Units*, Cambridge University Press, London, 1968.

National Bureau of Standards, *Annotated Accession List of Data Compilations*, NBS Technical Note 554, 1970.

National Bureau of Standards, *Description of the Magnetic Tape Version of the Bulletin of Thermodynamics and Thermochemistry*, NBS Technical Note 760, 1973.

National Bureau of Standards, *Selected Values of Chemical Thermodynamic Properties*, *NBS Circular* 500, 1952, currently being issued in revised form in six or seven parts as NBS Technical Note 270.

Perry, R. H., and D. W. Green, Eds., *Perry's Chemical Engineers' Handbook*, 6th Edition, McGraw-Hill, New York, 1984.

Reid, R. C., J. M. Prausnitz, and T. K. Sherwood, *The Properties of Gases and Liquids*, 3rd Edition, McGraw-Hill, New York, 1977.

Ross, L. W., "Data Sources for Calculations: Free Energy and Heat of Reaction," *Chemical Engineering*, October 10, 1966, p. 205 (list of sources).

Weast, R. C., Ed., *Handbook of Chemistry and Physics*, 66th Edition, CRC Press, Boca Raton, 1985.

Yaws, C. L., K. Y. Ki, and C. H. Kuo, "Physical and Thermodynamics Properties," series of articles appearing in issues of *Chemical Engineering* beginning June 10, 1974 and ending November 22, 1976.

PROBLEMS

1. A certain gasoline engine has an efficiency of 25%; that is, it converts into useful work 25% of the heat generated by burning a fuel. If the engine consumes 0.80 liters/h of a gasoline with a heating value of 3.5×10^4 kJ/liter, how much power does it put out? Express the answer both in kW and horsepower.

2. Consider an automobile with a mass of 5500 lb_m, braking to a stop from a speed of 55 miles/h.
 (a) How much heat (Btu) is generated by the friction of the braking process?
 (b) Suppose that throughout the United States, 100 000 000 such braking processes occur in the course of a given day. Calculate the average rate (MW) at which energy is being dissipated by the resulting friction.

3. The flow rate of liquid water through a 1-in. ID pipe is 3.00 gal/min. At what rate in ft·lb$_f$/s is kinetic energy transported by the water?

4. Air at 300°C and 130 kPa flows through a 7-cm ID pipe at a velocity of 42.0 m/s. Calculate the rate in J/s at which kinetic energy is being transported by the stream.

5. Suppose you pour a gallon of water on a yowling cat 10 ft below your bedroom window.
 (a) How much potential energy does the water lose?
 (b) How fast is the water traveling just before impact?
 (c) True or false: Energy must be conserved, therefore the kinetic energy of the water before impact must equal the kinetic energy of the cat after impact.

6. Methane enters a 3-cm ID pipe at 30°C and 10 bars with an average velocity of 5.00 m/s and emerges at a point 200 m lower than the inlet at 30°C and 9 bars. Calculate ΔE_k and ΔE_p (J/s) for the gas, assuming that the methane behaves as an ideal gas, where Δ signifies (final value − initial value).

7. You recently purchased a large plot of land in the Amazon jungle at an extremely low cost. You are quite pleased with yourself until you arrive there and find that the nearest source of electricity is 1500 miles away, a fact that your brother-in-law, the real estate agent, somehow forgot to mention. Since the local hardware store does not carry 1500-mile-long extension cords, you decide to build a small hydroelectric generator under a 75-m-high waterfall located nearby. The flow rate of the waterfall is 10^5 m^3/h, and you anticipate needing 750 kW·h/wk to run your lights, air conditioner, and television. Calculate the maximum power theoretically available from the waterfall, and see if it is sufficient to meet your needs.

8. Write and simplify the closed system energy balance equation for each of the following processes, and state whether nonzero heat and work terms are positive or negative. The solution of part (a) is given as an illustration.
 (a) The contents of a closed flask are heated from 25°C to 80°C.

 Solution

 $$Q + W = \Delta U + \Delta E_k + \Delta E_p$$

 $\quad\quad W = 0$ (no moving parts or generated currents)
 $\quad\quad \Delta E_k = 0$ (system is stationary)
 $\quad\quad \Delta E_p = 0$ (no height change)

 $$\boxed{Q = \Delta U}$$

 $$\boxed{Q > 0 \text{ (heat is added to the system)}}$$

 (b) The flask of part (a) is taken off the burner and cools back down to 25°C.
 (c) A chemical reaction takes place in a closed adiabatic (perfectly insulated) rigid container.
 (d) Repeat part (c), only suppose that the reactor is isothermal rather than adiabatic and that when the reaction was carried out adiabatically the temperature in the reactor increased.

9. If the volume of a gas increases by ΔV(liters) against a constant restraining pressure P(atm), the work transferred from the system (the gas) to its surroundings is $P\Delta V$ (liter·atm).

 (a) Calculate the energy conversion factor (J/liter·atm).

 (b) Suppose 200 calories are transferred to a gas in a balloon, and in expanding the balloon the gas does 3.62 liter·atm of work. Calculate ΔU(J) for this process.

 (c) Since energy is conserved, the 3.62 liter·atm of work done by the gas must have ended up as kinetic and/or potential and/or internal energy of something other than the gas. What became of it?

10. A cylinder with a movable piston contains 4.00 liters of a gas at 30°C and 5.00 bars. The piston is slowly moved to compress the gas to 8.00 bars.

 (a) Considering the system to be the gas in the cylinder and neglecting ΔE_p, write and simplify the energy balance equation. Do not assume that the process is isothermal in this part.

 (b) Suppose now that the process is carried out isothermally, and the compression work done on the gas equals 6.75 liter·bar. If the gas is ideal so that \hat{U} is a function only of T, how much heat (in joules) is absorbed from or released to (state which) the surroundings? (Use the gas constant table in the back of the book to determine the factor needed to convert liter·bars to joules.)

 (c) Suppose instead that the process is adiabatic, and that \hat{U} increases as T increases. Is the final system temperature greater than, equal to, or less than 30°C? (Briefly state your reasoning.)

11. A piston-fitted cylinder with a 5-cm ID contains 1.40 g of nitrogen. The mass of the piston is 5.00 kg, and a 20.00-kg weight rests on the piston. The gas temperature is 30°C, and the pressure outside the cylinder is 1.00 atm.

 (a) Prove that the absolute pressure of the gas in the cylinder is 2.26×10^5 Pa. Then calculate the volume occupied by the gas, assuming ideal gas behavior.

 (b) Suppose the weight is abruptly lifted, and the piston rises to a new equilibrium position. Further suppose that the process takes place in two steps: a rapid step in which a negligible amount of heat is exchanged with the surroundings, followed by a slow step in which the gas returns to 30°C. Considering the gas as the system, write the energy balance equation for step 1, step 2, and the overall process. In

all cases, neglect ΔE_p and ΔE_k. If \hat{U} varies proportionately with T, does the gas temperature increase or decrease in step 1? (Briefly explain your answer.)

(c) The work done by the gas equals the restraining force (the weight of the piston plus the force due to atmospheric pressure) times the distance traveled by the piston. Calculate this quantity, and use it to determine the heat released to or absorbed from (state which) the surroundings during the process.

12. The density of oxygen vapor at 150 K and 41.64 atm is 213.5 g/liter, and the specific internal energy of the vapor at this condition is listed in a table as 1706 J/mol.

 (a) Calculate the specific volume (liters/mol) of O_2 at 150 K and 41.64 atm using (i) the given density, (ii) the ideal gas law. What percentage error results from the use of the ideal gas law to calculate this quantity?

 (b) Calculate the specific enthalpy of O_2 in this state relative to the same reference state used to determine the given value of \hat{U}. (See Example 7.4-1.)

13. Saturated carbon dioxide vapor at 70°F and 853.4 psia has a specific volume of 0.08040 ft^3/lb_m and a specific enthalpy of 127.5 Btu/lb_m.

 (a) Calculate the percentage error that would result from the use of the ideal gas law to calculate \hat{V} for CO_2 at 70°F and 853.4 psia.

 (b) Calculate the specific internal energy of CO_2 in this state.

14. ,Values of the specific internal energy of bromine at three conditions are listed here.

State	$T(K)$	$P(bars)$	$\hat{V}(liters/mol)$	$\hat{U}(kJ/mol)$
1	300	0.310	0.0516	0.000
v	300	0.310	7.994	7.417
v	340	1.33	20.92	8.790

 (a) What reference state was used to generate the listed specific internal energies?

 (b) Calculate $\Delta\hat{U}(kJ/mol)$ for a process in which bromine vapor at 300 K is condensed at constant pressure. Then calculate $\Delta\hat{H}(kJ/mol)$ for the same process. (See Example 7.4-1.) Finally, calculate $\Delta H(kJ)$ for 5.00 mol of bromine undergoing the process.

 (c) Calculate $\Delta\hat{H}(kJ/mol)$ and $\Delta U(kJ)$ for a process in which bromine vapor at 300 K and 0.175 bar is heated to 340 K in a 5.00-liter container, assuming that \hat{U} is independent of pressure. Then calculate the heat (Q) required to achieve this temperature increase.

15. Prove that for an ideal gas, \hat{U} and \hat{H} are related as $\hat{H} = \hat{U} + RT$, where R is the gas constant. Then

 (a) Taking as given that the specific internal energy of an ideal gas is independent of the gas pressure, justify the claim that $\Delta\hat{H}$ for a process in which an ideal gas goes from (T_1, P_1) to (T_2, P_2) equals $\Delta\hat{H}$ for the same gas going from T_1 to T_2 at a constant pressure of P_1.

 (b) Calculate $\Delta H(cal)$ for a process in which the temperature of 2.5 mol of an ideal gas is raised by 50°C, resulting in a specific internal energy change of 3500 cal/mol.

16. You are performing an experiment to measure the specific internal energy of a gas relative to a reference state of 25°C and 1 atm (at which \hat{U} is arbitrarily set equal to 0). The gas is placed in a closed insulated 2.10-liter container at 25°C and 1 atm. A switch is

alternately closed and opened, causing a current to flow intermittently through an electrical heating coil in the chamber. The gas temperature, which is monitored with a calibrated thermocouple, increases while the circuit is closed and remains constant while it is open. A wattmeter reads 1.4 W when the circuit is closed; 90% of this power is transferred to the gas as heat. The thermocouple calibration curve is a straight line through the points $(T = 0°C, E = -0.249\ \text{mV})$ and $(T = 100°C, E = 5.27\ \text{mV})$, where E is the thermocouple potentiometer reading.

The following data are taken, where t represents the cumulative time during which the circuit was closed.

t(s)	0	30	60	90
E(mV)	1.13	2.23	3.34	4.44

(a) Which given item of information suggests that the chamber may be considered adiabatic? (*Note:* Simply saying the container is insulated does not guarantee that it is adiabatic.)

(b) Write the energy balance for the gas in the chamber and use it to calculate \hat{U}(J/mol) at each of the observed temperatures above 25°C, neglecting the work done on the gas by the stirrer. Express your solution as a table of \hat{U} versus T.

(c) What might the purpose of the stirrer be?

(d) What happens to the 0.14 W of power that do not go to heat the gas?

(e) A colleague points out to you that the calculated values of \hat{U} do not precisely correspond to the values at the calculated temperatures and 1 atm. You reply that he is quite correct, but it does not matter. Justify his statement, and state the basis of your reply. Suggest several ways to provide quantitative validation of your claim.

17. Simplify the open-system energy balance equation (Equation 7.4-15) for each of the following cases. State when possible whether nonzero heat and shaft work terms are positive or negative. The solution of part (a) is given as an illustration.

(a) Steam enters a rotary turbine and turns a shaft connected to a generator. The inlet and outlet steam ports are at the same height. Some energy is lost to the surroundings as heat.

Solution

$$Q + W_s = \Delta H + \Delta E_k + \Delta E_p$$

$$\Downarrow \Delta E_p = 0 \text{ (no height change)}$$

$$\boxed{\Delta H + \Delta E_k = Q + W_s}$$

$$\boxed{\begin{array}{l} Q \text{ is negative} \\ W_s \text{ is negative} \end{array}}$$

(b) A liquid stream is heated from 25°C to 80°C. The inlet and outlet pipes have the same diameter, and there is no change in elevation between these points.

(c) Water passes through the sluice gate of a dam and falls on a turbine rotor, which turns a shaft connected to a generator. The fluid velocity on both sides of the dam is negligible, and the water undergoes insignificant pressure and temperature changes between the inlet and outlet. (See Example 7.4-2.)

(d) Crude oil is pumped through a cross-country pipeline. The pipe inlet is 200 m higher than the outlet, the pipe diameter is constant, and the pump is located near the midpoint of the pipeline. Heat generated by friction in the line is lost through the wall.

(e) A chemical reaction takes place in a continuous reactor that contains no moving parts. Kinetic and potential energy changes from inlet to outlet are negligible.

18. Air flowing at a rate of 125 liters/min is heated from 25°C to 150°C at a constant pressure of 1.37 atm prior to entering a combustion furnace. The change in specific enthalpy associated with this transition is 870 cal/mol. Calculate the heat requirement in kW, assuming ideal gas behavior and that potential and kinetic energy changes are negligible.

19. A **Thomas flowmeter** is a device in which a known quantity of heat is added to a flowing stream and the flow rate of the stream is calculated from the measured temperature increase of the fluid. Suppose an electrical coil heater in such a device adds 1.50 kW of heat to a stream of nitrogen, raising its temperature from 30.0°C to 34.0°C at a constant absolute pressure of 1.10 atm. If the specific enthalpy of nitrogen is given by the formula

$$\hat{H}(\text{kJ/kg}) = 1.04[T(°C) - 25]$$

calculate the volumetric flow rate of the gas (liters/s) upstream of the heater (i.e., at 30°C and 1.1 atm).

20. The specific enthalpy of liquid n-hexane at 1 atm varies linearly with temperature, and equals 6.16 cal/g at 30°C and 30.8 cal/g at 50°C.

(a) Determine the equation that relates $\hat{H}(\text{cal/g})$ to $T(°C)$, and calculate the reference temperature on which the given enthalpies are based. Then derive an equation for $\hat{U}(T)(\text{cal/g})$ at 1 atm.

(b) Calculate the average rate at which heat must be removed to cool 20 kg of liquid n-hexane from 80°C to 20°C in 5 min.

21. Steam at 260°C and 7.00 bars (absolute) is expanded through a nozzle to 200°C and 4.00 bars. Negligible heat is lost from the nozzle to its surroundings. The approach velocity

of the steam is also negligible. The specific enthalpy of steam is 2974 kJ/kg at 260°C and 7 bars, and 2860 kJ/kg at 200°C and 4 bars. Use the open system energy balance equation to calculate the exiting steam velocity.

22. A flow diagram of a reactor feed mixer and preheater is shown here.

Reactant A
n_A = 100 kg/min
\hat{H}_A = 5 kJ/kg

Reactant B
n_B = 70 kg/min
\hat{H}_B = 3 kJ/kg

Reactor feed F
\hat{H}_F = 8 kJ/kg

Q (kW)

(a) Calculate ΔH for this process from Equation 7.4-14a.
(b) Under what circumstances is the number calculated in part (a) equal to the rate of heat input, Q?

23. The heart pumps blood at an average rate of 5 liters/min. The gauge pressure on the venous (intake) side is 0 mm Hg, and that on the arterial (discharge) side is 100 mm Hg. Energy is supplied to the heart as heat released by the absorption of oxygen in the cardiac muscles: 5 ml (STP) O_2/min are absorbed, and 20.2 kJ are released per ml of O_2 absorbed. Part of this absorbed energy is converted to flow work (the work done to pump blood through the circulatory system), and the balance is lost as heat transferred to the tissues surrounding the heart.
(a) Simplify Equation 7.4-12 for this system, assuming (among other things) that there is no change in internal energy from inlet to outlet.
(b) What percentage of the heat input to the heart (Q_{in}) is converted to flow work? (The answer may be thought of as the efficiency of the heart as a pump.)

24. Saturated steam at 100°C is heated to 300°C. Use the steam tables to determine (i) the required heat input (J/s) if a continuous stream flowing at 100 kg/s undergoes the process at constant pressure; (ii) the required heat input (J) if 100 kg undergo the process in a constant-volume container. What is the physical significance of the difference in the numerical values of these two quantities?

25. A fuel oil is burned in a boiler furnace, generating 556 kW of heat, of which 90% goes to produce saturated steam at 10 bars (absolute) from liquid water at 20°C. At what rate in m^3/h is steam produced?

26. Liquid water is fed to a boiler at 30°C and 15 bars and is converted at constant pressure to saturated steam. Use the steam tables to calculate $\Delta \hat{H}$(kJ/kg) for this process, and then calculate the heat input required to produce 15 000 m^3/h of steam at the exiting conditions. Assume that the kinetic energy of the entering liquid is negligible and that the steam is discharged through a 20-cm ID pipe.

27. You have been assigned to collect thermodynamic data for a new liquid product your company is about to begin manufacturing, and decide to use a continuous flow technique to generate a correlation of \hat{H} versus T. You wrap an electrical heating tape around a pipe, cover the tape with a thick layer of insulation, pump the liquid through the pipe

at the rate of 228 g/min, and adjust the power input to the heating tape with a variable resistor. For each resistance setting, you record the power input and the temperature of the liquid at the pipe outlet. You multiply the power input by a correction factor of 0.94 to determine the rate of heat input to the liquid. The entering liquid temperature remains at 25°C throughout the experiment.

The following data are taken:

T_{out}(°C)	Heat Input to the Liquid (W)
25.0	0.0
26.4	17.0
27.8	35.3
29.0	50.9
32.4	94.4

(a) Generate a table of \hat{H}(J/g) versus T(°C), taking 25°C and 1 atm as the reference state.

(b) Fit a line to the data (either graphically or by the method of least squares) to determine the coefficient b of an expression of the form $\hat{H} = b(T - 25)$.

(c) Estimate the heat required to raise 350 kg/min of the liquid from 20°C to 40°C.

(d) The correction factor of 0.94 accounts for the fact that the rate of energy input to the heating tape is somewhat greater than the rate of energy input to the liquid. Where does the additional energy go? (There are several answers.)

28. Saturated steam at 29.8 psig is to be used to heat a stream of ethane at constant pressure. The ethane enters the heat exchanger at 60°F and 20 psia at a rate of 27 890 ft³/min and is heated to 200°F. The steam condenses and leaves the exchanger as a liquid at 80°F. The specific enthalpy of ethane at 20 psia is 450 Btu/lb$_m$ at 60°F and 513 Btu/lb$_m$ at 200°F.

(a) How much heat (Btu/min) must be provided to heat the ethane from 60°F to 200°F?

(b) At what rate in ft³/min must steam be supplied to the heat exchanger?

(c) Should the heat exchanger be set up for cocurrent or countercurrent flow (see the diagram)? Explain. (*Hint:* Remember what the process is supposed to accomplish and the fact that heat always flows from a high temperature to a lower temperature).

Cocurrent flow Countercurrent flow

29. Superheated steam at 40 bars and 450°C flows at a rate of 250 kg/min to an adiabatic turbine, where it expands to 5 bars. The turbine develops 1500 kW. From the turbine the steam flows to a heater, where it is reheated isobarically to its initial temperature. Neglect kinetic energy changes.

(a) Write an energy balance on the turbine, and use it to determine the outlet steam temperature.

(b) Write an energy balance on the heater, and use it to determine the required heat input (kW) to the steam.

(c) Verify that an overall energy balance on the two-unit process is satisfied.

(d) Suppose the turbine inlet and outlet pipes both have diameters of 0.5 meter. Was it reasonable to neglect the change in kinetic energy for this unit?

30. During a period of relative inactivity, the average rate of transport of enthalpy by the metabolic and digestive waste products leaving the body minus the rate of enthalpy transport by the raw materials ingested and breathed into the body is approximately $\Delta H = -300$ kJ/h. Heat is transferred from the body to its surroundings at a rate given by

$$Q = hA(T_s - T_0)$$

where A is the body surface area (roughly 1.8 m^2 for an adult), T_s is the skin temperature (normally 34.2°C), T_0 is the temperature of the body surroundings, and h is a **heat transfer coefficient.** Typical values of h for the human body are

$$h = 8 \text{ kJ/m}^2 \cdot \text{h} \cdot °\text{C} \quad \text{(fully clothed, slight breeze blowing)}$$

$$h = 64 \text{ kJ/m}^2 \cdot \text{h} \cdot °\text{C} \quad \text{(nude, immersed in water)}$$

(Data taken from R. C. Seagrave, *Biomedical Applications of Heat and Mass Transfer*, Iowa State University Press, Ames, Iowa, 1971.)

(a) Consider the human body as a continuous system at steady state. Write the energy balance equation, making all appropriate simplifications and substitutions.

(b) Calculate the surrounding temperature for which the energy balance equation is satisfied (i.e., at which a person would feel neither hot nor cold) for a clothed person and for a nude person immersed in water.

(c) In terms of the preceding discussion, suggest why you feel colder on a windy day than on a day when the temperature is the same but there is no wind.

(d) When someone is engaged in physical exercise, would you expect the average value of ΔH to increase or decrease? In terms of the preceding discussion, suggest why you feel hot after running on a day when you felt perfectly comfortable at rest.

31. Two streams of liquid water are blended adiabatically. The first stream enters the blender at 30°C, and the second enters at 85°C. The ratio of the mass flow rate of the second stream to that of the first is 1.53:1. Write an energy balance for this process, and use it to calculate the specific enthalpy and hence (from the steam tables) the temperature of the outlet stream. (See Example 7.6-1.)

32. Steam produced in a boiler is frequently "wet"—that is, it is a saturated mixture of vapor and entrained liquid droplets. The **quality** of a wet steam is defined as the fraction of the mixture by mass that is vapor.

(a) The quality of a wet steam at 1 atm is 0.70. Use the steam tables to estimate the enthalpy of 150 kg of this steam.

(b) The specific enthalpy of a wet steam at 150.3°C is determined from the steam tables to be 2429 kJ/kg. Calculate the quality of the steam.

33. A wet steam at 5.0 bars with a quality of 0.85 (see Problem 32) is to be isothermally "dried" by evaporating the liquid mist. The flow rate of the dried steam is 500 m³(STP)/h.
 (a) Determine the temperature at which this operation occurs and the specific enthalpies of the wet and dry steams. (See Example 7.6-2.)
 (b) Calculate the heat input (kW) required for the evaporation process.

34. A **steam trap** is a device to purge steam condensate from a system without venting uncondensed steam. In one of the crudest trap types, the condensate collects and raises a float attached to a drain plug. When the float reaches a certain level, it "pulls the plug," opening the drain valve and allowing the liquid to discharge. The float then drops down to its original position and the valve closes, preventing uncondensed steam from escaping.
 (a) Suppose saturated steam at 25 bars is to be used to heat 100 kg/min of an oil with a heat capacity of 2.00 kJ/kg·°C from 135°C to 185°C. The steam condenses on the exterior of a number of tubes through which the oil is flowing. Condensate collects in the bottom of the exchanger and exists through a steam trap set to discharge when 1200 g of liquid are collected. How often does the trap discharge?
 (b) Especially when periodic maintenance checks are not performed, steam traps often fail to close completely, and so leak steam continuously. Suppose a process plant contains 1000 leaking traps (not an unrealistic supposition for some plants) operating at the conditions of part (a), and that on the average 10% additional steam must be fed to the condensers to compensate for the uncondensed steam venting through the leaks. Further, suppose that the cost of generating the additional steam is $1/10^6 Btu, where the denominator refers to the enthalpy of the leaking steam relative to liquid water at 20°C. Estimate the yearly cost of the leaks based on 24 h/day, 360 day/yr operation.

35. A turbine discharges 200 kg/h of saturated steam at 5.0 bars (absolute). It is desired to generate steam at 250°C and 5.0 bars by mixing the turbine discharge with a second stream of superheated steam at 300°C and 5.0 bars.
 (a) If 300 kg/h of the product stream are to be generated, how much heat must be added to the mixer?
 (b) If instead the mixing is carried out adiabatically, at what rate is the product steam generated?

36. Steam is contained in each of two compartments of an insulated chamber. Compartment A contains 1.5 kg of saturated steam at 1 atm, and compartment B contains steam at 2.00 bars (absolute) and 200°C. The barrier between the compartments is removed, and equilibrium is eventually reestablished. The final system internal energy is 2549 kJ/kg. Using the steam tables where necessary, calculate the mass of steam initially in compartment B. (*Note:* The flowchart for this closed system should resemble that of Example 7.6-1.)

37. Liquid water at 60 bars and 250°C passes through an adiabatic expansion valve, instantly forming a mixture of liquid and vapor at 1.2 bars. Determine the temperature of the mixture, and estimate the mass fraction of the vapor phase by writing an energy balance about the valve and neglecting ΔE_k.

38. The following diagram shows a simplified version of how a refrigerator works.

In (1) a **liquid receiver,** a liquid refrigerant (any one of a number of halogenated hydrocarbons such as CCl_2F_2) is contained at a high pressure and temperature. The liquid passes through (2) an **expansion valve,** where it flashes to a low pressure, cooling to its boiling point at this pressure and partially evaporating. The liquid–vapor mixture passes through (3) an **evaporator coil.** Air from the food storage area circulates over the coil, and the heat absorbed by the evaporating refrigerant in the coil causes the air to cool. The cold refrigerant vapor emerging from the coil passes to (4) a **compressor,** in which it is brought back to a high pressure, and in the process is raised to a high temperature. The hot vapor then passes through (5) a **condenser,** where it is cooled and condensed at constant pressure. The air that absorbs the heat given up by the condensing fluid is discharged outside the refrigerator, and the liquefied refrigerant returns to the liquid receiver.

Suppose Refrigerant R-12 (the standard name for CCl_2F_2) undergoes this cycle at a circulation rate of 40 lb_m/min, with the temperatures and pressures at the different points of the cycle being those shown on the flow diagram. Thermodynamic data for Refrigerant R-12 are as follows.

Saturated Fluid: $T = 5°F, \hat{H}_{liq} = 9.6$ Btu/lb_m, $\hat{H}_{vap} = 77.8$ Btu/lb_m
$T = 86°F, \hat{H}_{liq} = 27.8$ Btu/lb_m, $\hat{H}_{vap} = 85.8$ Btu/lb_m

Superheated Vapor: $T = 114°F, P = 93.3$ psig, $\hat{H}_{vap} \simeq 90$ Btu/lb_m

(a) Suppose the expansion valve operates adiabatically, and ΔE_k is negligible. Use an energy balance about the valve to calculate the fraction of the refrigerant that evaporates in this stage of the process.

(b) Calculate the rate in Btu/min at which heat is absorbed by the refrigerant that evaporates in the coil. (This is the useful cooling done by the system.)

(c) If the heat loss in the condenser is 2500 Btu/min, how much horsepower must the

compressor deliver to the system? (Use an overall energy balance to solve this problem.)

39. Three hundred liters/h of a 20 mole% C_3H_8–80% n-C_4H_{10} gas mixture at 0°C and 1.1 atm and 200 liters/h of a 40 mole% C_3H_8–60% n-C_4H_{10} mixture at 25°C and 1.1 atm are mixed and heated to 227°C at constant pressure. Calculate the heat requirement in kJ/h. (See Example 7.6-2.) Enthalpies of propane and n-butane are listed below. Assume ideal gas behavior.

$T(°C)$	Propane $\hat{H}(J/mol)$	Butane $\hat{H}(J/mol)$
0	0	0
25	1 772	2 394
227	20 685	27 442

40. Air saturated with water at 80°C and 1 atm is heated to 220°C and 1 atm. How much heat (kJ) is required per cubic meter (STP) of the humid air if the specific enthalpy of dry air is 0.418 kJ/mol at 80°C and 4.519 kJ/mol at 220°C. Assume that the enthalpy of water vapor at 80°C and 1 atm is that of saturated steam at 80°C.

41. Air at 35°C and 97% relative humidity is to be cooled to 18°C and fed into a plant area at a rate of 480 m³/min.

(a) Calculate the rate at which water condenses.

(b) Calculate the cooling requirement (in tons), assuming that the enthalpy of water vapor is that of saturated steam at the same temperature, and the enthalpy of dry air is given by the expression

$$\hat{H}(kJ/mol) = 0.0291[T(°C) - 25]$$

One ton of cooling corresponds to a heat transfer rate of 12 000 Btu/h.

42. A mixture containing 65 mole% acetone (Ac) and the balance acetic acid (AA) is separated in a continuous distillation column at 1 atm. A flowchart for the operation is as follows.

The overhead stream from the column is a vapor that passes through a condenser. The condensed liquid is divided into two equal streams: one is taken off as the overhead product and the other (the reflux) is returned to the column. The bottom stream from the column is a liquid that is partially vaporized in a reboiler. The liquid stream emerging from the reboiler is taken off as the bottoms product, and the vapor is returned to the column as boilup. Negligible heat is lost from the column, so that the only places in the system where external heat transfer takes place are the condenser and the reboiler.

Stream Composition Data

Feed—1: Liquid, 65.0°C; 65 mole% Ac, 35% AA

Overhead—2: Vapor, 63.0°C; 98 mole% Ac, 2% AA

Distillate—3:
Reflux—4: } Liquid, 56.8°C; 98 mole% Ac, 2% AA

Bottoms—5: Liquid, 98.7°C; 15.5 mole% Ac, 84.5% AA

Boilup—6: Vapor, 98.7°C; 54.4 mole% Ac, 45.6% AA

Thermodynamic Data

	\hat{H}(cal/mol)			
	Acetone (1 atm)		Acetic Acid (1 atm)	
$T(°C)$	\hat{H}_l	\hat{H}_v	\hat{H}_l	\hat{H}_v
56.8	0	7205	0	5723
63.0	205	7322	194	6807
67.5	354	7403	335	6884
98.7	1385	7946	1312	7420

(a) Taking 100 mol feed as a basis, calculate the net heating requirement (cal) for the process. (You may neglect heats of mixing, although doing so for dissimilar liquids such as acetone and acetic acid may introduce some error.)

(b) For the same basis, calculate the required heat input to the reboiler and the required heat removal from the condenser.

43. Saturated steam at 7.0 bars is blended with superheated steam at 10.0 bars and 300°C to produce 200 kg/h of superheated steam at 250°C. Kinetic and potential energy changes are negligible.

(a) Calculate the flow rates of the two feed streams, assuming that the process is adiabatic. (See Example 7.6-3.) State any additional assumptions you make.

(b) Recalculate the flow rates assuming a heat loss from the blender of 8500 kJ/h.

44. A **throttling calorimeter** is used to measure the quality of a wet steam (see Problem 32) flowing in a pipeline at 2.0 bars (absolute). A sidestream is bled from the pipeline through a throttling valve into a small chamber that is open to the atmosphere, and the temperature of the steam in the chamber is measured and found to be 108°C. The expansion through the valve may be considered adiabatic, and the steam in the chamber is dry. What is the quality of the steam in the pipeline?

45. A wet steam at 20 bars with a quality of 0.97 leaks through a defective steam trap (see Problems 32 and 34) and expands to a pressure of 1 atm. The process can be considered to take place in two stages: a rapid adiabatic expansion to 1 atm accompanied by complete evaporation of the liquid droplets in the wet steam, followed by cooling at 1 atm to ambient temperature. ΔE_k may be neglected in both stages.
 (a) Estimate the temperature of the superheated steam immediately following the rapid adiabatic expansion.
 (b) Someone looking at the steam trap would see a clear space just outside the leak, and a white plume forming a short distance away. (The same phenomenon can be observed outside the spout of a teakettle in which water is boiling.) Explain this observation. What would the temperature be at the point where the plume begins?

46. Eight ounces of a beverage in a glass at 65°F are to be cooled by adding ice and stirring. The enthalpy of the ice relative to liquid water at 32°F and the specific volume of the ice are $-150 \, \text{Btu/lb}_m$ and $0.01745 \, \text{ft}^3/\text{lb}_m$, respectively, and the properties of the beverage may be taken to be those of liquid water. Calculate the mass of ice that must melt to bring the liquid temperature to 40°F, neglecting heat losses to the surroundings.

47. A 25.0-g block of iron at 175°C is dropped into a liter of water in an insulated flask at 20°C and 1 atm. The specific internal energy of iron is given by the expression $\hat{U}(\text{cal/g}) = 4.13T(°C)$. Calculate the final temperature of the flask contents, assuming that the process is adiabatic, negligible evaporation occurs, negligible heat is absorbed by the flask wall, and the specific internal energy of liquid water at 1 atm and a given temperature is that of the saturated liquid at the same temperature.

48. Horatio Meshuggeneh has his own ideas of how to do things. For instance, when given the task of determining an oven temperature, most people would use a thermometer. Being allergic to doing anything most people would do, however, Meshuggeneh instead performs the following experiment. He puts a copper bar with a mass of 5.0 kg in the oven, and puts an identical bar in a well-insulated 20.0-liter vessel containing 5.0 liters of liquid water and the remainder saturated steam at 760 mm Hg. He waits long enough for both bars to reach thermal equilibrium with their surroundings, then quickly takes the first bar out of the oven, removes the second bar from the vessel, drops the first bar in its place, covers the vessel tightly, waits for the contents to come to equilibrium, and notes the reading on a pressure gauge built into the vessel. The value he reads is 50.1 mm Hg. He then uses the facts that copper has a specific gravity of 8.92, and a specific internal energy given by the expression $\hat{U}(\text{kJ/kg}) = 0.36T(°C)$, to calculate the oven temperature.
 (a) The Meshuggeneh assumption is that the bar can be transferred from the oven to the vessel without any heat being lost. If he makes this assumption, what oven temperature does Meshuggeneh calculate? How many grams of water evaporate in the process? (Assume ideal gas behavior in the vapor phase, and neglect the heat absorbed by the vessel wall—that is, assume that the heat lost by the bar is transferred entirely to the water in the vessel. Also, remember that you are dealing with a closed system once the hot bar goes into the vessel.)
 (b) In fact, the bar lost 8.3 kJ of heat between the oven and the vessel. What is the true oven temperature?

(c) The experiment just described was actually Meshuggeneh's second attempt. The first time he tried it, the final gauge pressure in the vessel was negative. What had he forgotten to do?

49. A perfectly insulated cylinder fitted with a leakproof frictionless piston with a mass of 30.0 kg and a face area of 400.0 cm^2 contains 7.0 kg of liquid water and a 3.0 kg bar of aluminum. The aluminum bar has an electrical coil imbedded in it, so that a known amount of heat can be added to it. Aluminum has a specific gravity of 2.70, and a specific internal energy given by the formula $\hat{U}(kJ/kg) = 0.94T(°C)$. The internal energy of liquid water at any temperature may be taken to be that of the saturated liquid at that temperature. Negligible heat is absorbed by the cylinder wall. Atmospheric pressure is 1.00 atm. The cylinder and its contents are initially at 20°C.

Suppose that 2680 kJ are added to the bar by means of the heating coil, and the contents of the cylinder are then allowed to equilibrate.

(a) Calculate the pressure of the cylinder contents throughout the process. Then determine whether the amount of heat added to the system is sufficient to vaporize any of the water.

(b) Determine the following quantities: (i) the final system temperature; (ii) the volumes (cm^3) of the liquid and vapor phases present at equilibrium; (iii) the vertical distance traveled by the piston from the beginning to the end of the process.

(c) Calculate an upper limit on the temperature attainable by the aluminum bar during the process, and state the condition that would have to apply for the bar to come close to this temperature.

Suggestion for part (b): Write an energy balance on the complete process, taking the cylinder contents to be the system. Note that the system is closed, and that work is done by the system when it moves the piston through a vertical displacement. The magnitude of this work is $W = -P\Delta V$, where P is the constant system pressure and ΔV is the change in system volume from the initial to the final state.

50. A rigid 5.00-liter vessel contains 4.00 liters of liquid water in equilibrium with 1.00 liter of water vapor at 25°C. Heat is added to the water by means of an immersed electrical coil. The volume of the coil is negligible. Use the steam tables to calculate the final temperature and pressure (bars) of the system and the mass of water vaporized (g) if 3164 kJ are added to the water and no heat is lost from the water to its surroundings. (*Note:* A trial-and-error calculation is required.)

***51.** A liquid mixture of benzene and toluene is to be separated in a continuous single-stage equilibrium flash tank.

The pressure in the unit may be adjusted to any desired value, and the heat input may be similarly adjusted to vary the temperature at which the separation is carried out. The vapor and liquid product streams both emerge at the temperature $T(°C)$ and pressure P(mm Hg) maintained in the vessel.

Assume that the vapor pressures of benzene and toluene are given by the Antoine equation, Table 6.1-1; that Raoult's law—Equation 6.4-1—applies; and that enthalpies of benzene and toluene liquid and vapor are linear functions of temperature. Specific enthalpies at two temperatures are given here for each substance in each phase.

$C_6H_6(l)$: $(T = 0°C, \ \hat{H} = 0 \text{ kJ/mol})$, $(T = 80°C, \ \hat{H} = 10.85 \text{ kJ/mol})$
$C_6H_6(v)$: $(T = 80°C, \ \hat{H} = 41.61 \text{ kJ/mol})$, $(T = 120°C, \ \hat{H} = 45.79 \text{ kJ/mol})$
$C_7H_8(l)$: $(T = 0°C, \ \hat{H} = 0 \text{ kJ/mol})$, $(T = 111°C, \ \hat{H} = 18.58 \text{ kJ/mol})$
$C_7H_8(v)$: $(T = 89°C, \ \hat{H} = 49.18 \text{ kJ/mol})$, $(T = 111°C, \ \hat{H} = 52.05 \text{ kJ/mol})$

(a) Write a subroutine to take as input a temperature $T(°C)$ and total pressure P (mm Hg) and to calculate the equilibrium compositions (mole fractions) of both the liquid and vapor phases. The program should first determine if $P < p_T^*(T)$ or $P > p_B^*(T)$, and stop execution if either condition applies. Test your program by proving that at 90°C and 652 mm Hg the liquid contains 39.9% benzene and the vapor contains 62.5% benzene.

(b) Write a program to take as input a temperature and pressure and to solve material and energy balances on the unit for the moles of liquid product, moles of vapor product, composition of each phase, and heat input required per mol of feed. The subroutine of part (a) should be incorporated into this program. Use the program to calculate the product yields and required heat input for the flash vaporization of an equimolar mixture of benzene and toluene with $T_F = 130°C$ at (i) $T = 90°C$, $P = 652$ mm Hg, and (ii) $T = 89°C$, $P = 652$ mm Hg.

52. A brine solution with a specific gravity of 1.03 flows through a channel with a variable cross section. Data taken at two axial positions in the channel are shown here.

* Computer problem.

	Point 1	Point 2
P_{gauge}	1.5×10^5 Pa	9.77×10^4 Pa
v	5 m/s	?

(a) Neglecting friction, calculate the velocity at Point 2. (See Example 7.7-1.)

(b) If the pipe diameter at Point 2 is 6 cm, what is the diameter at Point 1?

53. A **venturi meter** is a device to measure fluid flow rates, which in its operation resembles the orifice meter (Section 3.2b). It consists of a tapered constriction in a line, with pressure taps leading to a differential manometer at points upstream of the constriction and at the point of maximum constriction (the **throat**). The manometer reading is directly related to the flow rate in the line.

Suppose the flow rate of an incompressible fluid is to be measured in a venturi meter in which the cross sectional area at Point 1 is exactly four times that at Point 2.

(a) Derive the relationship between the velocities v_1 and v_2 at Points 1 and 2.

(b) Write the Bernoulli equation for the system between Points 1 and 2, and use it to prove that to the extent friction is negligible

$$P_1 - P_2 = \frac{15\rho V^2}{2A_1^2 g_c}$$

where P_1 and P_2 are the pressures at Points 1 and 2, ρ and V are the density and volumetric flow rate of the flowing fluid, and A_1 is the cross-sectional area of the pipe at Point 1.

(c) Suppose this meter is used to measure the flow rate of a stream of water, using mercury as the manometer fluid, and a manometer reading $h = 30$ cm is obtained. What is the volumetric flow rate of the water if the pipe diameter at Point 1 is 15 cm? (Recall the differential manometer equation, Equation 3.4-6.)

54. Methanol is contained in a large tank under a pressure of 3.5 bars (absolute). When a valve on the bottom of the tank is opened, the methanol drains freely through a 1-cm ID tube, whose outlet is 7 m below the surface of the methanol. The pressure at the outlet of the discharge pipe is 1 atm.

(a) Use the Bernoulli equation to estimate the methanol discharge velocity and flow rate in liters/min when the discharge valve is fully opened. Neglect the rate of fall of the methanol level in the tank. (See Example 7.7-2.)

(b) When the discharge valve is partially closed, the flow rate decreases, which means that $\Delta v^2/2g_c$ changes. However, the other two terms in the Bernoulli equation ($\Delta P/\rho$ and $g\,\Delta z/g_c$) remain the same. How do you account for this apparently contradictory result? (*Hint:* Examine the assumptions made in the derivation of the Bernoulli equation.)

55. Water is to be pumped from a lake to a ranger station on the side of a mountain (see figure). The flow rate is to be 100 gal/min, and the flow channel is a standard 1-in. Schedule 40 steel pipe (ID = 1.049 in.). A pump capable of delivering 8 hp ($= W_s$) is available. The friction loss F (ft·lb$_f$/lb$_m$) equals $0.041L$, where L(ft) is the length of the pipe. Calculate the maximum elevation, z, of the ranger station above the lake if the pipe rises at an angle of 30°.

56. Water from a reservoir passes over a dam through a turbine and discharges from a 70-cm ID pipe at a point 60 m below the reservoir surface. The turbine delivers 0.80 MW. Calculate the required flow rate of water in m³/min if friction is neglected. (See Example 7.7-3.) If friction were included, would a higher or lower flow rate be required? (*Note:* The equation you will solve in this problem has multiple roots. Find a solution less than 2 m³/s.)

57. Cloth bag filters are used to remove particulate matter from the gases passing out through several process and boiler stacks in a large industrial plant. The bags become clogged and must be replaced frequently. Since they are quite expensive, rather than

being discarded they are emptied, washed, and reused. In the washing process, a detergent solution with a specific gravity of 0.96 flows from a storage tank to a washing machine. The liquid effluent from the machine is pumped through a filter to remove dirt, and the cleaned detergent is recycled back to the storage tank.

Detergent flows from the storage tank to the washing machine by gravity feed at a rate of 600 liters/min. All pipes in the line have 4.0-cm ID. Friction losses are negligible in the line from the tank to the washing machine when the valve is fully opened, and $F = 72$ J/kg in the return line, which includes the pump and filter.

(a) Calculate the value of the height H (see figure) required to provide the desired flow rate of detergent into the washing machine when the valve is fully opened.

(b) Suppose the pump has an efficiency of 75%; that is, it delivers 75% of its rated value as shaft work. What must the rated value (kW) of the pump be to return 600 liters/min of detergent to the storage tank?

58. One thousand liters of a 95 wt% glycerol°5% water solution are to be diluted to 60% glycerol by adding a 35% solution pumped from a large storage tank through a 5-cm ID pipe at a steady rate. The pipe discharges at a point 20 m higher than the liquid surface in the storage tank. The operation is carried out isothermally and takes 13 min to complete. The friction loss (F of Equation 7.7-2) is 50 J/kg. Calculate the final solution volume and the shaft work in kW that the pump must deliver, assuming that the surface of the stored solution and the pipe outlet are both at 1 atm.

Data: $\rho_{H_2O} = 1.00$ kg/liter, $\rho_{gly} = 1.26$ kg/liter. (Use to estimate solution densities.)

Chapter Eight

Balances on Nonreactive Processes

In the energy balance problems of Chapter 7, all enthalpies and internal energies could be found in tables. In this chapter and Chapter 9 we present methods for evaluating $\Delta \hat{H}$ and $\Delta \hat{U}$ when tables of these properties are not available. Once these calculations have been performed for a process, the energy balance may be written and solved as before.

8.1 STATE PROPERTIES AND HYPOTHETICAL PROCESS PATHS

Even when you cannot find a table of enthalpies for a substance, you will often find data that allow you to calculate the enthalpy changes associated with certain processes; specifically,

1. Changes in pressure at constant temperature.
2. Changes in temperature at constant pressure.
3. Changes in phase at constant temperature and pressure.
4. Mixing or separation of two or more substances at constant temperature and pressure.
5. Chemical reaction at constant temperature and pressure.

To evaluate $\Delta \hat{H}$ for an arbitrary process, you may substitute a hypothetical sequence composed of such steps leading from the initial state to the final state. The sequence is referred to as a **process path.** Having done this, you may evaluate $\Delta \hat{H}$ for each step (the required data are presumably available), and *since \hat{H} is a state property*, calculate $\Delta \hat{H}$ for the entire process as the sum of the $\Delta \hat{H}$'s for the individual steps.

For example, suppose you wish to calculate $\Delta \hat{H}$ for a process in which ice at $-5°C$ and 1 atm is converted to steam at $300°C$ and 5 atm. One possible process path is shown in Figure 8.1-1. (Verify that each of the six steps in this path is one of the five basic types.) For the overall process,

$$\Delta \hat{H} = \sum_{i=1}^{6} \Delta \hat{H}_i$$

In the remainder of this chapter we discuss the evaluation of enthalpy and internal energy changes associated with changes in temperature and pressure, phase changes,

FIGURE 8.1-1

Hypothetical sequence of process steps

and mixing. Energy balances on processes involving chemical reactions will be discussed in Chapter 9.

TEST YOURSELF

1. How does the fact that \hat{H} is a state property simplify the calculation of $\Delta\hat{H}$ for an arbitrary change of state?

2. Construct appropriate paths for each of the following processes that consist of sequential steps of the five types listed at the beginning of Section 8.1.
 (a) Nitrogen at 20°C and 200 mm Hg is heated and compressed to 140°C and 40 atm. (*Solution:* Heat at 200 mm Hg from 20°C to 140°C, then compress isothermally from 200 mm Hg to 40 atm.)
 (b) Liquid acetone at 65°F and 1 atm is heated to form acetone vapor at 400°F and 1 atm.
 (c) Water at 30°C and 1 atm and NaOH at 25°C and 1 atm are mixed to form an aqueous solution of NaOH at 50°C and 1 atm.
 (d) O_2 and CH_4 are mixed at 170°C and 1 atm and react completely to form CO_2 and H_2O at 300°C and 1 atm. Assume that the enthalpy change for the reaction occurring at 25°C and 1 atm is known.

8.2 CHANGES IN PRESSURE AT CONSTANT TEMPERATURE

It has been observed experimentally that internal energy is nearly independent of pressure for solids and liquids, as is specific volume. *If the pressure of a solid or liquid changes but the temperature remains constant, you may therefore write* $\Delta\hat{U} \approx 0$ *and* $\Delta\hat{H} (= \Delta\hat{U} + \Delta P\hat{V}) \approx \hat{V}\Delta P.$

Both \hat{U} and \hat{H} can be shown to be independent of pressure for ideal gases. Consequently, *unless gases near or above their critical pressures are involved, it is normally*

safe to assume $\Delta \hat{U} \approx 0$ *and* $\Delta \hat{H} \approx 0$ *for a gas undergoing an isothermal pressure change.* (If enthalpy tables are available for the gas, there is of course no need to make this assumption.)

If you are required to evaluate $\Delta \hat{H}$ for a change in pressure of a gas under conditions such that the ideal gas law cannot be invoked, you must find either tabulated values of $\hat{H}(T, P)$ for the gas or a thermodynamic relation for changes in \hat{H} with changes in P at constant T. A good source of such relations is Chapter 5 of Reid, Prausnitz, and Sherwood.[1]

TEST YOURSELF

1. Which of the following assumptions appears reasonable for each of the isothermal processes to be described below? (i) $\Delta \hat{U} \approx 0$, $\Delta \hat{H} \approx 0$; (ii) $\Delta \hat{U} \approx 0$, $\Delta \hat{H} \neq 0$; (iii) Neither (i) or (ii).
 (a) H_2O (l, 1 atm) \rightarrow H_2O (1,200 atm), $T \equiv 25°C$
 (b) N_2 (g, 1 atm) \rightarrow N_2 (g, 1.2 atm), $T \equiv 25°C$
 (c) N_2 (g, 1 atm) \rightarrow N_2 (g, 200 atm), $T \equiv 25°C$
2. Consider the process

$$C_2H_6 \text{ (g, 25°C, 1 atm)} \rightarrow C_2H_6 \text{ (g, 25°C, 30 atm)}$$

How would you use the compressibility charts to determine whether it is reasonable to neglect $\Delta \hat{H}$ for this process?

8.3 CHANGES IN TEMPERATURE

8.3a Sensible Heat and Heat Capacities

The term **sensible heat** denotes heat that must be transferred to raise or lower the temperature of a substance or mixture of substances. The quantity of heat required to produce a specified temperature change in a system can be determined from the appropriate form of the first law of thermodynamics:

$$Q = \Delta U \text{ (Closed system)} \tag{8.3-1}$$

$$Q = \Delta H \text{ (Open system)} \tag{8.3-2}$$

(We have neglected kinetic and potential energy changes and work.) To determine the sensible heat Q for a heating or cooling process, you must therefore be able to evaluate ΔU or ΔH for the specified temperature change.

The specific internal energy of a substance depends strongly on temperature. If the temperature is raised or lowered in such a way that the system volume remains constant, the specific internal energy might vary as shown in the following plot.

[1] R. C. Reid, J. H. Prausnitz, and T. K. Sherwood, *The Properties of Gases and Liquids*, 3rd Edition, McGraw-Hill, New York, 1977.

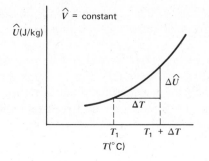

A temperature change ΔT from T_1 leads to a change $\Delta \hat{U}$ in specific internal energy. As $\Delta T \to 0$, the ratio $\Delta \hat{U}/\Delta T$ approaches a limiting value (i.e., the slope of the curve at T_1), which is by definition the **heat capacity at constant volume** of the substance, denoted by C_v.

$$C_v(T) = \left\{ \lim_{\Delta T \to 0} \frac{\Delta \hat{U}}{\Delta T} \right\} = \frac{d\hat{U}}{dT} \tag{8.3-3}$$

Since the plot of \hat{U} versus T is not generally a straight line, C_v (the slope of the curve) is a function of temperature.

The change in \hat{U} for a temperature rise from T to $T + dT$ at constant volume is from Equation 8.3-3

$$d\hat{U} = C_v(T)\,dT \tag{8.3-4}$$

and the change $\Delta \hat{U} = \hat{U}_2 - \hat{U}_1$ associated with a temperature change from T_1 to T_2 at constant volume is, accordingly,

$$\Delta \hat{U} = \int_{T_1}^{T_2} C_v(T)\,dT \tag{8.3-5}$$

Now, suppose both the temperature and the volume of a substance change. To calculate $\Delta \hat{U}$, you may break the process into two steps—a change in \hat{V} at constant T followed by a change in T at constant \hat{V}.

$$A(T_1, \hat{V}_1) \xrightarrow{\Delta \hat{U}_1} A(T_1, \hat{V}_2) \xrightarrow{\Delta \hat{U}_2} A(T_2, \hat{V}_2)$$
$$\underset{\Delta \hat{U}}{\underline{}}$$

Since \hat{U} is a state property, $\Delta \hat{U}$ (the desired quantity) equals $\Delta \hat{U}_1 + \Delta \hat{U}_2$. But for ideal gases and (to a good approximation) liquids and solids, \hat{U} depends only on T, so that since T is constant in step 1, $\Delta U_1 \approx 0$ for any substance but a nonideal gas. Moreover, since the second step is a change in temperature at constant volume, $\Delta \hat{U}_2$ is given by Equation 8.3-5. In summary, then, for a temperature change from T_1 to T_2

$$\boxed{\Delta \hat{U} = \int_{T_1}^{T_2} C_v(T)\,dT} \quad \begin{array}{l} \text{Ideal gas: exact} \\ \text{Solid or liquid: good approximation} \\ \text{Real gas: valid only if } \hat{V} \text{ is constant} \end{array} \tag{8.3-6}$$

EXAMPLE 8.3-1 Evaluation of an Internal Energy Change from a Tabulated Heat Capacity

Calculate the heat required to raise 200 kg of nitrous oxide from 20°C to 150°C. The constant volume heat capacity of N_2O in this temperature range is given by the equation

$$C_v \, (kJ/kg \cdot °C) = 0.855 + 9.42 \times 10^{-4}T$$

where T is in °C.

Solution

Assume ideal gas behavior. From Equation 8.3-6,

$$\Delta \hat{U}(kJ/kg) = \int_{20}^{150} (0.855 + 9.42 \times 10^{-4}T)\left(\frac{kJ}{kg \cdot °C}\right) dT(°C)$$

$$= 0.855T \Big]_{20}^{150} + \frac{9.42 \times 10^{-4}T^2}{2}\Big]_{20}^{150}$$

$$= (111 + 10.4) = 121 \, kJ/kg$$

The energy balance for this closed system is

$$Q = \Delta U = m(kg)\,\Delta \hat{U}(kJ/kg) = (200 \, kg)(121 \, kJ/kg) = \boxed{24\,200 \, kJ}$$

Question: Where was the assumption of ideal gas behavior used?

Next, let us suppose that we heat a substance at constant pressure and consider the resulting change in enthalpy. Like internal energy, enthalpy depends strongly on temperature. If $\Delta \hat{H}$ is the change in specific enthalpy resulting from a temperature increase at constant pressure from T to $T + \Delta T$, then as ΔT approaches zero the ratio $\Delta \hat{H}/\Delta T$ approaches a limiting value, defined as the **heat capacity at constant pressure** and denoted by C_p.

$$C_p(T) = \left\{ \lim_{\Delta T \to 0} \frac{\Delta \hat{H}}{\Delta T} \right\} = \frac{d\hat{H}}{dT} \tag{8.3-7}$$

Proceeding as before, we observe that the change in \hat{H} for a temperature change at constant pressure from T to $T + dT$ is

$$d\hat{H} = C_p(T)\,dT$$

and so for a change from T_1 to T_2 at constant pressure

$$\Delta \hat{H} = \int_{T_1}^{T_2} C_p(T)\,dT \tag{8.3-8}$$

For a process $A(T_1, P_1) \to A(T_2, P_2)$, we may construct a two-step process path

$$A(T_1, P_1) \xrightarrow{\Delta \hat{H}_1} A(T_1, P_2) \xrightarrow{\Delta \hat{H}_2} A(T_2, P_2)$$

$$\Delta \hat{H}$$

The first step is a change in pressure at constant temperature, the type of process described in Section 8.2. We saw in that section that

$$\Delta \hat{H}_1 = 0 \text{ (ideal gas)}$$
$$\approx \hat{V} \Delta P \text{ (solid or liquid)} \tag{8.3-9}$$

The second step is a change in temperature at constant pressure, so that $\Delta \hat{H}_2$ is given by Equation 8.3-8. Finally, since $\Delta \hat{H} = \Delta \hat{H}_1 + \Delta \hat{H}_2$ (why?), we obtain

$$\boxed{\Delta \hat{H} = \int_{T_1}^{T_2} C_p(T) \, dT} \quad \begin{array}{l} \text{Ideal gas: exact} \\ \text{Real gas: exact only if } P \text{ is constant.} \end{array} \tag{8.3-10a}$$

$$\boxed{\Delta \hat{H} = \hat{V} \Delta P + \int_{T_1}^{T_2} C_p(T) \, dT} \quad \text{Solid or liquid} \tag{8.3-10b}$$

Evaluating $\Delta \hat{H}$ for a nonideal gas undergoing a temperature and pressure change is best done using tabulated enthalpies. If none are available, a thermodynamic relation for variations of \hat{H} with P must be combined with Equation 8.3-8 to determine the enthalpy change; such relations can be found on pp. 263-267 of Reid, Prausnitz, and Sherwood.

TEST YOURSELF

1. By definition, what are C_v and C_p?
2. Suppose the formula

$$\Delta \hat{H} = \int_{T_1}^{T_2} C_p(T) \, dT$$

 is used to calculate the specific enthalpy change for a change in temperature *and pressure* undergone by (a) an ideal gas, (b) a highly nonideal gas, (c) a liquid. For which of these is the formula exact, and for which is it most likely to lead to a significant error?
3. If C_p for an ideal gas is 0.5 cal/g·°C (a constant), what is the enthalpy change in calories corresponding to a change from 10°C to 30°C undergone by 5 g of the gas?

CREATIVITY EXERCISE

Your task this time is to estimate the heat capacity of an unknown liquid. You have available a laboratory balance, a very well-insulated container, a sensitive thermometer that can measure liquid temperatures, and a thermocouple that can measure temperatures of solids. The container is a very poor conductor of heat, so that any heat added to or removed from its contents goes entirely into changing the temperature of those contents. If you need anything else (within reason), you can get it. Devise

as many ways as you can to estimate C_v, which you may assume is independent of temperature. (*Example:* Mix in the insulated flask a known mass, m_1, of your liquid at temperature T_1, and a known mass, m_2, of hot water at temperature T_2, and measure the final temperature T_f. Since you can calculate the heat lost by the water, $Q = m_2 C_{vw}(T_2 - T_f)$, and you know Q must also equal the heat gained by the other liquid, $m_1 C_{v1}(T_f - T_1)$, you can solve for C_v.)

8.3b Heat Capacity Formulas and Mean Heat Capacities

The heat capacities C_v and C_p are physical properties of materials, and are tabulated in standard references such as *Perry's Chemical Engineers' Handbook*.[2] They may be expressed in any units of energy per unit amount per unit temperature interval—for example, J/mol·K, or Btu/lb$_m$·°F. The term **specific heat** is sometimes used to denote a heat capacity expressed on a per-unit-mass basis.

Heat capacities are functions of temperature and are frequently expressed in polynomial form $(C_p = a + bT + cT^2 + dT^3)$. Values of the coefficients a, b, c, and d are given in Table B.2 of Appendix B for a number of substances, and listings for additional substances are given on pp. 3-129 to 3-146 of *Perry's Chemical Engineers' Handbook*.

When reading the coefficients of a heat capacity formula from Table B.2, do not mistake their orders of magnitude: If a value of 0.83 is read from the column labeled $b \cdot 10^2$, then the value of b must be two orders of magnitude *less* than 0.83, or $b = 0.83 \times 10^{-2}$.

Simple relationships exist between C_p and C_v in two cases:

$$\boxed{\textit{Liquids and solids: } C_p \approx C_v} \qquad \text{(8.3-11)}$$

$$\boxed{\textit{Ideal gases: } C_p = C_v + R} \qquad \text{(8.3-12)}$$

where R is the gas constant. (Try to prove the second relation.) The relationship between C_p and C_v for nonideal gases is complex and will not be discussed in this text.

EXAMPLE 8.3-2 Cooling of an Ideal Gas

Assuming ideal gas behavior, calculate the heat that must be transferred in each of the following cases.

1. A stream of nitrogen flowing at a rate of 100 mol/min is heated from 20°C to 100°C.

2. Nitrogen contained in a 5-liter flask at an initial pressure of 3 bars is cooled from 90°C to 30°C.

[2] R. H. Perry and D. W. Green, Eds., *Perry's Chemical Engineers' Handbook*, 6th Edition, McGraw-Hill, New York, 1984.

Solution

Neglecting kinetic energy changes, the energy balance equation for the open system of part (1) is $Q = \Delta H$, and that for the closed system of part (2) is $Q = \Delta U$. (Prove it.) The problem is therefore to evaluate ΔH and ΔU for the two specified processes.

1. From Table B.2, Appendix B, the heat capacity at constant pressure of N_2 is

$$C_p(\text{J/mol}\cdot^\circ\text{C}) = 29.00 + 0.2199 \times 10^{-2}T + 0.5723 \times 10^{-5}T^2 - 2.871 \times 10^{-9}T^3$$

where T is in $^\circ$C. Since we are assuming ideal gas behavior, the enthalpy change for the gas is independent of any pressure change that may occur, and hence, from Equation 8.3-10a,

$$\Delta\hat{H} = \int_{20}^{100} C_p(T)\,dT$$

$$\Downarrow$$

$$\Delta\hat{H}(\text{J/mol}) = 29.00T\Big]_{20}^{100} + 0.2199 \times 10^{-2}\frac{T^2}{2}\Big]_{20}^{100} + 0.5723 \times 10^{-5}\frac{T^3}{3}\Big]_{20}^{100}$$

$$-2.871 \times 10^{-9}\frac{T^4}{4}\Big]_{20}^{100}$$

$$= 2320 + 10.6 + 1.9 - 0.07 = 2332 \text{ J/mol}$$

Finally,

$$Q = \Delta H = n\,\Delta\hat{H}$$

$$= 100\,\frac{\text{mol}}{\text{min}}\,\Big|\,\frac{2332 \text{ J}}{\text{mol}} = \boxed{2.332 \times 10^5 \text{ J/min}}$$

2. To evaluate ΔU, we need the number of moles n, which may be calculated using the ideal gas law, and $\Delta\hat{U}$. To determine the latter quantity we need the constant volume heat capacity, which from Equation 8.3-12 is

$$C_v = C_p - R$$

$$\Big\| \quad C_p \text{ given above}$$

$$\Downarrow \quad R = (8.314 \text{ J/mol}\cdot\text{K})(1 \text{ K/1}^\circ\text{C})$$

$$C_v = 20.69 + 0.2199 \times 10^{-2}T + 0.5723 \times 10^{-5}T^2 - 2.871 \times 10^{-9}T^3$$

Calculate $\Delta\hat{U}$

$$\Delta\hat{U} = \int_{90}^{30} C_v(T)\,dT$$

$$\Downarrow$$

$$\Delta\hat{U} = 20.69T\Big]_{90}^{30} + 0.2199 \times 10^{-2}\frac{T^2}{2}\Big]_{90}^{30} + 0.5723 \times 10^{-5}\frac{T^3}{3}\Big]_{90}^{30}$$

$$-2.871 \times 10^{-9}\frac{T^4}{4}\Big]_{90}^{30}$$

$$= -1241 - 7.92 - 1.34 + 0.05 = -1.250 \times 10^3 \text{ J/mol}$$

Calculate n

At the initial condition (the only point at which we know P, V, and T)

$$n = PV/RT$$

$$= \frac{(3.00 \text{ bars})(5.00 \text{ liters})}{(0.08314 \text{ liter} \cdot \text{bar/mol} \cdot \text{K})(363 \text{ K})} = 0.497 \text{ mol}$$

Calculate Q

$$Q = \Delta U = n \Delta \hat{U}$$

$$= (0.497 \text{ mol})(-1.250 \times 10^3 \text{ J/mol}) = \boxed{-621 \text{ J}}$$

If the change in specific enthalpy of a substance that goes from T_1 to T_2 is $\hat{H}_2 - \hat{H}_1$, a **mean heat capacity**, \bar{C}_p, may be defined as

$$\bar{C}_p = \frac{\hat{H}_2 - \hat{H}_1}{T_2 - T_1} \tag{8.3-13}$$

Substituting for $\hat{H}_2 - \hat{H}_1$ from Equation 8.3-8 yields

$$\bar{C}_p(T_1 \rightarrow T_2) = \frac{\int_{T_1}^{T_2} C_p(T) \, dT}{T_2 - T_1} \tag{8.3-14}$$

Once C_p is known for specified change in temperature (e.g., from a table of mean heat capacities), you may calculate the corresponding enthalpy change as

$$\boxed{\Delta \hat{H} = \bar{C}_p \Delta T} \tag{8.3-15}$$

A simple multiplication thus replaces integration of $C_p \, dT$ in the calculation of $\Delta \hat{H}$. Note, however, that if C_p is not known and the calculation of $\Delta \hat{H}$ is to be done only once, there is little point in first determining \bar{C}_p from Equation 8.3-14, since the numerator of this equation is the desired quantity.

Tables 8.3-1 and 8.3-2 give mean heat capacities for the transitions of several common gases from a single base temperature T_{ref} to higher temperatures. You may use these tables to calculate the specific enthalpy of any of the tabulated species relative to the reference state as $\hat{H} = \bar{C}_p(T - T_{\text{ref}})$; you may also use the table to calculate $\Delta \hat{H}$ for a change between any two temperatures T_1 and T_2:

$$\boxed{\begin{aligned} \Delta \hat{H}(T_1 \rightarrow T_2) &= \hat{H}(T_2) - \hat{H}(T_1) \\ &= (\bar{C}_p)_{T_2}(T_2 - T_{\text{ref}}) - (\bar{C}_p)_{T_1}(T_1 - T_{\text{ref}}) \end{aligned}} \tag{8.3-16}$$

(The validity of this equation follows from the fact that \hat{H} is a state property. Do you see how?)

TABLE 8.3-1

Mean Heat Capacities of Ideal Combustion Gases: SI Units[a]

				\bar{C}_p (J/mol·°C)			
			Reference state: $P_{ref} = 1$ atm, $T_{ref} = 25°C$				
$T(°C)$	Air	O_2	N_2	H_2	CO	CO_2	H_2O
0	28.94	29.24	29.03	28.84	29.00	36.63	33.55
25	29.05	29.39	29.06	28.84	29.06	37.15	33.63
100	29.21	29.80	29.16	28.86	29.23	38.63	33.92
200	29.45	30.32	29.32	28.90	29.47	40.45	34.34
300	29.71	30.80	29.52	28.95	29.72	42.10	34.80
400	29.97	31.24	29.74	29.03	29.99	43.59	35.29
500	30.25	31.65	29.98	29.12	30.27	44.93	35.81
600	30.53	32.02	30.24	29.23	30.56	46.14	36.36
700	30.81	32.39	30.51	29.35	30.85	47.23	36.92
800	31.10	32.71	30.79	29.48	31.14	48.20	37.49
900	31.38	33.02	31.07	29.63	31.42	49.07	38.08
1000	31.65	33.30	31.34	29.78	31.70	49.85	38.66
1100	31.92	33.55	31.62	29.94	31.97	50.54	39.24
1200	32.18	33.79	31.88	30.12	32.23	51.18	39.81
1300	32.42	34.02	32.13	30.29	32.47	51.75	40.37
1400	32.65	34.23	32.37	30.47	32.69	52.28	40.91
1500	32.85	34.42	32.58	30.66	32.89	52.77	41.42

[a] Calculated from the heat capacity formulas of Table B.2.

Example: The enthalpy of oxygen at 700°C relative to O_2 at 25°C [or the enthalpy change for the process $O_2(25°C) \rightarrow O_2(700°C)$] is

$$\hat{H}_{O_2}(700°C) = (\bar{C}_p)_{700°C}(700°C - 25°C)$$

$$= \left(32.39 \frac{J}{mol·°C}\right)(675°C) = 21\ 860\ J/mol$$

EXAMPLE 8.3-3 Mean Heat Capacities

Fifteen kilogram-moles per second of air are cooled from 430 to 100°C. Calculate the required heat removal rate using tabulated mean heat capacities.

Solution

Assume ideal gas behavior, so that pressure changes (if there are any) do not affect $\Delta\hat{H}$. From Table 8.3-1,

$$(\bar{C}_p)_{25°C \rightarrow 100°C} = 29.21\ kJ/kmol·°C$$

$$(\bar{C}_p)_{25°C \rightarrow 430°C} = 30.05\ kJ/kmol·°C \text{ (by linear interpolation)}$$

$$\Downarrow \text{ Equation 8.3-16, } T_{ref} = 25°C$$

$$(\Delta\hat{H})_{430°C \rightarrow 100°C} = \left(29.21 \frac{kJ}{kmol·°C}\right)(100°C - 25°C) - (30.05)(430 - 25)$$

$$= -9980\ kJ/kmol$$

TABLE 8.3-2

Mean Heat Capacities of Ideal Combustion Gases: American Engineering Units[a]

$$\overline{C}_p(\text{Btu/lb-mole}\cdot{}^\circ\text{F})$$
Reference state: $P_{ref} = 1$ atm, $T_{ref} = 77{}^\circ\text{F}$

$T({}^\circ\text{F})$	Air	O_2	N_2	H_2	CO	CO_2	H_2O
32	6.942	6.989	6.938	6.893	6.931	8.755	8.019
77	6.943	7.024	6.944	6.895	6.945	8.879	8.038
100	6.949	7.041	6.948	6.895	6.952	8.941	8.049
200	6.979	7.114	6.967	6.898	6.981	9.203	8.100
300	7.010	7.184	6.988	6.902	7.013	9.451	8.154
400	7.042	7.251	7.011	6.908	7.045	9.686	8.211
500	7.075	7.315	7.036	6.915	7.079	9.909	8.271
600	7.110	7.377	7.063	6.923	7.113	10.12	8.334
700	7.145	7.437	7.092	6.933	7.149	10.32	8.399
800	7.181	7.494	7.123	6.944	7.186	10.51	8.467
900	7.217	7.548	7.155	6.956	7.223	10.68	8.536
1000	7.254	7.600	7.189	6.970	7.261	10.85	8.607
1100	7.291	7.650	7.233	6.984	7.299	11.01	8.680
1200	7.329	7.698	7.258	7.000	7.337	11.16	8.754
1300	7.367	7.744	7.295	7.016	7.376	11.30	8.830
1400	7.405	7.788	7.331	7.034	7.414	11.43	8.906
1500	7.442	7.830	7.368	7.052	7.453	11.55	8.983
1600	7.480	7.871	7.405	7.071	7.491	11.67	9.060
1700	7.517	7.909	7.443	7.091	7.529	11.78	9.138
1800	7.554	7.946	7.480	7.112	7.566	11.88	9.215
1900	7.590	7.982	7.516	7.133	7.602	11.98	9.293
2000	7.625	8.016	7.553	7.155	7.638	12.07	9.370

[a] Calculated from the heat capacity formulas of Table B.2.
Example: The enthalpy of water vapor at 1500°F relative to $H_2O(v)$ at 77°F is

$$\hat{H}_{H_2O(v)}(1500{}^\circ\text{F}) = \overline{C}_p(1500{}^\circ\text{F})(1500{}^\circ\text{F} - 77{}^\circ\text{F})$$

$$= \left(8.983\,\frac{\text{Btu}}{\text{lb-mole}\cdot{}^\circ\text{F}}\right)(1423{}^\circ\text{F}) = 12\,780\,\text{Btu/lb-mole}$$

The energy balance, neglecting kinetic energy changes, is

$$Q = \Delta H = n\Delta\hat{H}$$

$$= \frac{15\,\text{kmol}}{\text{s}}\left|\frac{-9980\,\text{kJ}}{\text{kmol}}\right|\frac{1\,\text{kW}}{1\,\text{kJ/s}} = \boxed{-1.497 \times 10^5\,\text{kW}}$$

Note: The mean heat capacities of the two preceding tables (and for that matter, the heat capacity formulas of Table B.2) apply strictly to heating or cooling at a constant pressure of 1 atm. The tabulated capacities may also be used for nonisobaric heating or cooling of ideal or nearly ideal gases. However, at pressures high enough for the gases to be highly nonideal, enthalpy tables or more precise heat capacity formulas should be used.

TEST YOURSELF

1. The gas constant R is approximately equal to 2 cal/mol·K. If C_p for a gas is 7 cal/mol·°C, estimate C_v. If C_p for a liquid is 7 cal/mol·°C, estimate C_v.
2. The mean heat capacity of a gas A is 1 cal/g·°C at 50°C, and 2 cal/g·°C at 100°C. The reference temperature on which these heat capacities are based is 25°C. Calculate $\Delta \hat{H}$(cal/g) for each of the following processes:
 (a) A(25°C) → A(50°C)
 (b) A(100°C) → A(25°C)
 (c) A(50°C) → A(100°C)

8.3c Estimation of Heat Capacities

The polynomial expressions for C_p in Table B.2 are based on experimental data for the listed compounds and provide a basis for accurate calculations of enthalpy changes. Several approximate methods for estimating heat capacities in the absence of tabulated formulas follow.

Kopp's rule is a simple empirical method for estimating the heat capacity of a solid or liquid at or near 20°C. According to this rule, C_p for a molecular compound is the sum of contributions (given in Table 8.3-3) for each atomic elements in the compound. For example, the heat capacity of solid calcium hydroxide, $Ca(OH)_2$, would be estimated from Kopp's rule as

$$(C_p)_{Ca(OH)_2} = (C_{pa})_{Ca} + 2(C_{pa})_O + 2(C_{pa})_H$$
$$= [26 + (2 \times 17) + (2 \times 9.6)](J/mol \cdot °C) = 79 \ J/mol \cdot °C$$

(The true value is 89.5.)

More accurate heat capacity estimation formulas for various types of gases and liquids are given in Chapter 5 of Reid, Prausnitz, and Sherwood, and several correla-

TABLE 8.3-3

Atomic Heat Capacities for Kopp's Rule[a]

	C_{pa}(J/g-atom·C)	
Element	Solids	Liquids
C	7.5	12
H	9.6	18
B	11	20
Si	16	24
O	17	25
F	21	29
P	23	31
S	26	31
All others	26	33

[a] D. M. Himmelblau, *Basic Principles and Calculations in Chemical Engineering*, 3rd Edition, Prentice-Hall, Englewood Cliffs, N.J.: 1974, p. 270.

tions are presented on pp. 3-275 of *Perry's Chemical Engineers' Handbook* and by Gold and Ogle.[3]

Suppose you wish to calculate the enthalpy change associated with a change in temperature undergone by a mixture of substances. Enthalpies and heat capacities of certain mixtures are tabulated in standard references. Lacking such data, you may use the following approximation:

Rule 1. *For a mixture of gases or liquids, calculate the total enthalpy change as the sum of the enthalpy changes for the pure mixture components.* You are in effect neglecting enthalpy changes associated with the mixing of the components; this is an excellent approximation for mixtures of gases and mixtures of similar liquids (such as aliphatic hydrocarbons), but it may be a poor one for dissimilar liquids. Enthalpies of mixing for systems of the latter type are discussed in detail in Section 8.5.

Rule 2. *For highly dilute solutions of solids or gases in liquids, neglect the enthalpy change of the solute.* The more dilute the solution, the better this approximation.

The calculation of enthalpy changes for the heating or cooling of a mixture of known composition may often be simplified by calculating a heat capacity for the mixture in the following manner:

$$C_{pm}(T) = \sum_{\substack{\text{all} \\ \text{mixture} \\ \text{components}}} y_i C_{pi}(T) \tag{8.3-17}$$

$$C_{pm} = \text{heat capacity of the mixture}$$
$$y_i = \text{mass or mole fraction of the } i\text{th component}$$
$$C_{pi} = \text{heat capacity of the } i\text{th component}$$

If C_{pi} and C_{pm} are expressed in molar units, then y_i must be the mole fraction of the ith component, and if the heat capacities are expressed in mass units, then y_i must be the mass fraction of the ith component. Once C_{pm} is known, $\Delta \hat{H}$ for a change in temperature from T_1 to T_2 may be calculated as

$$\Delta \hat{H} = \int_{T_1}^{T_2} C_{pm}(T) \, dT \tag{8.3-18}$$

Equation 8.3-17 is valid to the extent that enthalpies of mixing may be neglected.

EXAMPLE 8.3-4 Heat Capacity of a Mixture

Calculate the heat required to bring 150 mol/h of a stream containing 60% C_2H_6 and 40% C_3H_8 by volume from 0 to 400°C. Determine a heat capacity for the mixture as part of the problem solution.

[3] P. I. Gold and G. J. Ogle, "Estimating Thermochemical Properties of Liquids, Part 7—Heat Capacity," *Chem. Eng.*, April 7, 1969, p. 130.

Solution

The polynomial heat capacity formulas for ethane and propane given in Table B.2 are substituted into Equation 8.3-17 to yield

$$C_{pm}(\text{J/mol}\cdot{}^\circ\text{C}) = 0.600(49.37 + 13.92 \times 10^{-2}T - 5.816 \times 10^{-5}T^2 + 7.280 \times 10^{-9}T^3)$$
$$+ 0.400(68.03 + 22.59 \times 10^{-2}T - 13.11 \times 10^{-5}T^2 + 31.71 \times 10^{-9}T^3)$$
$$= 56.83 + 17.39 \times 10^{-2}T - 8.734 \times 10^{-5}T^2 + 17.05 \times 10^{-9}T^3$$

$$\Delta\hat{H} = \int_0^{400} C_{pm}\,dT = 34\,890 \text{ J/mol}$$

If potential and kinetic energy changes and shaft work are neglected, the energy balance becomes

$$Q = \Delta\hat{H} = n\,\Delta\hat{H} = \frac{150 \text{ mol}}{\text{h}} \left| \frac{34\,890 \text{ J}}{\text{mol}} \right. = \boxed{5.23 \times 10^6 \, \frac{\text{J}}{\text{h}}}$$

TEST YOURSELF

1. Estimate the heat capacity of solid calcium carbonate ($CaCO_3$) using Kopp's rule and Table 8.3-3.

2. Two kilograms of liquid n-hexane ($C_p = 2.5$ kJ/kg·°C) and 1 kg of liquid cyclohexane ($C_p = 1.8$ kJ/kg·°C) are mixed and heated from 20 to 30°C. Use the rule for liquid mixtures (Rule 1) given in this section to show that $\Delta H = 68$ kJ for this process. What is $\Delta\hat{H}$(kJ/kg mixture)?

3. A 0.100 wt% aqueous solution of sodium chloride is heated from 25°C to 50°C. Use the rule for solutions (Rule 2) given in this section to estimate $\Delta\hat{H}$(cal/g) for this process. C_p for water is 1 cal/g·°C.

4. The heat capacity of liquid water is 1 cal/g·°C, and that of ethanol is 0.54 cal/g·°C. Estimate the heat capacity of a mixture containing 50% ethanol and 50% water by mass.

8.3d Energy Balances on Single-Phase Systems

In Chapter 7, we saw that if the specific enthalpy of each component of each feed and product stream of a process is known, ΔH or ΔU for the process may be calculated and substituted in the energy balance equation. If tabulated enthalpies are available for the components, the calculation is straightforward; if not, it is necessary to choose a reference state (temperature, pressure, and state of aggregation) for each component, and calculate the specific enthalpies at the process temperatures using known or estimated heat capacities. (For mixtures of dissimilar liquids you may have to take into account heats of mixing.) The energy balance calculation then proceeds as in Chapter 7.

The choice of a reference state is arbitrary and may be made purely for computational convenience. If mean heat capacities are available for a component, choose the reference temperature on which they are based; otherwise, choose any process stream temperature as a reference, so that \hat{H} may be set equal to 0 at at least one point in the process.

For all but the simplest processes, a good procedure to make the energy balance calculations efficient is to prepare a table of component amounts and specific enthalpies for all input and output streams. As each quantity is calculated, its value is inserted in the table; when all values have been inserted, ΔH for the process may be calculated directly from the table.

For example, suppose 100 mol/h of a gas containing 60 mol% CH_4 and 40% H_2O at 350°C pass through a heat exchanger and cool down to 120°C, and that 10 mol/h of water enter the heat exchanger as a liquid at 25°C and leave as saturated steam at 1 atm.

To calculate the heat loss for this process unit, you must evaluate \hat{H} for each stream component and substitute into the energy balance equation ($Q = \Delta H$, assuming kinetic energy changes are negligible). This necessitates choosing reference conditions for CH_4 and H_2O, the two species involved in the process. Since the enthalpy of H_2O may be most easily determined from the steam tables, choose liquid water at the triple point (the reference condition for the steam tables) as the reference state of water; having nothing better to assume, choose methane gas at 350°C as the other reference state [so that $(\hat{H}_{CH_4})_{in} = 0$].

The input–output enthalpy table would appear as follows:

References: **CH_4(g, 350°C)**
 H_2O(triple point)

Substance	n_{in}	\hat{H}_{in}	n_{out}	\hat{H}_{out}	
CH_4	60	0	60		n in mol/h
H_2O(v, 350°C and 120°C)	40		40		\hat{H} in kJ/mol
H_2O(l, 25°C)	10		—	—	
H_2O(v, sat'd, 1 atm)	—	—	10		

Dashes by a substance indicate that the substance at the given condition does not appear as an input to or output from the process.

You must now evaluate the missing specific enthalpies as the enthalpy changes for the substances going from their reference states to the specified inlet and outlet

conditions. You would calculate $(\hat{H}_{CH_4})_{out}$ by integrating $(C_p)_{CH_4} dT$ from the reference state of methane (350°C) to the outlet state (120°C); look up the values of the water enthalpies in the steam tables; convert whatever units you must to express these enthalpies in kJ/mol; and insert the values in the table. You may then calculate

$$Q(kJ/h) = \Delta H = \sum n_{out}\hat{H}_{out} - \sum n_{in}\hat{H}_{in}$$

where the n's and \hat{H}'s are now conveniently summarized in the table.

It may at first seem like a waste of time to go through the formality of preparing a table rather than plunging right into the energy balance. However, when you begin to deal with relatively complex systems involving multiple inputs and outputs, and later with reactive systems, and still later, perhaps, with computer-aided material and energy balances, you will find the calculation efficiency that the table provides well worth the few seconds it takes to prepare it.

EXAMPLE 8.3-5 Energy Balance on a Gas Preheater

A stream containing 10% CH_4 and 90% air by volume is to be heated from 20 to 300°C. Calculate the required rate of heat input in kilowatts if the flow rate of the gas is 2.00×10^3 liters (STP)/min.

Solution

Basis: Given Flow Rate

Assume ideal gas behavior.

Recall that specifying the flow rate in liters (STP)/min does not imply that the feed gas is at standard temperature and pressure; this is simply an alternative way of giving the molar flow rate.

$$n = \frac{2000 \text{ liters (STP)}}{\text{min}} \left| \frac{1 \text{ mol}}{22.4 \text{ liters (STP)}} \right. = 89.3 \text{ mol/min}$$

Since there is only one input stream and one output stream, no material balances are needed, and we may proceed directly to the energy balance. Since we have assumed ideal gas behavior, we need not worry about any pressure changes that may be involved in the process, and since only mixtures of gases are involved, we may calculate \hat{H} for each component from the pure component heat capacities, neglecting heats of mixing (Rule 1 in Section 8.3c).

Let us choose the following reference conditions for the two substances involved in the process.

CH_4: gas, 20°C, 1 atm (so that $\hat{H}_{in} = 0$ for CH_4).

Air: gas, 25°C, 1 atm (so that \hat{H}_{in} and \hat{H}_{out} may be calculated directly from Table 8.3-1).

The input–output enthalpy table appears as shown here after all flow rates and specific enthalpies have been determined.

References: **CH_4(g, 20°C)**
air(g, 25°C)

Substance	n_{in}	\hat{H}_{in}	n_{out}	\hat{H}_{out}	
CH_4	8.93	0	8.93	12 090	n in mol/min
Air	80.4	-145	80.4	8 170	\hat{H} in J/mol

The calculations that led to the nonzero entries in this table are as follows:

$$n_{CH_4} = \frac{89.3 \text{ mol}}{\text{min}} \left| \frac{0.1 \text{ mol } CH_4}{\text{mol}} \right. = 8.93 \text{ mol } CH_4/\text{min}$$

$$n_{air} = 89.3 - 8.93 = 80.4 \text{ mol air/min}$$

$$(\hat{H}_{CH_4})_{out} = \int_{T_{ref}}^{T_{out}} (C_p)_{CH_4} \, dT$$

$$\left\Vert \begin{array}{l} \text{Substitute for } C_p \text{ from Table B.2} \\ T_{ref} = 20°C, \ T_{out} = 300°C \end{array}\right.$$

$$= \int_{20}^{300} (34.3 + 5.47 \times 10^{-2}T + 0.366 \times 10^{-5}T^2 - 11.0 \times 10^{-9}T^3) \, dT$$

$$= 12\,090 \text{ J/mol}$$

$$(\hat{H}_{air})_{in} = (C_p)_{20°C}(20 - T_{ref})°C$$
$$= (29.05 \text{ J/mol} \cdot °C)(-5°C) = -145 \text{ J/mol}$$

$$(\hat{H}_{air})_{out} = (\bar{C}_p)_{300°C}(300 - T_{ref})°C$$
$$= (29.71 \text{ J/mol} \cdot °C)(275°C) = 8170 \text{ J/mol}$$

Energy Balance

$$Q = \Delta H = \sum_{output} n_i \hat{H}_i - \sum_{input} n_i \hat{H}_i$$

$$\Downarrow$$

$$= [(8.93)(12\,090) + (80.4)(8170) - (8.93)(0) - (80.4)(-145)](J/\text{min})$$

$$= \frac{7.76 \times 10^5 \text{ J}}{\text{min}} \left| \frac{1 \text{ min}}{60 \text{ s}} \right| \frac{1 \text{ kW}}{1000 \text{ J/s}} = \boxed{12.9 \text{ kW}}$$

In the illustrative energy balance problems we have presented up to now the temperatures and states of aggregation of all input and output streams were specified, and the only unknown in the energy balance equation was the heat flow required to

achieve the specified conditions. You will also encounter problems in which the heat input is known but the temperature of an input or output stream is not. Example 8.3-6 illustrates the solution of such a problem.

EXAMPLE 8.3-6 Energy Balance on a Waste Heat Boiler

Methanol vapor leaves a synthesis unit at a temperature of 800°F and passes through a waste heat boiler in which it cools, and in doing so generates saturated steam at 70 psia. The water fed to the boiler enters at 70°F and 70 psia. If the molar feed

ratio of water to methanol is 0.2 and the boiler is assumed to operate adiabatically, what is the exit temperature of the methanol?

Data

The heat capacity of methanol vapor between 800°F and about 400°F is given by

$$C_p(\text{Btu/lb-mole} \cdot {}^\circ F) = 10.54 + 9.2 \times 10^{-3} T({}^\circ F)$$

From the steam tables in *Perry's Chemical Engineers' Handbook*, the enthalpy of liquid water at 70°F is 38.05 Btu/lb$_m$, and that of saturated steam at 70 psia is 1180 Btu/lb$_m$, both relative to liquid water at 32°F.

Solution

Basis: 1 lb-mole CH_3OH, 0.2 lb-mole H_2O.

Reference conditions: $H_2O(l)$ at 32°F, $CH_3OH(v)$ at 800°F.

Energy Balance

$$\Delta H = 0 \quad \text{(since the boiler operates adiabatically)}$$

$$n_{H_2O}(\hat{H}_{out} - \hat{H}_{in})_{H_2O} + n_{CH_3OH}(\hat{H}_{out} - \hat{H}_{in})_{CH_3OH} = 0$$

We know the numerical values of all quantities in this equation but $(\hat{H}_{out})_{CH_3OH}$, which equals

$$\int_{800}^{T} C_p(\text{Btu/lb-mole} \cdot {}^{\circ}\text{F}) \, dT({}^{\circ}\text{F})$$

$$= \int_{800}^{T} (10.54 + 9.2 \times 10^{-3} T) \, dT$$

$$= \left[10.54(T - 800) + \frac{9.2 \times 10^{-3}}{2} (T^2 - 800^2) \right] (\text{Btu/lb} \cdot \text{mole})$$

$$= 4.6 \times 10^{-3} T^2 + 10.54T - 11\,380$$

The energy balance equation becomes

$$\frac{0.2 \text{ lb-mole } H_2O}{} \bigg| \frac{18 \text{ lb}_m}{\text{lb-mole}} \bigg| \frac{(1180.6 - 38.05) \text{ Btu}}{\text{lb}_m}$$

$$+ (1 \text{ lb-mole CH}_3\text{OH})\underbrace{(4.6 \times 10^{-3} \, T^2 + 10.54T - 11\,380}_{\hat{H}_{\text{out}}} \underbrace{- \; 0)(\text{Btu/lb-mole})}_{\hat{H}_{\text{in}}} = 0$$

$$\Downarrow$$

$$4.6 \times 10^{-3} T^2 + 10.54T - 7267 = 0$$

Solving this quadratic equation yields $T = 555{}^{\circ}\text{F}$ and $T = -2846{}^{\circ}\text{F}$. Only the first solution is meaningful, so that the final result is

$$\boxed{T = 555{}^{\circ}\text{F}}$$

8.3e Numerical Integration of Tabulated Heat Capacities

You have now seen two ways to evaluate an expression of the type

$$\int_{T_1}^{T_2} C_p(T) \, dT \tag{8.3-19}$$

If a functional relation for $C_p(T)$ is available, such as one of the polynomials of Table B.2, the integration of (8.3-19) can be carried out analytically; and if tabulated mean heat capacities are available for the substance being heated or cooled, a simple multiplication replaces the integration.

Suppose, however, that the only information you have about C_p is its value at a series of temperatures that span the range from T_1 to T_2 [the limits of the integral of (8.3-19)]. The question is how to estimate the value of the integral from these data.

One way would of course be to plot C_p versus T, draw a curve by visual inspection through the points at which C_p is known, and estimate the integral graphically as the area under the curve from T_1 to T_2. This is a cumbersome procedure, however, even if you have access to a planimeter, a device that can calculate the area under a traced-out curve.

A better solution is to use one of many existing **quadrature formulas**—algebraic expressions that provide estimates of the integrals of tabulated data. Several such formulas are presented and illustrated in Appendix A.3; the use of one of them, **Simpson's rule,** is required for the integration of heat capacity data in several problems at the end of this chapter.

8.4 PHASE CHANGE OPERATIONS

Phase changes such as fusion and vaporization are usually accompanied by large changes in internal energy and enthalpy, and heat duties in phase change operations consequently tend to be substantial. The paragraphs that follow outline procedures for setting up and solving energy balances on such operations. The discussion will be limited to phase changes between liquid and vapor, solid and liquid, and solid and vapor; however, the principles to be discussed may be extended directly to other phase changes, such as between two solid phases.

8.4a Latent Heats

Consider a liquid at a fixed temperature, with its upper surface exposed to a gas. The liquid molecules move around with varying velocities, and hence varying kinetic energies, with the average kinetic energy being determined by the temperature. They also exert attractive forces on one another, which keep them in the relative closeness characteristic of the liquid state.

In the course of their random motion, liquid molecules frequently approach and strike the liquid surface. If the kinetic energy of such a molecule is great enough to overcome the attractive force exerted by the surrounding liquid at the point of impact, the molecule leaves the liquid and enters the vapor phase. This is the phenomenon of **vaporization.** The net rate of vaporization is the difference between the rate at which molecules escape from the liquid and the rate at which vapor molecules collide with the liquid surface and are captured.

Since the kinetic energies of vaporizing molecules are greater than the average kinetic energy of the liquid at the system temperature, the average energy of the residual liquid must be lowered by the vaporization. A lower average kinetic energy signifies a lower temperature. To maintain the liquid at its original temperature, therefore, an equivalent amount of energy must be added to the liquid as heat. From the energy balance equation, the heat added must result in a gain in internal energy and hence in enthalpy of the system. In short, *a gain in enthalpy is associated with the transition from liquid to vapor at a constant temperature.*

A similar consideration attends the transition from solid to liquid. The attractive forces between atoms in a solid are strong enough to lock the atoms in position relative to one another. However, while they may not undergo translational motion, the atoms may vibrate about their equilibrium positions with energies that increase with the system temperature.

If the temperature of a solid is raised at a constant system pressure, at some point the vibrational energies of the molecules become great enough to overcome the attractive forces exerted by adjacent molecules in the solid lattice, and melting (or fusion) takes place. Since fewer molecules of high vibrational energy are left in the solid after some fusion occurs, heat must be added to make up for the energy loss and hence to effect more fusion. The net result is a gain in enthalpy associated with the melting process at a fixed temperature. The same reasoning applies to sublimation, in which at a certain temperature the vibrational energies of some solid molecules are great enough to allow the molecules to escape into the vapor phase.

The enthalpy change associated with the transition of a unit amount (mass or moles) of a substance from one phase to another at constant temperature and pressure is known as the **latent heat** of the phase change. For example, the enthalpy change $\Delta \hat{H}$ for the transition from liquid water to steam at 100°C and 1 atm, which equals 40.6 kJ/mol, is by definition the **latent heat of vaporization** of water at this temperature and pressure. (The word "latent" is often omitted.)

Since condensation is the reverse process of vaporization, and enthalpy is a state property, the heat of condensation must be the negative of the heat of vaporization. Thus, the heat of condensation of water at 100°C and 1 atm must be -40.6 kJ/mol. Similarly, the heat of crystallization is the negative of the heat of fusion at the same temperature and pressure.

Latent heats for three phase changes are defined as follows:

1. **Latent heat of vaporization:** $\Delta \hat{H}_v$ = heat required to vaporize a unit amount of liquid at constant T and P.

2. **Latent heat of fusion:** $\Delta \hat{H}_m$ = heat required to melt a unit amount of solid at constant T and P.

3. **Latent heat of sublimation:** $\Delta \hat{H}_s$ = heat required to vaporize a unit amount of solid at constant T and P.

The latent heat of a phase change may vary considerably with the temperature at which the change occurs, but hardly varies at all with the pressure at the transition point. For example, the heat of vaporization of water at 25°C is 2443.9 J/g at $P = 23.75$ mm Hg and 2445.2 J/g at $P = 760$ mm Hg, while $\Delta \hat{H}_v$ for water at 100°C and 760 mm Hg is 2253.7 J/g. When using a tabulated latent heat, you must therefore be sure that the phase change in question takes place at the temperature for which the tabulated value is reported, but you may ignore moderate variations in pressure.

Tabulated values of latent heats, such as those in Table B.1, Appendix B, and on pp. 3-120 through 3-128 of *Perry's Chemical Engineers' Handbook*, usually apply to a substance at its normal boiling, melting, or sublimation point—that is, at a pressure of 1 atm. These quantities are referred to as **standard heats of phase change.**

EXAMPLE 8.4-1 Heat of Vaporization

At what rate in kilowatts must heat be transferred to a liquid stream of methanol at its normal boiling point to generate 1500 g/min of saturated methanol vapor?

Solution

From Table B.1, $\Delta \hat{H}_v = 35.3$ kJ/mol at $T_{bp} = 64.7$°C. The energy balance with potential and kinetic energy changes neglected is

$$Q = \Delta H = n \, \Delta \hat{H}_v$$

$$Q = \frac{1500 \text{ g}}{\text{min}} \left| \frac{1 \text{ mol}}{32.0 \text{ g CH}_3\text{OH}} \right| \frac{35.3 \text{ kJ}}{\text{mol}} \left| \frac{1 \text{ min}}{60 \text{ s}} \right| \frac{1 \text{ kW}}{1 \text{ kJ/s}} = \boxed{27.6 \text{ kW}}$$

Phase changes often occur at temperatures other than the temperature for which the latent heat is tabulated. When faced with this situation, you must select a hypothetical process path that permits the available data to be used.

Suppose, for example, that a substance is to be vaporized isothermally at 130°C, but the only available value of the heat of vaporization is at 80°C. A process path from the liquid at 130°C to the vapor at the same temperature must then be chosen that includes an isothermal vaporization step at 80°C: specifically, cool the liquid from 130°C to 80°C, vaporize the liquid at 80°C, and then heat the vapor back to 130°C. Summing the changes in enthalpy for each of these steps yields the change in enthalpy for the given process. (By definition, the calculated value is the latent heat of vaporization at 130°C.)

EXAMPLE 8.4-2 Vaporization and Heating

One hundred mol per hour of liquid n-hexane at 25°C and 7 bars are vaporized and heated to 300°C at constant pressure. Neglecting the effect of pressure on enthalpy, estimate the rate at which heat must be supplied.

Solution

An energy balance yields

$$Q = \Delta H \qquad (W_s = \Delta E_k = \Delta E_p = 0)$$

Therefore an evaluation of ΔH will yield the desired value of Q.

From Figure 6.1-4, the temperature at which the vapor pressure of n-hexane is 7 bars (104 psia) is approximately 295°F (146°C), and this is therefore the temperature at which the vaporization actually occurs. However, Table B.1, Appendix B, gives a value of $\Delta \hat{H}_v$ at the normal boiling point of n-hexane,

$$\Delta \hat{H}_v = 28.85 \text{ kJ/mol at } 69°C$$

We must therefore find a path that takes hexane from a liquid to a vapor at 69°C, rather than at the true vaporization temperature of 146°C.

As noted previously, the change in enthalpy associated with a process may be determined from any convenient path as long as the initial and final points of the chosen path correspond to those of the process. The diagram on the next page illustrates several possible paths from liquid hexane at 25°C to hexane vapor at 300°C.

If we knew $\Delta \hat{H}_v$ at 146°C, we would follow path $ABEH$ (the true process path) to evaluate the overall $\Delta \hat{H}$, or if we knew $\Delta \hat{H}_v$ at 25°C we would follow path CF, which would require only two calculations. Since we have the value at 69°C, we must follow path ADG.

$$n\text{-}C_6H_{14}(l) \xrightarrow{\Delta \hat{H}_A} n\text{-}C_6H_{14}(l) \xrightarrow{\Delta \hat{H}_D} n\text{-}C_6H_{14}(v) \xrightarrow{\Delta \hat{H}_G} n\text{-}C_6H_{14}(v)$$
$$\quad\; 25°C \qquad\qquad\quad 69°C \qquad\qquad\quad 69°C \qquad\qquad\quad 300°C$$

$\Delta \hat{H}_A$ and $\Delta \hat{H}_G$ are calculated by integrating tabulated heat capacity formulas for $C_6H_{14}(l)$ and $C_6H_{14}(v)$, respectively, and $\Delta \hat{H}_D$ is the known heat of vaporization.

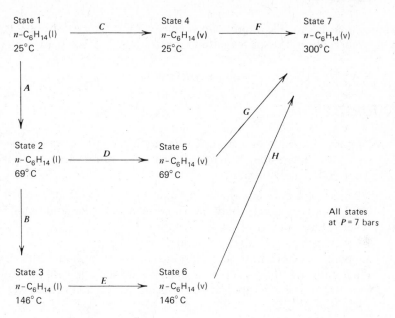

$$\Delta\hat{H}_A = \int_{25}^{69} (C_p)_{C_6H_{14}(l)}\, dT$$

$$\Downarrow C_p = 216.3 \text{ J/mol} \cdot {}^\circ\text{C}$$

$$\Delta\hat{H}_A = 9517 \text{ J/mol} = 9.517 \text{ kJ/mol}$$

$$\Delta\hat{H}_D = \Delta\hat{H}_v(69^\circ\text{C}) = 28.85 \text{ kJ/mol}$$

$$\Delta\hat{H}_G = \int_{69}^{300} (C_p)_{C_6H_{14}(v)}\, dT$$

$$\Downarrow C_p(\text{J/mol} \cdot {}^\circ\text{C}) = 137.44 + 0.408T - 2.39 \times 10^{-4}T^2 + 5.77 \times 10^{-8}T^3$$

$$\Delta\hat{H}_G = 47.1 \text{ kJ/mol}$$

For the overall process

$$Q = \Delta H = n(\text{mol/h})\,\Delta\hat{H}(\text{kJ/mol})$$

$$\Downarrow \Delta\hat{H} = \Delta\hat{H}_A + \Delta\hat{H}_D + \Delta\hat{H}_G$$

$$Q = \frac{100 \text{ mol}}{\text{h}} \, \bigg| \, \frac{(9.52 + 28.85 + 47.1) \text{ kJ}}{\text{mol}} = \boxed{8550 \text{ kJ/h}}$$

If a phase change takes place in a closed system, you must evaluate $\Delta\hat{U} = \Delta\hat{H} - \Delta P\hat{V}$ for the phase change to substitute into the energy balance equation. For phase changes such as fusion, which involve only liquids and solids, changes in \hat{V} are generally negligible compared to changes in \hat{H}, so that

$$\Delta\hat{U}_m \approx \Delta\hat{H}_m \tag{8.4-1}$$

For vaporization, $P\hat{V}$ for the vapor (which equals RT if ideal gas behavior may be assumed) is normally orders of magnitude greater than $P\hat{V}$ for the liquid, so that $\Delta PV \approx RT$, and

$$\Delta\hat{U}_v \approx \Delta\hat{H}_v - RT \tag{8.4-2}$$

TEST YOURSELF

1. If you are given a value of a heat of vaporization at 100°C and 1 atm, would you use it to estimate the enthalpy change for a vaporization at 100°C and 2 atm? For 200°C and 1 atm?

2. The enthalpies of a pure liquid and its vapor at 75°C and 1 atm are 100 J/mol and 1000 J/mol, respectively, both measured relative to the liquid at 0°C.
 (a) What is the enthalpy of the liquid at 0°C?
 (b) What is the heat of vaporization at 75°C?
 (c) Suppose you have heat capacity data for both the liquid and its vapor. What path would you follow to calculate the enthalpy change associated with 100 mol of the vapor at 400°C being cooled and condensed to form a liquid at 25°C?

3. The heat of fusion of zinc chloride at 556 K is $\Delta\hat{H}_m = 5500$ cal/mol, and the heat of vaporization of this substance at 1000 K is $\Delta\hat{H}_v = 28\,710$ cal/mol. Estimate $\Delta\hat{U}_m(556\text{ K})$ and $\Delta\hat{U}_v(1000\text{ K})$ for $ZnCl_2$. (Say $R = 2$ cal/mol·K.)

4. Reread the description of vaporization with which this section began. Then explain why the vapor pressure of a liquid (the pressure at which the rates of vaporization and condensation are equal at a given temperature) should increase with increasing temperature. (*Suggestion:* First consider the effect of T on the vaporization rate, then the effect of P on the condensation rate.)

8.4b Estimation and Correlation of Latent Heats

Reid, Prausnitz, and Sherwood review procedures for determining latent heats of vaporization, fusion and sublimation, and latent heat estimation methods are given on pp. 3-274 and 3-275 of *Perry's Chemical Engineers' Handbook*. Several of the methods given in these references are summarized below.

A simple formula for estimating a standard heat of vaporization ($\Delta\hat{H}_v$ at the normal boiling point) is **Trouton's rule:**

$$\Delta\hat{H}_v(\text{kJ/mol}) \begin{cases} \approx 0.088 T_b(\text{K}) \text{ (Nonpolar liquids)} \\ \approx 0.109 T_b(\text{K}) \text{ (Water, low molecular weight alcohols)} \end{cases} \tag{8.4-3}$$

where T_b is the normal boiling point of the liquid. Trouton's rule provides an estimate of $\Delta\hat{H}_v$ that is accurate to within 30%. Another formula that provides roughly 2% accuracy is **Chen's equation.**

$$\Delta\hat{H}_v(\text{kJ/mol}) = \frac{T_b[0.0331(T_b/T_c) - 0.0327 + 0.0297 \log_{10} P_c]}{1.07 - (T_b/T_c)} \tag{8.4-4}$$

where T_b and T_c are the normal boiling point and critical temperature in degrees Kelvin and P_c is the critical pressure in atmospheres.

A formula for approximating a standard heat of fusion is

$$\Delta \hat{H}_m (\text{kJ/mol}) \approx \begin{cases} 0.0092 T_m(\text{K}) \ (\text{Metallic elements}) \\ 0.025 T_m(\text{K}) \ (\text{Inorganic compounds}) \\ 0.050 T_m(\text{K}) \ (\text{Organic compounds}) \end{cases} \quad \textbf{(8.4-5)}$$

Latent heats of vaporization may be estimated from vapor pressure data by using the Clausius–Clapeyron equation, which was discussed in Section 6.1b.

$$\ln p^* = -\frac{\Delta \hat{H}_v}{RT} + B \quad \textbf{(8.4-6)}$$

Provided that $\Delta \hat{H}_v$ is constant over the range of temperatures encompassed by the vapor pressure data, the latent heat of vaporization may be determined from a plot of $\ln p^*$ versus $1/T$. (See Example 6.1-1.)

In many cases the latent heat of vaporization varies considerably with temperature, invalidating Equation 8.4-6. It is then necessary to use the **Clapeyron equation,** from which Equation 8.4-6 was derived. In the form of this equation expressed by Equation 6.1-2,

$$\frac{d(\ln p^*)}{d(1/T)} = \frac{\Delta \hat{H}_v}{R} \quad \textbf{(8.4-7)}$$

The heat of vaporization at a temperature T may be estimated from vapor pressure data by plotting $\ln p^*$ versus $1/T$, determining $[d(\ln p^*)/d(1/T)]$ at the temperature of interest as the slope of the tangent to the curve, and solving Equation 8.4-7 for $\Delta \hat{H}_v$. The slope may be determined graphically or by any of several numerical differentiation techniques outlined in texts on numerical analysis.

In Section 8.4a a procedure for calculating the latent heat of vaporization at any temperature from a known value at any other temperature was presented. The technique outlined is rigorous but time-consuming, and it requires heat capacity data that might not be available for the substance of interest. A useful approximation for estimating $\Delta \hat{H}_v$ at T_2 from a known value at T_1 is the **Watson correlation:**

$$\Delta \hat{H}_v(T_2) = \Delta \hat{H}_v(T_1) \left(\frac{T_c - T_2}{T_c - T_1} \right)^{0.38} \quad \textbf{(8.4-8)}$$

where T_c is the critical temperature of the substance.

EXAMPLE 8.4-3 Estimation of a Heat of Vaporization

The normal boiling point of methanol is 337.9 K, and the critical temperature of this substance is 513.2 K. Estimate the heat of vaporization of methanol at 200°C.

Solution

We first use Trouton's rule to estimate $\Delta \hat{H}_v$ at the normal boiling point, and then the Watson correlation to estimate $\Delta \hat{H}_v(473 \text{ K})$ from $\Delta \hat{H}_v(337.9 \text{ K})$.

Trouton's Rule:

$$\Delta \hat{H}_v(337.9 \text{ K}) = (0.109)(337.9) = 36.8 \text{ kJ/mol}$$

(The true value is 35.3 kJ/mol. Chen's equation yields 37.2 kJ/mol, so that in this unusual case Trouton's rule provides the better estimate.)

Watson Correlation:

Using the $\Delta \hat{H}_v$ estimated by Trouton's rule

$$\Delta \hat{H}_v(473 \text{ K}) = 36.8 \left(\frac{513.2 - 473}{513.2 - 337.9} \right)^{0.38} = \boxed{21.0 \text{ kJ/mol}}$$

The exact value is 19.8 kJ/mol.

TEST YOURSELF

How would you estimate $\Delta \hat{H}_v$ for a pure hydrocarbon at its normal boiling point under each of the following conditions?

1. You know only the normal boiling point.
2. You know the normal boiling point and the critical constants.
3. You have vapor pressure data over a range that encompasses $p^* = 1$ atm, and a semilog plot of p^* versus $1/T$ is a straight line.
4. Repeat 3, only suppose the plot is curved.
5. You know $\Delta \hat{H}_v$ at a temperature other than T_b, and do not know the heat capacities of the substance in its gaseous and liquid forms.
6. Repeat 5, only suppose that you have the heat capacity data.

8.4c Energy Balances on Processes Involving Phase Changes

When writing an energy balance on a process in which a component exists in two phases, you must choose a reference condition for that component by specifying both a phase and a temperature, and calculate the specific enthalpy of the component in all process streams relative to this condition. If the substance is a liquid at its reference state and a vapor in a process stream, \hat{H} may be calculated as outlined in Section 8.4a; that is, bring the liquid from the reference temperature to a point at which $\Delta \hat{H}_v$ is known, vaporize the liquid, bring the vapor to the stream temperature, and sum the individual enthalpy changes for the three steps.

EXAMPLE 8.4-4 Partial Vaporization of a Mixture

An equimolar liquid mixture of benzene (B) and toluene (T) at 10°C is fed continuously to a vessel in which the mixture is heated to 50°C. The liquid product is 40 mole% B, and the vapor product is 68.4 mole% B. How much heat must be transferred to the mixture per mol of feed?

Solution

Basis: 1 mol feed

We first determine V and L from material balances, and then Q from an energy balance. As before, we neglect the effects of pressure on specific enthalpy.

Overall mass balance: $\quad 1.00 = V + L$

Benzene balance: $\quad\quad 0.500 = 0.684V + 0.400L$ $\quad \Rightarrow \quad$ $V = 0.352$ mol, $L = 0.648$ mol

The energy balance for this process has the form $Q = \Delta H$. If enthalpies of mixing are neglected, the total enthalpy of each stream may be determined as the sum of the enthalpies of the individual stream components. An enthalpy table for this process appears as follow.

References: B(1, 10°C), T(1, 10°C)

Substance	n_{in}	\hat{H}_{in}	n_{out}	\hat{H}_{out}	
B(1)	0.500	0	0.259	5 332	
T(1)	0.500	0	0.389	6 340	n in mol
B(v)	—	—	0.241	37 520	\hat{H} in J/mol
T(v)	—	—	0.111	42 930	

The values of n_{out} were determined from the known mole fractions of benzene and toluene in the outlet streams and the calculated values of V and L. The inlet enthalpies both equal zero because of the choice of reference conditions. The heat capacity and latent heat data needed to calculate the outlet enthalpies are obtained from Tables B.1 and B.2.

B(l, 50°C): $C_6H_6(l, 10°C) \rightarrow C_6H_6(l, 50°C)$

$$\Delta\hat{H} = \int_{10}^{50} (C_p)_{C_6H_6(l)} \, dT = 5332 \text{ J/mol}$$

T(l, 50°C): $C_7H_8(l, 10°C) \rightarrow C_7H_8(l, 50°C)$

$$\Delta\hat{H} = \int_{10}^{50} (C_p)_{C_7H_8(l)} \, dT = 6340 \text{ J/mol}$$

B(v, 50°C): $C_6H_6(l, 10°C) \rightarrow C_6H_6(l, 80.26°C) \rightarrow C_6H_6(v, 80.26°C) \rightarrow C_6H_6(v, 50°C)$

$$\Delta \hat{H} = \int_{10}^{80.26} (C_p)_{C_6H_6(l)} \, dT + \Delta \hat{H}_v(80.26°C) + \int_{80.26}^{50} (C_p)_{C_6H_6(v)} \, dT$$

$$= 37\,520 \text{ J/mol}$$

T(v, 50°C): $C_7H_8(l, 10°C) \rightarrow C_7H_8(l, 110.8°C) \rightarrow C_7H_8(v, 110.8°C) \rightarrow C_7H_8(v, 50°C)$

$$\Delta \hat{H} = \int_{10}^{110.8} (C_p)_{C_7H_8(l)} \, dT + \Delta \hat{H}_v(110.8°C) + \int_{110.8}^{50} (C_p)_{C_7H_8(v)} \, dT$$

$$= 42\,930 \text{ J/mol}$$

Energy Balance

$$Q = \Delta H = \sum n_{out} \hat{H}_{out} - \sum n_{in} \hat{H}_{in}$$

$$\boxed{Q = 17\,660 \text{ J}}$$

CREATIVITY EXERCISE

A gas emerges from a stack at 1200°C. Rather than being released directly to the atmosphere, it can be passed through one or several heat exchangers, and the heat it loses can be put to use in a variety of ways. Think of as many uses of this heat as you can. (*Example:* During the winter, pass the gas through a series of radiators, thereby getting free heating.)

8.4d Psychrometric Charts

On a **psychrometric chart** (or **humidity chart**) several properties of a gas–vapor mixture are cross-plotted, providing a concise compilation of a large quantity of physical property data. The most common of these charts, that for the air–water system at 1 atm, is used extensively in the analysis of humidification, drying, and air-conditioning processes.

A psychrometric chart in SI units for the air–water system at 1 atm is shown in Figure 8.4-1, and a second chart in American engineering units is shown in Figure 8.4-2. A chart that covers a wider temperature range is given on p. 12-5 of *Perry's Chemical Engineers' Handbook.*

The ordinate of the psychrometric chart is the **absolute humidity** of the humid air, or the mass ratio of water vapor to dry air. Once you know this quantity, you may easily convert it to the mass fraction of water in the air, or to the mole ratio of water to air or the mole fraction of water. Curves showing the relative humidity $[p_{H_2O}/p_{H_2O}^*(T)]$ of humid air also appear on the chart. The curve for 100% relative humidity is referred to as the **saturation curve.**

The abscissa of the psychrometric chart—the **dry-bulb temperature**—is the temperature that would be determined by any of the devices discussed in Chapter 3,

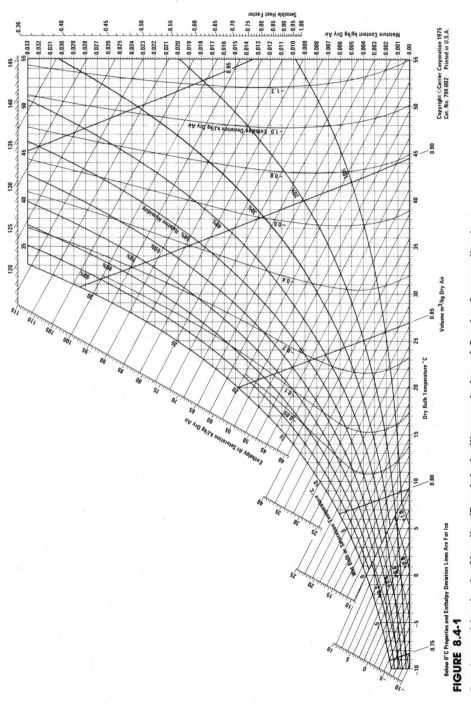

FIGURE 8.4-1

Psychrometric chart—SI units. (Reprinted with permission of Carrier Corporation.)

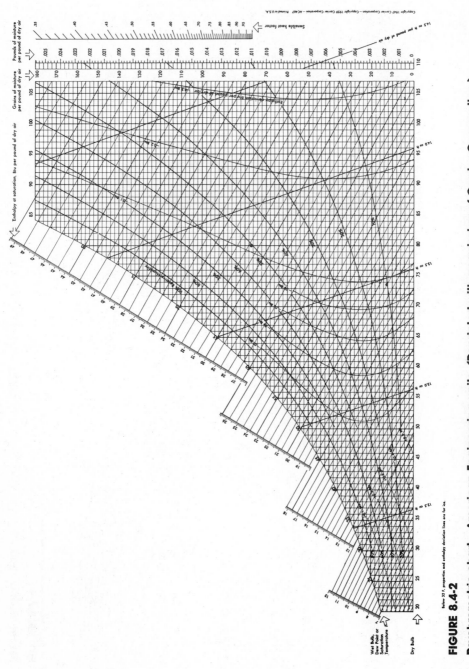

FIGURE 8.4-2

Psychrometric chart—American Engineering units. (Reprinted with permission of Carrier Corporation.)

including an ordinary thermometer. The other temperature plotted on the chart—the **wet-bulb temperature**—is best defined in terms of how it is measured.

Suppose a moist porous material (a **wick**) is wrapped around the bulb of a thermometer. If the thermometer is then placed in an unsaturated flowing stream of humid air and the wick is kept moist, evaporation of water from the wick causes a heat loss from the bulb, which causes a drop in the bulb temperature and hence in the thermometer reading. Eventually the bulb temperature stabilizes; at this point the reading of the thermometer corresponds to the wet-bulb temperature of the humid air.

The wet-bulb temperature of a gas–vapor mixture provides a measure of the mixture humidity. When the gas is saturated, no evaporation from the bulb occurs, and the wet-bulb and dry-bulb temperatures are the same; the lower the humidity, the greater the difference between the two temperatures.

The initial conditions (humidity and dry-bulb temperature) that lead to a single wet-bulb temperature in the experiment just described fall on a line on the psychrometric chart, called a **constant wet-bulb temperature line.** The constant wet-bulb temperature lines for air–water at 1 atm appear on Figures 8.4-1 and 8.4-2 as straight lines with negative slopes; the value of T_{wb} corresponding to a given line can be read at the intersection of the line with the saturation curve.

Lines that give the **humid volume** of air are also shown on the psychrometric chart. Notice that the units of \hat{V}_H are (m³ wet air)/(kg dry air); if you wish to determine the volume of a given mass of wet air, you must first determine the corresponding mass of dry air from the known absolute humidity, and then multiply this mass by \hat{V}_H.

Another variable shown on Figures 8.4-1 and 8.4-2 is the specific enthalpy of the air, relative to liquid water at 0°C and 1 atm and dry air at 1 atm and 0°C (Figure 8.4-1) or 0°F (Figure 8.4-2). The value of this quantity for saturated air can be read by extending the constant wet-bulb temperature line to the diagonal scale above the saturation curve. To calculate \hat{H} for unsaturated air, read the value for saturated air and add the **enthalpy deviation** obtained from the curves for this quantity shown on the chart. (See Example 8.4-5.)

The basis of the construction of the psychrometric chart is the Gibbs phase rule (Section 6.3a), which states that specifying a certain number of the intensive variables (temperature, pressure, specific volume, specific enthalpy, component mass or mole fractions, etc.) of a system automatically fixes the value of the remaining intensive variables. Humid air contains one phase and two components,[4] so that from Equation 6.2-1 the number of degrees of freedom is

$$F = 2 + 2 - 1 = 3$$

Specifying three intensive variables therefore fixes all other system properties. If the system pressure is fixed at 1 atm, then all other properties may be plotted on a two-dimensional plot, such as those shown in Figures 8.4-1 and 8.4-2.

[4] Since the components of dry air do not condense and are present in fixed proportion, dry air may be considered a single species (designated DA) in humidity calculations.

EXAMPLE 8.4-5 The Psychrometric Chart

Use the psychrometric chart to estimate (1) the absolute humidity, wet-bulb temperature, humid volume, dew point, and specific enthalpy of humid air at 41°C and 10% relative humidity, and (2) the amount of water vapor in 150 m³ of air at these conditions.

Solution

Following is a sketch of the psychrometric chart (Figure 8.4-1) showing the given state of the air:

1. Reading from the chart,

$$h_a \approx 0.0048 \text{ kg } H_2O/\text{kg DA}$$
$$T_{wb} = 19°C$$
$$\hat{V}(m^3/\text{kg DA}) \approx 0.895 \text{ (curve not shown)}$$

The dew point is the temperature at which the air with the given water content would be saturated at the same total pressure (1 atm), and is therefore located at the intersection of the horizontal constant absolute humidity line ($h_a \equiv 0.0048$) and the saturation curve, or

$$T_{dp} = 3°C$$

The specific enthalpy of saturated air at $T_{wb} = 19°C$ is 54.2 kJ/kg DA. Since the point corresponding to 41°C and 10% relative humidity falls roughly midway between the enthalpy deviation curves corresponding to -0.6 kJ/kg and -0.8 kJ/kg,

we may calculate \hat{H} as

$$\hat{H} = (54.2 - 0.7) \text{ kJ/kg DA}$$
$$\Downarrow$$
$$\boxed{\hat{H} = 53.5 \text{ kJ/kg DA}}$$

2. *Moles of humid air*

From Figure 8.4-1, the humid volume of the air is 0.897 m³/kg DA. We therefore calculate

$$\frac{150 \text{ m}^3}{} \left| \frac{1.00 \text{ kg DA}}{0.897 \text{ m}^3} \right| \frac{0.0048 \text{ kg H}_2\text{O}}{1.00 \text{ kg DA}} = \boxed{0.803 \text{ kg H}_2\text{O}}$$

The psychrometric chart can be used to simplify the solution of material and energy balance problems for constant pressure air–water systems, at the expense of some precision. Note the following points:

1. Heating or cooling humid air at temperatures above the dew point corresponds to horizontal movement on the psychrometric chart. The ordinate on the chart is the ratio (kg H_2O/kg dry air), which does not change as long as no condensation occurs.

2. If superheated humid air is cooled at 1 atm, the system follows a horizontal path to the left on the chart until the saturation curve (dew point) is reached; thereafter, the gas phase follows the saturation curve.

3. Since the psychrometric chart plots the mass ratio (kg H_2O/kg dry air) rather than the mass fraction of water, it is usually convenient to assume a quantity of dry air in a feed or product stream as a basis of calculation if the chart is to be used in the solution.

EXAMPLE 8.4-6 Material and Energy Balances on an Air Conditioner

Air at 80°F and 80% relative humidity is cooled to 51°F at a constant pressure of 1 atm. Use the psychrometric chart to calculate the fraction of the water that condenses, and the rate at which heat must be removed to deliver 1000 ft³/min of humid air at the final condition.

Solution

We begin by locating the point on Figure 8.4-2 that corresponds to 80°F and 80% relative humidity. The dew point, located by moving horizontally to the left to the saturation curve, is 73°F. The state of the gas phase at its final dry-bulb temperature of 51°F must therefore lie on the saturation curve.

Basis: 1 lb$_m$ dry air (DA)

(In assuming this basis, we are temporarily ignoring the specification of the volumetric flow rate at the outlet. After the process is balanced for the assumed basis, we will scale up to an outlet flow rate of 1000 ft³/min.)

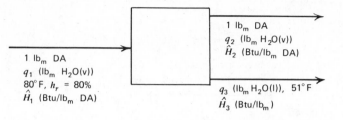

Note: In labeling the outlet gas stream, we have implicitly written a balance on dry air.

Point 1

$$80°F\brace 80\% \text{ RH} \xRightarrow{\text{Figure 8.4-2}} \begin{array}{l} h_a = 0.018 \text{ lb}_m \text{ H}_2\text{O/lb}_m \text{ DA} \\ \hat{H}_1 = 38.8 \text{ Btu/lb}_m \text{ DA} \end{array}$$

$$q_1 = \frac{1.0 \text{ lb}_m \text{ DA}}{} \left| \frac{0.018 \text{ lb}_m \text{ H}_2\text{O}}{\text{lb}_m \text{ DA}} \right. = 0.018 \text{ lb}_m \text{ H}_2\text{O}$$

Point 2

$$51°F\brace \text{Saturated} \xRightarrow{\text{Figure 8.4-2}} \begin{array}{l} h_a = 0.0079 \text{ lb}_m \text{ H}_2\text{O/lb}_m \text{ DA} \\ \hat{H}_2 = 20.9 \text{ Btu/lb}_m \text{ DA} \end{array}$$

$$q_2 = \frac{1.0 \text{ lb}_m \text{ DA}}{} \left| \frac{0.0079 \text{ lb}_m \text{ H}_2\text{O}}{\text{lb}_m \text{ DA}} \right. = 0.0079 \text{ lb}_m \text{ H}_2\text{O}$$

Balance on H_2O

$$q_1 = q_2 + q_3$$
$$\Big\| \begin{array}{l} q_1 = 0.018 \text{ lb}_m \\ q_2 = 0.0079 \text{ lb}_m \end{array}$$
$$q_3 = 0.010 \text{ lb}_m \text{ H}_2\text{O condensed}$$

Fraction H_2O Condensed

$$\frac{0.010 \text{ lb}_m \text{ condensed}}{0.018 \text{ lb}_m \text{ fed}} = \boxed{0.555}$$

Enthalpy of Condensate

Since the reference condition for water on Figure 8.4-2 is liquid water at 32°F, we must use the same condition to calculate \hat{H}_3.

$$\begin{array}{cc} \text{H}_2\text{O(l)} & \xrightarrow{\hspace{1cm}} \text{H}_2\text{O(l)} \\ 32°\text{F} & 51°\text{F} \end{array}$$

$$\Delta\hat{H} = \hat{H}_3 = 1.0 \frac{\text{Btu}}{\text{lb}_m \cdot °\text{F}} (51°\text{F} - 32°\text{F}) = 19.0 \text{ Btu/lb}_m \text{ H}_2\text{O}$$

Energy Balance

$$Q = \Delta H = \frac{1.0 \text{ lb}_m \text{ DA}}{} \left| \frac{20.9 \text{ Btu}}{\text{lb}_m \text{ DA}} + \frac{0.010 \text{ lb}_m \text{ H}_2\text{O(l)}}{} \right| \frac{19.0 \text{ Btu}}{\text{lb}_m \text{ H}_2\text{O}}$$

$$-\frac{1.0 \text{ lb}_m \text{ DA}}{} \left| \frac{38.8 \text{ Btu}}{\text{lb}_m \text{ DA}} = -17.7 \text{ Btu} \right.$$

(Observe carefully the units in this calculation.)

To calculate the cooling requirement for 1000 ft³/min of delivered air, we must first determine the volume of delivered air corresponding to our assumed basis and scale the calculated value of Q by the ratio (1000 ft³/min)/(V_{basis}). From the psychrometric chart, for humid air saturated at 51°F

$$\hat{V}_H = 13.0 \text{ ft}^3/\text{lb}_m \text{ DA}$$

$$\Downarrow$$

$$V_{\text{basis}} = \frac{1.0 \text{ lb}_m \text{ DA}}{} \left| \frac{13.0 \text{ ft}^3}{\text{lb}_m \text{ DA}} = 13.0 \text{ ft}^3 \right.$$

$$\Downarrow$$

$$Q = \frac{-17.7 \text{ Btu}}{} \left| \frac{1000 \text{ ft}^3/\text{min}}{13.0 \text{ ft}^3} = \boxed{-1360 \text{ Btu/min}} \right.$$

TEST YOURSELF

Air at 25°C and 1 atm has a relative humidity of 20%. Use the psychrometric chart to estimate the absolute humidity, the wet-bulb temperature, the dew point, the humid volume, and the specific enthalpy of the air.

8.4e Adiabatic Humidification

Drying and humidification operations, which involve the evaporation of a liquid into a gas–vapor mixture, are often carried out adiabatically, with the latent heat required to evaporate the liquid being provided by the sensible heat lost by the gas phase as it cools.

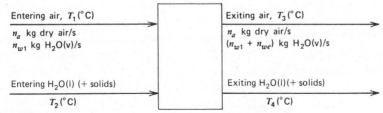

Entering air, T_1 (°C)

n_a kg dry air/s
n_{w1} kg H₂O(v)/s

Exiting air, T_3 (°C)

n_a kg dry air/s
$(n_{w1} + n_{we})$ kg H₂O(v)/s

Entering H₂O(l) (+ solids)

T_2 (°C)

Exiting H₂O(l)(+ solids)

T_4 (°C)

FIGURE 8.4-3
Flowchart of an adiabatic drying or humidification process

A flowchart is shown in Figure 8.4-3 of an adiabatic drying operation, in which a stream of warm humid air and a wet solid are brought into contact, and a quantity n_{we}(kg/s) of the water entering with the solid evaporates. If there are no solids involved, the process is humidification of air rather than drying of a solid, but the discussion that follows is still applicable.

You could now express the enthalpies of the stream components in terms of known heat capacities and the heat of vaporization of water, and write and solve material and energy balance equations on the unit to determine any unknown stream temperature or humidity. This is generally a straightforward but cumbersome procedure; however, if certain simplifying assumptions are made, the calculations may easily be performed with the aid of the psychrometric chart.

Let us assume:

1. $(C_p)_{air}$, $(C_p)_{H_2O}$, and $(\Delta \hat{H}_v)_{H_2O}$ are independent of temperature at the prevailing process conditions.

2. The enthalpy changes undergone by the unevaporated liquid water and the solid (if there is one) in going from T_2 to T_4 are negligible compared to the changes undergone by the air and the evaporated water.

3. The sensible heat required to raise liquid water from T_2 to T_3 is negligible compared to the heat of vaporization of water.

If the energy balance equation ($\Delta H = 0$) is written for the process of Figure 8.4-3 and these three assumptions are made, the simplified equation becomes

$$n_a(C_p)_{air}(T_3 - T_1) + n_{w1}(C_p)_{H_2O(v)}(T_3 - T_1) + n_{we}(\Delta \hat{H}_v)_{H_2O} = 0$$

$$\Downarrow$$

$$\frac{n_{we}}{n_a} = \frac{1}{(\Delta \hat{H}_v)_{H_2O}} \left[(C_p)_{air} + \frac{n_{w1}}{n_a}(C_p)_{H_2O(v)} \right](T_1 - T_3) \qquad \textbf{(8.4-9)}$$

Suppose now that the temperature T_1 and absolute humidity n_{w1}/n_a of the inlet air are specified, so that the state of the inlet air is fixed on the psychrometric chart. If we specify in addition the outlet air temperature $T_3(<T_1)$, then n_{we}/n_a may be calculated from Equation 8.4-9, and it may in turn be used to calculate the absolute humidity of the outlet air, $(n_{we} + n_{w1})/n_a$.

The outlet temperature and humidity determined in this manner are represented by a point on the psychrometric chart. If a lower value of T_3 is assumed, a higher outlet humidity would be calculated, yielding another point on the chart. The set of all such points for a specified T_1 and n_{w1}/n_a defines a curve on the chart, known as the **adiabatic saturation curve**. (See figure on next page.) *If the three stated assumptions are valid, the final state of air undergoing an adiabatic humidification must lie on the adiabatic saturation curve that passes through the inlet state on the pyschrometric chart.*

If the outlet temperature T_3 is low enough, the air leaves saturated with water. The temperature corresponding to this condition is called the **adiabatic saturation temperature** and is found at the intersection of the adiabatic saturation curve with the 100% relative humidity curve.

For most systems, the psychrometric chart would show a family of adiabatic saturation curves in addition to the families of curves shown on Figures 8.4-1 and 8.4-2. (See, for example, Figures 20-12 through 20-14 in *Perry's Chemical Engineers' Handbook*.) However, *for the air–water system at 1 atm, the adiabatic saturation curve through a given state coincides with the constant wet-bulb temperature line through that state, so that* $T_{as} = T_{wb}$.[5]

Let us summarize the procedure for using the psychrometric chart to carry out adiabatic humidification calculations for the air–water system.

1. *Assume that sensible heat changes undergone by the liquid water and by any solid present are negligible compared to the heat of vaporization of water.* If the water undergoes a substantial temperature change, you should not use the psychrometric chart, but rather write and solve the full energy balance equation.

2. *Locate the initial state of the air on the chart, and locate the final state on the constant wet-bulb temperature line that passes through the initial state.* Then perform whatever material and energy balance calculations are required.

EXAMPLE 8.4-7 Adiabatic Humidification

A stream of air at 30°C and 10% relative humidity is humidified in an adiabatic spray tower, which operates at $P \approx 1$ atm. The emerging air is to have a relative humidity of 40%.

[5] A detailed discussion of this coincidence is given in W. L. McCabe, J. C. Smith, and P. Harriott, *Unit Operations of Chemical Engineering*, 4th Edition, McGraw-Hill, New York, 1985, Ch. 23.

1. Determine the absolute humidity and the adiabatic saturation temperature of the entering air.

2. Use the psychrometric chart to calculate the rate at which water must be added to humidify 1000 kg/h of the entering air, and the temperature of the exiting air.

Solution

We assume that the sensible heat change undergone by the liquid in the spray tower is negligible compared with the heat of vaporization of water, so that the air follows an adiabatic saturation curve (constant wet-bulb temperature line) on the psychrometric chart.

1. Air at 30°C, 10% relative humidity

$$\Downarrow \text{ Figure 8.4-1}$$

$$h_a = 0.0026 \text{ kg H}_2\text{O/kg DA}$$

$$T_{wb} = T_{as} = 13.2°C$$

2. The state of the outlet air must lie on the $T_{wb} = 13.2°C$ line. From the intersection of this line with the curve for $h_r = 40\%$, the absolute humidity of the exit gas is determined to be 0.0063 kg H_2O/kg DA. The inlet (and outlet) flow rate of dry air, m_{DA}, is

$$m_{DA} = (1000 \text{ kg air/h})(1 \text{ kg DA}/1.0026 \text{ kg air}) = 997.4 \text{ kg DA/h}$$

The amount of water that must be evaporated, m_{H_2O}, may be calculated as the difference between the outlet and inlet water flow rates in the air stream.

$$m_{H_2O} = (997.4 \text{ kg DA/h})(0.0063 - 0.0026) \frac{\text{kg H}_2\text{O}}{\text{kg DA}}$$

$$= \boxed{3.7 \text{ kg H}_2\text{O/h}}$$

From Figure 8.4-1 the exit temperature is $\boxed{21.2°C}$.

TEST YOURSELF

1. (a) Under what conditions does the state (temperature and humidity) of a gas undergoing an adiabatic humidification follow a unique curve on the psychrometric chart, regardless of the entering liquid temperature?

 (b) Would this curve coincide with a constant wet-bulb temperature line if the gas is air and the liquid is water?

 (c) What if they were nitrogen and acetone?

2. Air at 26°C with a relative humidity of 10% undergoes an adiabatic humidification. Use Figure 8.4-1 to estimate the adiabatic saturation temperature of the air. If the exiting air has a dry-bulb temperature of 14°C, determine its absolute humidity, relative humidity, and specific enthalpy.

8.5 MIXING AND SOLUTION

When a solid or gas dissolves in a liquid or when two liquids are mixed, attractive forces are overcome between molecules of the feed materials and new forces exist between neighboring molecules of the product mixture or solution. A net absorption or release of energy accompanies these processes, resulting in internal energy and enthalpy differences between the pure components and the mixture.

We have so far neglected this effect, assuming that \hat{H} for a substance in a mixture equals \hat{H} for the pure substance. This is a good assumption for gas mixtures and for liquid mixtures of similar compounds, such as aliphatic hydrocarbons; significant enthalpy changes may accompany other types of mixing and solution processes, however. In this section we show how these changes may be determined and incorporated in energy balances.

8.5a Heats of Solution and Mixing

The **integral heat of solution** $\Delta\hat{H}_s(T, n)$ is defined as the change in enthalpy for a process in which 1 mole of a solute (gas or solid) is dissolved in n moles of a liquid solvent at a constant temperature T. As n becomes large, $\Delta\hat{H}_s$ approaches a limiting value known as the **heat of solution at infinite dilution.** The **heat of mixing** has the same meaning as the heat of solution when the process involves mixing two liquids rather than dissolving a gas or solid in a liquid.

Perry's Chemical Engineers' Handbook on pp. 3-157 through 3-159 gives heats of solution of various substances in water at either 18°C or "room temperature," which is roughly 25°C. *Caution:* The values given in the *Handbook* are *negatives* of the heats of solution $(-\Delta\hat{H}_s)$, although they are not identified as such explicitly.

As an illustration of how to use these data, suppose you wish to calculate ΔH for a process in which 2 mol of potassium cyanide (KCN) are dissolved in 400 mol of water at 18°C. First, calculate the moles of solvent per mole of solute:

$$n = 400/2 = 200 \text{ mol } H_2O/\text{mol KCN}$$

The value of $-\Delta\hat{H}_s$ (18°C, $n = 200$) is listed as -3.0 kcal/mol (meaning per mol of KCN dissolved). The total enthalpy change is therefore

$$\Delta H = n\,\Delta\hat{H}_s = \frac{+3.0 \text{ kcal}}{\text{mol KCN}} \left| \frac{2 \text{ mol KCN}}{} \right. = +6.0 \text{ kcal}$$

Table 8.5-1 lists values of the integral heats of solution at 25°C of HCl(g) and NaOH(s) in water, and the heat of mixing at 25°C of H_2SO_4(l) and water. Sources of data for other mixtures are given in the References for Chapter 7 of this text and on p. 3-160 of *Perry's Chemical Engineers' Handbook.*

Heats of solution such as those given in Table 8.5-1 may be used to determine directly the specific enthalpy of solutions at 25°C relative to the pure components at this temperature. Another common choice of reference conditions, however, is the pure solvent and an infinitely dilute solution at 25°C.

TABLE 8.5-1

Integral Heats of Solution and Mixing at 25°C

n(mol H$_2$O/mol solute)	$(\Delta \hat{H}_s)_{\text{HCl(g)}}$ kJ/mol HCl	$(\Delta \hat{H}_s)_{\text{NaOH(s)}}$ kJ/mol NaOH	$(\Delta \hat{H}_m)_{\text{H}_2\text{SO}_4}$ kJ/mol H$_2$SO$_4$
0.5	—	—	−15.73
1	−26.22	—	−28.07
1.5	—	—	−36.90
2	−48.82	—	−41.92
3	−56.85	−28.87	−48.99
4	−61.20	−34.43	−54.06
5	−64.05	−37.74	−58.03
10	−69.49	−42.51	−67.03
20	−71.78	−42.84	—
25	—	—	−72.30
30	−72.59	−42.72	—
40	−73.00	−42.59	—
50	−73.26	−42.51	−73.34
100	−73.85	−42.34	−73.97
200	−74.20	−42.26	—
500	−74.52	−42.38	−76.73
1 000	−74.68	−42.47	−78.57
2 000	−74.82	−42.55	—
5 000	−74.93	−42.68	−84.43
10 000	−74.99	−42.72	−87.07
50 000	−75.08	−42.80	—
100 000	−75.10	—	−93.64
500 000	—	—	−95.31
∞	−75.14	−42.89	−96.19

a From J. C. Whitwell and R. K. Toner, *Conservation of Mass and Energy*, pp. 344-346. Copyright © 1969 by McGraw-Hill, Inc. Used with permission of McGraw-Hill.

Consider, for example, a hydrochloric acid solution for which $n = 10$ moles H$_2$O/mole HCl. From Table 8.5-1, the specific enthalpy of this solution relative to pure HCl(g) and H$_2$O(l) at 25°C is $\Delta \hat{H}_s(n = 10) = -69.49$ kJ/mol HCl. Now, the enthalpy of the solution relative to H$_2$O(l) and a highly dilute solution of HCl (say $n = 10^6$ moles H$_2$O/mole HCl) is the enthalpy change for the process

$$\left\{ \begin{array}{l} 1 \text{ mol HCl} \\ 10^6 \text{ mol H}_2\text{O} \end{array} \right\} \xrightarrow{25°C} \left\{ \begin{array}{l} 1 \text{ mol HCl} \\ 10 \text{ mol H}_2\text{O} \end{array} \right\} + (10^6 - 10) \text{ mol H}_2\text{O(l)}$$

We may evaluate this enthalpy change using any convenient reference state—in particular, taking pure HCl(g) and H$_2$O(l) at 25°C as references. In view of the latter choice, \hat{H} for the $(10^6 - 10)$ moles of pure water equals zero, and the enthalpy change for the process is therefore

$$\Delta \hat{H} = \Delta \hat{H}_s(n = 10) - \Delta \hat{H}_s(n = \infty)$$
$$= -69.49 + 75.14 = 5.65 \text{ kJ/mol HCl}$$

In general, the enthalpy of a solution containing n moles H_2O/mole solute is

$$\boxed{\hat{H}_{\text{ref: pure solute and solvent @ 25°C}} = \Delta\hat{H}_s(n)}$$ (8.5-1)

$$\boxed{\hat{H}_{\text{ref: pure solvent and infinitely dilute solution @ 25°C}} = \Delta\hat{H}_s(n) - \Delta\hat{H}_s(\infty)}$$ (8.5-2)

Note again that these enthalpies are expressed per mole of *solute*, not per mole of solution.

TEST YOURSELF

The integral heat of solution of a solute A in water at 25°C is $-40\ \text{kJ/mol}$ A for $n = 10\ \text{mol}\ H_2O/\text{mol}$ A, and $-60\ \text{kJ/mol}$ A at infinite dilution.

1. What is the specific enthalpy (kJ/mol A) of an aqueous solution of A for which $n = 10\ \text{mol}\ H_2O/\text{mol}$ A relative to
 (a) pure H_2O and A at 25°C.
 (b) pure H_2O and an infinitely dilute aqueous solution of A?
2. If 5 mol of A are dissolved in 50 mol of H_2O at 25°C, how much heat is evolved or absorbed? (State which, observing that $Q = \Delta H$ for this process.)
3. How much heat is evolved or absorbed if the solution prepared in part 2 is poured into a large pond of water at 25°C?

8.5b Balances on Dissolution and Mixing Processes

When setting up an energy balance on a process that involves forming, concentrating, or diluting a solution, prepare an input–output enthalpy table considering the solution as a single substance. Because of the way heats of solution are tabulated, it generally proves to be convenient to calculate and list the products $n_i\hat{H}_i$ in the enthalpy table, rather than separating the amounts or flow rates (n_i) from the specific enthalpies (\hat{H}_i). (See Example 8.5-1.) If the solute by itself is involved in the process, it is convenient to choose the solute and solvent at 25°C (or at whatever temperature $\Delta\hat{H}_s$ is known) as references for enthalpy calculations, and if a dilute solution is input or output then an infinitely dilute solution and pure solvent are good choices as references.

To calculate the enthalpy of a solution at a temperature $T \neq 25°C$, first calculate its enthalpy at 25°C from tabulated heat of solution data using Equation 8.5-1 or 8.5-2, and then add the enthalpy change for the heating (or cooling) of the solution from 25°C to T. The enthalpy change for this step should be calculated from tabulated solution heat capacities if they are available (see, for example, pp. 3-145 and 3-146 of Perry's Chemical Engineers' Handbook for heat capacities of various aqueous solutions). Otherwise, follow the rules of thumb given in Section 8.3c of this chapter; use the average heat capacity of Equation 8.3-17 (for liquid mixtures) or neglect the enthalpy change of the solute (for dilute solutions).

EXAMPLE 8.5-1 Production of Hydrochloric Acid

Hydrochloric acid is produced by absorbing gaseous HCl in water. Calculate the heat that must be transferred to or from an absorption unit if HCl(g) at 100°C and $H_2O(l)$ at 25°C are fed to produce 1000 kg/h of 20 wt% HCl(aq) at 40°C.

Solution

It is advisable to determine the molar amounts or flow rates of the components of all feed and product solutions before drawing and labeling the flowchart. In this case

1000 kg/h 20 wt% HCl(aq)

$$n_{HCl} = \frac{1000 \text{ kg}}{h} \left| \frac{0.200 \text{ kg HCl}}{kg} \right| \frac{10^3 \text{ mol}}{36.5 \text{ kg HCl}} = 5480 \text{ mol HCl/h}$$

$$n_{H_2O} = \frac{1000 \text{ kg}}{h} \left| \frac{0.800 \text{ kg H}_2\text{O}}{kg} \right| \frac{10^3 \text{ mol}}{18.0 \text{ kg H}_2\text{O}} = 44\,400 \text{ mol H}_2\text{O/h}$$

Since $\Delta \hat{H}_s$ is known for HCl at 25°C and pure HCl(g) is involved in the process, we will choose pure HCl(g) and $H_2O(l)$ at 25°C as references.

References: HCl(g), H_2O(l) @ 25°C

Substance	$n_{in}\hat{H}_{in}$	$n_{out}\hat{H}_{out}$	
HCl(g)	1.19×10^4	—	$n\hat{H}$ in kJ/h
H_2O(l)	0	—	
HCl(aq)	—	-3.24×10^5	

Calculate $(n_{in}\hat{H}_{in})_{HCl(g)}$

$$HCl(g, 25°C) \rightarrow HCl(g, 100°C)$$

$$\Delta \hat{H} = \int_{25}^{100} C_p \, dT$$

\Downarrow C_p for HCl(g) from Table B.2

$$= 2.178 \text{ kJ/mol}$$

$$n_{in}\hat{H}_{in} = (5480 \text{ mol/h})(2.178 \text{ kJ/mol}) = 1.19 \times 10^4 \text{ kJ/h}$$

Calculate $(n_{out}\hat{H}_{out})_{HCl(aq)}$

$$n = (44\,400\ \text{mol H}_2\text{O})/(5480\ \text{mol HCl}) = 8.10$$

$$\text{HCl(g, 25°C)} + 8.10\ \text{H}_2\text{O(l, 25°C)} \xrightarrow{\Delta\hat{H}_1} \text{HCl(aq, 25°C)} \xrightarrow{\Delta\hat{H}_2} \text{HCl(aq, 40°C)}$$

$$\Delta\hat{H}_1 = \Delta\hat{H}_s(25°C, n = 8.1) \xrightarrow{\text{Table 8.5-1}} -67.4\ \text{kJ/mol HCl}$$

The heat capacity of hydrochloric acid is listed on p. 3-145 of *Perry's Chemical Engineers' Handbook* as a function of the mole fraction of HCl in the solution, which in our problem is

$$\frac{5480\ \text{mol HCl/h}}{(5480 + 44\,400)\ \text{mol/h}} = 0.110\ \text{mol HCl/mol}$$

$$\Downarrow$$

$$C_p = \frac{0.73\ \text{kcal}}{\text{kg}\cdot°\text{C}}\ \left|\ \frac{1000\ \text{kg solution}}{5480\ \text{mol HCl}}\ \right|\ \frac{4.184\ \text{kJ}}{\text{kcal}} = 0.557\ \frac{\text{kJ}}{\text{mol HCl}\cdot°\text{C}}$$

$$\Delta\hat{H}_2 = \int_{25}^{40} C_p\,dT = 8.360\ \text{kJ/mol HCl}$$

$$\Downarrow$$

$$n_{out}\hat{H}_{out} = \frac{5480\ \text{mol HCl}}{\text{h}}\ \left|\ \frac{(-67.4 + 8.36)\ \text{kJ}}{\text{mol HCl}}\right.$$

$$= -3.24 \times 10^5\ \text{kJ/h}$$

Energy Balance

$$Q = \Delta H = \sum n_{out}\hat{H}_{out} - \sum n_{in}\hat{H}_{in}$$

$$= -3.24 \times 10^5 - 1.19 \times 10^4 = \boxed{-3.36 \times 10^5\ \text{kJ/h}}$$

Thus, 336 000 kJ/h must be removed from the absorption unit.

8.5c Enthalpy-Concentration Diagrams

An **enthalpy-concentration diagram** is a plot of the specific enthalpy of a mixture versus the mole or weight percent of one component. It is an extremely efficient way of presenting enthalpy data, although its use is limited to two-component mixtures.

An enthalpy-concentration diagram for the NaOH–water system is shown in Figure 8.5-1. Notice that the enthalpies plotted are expressed per unit mass of solution (rather than of solute) and are calculated relative to pure water at 32°F and an infinitely dilute aqueous solution of NaOH at 68°F. Example 8.5-2 illustrates the use of this diagram in the solution of an energy balance problem.

EXAMPLE 8.5-2 Concentration of an Aqueous NaOH Solution

A 5-wt% NaOH solution at 80°F is to be concentrated to 23.1 wt% by evaporation of water. The concentrated solution and water vapor emerge from the evaporator at

230°F and 1 atm. Calculate the rate at which heat must be supplied to process 1000 lb_m/h of the feed solution.

Solution

Basis: Given feed rate of 5% solution

NaOH Balance

$$(0.05)(1000) = 0.231n_2 \Rightarrow n_2 = 216.5 \ lb_m/h$$

Total Mass Balance

$$1000 = n_1 + n_2 \xrightarrow{n_2 = 216.5} n_1 = 783.5 \ lb_m/h$$

Reference States for Energy Balance

$$H_2O(1, 32°F), \ NaOH(aq, 68°F, n = \infty)$$

From Figure 8.5-1: $\hat{H}_F = 45 \ Btu/lb_m$ (5% NaOH at 80°F)

$\hat{H}_L = 171 \ Btu/lb_m$ (23.1% NaOH at 230°F)

The enthalpy of water vapor at 230°F and 1 atm relative to liquid water at 32°F may be obtained from the steam tables in *Perry's Chemical Engineers' Handbook* as

$$\hat{H}_v = 1157.1 \ Btu/lb_m$$

Energy Balance

$$Q = \Delta H = n_1 \hat{H}_v + n_2 \hat{H}_L - 1000 \hat{H}_F$$
$$= (783.5)(1157.1) + (216.5)(171) - (1000)(45)$$
$$= \boxed{899 \ 000 \ Btu/h}$$

Compare the ease of this computation with that of Example 8.5-1. Having the enthalpy-concentration chart eliminates the need for all of the hypothetical heating, cooling, and isothermal mixing steps that would normally be required to evaluate the total enthalpy change for the process.

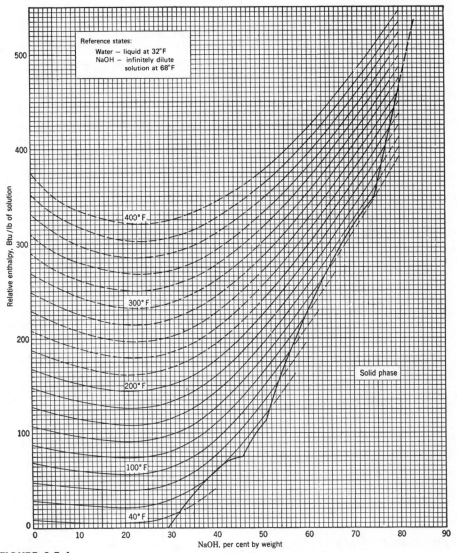

FIGURE 8.5-1

Enthalpy-concentration diagram for aqueous solutions of NaOH under a total pressure of 1 atm. The reference state for water is taken as liquid water at 32°F under its own vapor pressure. For sodium hydroxide, the reference state is that of an infinitely dilute solution at 68°F. (From G. G. Brown, et al., *Unit Operations*, © 1950, Figure 468. Reprinted by permission of John Wiley & Sons, Inc.)

Enthalpy-concentration diagrams are particularly useful for two-component systems in which vapor and liquid phases are in equilibrium. The Gibbs phase rule (Equation 6.2-1) specifies that such a system has $(2 + 2 - 2) = 2$ degrees of freedom; if as before we fix the system pressure, then specifying only one more intensive variable—the system temperature, or the mass or mole fraction of either component in either phase—fixes the value of all other intensive variables of both phases. An $\hat{H}-x$ diagram for the ammonia–water system at 1 atm is shown in Figure 8.5-2. The specific enthalpies of aqueous solutions and gaseous mixtures of ammonia and water are shown on separate curves on this figure.

Suppose the mass fraction of ammonia in a liquid solution of NH_3 and H_2O at 1 atm is specified to be 0.25. According to the phase rule, the system temperature and the mass fraction of NH_3 in the vapor phase are uniquely determined by these specifications. A **tie line** may therefore be drawn on the enthalpy-concentration chart from $x = 0.25$ on the liquid-phase curve to the corresponding point on the vapor curve, which is at $y = 0.95$, and the tie line may be labeled with the corresponding temperature, 100°F. Several tie lines constructed in this manner are shown in Figure 8.5-2; once drawn in, the lines may be used to determine directly the equilibrium composition and the specific enthalpy of each phase at a specified temperature.

EXAMPLE 8.5-3 Use of the Enthalpy-Concentration Chart for a Two-Phase System

An aqueous ammonia solution is in equilibrium with its vapor in a closed system at 160°F and 1 atm. The liquid phase accounts for 75% of the total mass of the system contents. Use Figure 8.5-2 to determine the weight percent of NH_3 in each phase and the enthalpy of the system per unit mass of the system contents.

Solution

The mass fractions of ammonia and specific enthalpies of each phase may be read from the intersections of the 160°F tie line with the vapor and liquid equilibrium curves on Figure 8.5-2.

Liquid Phase

$$\boxed{8\% \text{ NH}_3, 92\% \text{ H}_2\text{O}}\;; \hat{H}_L = 110 \text{ Btu/lb}_m$$

Vapor Phase

$$\boxed{64\% \text{ NH}_3, 36\% \text{ H}_2\text{O}}\;; \hat{H}_V = 855 \text{ Btu/lb}_m$$

Basis

$$1 \text{ lb}_m \text{ total mass} \Rightarrow 0.75 \text{ lb}_m \text{ liquid}, 0.25 \text{ lb}_m \text{ vapor}$$

$$\hat{H}(\text{Btu/lb}_m) = \frac{0.75 \text{ lb}_m \text{ liquid}}{\text{lb}_m} \left| \frac{110 \text{ Btu}}{\text{lb}_m} + \frac{0.25 \text{ lb}_m \text{ vapor}}{\text{lb}_m} \right| \frac{855 \text{ Btu}}{\text{lb}_m}$$

$$= \boxed{296 \frac{\text{Btu}}{\text{lb}_m}}$$

FIGURE 8.5-2

Enthalpy-concentration diagram for the ammonia–water system at 1 atm. (From G. G. Brown, et al., *Unit Operations*, ⓒ 1950, Figure 551. Reprinted by permission of John Wiley & Sons.)

If the overall composition of a two-phase two-component system is known, the fraction of the system that is liquid or vapor may easily be determined from the enthalpy-concentration diagram using the **lever rule,** which was first introduced in the discussion of phase diagrams in Chapter 6.

Suppose, for example, that a mixture of ammonia and water that is 40% NH_3 by mass is contained in a closed vessel at 140°F and 1 atm. Point A on Figure 8.5-2 corresponds to this set of conditions. Since this point lies between the vapor and liquid equilibrium curves, the mixture separates into two phases whose compositions are found at the extremities of the 140°F tie line (points B and C).

In general, if F, L, and V are the total mass of the mixture, the mass of the liquid phase and the mass of the vapor phase, respectively, and x_F, x_L, and x_V are the corresponding mass fractions of NH_3, then

$$\text{Total balance:} \quad F = L + V \qquad (8.5\text{-}3)$$

$$NH_3 \text{ balance:} \quad x_F F = x_L L + x_V V \qquad (8.5\text{-}4)$$

Substituting the expression of Equation (8.5-3) for F into Equation (8.5-4) and rearranging the result yields

$$\frac{L}{V} = \frac{x_V - x_F}{x_F - x_L} \qquad (8.5\text{-}5)$$

The tie line in question if isolated appears as shown below:

From the properties of similar triangles, the right side of Equation 8.5-5 equals the ratio of distances $\overline{AC}/\overline{AB}$. We have thus proved the following general rule: If A, B, and C are the points on a tie line corresponding to the total mixture, the liquid phase, and the vapor phase, respectively, and if F, L, and V are the corresponding masses, then the liquid-to-vapor mass ratio is

$$\frac{L}{V} = \frac{x_V - x_F}{x_F - x_L} = \frac{\overline{AC}}{\overline{AB}} \qquad (8.5\text{-}6)$$

This is the lever rule. It is also not difficult to prove that the mass fractions of the liquid and vapor phases are

$$\frac{L}{F} = \frac{x_V - x_F}{x_V - x_L} = \frac{\overline{AC}}{\overline{BC}} \qquad (8.5\text{-}7)$$

$$\frac{V}{F} = \frac{x_F - x_L}{x_V - x_L} = \frac{\overline{AB}}{\overline{AC}} \qquad (8.5\text{-}8)$$

Once you have located the total mixture on the chart from a specified set of feed conditions, it becomes a simple matter to determine the compositions, enthalpies, and relative proportions of each phase, calculations that would take a considerable amount of time in the absence of the chart.

EXAMPLE 8.5-4 Equilibrium Flash Vaporization

A 30% NH_3 solution at 100 psia is fed to a tank in which the pressure is 1 atm at a rate of 100 lb_m/h. The enthalpy of the feed solution relative to the reference conditions used to construct Figure 8.5-2 is 100 Btu/lb_m. The vapor composition is to be 89 wt% NH_3. Determine the temperature at which the process takes place, the mass fraction of NH_3 in the liquid product, the flow rates of the liquid and vapor product streams, and the rate at which heat must be transferred to the vaporizer.

Solution

Basis: 100 lb_m/h feed

From Figure 8.5-2,

$$x_V = 0.89$$
$$\Downarrow$$
$$\boxed{T = 120°F}$$
$$\boxed{x_l = 0.185 \ lb_m \ NH_3/lb_m}$$
$$\hat{H}_V = 728 \ \text{Btu/lb}_m$$
$$\hat{H}_L = 45 \ \text{Btu/lb}_m$$

From Equation 8.5-7

$$\frac{L}{F} = \frac{x_V - x_F}{x_V - x_L}$$

⇓

$$L = (100 \text{ lb}_m/\text{h})\frac{0.89 - 0.30}{0.89 - 0.185} = \boxed{84 \text{ lb}_m/\text{h liquid product}}$$

$$V = 100 - 84 = \boxed{16 \text{ lb}_m/\text{h vapor product}}$$

Energy Balance

$$Q = \Delta H = V\hat{H}_V + L\hat{H}_L - F\hat{H}_F$$

$$= (16)(728) + (84)(45) - (100)(100) = \boxed{5400 \text{ Btu/h}}$$

TEST YOURSELF

Use Figure 8.5-2 for the following calculations.

1. Estimate the temperature at which the mass fraction of NH_3 in the vapor phase of a two-phase system equals 0.85 and the corresponding liquid-phase NH_3 mass fraction.
2. What is the heat of vaporization of NH_3 at its normal boiling point?
3. If an NH_3–H_2O mixture whose overall composition is 50% NH_3–50% H_2O comes to equilibrium at 120°F, what fraction of the mixture is a vapor?

PROBLEMS

1. The specific internal energy of formaldehyde vapor at 1 atm and moderate temperatures is given by the formula

$$\hat{U}(\text{J/mol}) = 25.96T + 0.02134T^2$$

where T is in °C.
 (a) From the definition of the heat capacity at constant volume, derive a formula for $C_v(T)$, and use it to calculate C_v at 0°C and 100°C.
 (b) What reference temperature was used to generate the given expression for \hat{U}?
2. The heat capacity at constant volume of hydrogen sulfide at low pressures is

$$C_v(\text{J/mol}\cdot°\text{C}) = 25.2 + 1.547 \times 10^{-2}T - 3.012 \times 10^{-6}T^2$$

where T is in °C. Calculate the heat required to bring 3.00 liters of H_2S in a rigid vessel at an initial pressure of 3.00 atm from 25°C to 1000°C, retaining successively one term, two terms, and all three terms of the heat capacity formula. (See Example

8.3-1.) Calculate the percentage errors that result from retaining one and two terms of the formula, assuming that the full expression yields the correct result.

3. The heat capacity at constant pressure of hydrogen cyanide is given by the expression

$$C_p(J/mol \cdot °C) = 35.3 + 0.0291T(°C)$$

(a) Write an expression for the heat capacity at constant volume for HCN, assuming ideal gas behavior.

(b) Calculate $\Delta \hat{H}(J/mol)$ for the constant pressure process

$$HCN(25°C, 1 \text{ atm}) \rightarrow HCN(100°C, 1 \text{ atm})$$

(c) Calculate $\Delta \hat{U}(J/mol)$ for the constant volume process

$$HCN(25°C, 1 \text{ m}^3/kmol) \rightarrow HCN(100°C, 1 \text{ m}^3/kmol)$$

(d) If the process of part (b) were carried out in such a way that the initial and final pressures were each 1 atm but the pressure varied during the heating, the value of $\Delta \hat{H}$ would still be what you calculated assuming a constant pressure. Why is this so?

4. Use data in Table B.2 to calculate
 (a) The heat capacity of liquid benzene at 50°C.
 (b) The heat capacity of benzene vapor at 50°C.
 (c) The heat capacity of solid carbon at 50°C.
 (d) $\Delta \hat{H}(J/mol)$ for benzene vapor going from 100°C to 300°C.
 (e) $\Delta \hat{H}(J/mol)$ for solid carbon going from 100°C to 300°C.

5. Estimate the specific enthalpy of steam (kJ/kg) at 350°C and 100 bars relative to steam at 100°C and 1 atm using
 (a) The steam tables
 (b) Table B.2 and assuming ideal gas behavior
 What is the physical significance of the difference between the values of \hat{H} calculated by the two methods?

6. Two formulas for the heat capacity of CO are given here.

$$C_p(cal/mol \cdot °C) = 6.890 + 0.001436T(°C)$$
$$C_p(Btu/lb\text{-mole} \cdot °F) = 6.894 + 0.0007978T(°F)$$

Starting with the first formula, derive the second. (Recall Section 2.5, and remember that the temperature unit in the denominator of C_p refers to a temperature interval.)

7. Table B.2 lists values of the heat capacity of liquid methanol at two temperatures. Use the tabulated values to derive a linear expression for $C_p(T)$; then use the derived expression to calculate the heat required to raise the temperature of 20.0 kg of liquid methanol from 0°C to 64.7°C.

8. Chlorine gas is to be heated from 100°C and 1 atm to 200°C.
 (a) Calculate the heat input (kW) required to heat a stream of the gas flowing at 5.0 kmol/s at constant pressure.

(b) Calculate the heat input (kJ) required to raise the temperature of 5.0 kmol of chlorine in a closed rigid vessel from 100°C and 1 atm to 200°C. (*Suggestion:* Evaluate $\Delta \hat{U}$ directly from the result of the first calculation, so that you do not have to perform another integration.) What is the physical significance of the numerical difference between the values calculated in parts (a) and (b)?

(c) To accomplish the heating of part (b), you would actually have to supply an amount of heat to the vessel greater than the amount calculated. Why?

9. The heat capacity of liquid *n*-hexane is measured in a **bomb calorimeter.** Two liters of *n*-C_6H_{14} are placed in a well-insulated vessel at $T = 300K$. A small reaction flask (the bomb) is then immersed in the liquid. A combustion reaction known to release 16.73 kJ of heat takes place in the bomb, and the subsequent temperature rise of the system contents is measured and found to be 3.10K. In a separate experiment, it is found that 6.14 kJ of heat are required to raise the temperature of everything in the system except the *n*-hexane by 3.10K. Use these data to calculate $C_v(J/g \cdot K)$ for *n*-C_6H_{14} at $T \approx 300K$, and compare your result with a tabulated value.

10. The heat capacities of a substance have been defined as

$$C_v = \left(\frac{\partial \hat{U}}{\partial T}\right)_V, \quad C_p = \left(\frac{\partial \hat{H}}{\partial T}\right)_P$$

Use the defining relationship between \hat{H} and \hat{U} and the fact that \hat{H} and \hat{U} for ideal gases are functions only of temperature to prove that $C_p = C_v + R$ for an ideal gas.

11. Use tabulated mean heat capacities to calculate $\Delta \hat{H}$ for the following processes, all taking place at low pressures:
 (a) $N_2(25°C) \rightarrow N_2(700°C)$
 (b) $H_2(800°F) \rightarrow H_2(77°F)$
 (c) $CO_2(300°C) \rightarrow CO_2(1250°C)$
 (d) $O_2(970°F) \rightarrow O_2(0°F)$

12. Use tabulated mean heat capacity data to calculate the heat input rate in kilowatts required to raise 300 kg/min of carbon dioxide from 25°C to 450°C.

13. A stream of water vapor flowing at a rate of 250 mol/h is cooled from 700°C to 100°C at a constant pressure of 1 atm. Calculate the required cooling rate (kW) three ways:
 (a) From the steam tables
 (b) Using heat capacity data in Table B.2
 (c) Using mean heat capacity data in Table 8.3-1

14. Two hundred ft^3/h of air at 70°F and 1.2 atm are blown through ducts that pass through the interior of a large industrial motor. The air emerges at 500°F. Calculate the rate at which the air is removing heat generated by the motor, using tabulated mean heat capacity data.

15. Using data in Table B.2, calculate the mean heat capacity $C_p(J/mol \cdot °C)$ for the process

$$He(25°C, 1 \text{ atm}) \rightarrow He(500°C, 1 \text{ atm})$$

16. Calculate the mean heat capacity of water vapor (J/mol·°C) for the transition from 25°C and 1 atm to 750°C and 1 atm using

(a) The steam tables

(b) The heat capacity formula given in Table B.2

17. The heat capacity of oxygen at constant pressure in the range 0°C to 200°C is well represented by the formula

$$C_p(\text{J/mol} \cdot °\text{C}) = 29.83 + 0.0589T(°\text{C})$$

Following the development of the mean heat capacity concept in Section 8.3b, define a mean heat capacity at constant volume \bar{C}_v and calculate it for the process

$$O_2(0°\text{C}) \rightarrow O_2(200°\text{C})$$

using the given formula for C_p as the starting point. Assume ideal gas behavior.

18. Calculate the heat required to raise 50 kg of solid sodium carbonate (Na_2CO_3) from 10°C to 50°C using

(a) The true heat capacity of Na_2CO_3, which is 1.14 kJ/kg·°C.

(b) A heat capacity estimated by Kopp's rule. Calculate the percentage error in the latter calculation.

19. A liquid mixture of 30 wt% n-hexane and 70 wt% 2-methyl-1-pentanol ($C_6H_{14}O$) is cooled from 45°C to 20°C. Calculate the associated specific enthalpy change in J/g, using Kopp's rule to estimate any heat capacity for which a tabulated value cannot be found. State all assumptions you make.

20. A gas mixture contains one-third methane by volume and the balance oxygen at 350°C and 3.0 bars. Calculate the specific enthalpy of this stream in kJ/kg (not per kmol) relative to the pure components at 25°C and 1 atm. *State clearly all assumptions.*

21. Propane is to be burned with 20% excess oxygen. Before entering the furnace, the combined feed stream ($C_3H_8 + O_2$) is preheated from 77°F to 575°F. At what rate (Btu/h) must heat be supplied if the flow rate of the feed stream is 2.50×10^5 ft^3(STP)/h?

22. The radiant energy incident on the earth's surface on a sunny day is roughly 900 W/m². Collecting and focusing sunlight and using the focused beam to heat a fluid is an old idea, and as the economic and environmental costs of fossil fuel combustion mount, solar heating becomes an increasingly attractive technique.

Suppose a house is to be designed that will have a circulating forced air central heating unit, and solar energy is contemplated as a heat source (backed up with a conventional furnace to be used on cloudy days). If air is to be fed at a rate of 1000 m³/min at 30°C and 1 atm, and the air is to be heated to 55°C before being discharged into the living space, what is the required area of the solar collector plates? Assume 30% of the radiant energy incident on the plates is used to heat the air.

23. A natural gas containing 95 mole% methane and the balance ethane is burned with 25% excess air. The stack gas leaves the furnace at 900°C and 1 atm and contains no unburned hydrocarbons or carbon monoxide. The gas is cooled to 500°C in a heat exchanger. Taking as a basis of calculation 100 mol of the fuel gas fed to the furnace, calculate the amount of heat (kJ) that must be transferred from the gas in the heat exchanger to accomplish the indicated cooling.

24. Twenty liters of liquid n-propyl benzoate ($C_6H_5CO_2C_3H_7$, SG = 1.021) and 15 liters of liquid benzene are mixed and heated from 25°C to 75°C. Calculate the required heat input (kJ), using Kopp's rule when necessary. State all assumptions you make.

25. A stream containing propane gas enters the inner pipe of a double-pipe heat exchanger at 40°C and 250 kPa and exits from the pipe at 240°C and the same pressure. Super-heated steam at 300°C and 5.0 bars enters the outer (annular) pipe, flowing counter-currently to the propane, and exits as a saturated vapor at the same pressure. Calculate the feed ratio of the two streams, (m^3 steam fed/m^3 propane fed), neglecting heat losses from the heat exchanger to its surroundings.

26. Saturated steam at 300°C is used to heat a stream of methanol vapor from 65°C to 260°C in an adiabatic heat exchanger. The flow rate of the methanol is 50 liters(STP)/min, and the steam condenses and leaves the heat exchanger as a liquid at 90°C. Calculate the required flow rate of the steam in g/min.

27. A gas containing 8.5 mole% CO, 10.5% CO_2, 0.5% O_2, and 80.5% N_2 on a dry basis leaves a catalyst regeneration unit at 620°C and 1 atm with a dew point of 57°C. The flow rate of the gas is 28.5 m^3(STP)/h. A small amount of solid catalyst is entrained in the gas and is to be recovered in an electrostatic precipitator, but the gas must first be cooled to 425°C to prevent damage to the precipitator. The cooling will be accomplished by spraying water at 20°C into the gas. Calculate the required water feed rate (kg/h) neglecting the heat given up by the entrained solid particles as they cool.

28. All large chemical plants use a great deal of steam for heating, power generation, and process applications. The cost of fuel required to generate this steam in a power-plant boiler can be extremely high, and in many cases can make the difference between profit and loss for the plant. A common way to reduce the fuel costs associated with steam production is to pass hot gas streams produced in the plant through **waste heat boilers**—heat exchangers in which the heat lost by the cooling gases is used to produce steam from liquid water.

 Suppose a gas containing 40 mole% CO_2 and 60% H_2O at 1200°C and 830 mm Hg is about to be discharged to the atmosphere. If the gas is instead fed to a waste heat boiler and cooled to 300°C, and liquid water at 10 bars fed to the boiler is converted to saturated steam at the same pressure, how many kilograms of steam are produced per cubic meter of gas fed to the boiler? Assume the boiler operates adiabatically.

29. The air in a 55 m^3 room is to be cooled from 37°C, 1 atm, and 80% relative humidity to its dew point at 1 atm.
 (a) Calculate the masses of dry air and water vapor present in the room when the cooling has been accomplished.
 (b) Calculate the heat (kJ) that must be transferred to cool the quantity of air calculated in part (a), assuming $Q = \Delta H$ for this process.

30. On a cold winter day the temperature is 2°C and the relative humidity is 15%. You inhale air at an average rate of 5500 ml/min, and exhale a gas saturated with water at body temperature, roughly 37°C. If the mass flow rates of the inhaled and exhaled air excluding water are the same, the heat capacities of the water-free gases are each 1.05 J/g·°C, and water is ingested into the body as a liquid at 22°C, at what rate in J/day do you lose energy by breathing?

31. Ernest Acrophobe is a circus stunt man whose specialty is high diving. Faced with increasing competition from younger performers, he has decided to establish his superiority once and for all by diving from the top of a 250-ft tower into a half-empty barrel.

 The fateful day arrives, and Acrophobe climbs the ladder to the platform. Just as he is about to dive, he is struck with the thought that even if he survives the landing, the friction of his impact could heat the water sufficiently for him to be scalded to death. Unwilling to take unnecessary risks, he takes out his pocket calculator and estimates the final temperature of the water. He satisfies himself that the danger is negligible, and off he goes.

 Write an energy balance for this process from the moment before the dive to the moment after impact, and determine the final temperature, taking Acrophobe, the barrel, and the water as the system, and using the following data and somewhat questionable assumptions: (i) the barrel is 3 ft in diameter, contains water to a depth of 5 ft, and excluding the water weighs 100 lb_m; (ii) Acrophobe weighs 160 lb_m; (iii) the initial system temperature is 75°F, and the heat capacity of the system is 0.99 $Btu/lb_m \cdot °F$; (iv) all of the water remains in the barrel, and no heat is lost from the barrel after impact.

32. Seventy-five liters of liquid ethanol at 70°C and 65 liters of liquid water at 25°C are mixed in a closed, well-insulated vessel. Neglecting evaporation and the heat of mixing, calculate the final temperature of the mixture.

33. 110 g/s of liquid water at 25°C are sprayed into 500 liters(STP)/s of air initially at 500°C and 1 atm with a dew point of 30°C. Calculate the final temperature of the system after the water has vaporized in the gas, assuming that the process is adiabatic and takes place at a constant pressure of 1 atm. (*Suggestion:* Use the mean heat capacity tables for the gases and water vapor.)

34. In the manufacture of nitric acid, ammonia and preheated air are mixed to form a gas containing 10 mole% NH_3 at 600°C. (The ammonia is not preheated, since if it were it would decompose to N_2 and H_2.) The mixture is then catalytically oxidized to form NO_2, which is absorbed in water to form HNO_3. If ammonia enters the gas blending unit at 25°C at a rate of 520 kg/h, and heat is lost from the mixer to its surroundings at a rate of 7 kW, determine the temperature to which the air must be preheated. (Use the first two terms of the heat capacity formula for air of Table B.2, and see Example 8.3-5.)

35. Methane is burned completely with 70% excess air in a continuous combustion furnace. The air on its way to the furnace and the hot combustion products leaving the furnace pass through a heat exchanger in which the heat lost by the combustion products raises the temperature of the air from 25°C to 200°C.

If the combustion products leave the furnace at 800°C and the preheater operates adiabatically, at what temperature do the combustion products leave the preheater? (A trial-and-error solution is required.)

36. The heat capacity at constant pressure of a gas is determined experimentally at several temperatures, with the following results:

$T(°C)$	0	100	200	300	400	500	600
$C_p(J/mol·°C)$	33.5	35.1	36.7	38.4	40.2	42.0	43.9

 (a) Calculate the heat required to raise 300 kg/min of the gas from 0°C to 600°C, using Simpson's rule (Appendix A.3) to integrate the tabulated heat capacities.

 (b) Use the method of least squares (Appendix A.1) to derive a linear expression for $C_p(T)$ in the range 0 to 600°C, and use this expression to estimate once again the heat required to raise 300 kg/min of the gas from 0°C to 600°C. If the estimates differ, in which one would you have more confidence, and why?

***37.** As part of a design calculation, you must evaluate an enthalpy change for an obscure organic vapor that is to be cooled from 1800°C to 150°C in a heat exchanger. You search through all the standard references for tabulated enthalpy or heat capacity data for the vapor but have no luck at all, until you finally stumble on an article in the May 1922 *Serbian Journal of Obscure Organic Vapors* that contains a plot of $C_p(cal/g·°C)$ on a logarithmic scale versus $[T(°C)]^{1/2}$ on a linear scale. The plot is a straight line through the points ($C_p = 0.329$, $T^{1/2} = 7.1$) and ($C_p = 0.533$, $T^{1/2} = 17.3$).

 (a) Derive an equation for C_p as a function of T.

 (b) Suppose the relationship of part (a) turns out to be

$$C_p = 0.235 \exp[0.0473T^{1/2}]$$

and that you wish to evaluate

$$\Delta \hat{H}(cal/g) = \int_{1800}^{150} C_p \, dT$$

First perform the integration analytically; then write a computer program to do it using Simpson's rule (Appendix A.3). Have the program evaluate C_p at 11 equally spaced points from 150°C to 1800°C, estimate and print the value of $\Delta \hat{H}$, and repeat the calculation using 101 points. What can you conclude about the accuracy of the numerical calculation?

38. The heat of vaporization of ethylene glycol at its normal boiling point of 197.2°C is 56.9 kJ/mol. Calculate the rate (kW) at which heat must be removed to condense 175 kg $C_2H_6O_2$/min at 197.2°C and 1 atm.

39. Determine the enthalpy of *n*-pentane vapor at 200°C relative to the liquid at 20°C. Show clearly the process path you construct for this calculation.

40. Calculate the heat of vaporization of water (kJ/mol) at 25°C and low pressures from tabulated heat of vaporization (Table B.1) and heat capacities (Table B.2, Table 8.3-1), showing clearly the process path you construct for the calculation. Compare your answer with the value given in Table B.4. (See discussion following Example 8.4-1.)

41. Benzene vapor at 1000°F is cooled and converted to a liquid at 75°F in a continuous condenser. The condensate is drained into 55-gal drums, each of which takes 2.0 min to fill. Calculate the rate in Btu/h at which heat is being removed in the condenser.

* Computer problem.

42. In **gas adsorption** a vapor is transferred from a gas mixture to the surface of a solid. If the solid has an affinity for the vapor species and a large surface area, a substantial amount of vapor can be removed from a gas stream in this manner. An approximate but often useful way of analyzing adsorption is to treat it simply as condensation of vapor on a solid surface.

 Suppose a nitrogen stream at 35°C and 1 atm containing carbon tetrachloride with a 15% relative saturation is fed at a rate of 10.0 mol/min to a 6-kg bed of activated carbon. The temperature and pressure of the gas do not change appreciably from the inlet to the outlet of the bed, and there is no CCl_4 in the gas leaving the adsorber. The carbon can adsorb 40% of its own mass of carbon tetrachloride before becoming saturated. Neglecting the effect of temperature on the heat of vaporization of CCl_4, estimate the rate at which heat must be removed from the adsorber (kJ/min) to keep the process isothermal, and the time (min) it will take to saturate the bed.

43. If carbon dioxide is cooled at 1 atm, it condenses directly to a solid (**dry ice**) at $-78.4°C$. The heat of sublimation at this temperature is 6030 cal/mol. Calculate the heat removal rate (kW) required to produce 400 kg $CO_2(s)$/h at the sublimation temperature if CO_2 at 20°C is the feed.

44. Molten sodium chloride is to be used as a constant temperature bath for a high-temperature chemical reactor. Two-hundred kilograms of solid NaCl at 300K are charged into an insulated vessel, and a 3000-kW electrical heater is turned on, raising the salt to its melting point of 1073K and melting it.

 (a) The heat capacity (C_p) of solid NaCl is 50.41 J/mol·K at $T = 300K$, and 53.94 J/mol·K at $T = 500K$, and the heat of fusion of NaCl at 1073K is 30.21 kJ/mol. Use these data to determine a linear expression for $C_p(T)$ and to calculate $\Delta \hat{H}$(kJ/mol) for the transition of NaCl from a solid at 300K to a liquid at 1073K.

 (b) Write and solve the energy balance equation for this batch heating process to determine the required heat input in joules. (Recall the relationship between $\Delta \hat{U}$ and $\Delta \hat{H}$ for phase changes given in Section 8.3a.)

 (c) If 85% of the full power of 3000 kW goes into heating and melting the salt, how long does the process take?

45. Estimate the heat of vaporization of ethyl benzene at its normal boiling point using Trouton's rule and Chen's rule, and compare the results with a tabulated value of this quantity. Then estimate $\Delta \hat{H}_v$ at 100°C using the Watson correlation.

46. You are writing energy balances for a compound for which you cannot find heat capacity or latent heat data. All you know about the material are its molecular formula ($C_7H_{12}N$) and that it is a liquid at room temperature and has a normal boiling point of 200°C. Use this information to estimate the enthalpy of the vapor of this substance at 200°C relative to the liquid at 25°C. (Recall Section 8.3c.)

47. Use Equation 8.4-5 to estimate the heat of fusion of solid calcium chloride at its normal melting point, and compare the result with a tabulated value.

48. Estimate the heat of vaporization (kJ/mol) of benzene at a pressure of 100 mm Hg, using each of the following sets of data:

 (a) The heat of vaporization at the normal boiling point given in Table B.1, the

boiling point at 100 mm Hg as determined from the Antoine equation, and the Watson correlation.

(b) The boiling points at 50 mm Hg and 150 mm Hg as determined from the Antoine equation and the Clausius-Clapeyron equation.

(c) The heat of vaporization at the normal boiling point given in Table B.1, the boiling point at 100 mm Hg as determined from the Antoine equation, and heat capacity data given in Table B.2.

49. Humid air at 50°C and 1.0 atm with 2°C of superheat is fed to a condenser. Gas and liquid streams leave the condenser in equilibrium at 20°C and 1 atm. Calculate the mass of water condensed (kg) and the heat that must be transferred from the condenser (kJ) per cubic meter of air fed to the unit.

50. An air conditioner cools 8000 ft^3/min of humid air at 95°F and 98% relative humidity to 50°F. Calculate the rate of condensation of water in the unit and the cooling requirement in tons (1 ton = 12 000 Btu/h).

51. A gas stream containing n-hexane in nitrogen is fed to a condenser at 75°C and 3 atm (absolute), with a relative saturation of 90%. The product gas emerges at 0°C and 3 atm at a rate of 746.7 m^3/h. Calculate the percentage condensation of hexane (moles condensed/mol fed), and the rate (kW) at which heat must be transferred from the condenser.

52. A gas stream containing acetone in air flows from a solvent recovery unit at a rate of 300 ft^3/min at 250°F and 5 psig. A cooler–compressor is to be designed to bring the stream to 0°F and 5 atm (absolute), thereby condensing most of the acetone. To determine the composition of the feed stream, a 3.00-liter sample is taken at the condenser inlet and is cooled to a temperature at which all the acetone in the sample is recovered as a liquid. The liquid is poured into a flask, which is then weighed and found to have a mass of 4.973 g. The empty flask has a mass of 4.017 g.

(a) What is the mole fraction of acetone in the feed gas?

(b) At what rate in Btu/min must heat be removed in the cooler–compressor? (*Note:* $C_p = 0.124$ kJ/mol·°C for liquid acetone.)

53. A mixture of n-hexane vapor and air leaves a solvent recovery unit and flows through a 70-cm diameter duct at a velocity of 3.00 m/s. At a sampling point in the duct the pressure is 850 mm Hg, the temperature is 40°C, and the dew point of the sampled gas is 25°C. Calculate the temperature to which the gas must be cooled to recover 60% of the hexane as a liquid, and the required cooling rate in kW.

54. A sheet of cellulose acetate film containing 5 wt% liquid acetone enters an adiabatic dryer where 90% of the acetone evaporates into a stream of dry air flowing over the film. The film enters the dryer at $T_{f1} = 35$°C and leaves at T_{f2}(°C). The air enters the dryer at T_{a1}(°C) and 1.01 atm, and exits the dryer at $T_{a2} = 49$°C and 1 atm at a relative saturation of 40%. C_p may be taken to be 1.33 kJ/kg·°C for dry film, and 0.129 kJ/mol·°C for liquid acetone. Make a reasonable assumption regarding the heat capacity of dry air. The heat of vaporization of acetone may be considered independent of temperature. Take a basis of 100 kg film fed to the dryer for the requested calculations.

(a) Estimate the feed ratio (liters dry air(STP)/kg dry film).

(b) Derive an expression for T_{a1} in terms of the film temperature change, $(T_{f2} - 35)$, and use it to answer parts (c) and (d).

(c) Calculate the film temperature change if the inlet air temperature is 120°C.

(d) Calculate the required value of T_{a1} if the film temperature falls to 34°C, and that if it rises to 36°C.

(e) If you solved parts (c) and (d) correctly, you found that even though the air temperature is consistently higher than the film temperature in the dryer, so that heat is always transferred from the air to the film, the film temperature can drop from the inlet to the outlet. How is this possible?

55. The off-gas from a reactor in a process plant in the heart of an obscure Central American banana republic has been condensing and plugging up the vent line, causing a dangerous pressure buildup in the reactor. Plans have been made to pass the gas directly from the reactor into a cooling condenser, in which the gas will be brought to 25°C.

You have been called in as a consultant to aid in the design of this unit. Unfortunately, the chief (and only) plant engineer has disappeared after taking off for a weekend of bullfights, fiestas, and attempts to overthrow the government, and nobody else in the plant can tell you what the off-gas is (or what anything else is, for that matter).

However, a job is a job, and you set out to do what you can. You find an elemental analysis in the engineer's notebook indicating that the gas formula is $C_5H_{12}O$. On another page of the notebook, the off-gas flow rate is given as 235 m³/h at 116°C and 1 atm. You take a sample of the gas and cool it to 25°C, where it proves to be a solid. You then heat the solidified sample at 1 atm and note that it melts at 52°C and boils at 113°C. Finally, you make several assumptions and estimate the heat removal rate in kW required to bring the off-gas from 116°C to 25°C. What is your result?

56. A stream containing equimolar amounts of benzene and toluene at 60°C is fed to a continuous single-stage evaporator. Sixty percent of the benzene in the feed is vaporized. The vapor is analyzed and is found to contain 63.1 mole% benzene. Both the liquid and vapor product streams emerge at 150°C. Calculate the heat requirement for this process in kJ/kg feed. (See Example 8.4-4.)

57. Propane vapor emerging from a distillation column is fed to a condenser as a saturated vapor at 200 psia at a rate of 3.00×10^3 ft³/h and leaves the condenser as a saturated liquid. The condenser operates at a constant pressure of 200 psia. Cooling water enters the condenser coils at 80°F and may not undergo a temperature increase of more than 15°F. Neglecting the effects of pressure on heat capacity and heat of vaporization, calculate the rate (Btu/h) at which heat must be transferred (from the propane to the water) in the condenser and the minimum flow rate (lb_m/h) of cooling water. The heat capacity of water may be assumed constant at 1.0 Btu/$lb_m \cdot$°F.

58. An aqueous slurry at 50°C containing 20 wt% solids is fed to an evaporator, in which enough water is vaporized at 1 atm to produce a product slurry containing 35 wt% solids. Heat is supplied to the evaporator by saturated steam at 1.6 bars (absolute) condensing in an coil immersed in the liquid. There is no boiling point elevation due to the presence of the solids in the slurry. The heat capacity of the solids may be taken to be half that of liquid water.

(a) Calculate the steam requirement (kg/h) for a slurry feed rate of 1000 kg/h.

(b) **Vapor recompression** is often used in the operation of an evaporator. Suppose that the vapor (steam) generated in the evaporator described above is compressed to 1.6 bars and simultaneously heated so that no condensation occurs. The compressed steam and additional steam at 1.6 bars are then fed to the evaporator coil, in which isobaric condensation occurs. How much additional steam is required?

(c) What more would you need to know to determine whether or not vapor recompression is economically advantageous in this process?

59. A mixture that contains 46 wt% acetone (CH_3COCH_3), 27% acetic acid (CH_3COOH), and 27% acetic anhydride [$(CH_3CO)_2O$] is distilled at $P = 1$ atm. The feed enters the distillation column at $T = 348K$ at a rate of 15 000 kg/h. The distillate (overhead product) is essentially pure acetone, and the bottoms product contains 1% of the acetone in the feed.

The vapor effluent from the top of the column enters a condenser at 329K and emerges as a liquid at 303K. Half of the condensate is withdrawn as the overhead product, and the remainder is refluxed back to the column. The liquid leaving the bottom of the column goes into a steam-heated reboiler, in which it is partially vaporized. The vapor is returned to the column at a temperature of 398K, and the residual liquid, also at 398K, constitutes the bottoms product. A flowchart of the process and thermodynamic data for the process materials are shown here.

(a) Calculate the flow rates and compositions of the product streams.

(b) Calculate the condenser cooling requirement Q_c(kJ/h).

(c) Use an overall energy balance to determine the reboiler heating requirement Q_r(kJ/h). (Continued on next page.)

(d) If the reboiler heat is provided by the condensation of saturated steam at 10
bars(gauge), at what rate must steam be fed?

Thermodynamic Data (All temperatures are in degrees Kelvin)

Acetone: $C_{pl}(kJ/kg \cdot K) = 2.30$
$C_{pv}(kJ/kg \cdot K) = 0.459 + 3.15 \times 10^{-3}T - 0.790 \times 10^{-6}T^2$
$\Delta \hat{H}_v(329K) = 520.6 \text{ kJ/kg}$

Acetic acid: $C_{pl}(kJ/kg \cdot K) = 2.18$
$C_{pv}(kJ/kg \cdot K) = 0.688 + 1.87 \times 10^{-3}T - 0.411 \times 10^{-6}T^2$
$\Delta \hat{H}_v(391K) = 406.5 \text{ kJ/kg}$

Acetic anhydride: $C_{pl}(kJ/kg \cdot K) = ?$ (Estimate it—see Section 8.3c.)
$C_{pv}(kJ/kg \cdot K) = 0.751 + 1.34 \times 10^{-3}T - 0.046 \times 10^{-6}T^2$
$\Delta \hat{H}_v(413K) = ?$ (Estimate it—see Section 8.4b.)

60. A **double-effect evaporator** (two evaporators in series) is used to produce fresh water
from seawater containing 3.5 wt% dissolved salts. A flowchart for the process is shown
here.

Seawater enters the first effect at 300K at a rate of 5000 kg/h, and saturated steam
at a pressure of 4.00 bars is fed into a tube bundle in the first effect. The steam condenses
at 4.00 bars, and the condensate is withdrawn at the saturation temperature corre-
sponding to this pressure.

The heat given up by the steam condensing in the tubes causes water to evaporate
from the brine solution at the pressure of 0.60 bar maintained in the effect. The exiting
brine contains 5.5% salt. The steam generated in the first effect is fed to a tube bundle
in the second effect. The condensate from the bundle and the steam generated in the
second effect at a pressure of 0.20 bar constitute the fresh water produced in the process.

In solving the problems to be given, assume the brine solutions in both effects have
the physical properties of pure water, and that the effects operate adiabatically.

(a) Draw and label a flowchart of this process, giving the temperature and specific
enthalpy of each stream.

 (b) At what rate must steam be fed to the first effect?

 (c) What is the production rate of fresh water? What is the salt concentration (weight percent) of the final brine solution?

 (d) Why is it necessary that the pressure decrease from one effect to the next?

 (e) Suppose a single effect evaporator were used, operating at $P = 0.20$ bar. Calculate the feed rate of saturated steam at $P = 4.00$ bars that would be required to achieve the same production rate of fresh water. What more would you need to know to determine which process is more economical?

***61.** Seawater containing 3.5 wt% dissolved salts is to be desalinated in an adiabatic six-effect evaporator. (See Problem 60.) Backward feed is to be used: the seawater is fed to the last evaporator, and successively concentrated brine solutions flow counter-current to the direction of flow of steam from one effect to the next. Saturated steam at $P = 2$ bars is fed to the tube bundle in the first effect. The operating pressures in bars of the six effects are respectively 0.9, 0.7, 0.5, 0.3, 0.2, and 0.1. The brine leaving the first effect contains 30 wt% salt.

Following is a labeled diagram of the ith effect.

* Computer problem.

In terms of the variables defined in this diagram,

$$n_{L7} = 5000 \text{ kg/h}$$
$$x_{L7} = 0.035 \text{ kg salt/kg}$$
$$x_{L1} = 0.30 \text{ kg salt/kg}$$
$$n_{v0} = \text{feed rate of steam to first effect}$$

(a) Use a salt balance to calculate n_{L1}. Then use this result to determine how much fresh water is produced in the process.

(b) Prepare a table as follows.

	P (bars)	T (K)	n_L (kg/h)	x_L	\hat{H}_L (kJ/kg)	n_v (kg/h)	\hat{H}_v (kJ/kg)
Fresh steam	2.0		—	—	—		
Effect 1	0.9			0.30			
2	0.7						
3	0.5						
4	0.3						
5	0.2						
6	0.1						
(7)	1.0	300	5000	0.035		—	—

Fill in all *known* variable values (do not calculate any yet), including values obtained from the steam tables, *assuming that the physical properties of the brine solution are those of pure water.*

(c) Show that the following equations can be derived from balances:

$$x_{Li} = (n_L)_{i+1}(x_L)_{i+1}/n_{Li} \tag{1}$$

$$(n_v)_{i-1} = \frac{n_{vi}\hat{H}_{vi} + n_{Li}\hat{H}_{Li} - (n_L)_{i+1}(\hat{H}_L)_{i+1}}{(\hat{H}_v)_{i-1} - (\hat{H}_L)_{i-1}} \tag{2}$$

$$(n_L)_{i-1} = n_{Li} - (n_v)_{i-1} \tag{3}$$

(d) The missing entries in the table of part (b) may be determined by solving Equations 1 through 3 for all effects simultaneously, but this is a cumbersome task. Instead, a relatively simple trial-and-error method may be used, which consists of assuming a value for n_{v6}, calculating $n_{L6} = n_{L7} - n_{v6}$, and successively solving Equations 1 through 3 for x_{L6}, n_{v5}, n_{L5}, then x_{L5}, n_{v4}, n_{L4}, and so on, until finally x_{L1} is calculated. If the calculated value of x_{L1} is less than the known value of 0.3, a higher value of n_{v6} is assumed and the process is repeated; if $x_{L1} > 0.3$, a lower value of n_{v6} is assumed.

A reasonable first guess for n_{v6} is $\frac{1}{6}$ of the total rate of evaporation [determined in part (a)]. Termination occurs when the calculated value of x_{L1} falls between 0.298 and 0.302.

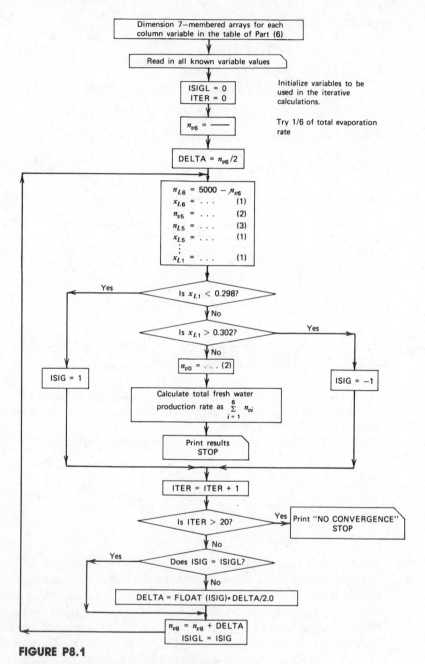

FIGURE P8.1

Flowchart for program of problem 59

A logic flowchart for this procedure is shown in Figure P8.1. Write a computer program to implement this procedure, and use it to determine the total rate of production of fresh water [compare with the result of part (a)], the required rate of steam flow to the first effect, and all variables in the table of part (b).

62. A liquid is placed in a well-insulated container, which is then sealed. Initially, the container and its contents (the liquid and pure nitrogen) are at 93°C and 1 atm; the liquid volume is 70 cm^3, and the gas volume is 3.00 liters. The liquid partially evaporates, and the system cools down and eventually comes to thermal equilibrium at 85°C with liquid still present. Physical property data for the liquid and its vapor are:

$$\Delta \hat{U}_v = 20 \text{ kcal/mol at } 90°C$$
$$(C_p)_{\text{liq}} = 20 \text{ cal/mol} \cdot °C$$
$$(C_p)_{\text{vap}} = 10 \text{ cal/mol} \cdot °C$$
$$(SG)_{\text{liq}} = 0.90$$
$$MW = 42$$

(a) Determine $(C_v)_{\text{liq}}$ and $(C_v)_{\text{vap}}$. (See Equations 8.3-11 and 8.3-12.)
(b) Draw and label a flowchart for this closed system process, and write and simplify the energy balance equation, assuming adiabatic operation.
(c) Use the energy balance to calculate the mass of liquid that evaporates, taking 4.97 cal/mol·°C as the heat capacity of nitrogen.
(d) Calculate the vapor pressure of the liquid at 85°C, assuming that the gas volume remains constant at 3.00 liters.

63. A small pharmaceutical firm plans to manufacture a new drug and has hired you as a consultant to design a condenser to remove the drug from a gas–vapor mixture. The mixture, which contains 20 mole% of the drug and the balance nitrogen, will be fed to the condenser at 510K and 1 atm at a rate of 3.5 liters/s. Of the drug fed to the unit, 90% must be condensed. No physical property data are available for the drug, and part of your job is to acquire the data needed to design the condenser. The company has sent you a large sample of the liquid drug for this purpose.

You acquire an insulated 2.000-liter container with a known heat capacity and a built-in electrical heating coil that can deliver a known heat input to the contents of the container. A calibrated thermocouple is used to measure the temperature in the vessel, and the pressure is measured with a mercury manometer.

You carry out a series of experiments on a day when atmospheric pressure is 763 mm Hg.

Experiment 1. Fill the container with the liquid, then seal and weigh.

$$\text{Mass of container} + \text{liquid} = 4.4553 \text{ kg}$$
$$\text{Mass of evacuated container} = 3.2551 \text{ kg}$$

Next, starting at each of two temperatures (T_0), add a fixed quantity of heat to the liquid, observe the final temperature (T_f), and subtract the heat absorbed by the container from the total heat input to determine the amount of the heat added to the liquid, Q_a.

$$T_0 = 283.0\text{K}, \ Q_a = 800.0 \text{ J} \Rightarrow T_f = 285.4\text{K}$$
$$T_0 = 330.0\text{K}, \ Q_a = 800.0 \text{ J} \Rightarrow T_f = 332.4\text{K}$$

You assume that the heat capacity is constant in each of these experiments in view of the small temperature changes.

Experiment 2. Pour a small quantity of the drug into the container, place the container in a liquid nitrogen bath to freeze the drug, evacuate all of the air, and seal the container. Weigh the container after it comes back to room temperature.

<div align="center">Mass of container + drug = 3.2571 kg</div>

Next, heat the sealed container until all of the liquid evaporates, and repeat Experiment 1.

$$T_0 = 363.0\text{K}, \ h_{\text{manometer}} = -500 \text{ mm}, \ Q_a = 1.30\text{J} \Rightarrow T_f = 366.9\text{K}$$
$$T_0 = 490.0\text{K}, \ h_{\text{manometer}} = -408 \text{ mm}, \ Q_a = 1.30\text{J} \Rightarrow T_f = 492.7\text{K}$$

Experiment 3. Fill approximately half the container with the drug, freeze, evacuate the air, and seal. Measure the pressure at several temperatures, verifying that liquid is present in the container at each temperature.

$$T = 315.0\text{K}, \ h_{\text{manometer}} = -564 \text{ mm}$$
$$T = 334.0\text{K}, \ h_{\text{manometer}} = -362 \text{ mm}$$
$$T = 354.0\text{K}, \ h_{\text{manometer}} = -2 \text{ mm}$$
$$T = 379.0\text{K}, \ h_{\text{manometer}} = +758 \text{ mm}$$

(a) Using the given data, determine the following physical properties of the drug: (i) liquid specific gravity, (ii) molecular weight, (iii) linear expressions for the heat capacities at constant volume (in J/mol·K) for both the liquid and vapor $[C_v = a + bT(\text{K})]$, (iv) linear expressions for C_p for both liquid and vapor, (v) a Clausius–Clapeyron expression for $p^*(T)$, (vi) the normal boiling point, and (vii) the heat of vaporization (in J/mol) at the normal boiling point.

(b) Calculate the required condenser temperature, assuming operation at 1 atm.

(c) Calculate the rate at which heat must be removed in the condenser, taking the heat capacity of nitrogen to be constant at 29.0 J/mol·K.

64. Freeze drying is a technique for dehydrating substances at low temperatures, thereby avoiding the degradation that may accompany heating. The material to be dried is cooled to a temperature at which all of the water present turns to ice. The frozen substance is then placed in a vacuum chamber and may also be subjected to radiant or microwave heating; the ice in the food sublimates, and the vapor is carried off by the vacuum pump.

Steaks are to be freeze-dried in a heated chamber at 1 torr (1 mm Hg). The steaks, which contain 72% water by mass, enter the chamber at $-26°\text{C}$ at a rate of 50 kg/min. Of the water entering with the steaks, 96% leaves as a vapor at 60°C; the remainder leaves as a liquid with the steaks at 50°C. Use the data given below and tabulated latent heat and enthalpy data for water to calculate the required heat input in kilowatts.

$$(C_p)_{\text{ice}} = 2.17 \text{ J/g·°C}$$
$$(C_p)_{\text{dry meat}} = 1.38 \text{ J/g·°C}$$

65. **Freeze concentration** is used to produce a fruit juice concentrate. A stream of fresh juice containing 12 wt% soluble solids in water at 20°C is combined with a recycle stream to form a preconcentrate, which is fed to a freezer. The mixture is cooled in the freezer to −7°C, thereby crystallizing 20 000 kg/h of ice. The slurry leaving the freezer, which contains 10 wt% ice, passes to a filter. The filtrate, which contains 45 wt% soluble solids, is removed as the process product. The remaining slurry, which contains all the ice and some concentrate (also containing 45% solids), is sent to an ice separator that cleanly removes all of the ice. The residual concentrate is the recycle stream that combines with the fresh feed to form the preconcentrate.

 (a) Determine the rates (kg/h) at which fresh fruit juice is fed and concentrate is produced, and the mass flow rate (kg/h) and solids concentration of the preconcentrate.

 (b) Calculate the cooling requirement (kW) for the freezer, assuming that the temperature of the recycled concentrate is 0°C and the heat capacity of all solutions is 4.0 kJ/kg·°C.

66. An equimolar liquid mixture of propane and n-butane enters a heat exchanger at 100°F and 250 psia and emerges as a vapor at 300°F and 250 psia. A high molecular weight liquid with a heat capacity $C_p = 0.600$ Btu/lb$_m$·°F is used as the heating fluid and undergoes a temperature drop from 600°F to 450°F in the heat exchanger. Of the heat given up by the fluid, 70% is transferred to the hydrocarbon mixture and the remainder is lost to the surroundings.

 (a) Assuming that the heat capacities of propane and butane vapors are independent of pressure, and that the specific enthalpies of the liquids of both species at any temperature are those of the pure saturated components at that temperature, calculate the enthalpy change in Btu/lb-mole undergone by the hydrocarbon mixture. (The necessary data may be found in Appendix B and in enthalpy tables in *Perry's Chemical Engineers' Handbook*.)

 (b) Calculate the required flow rate of the heating fluid if the entering flow rate of the hydrocarbon mixture is 200 lb$_m$/h.

67. A liquid mixture of benzene and toluene containing 50 wt% C_6H_6 at 90°C and 10.0 atm is fed into a flash tank at a rate of 550.0 liters/h. When the feed is exposed to the reduced pressure in the tank, a substantial quantity is vaporized. The temperature in the tank is maintained at 75°C by adding heat as needed. The liquid product contains 43.9 mole% benzene and is in equilibrium with the vapor product.

 (a) Using Raoult's law and the Antoine equation, calculate the total pressure of the system, the mole fraction of benzene in the vapor, and the molar flow rates of the liquid and vapor products.

 (b) Calculate the required heat input rate in kilowatts, using the following heat capacity formulas in your calculation.

$$C_6H_6(l): C_p(kJ/mol \cdot °C) = 0.144$$

$$C_6H_6(v): C_p(kJ/mol \cdot °C) = 0.0740 + 0.330 \times 10^{-3}T(°C)$$

$$C_7H_8(l): C_p(kJ/mol \cdot °C) = 0.176$$

$$C_7H_8(v): C_p(kJ/mol \cdot °C) = 0.0942 + 0.380 \times 10^{-3}T(°C)$$

68. A continuous adiabatic flash tank is to be used to separate a liquid mixture of two substances (A and B). The feed enters at temperature T_F and a high pressure and flashes to a low pressure, P, whereupon its temperature drops to T. For an assumed basis of 1 mol of feed, let

$$n_L, n_V = \text{moles of liquid product and vapor product}$$

$$x_L, x_V = \text{mole fractions of A in the liquid and vapor product}$$

$$p_A^*(T), p_B^*(T) = \text{vapor pressures of A and B}$$

$$T_{RA}, T_{RB} = \text{reference temperatures for enthalpy calculations}$$

$$\left. \begin{array}{l} \hat{H}_{AF}(T_F), \hat{H}_{AL}(T), \hat{H}_{AV}(T) \\ \hat{H}_{BF}(T_F), \hat{H}_{BL}(T), \hat{H}_{BV}(T) \end{array} \right\} \begin{array}{l} \text{specific enthalpies of A and B in} \\ \text{the feed, liquid product, and vapor} \\ \text{product, relative to } T_{RA} \text{ and } T_{RB} \end{array}$$

(a) Derive the following relations from Raoult's law and material and energy balances on the flash tank.

$$x_L = \frac{P - p_B^*(T)}{p_A^*(T) - p_B^*(T)} \tag{1}$$

$$x_V = x_L p_A^*(T)/P \tag{2}$$

$$n_L = \frac{x_V - x_F}{x_V - x_L} \tag{3}$$

$$n_V = 1 - n_L \tag{4}$$

$$n_L[x_L \hat{H}_{AL}(T) + (1 - x_L)\hat{H}_{BL}(T)] + n_V[x_V \hat{H}_{AV}(T) + (1 - x_V)\hat{H}_{BV}(T)]$$
$$- [x_F \hat{H}_{AF}(T_F) + (1 - x_F)\hat{H}_{BF}(T_F)] = 0 \tag{5}$$

*__(b)__ Write a program to read in and print out values of x_F, $T_F(°C)$, and P(mm Hg), and for each of the feed mixture components the Antoine equation coefficients, a reference temperature (e.g., 25°C), the liquid heat capacity (assume constant), normal boiling point, heat of vaporization, and coefficients of a two-term vapor heat capacity formula $C_p = a + bT$. Then have the program carry out a trial-and-error procedure to calculate and print out the flash tank temperature and the flow rates and compositions of the vapor and liquid product streams.

An inefficient but simple procedure is to assume a value of T (T_F is a reasonable first guess), solve equations 1 through 4 successively, and then evaluate the left side of equation 5 (which equals ΔH for the assumed temperature and must be positive when $T = T_F$). The value of T is then decreased by a small fixed amount, such as 0.5°C, and the procedure is repeated until the sign of ΔH changes from positive to negative from one T to the next. The true T (for which $\Delta H = 0$) must then lie between the two last assumed values and may be estimated by linear interpolation.

* Computer problem.

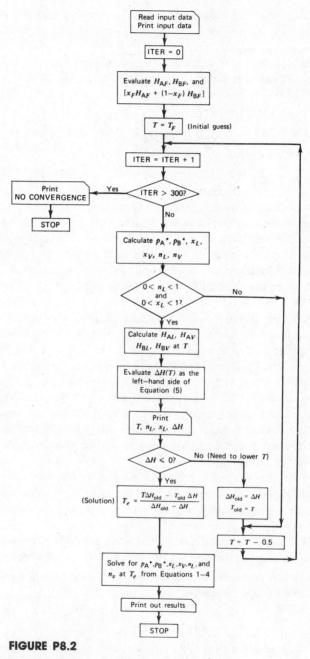

FIGURE P8.2

Flowchart for program of problem 68

A flowchart for a program to implement this procedure is shown in Figure P8.2. Write the program, and test it by estimating the flash tank temperature and product stream amounts (mol) and compositions (mole fractions) for the flash vaporization of one mol of an equimolar mixture of n-pentane (A) and n-hexane (B), if the feed temperature is 110°C and the tank pressure is 1.0 atm. Use the following heat capacity formulas:

$$C_5H_{12}(l): \; C_p(kJ/mol \cdot °C) = 0.195$$

$$C_5H_{12}(v): \; C_p(kJ/mol \cdot °C) = 0.115 + 0.341 \times 10^{-3}T(°C)$$

$$C_6H_{14}(l): \; C_p(kJ/mol \cdot °C) = 0.216$$

$$C_6H_{14}(v): \; C_p(kJ/mol \cdot °C) = 0.137 + 0.409 \times 10^{-3}T(°C)$$

69. A saturated vapor stream containing 10.9 mole% propane, 75.2% isobutane, and 13.9% normal butane passes from the top of a distillation column to a total condenser. Seventy-five percent of the condensate is returned to the column as reflux, and the remainder is removed as the overhead column product at a rate of 2500 kmol/h.

A decision must be made on whether to use a refrigerant or cooling water in the condenser. If the refrigerant is used, it will be fed to the condenser as a liquid and vaporized by the heat released by the condensing column vapor. The refrigerant pressure will be such that the vaporization takes place at $-6°C$, at which temperature $\Delta\hat{H}_v = 151$ kJ/kg. The other option calls for cooling water to be taken from a nearby river at an average summer temperature of 25°C. To avoid environmental problems, the temperature of the water returned to the river can be no greater than 34°C. With either system, the temperature of the condensate should be 6°C greater than the outlet temperature of the coolant, so that if the refrigerant is used the saturated condensate should have a temperature of 0°C, and if cooling water is used the saturated condensate should be at 40°C. The condenser pressure will be set to the minimum value needed to condense all of the vapor, which is to say the condensate will be at its bubble point temperature at the condenser pressure. Raoult's law may be used for all bubble point and dew point calculations (see Section 6.4c).

(a) Suppose the refrigerant is used for cooling. Estimate the condenser pressure P(mm Hg); the temperature $T_f(°C)$ of the vapor fed to the condenser, assuming that the vapor is at its dew point at pressure P; and the required coolant flow rate (kg/h).

(b) Repeat part (a) assuming that cooling water is fed to the condenser.

(c) What more would you need to know to be able to choose between the two options?

70. Formaldehyde is produced from methanol in a catalytic oxidation reactor, where the following reactions take place.

$$CH_3OH \rightarrow HCHO + H_2$$

$$2 H_2 + O_2 \rightarrow 2 H_2O$$

A stream of methanol is joined by a recycle stream, also methanol, and the combined stream is fed to the conversion reactor. Also entering the reactor are air (to oxidize the hydrogen produced in the methanol conversion reaction) and steam (to control the reactor temperature). The product gas from the reactor contains 19.9 mole% formaldehyde, 8.34% methanol, 30.3% nitrogen, 0.830% oxygen, 5.0% hydrogen, and 35.6% water vapor, and is at 600°C and 1 atm.

The following process is used to separate the formaldehyde from the unreacted methanol and noncondensable gases. The gases leaving the reactor are fed to a waste heat boiler, where they are cooled to 145°C, in the process generating saturated steam at 3.1 bars from saturated liquid water at the same pressure. The gases are cooled further to 100°C in a heat exchanger, where they come into thermal contact with cooling water fed at 30°C. To reduce scaling on the heat exchanger tubes, the temperature increase of the cooling water is limited to 15°C. The cooled gases are fed to an absorption column, where methanol and formaldehyde are absorbed in water. Pure water at 30°C is fed to the top of the column. The gas leaving the absorber is saturated with water vapor at 27°C and 1 atm and contains 200 parts of formaldehyde per million parts (by volume) of total gas. The aqueous solution leaving the bottom of the absorber at 88°C is fed to a distillation column that operates at 1 atm. The final product solution, which contains 37 wt% formaldehyde, 1% methanol, and the balance water, is removed from the reboiler at the bottom of the column, while pure methanol vapor emerges as the overhead product and is condensed at 1 atm. A portion of the condensate is refluxed to the top of the column, and the rest is recycled to the methanol conversion reactor. The reflux ratio, or ratio of methanol refluxed to methanol recycled to the reactor, is 2.5:1.

(a) Taking as a basis 100 mol of gas leaving the conversion reactor, draw and completely label a flowchart of this process. Then calculate the moles of fresh methanol feed, formaldehyde product solution, recycled methanol, and absorber off-gas, the kg of steam generated in the waste heat boiler, and the kg of cooling water fed to the heat exchanger between the waste heat boiler and the absorber. Finally, calculate the heat (kJ) that must be removed in the still overhead condenser, assuming that methanol enters as a saturated vapor at 1 atm and leaves as a saturated liquid at the same pressure.

(b) By what factor must all of the calculated quantities be multiplied to scale the flowchart to a production rate of 3.6×10^4 metric tons per year of formaldehyde solution, assuming that the process is on-stream 350 days per calendar year?

71. Use the psychrometric chart to estimate the absolute humidity, humid volume, specific enthalpy, wet bulb temperature, and dew point of air on a day when the temperature is 30°C and the relative humidity is 30%.

72. Humid air contains 1.95 wt% water vapor at 90°F. Use the psychrometric chart to estimate the relative humidity, wet bulb temperature, humid volume, dew point, and specific enthalpy of the air.

73. A **sling psychrometer** is a device to measure the humidity of air. A porous cloth (the **wick**) is wrapped around the bulb of a mercury thermometer, which is then whirled around in the air. As the water in the wick evaporates, the temperature of the thermometer bulb drops, and finally stabilizes at the wet bulb temperature of the air. The dry bulb temperature is read from a second thermometer mounted on the sling.

One summer day, the weather bureau reports a temperature of 33°C and a relative humidity of 40%. You wipe the sweat from your forehead and remark to a friend that you'd bet $5.00 the report is wrong, and the relative humidity is higher than 80%. He immediately puts a $5 bill on the table, accepting your challenge. You bring out your sling psychrometer, whirl it around, and read a dry bulb temperature of 35°C and a wet bulb temperature of 27°C. Who wins the bet?

74. Humid air is enclosed in a 2.00-liter flask at 40°C. The flask is slowly cooled. When the temperature reaches 25°C, drops of moisture become visible on the flask wall. Although the pressure in the flask changes when the temperature drops, it remains close enough to 1 atm for the psychrometric chart to provide a close representation of the behavior of the system throughout the process. Use the chart to solve the following problems.

(a) What were the relative humidity, absolute humidity, and wet bulb temperature of the air at 40°C?

(b) Calculate the mass of the water in the flask. (See Example 8.4-5.)

(c) Calculate the enthalpy change in joules undergone by the air in going from 40°C to 25°C.

(d) Write an energy balance for this closed system process, taking the wet air in the flask as the system, and use it to calculate the heat in joules that must be transferred from the air to accomplish the cooling. (Assume ideal gas behavior, so that $\hat{H} = \hat{U} + RT$.)

75. A wet filter cake passes through a continuous hot air dryer, in which 10 kg of water per minute evaporate. A stream of hot dry air enters the dryer at a rate of 400 kg/min. The temperature of the outlet gas is 50°C. The gas stream is then cooled to 20°C, causing a portion of the water to condense. The pressure is more or less constant at 1 atm.

(a) Use the psychrometric chart to estimate the specific enthalpy, wet bulb temperature, relative humidity, and dew point of the wet air at 50°C.

(b) Determine the absolute humidity and specific enthalpy of the wet air at 20°C.

(c) Use the results of parts (a) and (b) to calculate the rate of condensation of water.

(d) At what rate must heat be removed to carry out the cooling and condensation process?

76. You are ordering a central air conditioner for a small industrial plant. On a typical summer day, the outside air is at 87°F and 80% relative humidity. The air conditioner is to deliver 10 000 ft³/min of air at 70°F and 40% relative humidity.

(a) Use the psychrometric chart to estimate the rate at which water is condensed, the temperature to which the air must be cooled to condense water at this rate, and the net tons of cooling required (Q for the cooler $-Q$ for the reheater) if all of the air is brought in from the outside. (*Note:* 1 ton = 12 000 Btu/h.)

(b) Suppose that a recirculation system is used (as it would be in practice), as shown in the following diagram.

The recycle ratio (ft^3 recirculated/ft^3 exhaust) is $7:1$. Calculate the overall cooling requirement in tons. What percentage of the cooling requirement is saved by recirculating the air? Does this represent a net economic gain? If not, what additional factors contribute to the cost of the process with recirculation?

77. A gas stream leaves a combustion furnace at 800°F and atmospheric pressure, at a rate of 300 ft^3/min. The gas contains 7.1 mole% CO_2, 74.3% N_2, 4.3% O_2, and 14.3% H_2O. The gas is cooled to 200°F in a heat exchanger. The heat it loses goes into raising the temperature of air at 60°F and 30% relative humidity to 100°F. Calculate the standard cubic feet of air per hour that can be heated, using the psychrometric chart in your calculation.

78. Taking Figure 8.4-1 as a model and using tabulated heat capacity and latent heat data, construct a psychrometric chart for the system benzene–air at 1 atm. Find at least two points on each of the following curves on your chart, and sketch the curves.
 (a) The saturation curve between 0°C and 50°C.
 (b) The 50% relative humidity curve. Call the two points on this curve A and B.
 (c) The adiabatic saturation line through point A. Extend this curve to the enthalpy scale above the saturation curve, and give the value of H at this intersection.
 (d) The enthalpy deviation curve through point A.
 (e) The constant humid volume curve through point B.

79. Wet wood chips are dried in a continuous rotary dryer that operates at atmospheric pressure. The chips enter at 19°C with a water content of 40% and must leave with a moisture content of less than 15%. Hot air is fed to the dryer at a rate of 11.6 m^3(STP)/kg wet chips.

 To monitor the performance of the dryer by sampling the dried chips and determining their moisture content directly would be a cumbersome procedure, and almost impossible to automate. Instead, wet and dry bulb thermometers are mounted in both the inlet and outlet air lines, and the moisture content of the dried chips is determined by a material balance.

After the unit goes on stream, the inlet air dry bulb temperature is found to be 100°C, and the wet bulb temperature is low enough so that the moisture content of the air

may be neglected. The dry bulb temperature of the exiting air is found to be 38°C, and the wet bulb temperature is 29°C.

(a) Use the psychrometric chart to calculate the absolute humidity (kg H_2O/kg dry air) and specific enthalpy (kJ/kg dry air) of the outlet air stream. Then calculate the mass of water in the exiting air per kilogram of wet chips fed, assuming dry air has a molecular weight of 29.0.

(b) Calculate the moisture content of the emerging chips, and so determine whether the design specification of less than 15% H_2O has been achieved.

(c) If the unit is operating adiabatically and the heat capacity of the dry chips is 2.10 kJ/kg·°C, what is the exit temperature of the chips? (In estimating the specific enthalpy of the entering air, recall that the reference temperature for dry air used in constructing the psychrometric chart of Figure 8.4-1 is 0°C.)

80. Air at 45°C (dry bulb) and 10% relative humidity is to be humidified adiabatically to 60% relative humidity.

(a) Use the psychrometric chart to estimate the adiabatic saturation temperature of the air.

(b) Estimate the final temperature of the air, and the rate at which water must be added to humidify 15 kg/min of the entering air. (See Example 8.4-7.)

81. A solution of sugar in water is to be concentrated from 5 wt% sugar to 20%. The solution is fed continuously to a bubble column. Air at 45°C with a dew point of 4°C is bubbled through the column and emerges saturated. The humidification of the air may be considered adiabatic.

Use the psychrometric chart to solve the following problems:

(a) What are the absolute humidities of the entering and exiting air?

(b) How many kilograms of dry air must be fed per kilogram of the entering sugar solution? What is the corresponding volume of the entering wet air? (Use the chart for the latter problem.)

82. Air at 95°F with a dew point of 25°F enters a textile dryer at a rate of 250 ft³/min and leaves with a relative humidity of 90%.

(a) Use the psychrometric chart to determine the humid volume and absolute humidity of the entering air, and then use the results to determine the flow rate of *dry* air (lb_m/min) through the dryer.

(b) Assuming that the dryer operates adiabatically, calculate the final temperature of the air and the rate at which water is evaporated in the dryer.

83. Cold air at 20°F, 760 mm Hg pressure, and 70% relative humidity is conditioned by being passed through a bank of steam-heated coils, then through a water spray, and finally through a second set of steam-heated coils. In passing through the first coil bank, the air is heated to 75°F. The temperature of the water supplied to the spray chamber is adjusted to the wet bulb temperature of the air admitted to the chamber, so that the humidifying unit may be assumed to operate adiabatically. It is required that the air emerging from the conditioning unit be at 70°F and 35% relative humidity. Use Figure 8.4-2 to solve the following problems.
(a) What should be the temperature of the water supplied to the spray chamber?
(b) To secure the required final conditions, what must be the relative humidity of the air emerging from the spray chamber?
(c) What is the dry bulb temperature of the air emerging from the spray chamber?
(d) Calculate the number of pounds of water evaporated per cubic foot of original air.

84. **Spray cooling** is a technique for cooling and either humidifying or dehumidifying air by contacting it with a water spray.

The water leaving the tower is recirculated.

Two possible paths on the psychrometric chart for air initially at a temperature T_0 and a dew point T_{dp} undergoing spray cooling are shown in the following diagram.

If the air reaches saturation at a final temperature greater than T_{dp} (such as T_1), then the final air humidity $(h_a)_1$ is greater than the initial humidity $(h_a)_0$, so that the air is humidified as it cools. On the other hand, if the air approaches a final temperature less than T_{dp} (such as T_2), then the final air humidity $(h_a)_2$ is less than $(h_a)_0$, and the air is *dehumidified* as it cools. We thus arrive at the interesting conclusion that *you can remove water from air by spraying water into the air;* the only requirement is that the liquid water temperature must be less than the dew point of the entering air.

Use the psychrometric chart to solve the following spray cooling problems.

(a) A spray tower is used to cool air with dry bulb and wet bulb temperatures of 40°C and 18°C, respectively, to 20°C. No cooler is used in the recirculation loop, and the tower operation is such that the air follows an adiabatic saturation curve. How much water must be added as makeup per kg of dry air treated?

(b) One thousand pounds per hour of air at 85°F and 50% relative humidity are to be cooled to 50°F and dehumidified. The air is saturated as it emerges from the cooling tower. Liquid water emerges from the tower at 50°F; some is withdrawn, and the rest is cooled and recirculated. No heat is lost from the tower. Calculate the rate at which water must be withdrawn from the recirculation loop and the heat duty on the cooler in Btu/h. (*Suggestion:* Use an overall energy balance for the latter calculation.)

85. The integral heat of solution of ammonia in water at 1 atm is

$$\Delta \hat{H}_s(25°C, n = 2 \text{ mol } H_2O/1 \text{ mol } NH_3) = -78.2 \text{ kJ/mol}$$

Calculate the enthalpy change that accompanies the dissolution of 300 mol of NH_3 in 600 mol of water at 25°C and 1 atm.

86. Use Table 8.5-1 to determine the specific enthalpy (kJ/mol HCl) of hydrochloric acid containing 1 mole HCl/4 moles H_2O at 25°C relative to
(a) HCl(g) and H_2O(l) at 25°C.
(b) H_2O(l) and an infinitely dilute HCl solution at 25°C. (Note Equation 8.5-2.)

87. Sodium hydroxide is dissolved in enough water to make up a 15 mole% solution. If the NaOH and water are initially at 77°F (25°C), how much heat (Btu/lb product solution) must be removed for the solution also to be at 77°F. Assume the process is carried out at constant pressure, so that $Q = \Delta H$, and use Table 8.5-1 to evaluate $\Delta \hat{H}_s$.

88. A sulfuric acid solution is labeled 8N (where $1N = 1$ g-equivalent/liter, and 1 mol of H_2SO_4 contains two g-equivalents). The specific gravity of the solution is 1.230, and its heat capacity is 3.00 J/g·°C. Calculate the specific enthalpy of this solution at 50°C relative to pure H_2O and an infinitely dilute solution at 25°C.

89. You are about to dilute 2 mol of 100% sulfuric acid with enough water to produce a 30 mole% aqueous solution. The acid and water are initially at 25°C.
(a) How much heat would have to be removed to keep the final solution at 25°C?
(b) Suppose the flask has a mass of 150 g, and that the heat capacity of the flask and its contents is 3.30 J/g·°C. If the flask is sufficiently insulated to be considered adiabatic, what will be the final solution temperature?

90. An 8-molar hydrochloric acid solution (SG = 1.12, $C_p = 0.66$ cal/g·°C) is produced by absorbing HCl(g) in water. Liquid water enters the absorber at 25°C, and gaseous HCl

is fed at 130°C and 820 mm Hg (absolute). The hydrochloric acid emerges at 40°C. Estimate the volume (liters) of the gas that must be fed and the heat (kJ) that must be removed from the absorber per liter of product solution. State all assumptions you make.

91. Water is added to pure sulfuric acid in a well-insulated flask initially at 25°C and 1 atm. The final temperature of the product solution is to be 25°C, so that the water added must be chilled liquid ($T < 25°C$), ice, or a mixture of both. Taking as a basis of calculation 1 liter of each of the product solutions given below, determine the masses (g) of sulfuric acid, liquid water, and ice that must be mixed, and the initial temperature of the water or ice if either is added without the other. Assume $Q = \Delta H$ for the process. If you need to know the heat capacity of ice, take it to be half that of liquid water.

 (a) A 2.00 wt% H_2SO_4 solution (SG = 1.0104).

 (b) A 4.00 molar solution (SG = 1.231).

92. A 0.1 mole% caustic soda (NaOH) solution is to be concentrated in a continuous evaporator. The solution enters the unit at 25°C at a rate of 150 mol/min and is concentrated to 5 mole% at 50°C. Hot air at 200°C and 1.1 bars (absolute) is bubbled through the evaporator and leaves saturated with water at 50°C and 1 atm. Calculate the required volumetric flow rate of the entering air and the rate at which heat must be transferred to or from the unit. Assume that the heat capacity per unit mass of all liquid solutions is that of pure water.

93. Two hundred kilograms per hour of an aqueous solution containing 20 mole% sodium acetate ($NaC_2H_3O_2$) enters an evaporative crystallizer at 60°C. When the solution is exposed to the low pressure in the evaporator, 16.9% of the water evaporates, concentrating the remaining solution and causing crystals of sodium acetate trihydrate ($NaC_2H_3O_2 \cdot 3H_2O$) to precipitate. The product is an equilibrium mixture of crystals and a saturated aqueous solution containing 15.4 mole% $NaC_2H_3O_2$. The effluents (crystals, solution, and water vapor) are all at 50°C.

 (a) Calculate the feed rate to the crystallizer in kmol/h.

 (b) Calculate the production rate (kg/h) of trihydrate crystals and the mass flow rate (kg/h) of the liquid solution in which the crystals are suspended.

 (c) Estimate the rate (kJ/h) at which heat must be transferred to or from the crystallizer (state which), using the following physical property data:

$$(C_p)_{\text{all solutions}} = 3.5 \text{ kJ/kg} \cdot °C$$

$$(C_p)_{\text{crystals}} = 1.2 \text{ kJ/kg} \cdot °C$$

$$(C_p)_{H_2O(v)} = 32.4 \text{ kJ/kmol} \cdot °C$$

$$(\Delta \hat{H}_v)_{H_2O} = 4.39 \times 10^4 \text{ kJ/kmol}$$

Heat of solution of anhydrous sodium acetate:

$$\Delta \hat{H}_s(25°C) = -1.71 \times 10^4 \text{ kJ/kmol } NaC_2H_3O_2(s)$$

Heat of hydration: $NaC_2H_3O_2(s) + 3H_2O(l) \rightarrow NaC_2H_3O_2 \cdot 3H_2O(s)$

$$\Delta \hat{H}(25°C) = -3.66 \times 10^4 \text{ kJ/kmol } NaC_2H_3O_2$$

94. Ortho-phosphoric acid (H_3PO_4) is produced as a dilute aqueous solution that must be concentrated before further use. In one facility, 100 ton/day of a 28 wt% P_2O_5 solution [see part (a) of this problem] at 125°F is to be concentrated in a single evaporator to 42 wt% P_2O_5. Heat is supplied to the evaporator by condensing saturated steam at 27.5 psia. The evaporator is to operate at 3.7 psia, and there is a boiling point elevation of 37°F for the 42 wt% P_2O_5 solution in the evaporator (see Section 6.5c). The heat of solution of H_3PO_4 at 77°F may be taken to be -5040 Btu/lb-mole H_3PO_4 relative to $H_3PO_4(l)$ and $H_2O(l)$. The heat capacity of the 28% solution is 0.705 Btu/lb$_m$·°F, and that of the 42% solution is 0.583 Btu/lb$_m$·°F.

 (a) It is conventional for the compositions of phosphoric acid solutions to be expressed in terms of wt% P_2O_5. Write the stoichiometric equation for the formation of ortho-phosphoric acid ($MW = 98.00$) from phosphorus pentoxide ($MW = 141.96$), and use it to derive the expression

 $$\text{wt% } H_3PO_4 = 1.381(\text{wt% } P_2O_5)$$

 (b) Calculate the ratio (lb$_m$ water evaporated/lb$_m$ feed solution).

 (c) Suppose the water evaporated is subsequently condensed at a constant pressure of 3.7 psia. Determine the condensate flow rate in gal/min. How much heat (Btu/min) can be recovered through condensation of this water? At what temperature is this heat available? (To put it another way, if this heat were to be transferred to another stream, what is an upper bound on the temperature of that stream?)

 (d) How much steam (lb$_m$/h) must be supplied to the system to evaporate the required amount of water? Recast your answer in terms of lb$_m$ steam per lb$_m$ water evaporated.

95. Fifty milliliters of 100% H_2SO_4 at 25°C and 67.3 ml of liquid water at 15°C are mixed. The heat capacity of the product solution is 2.27 J/g·°C. Calculate the maximum temperature attainable by the product solution, and state the conditions that would have to be met for this temperature to be approached.

96. Suppose m_A g of species A [molecular weight M_a, heat capacity C_{pa}(J/g·°C)] at temperature T_{a0} (°C) and m_B g of species B (M_b, C_{pb}) at temperature T_{b0} are mixed adiabatically. The heat of mixing of A and B at 25°C is $\Delta\hat{H}_m(n)$ (J/mol A in solution), where $n = m_B M_A / m_A M_B$. The heat capacity of the product solution is C_{ps}(J/g·°C). All heat capacities may be considered independent of temperature.

 (a) Derive an expression for T_{max}, the highest temperature attainable by the product solution, in terms of the other quantities defined. State the conditions that would have to be met for this temperature to be approached.

 (b) Use your expression to estimate T_{max} for a process in which 100.0 g of sodium hydroxide at 25°C are added to 180.0 g of water at 40°C to form a product solution with a heat capacity of 3.3 J/g·°C.

97. A stirred tank is charged with n_b mol of a liquid B (molecular weight M_b). The space above the liquid, which has volume V_g (liters), is charged with a gas A (molecular weight M_a), at a pressure P_0(atm). The initial system temperature is T_0(K). The stirrer in the tank is then turned on, and A commences to dissolve in B. The dissolution continues

until the solvent is saturated with A at the final system temperature and pressure. The amount of heat lost from the mixer contents to their surroundings and the work input from the stirrer may be neglected, and the volume of the gas phase may be considered constant. A negligible amount of B evaporates.

The equilibrium solubility of A in B is governed by Henry's law at the conditions of interest; the following expression relates the molar A/B ratio in the liquid to the partial pressure of A in the gas phase in contact with the liquid at equilibrium.

$$r(\text{mol A/mol B}) = k_h p_A(\text{atm})$$

where

$$k_h = c + dT(\text{K})$$

The internal energy of solution of A in B at 298K is $\Delta \hat{U}_s(\text{kJ/mol A})$, and may be assumed independent of the solution concentration for the range of concentrations under consideration. The heat capacities at constant volume of pure A, pure B, and solutions of A in B are $C_{va}(\text{kJ/mol A·K})$, $C_{vb}(\text{kJ/mol B·K})$, and $C_{vs}(\text{kJ/g solution·K})$ and may be considered independent of temperature and (in the case of C_{vs}) composition. The gas phase behavior is well represented by the ideal gas law.

Let $T(\text{K})$ and $P(\text{atm})$ be the system temperature and pressure at equilibrium; n_{ao} the gram-moles of A charged to the mixer; and n_{av} and n_{al} the gram-moles of A in the vapor and liquid phases at equilibrium.

(a) Use material balances, the Henry's law relation, and the ideal gas law to derive expressions for n_{ao}, P, n_{av}, and n_{al} in terms of the final temperature, T, and known constants (n_b, V_g, P_0, T_0, M_a, M_b, c, and d). Then derive the following equation:

$$T = 298 + \frac{n_{al}(-\Delta \hat{U}_s) + (n_{ao}C_{va} + n_b C_{vb})(T_0 - 298)}{n_{av}C_{va} + (n_{al}M_a + n_b M_b)C_{vs}} \tag{1}$$

*(b) Write a computer program to perform the following tasks:

Read in values of n_b, T_0, P_0, V_g, c, d, $\Delta \hat{U}_s$, C_{va}, C_{vb}, and C_{vs}. Have the program terminate if $n_b \leq 0$.

Calculate n_{ao}.

Assume T (try $1.1T_0$ as a first guess).

Calculate P, n_{av}, and n_{al} from the equations you derived in part (a), then recalculate T from equation (1). Print out the values of T(assumed), P, n_{av}, n_{al}, and T(recalculated).

If the assumed and recalculated values of T are within 0.01K of each other, terminate. If they are not, and more than 15 iterations have been performed, terminate with an error message. Otherwise, repeat the previous step, using the recalculated value of T as the assumed value for this iteration. (Continued on next page.)

* Computer problem.

(c) Test your program with the following data.

$$n_b = 300 \text{ mol}, M_b = 18.0, C_{vb} = 0.0754 \text{ kJ/mol·K}$$
$$M_a = 35.0, C_{va} = 0.0291 \text{ kJ/mol·K}$$
$$C_{vs} = 4.2 \times 10^{-3} \text{ kJ/g·K}$$

$$\Delta \hat{U}_s = -74 \text{ kJ/mol A}$$
$$c = 1.54 \times 10^{-3}, d = -2.6 \times 10^{-3}$$
$$T_0 = 18°C$$

Run the program for the following two initial conditions: **(i)** $V_g = 15.0$ liters, $P_0 = 10.0$ atm, **(ii)** $V_g = 15.0$ liters, $P_0 = 50.0$ atm

(d) (*Optional exercise.*) Rewrite the program of part (b) to include the use of Wegstein's method (Appendix A.2) for convergence acceleration. Run the modified program with the data given in part (c), and compare the number of iterations required to achieve convergence with and without the accelerator for each of the two cases studied.

98. An aqueous solution containing 45 wt% NaOH at 80°F is diluted with liquid water at the same temperature. The mixing unit may be considered adiabatic.

(a) Use the enthalpy-concentration chart of Figure 8.5-1 to estimate the specific enthalpies of both the feed solution and the water.

(b) If the dilution ratio is 1 lb$_m$ H$_2$O/2 lb$_m$ feed solution, calculate the percentage by mass of NaOH in the product. Then from an energy balance on the process calculate \hat{H}(Btu/lb$_m$) of the final solution. Finally, use Figure 8.5-1 to determine the final solution temperature. (See Example 8.5-2.)

(c) Verify that the calculation of part (b) could have been done graphically using Figure 8.5-1 by drawing a line between the points corresponding to the two feed streams and locating the point for the product solution on the line, with the exact location being determined by the lever rule: if A, B, and C are the points corresponding to the feed solution, pure water, and the product solution, respectively, then

$$\frac{|AC|}{|BC|} = \frac{\text{mass of dilution water}}{\text{mass of feed solution}}$$

99. An aqueous caustic soda solution contains 6.0 wt% NaOH. Use the heat of solution data of Table 8.5-1 to estimate the enthalpy of this solution (Btu/lb$_m$ *solution*) at 200°F relative to pure water at 32°F and an infinitely dilute solution at 68°F. Assume that the heats of solution of Table 8.5-1 are applicable at 68°F, and that the heat capacity of a 6% solution is 0.94 Btu/lb$_m$·°F. Compare your result with the value of \hat{H} given in Figure 8.5-1. (*Suggestion:* Use a process path in which you form pure NaOH at 68°F from an infinitely dilute solution at this temperature; bring the water required to form the 6% solution from 32°F to 68°F; mix the water with the NaOH; and heat the resulting solution to 200°F.)

100. You have analyzed an aqueous ammonia solution and find that it contains 30 wt% NH$_3$.

(a) Use Figure 8.5-2 to determine the mass fraction of NH_3 in the vapor that would be in equilibrium with this solution in a closed flask at 1 atm and the corresponding system temperature.

(b) If the liquid phase in part (a) accounts for 90% of the total system mass, calculate the overall system composition and specific enthalpy using balances. (See Example 8.5-3.) Then verify graphically that the lever rule (p. 000) is satisfied.

101. An NH_3–H_2O mixture containing 60 wt% NH_3 is brought to equilibrium in a closed container at 140°F. The total mass of the mixture is 250 g. Use Figure 8.5-2 and the lever rule to determine the masses of ammonia and of water in each phase of the system.

102. An ammonia solution at a high pressure is flash-vaporized at a rate of 200 lb_m/h. The solution contains 0.70 lb_m NH_3/lb_m, and its enthalpy relative to $H_2O(1, 32°F)$ and $NH_3(1, -40°F)$ is -50 Btu/lb_m. Liquid and gas streams emerge from the unit at 1 atm and 80°F. Use Figure 8.5-2 to determine the mass flow rates and ammonia mass fractions of the vapor and liquid product streams and the rate (Btu/h) at which heat must be transferred to the vaporizer. (See Example 8.5-4.)

Chapter Nine

Balances on Reactive Processes

Large enthalpy changes—which may be positive or negative—are commonly associated with chemical reactions. If ΔH for a reactor is positive, heat must be added to keep the temperature (and hence the reaction rate) from dropping. The cost of providing this heat sometimes exceeds the cost of the reactants. On the other hand, if ΔH is negative, heat is emitted from the reactor; the heat may represent an asset or a liability, depending on how efficiently it can be transferred from the reactor and used elsewhere in the process. In either case, an absorbed or emitted heat of reaction can play a major role in the economics of a chemical process.

In this chapter we show how enthalpy changes that accompany chemical reactions may be determined from tabulated physical properties of the reactants and products, and how calculated enthalpies of reaction are incorporated in energy balances on reactive processes.

9.1 HEATS OF REACTION

The **heat of reaction** or **enthalpy of reaction** $\Delta \hat{H}_r(T, P)$ is the difference $H_{\text{products}} - H_{\text{reactants}}$ for a reaction that takes place under the following circumstances:

1. Stoichiometric quantities of the reactants are fed, and the reaction proceeds to completion.
2. The reactants are fed at temperature T and pressure P, and the products emerge at the same temperature and pressure.

For example, for the reaction

$$CaC_2(s) + 2\ H_2O(l) \rightarrow Ca(OH)_2(s) + C_2H_2(g) \qquad (9.1\text{-}1)$$

the heat of reaction at 25°C and a total pressure of 1 atm is

$$\Delta \hat{H}_r(25°C,\ 1\ \text{atm}) = -125.4\ \text{kJ/mol} \qquad (9.1\text{-}2)$$

Equations 9.1-1 and 9.1-2 together signify that if 1 mol of solid calcium carbide reacts completely with 2 mol of liquid water to form 1 mol of solid calcium hydroxide and 1 mol of acetylene, and the initial and final temperatures are each 25°C and the total pressure is constant at 1 atm, then $H_{\text{products}} - H_{\text{reactants}} = -125.4$ kJ. If the reaction is run under conditions such that the energy balance reduces to $Q = \Delta H$, then 125.4 kJ are emitted by the reaction system in the course of the reaction. (Recall that a negative Q implies flow of heat out of the system.)

422

The units of $\Delta\hat{H}_r$ often cause confusion. For example, if the heat of reaction for $2A + B \rightarrow 3C$ is reported to be -50 kJ/mol, you might ask "per mol of what?" This difficulty is avoided if you recall that the given $\Delta\hat{H}_r$ applies to stoichiometric quantities of each species: For $2A + B \rightarrow 3C$, for example, $\Delta\hat{H}_r = -50$ kJ/mol is simply a shorthand notation for

$$\Delta\hat{H}_r = \frac{-50 \text{ kJ}}{2 \text{ mol A reacted}} = \frac{-50 \text{ kJ}}{1 \text{ mol B reacted}} = \frac{-50 \text{ kJ}}{3 \text{ mol C produced}}$$

If you know, say, how much C was produced in a given reaction at a temperature T, you can multiply this quantity by the third of these ratios to determine the total enthalpy change. In more general terms, if v_A is the stoichiometric coefficient of a reactant or reaction product A, and n_A moles of A are consumed or produced at $T = T_o$, the total enthalpy change is

$$\Delta H = \frac{\Delta\hat{H}_r(T_o)}{v_A} n_A \tag{9.1-3}$$

Several definitions and properties of heats of reaction are summarized here.

1. The **standard heat of reaction** $\Delta\hat{H}_r^\circ$ is the heat of reaction when both the reactants and products are at a reference temperature and pressure, usually 25°C and 1 atm.

2. If $\Delta\hat{H}_r(T)$ is negative the reaction is said to be **exothermic** at temperature T, and if $\Delta\hat{H}_r(T)$ is positive the reaction is **endothermic** at T.

3. At low and moderate pressures, $\Delta\hat{H}_r$ is nearly independent of pressure.

4. The value of a heat of reaction depends on how the stoichiometric equation is written: For example, $\Delta\hat{H}_r^\circ$ for $A \rightarrow B$ is half of $\Delta\hat{H}_r^\circ$ for $2A \rightarrow 2B$. (This should seem reasonable to you if you look back at the definition of $\Delta\hat{H}_r$.)

5. The value of a heat of reaction depends on the states of aggregation (gas, liquid, or solid) of the reactants and products. For example:

$$CH_4(g) + 2\,O_2(g) \rightarrow CO_2(g) + 2\,H_2O(l): \Delta\hat{H}_r^\circ = -890.3 \text{ kJ/mol}$$
$$CH_4(g) + 2\,O_2(g) \rightarrow CO_2(g) + 2\,H_2O(g): \Delta\hat{H}_r^\circ = -802.3 \text{ kJ/mol}$$

The only difference between the reactions is that the water formed is a liquid in the first one and a vapor in the second. Since enthalpy is a state function, the difference between the two heats of reaction must be the enthalpy change associated with the vaporization of 2 mol of water at 25°C—that is, $2\Delta\hat{H}_v(25°C)$.

EXAMPLE 9.1-1

The standard heat of reaction for the combustion of gaseous n-butane is

$$C_4H_{10}(g) + \tfrac{13}{2}\,O_2 \rightarrow 4\,CO_2 + 5\,H_2O(l): \Delta\hat{H}_r^\circ = -2878 \text{ kJ/mol}$$

1. What is the standard heat of the reaction

$$2\ C_4H_{10}(g) + 13\ O_2 \rightarrow 8\ CO_2 + 10\ H_2O(l)$$

2. If 40 mol of CO_2 are produced, calculate the enthalpy change between reactants and products if both are at 25°C.

Solution

1.
$$\Delta \hat{H}_r^\circ = 2(-2878 \text{ kJ/mol}) = \boxed{-5756 \text{ kJ/mol}}$$

2. From Equation 9.1-3

$$\Delta H = \frac{-2878 \text{ kJ}}{4 \text{ mol } CO_2} \left| \frac{40 \text{ mol } CO_2}{} \right. = \boxed{-28\ 780 \text{ kJ}}$$

If a reaction takes place in a closed reactor at constant volume, the heat released or absorbed is determined by the change in internal energy between reactants and products. The **internal energy of reaction** $\Delta \hat{U}_r(T)$ is the difference $U_{\text{products}} - U_{\text{reactants}}$ if stoichiometric quantities of reactants react completely at temperature T.

Suppose a reaction occurs, and v_i is the stoichiometric coefficient of the ith gaseous reactant or product. If ideal gas behavior can be assumed and specific volumes of liquid and solid reactants and products are negligible compared with those of the gases, the internal energy of reaction is related to the heat of reaction by

$$\Delta \hat{U}_r(T) = \Delta \hat{H}_r(T) - RT \left(\sum_{\substack{\text{gaseous} \\ \text{products}}} v_i - \sum_{\substack{\text{gaseous} \\ \text{reactants}}} v_i \right) \tag{9.1-4}$$

For example, for the reaction

$$C_6H_{14}(l) + \tfrac{19}{2} O_2(g) \rightarrow 6\ CO(g) + 7\ H_2O(g)$$

the internal energy of reaction is, from Equation 9.1-4,

$$\Delta \hat{U}_r(T) = \Delta \hat{H}_r(T) - RT(6 + 7 - \tfrac{19}{2})$$
$$= \Delta H_r(T) - \tfrac{7}{2}RT$$

If there are no gaseous reactants or products, then to a good approximation $\Delta \hat{U}_r = \Delta \hat{H}_r$.

EXAMPLE 9.1-2 Evaluation of $\Delta \hat{U}_r$

The standard heat of the reaction

$$C_2H_4(g) + 2\ Cl_2(g) \rightarrow C_2HCl_3(l) + H_2(g) + HCl(g)$$

is $\Delta \hat{H}_r(25°C) = -420.8$ kJ/mol. Calculate $\Delta \hat{U}_r$ for this reaction.

Solution

From the stoichiometric equation

$$\sum v_i \text{ (product gases)} = 1 + 1 = 2$$
$$\sum v_i \text{ (reactant gases)} = 1 + 2 = 3$$

From Equation 9.1-4

$$\Delta \hat{U}_r = \Delta \hat{H}_r - RT(2 - 3)$$

$$= -420.8 \text{ kJ/mol} - \frac{8.314 \text{ J}}{\text{mol} \cdot \text{K}} \left| \frac{298 \text{ K}}{} \right| \frac{-1}{} \left| \frac{1 \text{ kJ}}{10^3 \text{ J}} \right.$$

$$= \boxed{-418.3 \text{ kJ/mol}}$$

TEST YOURSELF

1. What is a heat of reaction? A standard heat of reaction?
2. Suppose $\Delta \hat{H}_r^\circ$ is -40 kJ/mol for the reaction $2A \rightarrow B$.
 (a) What is the value of the ratio (kJ/mol A reacted)?
 (b) Is the reaction exothermic or endothermic at 25°C and 1 atm?
 (c) If the reactants and products are at the same temperature, must heat be added to or withdrawn from the reactor? (Assume that the energy balance equation is $Q = \Delta H$.)
 (d) If the reactor is adiabatic ($Q = 0$), would the products leave at a higher or a lower temperature than that of the entering reactants?
3. $C_6H_{14}(l) + \frac{19}{2}O_2 \rightarrow 6 CO_2 + 7 H_2O(l)$: $\Delta \hat{H}_r^\circ = -4163$ kJ/mol
 $C_6H_{14}(g) + \frac{19}{2}O_2 \rightarrow 6 CO_2 + 7 H_2O(l)$: $\Delta \hat{H}_r^\circ = -4195$ kJ/mol

 The standard state for the heats of reaction is 25°C and 1 atm. What is the physical significance of the difference between the two given values of $\Delta \hat{H}_r^\circ$?
4. Write the formula for $\Delta \hat{U}_r(T)$ in terms of $\Delta \hat{H}_r(T)$ for the reaction $A(g) + 2B(g) + C(l) \rightarrow D(g) + 2E(s)$.
5. Derive Equation 9.1-4.

9.2 MEASUREMENT AND CALCULATION OF HEATS OF REACTION: HESS'S LAW

A heat of reaction may be measured in a **calorimeter**—a closed reactor immersed in a fluid contained in a well-insulated vessel. The temperature rise or fall of the fluid can be measured and used to determine the heat generated or consumed by the reaction, and the value of $\Delta \hat{H}_r^\circ$ may then be calculated from the heat transferred and known reactant and product heat capacities.

There are serious limitations to this technique, however. Suppose, for example, you wish to determine $\Delta \hat{H}_r^\circ$ for the reaction

$$2 C(s) + O_2(g) \rightarrow 2 CO(g)$$

You could put 2 moles of carbon and 1 mole of oxygen together in a reactor, but you would never get 2 moles of carbon monoxide. If the reactants are at or near 25°C

or lower, nothing apparent would occur since the rate at which carbon and oxygen react is immeasurably low at this temperature. If on the other hand the mixture were heated to a temperature at which C and O_2 react at a measurable rate, the product would be either pure CO_2 or at best a mixture of CO and CO_2, making it impossible to determine the heat of the CO formation reaction alone.

However, you *can* carry out separately the reactions

1. $C + O_2 \rightarrow CO_2$: $\Delta\hat{H}_{r1}^{\circ} = -393.51 \text{ kJ/mol}$
2. $CO + \frac{1}{2}O_2 \rightarrow CO_2$: $\Delta\hat{H}_{r2}^{\circ} = -282.99 \text{ kJ/mol}$

and determine their heats of reaction experimentally. You may then construct a process path for the reaction

3. $C + \frac{1}{2}O_2 \rightarrow CO$: $\Delta\hat{H}_{r3}^{\circ} = ?$

Since \hat{H} is a state function,

$$\Delta\hat{H}_{r3}^{\circ} = \Delta\hat{H}_{r1}^{\circ} + (-\Delta\hat{H}_{r2}^{\circ}) = -393.51 + 282.99 = -110.52 \text{ kJ/mol}$$

You can therefore calculate the desired heat of reaction, which could not be measured, from two measurable heats of reaction.

This result could have been obtained more directly by treating the stoichiometric equations for reactions 1 and 2 as algebraic equations. If the equation for reaction 2 is subtracted from that for reaction 1, the result is

$$C + O_2 - CO - \tfrac{1}{2}O_2 \rightarrow 0 \text{ (zero)}$$
$$\Downarrow$$
$$C + \tfrac{1}{2}O_2 \rightarrow CO \text{ (reaction 3)}$$

The standard heat of reaction 3 may be calculated by applying the same operation to the heats of reactions 1 and 2—that is, $\Delta\hat{H}_{r3}^{\circ} = \Delta\hat{H}_{r1}^{\circ} - \Delta\hat{H}_{r2}^{\circ}$—confirming the result previously obtained.

The general statement of the validity of this procedure is called **Hess's law:** If the stoichiometric equation for reaction 1 can be obtained by algebraic operations (multiplication by constants, addition and subtraction) on stoichiometric equations for reactions 2, 3, ..., then the heat of reaction $\Delta\hat{H}_{r1}$ can be obtained by performing the same operations on the heats of reactions $\Delta\hat{H}_{r2}$, $\Delta\hat{H}_{r3}$,

EXAMPLE 9.2-1 Hess's Law

The standard heats of the following combustion reactions have been determined experimentally.

1. $C_2H_6 + \frac{7}{2} O_2 \rightarrow 2\, CO_2 + 3\, H_2O$: $\Delta \hat{H}_{r1}^{\circ} = -1559.8$ kJ/mol
2. $C + O_2 \rightarrow CO_2$: $\Delta \hat{H}_{r2}^{\circ} = -393.5$ kJ/mol
3. $H_2 + \frac{1}{2} O_2 \rightarrow H_2O$: $\Delta \hat{H}_{r3}^{\circ} = -285.8$ kJ/mol

Use Hess's law and the given heats of reaction to determine the standard heat of the reaction

4. $2C + 3\, H_2 \rightarrow C_2H_6$: $\Delta H_{r4}^{\circ} = ?$

Solution

You may easily verify that

$$(4) = 2 \times (2) + 3 \times (3) - (1)$$

so that from Hess's law

$$\Delta \hat{H}_{r4}^{\circ} = 2\,\Delta \hat{H}_{r2}^{\circ} + 3\,\Delta \hat{H}_{r3}^{\circ} - \Delta \hat{H}_{r1}^{\circ} = -84.6 \text{ kJ/mol}$$

This heat of reaction could not have been measured directly, since you cannot react carbon and hydrogen in such a way that ethane is the only reaction product.

TEST YOURSELF

1. What is Hess's law?
2. Suppose heats of reaction are measured experimentally for the following set of reactions.

$$2\,A + B \rightarrow 2C: \Delta \hat{H}_{r1}^{\circ} = -1000 \text{ kJ/mol}$$
$$A + D \rightarrow C + 3\,E: \Delta \hat{H}_{r2}^{\circ} = -2000 \text{ kJ/mol}$$

Use Hess's law to show that for

$$B + 6\,E \rightarrow 2\,D: \Delta \hat{H}_r^{\circ} = +3000 \text{ kJ/mol}$$

9.3 FORMATION REACTIONS AND HEATS OF FORMATION

A **formation reaction** of a compound is the reaction in which the compound is formed from its atomic constituents as they normally occur in nature (e.g., O_2 rather than O). The standard heat of such a reaction is the **standard heat of formation, $\Delta \hat{H}_f^{\circ}$**, of the compound.

Standard heats of formation for many compounds are listed in Table B.1 of this book and on pp. 3-147 through 3-154 of *Perry's Chemical Engineers' Handbook.*[1] For example, $\Delta \hat{H}_f^\circ$ for crystalline ammonium nitrate is given in Table B.1 as -365.14 kJ/mol, signifying

$$N_2(g) + 2\,H_2(g) + \tfrac{3}{2}O_2(g) \rightarrow NH_4NO_3(c): \Delta \hat{H}_r^\circ = -365.14\,\frac{kJ}{mol}$$

Similarly, for liquid benzene $\Delta \hat{H}_f^\circ = 48.66$ kJ/mol, or

$$6\,C(s) + 3\,H_2(g) \rightarrow C_6H_6(l): \Delta \hat{H}_r^\circ = +48.66\,\frac{kJ}{mol}$$

The standard heat of formation of an element is zero. (Why?)

It may be shown using Hess's law that *if v_i is the stoichiometric coefficient of the ith species participating in a reaction and $(\Delta \hat{H}_f^\circ$ is the standard heat of formation of this species, then the standard heat of the reaction is*

$$\boxed{\Delta \hat{H}_r^\circ = \sum_{\text{products}} v_i(\Delta \hat{H}_f^\circ)_i - \sum_{\text{reactants}} v_i(\Delta \hat{H}_f^\circ)_i} \qquad \textbf{(9.3-1)}$$

The standard heats of formation of all elements should be set equal to zero in this formula. The validity of Equation 9.3-1 is illustrated in the next example.

EXAMPLE 9.3-1 Determination of a Heat of Reaction from Heats of Formation

Determine the standard heat of reaction for the combustion of liquid *n*-pentane, assuming $H_2O(l)$ is a combustion product.

$$C_5H_{12}(l) + 8\,O_2(g) \rightarrow 5\,CO_2(g) + 6\,H_2O(l)$$

Solution

From Equation 9.3-1

$$\Delta \hat{H}_r^\circ = 5(\Delta \hat{H}_f^\circ)_{CO_2(g)} + 6(\Delta \hat{H}_f^\circ)_{H_2O(l)} - (\Delta \hat{H}_f^\circ)_{C_5H_{12}(l)}$$

$$\Downarrow \text{ Heats of formation from Table B.1}$$

$$\Delta \hat{H}_r^\circ = [5(-393.5) + 6(-285.84) - (-173.0)]\ kJ/mol$$

$$= \boxed{-3509\ kJ/mol}$$

To derive the formula for $\Delta \hat{H}_r^\circ$, we may write the stoichiometric equations for the formation reactions of the reactants and products:

1. $5\,C(s) + 6\,H_2(g) \rightarrow C_5H_{12}(l): \Delta \hat{H}_{r1}^\circ = (\Delta \hat{H}_f^\circ)_{C_5H_{12}(l)}$
2. $C(s) + O_2(g) \rightarrow CO_2(g): \Delta \hat{H}_{r2}^\circ = (\Delta \hat{H}_f^\circ)_{CO_2(g)}$
3. $H_2(g) + \tfrac{1}{2}O_2(g) \rightarrow H_2O(l): \Delta \hat{H}_{r3}^\circ = (\Delta \hat{H}_f^\circ)_{H_2O(l)}$

[1] R. H. Perry and D. W. Green, eds., *Perry's Chemical Engineers' Handbook*, 6th Edition, McGraw-Hill, New York, 1984.

The desired reaction,

4. $C_5H_{12}(l) + 8 O_2(g) \rightarrow 5 CO_2(g) + 6 H_2O(l)$: $\Delta \hat{H}_r^\circ = ?$

may be obtained as $5 \times (2) + 6 \times (3) - (1)$ (verify this), and the given formula for $\Delta \hat{H}_r^\circ$ then follows from Hess's law.

Techniques for estimating heats of formation of compounds with known molecular structures are given by Reid, Prausnitz, and Sherwood.[2]

TEST YOURSELF

1. The standard heat of the reaction

$$2 CO \rightarrow 2 C + O_2$$

is $\Delta \hat{H}_r^\circ = +221.0$ kJ/mol. Use this result to calculate the standard heat of formation of CO, and check your result with a tabulated value.

2. $\Delta \hat{H}_f^\circ$ is -28.64 kcal/mol for $C_3H_8(l)$, and -24.82 kcal/mol for $C_3H_8(g)$. What is the physical significance of the difference between these values?

3. Consider the reaction

$$CH_4 + 2 O_2 \rightarrow CO_2 + 2 H_2O(g)$$

Write the formula for $\Delta \hat{H}_r^\circ$ in terms of the standard heats of formation of the reactants and products.

9.4 HEATS OF COMBUSTION

The **standard heat of combustion** $\Delta \hat{H}_c^\circ$ of a substance is the heat of the reaction of that substance with oxygen to yield specified products, with both reactants and products at 25°C and 1 atm (the arbitrary but conventional reference state).

Table B.1 lists standard heats of combustion for a number of substances. The given values are based on the following assumptions: (1) All carbon in the fuel forms $CO_2(g)$, (2) all hydrogen forms $H_2O(l)$, (3) all sulfur forms $SO_2(g)$, (4) all nitrogen forms $N_2(g)$. The standard heat of combustion of liquid ethanol, for example, is given in Table B.1 as $\Delta \hat{H}_c^\circ = -1366.9$ kJ/mol; this signifies

$$C_2H_5OH(l) + 3 O_2 \rightarrow 2 CO_2(g) + 3 H_2O(l): \Delta \hat{H}_r(25°C, 1 \text{ atm}) = -1366.9 \text{ kJ/mol}$$

Additional heats of combustion are given on pp. 3-155 through 3-157 of *Perry's Chemical Engineers' Handbook*.

Standard heats of reactions that involve only combustible substances and combustion products can be calculated from tabulated standard heats of combustion, in

[2] R. C. Reid, J. M. Prausnitz, and T. K. Sherwood, *The Properties of Gases and Liquids*, 3rd Edition, McGraw-Hill, New York, 1977.

another application of Hess's law. A hypothetical reaction path may be constructed in which (1) all combustible reactants are burned with O_2 at 25°C, (2) the combustion products are combined to form the reaction products. Step (2) involves the reverse of the combustion reactions of the reaction products. Since both steps involve only combustion reactions, the total enthalpy change—which equals the desired heat of reaction—can be determined entirely from heats of combustion as

$$\Delta\hat{H}_r^\circ = \sum_{\text{reactants}} v_i(\Delta\hat{H}_c^\circ)_i - \sum_{\text{products}} v_i(\Delta\hat{H}_c^\circ)_i \qquad\qquad (9.4\text{-}1)$$

If any of the reactants or products are themselves combustion products $[CO_2, H_2O(l), SO_2, \dots)]$, their $\Delta\hat{H}_c^\circ$ terms in Equation 9.4-1 should be set equal to 0.

Note that this formula is similar to that used to determine $\Delta\hat{H}_r^\circ$ from heats of formation, except that in this case the product summation is subtracted from the reactant summation. The validity of this formula is illustrated in the next example.

EXAMPLE 9.4-1 Calculation of a Heat of Reaction from Heats of Combustion

Calculate the standard heat of reaction for the dehydrogenation of ethane

$$C_2H_6 \rightarrow C_2H_4 + H_2$$

Solution

From Table B.1, $(\Delta\hat{H}_c^\circ)_{C_2H_6} = -1559.9$ kJ/mol
$$(\Delta\hat{H}_c^\circ)_{C_2H_4} = -1411.0 \text{ kJ/mol}$$
$$(\Delta\hat{H}_c^\circ)_{H_2} = -285.84 \text{ kJ/mol}$$

From Equation 9.4-1, therefore,

$$\Delta\hat{H}_r^\circ = (\Delta\hat{H}_c^\circ)_{C_2H_6} - (\Delta\hat{H}_c^\circ)_{C_2H_4} - (\Delta\hat{H}_c^\circ)_{H_2} = \boxed{136.9 \text{ kJ/mol}}$$

As an illustration, let us demonstrate the validity of this formula using Hess's law. The combustion reactions are

1. $C_2H_6 + \frac{7}{2}O_2 \rightarrow 2 CO_2 + 3 H_2O$
2. $C_2H_4 + 3 O_2 \rightarrow 2 CO_2 + 2 H_2O$
3. $H_2 + \frac{1}{2}O_2 \rightarrow H_2O$

It is easy to show that

4. $C_2H_6 \rightarrow C_2H_4 + H_2$

is obtained as (1)–(2)–(3). The desired result follows from Hess's law.

One of the principal applications of Equation 9.4-1 is to determine heats of formation for combustible substances whose formation reactions do not occur naturally.

For example, the formation reaction of pentane

$$5 \; C(s) + 6 \; H_2(g) \to C_5H_{12}(l): \Delta \hat{H}_f^\circ = ?$$

cannot be carried out in a laboratory, but carbon, hydrogen, and pentane can all be burned and their standard heats of combustion determined experimentally. The heat of formation of pentane may then be calculated from Equation 9.4-1 as

$$(\Delta \hat{H}_f^\circ)_{C_5H_{12}(l)} = 5(\Delta \hat{H}_c^\circ)_{C(s)} + 6(\Delta \hat{H}_c^\circ)_{H_2(g)} - (\Delta \hat{H}_c^\circ)_{C_5H_{12}(l)}$$

CREATIVITY EXERCISE

When an exothermic reaction takes place, the energy liberated raises the temperature of the reactor contents unless something is done to offset the heating. Suppose such a reaction occurs in a batch reactor.

1. Think of as many reasons as you can why you might not want the reactor temperature to increase. (*Example:* The product might degrade or decompose at higher temperatures.)
2. Think of as many ways as you can to keep the reactor temperature from increasing as the reaction proceeds. (*Example:* Throw in some ice.)

9.5 ENERGY BALANCES ON REACTIVE PROCESSES

9.5a General Procedures

In Chapter 9, we outlined a method for setting up and solving energy balance problems for nonreactive processes. The procedure was to choose reference states for all species entering and leaving the system, calculate enthalpies of all stream components relative to these reference states, calculate

$$\Delta H = \sum_{\text{outlet}} n_i \hat{H}_i - \sum_{\text{inlet}} n_i \hat{H}_i \tag{9.5-1}$$

and substitute this value for ΔH in the energy balance equation. This procedure does not work without modifications if chemical reactions occur in the system, however, since the value of ΔH calculated from Equation 9.5-1 does not include enthalpy changes due to the reactions.

Several approaches to setting up energy balances on reactive systems are often adopted, which differ in the reference conditions used for enthalpy calculations. Suppose $\Delta \hat{H}_r^\circ$ for each reaction occurring in a system is known or calculable at a temperature T_0 (which is usually but not always 25°C). Following are two common choices of reference conditions and the procedures for calculating ΔH that correspond to each choice.

1. *Reference conditions: reactant and product species at T_0 in the state of aggregation for which $\Delta \hat{H}_r^\circ$ is known, and nonreactive species at any convenient temperature*

(such as the reactor inlet temperature, or the reference temperature of a mean heat capacity table).

If these reference conditions are chosen, the enthalpy change for the process may be determined by setting up and filling in a table of inlet and outlet stream component flow rates n_i and specific enthalpies \hat{H}_i, as was done in Chapter 8, and calculating

$$\Delta H = \frac{n_{AR}\,\Delta\hat{H}_r^\circ}{\nu_A} + \sum_{\text{outlet}} n_i\hat{H}_i - \sum_{\text{inlet}} n_i\hat{H}_i \tag{9.5-2}$$

where

A = any reactant or product

n_{AR} = moles of A produced or consumed in the process (not necessarily moles fed or moles present in the product)

ν_A = stoichiometric coefficient of A. Both n_{AR} and ν_A are positive numbers

If multiple reactions occur, a term of the form $n_A\,\Delta\hat{H}_r^\circ/\nu_A$ must be included in Equation 9.5-2 for each reaction. (Method 2 is generally easier to use in such a case.)

2. *Reference conditions: the elements that constitute the reactants and products at 25°C and the nonreactive molecular species at any convenient temperature.*

Again, a table of n_i's and \hat{H}_i's for all stream components is constructed, only now \hat{H}_i for a reactant or product is the sum of the heat of formation of the species at 25°C and any sensible and latent heats required to bring the species from 25°C to its inlet or outlet state. The overall enthalpy change for the process is then

$$\Delta H = \sum_{\text{outlet}} n_i\hat{H}_i - \sum_{\text{inlet}} n_i\hat{H}_i \tag{9.5-3}$$

The reaction term of Equation 9.5-2 is not required if elements are chosen as references, since this term is implicitly calculated when the heats of formation of the reactants are subtracted from those of the products in Equation 9.5-3.

For a process in which, say, CH_4 is burned with air to form CO_2 and H_2O, the reference states for the first method would be $CH_4(g)$, $O_2(g)$, $N_2(g)$, $CO_2(g)$, and $H_2O(l)$ at 25°C, and those for the second method would be $C(s)$, $H_2(g)$, $O_2(g)$, and $N_2(g)$ at 25°C.

The example that follows illustrates the first of these approaches.

EXAMPLE 9.5-1 Energy Balance about an Ammonia Oxidizer

The standard heat of reaction at 25°C and 1 atm for the oxidation of ammonia

$$4\,NH_3(g) + 5\,O_2(g) \rightarrow 4\,NO(g) + 6\,H_2O(g)$$

is $\Delta\hat{H}_r^\circ = -904.6$ kJ/mol. One hundred mol NH_3/h and 200 mol O_2/h at 25°C are fed into a reactor in which the ammonia is completely consumed. The product stream

emerges as a gas at 300°C. Calculate the heat transferred to or from the reactor, assuming operation at approximately 1 atm.

Solution

Basis: Given feed rates

Material balances lead to the flow rates shown here.

Q (kJ/h)

(Verify the product flow rates.) The enthalpy table appears as follows.

<p align="center">References: $NH_3(g)$, $O_2(g)$, $NO(g)$, $H_2O(g)$ at 25°C</p>

Substance	n_{in}	\hat{H}_{in}	n_{out}	\hat{H}_{out}	
NH_3	100	0	—	—	n in mol/h
O_2	200	0	75		\hat{H} in kJ/mol
NO	—	—	100		
H_2O	—	—	150		

Calculate Unknown Enthalpies

$$O_2(g, 300°C): \hat{H} = \bar{C}_p(300°C)[300°C - 25°C]$$
$$\Downarrow \text{ Table 8.3-1}$$
$$\hat{H} = (275°C)(30.80 \text{ J/mol·°C})$$
$$= 8470 \text{ J/mol} = 8.470 \text{ kJ/mol (Insert in enthalpy table)}$$

$$NO(g, 300°C): \hat{H} = \int_{25°C}^{300°C} (C_p)_{NO}\, dT$$
$$\Downarrow \text{ Table B.2}$$
$$\hat{H} = \int_{25}^{300} (29.50 + 0.8188 \times 10^{-2}T - 0.2925 \times 10^{-5}T^2$$
$$+ 0.3652 \times 10^{-9}T^3)\, dT$$
$$= 8453 \text{ J/mol} = 8.453 \text{ kJ/mol (Insert in table)}$$

$$H_2O(g, 300°C): \hat{H} = \bar{C}_p(300°C)[300°C - 25°C]$$
$$\Downarrow \text{ Table 8.3-1}$$
$$\hat{H} = (275°C)(34.80 \text{ J/mol·°C})$$
$$= 9570 \text{ J/mol} = 9.570 \text{ kJ/mol (Insert in table)}$$

Evaluate ΔH

Since 100 mol NH_3/h are consumed in the process, we may write Equation 9.5-2 as

$$\Delta H(kJ/h) = \frac{100 \text{ mol } NH_3 \text{ react}}{h} \bigg| \frac{-904.6 \text{ kJ}}{4 \text{ mol } NH_3} + \underbrace{\sum_{\text{out}} n_i \hat{H}_i - \sum_{\text{in}} n_i \hat{H}_i}_{\text{from table}}$$

$$= -19\,700 \text{ kJ/h}$$

Energy Balance

Neglecting kinetic energy changes

$$\boxed{Q = \Delta H = -19\,700 \text{ kJ/h}}$$

Thus, 19 700 kJ/h must be transferred *from* the reactor to maintain the product temperature at 300°C.

Suppose you know a heat of reaction at a temperature T_o and wish to determine $\Delta \hat{H}_r$ at some other temperature T. This calculation is performed exactly as in Example 9.5-1, taking as inputs and outputs stoichiometric quantities of the reactants and products at temperature T, and as reference states the states for which $\Delta \hat{H}_r(T_o)$ is known. For the reaction of Example 9.5-1, $\{n_{in}\}$ would be 4 mol NH_3(g) and 5 mol O_2(g), and $\{n_{out}\}$ would be 4 mol NO(g) and 6 mol H_2O(g).

The second possible choice of reference conditions—the atomic constituents of the reactants and products in their naturally occurring states—is often particularly convenient for processes that involve several simultaneous reactions. The next example illustrates this approach.

EXAMPLE 9.5-2 Energy Balance on a Methane Oxidation Reactor

Methane is oxidized with air to produce formaldehyde in a continuous reactor. A competing reaction is the combustion of methane to form CO_2.

1. $CH_4(g) + O_2 \rightarrow HCHO(g) + H_2O(g)$
2. $CH_4 + 2 O_2 \rightarrow CO_2 + 2 H_2O(g)$

A flowchart of the process for an assumed basis of 100 mol of methane fed to the reactor is shown here.

100 mol CH_4
25°C

100 mol O_2
376 mol N_2
100°C

Q

60 mol CH_4
30 mol HCHO(g)
10 mol CO_2
50 mol H_2O(g)
50 mol O_2
376 mol N_2
150°C

The pressure is low enough for ideal gas behavior to be assumed. If the methane enters the reactor at 25°C and the air enters at 100°C, how much heat must be withdrawn for the product stream to emerge at 150°C?

Solution

Basis: 100 mol CH₄ fed

Basis: 100 mol CH$_4$ fed

Since the component amounts of all streams are known, we may proceed directly to the energy balance. We choose as references the atomic species that form the reactants and products—that is, C, O$_2$, and H$_2$—at 25°C, and the nonreactive species—N$_2$—also at 25°C (the reference temperature of the mean heat capacity table, 8.3-1). The enthalpy table as it appears after all stream component enthalpies are calculated is shown here.

References: C, O$_2$, H$_2$, N$_2$ at 25°C

Substance	n_{in}	\hat{H}_{in}	n_{out}	\hat{H}_{out}	
CH$_4$	100	−74.85	60	−69.95	
O$_2$	100	2.235	50	3.758	n in mol
N$_2$	376	2.187	376	3.655	\hat{H} in kJ/mol
HCHO	—	—	30	−111.15	
CO$_2$	—	—	10	−388.6	
H$_2$O	—	—	50	−237.56	

Calculate Enthalpies

$O_2(100°C)$: $\hat{H} = \bar{C}_p(100°C)[100°C - 25°C]$

$\Downarrow \bar{C}_p$ from Table 8.3-1

$(\hat{H}_{O_2})_{in} = 2235$ J/mol $= 2.235$ kJ/mol

$O_2(150°C)$: $\hat{H} = \bar{C}_p(150°C)[150°C - 25°C] \xrightarrow{\text{Table 8.3-1}} (\hat{H}_{O_2})_{out} = 3.758$ kJ/mol

$N_2(100°C)$: $\hat{H} = 75\bar{C}_p(100°C) \xrightarrow{\text{Table 8.3-1}} (\hat{H}_{N_2})_{in} = 2.187$ lkJ/mol

$N_2(150°C)$: $\hat{H} = 125\bar{C}_p(150°C) \xrightarrow{\text{Table 8.3-1}} (\hat{H}_{N_2})_{out} = 3.655$ kJ/mol

$CH_4(25°C)$: We must form CH$_4$ from its elements at 25°C.

$$\hat{H} = (\Delta\hat{H}_f°)_{CH_4}$$

$\Downarrow \Delta\hat{H}_f°$ from Table B.1

$$(\hat{H}_{CH_4})_{in} = -74.85 \text{ kJ/mol}$$

$$CH_4(150°C): (\hat{H}_{CH_4})_{out} = (\Delta\hat{H}_f°)_{CH_4} + \int_{25}^{150} (C_p)_{CH_4}\, dT$$

$$= -74.85 + 4.90 = -69.95 \text{ kJ/mol}$$

$$HCHO(g, 150°C): \hat{H} = (\Delta\hat{H}_f°)_{HCHO(g)} + \int_{25}^{150} (C_p)_{HCHO}\, dT$$

$$= -115.90 + 4.75 = -111.15 \text{ kJ/mol}$$

$$CO_2(150°C): \hat{H} = (\Delta\hat{H}_f°)_{CO_2(g)} + [125\overline{C}_p(150°C)]$$

$$\left\|\begin{array}{l} \Delta\hat{H}_f° \text{ from Table B.1} \\ \overline{C}_p \text{ from Table 8.3-1} \end{array}\right.$$

$$(\hat{H}_{CO_2})_{out} = -393.5 + 4.94 = -388.6 \text{ kJ/mol}$$

$$H_2O(g, 150°C): \hat{H} = (\Delta\hat{H}_f°)_{H_2O(g)} + 125\overline{C}_p(150°C)$$

$$\left\|\begin{array}{l} \Delta\hat{H}_f° \text{ from Table B.1} \\ \overline{C}_p \text{ from Table 8.3-1} \end{array}\right.$$

$$\hat{H} = -241.83 + 4.27 = -237.56 \text{ kJ/mol}$$

Evaluate ΔH

From Equation 9.5-3,

$$\Delta H = \sum_{out} n_i\hat{H}_i - \sum_{in} n_i\hat{H}_i = -15\ 300 \text{ kJ}$$

Energy Balance

Neglecting kinetic energy changes

$$\boxed{Q = \Delta H = -15\ 300 \text{ kJ}}$$

What you are in effect doing when you use molecular species as references is constructing a hypothetical process path from the inlet to the outlet in which you bring the reactants from the inlet temperature to 25°C, react them isothermally, and bring the products up to the outlet temperature. You should be able to verify that Equation 9.5-2 for ΔH results when you sum these individual changes. Similarly, assuming the elements as reference states is tantamount to bringing the reactants to 25°C, decomposing them into their elements and forming the products at 25°C, and bringing the products up to the outlet temperature. As an exercise, verify that this procedure leads to Equation 9.5-3 for ΔH.

9.5b Processes with Unknown Outlet Conditions: Adiabatic Reactors

In the reactive systems we have looked at so far, the inlet and outlet conditions were specified and the required heat input could be determined from the energy balance equation.

In another important class of problems, the input conditions, heat input, and product composition are specified, and the outlet temperature is to be calculated. To solve such problems, you must evaluate the enthalpies of the products relative to the chosen reference states in terms of the unknown final temperature, and then substitute the resulting expressions into the energy balance equation ($Q = \Delta H$, or $\Delta H = 0$ for an adiabatic reactor) to calculate T_{outlet}.

EXAMPLE 9.5-3 Energy Balance on an Adiabatic Reactor

The dehydrogenation of ethanol to form acetaldehyde

$$C_2H_5OH(g) \rightarrow CH_3CHO(g) + H_2(g)$$

$$\Delta\hat{H}_r°(25°C) = +68.95 \text{ kJ/mol}$$

is carried out in an adiabatic reactor. Ethanol vapor is fed to the reactor at 300°C, and a conversion of 30% is obtained. Calculate the product temperature, using the following heat capacities.

$$C_2H_5OH(g): C_p = 0.110 \text{ kJ/mol} \cdot °C$$
$$CH_3CHO(g): C_p = 0.080 \text{ kJ/mol} \cdot °C$$
$$H_2(g): C_p = 0.029 \text{ kJ/mol} \cdot °C$$

Assume constant

Solution

Basis: 100 mol feed

Material balances lead to the information on the flowchart shown here.

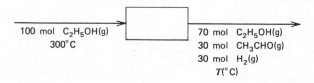

Since only one reaction occurs and the heat of reaction is known, we will choose the reactants and products at 25°C (rather than atomic species at this temperature) as references.

References: $C_2H_5OH(g)$, $CH_3CHO(g)$, $H_2(g)$
at 25°C

Substance	n_{in}	\hat{H}_{in}	n_{out}	\hat{H}_{out}
C_2H_5OH	100		70	
CH_3CHO	—	—	30	n in mol
H_2	—	—	30	\hat{H} in kJ/mol

Evaluate Inlet Enthalpy

$C_2H_5OH(g, 300°C)$:

$$\hat{H} = \int_{25}^{300} (C_p)_{C_2H_5OH(g)} \, dT$$
$$= (0.110 \text{ kJ/mol} \cdot °C)(300 - 25)°C$$
$$= 30.25 \text{ kJ/mol}$$
$$\sum_{in} n_i \hat{H}_i = (100 \text{ mol})(30.25 \text{ kJ/mol}) = 3025 \text{ kJ}$$

Evaluate Reaction Term at 25°C

$$\frac{\Delta \hat{H}_r^\circ}{v_A} n_A = \frac{+68.95 \text{ kJ}}{1 \text{ mol } CH_3CHO \text{ formed}} \left| \frac{30 \text{ mol } CH_3CHO \text{ formed}}{} \right.$$
$$= 2069 \text{ kJ}$$

Evaluate Outlet Enthalpies

In all cases,

$$\hat{H} = \int_{25}^{T} C_p \, dT \xrightarrow{\text{constant } C_p} C_p(T - 25)$$

$C_2H_5OH(g): \hat{H}(T) = 0.110(T - 25) \text{ kJ/mol}$

$CH_3CHO(g): \hat{H}(T) = 0.080(T - 25) \text{ kJ/mol}$

$H_2(g): \hat{H}(T) = 0.029(T - 25) \text{ kJ/mol}$

$$\Downarrow$$

$$\sum_{\text{out}} n_i \hat{H}_i = [(70.0)(0.110) + (30.0)(0.080) + (30.0)(0.029)](T - 25)$$

$$= (11.0T - 274) \text{ kJ}$$

Evaluate ΔH

From Equation 9.5-2,

$$\Delta H = \frac{\Delta \hat{H}_r^\circ}{\nu_A} n_A + \sum_{\text{out}} n_i \hat{H}_i - \sum_{\text{in}} n_i \hat{H}_i$$

$$= 2069 + (11.0T - 274) - 3025 = (11.0T - 1230) \text{ kJ}$$

Energy Balance

Neglecting kinetic energy changes

$$Q = \Delta H \xrightarrow[Q=0]{\text{adiabatic}} \Delta H = 11.0T - 1230 = 0 \Rightarrow T = \boxed{112°C}$$

This example was made simple by the assumption that the heat capacities were all constants, so that the energy balance became a linear equation in T. If $C_p = a + bT$, a quadratic equation must be solved, which is still not difficult; however, if $C_p(T)$ is a polynomial of order greater than one, a nonlinear equation results that must be solved by a trial-and-error technique.

Another class of problems involves processes for which the heat input and outlet temperature are specified, but the product composition is not. Solving such problems requires the simultaneous solution of material and energy balances, as the next example illustrates.

EXAMPLE 9.5-4 Simultaneous Material and Energy Balances

The ethanol dehydrogenation reaction of Example 9.5-3 is carried out with the feed entering at 300°C. The feed contains 90 mole% ethanol and the balance acetaldehyde. To keep the temperature from dropping too rapidly and hence quenching the reaction at a low conversion, heat is added to the reactor. You observe that when the heat addition rate is 1627 kJ per 100 mol of the feed gas, the outlet temperature is 253°C.

Calculate the fractional conversion of ethanol achieved in the reactor, using the heat capacity data given in Example 9.5-3.

Solution

Basis: 100 mol feed gas

$$C_2H_5OH(g) \rightarrow CH_3CHO(g) + H_2(g): \Delta\hat{H}_r^\circ = 68.95 \text{ kJ/mol}$$

There are three unknowns, and no three material balance equations will enable you to solve for n_1, n_2, and n_3 (try it and convince yourself). The problem must be solved by writing and simultaneously solving two material balances and an energy balance.

Balance on C

$$(90)(2) + (10)(2) = 2n_1 + 2n_2$$
$$\Downarrow$$
$$n_1 + n_2 = 100 \tag{1}$$

Balance on H

$$(90)(6) + (10)(4) = 6n_1 + 4n_2 + 2n_3$$
$$\Downarrow$$
$$6n_1 + 4n_2 + 2n_3 = 580 \tag{2}$$

Energy Balance

Reference: $C_2H_5OH(g)$, $CH_3CHO(g)$, $H_2(g)$ at 25°C

Substance	n_{in}	\hat{H}_{in}	n_{out}	\hat{H}_{out}	
$C_2H_5OH(g)$	90	30.3	n_1	25.1	n in mol
$CH_3CHO(g)$	10	22.0	n_2	18.2	\hat{H} in kJ/mol
$H_2(g)$	—	—	n_3	6.6	

$\hat{H} = \overline{C}_p(T - 25)$: Heat capacities from Example 9.5-3

$$(C_2H_5OH)_{inlet}: \hat{H} = (0.110 \text{ kJ/mol} \cdot °C)(300°C - 25°C) = 30.3 \text{ kJ/mol}$$
$$(CH_3CHO)_{inlet}: \hat{H} = (0.080)(275) = 22.0 \text{ kJ/mol}$$
$$(C_2H_5OH)_{outlet}: \hat{H} = (0.110)(253 - 25) = 25.1 \text{ kJ/mol}$$
$$(CH_3CHO)_{outlet}: \hat{H} = (0.080)(228) = 18.2 \text{ kJ/mol}$$
$$(H_2)_{outlet}: \hat{H} = (0.029)(228) = 6.6 \text{ kJ/mol}$$

$$Q = \Delta H$$

\Downarrow Equation 9.5-2

$$Q = \frac{(n_{H_2})_{out} \Delta \hat{H}_r^\circ}{v_{H_2}} + \sum_{out} n_i \hat{H}_i - \sum_{in} n_i \hat{H}_i$$

\Downarrow

$$1627 = \frac{n_3(68.95)}{1} + 25.1n_1 + 18.2n_2 + 6.6n_3 - (90.0)(30.3) - (10.0)(22.0)$$

\Downarrow

$$25.1n_1 + 18.2n_2 + 75.6n_3 = 4574 \qquad\qquad (3)$$

Solving Equations 1 through 3 simultaneously yields

$$n_1 = 59.0 \text{ mol } C_2H_5OH(g)$$
$$n_2 = 41.0 \text{ mol } CH_3CHO(g)$$
$$n_3 = 31.0 \text{ mol } H_2(g)$$

The fractional conversion of ethanol is

$$x = \frac{(n_{C_2H_5OH})_{in} - (n_{C_2H_5OH})_{out}}{(n_{C_2H_5OH})_{in}} = \frac{90.0 - 59.0}{90.0} = \boxed{0.344}$$

9.5c Thermochemistry of Solutions[3]

The enthalpy change associated with the formation of a solution from the solute elements and the solvent at 25°C is called the **standard heat of formation of the solution.** If a solution contains n moles of solvent per mole of solute, then

$$\boxed{(\Delta \hat{H}_f^\circ)_{solution} = (\Delta \hat{H}_f^\circ)_{solute} + \Delta \hat{H}_s^\circ(n)} \qquad\qquad (9.5\text{-}4)$$

where $\Delta \hat{H}_s^\circ$ is the heat of solution at 25°C (Section 8.5). From the definitions of $\Delta \hat{H}_f^\circ$ and $\Delta \hat{H}_s^\circ$, the dimensions of the heat of formation of the solution are (energy)/(mole of solute).

The standard heat of a reaction involving solutions may be calculated from heats of formation of the solutions. For example, for the reaction

$$2\,HNO_3(aq, n = 25) + Ca(OH)_2(aq, n = \infty) \rightarrow Ca(NO_3)_2(aq, n = \infty) + 2\,H_2O(l)$$

the standard heat of reaction is

$$\Delta \hat{H}_r^\circ = (\Delta \hat{H}_f^\circ)_{Ca(NO_3)_2(aq)} + 2(\Delta \hat{H}_f^\circ)_{H_2O(l)} - 2(\Delta \hat{H}_f^\circ)_{HNO_3(aq,\,n=25)} - (\Delta \hat{H}_f^\circ)_{Ca(OH)_2(aq)}$$
$$= -114.2 \text{ kJ/mol}$$

The last equation signifies that if a solution containing 2 mol of HNO_3 in 50 mol of $H_2O(n = 25)$ is neutralized at 25°C with 1 mol of $Ca(OH)_2$ dissolved in enough water

[3] Reviewing Sections 8.5a and 8.5b before reading this section might prove helpful.

so that the addition of more water would not cause a measurable enthalpy change ($n = \infty$), the resulting value of ΔH is -114.2 kJ.

If a standard heat of formation is tabulated for a solution involved in a reaction (as many are in Table B.1), the tabulated value may be substituted directly into the expression for $\Delta \hat{H}_r$; otherwise, $(\Delta \hat{H}_f^\circ)_{\text{soln}}$ must first be calculated by adding the heat of formation of the pure solute to the heat of solution.

EXAMPLE 9.5-5 Standard Heat of a Neutralization Reaction

1. Calculate $\Delta \hat{H}_r^\circ$ for the reaction

$$H_3PO_4(aq, n = \infty) + 3\,NaOH(aq, n = 50) \rightarrow Na_3PO_4(aq, n = \infty) + 3\,H_2O(l)$$

2. If 5.00 mol of NaOH dissolved in 250 mol of water are neutralized completely at 25°C with dilute phosphoric acid, what is the attendant enthalpy change?

Solution

1. $H_3PO_4(aq)$: $\Delta \hat{H}_f^\circ = -309.3$ kcal/mol $= -1294$ kJ/mol (from p. 3-150 of *Perry's Chemical Engineers' Handbook*).

$$NaOH(aq, n = 50)\text{: } \Delta \hat{H}_f^\circ = (\Delta \hat{H}_f^\circ)_{NaOH(s)} + \Delta \hat{H}_s^\circ(n = 50)$$

$\left\Vert\begin{array}{l}\text{Table B.1 } (\Delta \hat{H}_f^\circ) \\ \text{Table 8.5-1 } (\Delta \hat{H}_s^\circ)\end{array}\right.$

$$= (-426.6 - 42.51)\text{ kJ/mol} = -469.1\text{ kJ/mol}$$

$Na_3PO_4(aq)$: $\Delta \hat{H}_f^\circ = -471.9$ kcal/mol $= -1974$ kJ/mol (from p. 3-153 of *Perry's Chemical Engineers' Handbook*).

$H_2O(l)$: $\Delta \hat{H}_f^\circ = -285.8$ kJ/mol (from Table B.1)

$$\Delta \hat{H}_r^\circ = (\Delta \hat{H}_f^\circ)_{Na_3PO_4(aq)} + 3(\Delta \hat{H}_f^\circ)_{H_2O(l)}$$
$$- (\Delta \hat{H}_f^\circ)_{H_3PO_4(aq)} - 3(\Delta \hat{H}_f^\circ)_{NaOH(aq,\, n=50)}$$
$$= \boxed{-130.1\text{ kJ/mol}}$$

2. If 5 mol of dissolved NaOH are neutralized, then

$$\Delta H(25°C) = \frac{-130.1\text{ kJ}}{3.00\text{ mol NaOH}}\left|\frac{5.00\text{ mol NaOH}}{}\right. = \boxed{-217\text{ kJ}}$$

When you calculate ΔH for a reactive process as

$$\sum_{\text{products}} n_i \hat{H}_i - \sum_{\text{reactants}} n_i \hat{H}_i$$

and one of the reactant or product species is a solution, its specific enthalpy usually has the dimensions (energy/mole of solute), so that the corresponding value of n_i must be moles or molar flow rate of the *solute*, and not of the total solution. A complicating factor may be that while the heat of formation of a solution is always obtained in the desired units, solution heat capacities are normally based on a unit mass of the

total solution rather than of the solute. To calculate the specific enthalpy of a solution at a temperature T in (energy)/(mole of solute), you must first calculate m, the mass of solution corresponding to 1 mole of dissolved solute, and then add

$$m \int_{25°C}^{T} (C_p)_{\text{solution}} \, dT$$

to the standard heat of formation of the solution. The next example illustrates this calculation.

EXAMPLE 9.5-6　Energy Balance on a Neutralization Process

A 10 wt% aqueous solution of H_2SO_4 at 40°C is to be neutralized with a 20 wt% aqueous solution of NaOH at 25°C in a continuous reactor. At what rate in (kJ)/(kg H_2SO_4 solution) must heat be removed from the reactor if the product solution emerges at 35°C?

Solution

Basis: 1 kg H_2SO_4 solution

$$H_2SO_4(aq, 10\%) + 2\,NaOH(aq, 20\%) \rightarrow Na_2SO_4(aq) + 2\,H_2O(l)$$

This is a straightforward problem, but the number of intermediate calculations required to solve it might make it appear more difficult than it is. Let us summarize what must be done.

1. Solve for n_1, n_2, and n_3 by material balances.
2. Calculate the solvent-to-solute mole ratios of all solutions. (These quantities are needed to determine the solution enthalpies from tabulated heats of solution.)
3. Calculate the enthalpies of the solutions. (This will require additional composition calculations to enable the use of tabulated solution heat capacities.)
4. Write the energy balance equation, and solve it for the heat removal rate.

Observe that nothing here is really new, and, as we work our way through to the final result, recognize that most of the calculations are simply conversions of solution compositions from mass fractions to mole ratios back to mass ratios—conversions required by the nature of the available data for properties of solutions.

1. *Solve for n_1, n_2, and n_3 by material balances, and calculate the amount of water formed.*

 S balance:

$$\frac{100 \text{ g H}_2\text{SO}_4 \ \bigg| \ 32.0 \text{ g S}}{\ \bigg| \ 98.1 \text{ g H}_2\text{SO}_4} = \frac{n_2(\text{g Na}_2\text{SO}_4) \ \bigg| \ 32.0 \text{ g S}}{\ \bigg| \ 142 \text{ g Na}_2\text{SO}_4}$$

$$\Downarrow$$

$$n_2 = 145 \text{ g Na}_2\text{SO}_4$$

 Na balance: $(0.200n_1)(23/40) = (145)(46/142) \Rightarrow n_1 = 408 \text{ g NaOH(aq)}$
 Total mass balance: $1000 + 408 = 145 + n_3 \Rightarrow n_3 = 1263 \text{ g H}_2\text{O(l)}$
 Mass of product solution: $m = n_2 + n_3 = 1408 \text{ g}$

$$\text{Water formed by reaction} = \frac{145 \text{ g Na}_2\text{SO}_4 \text{ formed} \ \bigg| \ 1 \text{ mol} \ \bigg| \ 2 \text{ mol H}_2\text{O}}{\ \bigg| \ 142 \text{ g} \ \bigg| \ 1 \text{ mol Na}_2\text{SO}_4}$$

$$= 2.04 \text{ mol H}_2\text{O}$$

2. *Calculate solvent/solute mole ratios* (needed to determine heats of solution).

 $H_2SO_4(aq)$:

$$(900 \text{ g H}_2\text{O})/(18.0 \text{ g/mol}) = 50.0 \text{ mol H}_2\text{O}$$
$$(100 \text{ g H}_2\text{SO}_4)/(98.1 \text{ g/mol}) = 1.02 \text{ mol H}_2\text{SO}_4$$
$$\Downarrow$$
$$n = 50.0/1.02 = 49.0 \text{ mol H}_2\text{O/mol H}_2\text{SO}_4$$

 NaOH(aq):

$$(0.8 \times 408 \text{ g H}_2\text{O})/(18 \text{ g/mol}) = 18.1 \text{ mol H}_2\text{O}$$
$$(0.2 \times 408 \text{ g NaOH})/(40 \text{ g/mol}) = 2.04 \text{ mol NaOH}$$
$$\Downarrow$$
$$n = 18.1/2.04 = 8.90 \text{ mol H}_2\text{O/mol NaOH}$$

 $Na_2SO_4(aq)$:

$$(1263 \text{ g H}_2\text{O})/(18.0 \text{ g/mol}) = 70.2 \text{ mol H}_2\text{O}$$
$$(145 \text{ g Na}_2\text{SO}_4)/(142 \text{ g/mol}) = 1.02 \text{ mol Na}_2\text{SO}_4$$
$$\Downarrow$$
$$n = 70.2/1.02 = 68.8 \text{ mol H}_2\text{O/mol Na}_2\text{SO}_4$$

3. *Calculate ΔH.*

 This problem is made tricky by the fact that water is not just the solvent in the solutions involved, but is also formed as a reaction product. We will take as references the reactant and product solutions and evaluate ΔH as

$$\Delta H = n_A \Delta \hat{H}_r^\circ / v_A + \sum_{\text{out}} n\hat{H} + \sum_{\text{in}} n\hat{H}$$

where as usual n_A is the amount of species A reacted or formed by reaction, and v_A is the stoichiometric coefficient of that species. If we choose A to be H_2SO_4, then $n_A = 1.02$ mol and $v_A = 1$.

As indicated in Section 8.5b, it is convenient in solution chemistry calculations to tabulate the products $n\hat{H}$ rather than n and \hat{H} separately. The completed enthalpy table is shown below, followed by the calculations that led to the entries shown.

References: H_2SO_4(aq, $n = 49$), NaOH(aq, $n = 8.9$), Na_2SO_4(aq, $n = 69$) *at* 25°C

Substance	$n_{in}\hat{H}_{in}$	$n_{out}\hat{H}_{out}$	
H_2SO_4(aq)	57.8	—	
NaOH(aq)	0	—	$n\hat{H}$ in kJ
Na_2SO_4(aq)	—	58.9	

H_2SO_4(aq, $n = 49$, 40°C):

$$n\hat{H} = m \int_{25}^{40} C_p\, dT$$

$$= \frac{1000\text{ g}}{} \left| \frac{3.85\text{ J}}{\text{g}\cdot°\text{C}} \right| \frac{(40 - 25)°\text{C}}{} \left| \frac{1}{1000\text{ J}} \frac{\text{kJ}}{} \right. = 57.8\text{ kJ}$$

NaOH(aq, $n = 8.9$, 25°C): $n\hat{H} = 0$

Na_2SO_4(aq, $n = 69$, 35°C):

In the absence of better information, we will assume that the heat capacity of the solution is that of pure water, 4.184 J/g·°C.

$$n\hat{H} = m \int_{25}^{35} C_p\, dT$$

$$= \frac{1408\text{ g}}{} \left| \frac{4.184\text{ J}}{\text{g}\cdot°\text{C}} \right| \frac{(35 - 25)°\text{C}}{} \left| \frac{1}{1000\text{ J}} \frac{\text{kJ}}{} \right. = 58.9\text{ kJ}$$

The heats of formation of H_2SO_4(l) and NaOH(c) are given in Table B.1, and the heats of solution of these species are given in Table 8.5-1. *Perry's Chemical Engineers' Handbook* on p. 3-153 gives the standard heat of formation of Na_2SO_4(aq, $n = 1100$) as -330.82 kcal/mol $= -1384$ kJ/mol Na_2SO_4. In the absence of heat of solution data, we will assume that this value also applies to the solution for which $n = 69$ moles of water per mole of solute. The standard heats of formation of the species involved in the reaction

$$H_2SO_4(\text{aq}, n = 49) + 2\,\text{NaOH(aq}, n = 8.9) \rightarrow Na_2SO_4(\text{aq}) + 2\,H_2O(\text{l})$$

are obtained from Equation 9.5-4 (heat of formation of the solution equals heat of formation of the solute plus heat of solution) as

$$H_2SO_4(\text{aq}): \Delta\hat{H}_f° = [(-811.3) + (-73.3)] = -884.6\text{ kJ/mol } H_2SO_4$$
$$\text{NaOH(aq)}: \Delta\hat{H}_f° = [(-426.6) + (-41.5)] = -468.1\text{ kJ/mol NaOH}$$
$$Na_2SO_4(\text{aq}): \Delta\hat{H}_f° = -1384\text{ kJ/mol } Na_2SO_4$$
$$H_2O(\text{l}): \Delta\hat{H}_f° = -285.84\text{ kJ/mol } H_2O$$

and the standard heat of reaction is therefore

$$\Delta \hat{H}_r^\circ = (-1384)(1) + (-285.84)(2) - (-884.6)(1) - (-468.1)(2)$$
$$= -1349 \text{ kJ/mol}$$

4. *Energy balance.*

$$Q = \Delta H = (n_{H_2SO_4} \text{ reacting}) \Delta \hat{H}_r^\circ + \sum n_{out} \hat{H}_{out} - \sum n_{in} \hat{H}_{in}$$
$$= (1.02 \text{ mol})(-1349 \text{ kJ/mol}) + (58.9 - 57.8) \text{ kJ} = \boxed{-1375 \text{ kJ}}$$

When a strong acid or base is dissolved in water, it dissociates into ionic species; for example, dissolved NaOH exists as Na^+ and OH^- in a dilute solution. **Heats of formation of ions** may be determined from heats of solutions of such substances and may be used to calculate heats of formation of dilute solutions of highly dissociated materials. A good discussion of this topic and a table of heats of formation of ions is given by Hougen, Watson, and Ragatz.[4]

TEST YOURSELF

The heat of formation of A(s) is $(\Delta \hat{H}_f^\circ)_A = -100$ kJ/mol; heats of solution of A in a solvent B are $\Delta \hat{H}_s^\circ(n = 50 \text{ mol B/mol A}) = -10$ kJ/mol and $\Delta \hat{H}_s^\circ(n = \infty) = -15$ kJ/mol.

1. (a) What is the standard heat of formation of A(soln, $n = 50$) relative to B and the elements of A?
 (b) What is $\Delta \hat{H}_f^\circ$ for A(soln, $n = \infty$) relative to the same reference state?
2. (a) What is the enthalpy (kJ/mol A) of a solution of A in B at 25°C for which $n = 50$, relative to the elements of A at 25°C?
 (b) What is the enthalpy (kJ) of a solution containing 5 mol of A in 250 mol of B?

9.6 FUELS AND COMBUSTION

The use of heat generated by a combustion reaction to produce steam, which drives turbines to produce electricity, is presently the single most important commercial application of chemical reactions.

Most electrical power generated in the United States derives from the combustion of petroleum derivatives or natural gas. Unfortunately, the reserves of these materials in this country are finite, and at the present rate of domestic consumption they might run out in the next century. On the other hand, domestic coal reserves are sufficient to supply the nation's energy requirements for as much as 500 years, by some estimates.

[4] O. A. Hougen, K. M. Watson, and R. A. Ragatz, *Chemical Process Principles*, Part I, Wiley, New York, 1954, pp. 315-317.

The analysis of fuels and combustion reactions and reactors has always been an important activity for chemical engineers. It is critically important now in view of the need to develop a power generation technology based on available energy resources instead of fuels whose future availability is increasingly doubtful.

In this section, we review the properties of the fuels most often used for power generation, outline techniques for energy balances on combustion reactors, and discuss qualitatively various technical aspects of combustion reactions.

9.6a Fuels and Their Properties

Fuels burned in power plant furnaces may be solids, liquids, or gases. Some of the more common fuels are:

Solid Fuels: Principally coal (a mixture of carbon, water, noncombustible ash, hydrocarbons, and sulfur), coke (primarily carbon—the solid residue left after coal or petroleum is heated, driving off volatile substances and decomposing hydrocarbons), and to a small extent wood and solid waste (garbage).

Liquid Fuels: Principally hydrocarbons obtained by distilling crude oil (petroleum); also coal tars and shale oil. Considerable effort is being expended to extract shale oil from the plentiful shale deposits in the western United States. Extraction is possible now—the problem is to make it economical. There is also a growing interest in the use of alcohols obtained by fermenting grains (see Chapter 14) and in liquid fuels obtained from the catalytic liquefaction of coal.

Gaseous Fuels: Principally natural gas (80 to 95% methane, the balance ethane, propane, and small quantities of other gases); also light hydrocarbons obtained from petroleum or coal treatment (see Chapter 13), acetylene, and hydrogen (the latter two are relatively expensive to produce).

QUESTIONS FOR DISCUSSION

Coal contains primarily carbon and combustible hydrocarbons, but also contains substantial amounts of noncombustible ash, and as much as 5% sulfur by weight.

1. What becomes of the sulfur when coal is burned? What about the ash? (Suggest two possibilities for the ash.)
2. In view of the answer to the preceding question, why might coal be less desirable as a fuel than natural gas?
3. What might prompt a power company to use coal as its primary fuel despite its drawbacks relative to liquid or gaseous fuels?

The **heating value** (or **calorific value**) of a combustible material is the negative of the standard heat of combustion, usually expressed per unit mass of the material.

The **higher heating value** (or **total heating value** or **gross heating value**) is $-\Delta \hat{H}_c^\circ$ with $H_2O(l)$ as a combustion product, and the **lower heating value** (or **net heating value**) is the value based on $H_2O(v)$ as a product. Since $\Delta \hat{H}_c^\circ$ is always negative, the heating value is positive.

To calculate a lower heating value of a fuel from a higher heating value or vice versa, you must calculate the moles of water produced when a unit mass of the fuel is burned. If this quantity is designated n, then

$$HHV = LHV + n\,\Delta \hat{H}_v(H_2O, 25°C) \qquad (9.6\text{-}1)$$

(Try to prove this relationship from the definitions of LHV and HHV and Hess's law.) The heat of vaporization of water at 25°C is

$$\Delta \hat{H}_v(H_2O, 25°C) = 44.013 \text{ kJ/mol} \qquad (9.6\text{-}2a)$$
$$= 18\,934 \text{ Btu/lb-mole} \qquad (9.6\text{-}2b)$$

If a fuel contains a mixture of combustible substances, its heating value is

$$HV = \sum x_i(HV)_i \qquad (9.6\text{-}3)$$

where $(HV)_i$ is the heating value of the ith combustible substance. If the heating values are expressed in units of (energy)/(mass), then the x_i's are the mass fractions of the fuel components, while if the dimensions of the heating values are (energy)/(mole) then the x_i's are mole fractions.

EXAMPLE 9.6-1 Calculation of a Heating Value

A natural gas contains 85% methane and 15% ethane by volume. Calculate the higher heating value of this fuel in kJ/g from the standard heats of combustion of methane and ethane given below:

$$CH_4(g) + 2\,O_2(g) \rightarrow CO_2(g) + 2\,H_2O(g): \Delta \hat{H}_c^\circ = -802 \text{ kJ/mol}$$
$$C_2H_6(g) + \tfrac{7}{2}\,O_2(g) \rightarrow 2\,CO_2(g) + 3\,H_2O(g): \Delta \hat{H}_c^\circ = -1428 \text{ kJ/mol}$$

Solution

Since the heating value per unit mass of the fuel is desired, we will first calculate the composition on a mass basis.

$$1 \text{ mol fuel} \Rightarrow \begin{array}{l} 0.85 \text{ mol } CH_4 \Rightarrow 13.6 \text{ g } CH_4 \\ 0.15 \text{ mol } C_2H_6 \Rightarrow \underline{4.5 \text{ g } C_2H_6} \\ \phantom{0.15 \text{ mol } C_2H_6 \Rightarrow} 18.1 \text{ g total} \end{array}$$

Thus

$$x_{CH_4} = 13.6/18.1 = 0.751 \text{ g } CH_4/\text{g fuel}$$
$$x_{C_2H_6} = 1 - x_{CH_4} = 0.249 \text{ g } C_2H_6/\text{g fuel}$$

The higher heating values of the components are calculated from the given heats of combustion (which are the negatives of the lower heating values) as follows.

$$(HHV)_{CH_4} = (LHV)_{CH_4} + n_{H_2O}(\Delta\hat{H}_v)_{H_2O}$$

$$= \left[802 \frac{kJ}{mol\ CH_4} + \frac{2\ mol\ H_2O}{mol\ CH_4}\left(44.013\ \frac{kJ}{mol}\right)\right] \frac{1\ mol}{16\ g\ CH_4}$$

$$= 55.6\ kJ/g$$

$$(HHV)_{C_2H_6} = \left[1428 \frac{kJ}{mol\ C_2H_6} + 3\frac{mol\ H_2O}{mol\ C_2H_6}\left(44.013\ \frac{kJ}{mol}\right)\right] \frac{1\ mol}{30\ g\ C_2H_6}$$

$$= 52.0\ kJ/g$$

The higher heating value of the mixture is from Equation 9.6-3

$$HHV = x_{CH_4}(HHV)_{CH_4} + x_{C_2H_6}(HHV)_{C_2H_6}$$

$$= (0.751)(55.6) + (0.249)(52.0) = \boxed{54.7\ kJ/g}$$

Higher heating values for common solid, liquid, and gaseous fuels are tabulated in Section 9 of *Perry's Chemical Engineers' Handbook.* Representative values are given in Table 9.6-1. From the standpoint of heating value per unit mass, hydrogen is clearly the best fuel; however, it does not occur naturally in any appreciable quantities and the cost of producing it makes it less economical than the other fuels in Table 9.6-1.

TABLE 9.6-1

Typical Heating Values of Common Fuels

Fuel	Higher Heating Value	
	kJ/g	Btu/lb$_m$
Wood	17	7 700
Soft coal	23	10 000
Hard coal	35	15 000
Fuel oil, gasoline	44	19 000
Natural gas	54	23 000
Hydrogen	143	61 000

TEST YOURSELF

1. The standard heat of the reaction

$$n\text{-}C_4H_{10}(g) + \tfrac{13}{2}O_2 \rightarrow 4\ CO_2 + 5\ H_2O(g)$$

is -2658 kJ/mol. What is the lower heating value per mol of *n*-butane? The higher heating value?

2. A gas mixture contains 40% $H_2 (HHV = 143 \text{ kJ/g})$ and 60% CH_4 $(HHV = 55 \text{ kJ/g})$ by mass. Calculate the heating value of this mixture in kJ/g.

3. In 1986, bituminous coal cost roughly $150 per ton. What would natural gas have to cost ($/ton) to be as economical as coal on a $/Btu basis? (Use Table 9.6-1.)

9.6b Adiabatic Flame Temperature

When a fuel is burned, a considerable amount of energy is released. Some of this energy is transferred as heat through the reactor walls, and the remainder raises the temperature of the reaction products; the less heat transferred, the higher the product temperature. The highest achievable temperature is reached if the reactor is adiabatic. This temperature is called the **adiabatic flame temperature** (or **theoretical flame temperature.**)

The calculation of an adiabatic flame temperature follows the general procedure outlined in Section 9.5b. Unknown stream amounts or flow rates are first determined by material balances. Reference conditions are chosen, enthalpies of feed components are calculated, and enthalpies of product components are determined in terms of the product temperature T. Finally, ΔH for the process is evaluated and substituted into the energy balance equation ($\Delta H = 0$ for an adiabatic reactor), which is solved for T.

If n_f moles of a fuel are burned in an adiabatic reactor, the energy balance equation (Equation 9.5-2) becomes

$$\Delta H = n_f \Delta \hat{H}_c^\circ + \sum_{\text{output}} n_i \hat{H}_i(T_{ad}) - \sum_{\text{input}} n_i \hat{H}_i(T_{feed}) = 0$$

$$\Downarrow$$

$$\sum_{\text{output}} n_i \hat{H}_i(T_{ad}) = -n_f \Delta \hat{H}_c^\circ + \sum_{\text{input}} n_i \hat{H}_i(T_{feed}) \qquad (9.6\text{-}4)$$

where

$\Delta \hat{H}_c^\circ$ = heat of combustion of the fuel at 25°C

n_i = moles of the ith component in the feed or product

\hat{H}_i = specific enthalpy of the ith component relative to this component at 25°C in the state of aggregation for which $\Delta \hat{H}_c^\circ$ is specified

T_{ad} = product temperature, which by definition is the adiabatic flame temperature

The left side of Equation 9.6-4 is an algebraic expression in T_{ad}, whose form depends on how \hat{H}_i is evaluated. There are two ways to proceed with this calculation:

1. *Integrate tabulated heat capacity formulas from 25°C to T_{ad}.*

If the heat capacities are constants, a linear equation results (see Example 9.5-3); if C_p has the form $a + bT$, the equation is quadratic in T, and can be solved using the quadratic formula. However, if any heat capacity formula is second order or

higher, the energy balance equation is at least third order in T, and it must be solved by a trial-and-error technique such as Newton's rule (Appendix A.2).

2. *Use mean heat capacities from Table 8.3-1, and find by trial and error the value of T_{ad} that satisfies Equation 9.6-4.*

The next example illustrates the first of these approaches.

EXAMPLE 9.6-2 Calculation of an Adiabatic Flame Temperature

Liquid methanol is to be burned with 100% excess air. The engineer designing the furnace must calculate the highest temperature that the furnace walls will have to withstand so that an appropriate material of construction can be chosen. Perform this calculation, assuming that the methanol is fed at 25°C and the air enters at 100°C.

Solution

Basis: 1 mol CH₃OH

Assume complete combustion. From Table B.1,

$$CH_3OH(l) + \tfrac{3}{2}O_2 \rightarrow CO_2 + 2\,H_2O(l): \quad \Delta\hat{H}_c(25°C) = -726.6 \text{ kJ/mol}$$

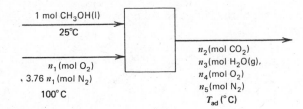

Calculate Component Amounts

$$(O_2)_{\text{theoretical}} = 1.5 \text{ mol}$$

$$n_1 = (2)(1.5 \text{ mol}) = 3 \text{ mol } O_2 \text{ fed}$$

$$\Downarrow$$

$$(3.76)(3) = 11.28 \text{ mol } N_2 \text{ fed}$$

Material balances yield

$$n_2 = 1.0 \text{ mol } CO_2$$
$$n_3 = 2.0 \text{ mol } H_2O$$
$$n_4 = 1.5 \text{ mol } O_2$$
$$n_5 = 11.28 \text{ mol } N_2$$

Calculate Feed Component Enthalpies

References: $CH_3OH(l)$, O_2, N_2 at 25°C

$CH_3OH(l, 25°C)$: $\hat{H} = 0$

Air (100°C):

$$0.21 \text{ mol O}_2 \xrightarrow{\Delta H = 0} 1 \text{ mol air} \to 1 \text{ mol air}$$
$$0.79 \text{ mol N}_2 \qquad\qquad 25°C \qquad\quad 100°C$$
$$25°C$$

$$\hat{H} = \bar{C}_p(100°C)[100°C - 25°C]$$
$$\Downarrow (\bar{C}_p)_{\text{air}} \text{ from Table 8.3-1}$$
$$\hat{H} = 2191 \text{ J/mol} = 2.191 \text{ kJ/mol}$$

Evaluate the Right Side of Equation 9.6-4

$$-n_f \Delta \hat{H}_c° + \sum_{\text{input}} n_i \hat{H}_i = -(1.00 \text{ mol CH}_3\text{OH})\left(-726.6 \frac{\text{kJ}}{\text{mol}}\right)$$

$$+(1.00 \text{ mol CH}_3\text{OH})\left(0 \frac{\text{kJ}}{\text{mol}}\right) + (14.28 \text{ mol air})\left(2.191\frac{\text{kJ}}{\text{mol}}\right)$$

$$= 757.9 \text{ kJ}$$

Calculate Enthalpy of Product Stream

References: $CO_2(g)$, $O_2(g)$, $N_2(g)$, and $H_2O(l)$ at 25°C. (Liquid water is assumed since this is the state for which $\Delta \hat{H}_c$ is known.)

From Table B.2, the heat capacities of the product gases in J/mol·°C in terms of $T(°C)$ are

$$(C_p)_{\text{CO}_2} = 36.11 + 4.233 \times 10^{-2}T - 2.887 \times 10^{-5}T^2 + 7.464 \times 10^{-9}T^3$$
$$(C_p)_{\text{H}_2\text{O(g)}} = 33.46 + 0.688 \times 10^{-2}T + 0.7604 \times 10^{-5}T^2 - 3.593 \times 10^{-9}T^3$$
$$(C_p)_{\text{O}_2} = 29.10 + 1.158 \times 10^{-2}T - 0.6076 \times 10^{-5}T^2 + 1.311 \times 10^{-9}T^3$$
$$(C_p)_{\text{N}_2} = 29.00 + 0.2199 \times 10^{-2}T + 0.5723 \times 10^{-5}T^2 - 2.871 \times 10^{-9}T^3$$

We could integrate each of these formulas from the reference temperature of 25°C to the unknown T_{ad} to calculate the \hat{H}_i's, and then substitute in $\sum n_i \hat{H}_i$; however, we can save some calculation time by summing before integrating. Substituting the values of n_i calculated previously and the tabulated heat capacities, and recalling that to calculate $(\hat{H})_{\text{H}_2\text{O(g)}}$ we must first vaporize the liquid at 25°C (using Equation 9.6-2a for $\Delta \hat{H}_v$), we obtain

$$\sum n_i C_{pi} = 473.8 + 9.826 \times 10^{-2}T + 4.178 \times 10^{-5}T^2 - 30.14 \times 10^{-9}T^3$$
$$\Downarrow$$
$$\sum_{\text{output}} n_i \hat{H}_i = n_3(\Delta \hat{H}_v)_{\text{H}_2\text{O}} + \int_{25}^{T_{\text{ad}}} \left(\sum n_i C_{pi}\right) dT$$

$$= 88026 + 473.8 T_{\text{ad}} + 4.913 \times 10^{-2}T_{\text{ad}}^2 + 1.393 \times 10^{-5}T_{\text{ad}}^3$$
$$-7.535 \times 10^{-9}T_{\text{ad}}^4 - 11\,845$$
$$\Downarrow \div 1000 \text{ J/kJ}$$
$$\sum_{\text{output}} n_i \hat{H}_i(\text{kJ}) = 0.4738 T_{\text{ad}} + 4.913 \times 10^{-5}T_{\text{ad}}^2 + 1.393 \times 10^{-8}T_{\text{ad}}^3$$

$$-7.535 \times 10^{-12}T_{\text{ad}}^4 + 76.18$$

Energy Balance

From Equation 9.6-4

$$\sum_{\text{output}} n_i \hat{H}_i = -n_f \Delta \hat{H}_c^\circ + \sum_{\text{input}} n_i \hat{H}_i = 757.9 \text{ kJ}$$

$$\Downarrow$$

$$7.535 \times 10^{-12} T_{\text{ad}}^4 - 1.393 \times 10^{-8} T_{\text{ad}}^3 - 4.913 \times 10^{-5} T_{\text{ad}}^2 - 0.4738 T_{\text{ad}} + 681.7 = 0$$

This quartic equation may be solved using Newton's rule (Appendix A.2) or by another trial-and-error procedure. A good first guess is obtained by dropping the quartic, cubic, and quadratic terms in the equation and solving the resulting linear equation ($-0.4738 T_{\text{ad}} + 681.8 = 0$) to estimate $T_{\text{ad}} \approx 1439°C$. The exact solution of the quartic equation is then eventually obtained as

$$\boxed{T_{\text{ad}} = 1256°C}$$

The furnace walls will therefore never be exposed to a temperature greater than 1256°C as long as the composition and temperature of the feed remain the same.

The adiabatic flame temperature is greatest when the fuel and air are fed in approximately stoichiometric proportions, and is much greater when pure oxygen rather than air is fed to the reactor.

TEST YOURSELF

1. What is the adiabatic flame temperature of a fuel?
2. Suppose T_{ad} is the adiabatic flame temperature calculated for a given fuel + air feed to a furnace. Give two reasons why the actual furnace temperature might be less than T_{ad}.
3. Why should the adiabatic flame temperature be much higher for a pure oxygen feed than for an air feed?

9.6c Flammability and Ignition

In this section and the one that follows, we discuss qualitatively what happens during the rapid chemical reaction between a fuel and oxygen. Along the way, we provide answers to the following questions:

1. What is a flame? Why are some flames blue and some yellow?
2. If you light a match in a mixture of methane and air that contains 10% CH_4 by volume, the mixture will burn explosively, but if the mixture contains 20% CH_4 nothing will happen. Why?

3. What is an explosion? What is the loud noise you hear when something explodes?

4. Hydrogen and oxygen react explosively to form water, yet if you mix these two gases in a flask, nothing happens. Why not?

We have so far in this text only considered the initial and final conditions in a chemical reactor, and not how long it may have taken to get from one to the other. When you study **chemical reaction kinetics,** you will learn that the rate of a reaction depends strongly on the reaction temperature; for many reactions, a temperature rise of only 10°C is enough to double the rate.

Suppose a mixture of methane and air containing 10 mole% CH_4 is heated by a central heat source (e.g., an electrical coil) at atmospheric pressure, beginning at room temperature. Although methane reacts with oxygen

$$CH_4 + 2\,O_2 \rightarrow CO_2 + 2\,H_2O$$

the reaction proceeds at an immeasurably low rate at room temperature, and to an observer nothing would seem to be happening in the reactor.

As the temperature increases, the rate of the oxidation reaction also increases, and measurable amounts of CO_2 and H_2O appear. However, if the heat source is turned off, the reactor temperature drops again—the rate at which heat is generated by the reaction alone is not enough to compensate for the rate at which heat is lost from the reaction zone.

However, if the temperature at any point in the reactor reaches about 640°C or higher, the rate of heat generation by the reaction exceeds the rate of heat loss from the reaction zone. The gas adjacent to this zone is then heated above 640°C, causing the zone of rapid reaction to spread. The temperature of the gas rapidly rises by several hundred or even a thousand degrees in a fraction of a second; even if the heating source is turned off, the rate of heat generation by the now rapidly occurring reaction is enough to maintain the system at its high temperature until the reactants are exhausted.

Combustion is defined as a rapid, high-temperature oxidation reaction. What happens in the reactor just described after the reaction rate accelerates dramatically is combustion, whereas the initial slow oxidation reaction between methane and oxygen to form CO_2 and H_2O, and other reactions between these species, such as the formation reaction of formaldehyde

$$CH_4 + O_2 \rightarrow HCHO + H_2O$$

are not combustion reactions.

The phenomenon whereby the rate of an oxidation reaction increases abruptly when the reaction mixture exceeds a certain temperature is called **ignition;** the temperature at which this occurs is called the **ignition temperature,** and the time between the instant when the mixture reaches the ignition temperature and the moment of ignition is the **ignition lag.** The ignition temperature and lag are shown here on a representative plot of the temperature of a fuel mixture that is being heated.

The value of the ignition temperature depends on a number of things for a given fuel, including the fuel-to-air ratio, the total pressure in the reactor, and even the reactor geometry. For any given fuel, there is a lower limit to this quantity called the **spontaneous ignition temperature.** Representative values of this quantity for stoichiometric fuel–air mixtures at 1 atm are 571°C for H_2, 633°C for CH_4, and 473°C for C_2H_6.

Ignition lags are typically 0.1–10 s in duration and decrease with increasing temperature above the spontaneous ignition temperature. Figures 9-7 and 9-8 on p. 9-17 of the fifth edition of the *Chemical Engineers' Handbook*[5] show how the ignition lag varies with temperature, pressure, and fuel-to-air ratio for several hydrocarbon gases.

We saw in Section 9.6b that the highest attainable temperature in a combustion reaction—the adiabatic flame temperature—depends on the fuel-to-air ratio, and we stated but did not prove that this upper temperature limit is a maximum when the fuel and oxygen are present in nearly stoichiometric proportion. If the mixture is either **rich** (fuel in excess) or **lean** (O_2 in excess), the adiabatic flame temperature decreases.

There exist two values of the mole percent of fuel in a reaction mixture—the **lower** or **lean flammability limit** and the **upper** or **rich flammability limit**—for which the adiabatic flame temperature equals the ignition temperature of the mixture. *A fuel–air mixture whose composition falls outside these limits is incapable of igniting or exploding, even if exposed to a spark or flame.* The composition range between the two flammability limits is called the **explosive range** of the mixture.

For example, the stoichiometric percentage of methane in a methane–air mixture is 9.5 mole%. (Prove it.) Experimentally, it is found that the lower flammability limit of CH_4–air mixtures at 1 atm is roughly 5% CH_4, and the upper flammability limit is approximately 15% CH_4. Thus, a CH_4–air mixture containing between 5% CH_4 and 15% CH_4 must be considered a fire or explosion hazard, while a mixture containing 3% CH_4 may be considered safe, and a mixture containing 18% CH_4 may also be considered safe as long as it is not brought into contact with additional oxygen.

Flammability limits of a number of hydrocarbon-air mixtures are listed in Table 9-20 on p. 9-19 of the *Chemical Engineers' Handbook.*[5] The given values apply to an initial temperature of roughly 25°C and a pressure of 1 atm.

[5] R. H. Perry and C. H. Chilton, eds., *Chemical Engineers' Handbook*, 5th Edition, McGraw-Hill, New York, 1973.

EXAMPLE 9.6-3 Ignition Temperature and Flammability Limits

Propane gas and air are to be mixed and fed to a combustion reactor. The combustion is to be initiated with a gas torch. Determine the minimum and maximum percentages of propane in the feed to the reactor and the minimum required temperature of the torch flame.

Solution

From Table 9-20 of the *Chemical Engineers' Handbook*,[5] a stoichiometric mixture of C_3H_8 in air contains 4.02% C_3H_8 by volume, and the lower and upper (lean and rich) flammability limits are 51% and 283% of stoichiometric, respectively. Therefore

Minimum mole% C_3H_8 for combustion: $0.51 \times 4.02\% = \boxed{2.05\% \ C_3H_8}$

Maximum mole% C_3H_8 for combustion: $2.83 \times 4.02\% = \boxed{11.4\% \ C_3H_8}$

The torch flame temperature must be at least as high as the spontaneous ignition temperature of a propane–air mixture, which from Table 9-20 is $\boxed{940°F}$.

If a liquid (or a volatile solid) is exposed to air, the vapor given off could form a combustible mixture with the air adjacent to it, and a spark or match lit in the vicinity of the liquid could cause the mixture to ignite or explode.

The **flash point** of a liquid is the temperature at which the liquid gives off enough vapor to form an ignitable mixture with the air above the liquid surface. The flash point of gasoline, for example, is roughly $-42°C$, and that of ethanol is $13°C$, so that these liquids constitute fire hazards at room temperature, while the flash points of fuel oils vary from $38°C$ to $55°C$, making the hazards associated with these materials considerably less.

TEST YOURSELF

1. Briefly define the following terms: (a) ignition, (b) ignition temperature, (c) ignition lag, (d) flammability limits of a fuel–air mixture, (e) flash point of a liquid.

2. The flammability limits of methane–air mixtures are 5% CH_4 and 14% CH_4 at 1 atm, and the spontaneous ignition temperature is $633°C$.
 (a) What would happen if a spark were struck in a methane–air mixture containing 10% CH_4? What about a mixture containing 20% CH_4?
 (b) If a methane–air mixture containing 20% CH_4 were heated to $700°C$, would the combustion reaction take place? What would happen if the heat source were turned off?
 (c) Pure methane is clearly not within the explosive range of methane–air mixtures, yet if pure methane is blown out of a cylinder into a room and a match is lit nearby, a flame is observed that persists after the match is withdrawn. How is this possible?

9.6d Flames and Detonations

Suppose a combustible gas–air mixture is contained in an open-ended tube, and a match or another ignition source is applied to one end of the tube. The gas mixture at this end is heated and eventually ignites. The intense heat generated by the combustion reaction raises the chemical species formed during the reaction to high energy states. When these species return to lower energy states, some of the energy they lose is given off in the form of light. The result is a visible **flame** accompanying the combustion.

Initially the flame is located at the end of the tube that was ignited. However, the heat of combustion quickly raises the adjacent unburned gas to its ignition point, causing the flame to "travel" toward the other end of the tube. At some point, the tube appears as follows.

The flame front moves in the direction of the unburned gases at a velocity called the **flame velocity,** which typically has a value of 0.3 to 1 m/s. The exact value of the flame velocity depends on a number of things, including the type of fuel, fuel-to-air ratio, initial temperature and pressure of the unburned gases, and the geometry of the combustion chamber. Table 9-20 on p. 9-19 of the *Chemical Engineers' Handbook*[5] gives maximum flame velocities for a number of combustible substances in air.

Suppose now that instead of being stationary in the tube, the combustion mixture is fed continuously into the bottom (as in a Bunsen burner), and the top is ignited. If the velocity with which the gases leave the tube equals the velocity with which the flame would travel down in a stationary gas, a **stationary flame** is achieved at the top. The wall of the tube lowers the flame velocity, so that the flame burns at the end but does not penetrate into the tube. (See figure on next page.)

If the gas flow rate is increased, the flame size and rate of heat generation both increase, since a larger quantity of gas is being burned. However, once the flow rate reaches a critical value, the flame can no longer travel back as fast as the combustion region is transported away from the burner. The gases in the combustion region become increasingly diluted with air, until the region finally falls outside the flammability limits and the fire is literally blown out.

On the other hand, if the gas flow rate to the burner tube is decreased, the gas velocity in the tube may become lower than the flame propagation velocity in the tube. The result is **flashback**—the flame travels back through the tube toward the fuel source. Flashback is extremely dangerous, and any flow system involving combustible gases must be designed to guarantee that the flow rate stays above the flame propagation velocity.

When combustion of a well-mixed fuel–air mixture occurs, the fuel rapidly reacts with oxygen to form a number of unstable intermediate species (such as oxygen and hydrogen atoms, and OH and HO_2 radicals), which then proceed through a complicated chain mechanism to form CO_2 and H_2O. Some of these species undergo transitions that cause them to emit radiation whose wavelength falls within the blue region of the visible spectrum. The result is that the flame appears blue.

On the other hand, when the fuel and air are not well mixed (such as when a pure hydrocarbon gas is burned as it emerges from a stack and mixes with atmospheric air), the combustion proceeds relatively slowly, and some of the hydrocarbon fuel decomposes to form elementary carbon and hydrogen before oxidation takes place. The heat of reaction is sufficient to raise the temperature to a point where the carbon particles glow incandescently. A yellow flame is the result.

Suppose ignition of a gas takes place in a confined or partially confined space. The large temperature rise in the combustion region causes a rapid buildup of pressure in this region. If the combustion is fast enough and the heat of reaction is high enough, the pressure buildup can create a sharply defined high-pressure front, or **shock wave,** which travels through the gas at a velocity well in excess of the flame propagation velocity in the gas. The shock wave rapidly compresses and ignites the gas as it passes through, giving the appearance of an instantaneous combustion. This is a **detonation.**

Even after the combustion reaction that gave rise to the detonation has consumed all the available fuel, the shock wave can persist for large distances, carrying with it considerable energy. The energy of even a small shock wave is sufficient to vibrate

the eardrums of anyone near the site of the detonation, producing the bang that always accompanies an explosion. The energy of a large shock wave may be sufficient to demolish a city.

TEST YOURSELF

You are the guest lecturer at a junior high school science fair. The following questions are put to you. How would you answer them in terms an intelligent 14-year-old might understand?

1. What is a flame?
2. What is a yellow flame? A blue flame?
3. What is a detonation? What is the loud noise you hear when something explodes? What is it that knocks buildings over when dynamite is set off? How does a gun work?
4. (This one from one of the science teachers who thought they should have asked *him* to give the lecture.) You say that hydrogen and oxygen react explosively to form water. Why is it that I can mix hydrogen and oxygen in a flask and nothing will happen? (*Hint:* Reread the beginning of Section 9.6c.)

PROBLEMS

1. The standard heat of the reaction

$$4\,NH_3(g) + 5\,O_2(g) \rightarrow 4\,NO(g) + 6\,H_2O(g)$$

 is

$$\Delta\hat{H}_r^\circ(25°C) = -0.9046\ kJ/mol$$

 (a) What is $\Delta\hat{H}_r^\circ$ for

$$NH_3(g) + \tfrac{5}{4}O_2(g) \rightarrow NO(g) + \tfrac{3}{2}H_2O(g)$$

 (b) What is $\Delta\hat{H}_r^\circ$ for

$$NO(g) + \tfrac{3}{2}H_2O(g) \rightarrow NH_3(g) + \tfrac{5}{4}O_2(g)$$

 (c) Calculate the enthalpy change associated with the consumption of 170 g $NH_3(g)$ at 25°C and 1 atm. (See Example 9.1-1.)

2. The standard heat of reaction for the combustion of liquid *n*-nonane to form CO_2 and $H_2O(l)$ at 25°C and 1 atm is $\Delta\hat{H}_r^\circ = -6124\ kJ/mol$, and that for the combustion of *n*-nonane vapor at the same conditions is $\Delta\hat{H}_r^\circ = -6171\ kJ/mol$.

 (a) If 3 mol/s of liquid nonane are consumed and the reactants and products are all at 25°C, calculate the required rate of heat input or output (state which) in kilowatts. (Assume $Q = \Delta H$ for this process).

 (b) What is the physical significance of the 47 kJ/mol difference between the two given values of $\Delta\hat{H}_r^\circ$? (Continued on next page).

(c) The second value of $\Delta \hat{H}_r^\circ$ applies to *n*-nonane vapor at 25°C and 1 atm, and yet the normal boiling point of *n*-nonane is 150.6°C. Can *n*-nonane exist as a vapor at 25°C?

3. Use the enthalpy data given below to determine $\Delta \hat{H}_r^\circ$ for

$$C_6H_{14}(g) + \tfrac{19}{2} O_2(g) \rightarrow 6 CO_2(g) + 7 H_2O(g)$$

Data: For the same reaction with $C_6H_{14}(l)$ as the reactant and $H_2O(l)$ as the product, the standard heat of reaction is -1.791×10^6 Btu/lb-mole. The heat of vaporization of hexane at 77°F (25°C) is 13 550 Btu/lb-mole and that of water is 18 934 Btu/lb-mole.

4. The standard heat of the reaction

$$CaC_2(s) + 5 H_2O(l) \rightarrow CaO(s) + 2 CO_2(g) + 5 H_2(g)$$

is $\Delta \hat{H}_r^\circ = +69.36$ kJ/mol.
(a) Calculate $\Delta \hat{U}_r^\circ$ for this reaction. (See Example 9.1-2.)
(b) Suppose you charge 120 g of CaC_2 and liquid water into a rigid container at 25°C, heat the container until the CaC_2 reacts completely, and cool the products back down to 25°C, condensing all the unconsumed water. What is the net amount of heat (kJ) that must be transferred from the reactor?

5. Use Hess's law to calculate the standard heat of the water–gas shift reaction

$$CO(g) + H_2O(g) \rightarrow CO_2(g) + H_2(g)$$

from each of the two sets of data given here.
(a) $CO(g) + H_2O(l) \rightarrow CO_2(g) + H_2(g)$: $\Delta \hat{H}_r^\circ = +1226$ Btu/lb-mole
 $H_2O(l) \rightarrow H_2O(g)$: $\Delta \hat{H}_v = +18\,935$ Btu/lb-mole
(b) $CO(g) + \tfrac{1}{2} O_2(g) \rightarrow CO_2(g)$: $\Delta \hat{H}_r^\circ = -121\,740$ Btu/lb-mole
 $H_2(g) + \tfrac{1}{2} O_2(g) \rightarrow H_2O(g)$: $\Delta \hat{H}_r^\circ = -104\,040$ Btu/lb-mole

6. Formaldehyde may be produced in the reaction between methanol and oxygen.

$$2 CH_3OH(l) + O_2(g) \rightarrow 2 HCHO(g) + 2 H_2O(l): \Delta \hat{H}_r^\circ = -326.2 \text{ kJ/mol}$$

The standard heat of combustion of hydrogen is

$$H_2(g) + \tfrac{1}{2} O_2(g) \rightarrow H_2O(l): \Delta \hat{H}_r^\circ = -285.8 \text{ kJ/mol}$$

Use these heats of reaction and Hess's law to determine the standard heat of the direct decomposition of methanol to form formaldehyde

$$CH_3OH(l) \rightarrow HCHO(g) + H_2(g)$$

7. Use tabulated heats of formation to determine the standard heats of the following reactions in kJ/mol, letting the stoichiometric coefficient of the first reactant in each reaction equal one.
(a) Nitrogen + oxygen react to form nitrogen dioxide (NO_2).
(b) Gaseous *n*-butane + oxygen react to form carbon monoxide + liquid water.
(c) Liquid *n*-pentane + oxygen react to form carbon dioxide + water vapor.
(d) Write the stoichiometric equations for the formation of the reactant and product compounds in part (c); then use Hess's law to derive the formula you used to calculate $\Delta \hat{H}_r^\circ$.

8. Trichloroethylene, a widely-used degreasing solvent for machine parts, is produced by a two-step reaction sequence. Ethylene is first chlorinated to yield tetrachloroethane, which is dehydrochlorinated to form trichloroethylene.

$$C_2H_4(g) + 2\ Cl_2(g) \rightarrow C_2H_2Cl_4(l) + H_2(g): \Delta\hat{H}_r^\circ = -385.76\ \text{kJ/mol}$$
$$C_2H_2Cl_4(l) \rightarrow C_2HCl_3(l) + HCl(g)$$

The standard heat of formation of trichloroethylene is -276.2 kJ/mol.

(a) Use the given data and tabulated standard heats of formation of ethylene and hydrogen chloride to calculate the standard heat of formation of tetrachloroethane and the standard heat of the second reaction.

(b) Use Hess's law to calculate the standard heat of the reaction

$$C_2H_4(g) + 2\ Cl_2(g) \rightarrow C_2HCl_3(l) + H_2(g) + HCl(g)$$

(c) If 300 mol/h of $C_2HCl_3(l)$ are produced according to the reaction of part (b) and the reactants and products are all at 25°C, how much heat is evolved or absorbed in the process? (Assume $Q = \Delta H$.)

9. The heat of combustion of 2,3,3-trimethyl pentane $[C_8H_{18}(l)]$ is measured in a constant-volume calorimeter. Two grams of this substance are burned with pure oxygen, and the heat released when the reactants and products $[CO_2(g)$ and $H_2O(g)]$ are all at 25°C is found to be sufficient to raise the temperature of 1.00 kg of liquid water by 21.34°C.

(a) Write an energy balance to show that the heat released in the calorimeter equals $n_{C_8H_{18}} \Delta\hat{U}_c^\circ$, and calculate $\Delta\hat{U}_c^\circ$ in kJ/mol. Then calculate $\Delta\hat{H}_c^\circ$. (See Example 9.1-2.)

(b) Use the results of part (a) to determine $\Delta\hat{H}_f^\circ$ for 2,3,3-trimethylpentane.

(c) Why would the heat of formation of 2,3,3-trimethylpentane probably be determined this way rather than directly from the formation reaction?

10. *n*-Butane is converted to isobutane in an isomerization reactor, which operates isothermally at 149°C.

(a) Calculate the heat of the isomerization reaction at 149°C, using tabulated standard heats of formation of normal butane and isobutane. (See Example 9.5-1 and the paragraph that follows it.)

(b) Suppose the feed to the reactor contains 93 mole% *n*-butane, 5% isobutane, and 2% HCl at 149°C, and a 40% conversion of *n*-butane is achieved. At what rate $[\text{kJ/m}^3(\text{STP})$ fed$]$ must heat be supplied to or removed from (state which) the reactor?

11. Normal heptane is dehydrocyclicized catalytically to toluene in a vapor-phase hydroforming process.

$$C_7H_{16} \rightarrow C_6H_5CH_3 + 4H_2$$

The reaction takes place in an isothermal reactor at 400°C and 1 atm. Pure heptane at 400°C is fed to the reactor, and a 35% conversion is achieved.

(a) Calculate the heat of the reaction at 400°C, using tabulated standard heats of formation of the reactants and products. (See Example 9.5-1 and the paragraph that follows it). The heat capacity of *n*-heptane vapor is $C_p(\text{J/mol}\cdot°\text{C}) = 150 + 0.60T(°\text{C})$.

(b) Estimate the rate (kJ/h) at which heat must be transferred to or from (state which) the reactor for a toluene production rate of 200 kg/h.

12. The reduction of hematite ore with coke (carbon) in a blast furnace to obtain pig iron is the first step in the production of most of the steel manufactured in the United States. The basic reaction is

$$Fe_2O_3(s) + 3\ C(s) \rightarrow 2\ Fe(s) + 3\ CO(g): \Delta\hat{H}_r^\circ = 2.111 \times 10^5\ \text{Btu/lb-mole}$$

although many other reactions simultaneously occur among the species present in the furnace. Estimate the heat requirements for a blast furnace per ton of iron produced, using the following data:

(a) Stoichiometric amounts of ore and carbon are fed at 77°F, reaction is complete, the iron emerges as a liquid at 2800°F, and CO emerges at 570°F.

(b) The heat capacity of Fe(s) in the temperature range of interest may be taken to be

$$C_p(\text{Btu/lb-mole}\cdot°\text{F}) = 5.90 + 1.50 \times 10^{-3}T(°\text{F})$$

Iron melts at 2794°F and has a heat of fusion of 6496 Btu/lb-mole. The heat capacity of molten iron is 8.15 Btu/lb-mole·°F.

13. The thermal decomposition of dimethyl ether

$$(CH_3)_2O(g) \rightarrow CH_4(g) + H_2(g) + CO(g)$$

is carried out in an isothermal 2.00-liter laboratory reactor at 600°C. The reactor is charged with pure dimethyl ether at a pressure of 350 mm Hg. After about 2 h, the reactor pressure is 875 mm Hg.

(a) Had the reaction proceeded to completion at the end of the 2-h period? If not, what percentage of the dimethyl ether decomposed?

(b) Use tabulated data for methane, hydrogen, and carbon monoxide, and the data given below for dimethyl ether, to calculate $\Delta\hat{H}_r(600°C)$ and $\Delta\hat{U}_r(600°C)$ for the decomposition reaction.

(c) How much heat was transferred to or from the reactor (state which) during the 2-h period of the reaction?

Data for $(CH_3)_2O(g)$:

$$\Delta\hat{H}_f^\circ(25°C) = -180.16\ \text{kJ/mol}$$
$$C_p(\text{J/mol}\cdot\text{K}) = 26.86 + 0.1659T(\text{K}) - 4.179 \times 10^{-5}T^2$$

14. Sulfur dioxide is oxidized to sulfur trioxide in a small pilot plant reactor. SO_2 and 100% excess air are fed to the reactor at 450°C. The reaction proceeds to a 65% conversion of SO_2, and the products emerge from the reactor at 550°C. The production rate of SO_3 is 100 kg/h. The reactor is surrounded by a water jacket, into which water at 25°C is fed. Calculate the minimum flow rate of the cooling water if the temperature rise of the water is to be kept below 15°C.

15. Ammonia is oxidized with air to form nitric oxide in the first step of the manufacture of nitric acid. Two principal reactions occur.

$$4\ NH_3 + 5\ O_2 \rightarrow 4\ NO + 6\ H_2O$$
$$2\ NH_3 + \tfrac{3}{2}O_2 \rightarrow N_2 + 3\ H_2O$$

A schematic representation of the process is as follows.

Calculate the heat transferred to or from the reactor, taking atomic species as references for enthalpy calculations. (See Example 9.5-2.)

16. Carbon monoxide at 25°C and steam at 150°C are fed to a reactor and react to form carbon dioxide and hydrogen. The product gas, which contains 40 mole% H_2, 40% CO_2, and 20% H_2O, emerges at 500°C at a rate of 2.5 SCMH (standard cubic meters per hour) and is cooled to 15°C and 1 atm, condensing a portion of the water. Calculate the percentage excess of steam fed to the reactor, the rate of condensation of the water, and the rates (kW) at which heat must be transferred from the reactor and the condenser.

17. Benzaldehyde is produced from toluene in the catalytic reaction

$$C_6H_5CH_3 + O_2 \rightarrow C_6H_5CHO + H_2O$$

Dry air and toluene vapor are mixed and fed to the reactor at 350°F and 1 atm. Air is supplied in 100% excess. Of the toluene fed to the reactor, 13% reacts to form benzaldehyde, and 0.5% reacts with oxygen to form CO_2 and H_2O. The product gases leave the reactor at 379°F and 1 atm. Water is circulated through a jacket surrounding the reactor, entering at 80°F and leaving at 105°F. During a 4-h test period, 29.3 lb_m of water are condensed from the product gases. (Total condensation may be assumed.) The standard heat of formation of benzaldehyde vapor is $-17\ 200$ Btu/lb-mole; the heat capacities of both toluene and benzaldehyde vapors are approximately 31 Btu/lb-mole·°F; and that of liquid benzaldehyde is 46 Btu/lb-mole·°F.

(a) Calculate the volumetric flow rates (ft^3/h) of the combined feed stream to the reactor and the product gas.

(b) Calculate the required rate of heat transfer from the reactor (Btu/h) and the flow rate of the cooling water (gal/min).

18. Ethanol is produced commercially by the hydration of ethylene

$$C_2H_4(g) + H_2O(g) \rightleftharpoons C_2H_5OH(g)$$

Some of the product is converted to diethyl ether in the undesired side reaction

$$2\ C_2H_5OH(g) \rightleftharpoons (C_2H_5)_2O(g) + H_2O(g)$$

The combined feed to the reactor contains 53.7 mole% C_2H_4, 36.7% H_2O, and the balance inerts, and enters the reactor at 310°C. The reactor operates isothermally at 310°C. An ethylene conversion of 5% is achieved, and the yield of ethanol (moles ethanol produced/mole ethylene consumed) is 0.900.

Data for diethyl ether:

$$\Delta \hat{H}_f^\circ = -272.8 \text{ kJ/mol } \textit{for the liquid}$$

$$\Delta \hat{H}_v = 26.05 \text{ kJ/mol. Assume independent of } T$$

$$C_p(\text{J/mol} \cdot {}^\circ\text{C}) = 89.45 + 0.4033T({}^\circ\text{C}) - 2.244 \times 10^{-4}T^2$$

(a) Calculate the reactor heating or cooling requirement in kJ/mol feed. (See Example 9.5-2.)

(b) Why would the reactor be designed to yield such a low conversion of ethylene? What processing step (or steps) would probably follow the reactor in a commercial implementation of this process?

19. The following gas-phase reactions take place in a continuous reactor.

$$A + B \rightarrow C$$

$$2\,C \rightarrow D + B$$

(a) Suppose the reactor feed contains A, B, and inerts, with mole fractions x_{A0}, x_{B0}, and x_{I0}, respectively. Letting f_A denote the fractional conversion of A (moles A consumed/mole A fed), and Ψ denote the yield of C (moles C produced/mole A consumed), prove that for a basis of 1 mol of feed, the moles of each species at the outlet are given by the following expressions.

$$n_A = x_{A0}(1 - f_A)$$

$$n_C = x_{A0}f_A\Psi$$

$$n_D = \tfrac{1}{2}(x_{A0}f_A - n_C)$$

$$n_B = x_{B0} - x_{A0}f_A + n_D$$

$$n_I = x_{I0}$$

*(b) Write a computer program to perform material and energy balances for the reactor. Have the program read in values of the following variables.

 (i) x_{A0}, x_{B0}, x_{I0}

 (ii) f_A and Ψ

 (iii) Feed and product temperatures, $T_f({}^\circ\text{C})$ and $T_p({}^\circ\text{C})$

 (iv) Standard heats of formation of A(g), B(g), C(g), and D(l), in kJ/mol

 (v) Heat of vaporization of D at 25°C, in kJ/mol

 (vi) Coefficients of heat capacity formulas $C_p = a + bT + cT^2$ for gaseous A, B, C, D, and I, where C_p has units of (kJ/mol·°C) and T is in °C.

The program should calculate and print out an enthalpy table like that of Example 9.5-2 and the heating or cooling requirement for the reactor. Test the program with the data and reactor specifications of Problem 18. Then run it again with the feed temperature changed to 125°C.

20. Ethylene oxide is produced by the catalytic oxidation of ethylene.

$$C_2H_4(g) + \tfrac{1}{2}O_2(g) \rightarrow C_2H_4O(g)$$

An undesired competing reaction is the combustion of ethylene to CO_2. The feed to a reactor contains 2 mol C_2H_4/mol O_2. The conversion in the reactor is 25%, and 0.70 mol

* Computer problem.

of C_2H_4O is produced per mol of C_2H_4 consumed. A multiple-unit process is used to separate the products: C_2H_4 and O_2 are recycled back to the reactor, $C_2H_4O(g)$ is sold as a product, and CO_2 and $H_2O(l)$ are discarded. The reactor inlet and outlet streams are each at 450°C, and the temperature of the makeup gases (fresh feed) and the unrecycled products is 25°C.

(a) Calculate the flow rate (kg/h) and the composition of the fresh feed needed to produce 1500 kg C_2H_4O/day.

(b) Calculate the net heat requirement (kW) for the entire process and that for the reactor alone.

Data for $C_2H_4O(g)$:

$$\Delta \hat{H}_f^\circ = -51.00 \text{ kJ/mol}$$
$$C_p(\text{J/mol·K}) = -4.69 + 0.2061 T(\text{K}) - 9.995 \times 10^{-5} T^2$$

21. Cumene ($C_6H_5C_3H_7$) is produced by reacting benzene with propylene in a fixed-bed catalytic reactor. [$\Delta \hat{H}_r^\circ (77°F) = -39\ 520$ Btu/lb-mole.] A liquid feed containing 75 mole% propylene and 25% n-butane and a second liquid stream containing essentially pure benzene are fed to the reactor. Fresh benzene and recycled benzene, both at 77°F, are mixed in a 1:3 ratio and passed through a heat exchanger where they are heated by the reactor effluent before being fed to the reactor. The reactor effluent enters this exchanger at 400°F and leaves at 200°F. The pressure in the reactor is sufficient to maintain the effluent stream as a liquid.

After being cooled in the heat exchanger, the reactor effluent is fed to a distillation column. All of the butane and unreacted propane are removed as overhead product from the column, and the cumene and unreacted benzene are removed as bottoms product and fed to a second distillation column where they are separated. The benzene leaving the top of the second column is the recycle that is mixed with the fresh benzene feed. The production rate of cumene is 1200 lb$_m$/h.

A flowchart and heat capacity data follow.

Process data:

Production rate = 1200 lb$_m$ of cumene/h

$$C_p(\text{Btu/lb}_m \cdot °\text{F}) = 0.57 \text{ for propylene}$$
$$= 0.55 \text{ for butane}$$
$$= 0.45 \text{ for benzene}$$
$$= 0.40 \text{ for cumene}$$

(a) Calculate the mass flow rates of the streams fed to the reactor, the molar flow rate and composition of the reactor effluent, and the molar flow rate and composition of the overhead product from the first distillation column.

(b) Calculate the temperature of the benzene stream fed to the reactor, and the required rate of heat addition to or removal from the reactor.

(c) Where in the given process data have approximations been made, or conditions given that probably could not be attained in practice?

22. Ethylbenzene is converted to styrene by the catalytic dehydrogenation reaction

$$C_8H_{10}(g) \rightarrow C_8H_8(g) + H_2: \Delta\hat{H}_r^\circ(600°C) = +124.5 \text{ kJ/mol}$$

A flowchart of a simplified version of the commercial process is shown here.

Fresh and recycled liquid ethylbenzene combine and are heated from 25°C to 500°C (A), and the heated ethylbenzene is mixed adiabatically with steam at 700°C (B) to produce the feed to the reactor at 600°C. (The steam suppresses undesired side reactions and removes carbon deposited on the catalyst surface). A once-through conversion of 35% is achieved in the reactor (C), and the products emerge at 560°C. The product stream is cooled (D) to 25°C, condensing essentially all of the water, ethylbenzene, and styrene, and allowing hydrogen to pass out as a recoverable byproduct of the process.

The water and hydrocarbon liquids are immiscible and are separated in a settling tank decanter (E). The water is vaporized and heated (F) and the steam is recycled to

mix with the ethylbenzene feed to the reactor. The hydrocarbon stream is fed to a distillation tower (G) (actually, a series of towers), in which the mixture is separated into essentially pure styrene and ethylbenzene, each at 25°C after cooling and condensation steps have been carried out. The ethylbenzene is recycled to the reactor preheater, and the styrene is taken off as a product.

(a) Calculate on the basis of 100 kg/h styrene produced the required fresh ethylbenzene feed rate, the flow rate of recycled ethylbenzene, and the circulation rate of water, all in mol/h. (Assume $P = 1$ atm.)

(b) Calculate the required rates of heat input or withdrawal in kJ/h for the ethylbenzene preheater (A), steam generator (F), and reactor (C).

(c) Suggest possible ways to improve the energy economy of this process.

Physical property data:

Ethylbenzene: $(C_p)_{liquid} = 182$ J/mol·°C
$(\Delta \hat{H}_v) = 36.0$ kJ/mol at 136°C
$(C_p)_{vapor} = 118 + 0.30T(°C)$ J/mol·°C

Styrene: $(C_p)_{liquid} = 209$ J/mol·°C
$(\Delta \hat{H}_v) = 37.1$ kJ/mol at 145°C
$(C_p)_{vapor} = 115 + 0.27T(°C)$ J/mol·°C

23. Formaldehyde is produced by decomposing methanol over a silver catalyst.

$$CH_3OH \rightarrow HCHO + H_2$$

To provide heat for this endothermic reaction, some oxygen is included in the feed to the reactor, leading to the partial combustion of the hydrogen produced in the methanol decomposition.

The feed to an adiabatic formaldehyde production reactor is obtained by bubbling a stream of air at 1 atm through liquid methanol. The air leaves the vaporizer saturated with methanol and contains 42% methanol by volume. The stream then passes through a heater, in which its temperature is raised to 145°C. To avoid deactivating the catalyst, the maximum temperature attained in the reactor must be limited to 600°C. For this purpose, saturated steam at 145°C is metered into the air–methanol stream, and the combined stream enters the reactor. A fractional methanol conversion of 70% is achieved in the reactor, and the product gas contains 5 mole% hydrogen. The product gas is cooled to 145°C in a waste heat boiler, in which saturated steam at 3.1 bars is generated from liquid water at 30°C. Several absorption and distillation units follow the waste heat boiler, and formaldehyde is ultimately recovered in an aqueous solution containing 37 wt% HCHO. The plant is designed to produce 36 metric kilotons of this solution per year, operating 350 days/yr.

(a) Calculate the operating temperature of the methanol vaporizer.

(b) Calculate the feed rate of steam to the reactor (kg/h), and the molar flow rate and composition of the product gas.

(c) Calculate the rate (kg/h) at which steam is generated in the waste heat boiler.

24. The synthesis of ethyl chloride is accomplished by reacting ethylene with hydrogen chloride in the presence of an aluminum chloride catalyst.

$$C_2H_4(g) + HCl(g) \xrightarrow{\text{Catalyst}} C_2H_5Cl(g): \Delta\hat{H}_r(0°C) = -64.5 \text{ kJ/mol}$$

Process data and a simplified schematic flowchart are given here.

Data: Reactor: adiabatic, outlet temperature = 50°C
Feed A: 100% HCl(g), 0°C
Feed B: 93 mole% C_2H_4, 7% C_2H_6, 0°C
Product C: Consists of 1.5% of the HCl, 1.5% of the C_2H_4, and all of the C_2H_6
 that enter the reactor
Product D: 1600 kg $C_2H_5Cl(l)$/h, 0°C
Recyle to reactor: $C_2H_5Cl(l)$, 0°C
C_2H_5Cl: $\Delta\hat{H}_v = 24.7$ kJ/mol (assume independent of T)
$(C_p)_{vapor}(kJ/mol \cdot °C) = 0.052 + 8.7 \times 10^{-5}T(°C)$

The reaction is exothermic, and if the heat of reaction is not removed in some way the reactor temperature could increase to an undesirably high level. To avoid this occurrence, the reaction is carried out with the catalyst suspended in liquid ethyl chloride. As the reaction proceeds, most of the heat liberated goes to vaporize the liquid, making it possible to keep the reaction temperature at or below 50°C.

The stream leaving the reactor contains ethyl chloride formed by reaction and that vaporized in the reactor. This stream passes through a heat exchanger where it is cooled to 0°C, condensing essentially all of the ethyl chloride and leaving only unreacted ethylene, HCl, and C_2H_6 in the gas phase. A portion of the liquid condensate is recycled to the reactor at a rate equal to the rate at which ethyl chloride is vaporized, and the rest is taken off as product. At the process conditions, heats of mixing and the influence of pressure on enthalpy may be neglected.

(a) At what rates (kmol/h) do the two feed streams enter the process?
(b) Calculate the composition (component mole fractions) and molar flow rate of product stream C.
(c) Write an energy balance around the reactor, and use it to determine the rate at which ethyl chloride must be recycled.

25. Nitric oxide and oxygen react to form nitrogen dioxide in a continuous adiabatic reactor. The feed contains NO and O_2 in stoichiometric proportion at 700°C, and a 70% conversion is achieved. Use the heat capacities given here to perform the requested calculations:

 Data: $(C_p)_{NO} = 37.7$ J/mol·°C
 $(C_p)_{NO_2} = 75.7$ J/mol·°C
 $(C_p)_{O_2} = 35.7$ J/mol·°C

 (a) Calculate the outlet temperature if the reaction is carried out in a continuous flow reactor. (See Example 9.5-3.)
 (b) Calculate the final reactor temperature if the reaction is carried out in a constant-volume batch reactor.

26. The equilibrium constant for the ethane dehydrogenation reaction,

$$C_2H_6(g) \rightleftharpoons C_2H_4(g) + H_2(g)$$

is defined as

$$K_p(\text{atm}) = \frac{x_{C_2H_4} x_{H_2}}{x_{C_2H_6}} P$$

where $P(\text{atm})$ is the total pressure and x_i is the mole fraction of the ith substance in an equilibrium mixture. The equilibrium constant has been found experimentally to vary with temperature according to the formula

$$K_p(T) = 7.28 \times 10^6 \exp[-17\,000/T(K)] \tag{1}$$

The heat of reaction at 1273K is $+145.6$ kJ/mol, and the heat capacities of the reactive species may be approximated by the formulas

$$
\left.
\begin{aligned}
(C_p)_{C_2H_4} &= 9.419 + 0.1147T(K) \\
(C_p)_{H_2} &= 26.90 + 4.167 \times 10^{-3}T(K) \\
(C_p)_{C_2H_6} &= 11.35 + 0.1392T(K)
\end{aligned}
\right\} \ (\text{J/mol·K})
$$

 Suppose pure ethane is fed to a continuous adiabatic reactor at 1273K and 1 atm, that the products emerge at $T_f(K)$ and 1 atm, and that the reactor is long enough for the outlet stream to be considered an equilibrium mixture of ethane, ethylene, and hydrogen.

 (a) Prove that the fractional conversion of ethane in the reactor is

$$f = \left(\frac{K_p}{1 + K_p}\right)^{1/2} \tag{2}$$

 (b) Write an energy balance on the reactor, and use it to prove that

$$f = \frac{1}{1 + \phi(T)} \tag{3}$$

 where

$$\phi(T) = \frac{\Delta\hat{H}_r(1273K) - \int_{T_f}^{1273} [(C_p)_{C_2H_4} + (C_p)_{H_2}]\,dT}{\int_{T_f}^{1273} (C_p)_{C_2H_6}\,dT} \tag{4}$$

Finally, substitute for $\Delta \hat{H}_r$ and the heat capacities in equation 4 to derive an explicit expression for $\phi(T)$.

(c)　We now have two expressions for the fractional conversion f: Equation 2 and Equation 3. If these expressions are equated, K_p is replaced by the expression of Equation 1, and $\phi(T)$ is replaced by the expression derived in part (b), the result is one equation in one unknown, T. Derive this equation, and transpose the right side to obtain an expression of the form

$$\psi(T) = 0 \qquad\qquad (5)$$

*(d)　Write a computer program to solve Equation 5 for T, and then to determine the final fractional conversion f. Begin by evaluating $\psi(T)$ at 1200K, 1150K, 1100K, and so on in 50K decrements until the value of ψ changes sign from one iteration to the next. Then use the regula-falsi algorithm (Appendix A.2) to estimate the solution, terminating when the estimated value of T changes by less than 0.01K from one iteration to the next. Have your program print out all values of T and the corresponding values of ψ, and build into the program a provision that if the number of regula-falsi iterations exceeds 20 the program will terminate with an error message.

27.　The synthesis of methanol from carbon monoxide and hydrogen is carried out in a continuous vapor-phase reactor at 5.00 atm (absolute). The feed contains CO and H_2 in stoichiometric proportion and enters the reactor at 25°C and 5.00 atm at a rate of 17.1 m^3/h. The product stream emerges from the reactor at 127°C. The rate of heat transfer from the reactor is 17.05 kW. Calculate the fractional conversion achieved and the volumetric flow rate (m^3/h) of the product stream. (See Example 9.5-4.)

28.　Coke can be converted into CO—a fuel gas—in the reaction

$$CO_2(g) + C(s) \rightarrow 2\, CO(g)$$

A coke that contains 84% carbon by mass and the balance noncombustible ash is fed to a reactor with a stoichiometric amount of CO_2. The coke is fed at 77°F, and the CO_2 enters at 400°F. Heat is transferred to the reactor in the amount of 5859 Btu/lb_m coke fed. The gaseous products and the solid reactor effluent (the ash and unburned carbon) leave the reactor at 1830°F. The heat capacity of the solid is 0.24 Btu/$lb_m \cdot$°F. Calculate the percentage conversion of the carbon in the coke.

29.　Carbon disulfide, a key component in the manufacture of rayon fibers, is produced in the reaction between methane and sulfur vapor over a metal oxide catalyst.

$$CH_4(g) + 4\, S(g) \rightarrow CS_2(g) + 2\, H_2S(g)$$
$$\Delta \hat{H}_r^\circ(700°C) = -274\ kJ/mol$$

Methane and molten sulfur, each at 150°C, are fed to a heat exchanger in stoichiometric proportion. Heat is exchanged between the reactor feed and product streams, and the sulfur in the feed is vaporized. The gaseous methane and sulfur leave the exchanger and pass through a second preheater in which they are heated to 700°C, the temperature at which they enter the reactor. Heat is transferred from the reactor at a rate of 41 kJ/mol

* Computer problem.

of feed. The reaction products emerge from the reactor at 800°C, pass through the heat exchanger, and emerge at 200°C with sulfur as a liquid. Use the heat capacity data given below to perform the following calculations.

(a) Estimate the fractional conversion achieved in the reactor.

(b) Estimate the quantity of heat (kJ/mol feed) that must be added to the feed in the second preheater to bring it to 700°C.

(c) Suggest a method to improve the energy economy of the process.

$$
\begin{aligned}
\textbf{\textit{Data:}}\quad & S(l): C_p = 29.4 \text{ J/mol} \cdot {}^\circ C \\
& S(g): C_p = 36.4 \text{ J/mol} \cdot {}^\circ C \\
& CH_4(g): C_p = 71.4 \text{ J/mol} \cdot {}^\circ C \\
& CS_2(g): C_p = 31.8 \text{ J/mol} \cdot {}^\circ C \\
& H_2S(g): C_p = 44.8 \text{ J/mol} \cdot {}^\circ C
\end{aligned}
$$

30. You are checking the performance of a reactor in which acetylene is produced from methane in the reaction

$$2\,CH_4(g) \rightarrow C_2H_2(g) + 3\,H_2(g)$$

An undesired side reaction is the decomposition of acetylene

$$C_2H_2(g) \rightarrow 2\,C(s) + H_2(g)$$

Methane is fed to the reactor at 1500°C at a rate of 10.0 mol CH_4/s. Heat is transferred to the reaction system at a rate of 975 kW. The product temperature is 1500°C, and the fractional conversion of methane is 0.600. Use the mean heat capacities given here for the transition from 20°C to 1500°C to calculate the product component flow rates and the yield of acetylene (mol C_2H_2 produced/mol CH_4 consumed).

$$
\begin{aligned}
& CH_4(g): C_p = 0.079 \text{ kJ/mol} \cdot {}^\circ C \\
& C_2H_2(g): C_p = 0.052 \text{ kJ/mol} \cdot {}^\circ C \\
& C(s): C_p = 0.022 \text{ kJ/mol} \cdot {}^\circ C \\
& H_2(g): C_p = 0.031 \text{ kJ/mol} \cdot {}^\circ C
\end{aligned}
$$

31. Hydrogen is produced in the steam reforming of propane:

$$C_3H_8(g) + 3\,H_2O(g) \rightarrow 3\,CO(g) + 7\,H_2(g)$$

The water–gas shift reaction also takes place in the reactor, leading to the formation of additional hydrogen:

$$CO(g) + H_2O(g) \rightarrow CO_2(g) + H_2(g)$$

The reaction is carried out over a nickel catalyst in the tubes of a shell-and-tube reactor. The feed to the reactor contains steam and propane in a 6:1 molar ratio at 125°C, and the products emerge at 800°C. The excess steam in the feed assures essentially complete consumption of the propane. Heat is added to the reaction mixture by passing a hot gas over the outside of the tubes that contain the catalyst. The gas is fed at a rate of 4.94 m^3/mol C_3H_8 fed, entering the unit at 1400°C and 1 atm and leaving at 900°C. The unit may be considered adiabatic.

Calculate the molar composition of the product gas, assuming that the heat capacity of the heating gas is 0.040 kJ/mol·°C.

32. Five liters of a 1.00-molar aqueous sulfuric acid solution (SG = 1.064) are stored at 25°C. Use data in Table B.1 and Table 8.5-1 to calculate the standard heat of formation of the solution in kJ/mol H_2SO_4 relative to the solute elements and water, and the total enthalpy of the solution relative to the same reference conditions.

33. Calculate the standard heat (kJ/mol) of the neutralization reaction between dilute hydrochloric acid and dilute sodium hydroxide solutions, taking the heat of solution of NaCl to be +4.87 kJ/mol. Then calculate the standard heat of the reaction between hydrogen chloride gas and solid sodium hydroxide to form solid sodium chloride and liquid water. What is the physical significance of the difference between the two calculated values?

34. One hundred cubic centimeters of a 10.0 mole% aqueous sulfuric acid solution (SG = 1.27) are to be titrated to neutrality at 25°C with a 3-molar caustic soda (sodium hydroxide) solution (SG = 1.13):

$$H_2SO_4(aq) + 2\,NaOH(aq) \rightarrow Na_2SO_4(aq) + 2\,H_2O(l)$$

(a) Calculate the required volume (cm^3) of the caustic soda solution.

(b) Calculate the standard heats of formation in kJ/mol of each of the solutions involved in this process, and the heat released if the reactants and products are all at 25°C. The heat of solution of sodium sulfate is −1.17 kJ/mol. Assume that $Q = \Delta H$ for the process. (See Example 9.5-5.)

35. Most of the world's chlorine and sodium hydroxide supply is obtained through the electrolysis of brine.

$$NaCl(aq) + H_2O(l) \rightarrow \tfrac{1}{2}H_2(g) + \tfrac{1}{2}Cl_2(g) + NaOH(aq)$$

(a) Determine the standard heats of formation of NaCl(aq) and NaOH(aq) and then the standard heat of the brine electrolysis reaction. The standard heat of solution of NaCl is $\Delta\hat{H}_s(n = \infty) = +4.87$ kJ/mol NaCl.

(b) In 1982 roughly 8400 metric kilotons of chlorine were produced in the United States. Calculate the energy in MW·h required for this production, assuming that all of the chlorine is produced by electrolysis at 25°C and that the required energy input equals ΔH for the process.

36. An aqueous solution containing 12.00 wt% NaOH (SG = 1.13) is used as a titration standard. An aliquot of an HCl solution with a mass of 5.000 g is titrated to neutrality with 6.06 cm^3 of the NaOH solution. The heat of solution of NaCl at 25°C may be taken to be 4.87 kJ/mol NaCl and $Q = \Delta H$ for the mixing process.

 (a) Calculate n(moles H$_2$O/mole solute) for both solutions.

 (b) Calculate the heat released if the reactants and products are all at 25°C.

 (c) If the process were adiabatic and the heat capacity of the product solution were that of pure water (4.184 J/g·°C), what would be the final solution temperature?

37. You have been given the task of determining the standard heat of hydration of calcium chloride hexahydrate; that is, the heat of the reaction

$$CaCl_2(s) + 6\ H_2O(l) \rightarrow CaCl_2 \cdot 6\ H_2O(s): \Delta\hat{H}_r^\circ = ?$$

You cannot carry out the hydration directly, so you resort to an indirect method. You first dissolve 1.00 mol of anhydrous CaCl$_2$ in 10.0 mol of water in a calorimeter and determine the standard heat of solution:

$$CaCl_2(s) + 10\ H_2O(l) \rightarrow CaCl_2(aq, n = 10): \Delta\hat{H}_s^\circ = -64.85\ \text{kJ/mol}$$

You next dissolve 1.00 mol of the hydrated salt in 4.00 mol water and determine

$$CaCl_2 \cdot 6\ H_2O(s) + 4\ H_2O \rightarrow CaCl_2(aq, n = 10): \Delta\hat{H}_r^\circ = +32.1\ \text{kJ/mol}$$

Use these results to calculate the desired heat of hydration. In addition, calculate the heat of formation of CaCl$_2$(aq, $n = 10$) relative to Ca(s), Cl$_2$(g), and H$_2$O(l) at 25°C.

38. A dilute aqueous solution of sulfuric acid at 25°C is used to absorb ammonia in a continuous reactor, thereby producing ammonium sulfate, a fertilizer.

$$2\ NH_3(g) + H_2SO_4(aq) \rightarrow (NH_4)_2SO_4(aq)$$

 (a) If the ammonia enters the absorber at 75°C, the sulfuric acid enters at 25°C, and the product solution emerges at 25°C, how much heat must be withdrawn from the unit per mol of (NH$_4$)$_2$SO$_4$ produced? (All needed physical property data may be found in Appendix B.)

 (b) How much energy is required to recover the ammonium sulfate from the solution, assuming both the solution and the separated products [(NH$_4$)$_2$SO$_4$(s) and H$_2$O(l)] are at 25°C?

39. A sulfuric acid solution (50 moles H$_2$O/mole H$_2$SO$_4$) is neutralized with a sodium hydroxide solution (20 moles H$_2$O/mole NaOH) in a continuous reactor. All reactants enter at 25°C. The standard heat of solution of sodium sulfate is -1.2 kJ/mol Na$_2$SO$_4$, and the heat capacities of all solutions (kJ/g·°C) may be taken to be that of pure liquid water.

 (a) Calculate the heat [kJ/kg H$_2$SO$_4$(aq)fed] that must be removed if the product solution emerges at 40°C.

 (b) Calculate the product solution temperature if the reactor is adiabatic.

40. Fifty milliliters of a 4-molar solution of sulfuric acid (SG = 1.23) are neutralized with a 12-molar solution of sodium hydroxide (SG = 1.37). Calculate the required volume of the sodium hydroxide solution and the final solution temperature if both feed solutions are at 25°C, the mixing is adiabatic, and the solution container absorbs a negligible quantity of heat. The heat capacity of the product solution may be taken to be that of pure liquid water.

41. Ammonia scrubbing is one of many processes for removing sulfur dioxide from flue gases. The gases are bubbled through an aqueous solution of ammonium sulfite, and the SO_2 reacts to form ammonium bisulfite

$$(NH_4)_2SO_3(aq) + SO_2(g) + H_2O(l) \rightarrow 2\,NH_4HSO_3(aq)$$

Subsequent process steps yield concentrated SO_2 and regenerate ammonium sulfite, which is recycled to the scrubber. The sulfur dioxide is either oxidized and absorbed in water to form sulfuric acid or reduced to elemental sulfur.

Flue gas from a power-plant boiler contains 0.30% SO_2 by volume, and enters a scrubber at a rate of 50 000 mol/h at 50°C. The gas is bubbled through an aqueous solution containing 10 mole% ammonium sulfite, which enters the scrubber at 25°C. The gas and liquid effluents from the scrubber both emerge at 35°C. The scrubber removes 90% of the SO_2 entering with the flue gas. The effluent liquid is analyzed and is found to contain 1.5 moles $(NH_4)_2SO_3$ per mole of NH_4HSO_3. The heat of formation of $(NH_4)_2SO_3(aq)$ at 25°C is -890.0 kJ/mol, and that of $NH_4HSO_3(aq)$ is -760 kJ/mol. The heat capacities of all liquid solutions may be taken to be 4.0 J/g·°C and that of the flue gas may be taken to be that of nitrogen. Evaporation of water may be neglected. Calculate the required rate of heat removal in kW.

42. Benzene is chlorinated in a catalytic reactor. The desired product is monochlorobenzene (C_6H_5Cl), but some of the product is further chlorinated to dichlorobenzene $(C_6H_4Cl_2)$. Reaction data are as follows:

Feed: $C_6H_6(l)$ and 25% excess chlorine (based on the monochlorination reaction), 25°C.

Products: $C_6H_5Cl(l)$, $C_6H_4Cl_2(l)$, $HCl(g)$, unreacted benzene and chlorine, 40°C.

Conversion and selectivity: 85% conversion of C_6H_6
12 kg C_6H_5Cl/kg $C_6H_4Cl_2$

Heat capacity of combined products: $C_p = 1.67$ J/g·°C

Heats of combustion to $CO_2(g)$, $HCl(aq)$, $H_2O(l)$:

$$C_6H_6(l): \Delta\hat{H}_c^\circ = -3268 \text{ kJ/mol}$$
$$C_6H_5Cl(l): \Delta\hat{H}_c^\circ = -3140 \text{ kJ/mol}$$
$$C_6H_4Cl_2(l): \Delta\hat{H}_c^\circ = -2812 \text{ kJ/mol}$$

(a) Calculate the heats of combustion of monochlorobenzene and dichlorobenzene taking $HCl(g)$ rather than $HCl(aq)$ to be the reaction product, and use the results to calculate the standard heats of the reactions

$$C_6H_6(l) + Cl_2(g) \rightarrow C_6H_5Cl(l) + HCl(g)$$
$$C_6H_6(l) + 2\,Cl_2(g) \rightarrow C_6H_4Cl_2(l) + 2\,HCl(g)$$

(b) Calculate the heat that must be transferred from the reactor per kilogram of monochlorobenzene formed.

43. A natural gas is analyzed and found to consist of 86.5% methane, 8.0% ethane, 2.0% propane, and 3.5% N_2 (noncombustible) by volume. Calculate the lower heating value of this fuel in kJ/mol and kJ/kg. (See Example 9.6-1.)

44. A fuel gas is made by mixing a stream of propane gas with a stream of *n*-butane gas in a mass ratio of 2:1. Calculate the mole fraction of propane in the fuel and the higher heating value of the fuel (kJ/mol).

45. A fuel gas is known to contain methane, ethane, and carbon monoxide. A sample of the gas is charged into an initially evacuated 2.000-liter vessel at 25°C and 2323 mm Hg (absolute). The vessel is weighed before and after being charged, and the mass difference is found to be 4.929 g. Next, the higher heating value of the gas is determined in a calorimeter to be 841.9 kJ/mol. Calculate the molar composition of the gas.

46. A gas containing 85 mole% methane and 15 mole% ethane at 25°C is burned in a constant volume reactor, and the products are brought back down to 25°C. Calculate the net heat (kJ) transferred from the reactor. (*Hint:* Recall Equation 9.1-4.)

47. The heating value of a fuel oil is to be measured in a constant-volume bomb calorimeter. The bomb is charged with oxygen and 0.00215 lb_m of the fuel, and is then sealed and immersed in an insulated container of water. The initial temperature of the system is 77°F. The fuel–oxygen mixture is ignited, and the fuel is completely consumed. The final calorimeter temperature is 87.06°F. The mass of the calorimeter, including the bomb and its contents, is 4.62 lb_m, and the average heat capacity of the system is 0.900 Btu/lb_m·°F. The water formed in the combustion may be assumed to be a liquid, and the difference between $\Delta \hat{U}_c$ and $\Delta \hat{H}_c$ may be neglected. Estimate the higher heating value of the fuel oil in Btu/lb_m.

48. An **ultimate analysis** of a coal is a series of operations that yields the percentages by mass of carbon, hydrogen, nitrogen, oxygen, and sulfur in the coal. The heating value of a coal is best determined in a calorimeter, but it may be estimated with reasonable accuracy from the ultimate analysis using the **Dulong formula.**

$$\text{HHV(kJ/kg)} = 33\,801(C) + 144\,158[(H) - 0.125(O)] + 9413(S)$$

where (C), (H), (O), and (S) are the mass fractions of the corresponding elements. The 0.125(O) term accounts for the hydrogen bound in the water contained in the coal.

(a) Derive an expression for the higher heating value (HHV) of a coal in terms of C, H, O, and S, and compare your result with the Dulong formula. Suggest a reason for the difference.

(b) A coal with an ultimate analysis of 75.8% C, 5.1% H, 8.2% O, 1.5% N, 1.6% S, and 7.8% ash (noncombustible) is burned in a power-plant boiler furnace, and all the sulfur in it (except for any sulfur in the ash) forms SO_2. Estimate the ratio (kg SO_2 formed)/(kJ heating value of the fuel). (This ratio must be below a specified value for the power plant to be in compliance with federal regulations regarding sulfur emissions.)

49. Methanol vapor is burned with excess air in a catalytic combustion chamber. Liquid methanol initially at 25°C is vaporized at 1.1 atm and heated to 100°C; the vapor is mixed with air that has been preheated to 100°C, and the combined stream is fed to the reactor at 100°C and 1 atm. The reactor effluent emerges at 300°C and 1 atm. Analysis of the product gas yields a dry-basis composition of 4.8% CO_2, 14.3% O_2, and 80.9% N_2.

(a) Calculate the percentage excess air supplied and the dew point of the product gas.

(b) Taking a basis of 1 g-mole of methanol burned, calculate the heat (kJ) needed to vaporize and heat the methanol feed, and the heat (kJ) that must be transferred from the reactor.

50. Methane is burned in a poorly designed combustion chamber. One thousand mol/h of methane are fed to the chamber at 25°C, along with the theoretically required amount of air preheated to 200°C. Calculate the heat transferred from the reactor if 90% of the methane is burned, the mole ratio of CO_2 to CO in the product gas is 10:1, and the combustion gases leave the furnace at 400°C.

51. A gaseous fuel containing methane and ethane is burned with excess air. The fuel enters the furnace at 25°C and 1 atm, and the air enters at 200°C and 1 atm. The stack gas leaves the furnace at 800°C and 1 atm and contains 5.321 mole% CO_2, 1.596% CO, 7.317% O_2, 12.239% H_2O, and the balance N_2.
 (a) Calculate the molar percentages of methane and ethane in the fuel gas and the percentage excess air fed to the reactor.
 (b) Calculate the heat (kJ) transferred from the reactor per cubic meter of fuel gas fed.
 (c) What would make you suspect that the data given in this problem were made up rather than being obtained in a real experiment?

52. A coal contains 73.0 wt% C, 4.7% H (not including the hydrogen in the coal moisture), 3.7% S, 6.8% H_2O, and 11.8% ash. The coal is burned at a rate of 50 000 lb_m/h in a power-plant boiler with air 50% in excess of that needed to oxidize the carbon in the coal to CO_2. The air and coal are both fed at 77°F and 1 atm. The solid residue from the furnace is analyzed and is found to contain 28.7 wt% C, 1.6% S, and the balance ash. The coal sulfur oxidized in the furnace is converted to $SO_2(g)$. Of the ash in the coal, 30% emerges in the solid residue and the balance is emitted with the stack gases as fly ash. The stack gas and solid residue emerge from the furnace at 600°F. The heating value of the coal is 18 000 Btu/lb_m.
 (a) Calculate the mass flow rates of all components in the stack gas and the volumetric flow rate of this gas. (Ignore the contribution of the fly ash in the latter calculation, and assume that the stack gas contains a negligible amount of CO.)
 (b) Assume that the heat capacity of the solid furnace residue is 0.22 Btu/lb_m·°F, that of the stack gas is the heat capacity per unit mass of nitrogen, and 35% of the heat generated in the furnace is used to produce electricity. At what rate in MW is electricity produced?
 (c) Calculate the ratio (heat transferred from the furnace)/(heating value of the fuel). Why is this ratio less than one?
 (d) Suppose the air fed to the furnace were preheated rather than being fed at ambient temperature, but that everything else (feed rates, outlet temperatures, and fractional coal conversion) were the same. What effect would this change have on the ratio calculated in part (c)? Explain. Suggest an economical way in which this preheating might be accomplished.

53. A gas stream consisting of n-hexane in methane is fed to a condenser at 60°C and 1.2 atm. The dew point of the gas (considering hexane as the only condensable component) is 55°C. The gas is cooled to 5°C in the condenser, recovering pure hexane as a liquid. The effluent gas leaves the condenser saturated with hexane at 5°C and 1.1 atm and is fed to a boiler

furnace at a rate of 207.4 liters/s, where it is burned with 100% excess air that enters the furnace at 200°C. The stack gas emerges at 400°C and 1 atm and contains no carbon monoxide or unburned hydrocarbons. The heat transferred from the furnace is used to generate saturated steam at 10 bars from liquid water at 25°C.

(a) Calculate the mole fractions of hexane in the condenser feed and product gas streams and the rate of hexane condensation (liters condensate/h).

(b) Calculate the rate at which heat must be transferred from the condenser (kW) and the rate of generation of steam in the boiler (kg/s).

54. In an industrial boiler, methane is burned completely with dry air to generate heat for the production of superheated steam. The hot combustion gases leave the furnace at 300°C and are then cooled to 150°C in a heat exchanger. The cooling medium in the exchanger is the air that is to be fed to the boiler furnace. The air enters the exchanger at 25°C. Water enters the boiler at 25°C and leaves as superheated steam at 17 bars and 250°C. The feed rate of methane is 450 000 mol/h at 25°C, and air is fed in 20% excess. Calculate

(a) The composition (component mole fractions) of the combustion gas leaving the furnace and, using the component heat capacities given below, the average molar heat capacity of the gas. (See Equation 8.3-17.)

(b) The temperature of the air leaving the preheater. (A trial-and-error solution is required.)

(c) The rate of steam production (kg/h).

55. A bituminous coal is burned with air in a boiler furnace. The coal is fed at a rate of 40 000 kg/h and has an ultimate analysis of 76 wt% C, 5% H, 8% O, negligible amounts of N and S, and 11% noncombustible ash, and a higher heating value of 25 700 kJ/kg. Air enters a preheater at 30°C and 1 atm with a relative humidity of 30%, exchanges heat with the hot flue gas, and enters the furnace. The composition of the flue gas on a dry basis is determined to be 7.8 mole% CO_2, 1.2% CO, 11.4% O_2, and the balance N_2. The flue gas emerges from the furnace at 260°C and is cooled to 150°C in the preheater. Non-combustible residue leaves the furnace at 450°C and has a heat capacity of 0.97 kJ/kg·°C.

(a) Prove that the air-to-fuel ratio is 16.1 standard cubic meters/kg coal and that the stack gas contains 4.6% H_2O by volume.

(b) Calculate the rate of cooling required to cool the flue gas from 260°C to 150°C and the temperature to which the air is preheated. (*Suggestion:* In setting up the energy balance on the air stream, assume a value of T, use the mean heat capacities of Table 8.3-1 to evaluate the heat gained by the air, and iterate until the calculated heat matches the known heat loss of the flue gas in the preheater.)

(c) If 60% of the heat generated in the furnace (Q) goes into producing saturated steam at 30 bars from liquid boiler feedwater at 50°C, at what rate (kg/h) is steam generated?

(d) Qualitatively, how would the answer to part (c) change if the air were not preheated?

56. Carbon monoxide is burned with excess air at 1 atm. The reactants are fed at 25°C and the adiabatic flame temperature is 1500°C.

(a) Calculate the percentage excess air fed to the reactor.

(b) If the percentage excess air were increased, how would the adiabatic flame temperature change, and why would it change in this manner?

57. Propane is burned completely in a continuous reactor with a stoichiometric amount of oxygen. Calculate the adiabatic flame temperature (see Example 9.6-2), assuming that the heat capacity of the product gas is 1.00 kJ/kg·°C and the gases are fed at 25°C, for each of the following two cases:

(a) The feed is propane and pure oxygen.

(b) The feed is propane and air. Explain the difference between the two results.

58. A natural gas containing 87 mole% CH_4 and the balance C_2H_6 is burned with 20% excess air in a boiler furnace. The gas enters the furnace at 298K, and the air is preheated to 423K. The heat capacities of the stack gas components may be assumed to have the following constant values:

$$CO_2: C_p = 50.0 \text{ J/mol·K}$$
$$H_2O(v): C_p = 38.5 \text{ J/mol·K}$$
$$O_2: C_p = 33.1 \text{ J/mol·K}$$
$$N_2: C_p = 31.3 \text{ J/mol·K}$$

Assuming complete combustion of the fuel, calculate the adiabatic flame temperature.

59. In the operation of a coating plant, acetone is used as a solvent, and the coated material is cured by evaporating the acetone into a stream of air. Because environmental standards do not allow discharging acetone into the atmosphere, a proposal to incinerate the stream is to be evaluated.

The proposed process uses two parallel reactors containing beds of solid particles. The air–acetone stream, which contains acetone and oxygen in stoichiometric proportion, enters one of the beds at 1500 mm Hg (absolute) at a rate of 1410 standard cubic meters per minute. The particles in the bed have been preheated and transfer heat to the gas. The mixture ignites when its temperature reaches 562°C, and combustion takes place rapidly and adiabatically. The combustion products then pass through and heat the particles in the second bed, cooling down to 350°C in the process. Periodically the

flow is switched so that the heated outlet bed becomes the feed gas preheater/combustion reactor and vice versa.

Use the following average values for C_p (kJ/mol·°C) in solving the problems to be given.

$$C_3H_6O: 0.126$$
$$O_2: 0.033$$
$$N_2: 0.031$$
$$CO_2: 0.050$$
$$H_2O: 0.038$$

(a) If the relative saturation of acetone in the feed stream is 12.2%, what is the stream temperature?

(b) Determine the composition of the gases after combustion, assuming that all of the acetone is converted to CO_2 and H_2O, and estimate the temperature of this stream.

(c) Estimate the rates (kW) at which heat is transferred from the inlet bed particles to the feed gas prior to combustion and from the combustion gases to the outlet bed particles. Suggest an alternative to the two-bed feed switching arrangement that would achieve the same purpose.

60. You have been given a sample of a fuel oil and asked to determine its heating value. You measure exactly 0.500 g of the fuel into a well-insulated reactor with a volume of 100.0 cm³, flush out the reactor interior with pure oxygen, and fill the reactor with oxygen to a gauge pressure of 23.45 atm. You seal the reactor and ignite the fuel. The reactor temperature quickly goes from its initial value of 25°C to a final value of 1091°C. You allow the gases to cool (but not enough for water to condense) and withdraw and analyze a sample. The analysis is 44.7 mole% O_2, 32.0% CO_2, 23.2% H_2O, and 0.14% SO_2.

Assume that $\Delta \hat{H}_c$ for the combustion reaction equals the lower heating value of the fuel; $C_v = 5.0$ J/g·°C for the product gas; the volume occupied by the fuel oil is negligible; the fuel contains negligible quantities of water and oxygen; and the reactor and noncombustible ash absorb 7% of the heat of reaction, with the balance going to heat the product gases.

(a) Calculate the ultimate analysis of the fuel oil—that is, the percentages by mass of C, H, S, and ash—and the lower heating value of the fuel oil in kJ/g.

(b) Use Table 9-10 on p. 9-10 of *Perry's Chemical Engineers' Handbook* to deduce the probable grade of the fuel oil (No. 1, 2, 3, 4, or 6).

61. Calculate the adiabatic flame temperature of liquid pentane at 25°C being burned with 30% excess oxygen (not air) fed at 75°C. Use heat capacity formulas from Table B.2 and Newton's rule (Appendix A.2) in the calculation. (See Example 9.6-2.)

62. Methane is burned completely with 25% excess air in a continuous adiabatic reactor. The feed gases enter the reactor at 25°C and 1 atm, and the combustion products emerge at a pressure of 1 atm. Methane is fed at a rate of 2000 liters/h. Calculate the volumetric flow rate and the dew point of the combustion products emerging from the reactor. In evaluating the adiabatic flame temperature, use the mean heat capacities of Table 8.3-1 for the products, and find T_{ad} by trial and error.

63. The adiabatic flame temperature of natural gas that contains methane, ethane, and propane is to be calculated from specified values of the following quantities:

$$y_{CH_4}, y_{C_2H_6}, y_{C_3H_8} = \text{mole fractions of fuel components}$$
$$T_f, T_a = \text{inlet temperatures of fuel and air, °C}$$
$$P_{xs} = \text{percent excess air}$$
$$y_{wo} = \text{mole fraction of water in the inlet air}$$

(a) For a basis of 1 g-mole of natural gas, calculate the gram-moles of each molecular species in the feed and product streams, assuming complete combustion and negligible CO formation. The answer should be expressed in terms of the variables given above.

(b) Given here are expressions for the specific enthalpies of the feed and product components, relative to their elements at 25°C.

$$\hat{H}_i(\text{kJ/mol}) = a_i + b_i T + c_i T^2 + d_i T^3 + e_i T^4, \; T \text{ in °C}$$

Substance (i)	a	b × 10²	c × 10⁵	d × 10⁸	e × 10¹²
CH_4 (1)	−75.72	3.431	2.734	0.122	−2.75
C_2H_6 (2)	−85.95	4.937	6.96	−1.939	1.82
C_3H_8 (3)	−105.6	6.803	11.30	−4.37	7.928
N_2 (4)	−0.7276	2.900	0.110	0.191	−0.7178
O_2 (5)	−0.7311	2.910	0.579	−0.2025	0.3278
$H_2O(v)$ (6)	−242.7	3.346	0.344	0.2535	−0.8982
Cl_2 (7)	−394.4	3.611	2.117	−0.9623	1.866

Derive the given expression for the specific enthalpy of methane from the heat capacity data in Table B.2. Then show that ΔH for the reactor is given by an expression of the form

$$\Delta H = \alpha_0 + \alpha_1 T + \alpha_2 T^2 + \alpha_3 T^3 + \alpha_4 T^4$$

where T is the product temperature, and

$$\alpha_0 = \sum_{i=4}^{7} (n_i)_{out} a_i - \sum_{i=1}^{3} (n_i)_{in} \hat{H}_i(T_f) - \sum_{i=4}^{6} (n_i)_{in} \hat{H}_i(T_a)$$

$$\alpha_1 = \sum_{i=4}^{7} (n_i)_{out} b_i \qquad \alpha_3 = \sum_{i=4}^{7} (n_i)_{out} d_i$$

$$\alpha_2 = \sum_{i=4}^{7} (n_i)_{out} c_i \qquad \alpha_4 = \sum_{i=4}^{7} (n_i)_{out} e_i$$

***(c)** Write a computer program to read in values of $y_{CH_4}, y_{C_2H_6},$ and $y_{C_3H_8}, T_f, T_a, P_{xs},$ and y_{wo}, and to solve the equation $\Delta H(T) = 0$ to determine the adiabatic flame temperature T to within 0.1°C. Use either Newton's rule (Appendix A.2) with an initial guess of 1000°C, or any other trial-and-error procedure you choose. Be

* Computer problem.

sure to build in a limit on the number of iterations in case the procedure does not converge.

(d) Test your program by calculating the adiabatic flame temperature of a natural gas containing 86% CH_4, 10% C_2H_6, and 4% C_3H_8 that enters the furnace at 40°C and is burned with 25% excess air that has been preheated to 150°C. Assume that on the day of the run the outside temperature is 27°C and the relative humidity is 87%.

64. Acetylene is produced by pyrolyzing (decomposing at high temperature) natural gas (predominantly methane).

$$2\,CH_4(g) \rightarrow C_2H_2(g) + 3\,H_2$$

The heat required to sustain this endothermic reaction is provided by feeding oxygen to the reactor and burning a portion of the methane to form primarily CO and some CO_2.

A simplified version of the process is as follows. A stream of natural gas, which for the purposes of this problem may be considered pure methane, and a stream containing 96.0 mole% oxygen and the balance nitrogen, are each preheated from 25°C to 650°C. The streams are combined and fed into an adiabatic converter, in which most of the methane and all of the oxygen are consumed, and the product gas is rapidly quenched to 38°C as soon as it emerges from the converter. The residence time in the converter is less than 0.01 s, low enough to prevent most but not all of the hydrocarbons from decomposing to form hydrogen and solid carbon particles (soot). Of the carbon in the feed, 5.67% emerges as soot.

The cooled effluent passes through a carbon filter in which the soot is removed. The clean gas is then compressed and fed to an absorption column, where it is contacted with a liquid solvent, dimethylformamide, or DMF (MW = 73.09). The off-gas leaving the absorber contains all of the hydrogen and nitrogen, 98.8% of the CO, and 95% of the methane in the gas fed to the column. The "lean" solvent fed to the absorber is essentially pure DMF; the "rich" solvent leaving the column contains all of the water and CO_2 and 99.4% of the acetylene in the gas feed. This solvent is analyzed and is found to contain 1.55 mole% C_2H_2, 0.68% CO_2, 0.055% CO, 0.055% CH_4, 5.96% H_2O, and 91.7% DMF.

The rich solvent goes to a stripping column, where it is heated at a reduced pressure, causing the dissolved gases to come out of solution. Two gas streams are taken off: one—the product gas—contains 99.1 mole% C_2H_2, 0.059% H_2O, and the balance CO_2, and the other—the stripper off-gas—contains methane, carbon monoxide, carbon dioxide, and water. The regenerated solvent is recirculated to the absorber.

A plant is designed to produce 5 metric tons/day of product gas. Calculate the following quantities.

(a) The required flow rates (SCMH) of the methane and oxygen feed streams.

(b) The molar flow rates (kmol/h) and compositions of the gas feed to the absorber, the absorber off-gas, and the stripper off-gas.

(c) The DMF circulation rate (kmol/h).

(d) The overall product yield (mol C_2H_2 in product gas/mol CH_4 in the feed to the

reactor), and the fraction that this quantity represents of its theoretical maximum value.

(e) The total heating requirement (kW) for the methane and oxygen feed preheaters.

(f) The temperature attained in the converter.

Chapter Ten

Computer-aided Balance Calculations

If you have worked your way through the preceding nine chapters, you are aware—maybe painfully so—that solving balance problems manually for even relatively simple systems can be a cumbersome, time-consuming, and error-laden task. Imagine now what it must be like to do it for, say, an entire petroleum refinery.

An alternative to this horrendous undertaking is to devise an **algorithm** (a systematic procedure) for performing the calculations, and then to program a computer to implement it. This technique is generally referred to as *flowsheet simulation*. (Flowsheet and flowchart are synonymous terms.) In this chapter, we introduce two approaches to flowsheet simulation that have been developed over the past two decades. In the first, termed the **sequential modular approach,** subroutines are written to solve the equations for each process unit, and a main program calls the subroutines in a sequence that more or less follows the process flowchart. In the second, known as the **equation-based approach,** all of the equations that govern the behavior of the individual units are collected and solved simultaneously.

Before discussing these techniques, we present some introductory material regarding what we have earlier referred to as process bookkeeping. You will recall that when you first learned how to approach material balance problems you were instructed to count the number of unknowns on a labeled flowchart, then to count the number of independent equations that could be written in terms of these unknowns. If the two numbers did not match, the problem was known to be either underdefined or overdefined; in either case, there was no point wasting time trying to find a solution until the discrepancy could be resolved. This procedure is an example of a general technique known as **degree-of-freedom analysis;** it constitutes an integral part of computer-aided balancing.

10.1 DEGREE-OF-FREEDOM ANALYSIS

The degrees of freedom of a system, n_{df}, is the number of variable values that must be specified before the remaining variables can be calculated. (This concept was first introduced in Section 6.2 in the context of the Gibbs phase rule.) If a system is described by n_e equations in n_v variables, then n_{df} equals $n_v - n_e$.

If $n_{df} = 0$ (e.g., if there are three equations in three unknowns), then the unknown variables can in principle be calculated. If the equations are all linear and independent, there is one and only one solution, while if any of the equations are nonlinear, there may be any number of solutions, including none (see Appendix A.2).

If $n_{df} > 0$—e.g., three equations in five unknowns, so that $n_{df} = 2$—then n_{df} variable values must be specified as part of the system definition. Once this has been done, the remaining unknown variables may be calculated from the system equations. Variables that are externally specified are called **design variables,** and those that are then calculated from the system equations are called **state variables.**

If $n_{df} < 0$, there are more equations than variables, and the process is overdefined; either you have written more equations than you are entitled to write (e.g., three material balances on a two-component system), or you have miscounted the variables (probably by not fully labeling the flowchart).

The remainder of this section illustrates degree-of-freedom analysis for increasingly complex system: first a single nonreactive process unit, then a single reactive unit, and finally a system of several interconnected units.

10.1a Single Nonreactive Process Unit

The number of unknown variables for a single unit is the sum of the unknown component amounts or flow rates for all inlet and outlet streams, plus all unknown stream temperatures and pressures, plus the rate of heat input. The equations available to determine these unknowns include material balances for each independent species, an energy balance, phase and chemical equilibrium relations, and additional specified relationships among the process variables.

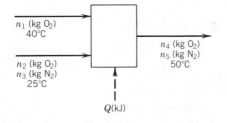

Let us first consider a simple heated mixer in which a stream of oxygen and nitrogen at 25°C is mixed with a stream of pure oxygen at 40°C, and the combined stream is heated to 50°C. A degree-of-freedom analysis on this simple process would proceed as follows:

$$6 \text{ variables } (n_1, \ldots, n_5, \text{ and } Q)$$
$$\underline{-3 \text{ relations (2 material balances, and 1 energy balance)}}$$
$$= 3 \text{ degrees of freedom}$$

Thus, three specifications would have to be provided in the problem statement, following which the system equations could be solved for all remaining unknowns. Specifying values of three of the variables would suffice, as would giving values for two of the variables and a relationship between the masses of the two feed streams. (Convince yourself—choose values for any three variables and mentally go through the calculation of the remaining three from the system equations.)

The following example illustrates a degree-of-freedom analysis for a process that involves vapor–liquid equilibria between effluent process streams.

EXAMPLE 10.1-1 Degree-of-Freedom Analysis of a Flash Vaporizer

A liquid mixture of *n*-hexane (HX) and *n*-heptane (HP) at a high pressure is abruptly exposed to a lower pressure. A portion of the mixture evaporates, yielding a vapor mixture relatively rich in hexane (the more volatile of the two feed components) and a residual liquid mixture relatively rich in heptane. The two product streams are in equilibrium at temperature T and pressure P; their compositions are related by Raoult's law (Section 6.4b).

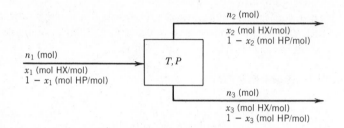

The relations among the system variables are as follows:

$$n_1 = n_2 + n_3 \qquad \text{(Total mole balance)} \qquad (1)$$

$$n_1 x_1 = n_2 x_2 + n_3 x_3 \qquad \text{(Hexane balance)} \qquad (2)$$

$$\left. \begin{array}{l} x_2 P = x_3 p^*_{\mathrm{HX}}(T) \\[4pt] (1 - x_2)P = (1 - x_3)p^*_{\mathrm{HP}}(T) \end{array} \right\} \qquad \text{(Raoult's law)} \qquad \begin{array}{l}(3)\\[4pt](4)\end{array}$$

$$\left. \begin{array}{l} \log[p^*_{\mathrm{HX}}(T)] = A_{\mathrm{HX}} - B_{\mathrm{HX}}/(T + C_{\mathrm{HX}}) \\[4pt] \log[p^*_{\mathrm{HP}}(T)] = A_{\mathrm{HP}} - B_{\mathrm{HP}}/(T + C_{\mathrm{HP}}) \end{array} \right\} \qquad \text{(Antoine equation)} \qquad \begin{array}{l}(5)\\[4pt](6)\end{array}$$

Perform a degree-of-freedom analysis on this system, and outline a solution procedure for the following sets of design variables.

1. $n_1, x_1, T,$ and x_3.
2. $n_2, x_2, n_3,$ and x_3.
3. $n_1, n_2, n_3,$ and T.

Solution

Ten variables—$n_1, n_2, n_3, x_1, x_2, x_3, T, P, p^*_{\mathrm{HX}},$ and p^*_{HP}—appear in these six equations. The system consequently has $10 - 6 = 4$ degrees of freedom. No more nor less than four of these variables must be specified in a problem statement, following which the remaining six state variables can be calculated from the given six equations. Notice how the difficulty of the calculation depends dramatically on which four variables are specified (i.e., which ones are the design variables).

1. *Design variables $n_1, x_1, T,$ and x_3*
 If values of these four variables are specified, the following relatively straight-forward solution procedure could be used to solve for the remaining (state) variables.

The first four equations each contain one unknown, and can be solved directly; the final two equations in two unknowns are linear, so they also can be solved without difficulty.

2. ***Design variables*** n_2, x_2, n_3, ***and*** x_3

For this choice of design variables, a relatively lengthy trial-and-error procedure is required to solve for the remaining variables. One such procedure is shown here.

Several of the methods outlined in Appendix A.2 can be used for the trial-and-error calculation, such as defining $F(T) = x_2(\text{calc}) - x_2(\text{specified})$, and using the regula-falsi method to find the value of T for which $F(T) \approx 0$.

3. **Design variables n_1, n_2, n_3, and T**
 Some choices of design variables lead to contradictions of the system equations, so they are not allowed. For example, choosing n_1, n_2, n_3 as design variables is forbidden; once any two of these variables are specified, the third may not be assigned a value independently, but must be calculated from Equation 1.

Counting variables and relations can be tricky. For example, consider a stream of air and water vapor that passes through a condenser in which 80% of the water and none of the air is condensed. The flowchart might appear as follows.

In doing the degree-of-freedom analysis on this unit, we count four variables (n_1, \ldots, n_4), and we might be tempted to subtract two balances (one for air and one for water), at least one additional relation (80% of the water is condensed), and possibly a second additional relation (none of the air is condensed), giving us either one or zero degrees of freedom.

We can also look at the flowchart and see by inspection how many degrees of freedom the system must have. It clearly cannot be zero: There is no way to calculate any of the variables without being given values for a single one of them. It also cannot be one: If you are given n_1, for example, you have no way of calculating n_2, n_3, and n_4. If, however, you are given values for both n_1 and n_2, you can determine n_4 as 80% of n_2, and then calculate n_3 as $n_2 - n_4$. The process therefore must have two degrees of freedom.

Where did our original degree-of-freedom analysis go wrong? We made two mistakes: When we did not label an air flow rate in the condensate, we implicitly used the information that all the air leaves in a single outlet stream, and when we gave the outlet flow rate of air the same label as the inlet flow rate, n_1, we implicitly wrote a balance on air. Since these two items of information have already been built into the flowchart, we cannot include them when itemizing the relations available to help solve for the unknown variables. The correct analysis is therefore as follows:

$$\begin{array}{l} 4 \text{ variables } (n_1, \ldots, n_4) \\ \underline{- 2 \text{ relations } (\% \text{ condensation, water balance})} \\ = 2 \text{ degrees of freedom} \end{array}$$

TEST YOURSELF

1. Consider the following set of equations:

$$x_1 + 2x_2 - x_3^2 = 0$$
$$5x_1 - x_2^3 + 4 = 0$$

 (a) How many degrees of freedom does this system possess?
 (b) Which choice of a design variable would lead to an easier procedure to solve for the state variables, x_1 or x_3?

2. A liquid stream containing species A, B, C, and D enters a vaporizer (flow rates n_1, \ldots, n_4 kg/s). Essentially all of the A, 25% of the B, and none of the D is vaporized. The C is distributed between the liquid and vapor products according to a known equilibrium relationship. How many degrees of freedom does this process possess?

10.1b Single Reactive Process Unit

Consider a chemical reactor in which the single reaction A → B takes place. A flowchart for this unit is shown here.

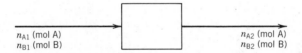

n_{A1} (mol A)
n_{B1} (mol B)

n_{A2} (mol A)
n_{B2} (mol B)

If there were no reaction, two degrees of freedom would be associated with this process (4 variables − 2 material balances = 2 degrees of freedom). Thus, if you were given, say, that $n_{A1} = 60$ and $n_{B1} = 40$, you could easily determine the values of the remaining two variables.

But there is reaction, and a consequent change in the degrees of freedom. You might again be given values for n_{A1} and n_{B1}, but now you would not be able to calculate n_{A2} and n_{B2}; their values would depend on how much A reacts. If, however, you are given the value of n_{A2} in addition to the first two given values, you could calculate the value of n_{B2}. (Try it: if $n_{A1} = 60$, $n_{B1} = 40$, and $n_{A2} = 20$, what is n_{B2}?) Thus, the reactive process has *three* degrees of freedom instead of two.

Note: If it were desired to calculate the required heat transfer to or from the process, a new variable (Q) would be labeled and an additional relation (the energy balance) would be counted, so that the degrees of freedom would still equal three.

 In general, *each independent chemical reaction occurring among the species in a process system adds one degree of freedom to the number that would be calculated in the absence of reactions.* To be able to solve the system equations for all unknown variables, you need an additional piece of information for each independent reaction.

It might be a specified value of one of the process variables, or a specified fractional conversion, yield, or selectivity, or an equilibrium relationship. The next example illustrates the required analysis.

EXAMPLE 10.1-2 Degree-of-Freedom Analysis of a Chemical Reactor

Methane is burned with 25% excess air. A methane conversion of 80% is achieved, and the molar ratio of CO_2 to CO in the product gas is 10:1. Perform a degree-of-freedom analysis on this process.

Solution

n_f (mol CH_4)

25% excess air

n_c mol O_2
3.76 n_c mol N_2

n_1 mol CH_4
n_2 mol O_2
n_3 mol N_2
n_4 mol CO_2
n_5 mol CO
n_6 mol H_2O

$$CH_4 + 2O_2 \rightarrow CO_2 + 2H_2O$$

$$CH_4 + (3/2)O_2 \rightarrow CO + 2H_2O$$

8 variables (n_f, n_o, n_1, ..., n_6)
−6 balances (CH_4, O_2, N_2, CO_2, CO, H_2O)
−3 additional relations (the specified percent excess air, methane conversion, and CO_2/CO selectivity)
+2 additional degrees of freedom for the two reactions
= 1 degree of freedom

One variable must therefore be specified for the system to be determinate. At this point, we observe that we did not follow our usual practice of specifying a basis of calculation. If we assign a value to, say, n_f (e.g., 100 mol CH_4), the system then has zero degrees of freedom, and the problem is therefore well-specified. The actual solution is straightforward. (Outline it mentally.)

In the previous section, we pointed out several mistakes that are commonly made in counting variables and relations for nonreactive processes. The same problems occur for reactive processes, along with several new ones.

Consider, for example, a reactor in which nitric oxide is oxidized completely to nitrogen dioxide with 10% excess oxygen:

$$2NO + O_2 \rightarrow 2NO_2$$

The flowchart of the reactor might be drawn and labeled as shown here.

n_1 (mol NO/s)
n_2 (mol O_2/s)

n_3 (mol NO_2/s)
n_4 (mol O_2/s)

The degree-of-freedom analysis might then proceed as follows:

$$
\begin{array}{l}
4 \text{ variables } (n_1,\ldots,n_4) \\
-3 \text{ balances } (NO, O_2, NO_2) \\
-2 \text{ additional relations } (10\% \text{ excess } O_2, 100\% \text{ conversion of NO}) \\
+1 \text{ chemical reaction} \\
\hline
=0 \text{ degrees of freedom}
\end{array}
$$

This result is clearly wrong, however: There is obviously no way to calculate n_1 through n_4 without being given any additional information. In fact, you can see by inspection that the unit has one degree of freedom: If you are given the value of n_1, you can calculate n_2 from the given percentage excess of oxygen, and then n_3 and n_4 from the specified fractional conversion of NO and the reaction stoichiometry.

Our mistake was overcounting the number of additional relations. When we labeled the flowchart and did not show a flow rate of NO in the exit stream, we implicitly used the information that the conversion of NO is 100%. We cannot count this information again as a separate relation to determine the remaining unknowns. When we eliminate this relation, the degree-of-freedom analysis yields the correct value of one.

TEST YOURSELF

Water is electrolyzed to form hydrogen and oxygen in a continuous electrolytic cell. Not all of the water that enters the cell is electrolyzed. How many degrees of freedom does this system possess?

10.1c Multiple-Unit Processes

A process that contains two or more interconnected units has fewer degrees of freedom than the sum of the degrees of freedom of the individual unconnected units. Suppose, for example, that an output stream from the first of two units is the feed stream to the second unit and that no reactions occur in either unit. (See Fig. 10.1-1.)

Let us first perform a degee-of-freedom analysis for unit A. We count nine variables (three for each stream entering or leaving the unit), and three equations (mass balances on methane, ethane, and ethylene), to obtain six degrees of freedom. The same result is obtained for unit B. Individually, then, there are 12 *local* degrees of freedom—local signifying that we are considering each unit as isolated, unconnected to the others.

However, the units are in fact linked, and in the above bookkeeping we have consequently overcounted variables: Specifically, the three variables associated with the product stream from unit A (n_3, x_3, and y_3) are identical to the three associated with the feed to unit B. The bookkeeping proceeds as follows.

$$
\begin{array}{rcl}
\text{System variables} = 18 \text{ local} - 3 \text{ overcounted} &=& 15 \\
\text{System equations} = 3 \text{ balances for each unit} &=& 6 \\
\hline
\text{Degrees of freedom} &=& 9
\end{array}
$$

FIGURE 10.1-1

Flowchart of a two-unit process

Thus, nine process stream variable values must be specified for the given system, at which point balances can be used to determine the remaining six variables.

In general, the number of degrees of freedom of a system of linked units equals the number of degrees of freedom of the individual units—that is, the total local degrees of freedom–minus the number of **ties** (variables common to two or more units, and hence overcounted), minus the number of any additional relations that were not counted in the individual unit analyses.

The procedure for performing a degree-of-freedom analysis on a process may be summarized as follows:

1. *Draw and completely label the flowchart.*

2. *Determine the local degrees of freedom for each process unit, mixing point, and stream-splitting point in the process.*

 The local degrees of freedom equals the number of unknown variables involved in the unit equations, minus the number of relations among these variables, plus the number of independent chemical reactions taking place in the unit. The relations may include material balances, an energy balance, equilibrium relations, and specifications of conversions, yields, selectivities, and mass ratios. If a relation involves variables belonging to units other than the one in question, however, count it in step 3 rather than here.

3. *Determine the degrees of freedom of the total process.*

 This quantity equals the sum of the local degrees of freedom, minus the number of relations among process variables that were not counted in step 2 (such as overall reactant conversions or percentage recoveries), minus the number of ties (overcounted variables) between units.

If the number of degrees of freedom determined in this manner equals zero, the process is well-specified, and all the unknown variables can in principle be determined. If n_{df} is greater than zero, then n_{df} variable values must be specified as design variables, and the remaining state variables may then be calculated.

The following example illustrates this procedure.

EXAMPLE 10.1-3 Degree-of-Freedom Analysis of a Multiple-Unit Process

Methanol may be produced from carbon monoxide and hydrogen by the reaction

$$CO + 2H_2 \rightarrow CH_3OH$$

The fresh feed, which contains CO and H_2 in stoichiometric proportion, enters the process at a rate of 2.2 m³/s at 25°C and 6.0 MPa, and combines adiabatically with a recycle stream. The combined stream is heated to 250°C, and the reactor effluent emerges at the same temperature. The effluent is cooled to 0°C at $P = 6.0$ MPa, partially condensing the product methanol. The gas leaving the condenser is saturated with methanol: 1% is taken off as a purge, and the remainder is recycled. An overall conversion of 98% is achieved. Ideal gas behavior may be assumed.

Carry out a degree-of-freedom analysis for this process, and show that the given information is sufficient to allow the calculation of the component flow rates for all streams, the required heat duties for all process units, and the temperature of the combined feed stream to the reactor preheater.

Solution

In labeling the flowchart, we have taken advantage of the fact that CO and H_2 remain in stoichiometric proportion throughout the process, which cuts down on the number of unknown variables that must be defined: If the flow rate of CO is n_i, the flow rate of H_2 must be $2n_i$. However, this also means that we cannot write independent balances on these two species. We must therefore allow for no more than *two* independent material balances on each process unit when we do the degree-of-freedom analysis, rather than the three that would normally be allowed when three species are involved in a process.

Mixing Point

$$
\begin{aligned}
&\text{6 variables } (n_0, n_1, n_2, n_3, n_6, T_a)\\
&-4 \text{ relations (2 material balances, 1 energy balance,}\\
&\qquad \text{and the ideal gas law for the fresh feed)}\\
\hline
&=2 \text{ local degrees of freedom}
\end{aligned}
$$

Heater

4 variables (n_1, n_2, T_a, Q_h)

-1 relation (The energy balance. We cannot count material balances, since in labeling the outlet flow rates the same as the inlet flow rates, we have implicitly used up the two allowed material balances.)

$= 3$ local degrees of freedom

Reactor

5 variables (n_1, n_2, n_3, n_4, Q_r)

-3 relations (2 material balances, 1 energy balance)

$+1$ chemical reaction

$= 3$ local degrees of freedom

Condenser

5 variables (n_3, n_4, n_5, n_6, Q_c)

-3 relations [1 material balance (Can you see why?), 1 energy balance, saturation condition at outlet]

$= 2$ local degrees of freedom

Purge Point

2 variables (n_3, n_6)

-0 relations (In labeling the chart, we have built in all allowed balances. If this is not clear to you, try writing a balance on, say, hydrogen, and see the result.)

$= 2$ local degrees of freedom

Process

12 local degrees of freedom $(2 + 3 + 3 + 2 + 2)$

-11 ties (n_1, n_2, and n_6 were each counted three times, n_3 was counted four times, and n_4 and T_a were each counted twice)

-1 additional relation (98% overall conversion)

$= 0$ degrees of freedom

The process is therefore well-specified.

The solution of the system equations for all unknown variables is straightforward. The ideal gas law applied to the fresh feed stream yields n_o. The specified overall conversion yields n_3 as 2% of n_o. The saturation condition at the condenser outlet combined with the calculated value of n_3 yields n_6, and an overall carbon balance yields n_5. Balances on CO and CH_3OH at the mixing point yield n_1 and n_2, and an energy balance for the same subsystem yields T_a. An energy balance on the heater then yields Q_h. A carbon balance on either the reactor or the condenser yields

n_4, and then energy balances on the reactor and the condenser yield Q_r and Q_c, respectively.

TEST YOURSELF

Consider the flowchart shown here for a nonreactive mixing–separation process.

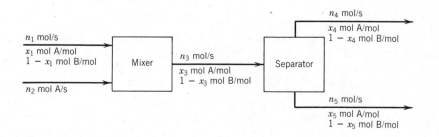

1. How many local degrees of freedom does each unit possess?
2. How many total degrees of freedom does the system possess?

10.1d Solution of Algebraic Equations

Once you have specified values for your design variables, you are faced with the task of solving the system equations for the state variables. For relatively simple processes, the manual techniques presented in previous chapters are applicable and should be used.

Unfortunately, like most things in life, chemical processes are seldom relatively simple. For example, most problems you have encountered in this text could eventually be reduced to the solution of linear algebraic equations.

$$0.21 n_{air} = 2015 \qquad \text{Solve for } n_{air}$$

$$\left.\begin{array}{l} V + B = 100 \\ 0.96V + 0.05B = 50 \end{array}\right\} \text{Solve for } V \text{ and } B$$

However, some could not. You encountered nonlinear equations, for example, when learning about gas laws:

$$\frac{PV}{RT} = 1 + \frac{B}{V} + \frac{C}{V^2} + \frac{D}{V^4} + \frac{E}{V^5} \qquad \begin{array}{l} \text{Given } P \text{ and } T, \\ \text{solve for } V \end{array}$$

and when studying vapor–liquid equilibrium

$$\log_{10} p = A + B/(C + T) \qquad \text{Given } T, \text{ solve for } p$$

and the first law of thermodynamics

$$\Delta H = a(T - 25) + b(T^2 - 25^2) + c(T^3 - 25^3) \qquad \text{Given } \Delta H, \text{ solve for } T$$

Solving sets of linear equations is tedious, but straightforward. Sets involving three or fewer unknowns can be solved using your favorite high school algebra technique— substitution, or addition and subtraction. For sets involving four or more unknowns, it is advisable to set the equations up in matrix form:

$$a_{11}x_1 + a_{12}x_2 + \ldots + a_{1n}x_n = b_1$$
$$a_{21}x_1 + a_{22}x_2 + \ldots + a_{2n}x_n = b_2$$
$$\vdots$$
$$a_{n1}x_1 + a_{n2}x_2 + \ldots + a_{nn}x_n = b_n$$

$$(10.1\text{-}1)$$

where the a's and b's are known constants. (For example, a_{23} signifies the coefficient of x_3 in the second equation.) If a variable does not appear in an equation, the corresponding a_{ij} equals zero. Most computers have built-in subroutines for solving sets of equations like this; you provide as input the values of the coefficients (a_{ij}) and right-hand sides (b_i), and the program determines the solution (x_1, \ldots, x_n).

Solving nonlinear equations, on the other hand, may or may not be straightforward. Doing so in all but the simplest cases involves trial-and-error, and there is usually no guarantee that you will be able to find a solution, or that a solution you find is the only possible solution, or even that a solution exists.

Some of the examples in the sections that follow and a number of the problems at the end of the chapter involve nonlinear equations. In some cases, the solutions will be easy to obtain; in others, they will not be. When you find yourself confronted with one in the latter category, refer to Appendix A.2, which discusses solution techniques for nonlinear equations.

10.2 SEQUENTIAL MODULAR SIMULATION

As we noted at the beginning of this chapter, there are two broad approaches to the automated solution of the balance equations for a process system: the **sequential modular approach** and the **equation-based approach.** This section outlines the first of these methods. The balance equations (and any other equations that may arise from physical considerations or process specifications) for each unit are written and solved. If there are no recycle streams, the calculation moves from one unit to another, until all units have been covered. If there is a **cycle** (the conventional term for a recycle loop in a process flowchart), a trial-and-error procedure is required: Values of one or more stream variables in the cycle are assumed; the balance equations for units in the cycle are solved, one unit at a time, until the values of the assumed variables are recalculated; and the procedure is repeated until the assumed and calculated values agree.

The object of the presentation that follows is not to enable you to go out and simulate complex processes—the treatment is too brief and general for that—but simply to give you a feeling for the structure of process simulation programs and calculations and to provide an idea of what such programs can do. For a more detailed examination of the features, applications, and theory of simulation programs, the references given at the end of Section 10.4 should be consulted.

10.2a Elements of a Modular Simulation

The first step in setting up a process for the sequential modular approach is to reconstruct the process flowchart in terms of **blocks** (process units or operations) and streams connecting them. Several types of blocks and names that might be given to them are as follows:

MIX	Mix several inlet streams adiabatically to form one product stream.
SPLIT	Split a single inlet stream into two or more product streams with the same composition and temperature.
COMPRESS	Raise the pressure of a gas by a specified amount.
PUMP	Raise the pressure of a liquid by a specified amount.
FLASH	Convert a liquid stream at one pressure to liquid and vapor streams in equilibrium at a lower pressure.
DISTILL EXTRACT CRYSTAL ABSORB	Simulate the separation processes of distillation, extraction, crystallization, and absorption.
REACT	Simulate a chemical reactor.

Two additional block types—convergence blocks and control blocks—have functions to be described later.

The simulation program contains a built-in subroutine corresponding to each block type. To simulate a process, you would write an executive program that would include a series of calls to these subroutines, resulting in the solution of the material and energy balance equations for the process, unit by unit.

For example, suppose two streams labeled S1 and S2 are mixed adiabatically to form a third stream S3. A MIX block might be used to simulate this operation. (What the subroutine for this block might look like is the subject of the next example.)

The step in the program that would perform the material and energy balances for this unit would be initiated by a statement equivalent to the Fortran statement

<div align="center">CALL MIX(M1,S1,S2,S3)</div>

M1 is a label, identifying the unit that has streams S1 and S2 as inputs and stream S3 as an output. In the subroutine, S1, S2, and S3 would each be arrays, the elements of which contain enough information to specify the flow rate, composition, and temperature of each stream. Prior to this statement in the program, the values of the elements of arrays S1 and S2 would have been either specified or determined as the

outputs of other blocks. The subroutine would calculate the component flow rates of stream S3 by solving material balance equations and (if required) the temperature of stream S3 by solving the energy balance equation.

EXAMPLE 10.2-1 Structure of a Mixing Block Subroutine

Two streams, S1 and S2, are mixed adiabatically to form a stream S3. Each stream may contain any of five components, and the feed streams may be at different temperatures. The heat capacities of all components are constants, and the heat of mixing may be neglected. Write a Fortran subroutine to calculate the component molar flow rates and temperature of the product stream from specified values of these quantities for the feed streams.

Solution

A degree-of-freedom analysis of this unit reveals 18 variables (six for each stream) and six equations (five material balances and an energy balance) for a net 12 degrees of freedom. The design variables are the 12 variables corresponding to the feed streams, leaving the component flow rates and temperature of the product stream as the state variables.

The material balance equations are trivial:

$$(S3)_i = (S1)_i + (S2)_i, \qquad i = 1, 5$$

The energy balance equation for this open adiabatic system is $\Delta H = 0$, or taking the temperature of stream S1 as the reference:

$$\left[\sum_{i=1}^{5} (S3)_i C_{pi} \right] (T_3 - T_1) - \left[\sum_{i=1}^{5} (S2)_i C_{pi} \right] (T_2 - T_1) = 0 \text{ (Verify!)}$$

$$\Downarrow$$

$$T_3 = T_1 + \frac{\left[\sum_{i=1}^{5} (S2)_i C_{pi} \right] (T_2 - T_1)}{\sum_{i=1}^{5} (S3)_i C_{pi}}$$

We will presume that the heat capacities have been specified in the main program and their values are transmitted to this subroutine through a common block in an array CP. The arrays S1, S2, and S3 each have six elements—the molar flow rates of the five components, and the stream temperature.

```
        SUBROUTINE MIX(M,S1,S2,S3)
C****   M IS A UNIT LABEL
C****   ARRAYS S1 AND S2 ARE INPUT, AND S3 IS OUTPUT
        DIMENSION S1(6),S2(6),S3(6)
        COMMON CP(5)
C****   SOLVE MATERIAL BALANCES
        DO 100J = 1,5
  100   S3(J) = S1(J) + S2(J)
C****   IF T1 = T2, SET T3 = T1 AND RETURN
        T1 = S1(6)
        T2 = S2(6)
        IF (T1.EQ.T2) GO TO 200
C****   EVALUATE SUMS FOR ENERGY BALANCES
        SUM2 = 0.0
        SUM3 = 0.0
        DO 150J = 1,5
        SUM2 = SUM2 + S2(J)*CP(J)
  150   SUM3 = SUM3 + S3(J)*CP(J)
C****   SOLVE ENERGY BALANCE FOR T3
        T3 = T1 + SUM2*(T2-T1)/SUM3
        S3(6) = T3
        RETURN
  200   S3(6) = T1
        RETURN
        END
```

The next example illustrates the structuring of a sequential modular process simulation using blocks of the types just described.

EXAMPLE 10.2-2 Simulation of an Acyclic Process

The flowchart shown here depicts a multistage separation process. Three liquid streams are mixed adiabatically; the product stream is pumped through a heater to a distillation

column, and the overhead still product is partially condensed to yield liquid and vapor products.

Using the blocks MIX (mix two streams to form a third), PUMP, HEAT, DISTILL, and CNDS, construct a block diagram for the simulation of this process.

Solution

Since three streams are mixed in the process but the MIX routine can only handle two feed streams, we need to use this block twice. In other respects, the block diagram looks like the process flowchart.

In the main simulation program, a series of statements resembling the following would be included (the first argument of each calling statement is the unit identifier):

```
CALL MIX(M1,S1,S2,SA)
CALL MIX(M2,S3,SA,S4)
CALL PUMP(P1,S4,S5,parameters)
CALL HEAT(H1,S5,S6,parameters)
CALL DISTILL(ST1,S6,S7,S8,parameters)
CALL CNDS(C1,S7,S9,S10,parameters)
```

The PUMP calculation would cause the pressure of the liquid stream to be raised by a specified amount; the routine might also calculate the required shaft work for this operation and could go as far as to specify the type and size of pump needed for the job. The HEAT routine would calculate the heat input required to achieve the required temperature change and might also perform heat exchanger design calculations.

The DISTILL subroutine would solve material and energy balance equations to determine the flow rates and compositions of both product streams and the heat requirement for the distillation column. Depending on the level of sophistication of the simulation program, the subroutine might then go on to determine details of the column design for the given separation.

The CNDS subroutine would proceed in a similar manner for the condenser calculations—first solving material and energy balance calculations to determine unknown flow rates and the heat duty, and then possibly proceeding to perform design calculations.

10.2b Cyclic Systems and the Convergence Block

Suppose the block diagram of a process has the following appearance.

Further suppose that you know the values of the variables of stream S1 and you wish to calculate the remaining stream variables.

The procedure outlined in the previous section will not work, since there is no way to get the calculation started. To solve the equations for block B1 you must know the variable values for both S1 and S5, but you cannot know S5 until you solve B3, for which you need S3 and hence the solution of B2, which in turn requires a knowledge of S2 and hence the solution of B1, and you are back where you began.

If the calculations were to be done by hand, overall system and subsystem balances would eventually yield n equations in n unknowns, and the equations could then in principle be solved for all the desired process variables. It would be difficult to write a general program to implement this method for an arbitrary process, however. Instead, the following iterative approach is used.

1. Assume values of the variables of a stream within the cycle (this is known as **tearing the cycle**).

2. Solve the system balance equations, working around the cycle from unit to unit until the tear stream variables are recalculated.

3. If the assumed and calculated tear stream variables agree within a specified tolerance, the solution is complete; if they do not, use the new values (or some combination of the new and old values) to initiate another swing around the cycle, and iterate in this manner until convergence is achieved.

Suppose, for example, that in the example just given we choose S5 as the tear stream. The flowchart might appear as shown here:

The calculation would begin with an assumption of values of the variables associated with stream S5. The equations for blocks B1, B2, and B3 could now be solved sequentially. The solution for B3 would yield a new set of values for stream S5. The

procedure would be repeated until the assumed and recalculated values of S5 agree to within a specified tolerance.

What you are doing with this procedure, if you think about it, is tantamount to solving an equation of the form $x = f(x)$, where $f(x)$ is the "function" that generates a new value of the tear stream variable (or variables) x by working around the cycle. Techniques described in Appendix A.2—successive substitution and Wegstein's algorithm—can be used to perform this calculation.

Most simulation programs have a block, called a **convergence block,** that performs such calculations using the bounded Wegstein algorithm. The output stream from this block contains the guessed set of tear stream variables, and the input stream contains the values calculated by working around the cycle. The block diagram for the simulated process would appear as follows.

It is conventional to show the convergence block using dashed lines, since it does not correspond to a physical operation in the process.

EXAMPLE 10.2-3 Simulation of a Cyclic Process

The gas-phase dehydrogenation of butane to butene

$$C_4H_{10} \rightarrow C_4H_8 + H_2$$

is carried out in a catalytic reactor, in which the overall conversion of butane is 93%. The reactor effluent is fed to a multiple-unit separation process. One product stream contains all the hydrogen, 95% of the butene, and 1% of the butane leaving the reactor. The second product stream is recycled to the reactor. It is desired to calculate the quantities of each component in each stream for an assumed basis of 100 mol of fresh butane feed.

1. Draw and label a flowchart of this process, and carry out a degree-of-freedom analysis for the system.

2. Choose tear stream variables, and convert the flowchart into a block diagram for a flowsheet simulation, using blocks MIX, REACT, DISTILL, and a convergence block CONVG.

Solution

1. **Basis: 100 mol fresh feed**

Three "units" comprise this process: the reactor, the separation process, and the mixing point, which in the process may be a stirred tank or simply a tee in a pipeline. Let us first calculate the local degrees of freedom associated with each unit, and then the total process degrees of freedom.

Mixing Point

$$4 \text{ unknowns } (n_{A1}, n_{B1}, n_{A4}, n_{B4})$$
$$\underline{-2 \text{ equations (material balances)}}$$
$$= 2 \text{ local degrees of freedom}$$

Reactor

$$5 \text{ unknowns } (n_{A1}, n_{B1}, n_{A2}, n_{B2}, n_{C2})$$
$$-3 \text{ equations (3 material balances)}$$
$$\underline{+1 \text{ chemical reaction}}$$
$$= 3 \text{ local degrees of freedom (See Section 10.1b.)}$$

Separator

$$8 \text{ unknowns}$$
$$-3 \text{ balances}$$
$$\underline{-2 \text{ additional relations (the splits of butane and butene)}}$$
$$= 3 \text{ local degrees of freedom}$$

Process

$$8 \text{ local degrees of freedom } (2 + 3 + 3)$$
$$-1 \text{ additional relation (the overall butane conversion)}$$
$$\underline{-7 \text{ ties } (n_{A1}, n_{B1}, n_{A2}, n_{B2}, n_{C2}, n_{A4}, \text{ and } n_{B4} \text{ were each counted twice})}$$
$$= 0 \text{ net degrees of freedom}$$

(We used the information that the recycle contains no hydrogen when we did not show a variable n_{C4} on the flowchart, so we should not count this fact as a fourth additional relation.) The problem can in principle be solved.

2. Before planning the solution strategy, let us observe the flowchart and confirm that we cannot proceed in a unit-to-unit manner without trial and error.

The mixing unit has two local degrees of freedom and four unknowns. We cannot solve for n_{A1} and n_{B1} until we know n_{A4} and n_{B4}. To determine these variables, we must solve the separation process balance equations—but we cannot do so unless we know n_{A2}, n_{B2}, and n_{C2}. To calculate *these* variables, we must

solve the reactor equations. Unfortunately, we cannot solve them either, since we must first know n_{A1} and n_{A2}, for which we must solve the mixer equations, and we are back to where we started.

The solution is to tear the cycle. We can tear it in any of three places: between the mixing point and the reactor, between the reactor and the separation process, or between the separation process and the mixing point. The first and third choices involve the trial-and-error determination of two variables each, while the second one involves three variables. The fewer variables you have to search for by trial and error, the better off you generally are. Let us, therefore, choose the recycle stream as the tear stream. (The feed to the reactor would do equally well.)

The block diagram for the simulation now appears as follows.

The program would take as input a guessed pair of values for the tear stream variables n_{A4} and n_{B4} (stream S4A) and might contain the following sequence of statements.

```
CALL MIX(M1,S0,S4A,S1)
CALL REACT(R1,S1,S2,parameters)
CALL SEPAR(SP1,S2,S3,S4R,parameters)
CALL CONVG(C1,S4R,S4A)
```

(The first unit called is always the one following the tear stream.) The call of SEPAR would result in the recalculation of n_{A4} and n_{B4} from the separation process balances (stream S4R). CONVG would compare these values with the ones initially assumed. If they agree to within a specified tolerance, the calculation would be terminated; otherwise, the new values would be used to begin another journey around the cycle. The nature of the iterative procedure performed by CONVG is discussed in greater detail in Appendix A.2.

Networks of recycle loops are commonly encountered in large processes, and a suitable choice of a tear stream may minimize the number of iterations required to solve the balance equations of such systems. For example, consider the first block diagram on the next page. There are two cycles in this process: S2-S3-S4 and S3-S5-S7. To solve the system equations you could, for example, tear both S4 and S7, which would require the inclusion of two convergence blocks and hence the simultaneous

solution of two iterative loops; however, you can instead tear one stream common to both cycles (S3), thereby probably (but not necessarily) decreasing the computation time required to achieve the solution.

On the other hand, you will commonly encounter systems where you have no choice but to tear two or more streams, as in the next process.

There are three cycles: S2-S3-S4-S5, S7-S9-S11, and S3-S4-S6-S7-S8, and no single stream that if torn would permit the solution of all of the system equations. For instance, if you tear S3 you could work your way around the first cycle to unit M2 but you would be stuck there for lack of knowledge of S8, and you would be similarly stuck at unit M3 in the third cycle since you would not know S11.

One of many possible ways to analyze this system is to tear S4 and S7 (so that at least one stream in each cycle is torn). You must call the block programs in an order for which they can be solved. Remember that S1, S4B, and S7B are presumably known, and convince yourself that the following program would in principle work.

CALL DISTILL (D1, S4B, S5, S6, parameters)	(Calculates S5, S6)
CALL MIX (M1, S1, S5, S2)	(Calculates S2)
CALL SPLIT (SP1, S7B, S8, S9, parameters)	(Calculates S8, S9)
CALL MIX (M2, S2, S8, S3)	(Calculates S3)
CALL REACT (R1, S3, S4A, parameters)	(Calculates S4A)
CALL CONVG (C1, S4A, D1, S4B, parameters)	(Forces convergence of S4)
CALL DISTILL (D2, S9, S10, S11, parameters)	(Calculates S10, S11)
CALL MIX (M3, S6, S11, S7A)	(Calculates S7A)
CALL CONVG (C2, S7A, D1, S7B)	(Forces convergence of S7)

Other choices of tear streams could be made, and some simulation programs would permit you to use a single convergence block to force convergence of both tear streams simultaneously.

A rapidly growing body of literature deals with the analysis of process flowcharts to determine systematically how many streams must be torn and in which order to carry out the subsequent calculations. We will not discuss these matters here; a good review of the subject is provided by Westerberg et al.[1]

10.2c The Control Block

In the process simulations we have discussed so far, the direction of flow of information corresponded to the direction of flow of the process streams—from feeds to products and around cycles. This mode of information flow is appropriate if the object is to calculate the output of a process for a given input and set of process parameters (temperatures, pressures, etc.); however, it is often the case that a desired output is specified and input or process unit parameters required to achieve this output are to be calculated. A feature of process simulation programs called a **control block** is used for calculations of this type.

Consider, for example, a single adiabatic flash evaporator for which the flow rate, composition, temperature, and pressure of the feed stream are known. If the pressure in the evaporator is specified, a single call to the FLASH subroutine results in the calculation of the component flow rates of both product streams, as well as the equilibrium temperature of the unit. Suppose, however, that one of the product component flow rates is specified, and you wish to determine the value of P required to achieve the attendant separation. The FLASH program by itself is incapable of performing this calculation; what is done is to set up an artificial cycle, using a subroutine that varies the value of P until the calculated output variable agrees with the specified value. The *control block* is the entity that performs this manipulation.

[1] Reference at the end of Section 10.4.

Suppose the first element of stream S2 is the specified component flow rate. The block diagram for the simulation program would be as shown here. The execution of this cycle resembles that of a convergence block cycle: An initial guess for P is read in; FLASH is called; the calculated value of S2(1) is compared with the desired value; and if agreement is not achieved, the value of P is adjusted and the cycle is executed repeatedly until convergence is achieved.

The control variable [S2(1)] may be an element of the product stream of the unit for which the parameter is being adjusted, as in this example, or it may be an element of a stream well downstream of this unit. General simulation programs provide control blocks that can manipulate several parameters to adjust the values of several stream variables, providing considerable flexibility for process design.

An example of a sequential modular simulation for a relatively large process is given in Example 10.3-3, following a discussion of the second broad approach to flowsheet simulation.

10.3 EQUATION-BASED SIMULATION

The sequential modular approach solves the system equations in blocks corresponding to the unit operations that make up the process. The block diagram for the process looks very much like the traditional process flowchart. Since engineers are accustomed to viewing chemical processes as sequences of unit operations, they tend to feel comfortable with this approach to flowsheet simulation.

In the **equation-based approach,** the equations for all units are collected and solved simultaneously. The natural decomposition of the system into its constituent unit operations is therefore lost. Moreover, the simultaneous solution of large numbers of equations, some of which may be nonlinear, can be a cumbersome and extremely time-consuming problem, even for a powerful computer. For all these reasons, most existing flowsheet simulation programs are based on the sequential modular approach.

However, there are disadvantages associated with the sequential method. The block subroutines for simulating specific unit processes are designed to operate in a forward calculation mode only: That is, they calculate product stream variables from given values of feed stream variables and process conditions. Frequently, the problems the process engineer wants to solve fall into either of two other categories: (1) given process conditions and product stream variables, calculate feed stream variables; (2) given feed and product stream variables, calculate process conditions. In both cases, iterative calculations using control blocks are required to solve the problems using a sequential modular program.

This particular difficulty vanishes if the system equations are simply collected and solved for all unknown variables. In recent years several powerful equation-solving algorithms have been developed that make the equation-based approach competitive with the sequential modular approach, and many researchers in the field believe that as this trend continues, the former approach will replace the latter one as the standard method for flowsheet simulation. (Engineers are also working on so-called "simultaneous modular" methods, which combine features of both sequential modular and equation-based approaches. We will not deal with these refinements here, however.)

The following example illustrates the equation-based approach.

EXAMPLE 10.3-1 Simulation and Design of a Two-Column Separation Process

A stream containing 35 wt% benzene(B), 50% toluene(T), and the balance xylene(X) is fed to a distillation column. The overhead product from the column contains 91.4 wt% benzene and 8.3% toluene. The bottoms product is fed to a second column. The overhead product from the second column contains 4.25 wt% benzene and 91.6% toluene. Of the toluene fed to the process, 10.0% is recovered in the bottoms product from the second column, and 93.3% of the xylene fed to the process is recovered in the same stream.

1. Perform a degree-of-freedom analysis on the process, and outline a manual calculation procedure to determine the compositions of all streams.

2. Construct a set of n equations in the n unknown stream variables identified in part 1.

Solution

Basis: 100 kg feed

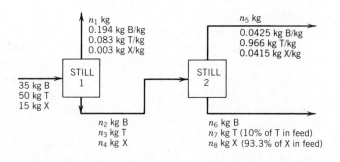

1. **Still 1**

$$
\begin{array}{l}
4 \text{ variables } (n_1, \ldots, n_4) \\
-3 \text{ equations (balances on B, T, and X)} \\
\hline
=1 \text{ local degree of freedom}
\end{array}
$$

Still 2

$$7 \text{ variables } (n_2, \ldots, n_8)$$
$$\underline{-3 \text{ equations (balances)}}$$
$$=4 \text{ local degrees of freedom}$$

Process

$$5 \text{ local degrees of freedom } (4 + 1)$$
$$-3 \text{ ties } (n_2, n_3, n_4)$$
$$-2 \text{ additional relations (fractional recoveries}$$
$$\underline{\text{of toluene and xylene in column 2 bottoms)}}$$
$$=0 \text{ net degrees of freedom}$$

The process is therefore properly specified and can in principle be solved.

A manual solution is easily outlined. The specified recoveries of toluene and xylene in the bottoms from the second column allow the immediate determination of n_7 and n_8, leaving three unknowns—n_1, n_5, and n_6—in the overall process system. Overall toluene and xylene balances provide two equations in two unknowns—n_1 and n_5—and an overall mass balance then yields n_6. Finally, benzene, toluene, and xylene balances on either still 1 or still 2 yield solutions for n_2, n_3, and n_4.

2. The system equations are given below. We first write the allowed balances for still 1, then those for still 2, and finally the additional process specifications.

Still 1 Balances

$$\text{B:} \quad 35.0 \text{ kg} = 0.914n_1 + n_2 \tag{1}$$
$$\text{T:} \quad 50.0 \text{ kg} = 0.083n_1 + n_3 \tag{2}$$
$$\text{X:} \quad 15.0 \text{ kg} = 0.003n_1 + n_4 \tag{3}$$

Still 2 Balances

$$\text{B:} \quad n_2 = 0.0425n_5 + n_6 \tag{4}$$
$$\text{T:} \quad n_3 = 0.916n_5 + n_7 \tag{5}$$
$$\text{X:} \quad n_4 = 0.0415n_5 + n_8 \tag{6}$$
$$10\% \text{ T recovery:} \quad n_7 = 0.100(50.0 \text{ kg}) = 5.00 \text{ kg} \tag{7}$$
$$93.3\% \text{ X recovery:} \quad n_8 = 0.933(15.0 \text{ kg}) = 14.0 \text{ kg} \tag{8}$$

We thus have eight equations in eight unknowns. Moreover, they are linear equations, so they can be written in the form of Equation 10.1-1 and solved by standard matrix techniques, using programs built into most computers.

Although you do not need a computer to solve this particular set of equations, the same method can be used to collect similar sets of equations corresponding to much more complex processes—processes that do not easily yield to manual solution techniques.

The next example illustrates the equation-based approach for a system that involves nonlinear relations and recycle.

EXAMPLE 10.3-2 Simulation of an Equilibrium Reaction/Separation Process

Ethane is dehydrogenated to ethylene and acetylene in the following pair of catalytic reactions:

$$C_2H_6 \rightleftharpoons C_2H_4 + H_2 \qquad [A \rightleftharpoons B + D]$$
$$C_2H_6 \rightleftharpoons C_2H_2 + 2H_2 \qquad [A \rightleftharpoons C + 2D]$$

The reactions take place at 977°C and 1 atm and proceed to a point such that the product gas composition satisfies the following equilibrium conditions:

$$\frac{y_B y_D}{y_A} = 3.75 \qquad \frac{y_C y_D^2}{y_A} = 0.135$$

where y_i denotes mole fraction. The product gas goes to a complex separation process that separates 95% of the unreacted ethane from the ethylene, acetylene, and hydrogen and recycles the separated ethane to the reactor.

Perform a degree-of-freedom analysis on this process, set up the equations for all unknown stream variables, and outline a method of solution.

Solution

Basis: 100 mol Ethane Fed to the Reactor

The degree-of-freedom analysis proceeds in the customary manner.

Mixing Point

$$2 \text{ variables } (n_0, n_6)$$
$$\underline{-1 \text{ balance (ethane)}}$$
$$= 1 \text{ local degree of freedom}$$

Reactor

$$4 \text{ variables } (n_1, \ldots, n_4)$$
$$-4 \text{ balances (ethane, ethylene, acetylene, hydrogen)}$$
$$-2 \text{ equilibrium relations}$$
$$\underline{+2 \text{ chemical reactions}}$$
$$= 0 \text{ local degrees of freedom}$$

Separation Process

6 variables (n_1, \ldots, n_6)
- 1 balance (Ethane. We implicitly used up the balances on ethylene, acetylene, and hydrogen when we labeled the outlet flow rates of these species n_2, n_3, and n_4)
- 1 additional relation (95% of the ethane fed to the separator leaves with the product)

= 4 local degrees of freedom

Process

5 local degrees of freedom $(1 + 0 + 4)$
- 5 ties (n_1, \ldots, n_4, n_6)

= 0 net degrees of freedom

The system equations are as follows.

Mixing Point

$$n_0 + n_6 = 100 \tag{1}$$

Reactor

It is convenient to analyze the reactor effluent in terms of the extents of the two reactions (see Section 4.6b). From the two given stoichiometric equations,

$$n_1 = 100 - \xi_1 - \xi_2 \tag{2}$$

$$n_2 = \xi_1 \tag{3}$$

$$n_3 = \xi_2 \tag{4}$$

$$n_4 = \xi_1 + 2\xi_2 \tag{5}$$

$$n_{tot} = n_1 + n_2 + n_3 + n_4 = 100 + \xi_1 + 2\xi_2 \tag{6}$$

The mole fraction of the ith component in the reactor effluent is n_i/n_{tot}. The mole fractions of all four species can be expressed in terms of ξ_1 and ξ_2 using equations (2) through (6), and the resulting expressions can be substituted into the given equilibrium relations to yield the following:

$$\frac{\xi_1(\xi_1 + 2\xi_2)}{(100 - \xi_1 - \xi_2)(100 + \xi_1 + 2\xi_2)} = 3.75 \tag{7}$$

$$\frac{\xi_2(\xi_1 + 2\xi_2)^2}{(100 - \xi_1 - \xi_2)(100 + \xi_1 + 2\xi_2)^2} = 0.135 \tag{8}$$

Separation Process

Ethane split: $n_5 = 0.05n_1$ $\qquad\qquad$ (9)

Ethane balance: $n_1 = n_5 + n_6$ $\qquad\qquad$ (10)

We now have 10 equations in 10 unknowns $(n_0, \ldots, n_6, n_{tot}, \xi_1, \text{ and } \xi_2)$. Once the equations are solved, any desired quantities may be determined, such as the component mole fractions in the reactor effluent, the overall and single-pass conversions, yields and selectivities, and required heat duties on the reactor and separation process.

Examining the 10 equations, we see that two are nonlinear [equations (7) and (8)] and the remainder are linear. The calculation proceeds as follows.

Solve equations (7) and (8) simultaneously for ξ_1 and ξ_2 (Trial and error—see below)

$$n_1 = 100 - \xi_1 - \xi_2 \quad \text{(2)}$$
$$n_2 = \xi_1 \quad \text{(3)}$$
$$n_3 = \xi_2 \quad \text{(4)}$$
$$n_4 = \xi_1 + 2\xi_2 \quad \text{(5)}$$

$$n_5 = 0.05 n_1 \quad \text{(9)}$$
$$n_6 = n_1 - n_5 \quad \text{(10)}$$

$$n_0 = 100 - n_6 \quad \text{(1)}$$

The only problem that remains is the simultaneous solution of equations (7) and (8). Referring to Section A.2h in Appendix A, we see that the Newton-Raphson method can be used to solve this problem. We first rewrite each of the two equations in the form $f(\xi_1, \xi_2) = 0$ by multiplying through by the denominators of the left sides and bringing all terms to the left of the equal sign. Equation (7) becomes

$$f_1(\xi_1, \xi_2) = \xi_1(\xi_1 + 2\xi_2) - 3.75(100 - \xi_1 - \xi_2)(100 + \xi_1 + 2\xi_2) = 0$$

and from equation (8)

$$f_2(\xi_1, \xi_2) = \xi_2(\xi_1 + 2\xi_2)^2 - 0.135(100 - \xi_1 - \xi_2)(100 + \xi_1 + 2\xi_2)^2 = 0$$

The roots of these two equations may then be found following the procedure outlined in Section A.2h.

The final example illustrates the analysis of a multiple-unit process with several internal cycles, using both the sequential modular and equation-based approaches.

EXAMPLE 10.3-3 Simulation of an Ammonium Nitrate Plant[2]

Ammonium nitrate, a constituent of many fertilizers, is manufactured in the reaction of ammonia and aqueous nitric acid.

$$NH_3(g) + HNO_3(aq) \rightarrow NH_4NO_3(l)$$

[2] This example is based on a process description in L. B. Andersen and L. A. Wenzel, *Introduction to Chemical Engineering*, New York, McGraw-Hill, 1961.

A labeled flowchart is shown in Figure 10.3-1. The principal features of the process are summarized in the paragraphs that follow.

The nitric acid fed to the process is a 59.5 wt% solution, which enters a charge tank at a rate of 10 970 kg/h. Also fed to the tank are two recycle streams coming from different parts of the process. Both of these streams contain ammonium nitrate, and one also contains dissolved ammonia.

A liquid stream from the charge tank is heated from 24°C to 149°C and fed to the reactor, along with a stream of ammonia vapor at 108°C and 4.5 bars. The total ammonia fed to the reactor in both feed streams is 5% in excess of the amount needed to react completely with the nitric acid in the feed. At the reactor operating conditions, the ammonium nitrate is formed as liquid droplets and most of the water in the acid is vaporized. The reaction goes to completion.

The reactor effluent—a gas–liquid mixture—leaves the reactor at a temperature of 238°C and flows into a cyclone separator. The tangential flow of the mixture generates a centrifugal force on the liquid droplets, which impels them to the wall. The droplets adhere to and flow down the wall to the base of the unit where they are contacted with hot air, which vaporizes essentially all of the water and a small amount of the nitrate, leaving the remaining nitrate in a molten condition. The mass ratio of air to reactor effluent entering the cyclone is 0.045:1. Before entering the cyclone, the air is heated from 24°C to 205°C by exchange of heat with the gases leaving the cyclone.

The molten nitrate stream leaves the separator at 199°C and is air-cooled on a slowly moving belt to a temperature at which it all solidifies. The solid nitrate is then ground and screened. All but the smallest particles (the "fines") go through a coating process, and then to bagging and shipping. The fines, which account for 16.4% of the nitrate fed to the grinding mill, are recycled to a tank where they are redissolved in acid and recycled to the charge tank.

The gas stream leaving the cyclone contains hot air, the excess ammonia, water evaporated from the nitric acid solution in the reactor and from the collected liquid in the cyclone, and 3% of the ammonium nitrate in the reactor effluent. The stream leaves the separator at 233°C, passes through the air preheater, and enters a partial condenser where some of the water and ammonia and essentially all of the nitrate are condensed. The equilibrium relationship between the compositions of the vapor and liquid streams leaving this unit may be expressed in the form

$$y_{H_2O} = f_1(x_{H_2O}, x_{NH_3})$$
$$y_{NH_3} = f_2(x_{H_2O}, x_{NH_3})$$

where y and x denote mole fractions in the gas and liquid phases, respectively. The gas stream leaving the partial condenser passes to a second condenser, which removes the remaining water and ammonia, leaving essentially pure air to be discharged to the atmosphere.

It is desired to calculate the flow rates n_0, \ldots, n_{17} (see Figure 10.3-1), the temperature of the gas entering the partial condenser (T_a), and the heat duties on the acid heater (Q_h) and the reactor (Q_r).

FIGURE 10.3-1
Flowchart of an ammonium nitrate production process

1. Carry out a degree-of-freedom analysis on the process.

2. Set up a sequential modular simulation of the process, using the following blocks:

 MIX Mix two or more streams of known composition adiabatically to form a single outlet stream. Calculate the component flow rates and temperature of the outlet stream.

 REACT Take as input a stream of reactants at a specified temperature, and carry the reaction to completion at a specified outlet temperature.

 SEPR Separate a feed stream with specified component flow rates into two outlet streams. The fractions of each component that go into the first outlet stream are specified.

 CONDNS Carry out a partial condensation on a multicomponent feed stream.

 EXCH Exchange heat between two streams. The flow rates of both streams are specified, as are the inlet and outlet temperatures of one stream and the inlet temperature of the second.

 HEAT Raise or lower the temperature of a stream by a specified amount.

 CONVG Perform a bounded Wegstein iteration to determine the value of one or more tear stream variables.

 CALC Perform a simple calculation of a specified stream variable from input stream variable values (a user-written subroutine).

3. Write a set of equations for all unknown variables; that is, set up the problem in a form suitable for solution using an equation-based algorithm.

Solution

1. *Degree-of-freedom analysis*
 We will present the analysis with a minimum of comment. Try to justify to yourself the given number of relations for each process unit.

 Reactor

 > 8 variables (n_0, \ldots, n_6, Q_r)
 > -4 balances (ammonia, water, nitrate, energy) (We do not include a nitric acid balance, since we did not label an outlet flow rate for this species)
 > -2 additional relations (% excess ammonia, complete conversion)
 > $+1$ chemical reaction
 > $\overline{}$
 > $=3$ local degrees of freedom

 Cyclone

 > 8 variables (n_4, \ldots, n_{11})
 > -3 balances (ammonia, water, nitrate) (Why not air?)
 > -2 additional relations (air/reactor effluent mass ratio, nitrate split)
 > $\overline{}$
 > $=3$ local degrees of freedom

Grinders

> 3 variables (n_{11}, \ldots, n_{13})
> -1 balance (nitrate)
> -1 additional relation (% fines)
> $= 1$ local degree of freedom

Preheater

> 5 variables $(n_7, \ldots, n_{10}, T_a)$
> -1 balance (energy—Why no material balances?)
> $= 4$ local degrees of freedom

Partial Condenser

> 9 variables $(n_7, \ldots, n_{10}, n_{14}, \ldots, n_{17}, T_a)$
> -2 balances (ammonia, water)
> -2 additional relations (equilibrium relations for ammonia and water)
> $= 5$ local degrees of freedom

Final Condenser

> 3 variables (n_7, n_{16}, n_{17})
> -0 balances (Why?)
> $= 3$ local degrees of freedom

Fines Dissolving Tank

> 1 variable (n_{12}) (We are ignoring n_{acid},
> for reasons to be discussed below)
> -0 balances
> $= 1$ local degree of freedom

Acid Charge Tank

> 7 variables $(n_1, n_2, n_3, n_{10}, n_{12}, n_{14}, n_{15})$
> -3 balances (ammonia, water, nitrate)
> $= 4$ local degrees of freedom

Acid Heater

> 4 variables (n_1, n_2, n_3, Q_h)
> -1 balance (energy)
> $= 3$ local degrees of freedom

Process

> 27 local degrees of freedom
> -27 ties (Referring to Figure 10.3-1, count them)
> $= 0$ net degrees of freedom

The process is therefore determinate.

In our analysis of the fines dissolving tank, we paid no attention to n_{acid}. The reason is that from the standpoint of the requested process analysis, the value of this flow rate is both indeterminate and immaterial. Whatever the quantity of acid that flows to the fines dissolving tank from the acid charge tank, the same quantity flows back. Since the value of n_{acid} has no bearing on any of the other material flows and we have no way of calculating this value from the given information about the process, we may as well ignore it when setting up the flowsheet simulation.

2. ***Sequential modular simulation***

Examination of the flowchart of Figure 10.3-1 shows that there are two cycles in the process. The first includes the reactor, cyclone, preheater, partial condenser, acid charge tank, and acid heater, and the second includes the reactor, cyclone, nitrate cooler, grinders, fines dissolving tank, acid charge tank, and acid heater. We can tear both cycles at once between the charge tank and the acid heater, between the heater and the reactor, or between the reactor and the cyclone. The stream between the reactor and the cyclone involves three variables (n_4, n_5, and n_6), one of which may immediately be eliminated by virtue of the specified percentage excess of ammonia in the reactor feed. Since no other potential tear stream involves fewer variables, we will choose this location as the tear point.

Figure 10.3-2 shows a block diagram of the simulation. Note the following features of the diagram.

1. Blocks are not included for the fines dissolver and the nitrate cooler, since both of these units serve simply to pass along a stream of material from one unit to another. If energy balances on these units were required, we would then include blocks for them.

2. Two calculation blocks are included. In the first (C1), the value of the ammonia fresh feed rate (n_0) will be calculated from the known value of the nitric acid feed rate, the calculated quantity of recycled ammonia (n_1), and the specified percentage excess ammonia in the reactor feed (5%). In the second block (C2), the flow rate of air (n_7) will be calculated from the known mass flow rate of the reactor effluent ($n_4 + n_5 + n_6$) and the specified mass ratio of air to reactor effluent (0.045).

3. The SEPR block must be used to simulate the cyclone. However, since this block only accepts a single input stream, the reactor effluent and air feed must first be combined in a MIX operation, even though there is no physical mixing stage in the process.

The simulation would begin with a guess at the tear stream variables (S5B). The subroutine calling sequence for the simulation would then be as follows.

CALL CALC2(C1,S5B,S3)

Calculates n_7, hence streams S3, S4, and S11

CALL MIX(M1,S5B,S4,S6)
CALL SEPR(SP1,S6,S7,S13, parameters)

FIGURE 10.3-2
Block diagram for sequential modular simulation of ammonium nitrate process

Takes as input the flow rates of air (n_7), ammonia (n_4), ammonium nitrate (n_5), and water (n_6) entering the cyclone, and the known splits of air, ammonia, water, and nitrate, and calculates the component flow rates of streams S7 and S13 (n_8, \ldots, n_{11}).

CALL SEPR(SP2,S13,S14,S15)

Calculates n_{12} and n_{13} from the specified fraction of fines in the grinder product.

CALL EXCH(E1,S3,S4,S7,S8, parameters)

Calculates T_a from the known stream flow rates and the three other stream temperatures.

CALL CONDS(CD1,S8,S9,S10, parameters)

Calculates n_{14}, \ldots, n_{17} from the known flow rates of stream S8 and the equilibrium relations for ammonia and water. If the temperatures of S9 and S10 were specified, the subroutine could also determine the required heat duty on the condenser.

At this point in the calculation, we could call SEPR for unit SP3. However, since this calculation has no bearing on the convergence of the iterative cyclic calculations, we may as well hold off on it and do it only after the iterative procedure has converged.

CALL MIX(M2,S1,S10,S14,S16)

Calculate n_1, n_2, and n_3. Notice that in previous flowsheet simulation examples, our MIX block could only handle two inlet streams. If that were the case here, we would need two block calls—one to mix S1 and S14, and the next to mix the product of the first stage with S10.

CALL CALC(C1,S16,S2)

Calculate n_0 from the known value of the nitric acid feed rate, the current value of the ammonia recycle rate (n_1), and the specified value of 5% excess ammonia in the reactor feed.

We will bypass the call of HEAT for the acid heater at this point, for the same reason we did not call SEPR for the final condenser. This calculation is not needed to complete the cycle, and so we need only do it after the cyclic calculation has converged.

CALL MIX(M3,S2,S17,S18)

Calculates the combined feed to the reactor.

CALL REACT(R1,S18,S5A, parameters)

Calculates the reactor effluent (S5A) flow rates n_4, n_5, and n_6 and the required heat duty on the reactor, Q_r.

CALL CONVG(CN1,S5A,S5B)

Compares the estimated (S5B) and recalculated (S5A) values of the tear stream variables. If they do not agree within a specified tolerance, the program generates a new set of estimated values using the bounded Wegstein algorithm and goes back to the first step [CALL MIX(M1, . . .)] to begin a new cycle. If convergence is achieved, the program proceeds to do the remaining calculations.

CALL SEPR(SP3,S9,S11,S12)

Calculates the flow rates of two of the process product streams. If temperatures of S9, S11, and S12 were specified, an energy balance could be performed on this unit as well.

CALL HEAT(H1,S16,S17)

Calculates the required heat duty on the acid heater, Q_h.

3. **_Equation-Based Simulation_**

The system equations are presented below. Verify that each set of equations is consistent with the degree-of-freedom analysis for that unit.

Reactor

Nitric acid fed (and reacted) = (6527/63.0) = 103.6 kmol/h

$$Excess\ ammonia: \frac{(n_0 + n_1)}{17.0} = 1.05 \times 103.6\ \text{kmol/h}$$

$$\Downarrow$$

$$n_0 + n_1 = 1849\ \text{kmol/h} \tag{1}$$

Water balance: $n_2 = n_6$ (2)

Nitrate balance: $n_5 = 103.6$ kmol/h formed (80.0 kg/kmol)

$$\Downarrow$$

$$n_5 = 8288\ \text{kmol/h}$$

Ammonia balance: $n_4 = 1849$ kg/h fed $-$ (103.6 kmol/h react)(17.0 kg/kmol)

$$\Downarrow$$

$$n_4 = 88\ \text{kg/h}$$

Energy balance: $Q_r = \sum_{out} n_i \hat{H}_i - \sum_{in} n_i \hat{H}_i + [103.6 \text{ kmol HNO}_3/\text{h}]\Delta H_r^\circ$ (3)

The specified enthalpies of the feed and product components would be calculated in kJ/kg, relative to the components at 25°C (liquid nitric acid, gaseous ammonia, and solid ammonium nitrate), and the standard heat of reaction would be determined in kJ/kmol.

Centrifuge

Air feed: $n_7 = 0.045(n_4 + n_5 + n_6)$

$\qquad \Downarrow n_4 = 88, n_5 = 8288$

$\qquad n_7 = 0.045n_6 + 377$ (4)

Ammonia balance: $n_8 = n_4 = 88 \text{ kg/h}$

Water balance: $n_9 = n_6$ (5)

Nitrate vaporization: $n_{10} = 0.03n_5 = 249 \text{ kg/h}$

Nitrate balance: $n_{11} = n_5 - n_{10} = 8288 - 249 = 8039 \text{ kg/h}$

Grinders

Fines: $n_{12} = 0.164n_{11} = 1318 \text{ kg/h}$

Nitrate balance: $n_{13} = n_{11} - n_{12} = 6721 \text{ kg/h}$

Air Preheater—energy balance

$$\left[\sum_i n_i C_{pi}\right](233 - T_a) = n_7 C_{p7}(205 - 24)$$ (6)

Partial Condenser

$$y_{H_2O} = n_{16}/(n_7 + n_{16} + n_{17})$$ (7)

$$y_{NH_3} = n_{17}/(n_7 + n_{16} + n_{17})$$ (8)

$$x_{H_2O} = n_{15}/(n_{10} + n_{14} + n_{15})$$ (9)

$$x_{NH_3} = n_{14}/(n_{10} + n_{14} + n_{15})$$ (10)

$$y_{H_2O} = f_1(x_{H_2O}, x_{NH_3})$$ (11)

$$y_{NH_3} = f_2(x_{H_2O}, x_{NH_3})$$ (12)

$$\text{NH}_3 \text{ balance: } n_8 = n_{14} + n_{17}$$ (13)

$$\text{H}_2\text{O balance: } n_9 = n_{15} + n_{16}$$ (14)

Acid Charge Tank

NH₃ balance: $n_{14} = n_1$ (15)

H₂O balance: $4443 + n_{15} = n_2$ (16)

Nitrate balance: $n_{10} + n_{12} = n_3$

$\qquad \| \quad n_{10} = 249$

$\qquad \Downarrow \quad n_{12} = 1318$

$\qquad \quad n_3 = 1567 \text{ kg/h}$

Acid Heater—energy balance

$$Q_h = \left[\sum_{\text{all species}} n_i C_{pi} \right] (149°\text{C} - 24°\text{C}) \tag{17}$$

At this point we have 17 equations in 17 unknowns: n_0, n_1, n_2, n_6, n_7, n_9, n_{14} to n_{17}, T_a, Q_r, Q_h, and the four mole fractions defined by equations 6 through 9. The system of equations may be simplified immediately by substituting n_6 wherever n_2 appears (from equation 2) and whenever n_9 appears (equation 4), and by replacing n_{14} with n_1 (equation 14). The remaining set of 14 equations in 14 unknowns could then in principle be solved by an equation-based algorithm to yield the remaining flow rates, the unknown stream temperature, and the two unknown heat duties.

10.4 COMMERCIAL SIMULATION PACKAGES

A large number of programs have been written to simulate process flowsheets using the sequential modular approach. Two widely used programs are FLOWTRAN, developed by the Monsanto Company, and ASPEN, developed at the Massachusetts Institute of Technology. Simulation programs are also available from several companies that provide time-shared computer facilities.

Besides having provisions for simulating most of the principal types of process equipment, the more extensive of these programs have routines that generate physical properties of hundreds of chemical species, either from built-in data tables or by using estimation formulas. Included in the stored data are molecular weights, normal boiling and melting points, liquid densities, critical constants, heat capacities, latent heats, solubilities, and vapor pressure estimation formula constants.

The simulation program is written so that a block subroutine that requires a physical property automatically retrieves the property from the data bank. In addition, it is possible for the user to have direct access to the data, and it is not uncommon for a simulation program to be used just to generate physical properties of materials, without actually carrying out a simulation.

The design of a process unit to achieve a desired output from a specified input normally includes a calculation of the unit size, which is the principal determinant of the installation and operating costs of the unit. A number of general estimation formulas have been proposed to predict the cost of a specific type of unit from a given size-dependent parameter (e.g., volume, heat transfer area, or power rating of a pump). These formulas have been built into some simulation programs. The normal procedure for using them is to solve all system material and energy balances in the manner we have described, and after all convergence and control block cycles have converged, to call the subroutines that determine the sizes and costs of the units.

Equation-based flowsheet simulators have not reached the level of commercialization of their sequential modular counterparts as of the time of this writing, and the development of efficient algorithms is the subject of much continuing research.

Sources of additional information about computer simulation of chemical processes are the following.

Reklaitis, G. V., *Introduction to Material and Energy Balances*, Wiley, New York, 1983.

Seader, J. D., W. D. Seider, and A. C. Pauls, *Flowtran Simulation—An Introduction*, Monsanto Company, St. Louis, 1974.

Westerberg, A. W., H. P. Hutchison, R. L. Motard, and P. Winter, *Process Flowsheeting*, Cambridge University Press, London, 1979.

PROBLEMS

1. Draw and label a flowchart and determine the number of degrees of freedom for each of the given systems. Give a feasible set of design variables, and, if possible, an infeasible set. The solution to part (a) is given as an example.

 (a) An aqueous solution of sulfuric acid is diluted with pure water in an adiabatic mixer. The two feed streams are at the same temperature.

 Solution

 n_1 [kg H_2SO_4 (aq)] @ T_0 (°C)

 x_1 (kg H_2SO_4/kg)

 n_2 [kg H_2O(l)] @ T_0 (°C)

 n_3 (kg) @ T (°C)

 x_3 (kg H_2SO_4/kg)

 $$\begin{array}{l} \text{7 variables } (n_1, x_1, n_2, n_3, x_3, T_0, T) \\ \underline{-3 \text{ relations (2 material balances and 1 energy balance)}} \\ = 4 \text{ degrees of freedom} \end{array}$$

 One feasible set of design variables (there are others) is

 $$\{n_1, x_1, n_2, T_0\}$$

 If you are given values of these variables you can calculate n_3 (total mass balance), x_3 (H_2SO_4 balance), and T (energy balance). An infeasible set is

 $$\{n_1, n_2, n_3, T_0\}$$

 Once n_1 and n_2 have been specified, n_3 is fixed by a total material balance and may not be independently assigned a value.

 (b) A natural gas containing methane, ethane, and propane is mixed with preheated air, and the mixture is heated to 200°C.

 (c) A stream containing hexane vapor in nitrogen is cooled at constant pressure, condensing a portion of the hexane.

2. Benzene (B) and chlorobenzene (C) are being separated in a distillation column. Vapor and liquid streams, each containing both species, are fed to one of the trays of the column,

and liquid and vapor streams are taken off the tray. The tray functions as an **ideal stage** (see Problem 6-51): The effluent streams are in equilibrium at temperature T and pressure P, with compositions related by Raoult's law, Equation (6.4-1).

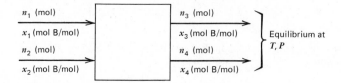

Calculate the number of degrees of freedom. Then specify sets of design variables for which the solution for the remaining state variables would be (a) straightforward, (b) iterative, (c) impossible.

3. Barium sulfate is reduced with carbon as part of the manufacture of lithopone, a white pigment:

$$2\,BaSO_4(s) + 4C(s) \rightarrow 2\,BaS(s) + 4CO_2(g)$$

A mixture of crushed barite (an ore containing x_B kg $BaSO_4$/kg barite) and coal (containing x_C kg C/kg coal) at temperature $T_0(K)$ is fed to a furnace. Essentially all of the barium sulfate in the feed is converted. As part of the furnace design, you must specify for a basis of 100 kg barite ore the masses of coal fed and of carbon, barium sulfide, carbon dioxide, and other solids emerging from the reactor, as well as the product temperature and required heat input to the furnace.

(a) Draw and label a flowchart, and calculate the degrees of freedom of the process.
(b) Outline the solution procedure (i.e., state in order the relations you would write and the variables you would determine) if you are given values of x_B, x_C, T_0, the percent excess coal fed, and the product temperature. If the solution involves iteration or is impossible, say so.
(c) Repeat part (b), assuming that the specified variables are x_B, x_C, T_0, the mass of BaS produced, and the heat input to the furnace.
(d) Repeat part (b), assuming that the specified variables are x_B, x_C, T_0, the percent excess coal fed, and the heat input to the furnace.

4. Ethyl alcohol can be bacterially oxidized to acetic acid in the following two-step fermentation sequence:

$$2\,C_2H_5OH + O_2 \rightarrow 2\,CH_3CHO + 2\,H_2O$$
$$2\,CH_3CHO + O_2 \rightarrow 2\,CH_3COOH$$

If the alcohol-containing feedstock is wine, cider, or a malt solution, the resulting solution is vinegar.

An aqueous solution containing ethyl alcohol in water is fermented to produce dilute acetic acid. The feed mixture (the ethanol solution and the bacteria that make the fermentation occur) and air are fed at a temperature T_0. The product solution contains ethanol, acetaldehyde (CH_3CHO), acetic acid, and water. All liquid and gaseous effluents are at temperature T. The variables involved in the process are n_f (mol feed solution),

x_{ef} (mol ethanol/mol feed solution), n_{air} (mol air fed), P_{xs} (percent excess air), n_e, n_{ah}, n_{aa}, n_w (gram-moles of ethanol, acetaldehyde, acetic acid, and water, respectively, in the product mixture), n_{ox}, n_n (gram-moles of oxygen and nitrogen, respectively, emerging from the reactor), T_0, T, and Q (kJ heat removed).

(a) Calculate the number of degrees of freedom of the process. How would the answer differ if the reactor were adiabatic?

(b) Outline the solution procedure to determine all unknown variables from given values of n_f, x_{ef}, P_{xs}, n_e, n_{ah}, T_0, and T.

(c) Specify a set of design variables that would entail an iterative solution for the remaining state variables.

(d) Specify an infeasible set of design variables.

5. Propylene is hydroformylated to butyraldehyde and *n*-butanol in the following reaction sequence:

$$C_3H_6 + CO + H_2 \rightarrow C_3H_7CHO \text{ (butyraldehyde)}$$
$$C_3H_7CHO + H_2 \rightarrow C_4H_9OH \text{ (}n\text{-butanol)}$$

Liquid propylene, gaseous carbon monoxide and hydrogen, and a soluble cobalt catalyst are fed to a high-pressure catalytic reactor. The reactor effluent goes to a flash tank, where all of the solution constituents are vaporized except the catalyst, which is recycled to the reactor. The reaction products are separated from unconsumed reactants in a multiple-unit process, and the product stream, which contains both butyraldehyde and *n*-butanol, is subjected to additional hydrogenation with excess hydrogen, converting all of the butyraldehyde to butanol.

(A more detailed flowchart is shown on p. 769 of G. T. Austin, *Shreve's Chemical Process Industries*, 5th Edition, McGraw-Hill, New York, 1984.)

Calculate the number of degrees of freedom for each subprocess (the reactor, flash tank, separation process, and hydrogenator), and then the number for the entire process. (See Example 10.3-1.)

6. Normal butane is catalytically isomerized to isobutane in a stirred-tank reactor. A fresh feed stream containing pure *n*-butane is combined with a recycle stream containing *n*-butane and isobutane, and the combined stream is fed to the reactor, where some but not all of the *n*-butane in the feed is converted. The reactor effluent is pumped to a distillation column. The overhead distillation product contains primarily isobutane and a small amount of *n*-butane. The bottoms product, which also contains both species, is the stream recycled to the reactor.

(a) Determine the number of degrees of freedom associated with the feed mixer, the reactor, the still, and the entire process.

(b) Suppose a single-pass conversion of 35.0% is attained in the reactor, the overhead still product contains 88.5 mole% isobutane, and the recycle stream contains 85.0 mole% n-butane. Calculate the overall conversion, the recycle ratio (mol recycle/ mol fresh feed), and the gram-moles of overhead still product produced per 100 mol fresh feed.

(c) Solve the problem of part (b) again, only this time do it as it might be done by a sequential modular simulation of the process. Take a basis of calculation of 100 mol fresh feed, and choose n_r (mol recycle) as the tear steam variable. Begin by assuming $n_r = 100$ mol, and go through three successive substitution iterations (Appendix A.2). What is the percentage error in the third calculated value of n_r?

(d) Repeat part (c), only perform a Wegstein acceleration step (Appendix A.2) after the first two successive substitution iterations. What is the percentage error in the last calculated value of n_r?

In the next seven problems, you will be asked to write the computer codes for several different flowsheet simulation modules. In each case, the approach outlined in Example 10.2-1 for the mixing module should be followed, except that it will not be necessary to include a block identifier as a subroutine argument.

7. Write the code for a module called SPLIT, which takes a feed stream (SF) and splits it into two streams (S1 and S2), each having the composition and temperature of the feed stream. A fraction X1 of the feed stream becomes stream S1, and a fraction (1-X1) becomes S2. Up to seven components may be present in the feed stream, so that each stream has associated with it eight attributes: seven component molar flow rates, and the stream temperature. The arguments of the subroutine should be SF, S1, and S2 (each eight-membered arrays), X1, and N, the number of components in the feed stream (any number from 1 to 7). X1, N, and the elements of SF are input variables, and the elements of S1 and S2 should be calculated by the subroutine.

Test your code for a feed stream flowing at a rate of 150 mol/h at $T = 315K$, containing 35 mole% n-octane (A), 57% iso-octane (B), and the balance a chemically inert species (C), which is to be split into two streams in a 60:40 ratio. Write and run a calling program that defines the input variables, calls the module subroutine, and prints out the attributes of S1 and S2.

8. You are to write the code for a two-component flash vaporizer module that converts a liquid feed stream (SF) containing two species at a temperature $TF(K)$ to liquid and vapor product streams (SL, SV) in equilibrium at a temperature $T(K)$ and a pressure $P(mm\ Hg)$ lower than that of the feed stream. The compositions of the liquid and vapor product streams are related by Raoult's law (Equation 6.4-1), and the component vapor pressures are correlated with temperature by the Antoine equation, Table 6.1-1.

(a) Write the system mass and energy balances and equilibrium relations in terms of the following variables.

NF,NL,NV	The moles of feed, liquid product, and vapor product.
XF1,XL1,XV1	The mole fractions of the first of the two species in the feed and the liquid and vapor product streams.
TF,T,P	Feed stream temperature, vaporizer temperature, and vaporizer pressure.
PV1,PV2	Vapor pressures of the two species at temperature T.
A1,B1,C1 A2,B2,C2	$\begin{cases}\text{Antoine equation constants for the two species involved in the}\\\text{process (transmitted to the subroutine via a COMMON statement).}\end{cases}$
CP1 CP2	$\begin{cases}\text{Liquid-phase heat capacities (kJ/kmol·K) of the two species (trans-}\\\text{mitted via COMMON). Assume independent of temperature.}\end{cases}$
HV1 HV2	$\begin{cases}\text{Heats of vaporization (kJ/mol) of the two species (transmitted via}\\\text{COMMON). Assume independent of temperature.}\end{cases}$
Q	The required heat input (kW) to the vaporizer.

Show that the system has five degrees of freedom. Then outline a procedure to solve for all outlet stream variables from specified values of the feed-stream component molar flow rates, *TF*, *T*, and *P*.

(b) Write the code for a module subroutine called FLASH2 to implement the procedure outlined in part (a). The input variables are the attributes of SF (two component flow rates and the temperature), *T*, and *P*. The output variables are the attributes of SL and SV, and Q.

(c) Test your code for a feed stream containing 30.0 mole% benzene and 70 mole% toluene flowing at a rate of 1.00 mol/s that is to be flashed from 385K and a high pressure to 363K and 512 mm Hg. The heat capacities of liquid benzene and toluene may be taken to be 160 kJ/kmol·K and 190 kJ/kmol·K, respectively, and the heats of vaporization may be taken to be those at the normal boiling points of the two species. Write and run a calling program that defines the attributes of SF and other input parameters, calls the module subroutine, and prints out the attributes of the product streams and the required heat input.

9. You are to write the code for a flash vaporizer module that converts a liquid feed stream (SF) containing N species at a high pressure to liquid and vapor product streams (SL, SV) in equilibrium at a temperature T(K) and a pressure P(mm Hg) lower than that of the feed stream. The compositions of the liquid and vapor product streams are related by Raoult's law (Equation 6.4-1), and the component vapor pressures are correlated with temperature by the Antoine equation, Table 6.1-1.

(a) Write the system mass and energy balances and equilibrium relations in terms of the following variables.

NF,NL,NV	The moles of feed, liquid product, and vapor product.
XF(I) XL(I) XV(I)	$\begin{cases}\text{The mole fractions of the }i\text{th component in the feed, liquid product,}\\\text{and vapor product, where I runs from 1 to }(N-1).\end{cases}$

TF,T,P Feed temperature, vaporizer temperature, and pressure, respectively.

PV(I) Vapor pressures of the N species at temperature T, where I runs from 1 to N.

A(I)
B(I) { Antoine equation constants for the N species involved in the process (transmitted to the subroutine via a COMMON statement).
C(I)

CP(I) Liquid-phase heat capacities (kJ/kmol·K) of the N species (transmitted via COMMON). Assume independent of temperature.

HV(I) Heats of vaporization (kJ/mol) of the N species (transmitted via COMMON). Assume independent of temperature.

Q The required heat input (kW) to the vaporizer.

Show that the system has $(N + 3)$ degrees of freedom, counting as the system variables three stream flow rates, $3(N - 1)$ mole fractions, N vapor pressures, TF, T, P, and Q. Then work out a trial-and-error procedure for determining the product stream flow rates and mole fractions and Q from specified values of TF, T, P, and the feed stream molar flow rate and component mole fractions.

(b) Write a module subroutine FLASHN to implement the procedure outlined in part (a) for a feed stream containing up to seven components. The arguments of the subroutine should be N, SF, SV, SL, P, and Q. The input variables are SL(N+1) and SV(N+1) (both of which equal T), P, and the attributes of SF (N molar flow rates and the feed temperature), and the outlet variables are Q and the remaining attributes of SV and SL.

(c) Test your code for a feed stream containing 34.8 mole% n-pentane, 30.0 mole% n-hexane, and the balance n-heptane flowing at a rate of 1.00 mol/s which is to be flashed from 363K and a high pressure to 338K and 611 mm Hg. The heat capacities of liquid pentane, hexane, and heptane (in kJ/kmol·K) may be taken to be 188, 216, and 213, respectively, and the heats of vaporization may be taken to have their values at the normal boiling points of these substances. Write and run a calling program that defines the attributes of SF and other input parameters (including the heat capacities and heats of vaporization), calls the module subroutine, and prints out the attributes of the product streams and the required heat input.

10. You are to write the code for a partial condenser module that converts a vapor feed stream (SF) containing a single condensable species at temperature TF(K) to liquid and vapor product streams (SL, SV) in equilibrium at a temperature T(K). The process takes place at a constant pressure P(atm). The compositions of the liquid and vapor product streams are related by Raoult's law (Equation 6.4-1), and the component vapor pressures are correlated with temperature by the Antoine equation, Table 6.1-1.

The system variables are as follows.

NF,NL,NV mol/s of feed, liquid product, and vapor product.

XF { Mole fractions of the condensable substance in the feed and the vapor
XV product.

TF,T,P Feed temperature (K), condenser temperature (K), and pressurre (atm), respectively.

PV Vapor pressure (mm Hg) of the condensable substance at temperature T.

A,B,C Antoine equation constants for the condensable substance (transmitted to the subroutine via a COMMON statement).

CPL Liquid-phase heat capacity (kJ/mol \cdot K) of the condensable substance transmitted via COMMON). Assume independent of temperature.

HV Heat of vaporization (kJ/mol) of the condensable substance (transmitted via COMMON). Assume independent of temperature.

CPV $\begin{cases} \text{Heat capacities (kJ/mol} \cdot \text{K) of the condensable vapor and the non-} \\ \text{condensable gas (transmitted via COMMON). Assume independent of} \\ \text{temperature.} \end{cases}$
CPG

Q Required heat input (kW) to the vaporizer.

(a) Show that the system has five degrees of freedom, counting as the system variables three stream flow rates, two mole fractions, one vapor pressure, TF, T, P, and Q.

(b) The input variables to the module are to be the molar flow rates of the condensable and noncondensable feed-stream constituents [SF(1) and SF(2)], the feed temperature (SF(3) = TF), the operating pressure of the condenser (P), and the fraction YC of the condensable species in the feed that is to be condensed. (The heat capacities and heat of vaporization must also be supplied to the subroutine.) The output variables are the molar flow rate of the liquid product stream [SL(1)], the molar flow rates of the vapor product stream constituents [SV(1) and SV(2)], the operating temperature of the condenser [SL(2) = SV(3) = T], and the required heat removal rate (Q). Outline the required calculations.

(c) Write a module subroutine CNDNS to implement the procedure outlined in part (b). The subroutine arguments should be SF, SV, SL, P, YC, and Q. Input variables are YC, P, and the attributes of SF, and the outlet variables are Q and the attributes of SV and SL.

(d) Test your code for a feed stream flowing at 1 mol/s containing methyl alcohol in air at 60°C and 1 atm with a relative saturation of 85%, from which 90% of the methanol is to be removed by condensation at constant pressure. The heat capacities in kJ/mol·K of liquid methanol and methanol vapor may be taken to be 0.078 and 0.050. Write and run a calling program that defines the attributes of SF and other input parameters, calls the module subroutine, and prints out the required condenser temperature, the flow rates and compositions of the liquid and vapor product streams, and the required rate of heat removal.

11. You are to write the code for a module to perform material and energy balance calculations on a single-phase reactor in which a single reaction takes place. The reaction has the general form

$$v_1 A_1 + v_2 A_2 + v_3 A_3 + \ldots + v_m A_m = 0$$

In this equation A_i is the ith reactant or product and v_i is the stoichiometric coefficient of this species; v_i is negative for reactants and positive for products. It is also convenient to define v_i for each inert species in the feed to the reactor, assigning it a value of 0.

The input to the module is the feed stream flow rate, composition, and temperature, the fractional conversion of one of the reactants, and the product stream temperature. The module is to calculate the product stream component flow rates and the required heat transfer to the reactor.

The module equations should be written in terms of the following variables:

N	The total number of reactive and inert species present in the feed and product streams.
SF(I) I = 1, . . . ,N + 1	The molar flow rates of the feed stream components (mol/s) and the feed stream temperature (K).
SP(I), I = 1, . . . ,N + 1	The molar flow rates of the product stream components (mol/s) and the product stream temperature (K).
NU(I), I = 1, . . . ,N	The stoichiometric coefficients of all species ($-$ for reactants, $+$ for products, 0 for inerts).
X	The fractional conversion of one of the reactants
IX	The number of the reactant for which the fractional conversion is specified.
CP(I), I = 1, . . . ,N	The heat capacities of all species, kJ/mol·K (assume constant—transmit through COMMON).
HF(I), I = 1, . . . ,N	The standard heats of formation of all species, kJ/mol (transmit through COMMON).
Q	The heat input to the reactor, kW. (If Q is negative, heat must be transferred from the reactor.)

(a) Write the equations you would use to calculate Q and the first N components of SP from specified values of all of the other variables defined above.

(b) Write the code for a module called REACTS to implement the procedure outlined in part (a). The arguments of the subroutine should be SF, SP, NU, N, X, IX, and Q, and the arrays CP and HF should be transmitted through COMMON. All variables but Q and the first N attributes of SP should be considered input variables; Q and SP(1), . . . ,SP(N) should be calculated by the subroutine.

(c) Test your code for a reactor in which propane is burned with 20% excess air at a rate of 270 standard cubic meters per hour of propane. The combined feed stream enters the reactor at 423 K, and the stack gas leaves at 1050 K. A percentage propane conversion of 90% is achieved in the reactor, and no CO is contained in the stack gas.

Write and run a calling program to define the input variables, call the module subroutine, and print out the results. Number the species involved in the process as follows: $1-C_3H_8$, $2-O_2$, $3-N_2$, $4-CO_2$, $5-H_2O$. Thus, for example, NU(1) = -1.0, NU(2) = -5.0, SF(1) = 3.348, SF(3) = 75.54, SF(6) = 423, and SP(6) = 1050. (You will have to verify these values as part of the problem solution.) Take the heat capacity of each species to be its value at 700K as calculated from Table B.2 (so that, for example, CP(1) = 0.1431 kJ/mol·K).

12. You are to write the code for a module to perform material and energy balance calculations on a single-phase adiabatic reactor in which a single reaction takes place. The

reaction has the general form

$$v_1 A_1 + v_2 A_2 + v_3 A_3 + \ldots + v_m A_m = 0$$

In this equation A_i is the ith reactant or product and v_i is the stoichiometric coefficient of this species; v_i is negative for reactants and positive for products. It is also convenient to define v_i for each inert species in the feed to the reactor, assigning it a value of 0. The inputs to the module are the feed-stream flow rate, composition, and temperature, and the fractional conversion of one of the reactants. The module is to calculate the product stream component flow rates and temperature.

The module equations should be written in terms of the following variables:

N	The total number of reactive and inert species present in the feed and product streams.
SF(I), I = 1, . . . ,N+1	The molar flow rates of the feed stream components (mol/s) and the feed stream temperature (K).
SP(I), I = 1, . . . ,N+1	The molar flow rates of the product stream components (mol/s) and the product stream temperature (K).
NU(I), i = 1, . . . ,N	The stoichiometric coefficients of all species ($-$ for reactants, $+$ for products, 0 for inerts) (transmit through COMMON).
X	The fractional conversion of one of the reactants.
IX	The number of the reactant for which the fractional conversion is specified.
ACP(I), I = 1, . . . ,N BCP(I), I = 1, . . . ,N CCP(I), I = 1, . . . ,N DCP(I), I = 1, . . . ,N	$\begin{cases} \text{Coefficients of the heat capacity formula } C_p(\text{kJ/mol} \cdot {}^\circ\text{C}) = \\ a + bT + cT^2 + dT^3 \text{ (transmit through COMMON).} \end{cases}$
HF(I), I = 1, . . . ,N	The standard heats of formation of all species, kJ/mol (transmit through COMMON).

(a) Write the equations you would use to calculate the attributes of SP from specified values of all of the other variables presented. Use Newton's rule (Appendix A.2) to solve the 4th-order equation for the reactor temperature.

(b) Write the code for a module called REACTAD to implement the procedure outlined in part (a). The arguments of the subroutine should be SF, SP, NU, N, X, and IX, and the arrays CP and HF should be transmitted through COMMON. All arguments but the attributes of SP should be considered input variables.

(c) Test your code for a reactor in which carbon monoxide is oxidized with 25% excess air at 1 atm to form carbon dioxide. The combined feed stream enters the reactor at 650°C at a rate of 23.0 kmol/h, and a CO conversion of 45% is achieved.

Write and run a calling program to define the input variables, call the module subroutine, and print out the results. Number the species involved in the process as follows: 1—CO, 2—O_2, 3—N_2, 4—CO_2. Thus, for example, NU(1) = -2.0, NU(2) = -1.0, SF(1) = 1.607, and SF(3) = 3.777. (You will have to verify these values as part of the problem solution.)

13. You are to write the code for a convergence module that can deal with one to three tear-stream variables using the bounded Wegstein algorithm, as outlined in Appendix A.2. The object is to determine the values of one, two, or three of the variables x_1, x_2, and x_3 that satisfy the relations

$$x_1 = f_1(x_1, x_2, x_3)$$
$$x_2 = f_2(x_1, x_2, x_3)$$
$$x_3 = f_3(x_1, x_2, x_3)$$

where f_1, f_2, and f_3 may be simple algebraic functions, or they may represent the composite set of calculations involved in proceeding around a cycle of unit operations to recalculate tear-stream variables. The calculation consists of assuming values for the tear-stream variables $[x_{ia}, i = 1, \ldots, 3]$; computing $[x_{ic} = f_i(x_{1a}, x_{2a}, x_{3a}), i = 1, \ldots, 3]$; determining whether x_{ia} and x_{ic} are within a specified tolerance of each other for each i; and if they are not, using the bounded Wegstein algorithm to estimate the values of x_{1a}, x_{2a}, and x_{3a} to be used in the next iteration.

The module code should be written in terms of the following variables.

N	Number of tear-stream variables (1, 2, or 3).
XA(I), I=1, . . . ,N	Assumed values of x_i (initially guessed, subsequently calculated by the Wegstein algorithm).
XC(I), I=1, . . . ,N	Recalculated values of x_i obtained from the functions f_i (or by proceeding around a cycle).
KMAX	Maximum number of iterations to be allowed.
EPS	Relative convergence tolerance (see Equation A.2-8). The difference between XA(I) and XC(I) divided by the value of XC(I) must be less than EPS for convergence to be declared.
IPR	A printout parameter. The code should be written such that if IPR=1 the values of x_i are printed after each iteration, while if IPR = 0 the printing is suppressed.

(a) Write the code for a module subroutine called CONVG that has as its arguments XA, XC, N, KMAX, EPS, and IPR. The values of N, KMAX, EPS, IPR, and the elements of XA are input variables, and the final estimated values of x_i are to be returned in array XC. The code should execute the procedure outlined in Section A.2f separately on each of the N variables, up to a maximum of KMAX iterations. If this limit is reached and convergence has not been attained, an error message should be printed and the program terminated.

In a real flowsheet simulator, the code would be written so that the module could be inserted as a block in the flowsheet (see, for example, the flowsheet block diagram on p. 502, and the tear-stream variables XC(I) would then be recalculated from assumed values of XA(I) by proceeding around the cycle. To simplify the coding in this problem, however, have the subroutine call a function-generating

subroutine of the form

CALL FUNCGEN(N,XA,XC)

You will need to write Subroutine FUNCGEN to evaluate

$$XC(1) = f_1[XA(1),XA(2),XA(3)]$$

and similarly for XC(2) and XC(3).

(b) Apply your code to the problem of Example A.2-2. Build in an upper limit of 20 iterations (KMAX = 20).

14. If you have access to a commercial flowsheet simulation program such as FLOWTRAN or ASPEN, use it to simulate the ammonium nitrate manufacturing process described in Example 10.3-3.

Chapter Eleven

Balances on Transient Processes

A system is said to be in a transient (or unsteady-state) condition if the value of any system variable changes with time. Batch and semibatch process systems are inherently transient. Continuous systems are always transient when they are being started up and shut down, and they may become transient during normal operation due to imposed changes in operating conditions or to unplanned fluctuations in the values of input or control variables.

The principles of deriving balances on transient systems are basically those developed in Chapters 4 and 7; the essential difference is the existence of a nonzero accumulation term in the balance equation. The accumulation term is generally a derivative, however, so that the techniques required to solve transient balances are more complex than those we have developed in previous chapters.

11.1 THE GENERAL BALANCE EQUATION . . . AGAIN

In Section 4.2, the general balance equation was given as

$$\text{Accumulation} = \text{input} + \text{generation} - \text{output} - \text{consumption} \qquad \textbf{(4.2-1)}$$

Two forms of this equation were discussed: differential balances, which are applicable at an instant of time, and integral balances, which account for a finite time period. We examine in this section the nature of the relationship between these two types of balances; in doing so, we belatedly show why they are called differential and integral.

11.1a Differential Balances

Suppose a species A is involved in a process. Let q_{in}(kg/s) and q_{out}(kg/s) be the rates at which A enters and leaves the process by crossing the boundaries, and let r_{gen}(kg/s) and r_{cons}(kg/s) be the rates of generation and consumption of A within the system by chemical reaction. Any or all of the variables q_{in}, q_{out}, r_{gen}, and r_{cons} may vary with time.

Let us now write a balance on A for a period of time from t to $t + \Delta t$, supposing that Δt is small enough for the quantities q_{in}, q_{out}, r_{gen}, and r_{cons} to be considered constant. (Since we will eventually let Δt approach 0, this assumption is not restrictive.) The terms of a balance on A are easily calculated.

532

$$\text{Input (kg)} = q_{in}(kg/s)\,\Delta t(s)$$
$$\text{Output} = q_{out}\,\Delta t$$
$$\text{Generation} = r_{gen}\,\Delta t$$
$$\text{Consumption} = r_{cons}\,\Delta t$$

We also suppose that the mass of A in the system changes by an amount Δm(kg) during this small time interval; this by definition is the accumulation of A in the system. Then, from the balance equation, (4.2-1),

$$\Delta m = (q_{in} + r_{gen} - q_{out} - r_{cons})\,\Delta t \qquad\qquad \textbf{(11.1-1)}$$

If now we divide by Δt and then let Δt approach 0, the ratio $\Delta m/\Delta t$ becomes the derivative of m with respect to t (dm/dt), and consequently the balance equation becomes

$$\boxed{\frac{dm}{dt} = q_{in} + r_{gen} - q_{out} - r_{cons}} \qquad\qquad \textbf{(11.1-2)}$$

This is the general differential balance equation: m is the amount of the balanced quantity in the system, and the four terms on the right side are rates that may vary with time.

If Equation 11.1-2 is applied to a continuous system at steady state, the quantity m must be a constant, its time derivative therefore equals zero, and the equation reduces to the familiar

$$\text{Input} + \text{generation} = \text{output} + \text{consumption} \qquad\qquad \textbf{(4.2-2)}$$

However, as long as any term varies with time, the derivative on the left side of Equation 11.1-2 remains part of the equation. We thus conclude that *the balance equation for an unsteady-state system at an instant of time is a differential equation* (hence the term differential balance).

Equation 11.1-2 is an ordinary first-order differential equation. Before it can be solved to yield an expression for $m(t)$, a **boundary condition** must be provided—a specified value of the dependent variable (m) at some value of the independent variable (t). Frequently, the value of m at time $t = 0$ (an "initial condition") is specified. The complete balance equation would be Equation 11.1-2 followed by

$$t = 0, m = \ldots$$

or simply

$$m(0) = \ldots$$

When you analyze a transient system, your analysis is not complete unless every differential equation you derive is accompanied by a boundary condition similar to one of those just given.

EXAMPLE 11.1-1 Differential Balances on a Chemical Reactor

A continuous stirred-tank reactor is used to produce a compound R in the liquid-phase reaction A → R. Feed enters the reactor at a rate ϕ_0(liters/s); the concentration of the reactant in the feed is C_{A0}(mol A/liter). The volume of the tank contents is V(liters). The vessel may be considered perfectly mixed, so that the concentration of A in the product stream equals that in the tank. For this process the rate of consumption of A equals kC_A (mol/s·liter of reaction volume). All fluids (the feed, the tank contents, and the product) may be taken to have the same density, ρ(g/liter).

Write differential balances on total mass and on moles of A, expressing the balances in terms of the variables shown on the following diagram:

Solution

Basis: Given Quantities

Total Mass Balance

$$\text{Accumulation} = \text{input} - \text{output}$$

$$\text{Mass in reactor: } m(\text{g}) = V(\text{liters})\rho(\text{g/liter})$$

$$\Downarrow$$

$$\text{Accumulation (g/s)} = \frac{dm}{dt} = \frac{d(\rho V)}{dt} = \rho\frac{dV}{dt} \text{ (since } \rho \text{ is constant)}$$

$$\text{Input (g/s)} = \phi_0(\text{liters/s})\rho(\text{g/liter})$$

$$\text{Output (g/s)} = \phi(\text{liters/s})\rho(\text{g/liter})$$

$$\Downarrow$$

$$\rho\frac{dV}{dt} = \phi_0\rho - \phi\rho$$

$$\Downarrow \text{ cancel } \rho$$

$$\boxed{\begin{array}{c} dV/dt = \phi_0 - \phi \\ t = 0, V = V_0 \end{array}}$$

where V_0 is the volume of the initial tank contents. (*Question:* If $\phi_0 = \phi$, what does the mass balance tell you?)

Balance on A

$$\text{Accumulation} = \text{input} - \text{output} - \text{consumption}$$

$$\text{Moles of A in the reactor} = V(\text{liters})C_A(\text{mol/liter})$$

$$\Downarrow$$

$$\text{Accumulation (mol A/s)} = \frac{d(VC_A)}{dt}$$

$$\text{Input (mol A/s)} = \phi_0(\text{liters/s})C_{A0}(\text{mol A/liter})$$

$$\text{Output (mol A/s)} = \phi(\text{liters/s})C_A(\text{mol A/liter})$$

$$\text{Consumption (mol A/s)} = kC_A(\text{mol A/s·liter})V(\text{liters})$$

$$\Downarrow$$

$$\boxed{\begin{array}{c} \dfrac{d(VC_A)}{dt} = \phi_0 C_{A0} - \phi C_A - kC_A V \\[2ex] t = 0, \ C_A = C_A(0) \end{array}}$$

where $C_A(0)$ is the concentration of A in the initial tank contents. How you would proceed to solve this equation for the output concentration $C_A(t)$ depends on how the quantities ϕ_0, ϕ, and C_{A0} vary with time.

11.1b Integral Balances

Let us reconsider the form of the differential balance given in Equation 11.1-2.

$$\frac{dm}{dt} = q_{in} + r_{gen} - q_{out} - r_{cons} \tag{11.1-2}$$

The equation may be rewritten as

$$dm = q_{in}\,dt + r_{gen}\,dt - q_{out}\,dt - r_{cons}\,dt$$

and integrated from an initial time t_0 to a later time t_f, to obtain

$$\int_{t_0}^{t_f} dm = m(t_f) - m(t_0) = \int_{t_0}^{t_f} q_{in}\,dt + \int_{t_0}^{t_f} r_{gen}\,dt - \int_{t_0}^{t_f} q_{out}\,dt - \int_{t_0}^{t_f} r_{cons}\,dt \tag{11.1-3}$$

This is the integral balance equation. The left side is the accumulation of the balanced quantity in the system between t_0 and t_f. The term $(q_{in}\,dt)$ is the amount of the balanced quantity that enters the system in the infinitesimal interval from t to $(t + dt)$, so that the integral

$$\int_{t_0}^{t_f} q_{in}\,dt$$

is the total amount that enters between t_0 and t_f. Similar reasoning can be applied to the other terms, leading to the conclusion that Equation 11.1-3 is simply another statement of the general balance equation

$$\text{Accumulation} = \text{input} + \text{generation} - \text{output} - \text{consumption} \quad \textbf{(11.1-4)}$$

only now each term represents an amount of the balanced quantity rather than a rate. It is not difficult to show that this general equation 11.1-4 reduces to the usual form of the steady-state balance equation for continuous systems. (Show it—let $m_{\text{final}} = m_{\text{initial}}$, assume all q's and r's are constant, and simplify the equation.) For a closed (batch) system, if the balanced quantity is mass (rather than energy), $q_{\text{in}} = q_{\text{out}} \equiv 0$, and the equation may be written

$$m_{\text{initial}} + \int_{t_0}^{t_f} r_{\text{generation}} \, dt = m_{\text{final}} + \int_{t_0}^{t_f} r_{\text{consumption}} \, dt$$

or once again

$$\text{Initial input} + \text{generation} = \text{final output} + \text{consumption} \quad \textbf{(4.2-3)}$$

This is the form of the integral balance equation given in Chapter 4 for a closed system.

The procedure to follow when writing a balance on a transient system is to derive the differential balance equation, integrate it between initial and final times, and solve for the unknown quantity in the resulting equation.

EXAMPLE 11.1-2 Water Balance on a City Reservoir

The water level in a municipal reservoir has been decreasing steadily during a dry spell, and there is concern that the drought could continue for another 60 days. The local water company estimates that the consumption rate in the city is approximately 10^7 liters/day. The State Conservation Service estimates that rainfall and stream drainage into the reservoir coupled with evaporation from the reservoir should yield a net water input rate of $10^6 \exp(-t/100)$ liters/day, where t is the time in days from the beginning of the drought, at which time the reservoir contained an estimated 10^9 liters of water.

1. Write a differential balance on the water in the reservoir.
2. Integrate the balance to calculate the reservoir volume at the end of 60 days of continued drought.

Solution

1. We will write a balance on the mass $m(\text{kg})$ of water in the reservoir, but will express the equation in terms of volumes to make use of the given data, using the relationship $m(\text{kg}) = \rho(\text{kg/liter}) V(\text{liters})$. The differential balance equation is

as follows:

$$\frac{dm}{dt} = q_{in} + r_{gen} - q_{out} - r_{cons} \quad \text{(each term in kg/day)}$$

$$\left\| \begin{array}{l} \dfrac{dm}{dt} = \dfrac{d}{dt}(\rho V) = \rho \dfrac{dV}{dt} \quad \text{(since } \rho \text{ is constant)} \\[2mm] q_{in} = \rho(10^6 e^{-t/100}) \\[2mm] q_{out} = \rho(10^7) \\[2mm] r_{gen} = r_{cons} = 0 \quad \text{(water is not produced or consumed in the reservoir)} \\[2mm] \text{cancel } \rho \end{array} \right.$$

$$\boxed{\begin{array}{l} \dfrac{dV(t)}{dt} = 10^6 \exp(-t/100) - 10^7 \\[3mm] t = 0, \ V = 10^9 \end{array}}$$

2. We now separate variables and integrate the differential balance equation from $t = 0$ to $t = 60$ days.

$$\int_{V(0)}^{V(60)} dV = \int_0^{60} [10^6 \exp(-t/100) - 10^7] \, dt$$

$$\Downarrow$$

$$V(60 \text{ days}) - V(0) = \int_0^{60} 10^6 e^{-t/100} \, dt - \int_0^{60} 10^7 \, dt$$

$$\Downarrow V(0) = 10^9 \text{ liters}$$

$$V(60 \text{ days}) = 10^9 - 10^6 (10^2) e^{-t/100} \Big]_0^{60} - 10^7 \, t \Big]_0^{60}$$

$$= \boxed{4.45 \times 10^8 \text{ liters}} \quad \text{(Verify)}$$

TEST YOURSELF

A liquid, A, is poured at a rate of 10 kg/h into a tank containing a second liquid, B. The two substances react and A is consumed, but at a lower rate than that at which it is added. Liquid is withdrawn from the reaction vessel at a rate of 10 kg/h.

1. Which terms of the general balance equation

$$\text{Accumulation} = \text{input} + \text{generation} - \text{output} - \text{consumption}$$

 do *not* drop out in each of the following balances on the reaction vessel?
 (a) total mass
 (b) moles of A
 (c) moles of B

2. Write a differential mass balance on the system, letting $m(t)$ be the total mass of the system contents.

11.2 MATERIAL BALANCES

Two of the key steps in the formulation and solution of differential mass balances are expressing mathematically the accumulation term in the balance equation and integrating the resulting differential equation.

As shown in Section 11.1, the rate of mass accumulation can be expressed as the derivative with respect to time of the mass of the balanced quantity in the system. For example, suppose at an instant in time the contents of a system have a total mass m and a $CaSO_4$ mass fraction x. Rates of accumulation of total mass and of the mass of $CaSO_4$ may then be expressed as follows.

$$\textit{Rate of Total Mass Accumulation} = \frac{dm}{dt}$$

$$\textit{Rate of CaSO}_4 \textit{ Accumulation} = \frac{d(xm)}{dt} = x\frac{dm}{dt} + m\frac{dx}{dt}$$

Example 11.2-1 further illustrates the procedure for solving differential material balances on transient systems.

EXAMPLE 11.2-1 Material Balances on a Storage Tank

A 250-m^3 fuel storage tank that contains pure methane is to be switched to propane. Propane is fed to the tank at a rate of 5.00 m^3/min, and gas is withdrawn from the tank at the same rate. The contents of the tank may be considered uniform, so that the composition of the outlet stream is that of the tank contents. All gases are at 27°C and 1 atm. How long will it take to flush out 99% of the methane originally in the tank?

Solution

Let n be the moles of gas in the tank, and q the molar flow rate of the input and output gases.

5.00 m^3 C_3H_8 (g)/min @ 27°C, 1 atm
q_{in} kmol C_3H_8/min

250 m^3
n (kmol)
x (kmol CH_4/kmol)

5.00 m^3/min @ 27°C, 1 atm
q_{out} (kmol/min)
x (kmol CH_4/kmol)

Assuming ideal gas conditions,

$$q_{in} = q_{out} = \frac{5.00 \text{ m}^3}{\text{min}} \left| \frac{273 \text{ K}}{300 \text{ K}} \right| \frac{1 \text{ kmol}}{22.4 \text{ m}^3(\text{STP})} = 0.203 \text{ kmol/min}$$

$$n = (250 \text{ m}^3)(273/300)(1.00 \text{ kmol}/22.4 \text{ m}^3) = 10.2 \text{ kmol}$$

Since $q_{in} = q_{out}$, a total mole balance leads to the conclusion that n is constant.

When 99% of the methane has been flushed out of the tank, the methane mole fraction in the remaining gas must be 0.01. (Why?) We will write and integrate a balance on methane to calculate the time required for this state to be achieved.

CH$_4$ Balance

The amount of CH$_4$ in the tank at any time is n(kmol)x(kmol CH$_4$/kmol). The balance is

$$\text{Accumulation} = -\,\text{output}$$

$$\Downarrow$$

$$\frac{d}{dt}(nx) = -qx$$

$$\Big\Downarrow \begin{array}{l} n = 10.2 \\ q = 0.203 \end{array}$$

$$\frac{dx}{dt} = -0.0200x$$

$$t = 0,\, x = 1.00$$

This is the balance equation. To solve it, we separate the variables and integrate from the initial condition ($t = 0$, $x = 1$) to the final state (t_f, $x = 0.01$)

$$\frac{dx}{x} = -0.02\,dt$$

$$\Downarrow$$

$$\int_{1}^{0.01} \frac{dx}{x} = \int_{0}^{t_f} (-0.02)\,dt$$

$$\Downarrow$$

$$\ln x \Big]_{1}^{0.01} = -0.02t \Big]_{0}^{t_f}$$

$$\Downarrow$$

$$\ln\left(\frac{0.01}{1}\right) = -0.02t_f$$

$$\Downarrow$$

$$\boxed{t_f = 230 \text{ min}}$$

If a chemical reaction occurs in a process, the balance on a reactive species must include a term for the rate of generation or consumption of the product or reactant being balanced.

EXAMPLE 11.2-2 Material Balance on a Batch Reactor

A reaction with stoichiometry A → B is carried out in a constant-volume isothermal batch reactor. The rate of consumption of A, r_A, is directly proportional to the molar

concentration of A in the reactor and to the reactor volume:

$$r_A(\text{mol/s}) = 0.200V(\text{liters})C_A(\text{mol/liter})$$

(The *rate constant*, 0.200, has units of s^{-1}.) The initial concentration of A in the reactor is 0.100 mol/liter. Write a differential balance on A and use it to calculate the time required to achieve a conversion of 90%.

Solution

The amount of A in the reactor at any time is $n_A = VC_A$. The balance on A is

$$\text{Accumulation} = -\text{consumption}$$

$$\Downarrow$$

$$\frac{d(VC_A)}{dt} = -0.200VC_A$$

$$\Downarrow V = \text{constant}$$

$$\boxed{\begin{array}{c} \dfrac{dC_A}{dt} = -0.200C_A \\[2mm] t = 0, \; C_A = 0.100 \end{array}}$$

The final concentration of A must be 10% of the initial concentration, or $C_{A0}/10 = 0.01$. The balance equation may easily be integrated by separating the variables and integrating from $(t = 0, \, C_A = 0.1)$ to $(t_f, \, C_A = 0.01)$.

$$\frac{dC_A}{C_A} = -0.200 \, dt$$

$$\Downarrow$$

$$\int_{C_{A0}}^{C_{Af}} \frac{dC_A}{C_A} = -0.200 \int_0^{t_f} dt$$

$$\Downarrow$$

$$\ln \frac{C_{Af}}{C_{A0}} = -0.200 t_f$$

$$\Downarrow C_{Af}/C_{A0} = 0.100$$

$$t_f = -\frac{\ln(0.100)}{0.200(s^{-1})} = \boxed{11.5 \text{ s}}$$

The third example in this section presents a situation where several balances must be solved simultaneously to achieve a desired result. Observe how we keep track of unknowns and relations between them in the course of the solution, just as was done in balances on complex steady-state systems.

EXAMPLE 11.2-3 Batch Distillation

A batch (or, more precisely, semibatch) distillation unit is charged with 100 mol of a 60 mole% benzene–40 mole% toluene mixture. At any given instant the vapor leaving

the still may be considered to be in equilibrium with the remaining liquid. The benzene mole fraction in the exit vapor, y, and the benzene mole fraction in the remaining liquid, x, are related by the expression

$$y = \frac{2.6x}{1 + 1.6x}$$

Derive an equation relating the amount of liquid remaining in the still to the mole fraction of benzene in this liquid.

Solution

We will choose the liquid remaining in the still as our system. Note that among the labeled quantities on the flow chart, L is a quantity of liquid, while V is a vapor flow rate.

Since there are two components in this nonreactive system, we may write two independent balances—one on total moles and one on moles of benzene.

Total Balance

$$\text{Accumulation} = -\text{output}$$
$$\Downarrow$$
$$\frac{dL}{dt} = -V \tag{1}$$

Benzene Balance

$$\text{Accumulation} = -\text{output}$$
$$\Downarrow$$
$$\frac{d(xL)}{dt} = -yV$$
$$\Downarrow$$
$$L\frac{dx}{dt} + x\frac{dL}{dt} = -yV \tag{2}$$

From the equilibrium expression

$$y = 2.6x/(1.0 + 1.6x) \tag{3}$$

There are four dependent variables in this system—$L(t)$, $V(t)$, $x(t)$, and $y(t)$–and we wish to derive a relationship between L and x. We must therefore use two of the available relations—Equations 1, 2, and 3—to express V and y in terms of L and x, and substitute the resulting expressions for V and y into the third equation to obtain the desired relationship.

From Equation 1, $V = -dL/dt$, and Equation 3 directly expresses y as a function of x. Substituting for V and y in Equation 2, we obtain

$$L\frac{dx}{dt} + x\frac{dL}{dt} = \left(\frac{2.6x}{1.0 + 1.6x}\right)\frac{dL}{dt}$$

$$\Downarrow$$

$$L\frac{dx}{dt} = \left(\frac{2.6x}{1.0 + 1.6x} - x\right)\frac{dL}{dt}$$

Since there is only one independent variable, t, in the system, we may treat dt as though it were an algebraic variable and cancel it out, and then bring all the terms involving x to one side of the equation and those involving L to the other.

$$\frac{dL}{L} = \frac{dx}{\left(\dfrac{2.6x}{1.0 + 1.6x} - x\right)} = \frac{1.0 + 1.6x}{1.6x(1 - x)}\,dx$$

Finally, we obtain an explicit relationship between L and x by integrating from the initial condition ($L = 100$ mol, $x = 0.6$ mol B/mol) to an arbitrary final state (L, x).

$$\int_{100}^{L}\frac{dL}{L} = \int_{0.6}^{x}\frac{1.0 + 1.6x}{1.6x(1 - x)}\,dx$$

The integral on the left side is easily evaluated; the right side requires integration by partial fractions or a table of integrals. The result is

$$\ln L\Big]_{100}^{L} = 0.625\ln x\Big]_{0.6}^{x} - 1.625\ln(1 - x)\Big]_{0.6}^{x}$$

$$\Downarrow$$

$$\ln\left(\frac{L}{100}\right) = \ln\left[\left(\frac{x}{0.6}\right)^{0.625}\left(\frac{1-x}{0.4}\right)^{-1.625}\right]$$

$$\Downarrow$$

$$\boxed{L = 100\left(\frac{x}{0.6}\right)^{0.625}\left(\frac{1-x}{0.4}\right)^{-1.625}}$$

A plot of x, the mole fraction of benzene in the liquid, versus L, the moles of liquid remaining, appears in the diagram on the next page. The plot shows that the liquid remaining in the still becomes progressively leaner in benzene; the last remaining drop ($L \approx 0$) is almost pure toluene.

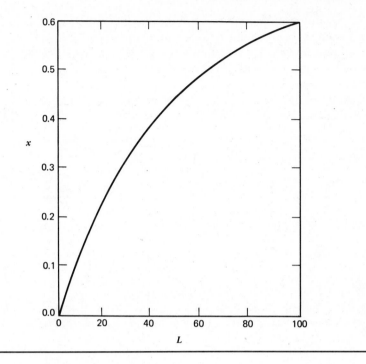

11.3 ENERGY BALANCES ON SINGLE-PHASE NONREACTIVE PROCESSES

The general energy balance equation has the form

$$\text{Accumulation} = \text{input} - \text{output} \tag{11.3-1}$$

since energy can neither be generated nor consumed.

Suppose $E_{sys}(t)$ is the total energy (internal + kinetic + potential) of a system, and m_{in} and m_{out} are the mass flow rates of the system input and output streams. (If the system is closed, these quantities each equal zero.) Proceeding as in the development of the transient mass balance equation, we apply the general energy balance equation (11.3-1) to the system in a small time interval from t to $(t + \Delta t)$, during which time the properties of the input and output streams remain approximately constant. The terms of the equation are as follows:

$$\text{Accumulation} = \Delta E_{sys} = \Delta U_{sys} + \Delta E_{k,sys} + \Delta E_{p,sys}$$

$$\text{Input} = m_{in}\left(\hat{H}_{in} + \frac{v_{in}^2}{2g_c} + \frac{g}{g_c}\, z_{in} \right) \Delta t + Q\,\Delta t + W_s\,\Delta t$$

$$\text{Output} = m_{out}\left(\hat{H}_{out} + \frac{v_{out}^2}{2g_c} + \frac{g}{g_c}\, z_{out} \right) \Delta t$$

where the quantities in parentheses are the specific enthalpies, kinetic energies, and potential energies of the input and output streams, and Q and W_s are the rates of transfer of heat and shaft work. (See Section 7.4c.) Any or all of the variables m, \hat{H}, v, z, Q and W_s may vary with time.

If we now substitute the given expressions for accumulation, input, and output into Equation 11.3-1, divide by Δt, and let Δt approach zero, we obtain the general differential energy balance equation.

$$\frac{dU_{sys}}{dt} + \frac{d(E_{k,sys})}{dt} + \frac{d(E_{p,sys})}{dt} = m_{in}\left(\hat{H}_{in} + \frac{v_{in}^2}{2g_c} + \frac{g}{g_c}z_{in}\right)$$

$$- m_{out}\left(\hat{H}_{out} + \frac{v_{out}^2}{2g_c} + \frac{g}{g_c}z_{out}\right) + Q + W_s \quad \textbf{(11.3-2)}$$

If there are several input and output streams, a term of the form

$$m\left(\hat{H} + \frac{v^2}{2g_c} + \frac{g}{g_c}z\right)$$

must be included in Equation 11.3-2 for each stream.

It is quite difficult to solve Equation 11.3-2 unless a number of simplifications are made. We will restrict our consideration to systems that satisfy the following conditions:

1. The system has at most a single input stream and a single output stream, each with the same mass flow rate.

$$m_{in} = m_{out} = m \quad \textbf{(11.3-3)}$$

A consequence of this assumption is that the mass of the system contents does not change with time.

2. Kinetic and potential energy changes in the system and between the inlet and outlet streams are negligible.

$$\frac{dE_{k,sys}}{dt} \approx \frac{dE_{p,sys}}{dt} \approx 0 \quad \textbf{(11.3-4)}$$

$$m\left(\frac{v_{in}^2}{2g_c} - \frac{v_{out}^2}{2g_c}\right) \approx 0 \quad \textbf{(11.3-5)}$$

$$m\left(\frac{g}{g_c}z_{in} - \frac{g}{g_c}z_{out}\right) \approx 0 \quad \textbf{(11.3-6)}$$

Under these conditions, Equation 11.3-2 simplifies to

$$\frac{dU_{sys}}{dt} = m(\hat{H}_{in} - \hat{H}_{out}) + Q + W_s \quad \textbf{(11.3-7)}$$

If the equation is applied to a closed system, $m \equiv 0$ and the term W_s must be replaced by the total work W.

Equation 11.3-7 is simple in appearance, but its solution is still generally difficult to obtain. If, for example, the composition or temperature of the system contents varies with position in the system, it is difficult to express the total internal energy U_{sys} in terms of measurable quantities, and a similar problem occurs if phase changes or chemical reactions take place in the course of the process. To illustrate the solution of energy balance problems without becoming too involved in thermodynamic complexities, we will impose the following additional restrictions.

3. The temperature and composition of the system contents do not vary with position within the system. In consequence, the outlet stream and the system contents must be at the same temperature, or

$$T_{out} = T_{sys} \qquad (11.3\text{-}8)$$

4. No phase changes or chemical reactions take place within the system; \hat{U} and \hat{H} are independent of pressure; and the mean heat capacities C_v and C_p of the system contents (and of the inlet and outlet streams) are independent of composition and temperature, and hence unchanging with time. Then if T_r is a reference temperature at which \hat{H} is defined to be zero, and M is the mass (or number of moles) of the system contents,

$$U_{sys} = M\hat{U}_{sys} = M[\hat{U}(T_r) + C_v(T_{sys} - T_r)]$$
$$\Downarrow M, \hat{U}(T_r), \text{ and } C_v \text{ are constant}$$
$$\frac{dU_{sys}}{dt} = MC_v \frac{dT_{sys}}{dt} \qquad (11.3\text{-}9)$$

$$\hat{H}_{in} = C_p(T_{in} - T_r) \qquad (11.3\text{-}10)$$
$$\hat{H}_{out} = C_p(T_{out} - T_r)$$
$$\Downarrow \text{ Equation } 11.3\text{-}8$$
$$\hat{H}_{out} = C_p(T_{sys} - T_r) \qquad (11.3\text{-}11)$$

Finally, we may substitute the expressions of Equations 11.3-3 through 11.3-11 into the general energy balance equation, (11.3-2), to obtain for an open system

Open System:
$$\boxed{MC_v \frac{dT_{sys}}{dt} = mC_p(T_{in} - T_{sys}) + Q + W_s} \qquad (11.3\text{-}12)$$

(Verify this result for yourself.) For a closed system, the equation is

Closed System:
$$\boxed{MC_v \frac{dT_{sys}}{dt} = Q + W} \qquad (11.3\text{-}13)$$

To summarize, the conditions under which Equations 11.3-12 and 11.3-13 are valid are: (1) negligible kinetic and potential energy changes, (2) no accumulation of mass in the system, (3) pressure independence of \hat{U} and \hat{H}, (4) no phase changes or chemical

reactions, (5) a spatially uniform system temperature. Any or all of the variables T_{sys}, T_{in}, Q, and W_s (or W) may vary with time, but the system mass, M, the mass through-put rate, m, and the heat capacities, C_v and C_p, must be constants.

The following example illustrates the derivation and solution of an energy balance on a closed system that satisfies these restrictions.

EXAMPLE 11.3-1 Startup of a Batch Reactor

A well-stirred batch reactor wrapped in an electrical heating mantle is charged with a liquid reaction mixture. The reactants must be heated from an initial temperature of 25°C to 250°C before the reaction can take place at a measurable rate. Use the data given below to determine the time required for this heating to take place.

$$\text{Reactants: Mass} = 1.50 \text{ kg}$$
$$C_v = 0.900 \text{ cal/g} \cdot °C$$
$$\text{Reactor: Mass} = 3.00 \text{ kg}$$
$$C_v = 0.120 \text{ cal/g} \cdot °C$$
$$\text{Heating rate: } Q = 500.0 \ W$$

Negligible reaction and no phase changes during heating
Negligible energy added to the system by the stirrer

Solution

We first note that the conditions of validity of the simplified closed system energy balance equation, Equation 11.3-13, are all satisfied (verify this); moreover, since the system has constant volume and the energy input due to the stirrer is presumed negligible, $W \approx 0$. The equation therefore becomes

$$MC_v \frac{dT_{sys}}{dt} = Q$$

$$t = 0, \ T_{sys} = 25°C$$

The task is now to integrate this equation from the initial state of the system ($t = 0$, $T_{sys} = 25°C$) to the final state ($t = t_f$, $T_{sys} = 250°C$), and to solve the integrated equation for the heating time t_f. Rearranging the equation,

$$MC_v \, dT_{sys} = Q \, dt$$
$$\Downarrow \text{ Integrate}$$
$$\int_{25}^{250} MC_v \, dT_{sys} = \int_0^{t_f} Q \, dt$$
$$\Downarrow Q, M, \text{ and } C_v \text{ are constant}$$
$$MC_v(250 - 25) = Qt_f$$
$$\Downarrow$$
$$t_f = \frac{225 \, MC_v}{Q}$$

The heat capacity of the system is obtained from Equation 8.3-17 as

$$C_v = \frac{M_{\text{reactants}}}{M} (C_v)_{\text{reactants}} + \frac{M_{\text{reactor}}}{M} (C_v)_{\text{reactor}}$$

$$\Downarrow$$

$$MC_v = (1500 \text{ g}) \left(0.900 \frac{\text{cal}}{\text{g} \cdot {}^\circ\text{C}} \right) + (3000 \text{ g}) \left(0.120 \frac{\text{cal}}{\text{g} \cdot {}^\circ\text{C}} \right)$$

$$= (1710 \text{ cal}/{}^\circ\text{C})(4.184 \text{ J/cal}) = 7150 \text{ J}/{}^\circ\text{C}$$

The final result is

$$t_f = \frac{225 \, MC_v}{Q}$$

$$\left\Vert \begin{array}{l} MC_v = 7150 \text{ J}/{}^\circ\text{C} \\ Q = 500 \text{ W} = 500 \text{ J/s} \end{array} \right.$$

$$t_f = \frac{7150(\text{J}/{}^\circ\text{C})}{500(\text{J/s})} (225{}^\circ\text{C}) = 3220 \text{ s} = \boxed{53.7 \text{ min}}$$

Question: Which restriction on the energy balance equation (11.3-13) would probably be violated if the reactants were not stirred?

The final example illustrates the application of the transient energy balance equation to a continuous system.

EXAMPLE 11.3-2 Transient Behavior of an Air-Cooling System

An engine that, when operating, generates heat at a constant rate $Q_{\text{gen}} = 8530$ Btu/min, is cooled with air.

The air in the engine housing is circulated rapidly enough for its temperature to be considered uniform and equal to the temperature of the outlet air. Air passes through the engine housing at a rate of 6.00 lb-mole/min, entering at a temperature of 65°F, and an average of 0.200 lb-mole of air is contained within the engine housing. (We will neglect the variation of this quantity with the changing temperature of the gas.)

Heat is lost from the housing to its surroundings at a rate

$$Q_l(\text{Btu/min}) = 33.0(T - 65°\text{F})$$

Suppose the engine is started with the inside air temperature equal to 65°F.

1. Calculate the steady-state air temperature if the engine runs continuously for an indefinite period of time, assuming

$$C_v = 5.00 \text{ Btu/lb-mole·°F}$$

2. Derive a differential equation for the variation of the outlet temperature with the time from startup, and solve it.

Solution

System = the air within the engine housing.

1. The steady-state energy balance equation may be obtained by setting dT/dt equal to zero in Equation 11.3-12.

$$0 = mC_p(T_{in} - T_{out}) + Q + W_s$$

$T_{in} = 65°\text{F (given)}$
$T_{out} = T_s(°\text{F})$ (steady-state outlet temperature)
$W_s = 0$ (no moving parts)
$Q = Q_{gen} - Q_l$

$$mC_p(T_s - 65.0) = 8530 - 33.0(T_s - 65.0)$$

Assuming ideal gas behavior

$$C_p = C_v + R = (5.00 + 1.99) \text{ Btu/lb-mole·°F (Equation 8.3-12)}$$

$$mC_p = \frac{6.00 \text{ lb-mole}}{\text{min}} \left| \frac{6.99 \text{ Btu}}{\text{lb-mole·°F}} \right. = 41.9 \frac{\text{Btu}}{\text{min·°F}}$$

The energy balance equation then becomes

$$41.9(T_s - 65.0) = 8530 - 33.0(T_s - 65.0)$$

$$\Downarrow$$

$$\boxed{T_s = 179°\text{F}}$$

2. The unsteady-state balance equation (11.3-12) is, for our system,

$$MC_v \frac{dT}{dt} = mC_p(65 - T) + Q_{gen} - Q_l$$

$M = 0.2$ lb-mole
$C_v = 5$ Btu/lb-mole·°F
$mC_p = 41.9$ Btu/min·°F [from part 1]
$Q_{gen} = 8530$ Btu/min
$Q_l = 33.0(T - 65)$(Btu/min)

$$\boxed{\begin{array}{l} \dfrac{dT}{dt} = -74.9T + 13\ 400 \\[2mm] t = 0, \ T = 65°\text{F} \end{array}}$$

You can check this result by setting $dT/dt = 0$ and solving the resulting equation for the steady-state value of T; the result is $(13\,400/74.9) = 179°F$, in agreement with the result of part 1.

The solution of the equation is obtained by separating variables and integrating:

$$\int_{65}^{T} \frac{dT}{13\,400 - 74.9T} = \int_{0}^{t} dt$$

$$\Downarrow$$

$$-\frac{1}{74.9} \ln{(13\,400 - 74.9T)} \Bigg]_{65}^{T} = t$$

$$\Downarrow$$

$$\ln{(13\,400 - 74.9T)} = -74.9t + \ln{(8352)}$$

$$\Downarrow$$

$$13\,400 - 74.9T = 8352 \exp{(-74.9t)}$$

$$\Downarrow$$

$$\boxed{T = 179 - 112 \exp{(-74.9t)}}$$

A check on the solution may be obtained by substituting $t = 0$ and verifying that T equals the specified initial value, $65°F$.

If you evaluated T from this equation for a number of values of t, a plot would appear as follows:

The plot starts from the initial condition of $65°F$ and asymptotically approaches the steady-state value of $179°F$; it takes 3.9 s for the temperature to reach $178°F$, one degree short of its final value.

TEST YOURSELF

1. Under what conditions is the simplified form of the energy balance equation, (11.3-12), valid?

2. Suppose the heat capacity C_v is a function of T, which varies with time. Where would the derivation of Equation 11.3-12 break down?

PROBLEMS

1. A solution containing hydrogen peroxide with a mass fraction x_{p0} (kg H_2O_2/kg solution) is added to a storage tank at a steady rate q_0(kg/h). During this process, the liquid level reaches a corroded spot in the tank wall, and a leak develops. As the filling continues, the leak rate q_1(kg/h) becomes progressively worse. Moreover, the peroxide begins to decompose once it is in the tank, at a rate

$$r_d(kg/h) = km_p$$

 where m_p(kg) is the mass of peroxide in the tank.

 Let $m(t)$ be the total mass of solution in the tank at time t, where $t = 0$ is the moment the liquid level reaches the corroded spot, and let m_0 and m_{p0} be the total liquid mass and mass of peroxide, respectively, in the tank at $t = 0$. Write differential balances on the total tank contents and on the peroxide in the tank, and provide initial conditions. Your solution should involve only the quantities q_0, q_1, x_{p0}, k, m, m_0, m_p, m_{p0}, and t. Assume that the peroxide concentration does not vary with position in the tank.

2. A **ball mill** is a particle size-reduction device used for crushing and grinding (**comminution**) operations, which consists of a horizontal rotating drum partially filled with steel balls. A mineral ore, coal, or another solid material is fed into the mill; the ore and steel balls are lifted by the mill's rotation and then fall, causing the ore to be broken down into smaller particles.

 Suppose a copper ore is fed to a ball mill at a steady rate Q(kg/h), crushed ore is withdrawn at the same rate, and at an initial time $t = 0$ the total mass of ore in the mill is M(kg). Further suppose that a series of 10 particle size ranges is defined, and that x_i is the mass fraction of particles in the ith size range, where $i = 1$ represents the largest size range and $i = 10$ represents the smallest range.

 (a) Write a differential balance on the total mass of ore in the mill, and use it to prove that M does not change with time.

 (b) Suppose that the rate at which particles are broken out of the ith size range is

$$r_i(kg/s) = k_i m_i$$

 where m_i is the mass of particles in this size range. Further suppose that of the particles broken out of size range i in a differential time interval, a fraction b_{ij} go into size range j. Assume that the size distribution of particles is uniform throughout the mill and equals that of the product, and let x_{if} equal the mass fraction of the feed that falls in the ith size range. Show that a mass balance on the jth size fraction in the tank yields after some rearrangement

$$dx_j/dt = (Q/M)(x_{jf} - x_j) - k_j x_j + \sum_{i=1}^{j-1} k_i x_i b_{ij}$$

3. Methanol is withdrawn from a 5.00-m³ storage tank at a rate that on the average increases linearly with time. At one point the tank contains 750 kg of the liquid, and the withdrawal rate is 750 kg/h. Five hours later the withdrawal rate has increased to 1000 kg/h. Methanol is added to the tank to replenish the supply at an average rate of 1200 kg/h.

(a) Calculate an expression for the average withdrawal rate $q_w(t)$(kg/h), letting $t = 0$ signify the time at which $q_w = 750$ kg/h, and incorporate it into a differential mass balance on the methanol in the tank.

(b) How much methanol will be in the tank at $t = 5$ h? (See Example 11.2-1.)

(c) How long will it take for the methanol level in the tank to reach its highest value, and what percentage of the tank volume will be filled at this point?

(d) Calculate the time it will take to empty the tank.

4. An 8.00-ft^3 compressed air tank is being filled. Before the filling begins, the tank is open to the atmosphere. The reading on a Bourdon gauge mounted on the tank increases linearly from an initial value of 0.0 to 100 psi after 15 s. The temperature is constant at 72°F, and atmospheric pressure is 1 atm.

(a) Let $n(t)$ equal the lb-moles of air in the tank at any time, and let q(lb-moles/s) be the rate at which air is added. Write a differential balance on the air in the tank in terms of n and q.

(b) Calculate q, and integrate the balance equation to determine the lb-moles of *oxygen* in the tank after 2 min.

5. A gas storage tank with a floating roof receives a steady input of 540 m^3/h of a natural gas. The rate of withdrawal of gas from the tank, Q_w, varies more or less randomly during the day and is recorded at 10-min intervals. At 8:00 A.M. one morning the volume of stored gas is 3.00×10^3 m^3. The withdrawal rate data for the next 4 h are as follows:

Hour beginning at	$Q_w(m^3/min)$
8:00	11.4, 11.9, 12.1, 11.8, 11.5, 11.3
9:00	11.4, 11.1, 10.6, 10.8, 10.4, 10.2
10:00	10.2, 9.8, 9.4, 9.5, 9.3, 9.4
11:00	9.5, 9.3, 9.6, 9.6, 9.4, 9.9
12:00	9.8

The temperature and pressure of the inlet, stored, and outlet gases are equal and nearly constant throughout the given time period.

(a) Write a differential balance on the gas in the tank, and prove that when integrated it yields the following equation for the gas volume:

$$V(t) = 3.00 \times 10^3 + 9.00t - \int_0^t Q_w \, dt$$

where t(min) is the time elapsed since 8:00 A.M.

(b) Calculate the stored gas volume at noon, using Simpson's rule (Appendix A.3) to evaluate the integral.

(c) Although a running estimate of the tank volume is important to have, in practice it would probably not be obtained in the manner indicated. How would it be obtained? What would you infer if the value estimated in part (b) is greater than that obtained by the more accurate method?

*(d) Write and run a computer program to read in the data for $Q_w(t)$, and to estimate and print out the gas volume at each time for which Q_w is recorded, using the

* Computer problem.

trapezoidal rule (Appendix A.3) to evaluate the integral over each successive 10-min interval. What is the percentage difference between the volumes at noon estimated by Simpson's rule and the trapezoidal rule?

6. Water is added at varying rates to a 300-liter holding tank. When a valve in a discharge line is opened, water flows out at a rate proportional to the height and hence to the volume V of water in the tank. The flow of water into the tank is slowly increased, and the level rises in consequence, until at a steady input rate of 60.0 liters/min the level just reaches the top but does not spill over. The input rate is then abruptly decreased to 20.0 liters/min.

 (a) Write the equation that relates the discharge rate, q(liters/min), to the volume of water in the tank, V(liters), and use it to calculate the steady-state volume when the input rate is 20 liters/min.

 (b) Write a differential balance equation on the water in the tank for the period from the moment the input rate is decreased to the attainment of steady state, expressing it in the form $dV/dt = \ldots$. Use the equation to verify the steady-state value of V calculated in part (a).

 (c) Separate variables and integrate the balance equation to derive an expression for $V(t)$. Use this result to calculate the time in minutes required for the volume to decrease to within 1% of its steady-state value. What does the equation indicate about the time required to reach steady-state?

7. The production supervisor of a small pharmaceutical firm has observed a decreasing demand for potassium regurgitol (PRG) over a 2-month period, and since the plant manager has been throwing up the low sales of this product at the weekly staff meetings, the supervisor decides to discontinue its production immediately. On the day of this decision, the inventory of PRG is 18 000 kg. Based on the orders on hand, the manager projects the following weekly demand for the next six weeks:

Week	1	2	3	4	5	6
Demand D(kg/wk)	2385	1890	1506	1196	950	755

 (a) Use a semilog plot of the projected demand figures to derive an equation for D as a function of t(weeks) from the present time.

 (b) Write a differential balance on the inventory I(kg) of PRG, and integrate it to determine I as a function of t.

 (c) If the demand continues to follow the projected trend of the next six weeks, how much PRG will eventually have to be discarded?

8. A ventilation system has been designed for a large laboratory with a volume of 1100 m³. The volumetric flow rate of ventilation air is 700 m³/min @22°C and 1 atm; the latter two values may also be taken as the temperature and pressure of the room air. A reactor in the laboratory is capable of emitting as much as 1.50 mol of sulfur dioxide into the room if a seal ruptures. An SO_2 mole fraction in the room air greater than 1.0×10^{-6} (1 ppm) constitutes a health hazard.

 (a) Suppose the reactor seal ruptures at a time $t = 0$, and the maximum amount of SO_2 is emitted almost instantaneously. Assuming that the air flow is sufficient to make the room air composition spatially uniform, write a differential SO_2 balance,

letting n = total moles in the room and x = mole fraction of SO_2. (See Example 11.2-1.)

(b) Integrate the balance to calculate (i) the concentration of SO_2 (mol/liter) in the room 2 min after the rupture occurs, and (ii) the time required for the SO_2 concentration in the room to reach the level of safety.

(c) What might be done to decrease the environmentally hazardous nature of the reactor?

9. Ninety kilograms of sodium nitrate are dissolved in 110 kg of water in a well-mixed tank. When the dissolution is complete, pure water is fed to the tank at a constant rate q(kg/min), and solution is withdrawn from the tank at the same rate.

 (a) Write an expression for the total mass of solution in the tank at any time.

 (b) Write a balance on the sodium nitrate in the tank, and integrate it to determine the mass fraction of $NaNO_3$ in the vessel as a function of q and the time, t(min), from the moment the input and output flows commence.

 (c) If the throughput rate is $q = 100$ kg/min, how long will it take to flush out 90% of the sodium nitrate originally in the tank? How long to flush out 99%? 99.9%?

10. A tracer is used to characterize the degree of mixing in a continuous stirred tank. Water enters and leaves the mixer at a rate of Q(m^3/min). Scale has built up on the inside walls, so that the effective volume V(m^3) of the tank is unknown. At a time $t = 0$, a mass m_0(kg) of the tracer is injected into the tank and the tracer concentration in the outlet stream C(kg/m^3) is monitored. The injection is rapid enough so that all of the tracer may be considered to be in the tank at $t = 0$.

 (a) Write a differential balance on the tracer in the tank in terms of V, C, and Q, assuming that the tank contents are perfectly mixed.

 (b) Integrate the balance to prove that

$$C(t) = (m_0/V) \exp(-Qt/V)$$

 (c) Suppose the flow rate through the mixer is $Q = 30$ m^3/min and that the following data are taken:

Time from injection, t(min)	1	2	3	4
$C \times 10^3$ (kg/m^3)	0.223	0.050	0.011	0.0025

 (For example, at $t = 1$ min, $C = 0.223 \times 10^{-3}$ kg/m^3). Verify graphically that the tank is functioning as a perfect mixer, and determine the effective volume V(m^3).

11. A 40-ft^3 oxygen tent initially contains air at 68°F and 14.7 psia. At $t = 0$ a gas mixture containing 35% O_2 by volume and the balance N_2 is fed to the tent at 68°F and 1.3 psig at a rate of 60 ft^3/min, and gas is withdrawn from the tent at 68°F and 14.7 psia at a molar flow rate equal to that of the feed gas.

 (a) Write a differential balance on the oxygen in the tent, assuming the tent contents are uniform. Write the balance in terms of the total moles in the tent and x, the mole fraction of O_2 in the exit gas.

 (b) Integrate the balance to determine the time required for the mole fraction of oxygen in the tent to reach 0.27.

12. A radioactive isotope decays at a rate proportional to its concentration. If the concentration of an isotope is C(mg/liter), then its rate of decay may be expressed as

$$r_d(\text{mg/liter} \cdot \text{s}) = kC$$

where k is a constant.

(a) A solution of a radioisotope whose concentration is C_0(mg/liter) is placed in a closed vessel of volume V(liters). Write a balance on the isotope in the vessel, and integrate it to prove that the **half-life** $t_{1/2}$ of the isotope—by definition, the time required for the isotope concentration to drop to half of its initial value—equals $(\ln 2)/k$.

(b) The half-life of ^{256}Mn is 2.6 h. A batch of this isotope that was used in a radiotracing experiment has been collected in a holding tank. The radiation safety officer declares that the activity (which is proportional to the isotope concentration) must decay to 1% of its present value before the solution can be discarded. How long will this take?

13. A chemical reaction with stoichiometry $A \rightarrow$ products is said to follow an **nth-order rate law** if A is consumed at a rate proportional to the nth power of its concentration in the reaction mixture. If r_A is the rate of consumption of A per unit reactor volume, then

$$r_A(\text{mol/liter} \cdot \text{s}) = kC_A^n$$

where C_A(mol/liter) is the reactant concentration, and the constant of proportionality k is the reaction **rate constant**. A reaction that follows this law is referred to as **nth-order.** The rate constant is a strong function of temperature, but is independent of the reactant concentration.

(a) Suppose a first-order reaction ($n = 1$) is carried out in an isothermal batch reactor of constant volume V. Write a material balance on A, and integrate it to derive the expression

$$C_A = C_{A0} \exp(-kt)$$

where C_{A0} is the concentration of A in the reactor at $t = 0$.

(b) The gas-phase decomposition of sulfuryl chloride

$$SO_2Cl_2 \rightarrow SO_2 + Cl_2$$

is thought to follow a first-order rate law. The reaction is carried out in a constant-volume isothermal batch reactor, and the concentration of SO_2Cl_2 is measured at several reaction times, with the following results:

t(min)	4.0	20.2	40.0	60.0	120.0	180.0
C_A(mol/liter)	0.0158	0.0152	0.0144	0.0136	0.0116	0.0099

Use these results to verify the proposed rate law graphically, and determine the rate constant k. (Give both the value and units of k.)

14. A gas-phase decomposition reaction with stoichiometry $2A \rightarrow 2B + C$ follows a second-order rate law (see Problem 13):

$$r_d(\text{mol/m}^3 \cdot \text{s}) = kC_A^2$$

where C_A is the reactant concentration in mol/m^3. The rate constant k varies with the reaction temperature according to the **Arrhenius law**

$$k(m^3/mol \cdot s) = k_0 \exp(-E/RT)$$

where $k_0(m^3/mol \cdot s) = $ the **pre-exponential factor**
 $E(J/mol) = $ the reaction **activation energy**
 $R = $ the gas constant
 $T(K) = $ the reaction temperature

(a) Suppose the reaction is carried out in a batch reactor of constant volume $V(m^3)$ at a constant temperature $T(K)$, beginning with pure A at a concentration C_{A0}. Write a differential balance on A, and integrate it to obtain an expression for $C_A(t)$ in terms of C_{A0} and k.

(b) Let $P_0(atm)$ be the initial reactor pressure. Prove that $t_{1/2}$, the time required to achieve a 50% conversion of A in the reactor, equals RT/kP_0, and derive an expression for $P_{1/2}$, the reactor pressure at this point, in terms of P_0. Assume ideal gas behavior.

(c) The decomposition of nitrous oxide (N_2O) to nitrogen and oxygen is carried out in a 5.00-liter batch reactor at a constant temperature of 1015 K, beginning with pure N_2O at several initial pressures. The reactor pressure $P(t)$ is monitored, and the times ($t_{1/2}$) required to achieve 50% conversion of N_2O are noted.

$P_0(atm)$	0.135	0.286	0.416	0.683
$t_{1/2}(s)$	1060	500	344	209

Use these results to verify that the N_2O decomposition reaction is second-order, and determine the value of k at $T = 1015$ K.

(d) The same experiment is performed at several other temperatures at a single initial pressure of 1.00 atm, with the following results.

$T(K)$	900	950	1000	1050
$t_{1/2}(s)$	5464	1004	219	55

Use a graphical method to determine the Arrhenius law parameters (k_0 and E) for the reaction.

(e) Suppose the reaction is carried out in a batch reactor at $T = 980$ K, beginning with a mixture at 1.20 atm containing 70 mole% N_2O and the balance a chemically inert gas. How long (minutes) will it take to achieve a 90% conversion of N_2O?

15. In an enzyme-catalyzed reaction with stoichiometry $A \rightarrow B$, A is consumed at a rate given by an expression of the **Michaelis-Menten form:**

$$r_A(mol/liter \cdot s) = \frac{k_1 C_A}{1 + k_2 C_A}$$

where $C_A(mol/liter)$ is the reactant concentration, and k_1 and k_2 depend only on temperature.

(a) The reaction is carried out in an isothermal batch reactor of constant volume $V(liters)$, beginning with pure A at a concentration C_{A0}. Write a balance on A,

and integrate it to obtain an expression for the time required to achieve a specified concentration C_A.

(b) Use the expression of part (a) to devise a graphical method of determining k_1 and k_2 from data for C_A versus t. Your plot should involve fitting a straight line and determining the two parameters from the slope and intercept of the line. (There are several possible solutions.) Then apply your method to determine k_1 and k_2 for the following data taken in a 2.00-liter reactor, beginning with A at a concentration $C_{A0} = 5.00$ mol/liter.

t(s)	60.0	120.0	180.0	240.0	480.0
C_A(mol/liter)	4.484	4.005	3.561	3.154	1.866

16. Phosgene ($COCl_2$) is formed by CO and Cl_2 reacting in the presence of activated charcoal:

$$CO + Cl_2 \rightarrow COCl_2$$

At $T = 303.8$ K in the presence of 1 g of charcoal, the rate of formation of phosgene is

$$R_f(\text{mol/min}) = \frac{8.75\, C_{CO}\, C_{Cl_2}}{(1 + 58.6 C_{Cl_2} + 34.3 C_{COCl_2})^2}$$

where C denotes concentration in mol/liter.

(a) Suppose the charge to a 3.00-liter batch reactor is 1.00 g of charcoal and a gas containing 60 mole% CO and 40 mole% Cl_2, and that the initial reactor conditions are 303.8 K and 1 atm. Calculate the initial concentrations (mol/liter) of both reactants, neglecting the volume occupied by the charcoal. Then, letting $C_P(t)$ be the concentration of phosgene at an arbitrary time t, derive relations for C_{CO} and C_{Cl_2} in terms of C_P.

(b) Write a differential balance on phosgene, and show that it simplifies to

$$\frac{dC_P}{dt} = \frac{2.92(0.02407 - C_P)(0.01605 - C_P)}{(1.941 - 24.3 C_P)^2}$$

Provide an initial condition for this equation.

(c) Starting with the equation of part (b), derive an integral expression for the time required to achieve a 75% conversion of the limiting reactant. (Your solution should have the form $t =$ a definite integral.)

*(d) The integral you derived in part (c) can be evaluated analytically; however, more complex rate laws than the one given for the phosgene formation reaction would yield an integral that must be evaluated numerically. The procedure is to evaluate the integrand at a number of points between the limits of integration and to use a quadrature formula such as the trapezoidal rule or Simpson's rule (Appendix A.3) to estimate the value of the integral.

Write a computer program to evaluate the integrand of the integral of part (c) at n_p equally spaced points between and including the limits of integration, where n_p is an odd number, and then to evaluate the integral using Simpson's rule. Perform the calculation for $n_p = 5$, 21, and 51, and compare the results with the exact value of the integral.

* Computer problem.

17. A gas that contains CO_2 is contacted with liquid water in an agitated batch absorber. The equilibrium solubility of CO_2 in water is given by Henry's law (Section 6.4b)

$$C_A = p_A/H_A$$

where

$$C_A(\text{mol/cm}^3) = \text{concentration of } CO_2 \text{ in solution}$$

$$p_A(\text{atm}) = \text{partial pressure of } CO_2 \text{ in the gas phase}$$

$$H_A = \text{Henry's law constant}$$

The rate of absorption of CO_2 (that is, the rate of transfer of CO_2 from the gas to the liquid per unit area of gas–liquid interface) is given by the expression

$$r_a(\text{mol/cm}^2 \cdot \text{s}) = k(C_A^* - C_A)$$

where

$$C_A = \text{actual concentration of } CO_2 \text{ in the liquid}$$

$$C_A^* = \text{concentration of } CO_2 \text{ that would be in equilibrium with}$$
$$\text{the } CO_2 \text{ in the gas phase } (C_A^* = p_A/H_A)$$

The gas phase is at a total pressure $P(\text{atm})$ and contains y_A (mol CO_2/mol gas), and the liquid phase initially consists of $V(\text{cm}^3)$ of pure water. The agitation of the liquid phase is sufficient for its composition to be considered spatially uniform, and the amount of CO_2 absorbed is low enough for P, V, and y_A to be considered constant throughout the process.

(a) Write a differential balance on CO_2 in the liquid phase, and integrate it to derive the relation

$$C_A(t) = C_A^*[1 - \exp(-kSt/V)]$$

(b) Suppose the system pressure is 20 atm, the liquid volume is 5 liters, the tank diameter is 10 cm, the gas contains 30% CO_2 by volume, the Henry's law constant is 9230 atm/(mol/cm^3), and the absorption rate coefficient k equals 0.020 cm/s. Calculate the time required for C_A to reach 0.62 mol/liter if the gas-phase properties remain essentially constant.

18. A liquid-phase chemical reaction with stoichiometry $A \rightarrow B$ takes place in a semicontinuous reactor. The rate of consumption of A per unit volume of the reactor is given by the first-order rate expression (see Problem 13)

$$r_A(\text{mol/liter} \cdot \text{s}) = kC_A$$

where $C_A(\text{mol/liter})$ is the reactant concentration. The tank is initially empty. Beginning at a time $t = 0$, a solution containing A at a concentration $C_{A0}(\text{mol/liter})$ is fed to the tank at a steady rate $\phi(\text{liters/s})$.

(a) Write differential balances on the total volume of the tank contents, V, and on the moles of A in the tank, n_A. Provide initial conditions for each equation. Your equations should contain only the variables n_A, V, and t, and the constants ϕ and C_{A0}. Simplify the second equation to determine an expression for the steady-state value of n_A.

(b) Integrate each equation to derive expressions for $V(t)$ and $n_A(t)$. Use the latter solution to verify the expression for the steady-state value of n_A derived in part (a).

19. A liquid mixture containing 70 mole% n-pentane and the balance n-hexane is vaporized in a semicontinuous still. The equilibrium relationship between the mole fraction of C_5H_{12} in the liquid (x) and that in the vapor (y) is of the form

$$y = \frac{ax}{x + b}$$

(a) When $x = 1$, what must y equal? (Think of the definitions of these quantities.) Use the answer to derive a relationship between a and b.

(b) Use Raoult's law (Equation 6.4-1) and the Antoine equation (Table 6.1-1) to calculate the mole fraction of pentane in the vapor phase in equilibrium with the 70 mole% pentane–30 mole% hexane mixture at the initial system temperature of 46°C. Then calculate the coefficients a and b of the xy relationship.

(c) Derive an equation relating L, the amount of liquid left in the still, to x_f, the mole fraction of C_5H_{12} in this liquid, assuming that the values of a and b do not vary with conditions in the system and that the vapor and liquid phases are constantly in equilibrium with each other. (See Example 11.2-3.)

(d) Calculate x_f when 75% of the liquid has been vaporized. (Use either a graphical technique or a trial-and-error procedure like Newton's rule—Appendix A.2.)

20. A kettle containing 2 liters of water at a temperature of 18°C is placed on an electric stove and begins to boil in four minutes.

(a) Write an energy balance on the water, neglecting evaporation that may occur during the heating.

(b) What is the average rate (W) at which heat is added to the water during this period? (See Example 11.3-1.)

(c) The rate of heat output from the stove burner differs significantly from the heating rate calculated in part (b). In which direction, and why?

21. An electrical coil is used to heat 20.0 kg of water in a closed adiabatic vessel. The water is initially at 25°C and 1 atm. The coil delivers a steady 2.00 kW of power to the vessel and its contents.

(a) Write a differential energy balance on the water, assuming that 97% of the energy delivered by the coil goes into heating the water. What happens to the other 3%?

(b) Integrate the equation of part (a) to derive an expression for the water temperature as a function of time.

(c) How long will it take for the water to reach the normal boiling point? Will it boil at this temperature? Why not?

22. An iron bar 2 cm × 3 cm × 10 cm at a temperature of 95°C is dropped into a barrel of water at 25°C. The barrel is large enough so that the water temperature rises negligibly as the bar cools. The rate at which heat is transferred from the bar to the water is given by the expression

$$Q(J/\text{min}) = UA(T_b - T_w)$$

where

$U(= 0.050 \text{ J/min} \cdot \text{cm}^2 \cdot \text{°C})$ is a **heat-transfer coefficient**

$A(\text{cm}^2)$ is the exposed surface area of the bar

T_b, T_w are, respectively, the surface temperature of the bar and the temperature of the water.

The heat capacity of the bar is 0.460 J/g·°C. Heat conduction in iron is rapid enough for the temperature $T_b(t)$ to be considered uniform throughout the bar.
(a) Write an energy balance on the bar.
(b) Determine the final steady-state temperature of the bar.
(c) Derive an expression for $T_b(t)$, and calculate the time required for the bar to cool to 30°C.

23. A steam coil is immersed in a stirred heating tank. Saturated steam at 7.50 bars condenses within the coil, and the condensate emerges at its saturation temperature. A solvent with a heat capacity of 2.30 kJ/kg·°C is fed to the tank at a steady rate of 12.0 kg/min and a temperature of 25°C, and the heated solvent is discharged at the same flow rate. The tank is initially filled with 760 kg of solvent at 25°C, at which point the flows of both steam and solvent are commenced. The rate at which heat is transferred from the steam coil to the solvent is given by the expression

$$Q(\text{kJ/min}) = UA(T_{\text{steam}} - T)$$

where UA (the product of a heat transfer coefficient and the area through which the heat is transferred) equals 11.5 kJ/min·°C. The tank is well stirred, so that the temperature of the contents is spatially uniform and equals the outlet temperature.
(a) Write a differential energy balance on the tank contents.
(b) Simplify the balance to derive the steady-state outlet solvent temperature.
(c) Integrate the balance equation to calculate the time required to heat the solvent to an arbitrary temperature $T_f(\text{°C})$. (See Example 11.3-2.)
(d) Calculate the solvent temperature after 40 min.

24. At 9:30 A.M. an open flask containing 350 g of liquid benzene at 20°C is put on a burner, and the burner is set so that the heat input rate to the benzene is roughly constant at 40.2 W. The heat capacity of liquid benzene may be taken to be constant at 1.77 J/g·°C.
(a) Calculate the time at which the benzene temperature reaches 40°C. (Truncate the solution to the nearest minute.) Neglect evaporation of benzene during the heating.
(b) Calculate the quantity of benzene left in the flask at 10:10 A.M., assuming that once the benzene starts boiling, the rate of heat input to the flask equals the rate of vaporization (g/s) times the heat of vaporization.

25. A steam radiator is used to heat a 60-m³ room. Saturated steam at 3.0 bars condenses in the radiator, and emerges as a liquid at the saturation temperature. Heat is lost from the room to the outside at a rate

$$Q_0(\text{kJ/h}) = 30.0(T - T_0)$$

where $T(°C)$ is the room temperature and $T_0 = 0°C$ is the outside temperature. At the moment the radiator is turned on, the temperature in the room is 10°C.

(a) Let $m_s(kg/h)$ denote the rate at which steam condenses in the radiator, and $n(kmol)$ the quantity of air in the room. Write a differential energy balance on the room air, assuming that n remains constant at its initial value, and evaluate all numerical coefficients. Take the heat capacity of air (C_v) to be constant at 20.8 J/mol·°C.

(b) Write the steady-state energy balance on the room air, and use it to calculate the steam rate required to achieve a constant room temperature of 24°C.

(c) Integrate the transient balance to calculate the time required to achieve a temperature of 23°C, assuming that the steam rate is that calculated in part (b).

26. An immersed electrical heater is used to raise the temperature of a liquid from 20°C to 60°C in 20.0 min. The mass of the liquid and the container is 250 kg, and the mean heat capacity of the system is 4.00 kJ/kg·°C. The liquid decomposes explosively at 85°C.

At 10:00 A.M. a batch of liquid is poured into the vessel and the operator turns on the heater and goes for a coffee break. Ten minutes later, his supervisor walks by and looks at the strip chart record of the power input. This is what she sees.

The supervisor immediately shuts off the heater and charges off to pass on to the operator several brief observations that come to her mind.

(a) Calculate the required steady power input in kW, neglecting energy losses from the container.

(b) Write and integrate using Simpson's rule (Appendix A.3) an energy balance on the system to calculate the system temperature at the moment the heater is shut off. Use the following data from the recorder chart.

$t(s)$	0	30	60	90	120	150	180	210	240	270	300
$Q(kW)$	33	33	34	35	37	39	41	44	47	50	54

$t(s)$	330	360	390	420	450	480	510	540	570	600
$Q(kW)$	58	62	66	70	75	80	85	90	95	100

(c) Suppose that if the heat had not been shut off, Q would have continued to rise linearly at a rate of 10 kW/min. At what time would everyone in the plant realize that something was wrong?

PART FOUR
CASE STUDIES

Chapter Twelve

Production of Soda Ash from Brine[1]

Sodium carbonate is produced in the United States at a rate of about 8 million tons per year, which represents almost one-third of the world production rate of this material. It is used in the manufacture of glass, as a chemical intermediate, and as an ingredient in the production of soaps, detergents, wood pulp, and paper. Sodium carbonate is commonly called **soda ash.**

Soda ash can be synthesized from calcium carbonate by the Solvay process[2] or recovered from naturally occurring brines or ores. Modern environmental regulations and increased energy costs have made the latter method dominant. In the United States, two major sources of naturally occurring sodium carbonate are used in the production of soda ash: a brine from Searles Lake in California, and an ore from Green River, Wyoming. As much as 1.3 million tons per year of soda ash, representing approximately 16% of the total United States production, can be produced from the Searles Lake brine.

Your company has been asked to help with the commercial development of a naturally occurring brine whose composition is given in Table 12-1. Physical and chemical properties that may be useful are given in Table 12-2. Initial estimates of the brine utilization rate call for 38 m^3/min of the brine to be pumped to the plant from a subterranean lake. The lake extends through porous structure to within 10 m of the surface of the earth, although the brine intake will be at a depth of approximately 100 m. The project group you lead has been assigned the task of completing a preliminary process flowsheet that will be used to develop a prospectus for potential investors. As part of your effort, you have distributed to each person in your group the following information and list of questions.

PROCESS DESCRIPTION

The production of soda ash from a brine requires that the sodium carbonate be separated from water and other brine constituents. This can be accomplished by reacting the sodium carbonate with carbon dioxide to form sodium bicarbonate.

$$Na_2CO_3(aq) + CO_2(g) + H_2O(l) \rightarrow 2\,NaHCO_3(aq) \tag{12-1}$$

[1] This case study was prepared with the assistance of T. F. Canning of Kerr-McGee Chemical Corporation.

[2] For an introduction to the Solvay process, consult D. F. Rudd, G. J. Powers, and J. J. Siirola, *Process Synthesis*, Prentice-Hall, Englewood Cliffs, N.J., 1973.

TABLE 12-1

Composition of Brine

Component	Wt%
Sodium carbonate	6.2
Sodium bicarbonate	0.6
Sodium chloride	19.0
Sodium sulfate	4.0
Water	70.2
	100.0

Sodium bicarbonate has a limited solubility in water, which can be decreased further by reducing the solution temperature. As sodium bicarbonate is produced, it saturates the solution and begins to crystallize. Crystals can be separated by filtration from the liquor, washed to remove residual liquor, and sent to a sodium carbonate recovery unit.

Soda ash is recovered from the sodium bicarbonate by **calcination**—heating the recovered crystals and driving off both carbon dioxide and water. (See Reaction 12-2.)

TABLE 12-2

Process Data

Specific gravities	
Brine	1.25
Sodium bicarbonate filtrate	1.21
Monohydrate mother liquor	1.32
Heat capacities (cal/g · °C)	
Brine	0.77
$NaHCO_3(s)$	0.25
Monohydrate mother liquor	0.85
$Na_2CO_3 \cdot H_2O(s)$	0.30
$Na_2CO_3(s)$	0.28
Sodium bicarbonate filtrate	0.81
Heats of crystallization	
$NaHCO_3$	4.1 kcal/mol $NaHCO_3$
Na_2CO_3	-7.9 kcal/mol Na_2CO_3
$Na_2CO_3 \cdot H_2O$	-4.8 kcal/mol $Na_2CO_3 \cdot H_2O$
Heats of reaction (kcal/mol)	
Reaction 12-1 at 60°C	6.61
Reaction 12-3 at 25°C	13.8
Solubilities	
Sodium bicarbonate in brine	1.9 wt% at 20°C
	2.3 wt% at 60°C
Sodium carbonate monohydrate	
(wt% Na_2CO_3) in water	33.2 wt% at 35.4°C
	30.8 wt% at 109°C

$$2\,NaHCO_3(s) \rightarrow Na_2CO_3(s) + H_2O(g) + CO_2(g) \tag{12-2}$$

The resulting sodium carbonate, which is known as **light ash** because of its low bulk density, can be marketed or upgraded to a higher value product. The glass industry, for example, needs soda ash with a density and particle size comparable to those of sand and limestone. To obtain a soda ash with the desired properties, the light ash is bleached and fed to a crystallizer where sodium carbonate monohydrate ($Na_2CO_3 \cdot H_2O$) crystallizes. The monohydrate crystals are separated from the solution (**mother liquor**) in a centrifuge and dried in steam tube driers, where the monohydrate salt is converted to anhydrous sodium carbonate (**dense ash**).

In the proposed process, brine at 25°C is pumped from the underground source through three successive heat exchangers, each of which is associated with a bicarbonate crystallizer. In each of these crystallizers, liquid is drawn from the bottom of the vessel, pumped through a heat exchanger where it exchanges heat with the brine, and returned to the crystallizer. The brine goes first to the exchanger on the third crystallizer, then to that on the second crystallizer, and finally to that on the first crystallizer.

After passing through the heat exchangers, the brine is fed to a precarbonation column where it is contacted countercurrently with spent gas from the carbonation reactors. This step provides additional utilization of carbon dioxide from the gas leaving the carbonators, thereby reducing the carbon dioxide lost from the process. On leaving the precarbonator, the brine enters the first of two carbonation reactors in series. The carbonation reactors are operated adiabatically, and the temperature of the liquid leaving the precarbonator is adjusted so that the temperature of the second carbonation reactor will be 60°C. Both carbonation reactors operate at 2 bars absolute, and they may be considered well-mixed (that is, the contents of the reactor are uniform, and the liquid exiting the reactor has the same composition and temperature as the liquid in the reactor). Solubilities of CO_2 in water are given in Table 12-3.

Reaction kinetic data found in the literature indicate that each reactor should have a volume that gives a slurry residence time (slurry volume/slurry volumetric flow rate) of approximately 45 min. Under these conditions, 85% of the CO_2 fed is absorbed in the carbonators, and 5% is absorbed in the precarbonator. The gas fed to the carbonators is divided such that equal amounts go to each carbonator, and approximately the same amount of CO_2 is absorbed in each of the two carbonation stages. The gas streams exiting the carbonators are combined and fed to the precarbonator. The three-phase mixture in the carbonation reactors (gas–liquid–solid) occupies approximately 1.25 times the volume that would be occupied by the solid–liquid slurry

TABLE 12-3

Solubilities (S) of CO_2 in Water (mol/liter) at a Partial Pressure of 1 atm as a Function of Temperature (T) in °C

T	15	25	35	45	55	65	75	85	100
S	0.0455	0.0336	0.0262	0.0215	0.0175	0.0151	0.0120	0.0090	0.0065

alone. Moreover, to minimize entrainment of liquid with the gas exiting the carbonation reactors, approximately 20% of the reactor volume is to be free space above the three-phase mixture.

The sodium bicarbonate slurry (mixture of solids and liquid) leaves the carbonation reactors and is fed to the three-stage crystallization process. As described earlier, the brine fed to the process is used to cool the slurry at each of the crystallizers. The first crystallizer is operated at 49°C, the second at 38°C, and the third at 27°C.

The slurry of crystals and liquor exiting the third crystallizer is sent to a filtration system where sodium bicarbonate crystals are separated from the liquor. The filter cake, which contains 91 wt% solid crystals and 9 wt% solution, is subjected to a staged washing process to remove impurities (NaCl and Na_2SO_4) present in the original brine. The cake from each filtration stage is mixed with a wash liquor that is 85% saturated with sodium bicarbonate at 27°C and refiltered. The volume of the wash liquor is twice that of the solution entrained in the filter cake. Because the wash liquor is 85% saturated, negligible quantities of $NaHCO_3$ are redissolved during the washing process. At each stage of the washing process, 55% of the solution remaining in the filter cake from the previous filtration is purged and replaced by an equal amount of the bicarbonate solution. The filtrate and the wash liquor can be pumped back into the ground or to a lake for reconcentration through solar evaporation.

Following the washing process, sodium bicarbonate crystals are recovered from the filter and heated in dryers to remove water and CO_2. This process of calcination is accomplished by indirect heating of the solids with saturated steam at 32 bars absolute. The sodium carbonate crystals (light ash) and exiting gases are at 180°C. The off-gas contains the moisture and carbon dioxide generated during calcination. As the dryer is not perfectly sealed, air at 27°C is drawn into the system, and the off-gas contains 88 moles of CO_2 for every 12 moles of air. Of the off-gas, approximately 10% is lost to the surroundings, and the remaining 90% is compressed to 5.0 bars and recycled to the carbonation reactors. Before being fed to the carbonators, the off-gas is cooled to 60°C and blended with fresh, pure CO_2, which is supplied by a plant that recovers and purifies CO_2 from boiler off-gas.

The hot light ash leaving the dryer is sent to a bleaching facility where discoloring materials present in trace amounts are burned off. The hot, bleached light ash is fed to a crystallizer operating at 88°C, where it is converted to sodium carbonate monohydrate ($Na_2CO_3 \cdot H_2O$). The operation of the crystallizer is adjusted to maintain a solids content of 35 kg of monohydrate crystals per 100 kg of slurry. Slurry from the crystallizer is fed to a centrifuge system that reduces the amount of solution retained by the crystals to 5 wt% of the total mass of wet solids. Fresh water is added to the mother liquor recovered from the centrifuge, and the mixture is recycled to the crystallizer. The monohydrate crystals are then sent to a dryer where free and combined water are driven off and dense anhydrous soda ash is recovered.

$$Na_2CO_3 \cdot H_2O(s) \rightarrow Na_2CO_3(s) + H_2O(g) \qquad \text{(12-3)}$$

Heat is supplied to the steam-tube dryer by saturated steam at 32 bars, and gases and dense ash exit the dryer at 180°C.

PROBLEMS

In Problems 1 through 21, you may assume that the conversion of sodium carbonate to sodium bicarbonate is 100%; in Problems 1 through 14, assume that negligible moisture exits with the gas leaving the precarbonator. Unless otherwise indicated, all steam used for heating condenses completely, and the condensate is removed from the heat exchanger as a saturated liquid.

1. Construct a detailed flowchart of the process, showing known compositions, flow rates, and temperatures. Use this chart to keep track of your calculations and as a basis for a final flowchart to be prepared when you have finished the case study.

2. Why is sodium carbonate converted to sodium bicarbonate in the process, only to be reoverted later back to sodium carbonate?

3. Obtain a recent value of the selling price for soda ash and estimate total sales in the United States.

4. Estimate the rate (kg/h) at which sodium bicarbonate exits the second carbonation reactor in both crystalline and solution form. Assume the filtrate leaving the first filtration stage is saturated with sodium bicarbonate and estimate the crystal production rate in kg/h.

5. What are the production rates (kg/h) of light ash and dense ash that can be expected from this process? Convert your answers to metric tons per year. What is the yield of sodium carbonate as a percentage of that fed to the process?

6. At what rate (kmol/h) is CO_2 generated in the reaction of sodium bicarbonate to soda ash? At what rate (kmol/h) must fresh CO_2 be supplied to the process?

7. Determine the rate (kmol/h) at which carbon dioxide is absorbed in each of the carbonators and in the precarbonator. You may assume that all of the CO_2 absorbed reacts with sodium carbonate. What is the feed rate (kmol/h) of carbon dioxide to the carbonators? What is the composition of the feed gas to the carbonators?

8. Suppose that the fresh CO_2 is produced by the combustion of coal, whose ultimate composition is given in Table 12-4. Estimate the rate at which the coal must be burned to completion to generate the required amount of carbon dioxide. Determine the rate at

TABLE 12-4

Ultimate analysis of coal used in steam generation

Element	Wt%
Carbon	52.5
Hydrogen	4.8
Oxygen	18.3
Nitrogen	1.2
Sulfur	0.6
Ash	22.6

which SO_2 is emitted from the combustion process in kilograms per metric ton of coal and in metric tons per year.

9. Determine the volume of the carbonation reactors. Note that entrainment of liquid in the gas exiting the reactor is directly related to the velocity of that gas. Calculate the velocity (m/s) of the gas flowing through the free space above the three-phase mixture in the reactor as a function of reactor diameter.

10. Estimate the composition of the liquor or slurry streams leaving the precarbonator and each carbonator. Using the solubility data for sodium bicarbonate, indicate whether each of these streams is a liquid or a slurry. If your calculations show that a stream is a slurry, determine the solids fraction and the composition of the liquid portion of the stream.

11. Prepare plots of weight fractions of NaCl and Na_2SO_4 in liquor adhering to the sodium bicarbonate crystals as a function of the number of wash stages. How many stages are required to reduce the sodium chloride content of the filter cake to less than 0.12 kg NaCl per 100 kg $NaHCO_3$? Determine the amount of wash liquor required to wash the sodium bicarbonate crystals, expressing your answer as kg solution/kg $NaHCO_3$ crystals. Suggest possible sources of the wash liquor. How is the production rate of soda ash changed if some of the sodium bicarbonate is used to produce the 85%-saturated wash liquor? Obtain a revised estimate of the required feed of fresh CO_2.

12. Determine the feed rate (kg/h) of fresh water to the sodium carbonate monohydrate crystallizer, assuming no losses by evaporation.

13. Estimate the requirements (kW) for pumping the brine from its subterranean source to storage tanks whose height is 10 m. Assume the pumps have an efficiency of 55%, that the friction loss is 10% of the potential energy change, and that kinetic energy changes can be neglected. (*Hint:* Perform a mechanical energy balance, taking the first point to be at the upper surface of the underground lake at 1 atm and the second point at the upper surface of the storage tank. See Section 7.7.)

14. Determine the temperature of the brine as it leaves each of the heat exchangers in the three-stage crystallization operation.

15. At what rate (kg/h) does water leave in the gas discharged from the precarbonator if the gas is saturated at the temperature of the liquid entering this vessel?

16. As indicated, the carbonation reactors are operated so that the slurry leaving the second unit is at 60°C. Recalling that the precarbonator and each carbonation reactor function adiabatically, to what temperature must the feed to the first carbonation reactor be adjusted? At what rate must heat be added to or removed from the stream leaving the precarbonator before it is fed to the first carbonation reactor? What is the temperature of the product from the first carbonation reactor? Determine the amount of steam or cooling water required using the following description of the utilities: If heat addition is required, assume that the heat is supplied by saturated steam at 3 bars absolute; if cooling is required, use cooling water at 30°C and allow a temperature increase in the water of 15°C.

17. Suppose both carbonation reactors were to operate at 60°C. To what temperature would the liquid feed to each have to be adjusted? At what rate would heat have to be removed from or added to these feed streams to obtain the required temperatures?

18. It has been suggested that direct firing of the dryer with hot combustion gases could be used in place of calcination by heating with steam. Consider the composition of the combustion gases obtained in burning coal, and give a reason for rejecting this suggestion.

19. Estimate the amount of 32-bar saturated steam (kg/h) required to operate the dryer in which sodium bicarbonate is converted to light ash, and that needed to operate the dryer in which sodium carbonate monohydrate is converted to dry anhydrous product.

20. Use the Dulong formula from Problem 48 in Chapter 9 to estimate the higher heating value of the coal to be used in the plant. Assume 10% excess air in the furnace, and suppose that 80% of the lower heating value is used to generate saturated steam at 32 bars from saturated liquid at the same pressure. Determine the temperature of the stack gases from the furnace and the amount of coal that must be burned to provide all of the steam required to operate the dryers. Present your answer in both kilograms per hour and tons per year. The ash heat capacity is 0.22 cal/g·°C.

21. Is it possible to meet process CO_2 requirements by recovering carbon dioxide from the off-gas produced in the combustion of coal?

Additional Problems for Study

22. There is concern regarding accumulation of impurities, especially NaCl, in the monohydrate crystallizer. Sodium chloride is carried by the sodium bicarbonate crystals from the washing operation to the dryer, and then from the dryer to the monohydrate crystallizer by unrefined soda ash. The refined soda ash is to contain less than 0.2 kg NaCl per 100 kg Na_2CO_3. Estimate the rate at which NaCl enters the monohydrate crystallizer, and propose a process modification that will control the concentration of NaCl in the recycled mother liquor.

23. The Solvay process for manufacturing soda ash is used in only a few plants in the United States. Review this process and suggest reasons for the competitive advantage of natural soda ash over synthetic soda ash.

24. Using the information in Chapter 23 of *Perry's Chemical Engineers' Handbook*[3] as a guide, evaluate the corrosiveness of mixtures of CO_2 in brine and propose materials of construction for the carbonation reactors. Repeat this process for the heat exchangers and crystallizers.

25. Test the assumption of complete conversion of sodium carbonate to sodium bicarbonate in the carbonation reactors using the following expression[4]:

$$\frac{X^2 C^{1.29}}{S p_{CO_2}(1 - X)(185 - T)} = 10$$

[3] R. H. Perry and D. W. Green, Eds., *Perry's Chemical Engineers' Handbook*, 6th Edition, McGraw-Hill, New York, 1984.

[4] C. R. Harte, Jr., E. M. Baker, and H. H. Purcell, *Ind. Eng. Chem.* 25:528, 1933.

where C is total sodium concentration (normality), p_{CO_2} is partial pressure of carbon dioxide (atm), S is solubility of CO_2 in water (mol/liter) at 1 atm partial pressure, T is temperature (°C), and X is fraction of the sodium in the bicarbonate form.

26. What effect does incomplete conversion of sodium carbonate to bicarbonate have on the process yield? If the yield of light ash is substantially reduced through incomplete conversion of sodium carbonate to bicarbonate, suggest possible process modifications that would increase the conversion.

Chapter Thirteen

The Production of Chemicals from Coal[1]

Coal has been considered a major component of U.S. energy reserves for a long time, and more recently it has become an important raw material for the production of chemicals.[2] The worldwide abundance of coal, as reflected in Table 13-1, and its proximity to sites where it can be used, have made it an attractive alternative to more costly or less strategically located feedstocks. Most approaches to the conversion of coal to chemicals involve either coal **gasification** or coal **liquefaction**, processes that convert solid coal to fluids. The fluids are more easily handled and processed than the original coal; they can be used as fuels on-site or after pipeline transmission, or they can be converted to chemical intermediates. A secondary but important aspect of these conversion processes is that potentially harmful materials in coal are more easily separated from the fluids produced by gasification or liquefaction than from either the solid coal or the off-gases from its combustion.

The most common coal gasification process involves the reaction of carbon, the primary constituent of coal, with steam to produce carbon monoxide and hydrogen.

$$C + H_2O \rightarrow CO + H_2 \tag{13-1}$$

The carbon monoxide and hydrogen can be separated from the other compounds exiting the gasifier and used either as a fuel or a feedstock to a chemical manufacturing process. Simultaneously, oxygen fed to the reactor along with steam reacts with carbon to form carbon dioxide.

$$C + O_2 \rightarrow CO_2 \tag{13-2}$$

In addition, there is significant reaction between water and carbon monoxide in the gasifier via the **water–gas shift** reaction.

$$CO + H_2O \rightleftarrows CO_2 + H_2 \tag{13-3}$$

Although Reactions 13-1, 13-2, and 13-3 are the primary reactions in coal gasification, there are important secondary reactions. For example, the hydrogen-rich atmosphere in the gasification reactor causes the formation of hydrogen sulfide from the sulfur usually found in coal and of small amounts of methane.

$$S + H_2 \rightarrow H_2S \tag{13-4}$$

$$C + 2H_2 \rightarrow CH_4 \tag{13-5}$$

[1] This case study was prepared with the assistance of Victor H. Agreda of the Tennessee Eastman Co.

[2] H. W. Coover Jr. and R. C. Hart, "A Turn in the Road: Chemicals from Coal," *CEP*, *78*(4), 72 (1982).

TABLE 13-1

Distribution of Proven Coal Reserves in Billions of Tons[a]

Country	Reserves
Soviet Union	273
United States	182
Peoples Republic of China	101
German Federal Republic	40
Australia	24
India	12
Republic of South Africa	11
Other	55
	698

[a] From R. F. Probstein and R. E. Hicks, *Synthetic Fuels*, McGraw-Hill, New York, 1982.

Various hydrocarbon feedstocks other than coal can be used for production of chemicals, including crude oil, natural gas, and wood. Each of these raw materials yields a **synthesis gas** containing carbon monoxide and hydrogen, much as described above. As the hydrogen-to-carbon ratio in the synthesis gas differs from one feedstock to another, the differences may be used in selecting a feedstock to produce a specific product. For example, natural gas reforming yields a synthesis gas with a high hydrogen content that can be used as a raw material for manufacturing hydrogen-rich products such as ammonia. Gasification of coal, on the other hand, provides a gas with a high carbon monoxide content that can be used as a raw material for producing species with lower hydrogen contents such as acetic anhydride ($C_4H_6O_3$). The most efficient use of raw material is generally achieved when a feedstock contains hydrogen and carbon monoxide in roughly the proportions required to form the desired chemical product. Of course, there are other factors to consider in choosing a feedstock, such as the projected stability in price and supply of the feedstock and the location of the feedstock in relation to the plant.

Your company has plans to build a plant to convert coal to acetic anhydride, a chemical vital to the company's product line. The plant is to produce 160 000 metric tons of acetic anhydride per year. It is anticipated that the completed process will operate continuously 350 days/yr. The composition of coal from the nearest deposit is given in Table 13-2. Judging from the sulfur content, production of significant

TABLE 13-2

Composition of Coal to be Used as Feed to the Gasifier[a]

Element	C	O	H	S	N	Ash
Weight Percent	52.5	18.3	4.8	0.6	1.2	22.6

[a] The coal contains 10.5% moisture. The hydrogen and oxygen percentages given here include the masses of these elements in the coal moisture.

quantities of H_2S is expected. Since H_2S poisons (deactivates) catalysts used in the conversion of hydrogen and carbon monoxide into acetic anhydride, it must be removed from the gas leaving the gasifier. It is your responsibility to assist with the setting of design variables necessary for sizing equipment.

PROCESS DESCRIPTION

Before it can be fed to the gasification reactor, coal must be ground to meet specifications on particle size. The grinding is accompanied by the generation of coal dust that can represent a loss of raw material and a hazard to both personnel and equipment. This problem is minimized by recognizing that one of the primary reactions in coal gasification is between coal and water, and that grinding and separating the resulting coal particles by size can be performed on a coal–water mixture as well as on dry coal. The ratio of dry coal to water in the feed to the gasifier is to be 2 kg coal/kg water, and this ratio is set with the water added to the grinding and classification operations.

The reactor in which gasification is to be carried out is an **entrained flow** gasifier. A coal–water slurry is fed to the gasifier along with high-pressure oxygen. Both feed streams may be assumed to be at 25°C. The reactor is operated adiabatically at 6300 kPa absolute. At these conditions, only trace quantities of hydrocarbon byproducts other than methane are present in the gas leaving the gasifier, and essentially all of the reactive carbon, hydrogen, and sulfur in the coal are reacted. The gas and molten ash (**slag**) leaving the gasifier are fed to a wash tank where the gas is bubbled through water and the slag is solidified. The gas leaving the wash tank is fed to a scrubber where it contacts a water spray. Small amounts of residual particulate matter are removed from the gas, which is also cooled from 220°C to 170°C. The water with which the gas is contacted has been pumped from the wash tank, through a heat exchanger, and into the scrubber. The gas–liquid mixture formed in the scrubber is separated into saturated gas and liquid streams at 170°C and 5960 kPa absolute, and the water is returned to the wash tank. Solids collected in the wash tank are allowed to settle, and are then removed along with the water condensed from the gas leaving the gasifier. The composition of the gas leaving the gasifier is given in Table 13-3.

The ratio of hydrogen to carbon monoxide in the gas leaving the scrubber must be adjusted to satisfy stoichiometric requirements associated with downstream reactions. The adjustment is made by contacting a portion of the gas leaving the scrubber with a catalyst that increases the rate at which the water–gas shift reaction occurs. The shift reactor consists of a fixed bed of cobalt-molybdenum catalyst pellets and operates at 300°C and 5615 kPa absolute.

The product from the water–gas shift reactor and the unreacted portion of the synthesis gas stream obtained from the gasifier are blended and cooled by generating steam in waste-heat boilers. Water is removed completely from the cooled gas in a molecular sieve drier. The dry gas is then fed to an **acid gas removal system** where all of the sulfur-containing compounds are separated and sent to a sulfur recovery

reactor, and 99.9% of the CO_2 is separated and either vented or recovered for use elsewhere in the plant.

The acid gas removal system consists of two absorption columns, both of which operate at 5000 kPa absolute, and a solvent regeneration section. The gas enters the first absorption column and contacts refrigerated methyl alcohol flowing counter-currently to the gas. In the column, some CO_2 and essentially all of the H_2S are absorbed in the methyl alcohol. The gas leaving this column is sent to a second absorption column where most of the remaining CO_2 is absorbed in additional methyl alcohol. The gas exiting the second column is at $-30°C$ and contains 0.1% of the CO_2 that entered the first absorber with the crude coal gas.

There are two separate operations in the regeneration section of the acid gas removal system. In the first, the methyl alcohol from the first absorber is flashed (abruptly exposed to a lower pressure, causing a significant fraction of the absorbed gases to come out of solution). Essentially all of the absorbed CO_2 and H_2S are recovered in the off-gas, and the regenerated methyl alcohol is recirculated to the first absorber. The off-gas from the flash tank is a mixture consisting of 15 mole% H_2S and 85 mole% CO_2. In the second operation, methyl alcohol from the second absorber is fed to a stripping column where it flows downward to a reboiler. Indirect steam heating (heat transferred through a heat transfer surface) is used to vaporize a por-tion of the methyl alcohol, which is returned to the column and strips the CO_2 from the liquid methyl alcohol. The regenerated liquid methyl alcohol is cooled and re-circulated to the second absorber.

Sulfur emissions are potential health hazards, and safe disposal of the H_2S sepa-rated from the product gas is an important provision of the proposed plant. To accomplish this, the hydrogen sulfide is converted into elemental sulfur in a **Claus process.** In this process, the H_2S-rich gas discharged from the flash tank is split, with one-third going to a furnace where the hydrogen sulfide is burned at 1 atm with a stoichiometric amount of air to form SO_2.

$$H_2S + 1.5\,O_2 \rightarrow SO_2 + H_2O \tag{13-6}$$

The hot gases leave the furnace and are cooled prior to being mixed with the re-mainder of the H_2S-rich gas from the flash tank. The mixed gas is then fed to a catalytic reactor where hydrogen sulfide and SO_2 react to form elemental sulfur.

$$2\,H_2S + SO_2 \rightarrow 2\,H_2O + 3\,S \tag{13-7}$$

The gases leave the reactor at 380°C and are cooled to 130°C by generating steam at 205 kPa from saturated liquid at 205 kPa. Liquid sulfur is condensed and recovered as the hot gases are cooled at atmospheric pressure.

TABLE 13-3

Dry-Basis Composition of Syngas Exiting Gasifier

Component	CO	H_2	CO_2	N_2	CH_4	H_2S
mole %	47.95	37.25	13.85	0.6	0.1	0.25

As an intermediate step in the production of acetic anhydride, carbon monoxide and carbon dioxide are reacted with hydrogen to produce methanol.

$$CO + 2\,H_2 \rightleftharpoons CH_3OH \tag{13-8}$$

$$CO_2 + 3\,H_2 \rightleftharpoons CH_3OH + H_2O \tag{13-9}$$

These reactions are carried out at 200°C and 4925 kPa absolute. The feed to the methanol reactor is formed by combining recycle from the reactor with a fresh feed stream containing 5% excess hydrogen, based on complete conversion of CO and CO_2 to methanol. The composition of the gas leaving the acid gas removal system is adjusted to meet the fresh feed requirement on excess hydrogen. This is accomplished by splitting the gas from the first absorption column into two streams, one of which is sent to a low-temperature separation unit while the second is blended with the hydrogen-rich stream exiting this unit. The separation is designed to condense 95% of the inlet carbon monoxide and to operate at 5000 kPa.

Reactions 13-8 and 13-9 do not go to completion, and conversions of CO and CO_2 must be estimated based on equilibrium constants available from the literature. The equilibrium constant K_8 for Reaction 13-8 is given here in terms of the partial pressures of reactants and products in kPa and system temperature in degrees Kelvin[3]

$$K_8 = \frac{p_M}{p_{CO}p_H^2} = 1.390 \times 10^{-4} \exp\left[21.225 + (9143.6/T)\right.$$

$$\left. - 7.492 \ln T + 4.076 \times 10^{-3}T - 7.161 \times 10^{-8}T^2\right] \tag{13-10}$$

where the subscripts M and H represent methanol and H_2, respectively. Note that Reaction 13-9 can be represented as the sum of Reaction 13-8 and the reverse of the water–gas shift reaction

$$H_2 + CO_2 \rightleftharpoons CO + H_2O \tag{13-11}$$

The equilibrium constant for Reaction 13-11 is given by the equation

$$K_{11} = \frac{p_{CO}p_W}{p_{CO_2}p_H} = 3.330 \exp\left[13.148 - (5639.5/T) - 1.077 \ln T\right.$$

$$\left. - 5.44 \times 10^{-4}T + 1.125 \times 10^{-7}T^2 + (49\,170/T^2)\right] \tag{13-12}$$

where the subscript W represents water.[3] The stream leaving the reactor is sent to a multiple-unit process that separates methanol and water from the noncondensables, and then separates the methanol and water. The stream of noncondensables, which consists of carbon dioxide, carbon monoxide, hydrogen, methane, and nitrogen, is sent to a multiple-unit process that removes all of the nitrogen and methane and some of the hydrogen. The gas that remains contains the quantity of hydrogen stoichiometrically required to convert all of the CO and CO_2 to methanol. This stream is recycled to join the fresh feed to the reactor.

[3] L. E. Wade, R. B. Gengelbach, J. L. Taumbley, and W. L. Hallhauer, "Methanol," *Kirk-Othmer Encyclopedia of Chemical Technology*, 3rd ed., Vol. 15, John Wiley, New York, 1981, pp. 398-415.

Finally, acetic anhydride is produced in two successive reactions. Methyl acetate is first synthesized by reacting the methanol with acetic acid obtained from another manufacturing unit.

$$CH_3OH + CH_3COOH \rightleftarrows CH_3COOCH_3 + H_2O \qquad (13\text{-}13)$$

The feed to the reactor contains methanol and acetic acid in stoichiometric proportion. Chemical equilibrium is achieved in the reactor. At the reaction conditions, the equilibrium constant K_{13} in the following relationship is 5.2:

$$K_{13} = x_{MA}x_W/x_M x_A \qquad (13\text{-}14)$$

where x_i is the mole fraction of i in the product solution, and the subscripts MA and A represent methyl acetate and acetic acid respectively. The product from the reactor is sent to a facility that separates the reactants from the products and methyl acetate from water. The reactants are recycled to the reactor. The methyl acetate and carbon monoxide from the low temperature separation unit are fed to a second reactor in stoichiometric proportion, and the two substances react completely to form acetic anhydride.

$$CH_3COOCH_3 + CO \rightarrow CH_3COOCOCH_3 \qquad (13\text{-}15)$$

PROBLEMS

In solving the problems that follow, cite all sources of data not given in the process description and clearly state all assumptions.

1. Construct a detailed flowchart of the process, showing known compositions, flow rates, pressures, and temperatures. Use this chart to keep track of your calculations and as a basis for a final flowchart to be prepared when you have finished the case study.

2. Assume there are no side reactions in the synthesis of methanol, methyl acetate, and acetic anhydride, and calculate the molar flow rates of reactants (kmol/h) in the fresh feeds to each of the reactors. Where fresh feed must be supplied from a source external to the process, calculate the feed rates of these materials in kg/h.

3. Estimate the amount of oxygen fed to the gasification reactor in kmol and standard cubic meters per kg of coal fed to the process.

4. Suppose that air, rather than pure oxygen, were fed to the reactor. What would be the flow of air in kmol and standard m^3 per kg of coal fed to the process? What would be the composition of the gas leaving the gasifier under these conditions? Suggest reasons based on these calculations for the use of O_2 instead of air.

5. If all of the sulfur in the coal reacts to form H_2S, determine the air requirements for the Claus plant in kmol and standard m^3 per kg of coal fed to the process.

6. Estimate the mole fraction of water in the gas exiting the scrubber. What is the rate (kg/kg coal) at which water exits the scrubber in the product gas?

7. Determine the fraction of the gas leaving the scrubber that must be sent to the water-gas shift reactor. Calculate the composition of the stream resulting from mixing the product from the water-gas shift reactor and the stream bypassing the reactor.

8. Determine the coal feed rate (kg/h), and calculate the rates at which oxygen is fed to the gasifier (kmol/h) and air is fed to the Claus plant (kmol/h).

9. Estimate the composition of the product gas leaving the acid gas removal system. What fraction of this gas is sent to the cryogenic separator? What is the composition of the gas resulting from blending the H_2-rich product from the cryogenic separator with the stream bypassing the separator?

10. Using the equilibrium constants for Reactions 13-8 and 13-11, determine the composition of the effluent from the methanol synthesis reactor. What is the single-pass conversion of carbon monoxide? (A trial-and-error solution is required.)

11. Based on the equilibrium constant given for Reaction 13-13, calculate the composition of the product stream leaving the methyl acetate synthesis reactor and the single-pass conversion of methanol to methyl acetate. Determine the recycle rates of methanol and acetic acid.

12. Propose a modification of the methyl acetate reactor that would involve recycle of primarily one reactant. (*Hint:* Consider the effect of feeding an excess of one of the reactants on product composition.) Estimate the recycle rate for the proposed modification. Summarize features of this mode of operation that could make it more economical than the mode requiring recycle of both reactants.

13. Estimate the temperature to which the $CO-H_2$ stream must be cooled to condense 95% of the inlet CO, assuming ideal gas behavior and that CO is the only condensable component in the system. (The dependence of the vapor pressure of CO on temperature is given in Table 3-7 of *Perry's Chemical Engineers' Handbook.*[4]) Determine how well the assumption of ideal gas behavior is satisfied by using the temperature calculated in the first part of this question and the system pressure to estimate a compressibility factor for the product gas from the condenser using Kay's rule.

14. What fraction of the elemental sulfur leaving the Claus plant reactor is condensed? (The dependence of the vapor pressure of sulfur on temperature is given in Table 3-7 of *Perry's Chemical Engineers' Handbook.*) A suggestion has been made to burn the residual sulfur in the gas stream exiting the sulfur condenser, forming SO_2, and discharging this material into the atmosphere. Estimate the emission rate of SO_2 (kg/h) if this suggestion is followed.

15. The plant has a methanol storage tank with a capacity of 2 000 000 liters. The average temperature of the liquid in the tank is 40°C, and the tank is padded with nitrogen at a pressure of 1 atm (absolute) to reduce explosion hazards. The vapor space in the tank can be assumed saturated with methanol. The volume of methanol stored in the tank increases from 800 000 to 1 200 000 liters twice a day, and in the process a portion of

[4] R. H. Perry and D. W. Green, Eds., *Perry's Chemical Engineers' Handbook*, 6th edition, McGraw-Hill, New York, 1984, p. 3-47.

the vapor space is displaced by liquid methanol. What would be the methanol vapor emission to the atmosphere (kg/day) if there were no emission control devices? Propose two different schemes to reduce vapor emissions.

16. Determine the relationship between flow rate and temperature of the water fed to the scrubber. Is the rate at which heat must be removed from the water as it is pumped from the wash tank to the scrubber affected by the flow rate of water? Explain.

17. Assume that the temperature of the gases and slag leaving the gasifier is 1100°C. Estimate the total rate at which heat must be removed from the wash tank–scrubber system. (*Note:* The actual temperature of such streams in commercial entrained-flow gasifiers is closer to 1350°C.) The following properties of ash and slag may be used: for ash, $C_p(kJ/kg \cdot °C) = 0.604 + 0.586 \times 10^{-3} T(K)$, $\Delta\hat{H}_m = 710 \ kJ/kg$; for slag, $C_p(kJ/kg \cdot °C) = 0.767 + 0.744 \times 10^{-3} T(K)$.

18. The portion of the gas from the scrubber that is not sent to the water–gas shift reactor and the product gas from the water–gas shift reactor are used to generate steam in waste-heat boilers. Steam can be generated and used at three absolute pressures—2858 kPa, 790 kPa, and 205 kPa. The relative worths of steam at these pressures are 1.3, 1.2, and 1.0. Given that the gas fed to the acid gas removal system must be cooled to 100°C, and that gas fed to a boiler can be cooled to within 5°C of the temperature of the steam it generates, propose a bank of waste heat boilers that maximizes the value of the generated steam. How much of each steam is generated in the optimum configuration? Why do the respective steams have different values?

19. How much low pressure steam (205 kPa) in kg/h can be generated by heat recovered in the Claus plant?

20. Determine the standard heats of Reactions 13-1 and 13-2. Discuss the purpose of feeding oxygen to the reactor.

21. Assuming no heat is lost to the surroundings, estimate the temperature of the gas and slag leaving the gasifier. (*Hint:* Choose the elements at 25°C and 1 atm as reference states, and estimate $\Delta\hat{H}_f^\circ$ of coal from the higher heating value obtained using the Dulong formula, which is given on p. 474.)

22. At what rate (kW) is heat transferred to or from (state which) the methanol synthesis reactor to maintain its operation at 200°C? You may assume that the recycle stream is at 40°C, and the fresh feed is at -20°C.

Additional Problems

23. In addition to the components shown in Table 13-2, coal contains trace impurities such as vanadium, arsenic, and mercury. Where would you expect these components to end up in your flowsheet? Could they be expected to cause any problems? Explain.

24. If there were no storage tanks between the different sections of the process, all of the units would have to shut down on loss of gasifier synthesis gas. However, one of the plants could continue to operate, independent of syngas availability, if intermediate liquid

storage tanks were available. Identify this plant, and size the feed and product tanks so that the plant can operate for 48 hours in the event of a gasifier shutdown. Assume that it is not economical to store liquefied syngas, and state all other assumptions.

25. Given that several chemicals are synthesized in this process, what additional products should the company consider producing? For example, what are some of the useful chemicals that can be synthesized from methanol, other than methyl acetate?

26. Suppose that the plant under discussion were to be called upon to produce the acetic acid used in the synthesis of methyl acetate. How would this affect the design of the plant? How does the use of one of the intermediates in the production of acetic anhydride affect the design of the entire process? What are the other options for manufacturing acetic acid?

27. Determine the price of all raw materials used in the synthesis of acetic anhydride, and calculate the ratio of product worth to cost of raw materials.

Chapter Fourteen

Fermentation of Grain to Produce Ethyl Alcohol[1]

Ethyl alcohol can be produced from a variety of sources, including fermentation of carbohydrates derived from starch crops such as grain and potatoes, from sugar crops such as cane and beets, from cellulosic agricultural residues such as bagasse and corn stalks, and from wood and wood by-products. It can also be produced by hydration of ethylene or from synthesis gas containing hydrogen and carbon monoxide (see Chapter 13). In a number of countries, such as Brazil, considerable progress has been made in using ethanol to supplement gasoline. In addition, the abundance of grains produced in the United States and other countries has enhanced the attractiveness of converting a portion of these grains to ethyl alcohol for blending with gasoline as an octane enhancer, or for use as a feedstock in the synthesis of other chemicals.

Your company is considering a bid to purchase an ethanol production plant owned by a smaller company, and you have been given the assignment of reviewing the process. This review is to include a quantitative study of the operation and should develop detailed information regarding key process variables. Although the study need not be concerned with economics, consideration should be given to the energy requirements of the process. Suggestions for alternative processing schemes should be included in the final report.

In the process to be described, a portion of the starch in corn is converted to ethyl alcohol by two biological processes: **saccharification** and **fermentation.** In saccharification, the polymeric structure of starch (a polysaccharide) is hydrolyzed in the presence of the enzymes (biological catalysts) α-amylase and amyloglucosidase. The primary products of hydrolysis are maltose (a disaccharide consisting of two glucose units) and oligomers consisting of several glucose units.

$$\underset{\text{Starch}}{-(C_6H_{10}O_5)_x} \xrightarrow{\text{Enzymes}} \underset{\text{Maltose}}{C_{12}H_{22}O_{11}} + H_2O + \underset{\text{Oligomers}}{-(C_6H_{10}O_5)_y}$$

The fermentation process is based on the growth of a yeast culture that converts maltose to ethyl alcohol and carbon dioxide:

$$C_{12}H_{22}O_{11} + H_2O \,(+\text{yeast}) \rightarrow 4\,C_2H_5OH + 4\,CO_2 \,(+\text{yeast} + H_2O)$$

[1] Reviewed by Raphael Katzen of Raphael Katzen Associates. For a more complete description of a process for the synthesis of ethanol from grain, consult Raphael Katzen Associates, "Grain Motor Fuel Alcohol Technical and Economic Assessment Study," Report to the U.S. Department of Energy, NTIS HCP/J6639-01, June 1979.

580

TABLE 14-1

Properties of 199-Proof Ethyl Alcohol

Specific gravity at 60°F	0.7962
Composition	
Vol% ethyl alcohol	99.50
Wt% ethyl alcohol	99.19
Mole% ethyl alcohol	97.95

TABLE 14-2

Composition of Corn (Grain)[a]

Component	Wt%
Starch	72[b]
Protein	9[b]
Oil	3[b]
Fiber and ash	16[b]
Moisture	15.5

[a] One bushel of corn is equivalent to 56 lb_m of corn.
[b] Moisture-free basis.

As the yeast culture grows, 0.0794 lb_m of yeast is produced for every lb_m ethyl alcohol formed, and 0.291 lb_m water is produced for every lb_m of yeast formed. The residue from the alcohol recovery and dehydration operation can be further processed to recover distillers' dried grains and solubles (**DDGS**). The yeast solids become a part of the by-product grains, and the sale of DDGS as animal feed improves process economics and minimizes waste disposal problems.

The alcohol plant under consideration is designed to produce 50 million gallons of 199-proof (99.5 vol%) ethyl alcohol per year. Existing technology provides a yield of 2.57 gal of alcohol from a bushel of corn. (See Tables 14-1 and 14-2 for properties of 199-proof ethyl alcohol and corn.) The process is to be operated continuously, with the exception of the fermentation and fungal amylase sections of the plant, which must be operated in a batchwise fashion to allow for frequent sterilization of equipment. The nature of the plant dictates that it operate at least 330 days per year.

PROCESS DESCRIPTION

Corn is milled and the resulting meal is conveyed to a mixing tank where it is blended with recycled condensates and water to produce a mash. The total water input to this tank is controlled to produce 22 gal of mash per bushel of corn input. The specific gravity of the mash is approximately 1.1. From the mixing tank, the slurry flows to a precook vessel, which is maintained at 145°F. The condensates, which have been added hot to the mixing tank, yield a mash at 100°F. Live saturated steam at 15 psig is added to the precook vessel to increase the mash temperature to 145°F. (Live steam is steam injected directly into a process vessel.)

Mash from the precooking vessel is heated to 230°F by condensing saturated steam at 15 psig, and then to 320°F with saturated 150-psig live steam. The mash is then sent to a cooking tank and held there long enough for the starch structure to be broken down in preparation for the saccharification reactions. The cooked mash is flashed to 15 psig to produce saturated steam and a concentrated mash, which is cooled to 145°F by flashing to a vacuum. The flash-tank vacuum is maintained by drawing the vapor into a condenser where most of the steam is condensed and passing the remaining uncondensed vapor to a steam ejector. (Steam ejectors are described in *Perry's Chemical Engineers' Handbook.*[2]) Steam at 150 psig is fed to the ejector along with the uncondensed vapor, and the resulting mixture is sent to a condenser operating at 15 psig. The ratio of vapor drawn from the flash condenser to 150-psig steam fed to the ejector is 0.04 lb_m vapor/lb_m steam. Approximately 30 lb_m/h of vapor is drawn into the ejector from the flash condenser. All condensates from these units are pumped back to the mash mixing tank.

Mash from the vacuum flash is mixed with a small amount of fungal amylase, and the mixture is sent to the saccharification reactors where the starch is converted at 140°F to fermentable sugars. Thin stillage (s.g. = 1.1) recycled from a downstream unit is added to the liquor from the saccharification reactors to lower the pH, provide yeast nutrients, and obtain a final mash volume of 25 gal per bushel of corn fed to the process. The volume of the thin stillage added is 16% of the final mash volume going to the fermenters. The mixture of converted mash and thin stillage is cooled in heat exchangers to 100°F by cooling-tower water at 85°F, and then to 85°F by well water at 60°F. The temperature increases of both cooling water streams are limited to 25°F.

The cooled mash is fed to the fermentation reactors, which are operated in a batch mode. The reactor network consists of eight vessels, each having a capacity of 550 000 gal. The vessels are arranged in sets of two with one heat exchanger and circulation pump for each set of fermenters. Liquid loading is 90% of vessel capacity. Cooling is needed for only about 24 h out of the 48-h fermentation cycle, which makes it possible for one exchanger to service two fermenters. The fermenters are filled on an 8-h cycle; in other words, the flow rate of mash is sufficient to fill a single reactor every eight hours.

About 300 lb_m of yeast is added to each fermentation batch.[3] During fermentation, heat is released by the exothermic conversion of sugars to ethanol and carbon dioxide, and the batch temperature is allowed to increase from 85°F to 95°F. Removal of heat from the mash as it is recirculated through the heat exchangers is used to prevent the temperature from increasing above 95°F. Well water at 60°F is the cooling fluid used in these exchangers, and each fermenter requires a flow of about 2400 gal/min during the peak period. The temperature increase of the well water is limited to 25°F.

[2] R. H. Perry and D. W. Green, Eds., *Perry's Chemical Engineers' Handbook*, 6th Edition, McGraw-Hill, New York, 1984, pp. 6-31 through 6-32.

[3] For additional information on the fermenter operation see J. E. Bailey and D. F. Ollis, *Biochemical Engineering Fundamentals*, 2nd Ed., McGraw-Hill, New York, 1986, Chs. 9, 11, and 12.

The fermentation products are sent to a storage vessel from which dilute alcohol is pumped at 90°F and 1 atm and heated to 280°F by condensing 150-psig saturated steam. The dilute alcohol is then flashed to separate essentially all of the carbon dioxide. Water and alcohol vaporized with the carbon dioxide are condensed and returned to the flash drum. The liquid leaving the flash drum is at 250°F and consists of 7.1 wt% alcohol, 6.9 wt% soluble and suspended solids, and water. This liquid is fed to a recovery system consisting of two distillation columns and one stripping column.

Feed enters each distillation column at a location between the top and bottom, and two product streams are withdrawn. A vapor stream is removed from the top of the column and condensed. A portion of the condensate (**reflux**) is returned to the top of the column, and the remainder is taken as overhead product or **distillate.** A liquid stream leaves the bottom of the column and goes to a reboiler that vaporizes a portion of the liquid. The generated vapor is returned to the column as **boilup,** and the remaining liquid is taken as product and is referred to as **bottoms.** In the stripping column (**stripper**), the liquid feed is introduced at the top of the column, and two product streams are withdrawn. One of the product streams is a vapor leaving the top of the column and the other is a liquid that leaves the bottom of the column and is fed to a reboiler. The reboiler generates a vapor stream, which is returned to the stripper as boilup, and a liquid bottoms product.

The dilute liquid from the flash tank is fed to the first distillation column, which is operated at 50 psig and is known as the **ethanol concentrator.** Heat is supplied to the ethanol concentrator by condensing approximately $110\,000\,lb_m/h$ of 150-psig saturated steam in the reboiler. Condensate from the reboiler is saturated at 150 psig. The distillate and reflux are saturated liquids at 250°F and contain 95 vol% (190-proof) ethyl alcohol. The bottoms from the column are at a temperature of about 305°F, and contain all of the solids fed to the column and approximately 0.02 wt% alcohol on a solids-free basis.

An **azeotrope** containing two or more components is a mixture whose bubble point is greater or less than all pure-component bubble points.[4] If an azeotrope is encountered in distillation, its composition represents a limit to the separation or concentration that can be achieved. In the system at hand, the distillate from the ethanol concentrator has a composition approaching that of the azeotrope formed between ethanol and water, 89 mole% ethanol and 11% water. Further purification of the ethyl alcohol in this stream can be accomplished only by **breaking** the azeotrope, which is done in this case by adding pentane as an **entrainer** to the second distillation column (**azeotrope column**). The presence of pentane causes the formation of an ethanol–water–pentane azeotrope that is removed as a vapor from the top of the second distillation column. On condensation, this azeotrope separates into two liquid phases, one rich in pentane and the other rich in water. The two-phase condensate is collected in a vessel (**decanter**) whose design allows the two liquid phases to be withdrawn separately. The pentane-rich phase, which contains 76.5 mole% pentane, 8.3%

[4] For further information on azeotropes consult R. H. Perry and D. W. Green, Eds., *Perry's Chemical Engineers' Handbook*, 6th Edition, McGraw-Hill, New York, 1984, pp. 13-59 to 13-68.

TABLE 14-3

Composition of Flue Gas Fed to Dryer

Component	Vol%
Carbon dioxide	12.6
Water	9.3
Nitrogen	73.3
Oxygen	4.5
Sulfur dioxide	0.3

ethanol, and 15.2% water, is mixed with the condensed overhead from the stripping column, and the mixture is returned to the azeotrope column as reflux. The water-rich phase contains 75.0 mole% water and 25.0% ethanol. The molar ratio of flow rates from the decanter is 12.1 moles of pentane-rich liquid per mole of water-rich liquid. Sufficient pentane has been added to the system to give a ratio of 2.48 moles of pentane in the reflux per mole of feed to the azerotrope column.[5] The azeotrope column operates at 3.3 atm and produces a bottoms that is 99.5 vol% (199-proof) ethyl alcohol and 0.5% water. This product stream is cooled to 100°F and sent to storage.

The water-rich distillate from the azeotrope column, which is at 151°F, is sent to a stripper operating at 2 atm that separates residual alcohol and trace quantities of pentane from water. The bottoms from the stripper contains trace amounts of ethyl alcohol and pentane and is sent to a waste-treatment plant. Vapor leaves the stripping column at 208°F, and is condensed and returned to the azeotrope column as part of the reflux.

After the bottoms from the ethanol concentrator has been flashed to atmospheric pressure, the resulting liquor is divided by centrifugation into two fractions: a thin stillage containing 4 wt% total solids and a **wet cake** containing about 30 wt% total solids. As described earlier, part of the thin stillage is recycled to join the mash leaving the saccharification reactor. The remaining thin stillage is heated from 165°F to 208°F and is then sent to an atmospheric evaporator to produce a **syrup** with a solids content of 40 wt%.

The wet cake is mixed with the syrup and recycled dried grains, which have a moisture content of 10 wt%. The recycle of dried grains is adjusted so that the resulting mixture of wet cake, syrup, and recycled dry grains has a moisture content of about 30 wt%. The wet grains mixture is fed at 100°F to a dryer where it is contacted with hot flue gas from a boiler. The hot gas, whose composition is given in Table 14-3, enters the dryer at about 600°F, and gas and dried grains leave the dryer at 190°F. The dried grains, which contain 10 wt% moisture, are split into two portions, with one portion being recycled to regulate the moisture content of the feed to the dryer and the remainder being sent to storage and shipping.

[5] The composition of the ternary, heterogeneous azeotrope and the ratio of pentane-rich reflux to water-rich product are based on data in C. Black, "Distillation Modeling of Ethanol Recovery and Dehydration Processes for Ethanol and Gasohol," *CEP*, *76*(9), 78 (1980).

PROBLEMS

The heat capacity of mash may be assumed to be the same as that of water. The specific gravity of the mash prior to fermentation is 1.1; after fermentation it is 1.05. The heat capacity of grains is 0.25 $Btu/lb_m \cdot °F$. In your solutions to the following problems, cite all sources of data not given in the process description, and clearly state all assumptions.

1. Construct a flowchart of the process and a separate diagram for one of the fermentation batteries containing two reaction vessels, one pump, and one heat exchanger. Calculate and fill in the component flow rates (lb_m/h) and temperatures of each stream on the chart to the extent possible from the given information.

2. Operation of the fermentation reactors in a batch mode adds considerable complexity to the process. To assist with estimating the personnel requirements and options for automatic control for the fermentation plant, develop a schedule for switching the operation of the eight fermenters to provide for continuous feed to the section and appropriate circulation of mash through the heat exchangers.

3. Calculate the feed rate of corn in lb_m/h. Estimate the acreage required to supply this plant with corn. (*Note:* According to the U.S. Department of Agriculture, *Crop Production*, 11/9/78, p. A-3, 101.2 bushels of corn are harvested from an acre.)

4. Calculate the molar flow rate (lb-moles/h) of ethanol product.

5. How much water (combined fresh water and condensates) must be added to the mixing tank to obtain the desired mash volume?

6. The CO_2 produced as a by-product from the fermentation reactor may be useful if it is recovered in sufficient quantity. Estimate the production rate of CO_2 in pound-moles per hour and standard cubic feet per minute.

7. Using Raoult's law, estimate the temperature of the bottoms leaving the azeotrope column and the stripper.

8. What fraction of the dried grains must be recycled to control the moisture content of the grains entering the dryer? What production rate of dried grains can be expected from this process?

9. What is the minimum pressure at which the cooking vessel must operate? What is the pressure to which the mash must be flashed to provide a resultant liquor temperature of 145°F? At what rate (lb_m/h) is water vaporized in this step?

10. Determine the rate (lb_m/h) at which 15-psig steam must be supplied to heat the mash from the precooking vessel to 230°F. How much live steam must be injected to raise the temperature of this stream to 320°F? It has been suggested that a portion of these steam requirements could be met by the vapor from the flash occurring at 15 psig. Is this feasible? If so, what fraction of the required 15-psig steam can be supplied by this means?

11. At what rates must heat be removed to condense the vapors from each of the distillation columns and the stripper? Suppose cooling water at 85°F is available and the temperature increase of this water is limited to 25°F. At what rates must water be supplied to each condenser?

12. Calculate the flow rates of cooling-tower water and well water required throughout the process. Summarize your calculations by constructing a table containing the following information for each type of cooling water: location of heat exchanger, flow rate (lb_m/h) of process stream, inlet and outlet temperatures of process stream, flow rate of cooling water (lb_m/h and gal/min).

13. Given that the overhead product and reflux from the ethanol concentrator are saturated liquids at the given conditions, estimate the ratio of reflux flow rate to distillate flow rate. What effect does this ratio have on the operating cost of the column?

14. Suppose that saturated 150-psig steam is used to provide heat to raise the temperature of the feed to the recovery system to the specified value and to generate boilup for the ethanol concentrator, the azeotrope column, and the stripper. Determine the steam requirements (lb_m/h) for each of these units, assuming that the steam condensate is saturated at 150 psig.

15. What portion of the heat required to raise the temperature of the feed to the recovery system to 280°F can be supplied by the vapor generated on flashing the liquid stream leaving the reboiler of the ethanol concentrator? What would be the advantage of carrying out the flash in several stages? Illustrate by subjecting the bottoms from the ethanol concentrator to a three-stage flash, with each stage reducing the pressure by 16.7 psi. Determine the fraction of the heat required to raise the temperature of the feed to the recovery system to 280°F that can be provided by condensing the vapors generated by the staged flashing. You may neglect the effect of solids on the vapor pressure of water.

16. The efficient use of energy is an important consideration in the economics of the process. Note that the recovery system has several condensers and reboilers, in addition to the units used to heat the feed to the system. A suggestion has been made to reduce the steam requirements by using the vapor from the ethanol concentrator to provide heat to the reboilers on the azeotrope column and the stripper. Furthermore, it has been proposed that the vapors from the azeotrope column and the stripper be condensed by providing heat to the feed to the recovery system. Determine the feasibility of these suggestions and evaluate the savings in steam (lb_m/h) resulting from such steps.

17. Estimate the total amount of heat (Btu) that must be removed from a single batch fermentor for the temperature to rise as indicated and then to level off at 95°F. (The higher heating value of maltose is 5649.1 kJ/mol.)

18. Use the heat of reaction for fermentation and the stated maximum heat removal rate to estimate the rate of reaction (lb-moles ethanol formed per min) at the peak period of heat removal. Neglect the formation of yeast in estimating the heat of reaction.

19. At what rate must the hot gas be fed to the drier to provide the required drying of grains? If this gas is to be obtained from the combustion of Illinois No. 6 coal with 30% excess air, how much coal is required?

Additional Problems

20. The use of pentane as an entrainer requires that the azeotrope column and stripper operate at elevated pressures. Benzene has been used as an entrainer in other situations,

TABLE 14-4

Fusel Oil Formation in Sugar Fermentation

4.47 gal of fusel oil formed per 1000 gal of
ethyl alcohol
Average molecular weight = 82.6
Specific gravity at 60°F = 0.814

Component	Wt%
n-Propyl alcohol	10.64
Isobutyl alcohol	11.09
Active amyl alcohol	21.27
Isoamyl alcohol	54.27
Hexyl alcohol	2.73

3.94 lb_m carbon dioxide formed in the production of
1 lb_m fusel oil.

and has the advantage of allowing operation of the azeotrope column at atmospheric pressure. Using information available in the literature,[6] compare the equipment requirements of the recovery system using these two entrainers.

21. Fungal amylase is blended with the mash leaving the vacuum flash, and then a portion of this stream is diverted to the fungal amylase plant. Explain the reason for such an arrangement.

22. Small amounts of **fusel oils** are formed during the fermentation of maltose to ethanol and carbon dioxide. Information on fusel oils is given in Table 14-4. These materials are more volatile than water and less volatile than the water–ethyl alcohol azeotrope. As a consequence, they would accumulate at an intermediate location in the ethanol concentrator if they were not purged from the system. Estimate the rate at which these materials are produced, and suggest a method for purging them from the system.

23. A number of alternative processes for the synthesis of ethanol use nonbiological feedstocks. Summarize the chemistry of these processes.

[6] C. J. King, *Separation Processes*, 2nd ed., McGraw-Hill, 1980, pp. 344-349.

Appendix A

Computational Techniques

This appendix introduces several mathematical concepts and methods that have widespread applicability in the analysis of chemical processes. The presentation presumes a knowledge of elementary calculus, but not of linear algebra or numerical analysis. The student who wishes a broader or deeper treatment of the subjects discussed is advised to refer to a numerical analysis reference. Carnahan, Luther, and Wilkes,[1] for example, discuss many of the techniques treated in this appendix and provide Fortran programs to implement some of them.

APPENDIX A.1 THE METHOD OF LEAST SQUARES

In this section we outline a statistical technique for fitting a straight line to y versus x data. You should be aware, however, that in doing so we are barely scratching the surface of the field of statistical analysis; we will not discuss techniques for fitting functions of several variables, for determining quantitatively the uncertainties associated with a fit, and for comparing alternative functions used to fit a given data set.

Suppose you measure y at four values of x, plot the data on a graph of y versus x, and draw a line through the data points.

If the line you draw is $y = ax + b$, then at an abscissa point x_i ($i = 1, 2, 3$ or 4) the measured value of y is y_i, and the corresponding value of y on the line is $ax_i + b$. The vertical distance d_i from the ith data point to the line (called the ith **residual**) is, therefore,

$$d_i = y_i - (ax_i + b), \qquad i = 1, 2, 3, 4 \tag{A.1-1}$$

[1] B. Carnahan, H. A. Luther, and J. O. Wilkes, *Applied Numerical Methods*, Wiley, New York, 1969.

If d_i is positive then the ith data point must be above the line (why?), if d_i is negative the point is below the line, and if d_i equals zero the line passes through the point. A line is said to fit the data well if the values of most of the residuals are close to zero.

There are several ways to determine the line that best fits a set of data, which differ primarily in their definitions of "best." The most common method is the **method of least squares.**

Suppose there are n plotted points $(x_1, y_1), (x_2, y_2), \ldots, (x_n, y_n)$, so that a line $y = ax + b$ drawn through the points yields a set of n residuals d_1, d_2, \ldots, d_n. According to the method of least squares, *the best line through the data is the one that minimizes the sum of the squares of the residuals.*[2] The task is therefore to find the values of a and b that minimize

$$\phi(a, b) = \sum_{i=1}^{n} d_i^2 = \sum_{i=1}^{n} (y_i - ax_i - b)^2 \qquad \text{(A.1-2)}$$

You can obtain expressions for the best values of a and b in terms of known quantities by differentiating the equation for ϕ (Equation A.1-2) with respect to both a and b, setting the derivatives equal to zero, and solving the resulting algebraic equations for a and b. (See Problem 38 in Chapter 2.) The results of these calculations are as follows. If we define

$$s_x = \frac{1}{n} \sum_{i=1}^{n} x_i \qquad s_{xx} = \frac{1}{n} \sum_{i=1}^{n} x_i^2$$

$$s_y = \frac{1}{n} \sum_{i=1}^{n} y_i \qquad s_{xy} = \frac{1}{n} \sum_{i=1}^{n} x_i y_i \qquad \text{(A.1-3)}$$

then

1. Best line: $y = ax + b$

$$\text{Slope:} \quad a = \frac{s_{xy} - s_x s_y}{s_{xx} - (s_x)^2} \qquad \text{(A.1-4)}$$

$$\text{Intercept:} \quad b = \frac{s_{xx} s_y - s_{xy} s_x}{s_{xx} - (s_x)^2} \qquad \text{(A.1-5)}$$

2. Best line through the origin: $y = ax$

$$\text{Slope:} \quad a = \frac{s_{xy}}{s_{xx}} = \frac{\sum x_i y_i}{\sum x_i^2} \qquad \text{(The intercept equals 0.0.)} \qquad \text{(A.1-6)}$$

Once you determine a and b, you should plot the line $y = ax + b$ on the same graph as the data to get an idea of how good the fit is.

[2] We could also choose the best line as the one that minimizes the sum of the absolute values of the residuals, or the sum of the fourth powers of the residuals. Using the squares simply provides computational convenience.

EXAMPLE A.1-1 The Method of Least Squares

Two variables, P and t, are related by the equation

$$P = \frac{1}{mt^{1/2} + r}$$

The following data are taken:

P	0.279	0.194	0.168	0.120	0.083
t	1.0	2.0	3.0	5.0	10.0

Calculate m and r using the method of least squares.

Solution

The equation may be rewritten in the form

$$\frac{1}{P} = mt^{1/2} + r$$

so that a plot of $1/P$ versus $t^{1/2}$ should be a line with slope m and intercept r. From the tabulated data,

$y = 1/P$	3.584	5.155	5.952	8.333	12.048
$x = t^{1/2}$	1.0	1.414	1.732	2.236	3.162

$$\frac{1}{P} = mt^{1/2} + r$$

$$\Downarrow \; y = 1/P, \; x = t^{1/2}$$

$$y = mx + r$$

Evaluate the quantities of Equation A.1-3:

$$s_x = \tfrac{1}{5}(1.000 + 1.414 + 1.732 + 2.236 + 3.162) = 1.909$$
$$s_y = 7.014$$
$$s_{xx} = 4.200$$
$$s_{xy} = 15.582$$

Hence, from Equation A.1-4,

$$\textit{Slope: } m = \frac{s_{xy} - s_x s_y}{s_{xx} - (s_x)^2} = 3.94$$

and from Equation A.1-5

$$\textit{Intercept: } r = \frac{s_{xx} s_y - s_{xy} s_x}{s_{xx} - (s_x)^2} = -0.517$$

so that the final result is

$$\boxed{P = \frac{1}{3.94 t^{1/2} - 0.517}}$$

A good check on the results would be to plot $1/P$ versus $t^{1/2}$, showing both the data points and the line

$$\frac{1}{P} = 3.94t^{1/2} - 0.517$$

If the choice of this function to fit the given data was reasonable and if no mistakes were made in the calculation, the data points should be scattered about the line. This is in fact the case, as shown in the following diagram.

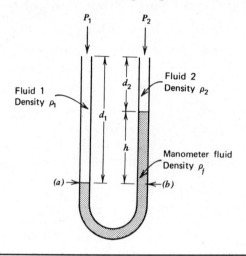

TEST YOURSELF

1. A line $y = 3x + 2$ has been fit to a set of data including the points $(x = 1, y = 4)$ and $(x = 3, y = 13)$. What are the residuals at these two points?

2. What is the definition of the best line through a set of data points that forms the basis of the method of least squares?

3. Could the method of least squares be used to fit a straight line to data points that fall on a distinct curve? (*Corollary:* Does the line that "best" fits the data necessarily fit the data well?)

4. An alternative to the method of least squares might be to minimize the sum of the residuals rather than the sum of squares of the residuals. What would be wrong with this method?

APPENDIX A.2 ITERATIVE SOLUTION OF NONLINEAR EQUATIONS

A.2a Linear and Nonlinear Equations

In Section 5.4 we presented the Van der Waal's equation of state.

$$(P + a/\hat{V})(\hat{V} - b) = RT$$

and observed that solving this equation to calculate P for a given \hat{V} and T is easy, while solving for \hat{V} for specified values of P and T is relatively difficult.

What makes an equation easy or difficult to solve is its **linearity** or **nonlinearity** in the unknown variable. Equations that contain unknown variables raised to the first power only (x, but not x^2 or $x^{1/2}$), and that do not contain products (xy) or transcendental functions ($\sin x$, e^x) of unknown variables, are called **linear equations.** Equations that do not satisfy these conditions are called **nonlinear equations.**

For example, if a, b, and c are constants, and x, y, and z are variables,

$$ax + by = c \text{ is linear}$$
$$ax^2 = by + c \text{ is nonlinear (contains } x^2)$$
$$x - \ln x + b = 0 \text{ is nonlinear (contains } \ln x)$$
$$ax + by = cx \text{ is linear}$$

Linear equations that contain a single unknown variable have one and only one solution (one **root**).

$$7x - 3 = 2x + 4 \Rightarrow x = 1.2$$

$$\left.\begin{array}{l} P\hat{V} = RT \\ P = 3, R = 2, T = 300 \end{array}\right\} \Rightarrow \hat{V} = RT/P = (2)(300)/(3) = 200$$

However, nonlinear equations that contain a single unknown variable may have any number of real roots (and they may also have imaginary and complex roots). For example,

$$x^2 + 1 = 0 \quad \text{has no real roots}$$
$$x^2 - 1 = 0 \quad \text{has two real roots } (x = +1, \text{ and } x = -1)$$
$$x - e^{-x} = 0 \quad \text{has one real root } (x = 0.56714\ldots)$$
$$\sin x = 0 \quad \text{has an infinite number of real roots } (x = 0°, 180°, 360°, \ldots)$$

The roots of some nonlinear equations, such as the second of the equations given above, can be obtained directly using simple algebra, but most nonlinear equations must be solved using an iterative or trial-and-error technique.

A single equation containing several variables may be linear with respect to some variables and nonlinear with respect to others. For example,

$$xy - e^{-x} = 3$$

is linear in y and nonlinear in x. If x is known, the equation may easily be solved for y, while the solution for x from a known value of y is much harder to obtain. Another example is provided by the three-term virial equation of state:

$$P\hat{V} = RT\left[1 + \frac{B(T)}{\hat{V}} + \frac{C(T)}{\hat{V}^2}\right]$$

where B and C are functions of temperature. This equation is linear in P and nonlinear in \hat{V} and T. It is consequently easy to solve for P from given values of T and \hat{V} and difficult to solve for either \hat{V} or T from given values of the other two variables.

Most of the problems you are called on to solve in this text reduce to one or two linear equations in as many unknowns. The hard part of the problems, if there is a hard part, is deriving the equations; solving them is a matter of simple algebra. However, many process problems involve nonlinear equations. Techniques for solving such problems are the subject of this appendix.

TEST YOURSELF

Classify the following single-variable equations as linear or nonlinear, considering a, b, and c constants.

1. $3x + 17 = 23x - 12$
2. $3x = a(\ln x) + b$
3. $x \exp(x) = 14$
4. $axy - bx^2 = cy/x$
 (a) x is known
 (b) y is known
5. $14x \cos(y) - 8/z = 23$
 (a) x and y are known
 (b) x and z are known
 (c) y and z are known

A.2b Graphical Solution

In this and the next several sections, we will discuss methods for solving one nonlinear equation in one unknown. Extensions to multivariable problems will be presented in Section A.2h.

Suppose you have to solve an equation of the form $f(x) = 0$—that is, find the root or roots of the function $f(x)$. (Any equation can be written in this form by bringing every term to the left side. For example, $x = e^{-x}$ becomes $f(x) = x - e^{-x} = 0$.) A particularly obvious solution technique is to plot $f(x)$ versus x and to locate by graphical interpolation the point at which the curve crosses the axis.

There are several problems with this technique. It is basically a manual method; it is relatively slow; and it is not very precise. Its principal advantage is that it enables you to see how f varies with x, which is particularly useful when you are dealing with functions that have several roots.

The points x_1, x_2, and x_3 are all roots of the equation $f(x) = 0$. The computational techniques we will describe shortly would locate one or another of these roots, depending on the initial guess; however, for all but simple polynomial functions a plot is the only convenient method to detect the existence and approximate location of multiple roots. Therefore, unless you know that only one root exists, or you know the approximate location of each root you need to determine, a good procedure is to plot f versus x and use the plot to determine initial estimates for more precise root-finding methods.

TEST YOURSELF

1. Use a graphical argument to justify the statement that a linear function has only one root.

2. How many roots does the function $f(x) = x - \exp(-x)$ have? (*Suggestion:* Sketch plots of $f_1(x) = x$ and $f_2(x) = \exp(-x)$ versus x, and use these plots to obtain your answer.)

A.2c Regula-falsi Method

In this and the next subsection, we outline algorithms for finding roots of single-variable equations of the form $f(x) = 0$. The first procedure, termed the **regula-falsi method,** is appropriately used when an analytical expression for the derivative of f with respect to x is not available—as, for example, when $f(x)$ is obtained as the output of a computer program for an input value of x. The algorithm is as follows:

1. Find a pair of values of x—x_n and x_p—such that $f_n = f[x_n] < 0$ and $f_p = f[x_p] > 0$.

2. Estimate the value of the root of $f(x)$ from the following formula:

$$x_{\text{new}} = \frac{x_n f_p - x_p f_n}{f_p - f_n} \tag{A.2-1}$$

and evaluate $f_{\text{new}} = f[x_{\text{new}}]$.

3. Use the new point to replace one of the original points, keeping the two points on opposite sides of the x axis. If $f_{\text{new}} < 0$, replace the old x_n and f_n with x_{new} and f_{new}. If $f_{\text{new}} > 0$, replace x_p and f_p with x_{new} and f_{new}. (If $f_{\text{new}} = 0$, you have of course found the root and need go no further.)

4. See if the new x_n and x_p are close enough for convergence to be declared (see Section A.2g). If they are not, go back to step 2.

What you are doing with this procedure is the algebraic equivalent of drawing a straight line between the two points $[x_n, f_n]$ and $[x_p, f_p]$ on a plot of f versus x and using the intersection of this line with the x axis as the next estimate of the root.

Successive pairs of points determined in this manner clearly approach each other. The procedure terminates when x_n and x_p are close enough together to satisfy a specified convergence criterion. Possible convergence criteria are suggested in Section A.2g.

Once you have found a root, you have no guarantee that it is the only one. For all but relatively simple algebraic functions, the safest procedure is to evaluate $f(x)$ at enough points over the range of interest to either establish or refute the existence of multiple roots.

TEST YOURSELF

1. Suppose a "black box" computer program gives values of a function $f(x)$ for specified values of x. Unknown to the programmer, the function is

$$f = 4 - (x - 2)^2.$$

 (a) What are the roots of this function? (You should be able to do it by inspection.)
 (b) Suppose the programmer tries values $x_p = 3$ and $x_n = 5$. If he uses the regula-falsi method, what will the next x_n-x_p pair be? To which root will the method ultimately converge?

2. Derive Equation A.2-1.

A.2d Newton's Rule

The next algorithm for finding the root of a function $f(x)$ is **Newton's rule.** It is considerably more efficient than regula-falsi, but only for functions for which the derivative $f'(x) = df/dx$ can be evaluated analytically. The formula for proceeding

from one estimate of the root to the next is

$$x_{k+1} = x_k - \frac{f_k}{f'_k}$$ (A.2-2)

As always, you begin by estimating a value of the root, x_0. Successive estimates are then generated from Equation A.2-2, with a test for convergence (Section A.2g) being applied after each estimate is obtained.

The easiest way to understand how Newton's rule works is graphically. Suppose the plot of f versus x appears as follows:

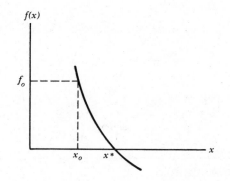

Although it may not be obvious at first glance, Newton's rule is equivalent to choosing a value of x_o and calculating $f_o = f(x_o)$, drawing a line tangent to the curve at (x_o, f_o), and using the intersection of this line with the x axis as the next estimate (x_1). As the following diagram shows, the successive values of x generated in this manner (x_1, x_2, x_3, \ldots) may converge on the root x^*, although this convergence is not guaranteed.

The formula for each estimate (x_{k+1}) in terms of the previous estimate (x_k) may easily be derived. The graphical representation of one step of the procedure is shown on the next page. The slope of the tangent line is $(df/dx)_{x_k} = f'_k$; however, two known points on this line are $(x_{k+1}, 0)$ and (x_k, f_k), so that the slope is also equal to

$(0 - f_k)/(x_{k+1} - x_k)$. Equating these two expressions for the slope yields

$$f'_k = \frac{-f_k}{x_{k+1} - x_k}$$

The solution of this equation for x_{k+1} is Newton's rule, Equation A.2-2

$$x_{k+1} = x_k - \frac{f_k}{f'_k}$$

EXAMPLE A.2-1 Newton's Rule

Determine the root of the equation $x = e^{-x}$ using Newton's rule.

Solution

$$f(x) = x - e^{-x}$$
$$f'(x) = df/dx = 1 + e^{-x}$$

When $x = 0$, $f(x)$ is negative, while when $x = 1$, $f(x)$ is positive (verify by inspection); the root x^* must therefore be between 0 and 1. Try $x_o = 0.2$ as a first guess.

First iteration: $x_o = 0.2$

$$\Downarrow$$
$$f(x_o) = 0.2 - e^{-0.2} = -0.6187$$
$$\Downarrow$$
$$f'(x_o) = 1 + e^{-0.2} = 1.8187$$
$$\Downarrow$$
$$x_1 = x_o - f(x_o)/f'(x_o) = 0.5402$$

Second iteration: $x_1 = 0.5402$

$$\Downarrow$$
$$f(x_1) = 0.5402 - e^{-0.5402} = -0.0424$$
$$\Downarrow$$
$$f'(x_1) = 1 + e^{-0.5402} = 1.5826$$
$$\Downarrow$$
$$x_2 = x_1 - f(x_1)/f'(x_1) = 0.5670$$

Third iteration: $x_2 = 0.5670$

$$\Downarrow$$
$$f(x_2) = 0.5670 - e^{-0.5670} = 2.246 \times 10^{-4}$$
$$\Downarrow$$
$$f'(x_2) = 1 + e^{-0.5670} = 1.5672$$
$$\Downarrow$$
$$x_3 = x_2 - f(x_2)/f'(x_2) = 0.56714$$

The successive estimates of x^* are therefore

$$0.2 \Rightarrow 0.5402 \Rightarrow 0.5670 \Rightarrow 0.56714$$

This is clearly a converging sequence; depending on how much precision you require, you might stop here or carry out one or two additional iterations. Let us stop here, and say that $x^* \approx 0.56714$.

TEST YOURSELF

1. Could you use Newton's rule to find a root of the equation $x^2 - 3x - 3 = 0$? Would you use it? Why not?

2. Suppose you wish to find a root of $f(x) = 0$, where $f(x)$ versus x appears as shown here.

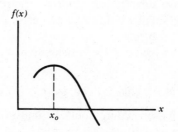

What would happen if you took as a first guess the value of x_o shown? (What would happen shows that Newton's rule does not necessarily work if a poor choice of x_o is made.)

A.2e Successive Substitution and Modified Successive Substitution

Problems involving the solution of nonlinear equations can often be expressed in the form

$$x = f(x)$$

where $f(x)$ is a nonlinear function. (*Example:* $x = e^{-x}$.) As shown in Chapter 10, balance equations for multiple unit processes with recycle often fall into this category: x would be the assumed value of a tear stream variable, and $f(x)$ would be the value generated by calculating around the cycle.

The simplest and most obvious solution method is **successive substitution.** An initial estimate, $x^{(0)}$, is selected; $f(x^{(0)})$ is calculated; and the calculated value is used as the next estimate of the root. The formula is

$$x^{(i+1)} = f(x^{(i)}) \tag{A.2-3}$$

The procedure is repeated until the specified convergence criterion is satisfied.

Sometimes successive substitution works very well, converging in a few steps. Three unsatisfactory convergence patterns are also observed from time to time, however. In the first, the successive estimates oscillate about a central value:

$$37.6, 2.3, 36.8, 2.6, 34.4, 2.9, \ldots$$

The root lies somewhere between 3 and 30, but the procedure will clearly take a large number of iterations to get there.

The problem here is that successive substitution generates overly large steps. Instead of jumping all the way from 37.6 to 2.3, as successive substitution dictates, we should only go part of the way from the first to the second value to obtain our estimate of $x^{(1)}$. To do this, we may use **modified successive substitution** (also called damped successive substitution). The formula is

$$x^{(i+1)} = x^{(i)} + p\{f(x^{(i)}) - x^{(i)}\} \tag{A.2-4}$$

where p, the **damping parameter,** is a number between 0 and 1. If $p = 1$, the procedure reduces to pure successive substitution, while as p approaches zero, the size of the step becomes smaller and smaller. A few trial-and-error iterations should yield a good value of p for a specific problem.

The second case of slow convergence in successive substitution involves a creeping progression, such as

$$151.7, 149.5, 147.4, 145.6, 143.8, \ldots$$

Again, it appears that the procedure could be converging to a solution, but it is equally clear that it is in no hurry to get there.

The remedy for this problem is to **accelerate** the convergence procedure—to jump over many of the intermediate solutions to which continued successive substitution would lead. The next section outlines Wegstein's method, one of the most commonly used acceleration algorithms.

The third unsatisfactory convergence pattern is instability. For example, if successive substitution yields a sequence like

$$1.0, 2.5, -6.8, 23.5, 97.0, \ldots$$

then successive substitution will clearly not work, no matter how many iterations are attempted. A better first estimate might yield a convergent sequence, or the problem might be intrinsically unstable and must be restructured or solved by a different technique. Texts on numerical analysis outline stability conditions for nonlinear equation solution algorithms; their consideration is beyond the scope of this appendix.

TEST YOURSELF

1. For each of the following sequences of successive estimates of a root, indicate whether successive substitution seems to be adequate, or whether you would resort to modified successive substitution or an acceleration method (state which).
 (a) 165, 132, 163, 133, 162, 133, ...
 (b) 43, 28, 26, 26.7, 26.71, ...
 (c) 21.0, 21.2, 21.4, 21.59, 21.79, ...
2. Suppose $x^{(i)} = 14.0$, $f(x^{(i)}) = 13.0$, and you are using modified successive substitution with $p = 0.4$. What is your next estimate of the root?

A.2f Bounded Wegstein Algorithm

The procedure outlined in this section encompasses successive substitution and modified successive substitution as special cases, and in addition provides acceleration capability. It is as follows.

1. Begin by choosing $x^{(1)}$. Calculate $f(x^{(1)})$, and let $x^{(2)} = f(x^{(1)})$. (That is, carry out a successive substitution step.) Let $k = 2$.
2. Calculate $f(x^{(k)})$.
3. Check for convergence. If $x^{(k)}$ and $f(x^{(k)})$ are close enough together to meet the convergence criterion, terminate the procedure. If convergence is not achieved, calculate

$$w = \frac{f(x^{(k)}) - f(x^{(k-1)})}{x^{(k)} - x^{(k-1)}} \qquad \text{(A.2-5a)}$$

$$q = w/(w - 1) \qquad \text{(A.2-5b)}$$

4. If $q > q_{max}$, reset q to q_{max}
 $q < q_{min}$, reset q to q_{min}

 where

$$q_{min} = -5 \qquad \text{and} \qquad q_{max} = +0.5$$

This step imposes limits on the size of the acceleration step and leads to the designation of this procedure as the **bounded** Wegstein algorithm.

5. Calculate

$$x^{(k+1)} = qx^{(k)} + (1 - q)f(x^{(k)}) \qquad \text{(A.2-6)}$$

6. Increase k by 1 and go back to step 2.

It is not difficult to show that Wegstein's method is tantamount to generating two points on the curve of $f(x)$ versus x and determining as the next estimate the intersection of the line between these points and the 45° line [at which $x = f(x)$].

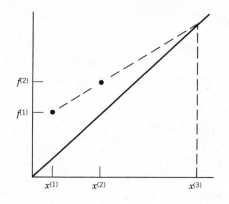

If you examine Equation A.2-6, you will see that if the parameter q equals zero, the procedure reduces to successive substitution; if q is between zero and 1 the procedure is modified successive substitution; and if q is negative the procedure involves acceleration. If the procedure seems to be unstable, raising the value of q_{min} (i.e., making it less negative) might improve matters; if it is converging very slowly and monotonically, you might lower q_{min}; and if it is converging in an oscillatory manner, try raising q_{max}.

TEST YOURSELF

An effort to solve the equation $x = f(x)$ begins with two successive substitution steps:

$$x = 2.00 \rightarrow f(x) = 2.30$$
$$x = 2.30 \rightarrow f(x) = 2.45$$

1. Calculate the next value of x using the Wegstein algorithm.
2. Sketch a plot of $f(x)$ versus x, showing the two given points, and show graphically that the value of x you calculated is approximately correct.

A.2g Convergence Criteria

A problem common to all iterative computational methods is knowing when to quit. An iterative method rarely yields a precise root, but rather it gives successive approximations that (if the method converges) approach the root more and more closely. Whether you are doing the calculation by hand or writing a program to do it, you must specify how close is close enough.

Suppose that our task is to find a root of the function $f(x)$ or a solution of the equation $x = f(x)$, and that we are using a method that gives successive estimates $x^{(1)}$, $x^{(2)}$, etc. Several different criteria may be used to determine when to terminate the procedure. Two of the most common ones are given below. In each case, if the given inequality is satisfied, the final estimate of the root is $x^{(i)}$.

$$\left| x^{(i)} - x^{(i-1)} \right| < \varepsilon \qquad \text{(A.2-7)}$$

$$\frac{\left| x^{(i)} - x^{(i-1)} \right|}{x^{(i)}} < \varepsilon \qquad \text{(A.2-8)}$$

where ε, the **convergence tolerance,** is a small number, chosen to be orders of magnitude less than the value of x. The smaller the value of ε, the more accurate will be the final estimate of the root, but the more steps it will take to get there.

Equation A.2-7 is an **absolute convergence criterion.** If the value of x changes by less than ε from one iteration to the next, the procedure is terminated and the last value of x is taken to be the desired root. The strictness of this criterion for a given ε depends on the magnitude of the estimated values. If $\varepsilon = 0.01$, for example, and successive estimates of x are 358 234.5 and 358 234.6, the procedure would not be terminated, even though the estimates are undoubtedly close enough for any realistic purpose. On the other hand, successive estimates of 0.003 and 0.006 would lead to termination for the same value of ε, despite the fact that the two estimates differ by a factor of two.

Equation A.2-8 is a **relative convergence criterion,** which avoids this difficulty. If a relative convergence criterion is used, a value of $\varepsilon = 0.01$ specifies that the procedure will be terminated when the value of x changes from one iteration to the next by less than 1%, regardless of the magnitude of that value. A relative convergence criterion is used for most root-finding calculations.

TEST YOURSELF

1. Suppose you wish to declare convergence when the estimated value of a root changes by less than one part in a million from one iteration to the next. Which convergence criterion would you use, and what would the value of ε be?
2. Two successively generated estimates of a root are 10.001 and 10.000. What are the maximum values of ε that would result in 10.000 being the final estimate if (a) an absolute convergence criterion and (b) a relative convergence criterion is used.
3. Suppose the true value of a root is 0.000. Which (if either) convergence criterion would lead to problems for this special case?

A.2h Multivariable Root-finding Algorithms

Solving n nonlinear equations in n unknowns is a difficult problem, to say the least, and a general treatment is well beyond the scope of this text. In this section we present several approaches without proof or much explanation. For additional details, standard texts on numerical analysis should be consulted, such as the one by Carnahan, Luther, and Wilkes.

Three methods that can be used to find the values of x_1, \ldots, x_n that satisfy n simultaneous equations are extensions of methods given previously for single-variable problems. They are (1) successive substitution; (2) the bounded Wegstein algorithm,

and (3) the Newton-Raphson method (a multivariable extension of Newton's rule). The example that follows the presentation of the last-named technique illustrates all three algorithms.

Successive Substitution Suppose the equations can be cast in the form:

$$x_1 = f_1(x_1, x_2, \ldots, x_n)$$
$$x_2 = f_2(x_1, x_2, \ldots, x_n)$$
$$\vdots$$
$$x_n = f_n(x_1, x_2, \ldots, x_n)$$

(A.2-9)

(A cycle with n tear-stream variables falls into this category.) The successive substitution method consists of assuming values for each of the n unknown variables, evaluating the functions f_1, \ldots, f_n, and using the calculated values as the next estimates of the variables. The procedure is terminated when all variable values meet a specified convergence criterion. For example, if $x_i^{(k)}$ is the value of the ith variable at the kth iteration, the procedure might be terminated when

$$\frac{|x_i^{(k)} - x_i^{(k-1)}|}{x_i^{(k)}} < \varepsilon, \qquad i = 1, 2, \ldots, n$$

(A.2-10)

This approach is obvious, simple, and generally inefficient. The greater the number of variables, the longer it takes for the procedure to converge, if it converges at all. It is generally preferable to use Wegstein's method or the Newton-Raphson method, according to whether or not the partial derivatives of the functions f_1, \ldots, f_n can be evaluated analytically. (Use Newton-Raphson if they can be, otherwise try Wegstein, but don't be too surprised if it doesn't converge.)

Bounded Wegstein Algorithm If the equations to be solved have the form of Equation A.2-9 [i.e., $x_i = f_i(x_1, x_2, \ldots, x_n)$], guess values for all n variables and apply the procedure of Section A.2f separately to each variable. Terminate when the convergence criteria are satisfied for all variables.

This procedure will work reasonably well if the generating function f_1 depends almost entirely on x_1, f_2 depends only on x_2, and so on (i.e., if there is little interaction among the variables). If this is not the case, convergence will generally be very difficult to achieve.

Newton-Raphson Method Suppose now that the equations to be solved take the form

$$g_1(x_1, x_2, \ldots, x_n) = 0$$
$$g_2(x_1, x_2, \ldots, x_n) = 0$$
$$\vdots$$
$$g_n(x_1, x_2, \ldots, x_n) = 0$$

(A.2-11)

(If the equations are of the form $x_i = f_i(x_1, \ldots, x_n)$, then g_i can be defined as $x_i - f_i$ to convert the equations to the required form.) The Newton-Raphson method is as

follows:

1. Guess values for all n variables, $x_1^{(1)}, \ldots, x_n^{(1)}$. Let $k = 1$ (the number of the iteration).
2. Evaluate the functions

$$g_i^{(k)} = g_i[x_1^{(k)}, \ldots, x_n^{(k)}] \qquad \text{(A.2-12)}$$

and the partial derivatives of each function with respect to each variable:

$$a_{ij} = (\partial g_i / \partial x_j) \text{ at } x_1^{(k)}, \ldots, x_n^{(k)} \qquad \text{(A.2-13)}$$

3. Let $d_i = x_i^{(k+1)} - x_i^{(k)}$. Solve the following set of linear equations for d_i:

$$
\begin{aligned}
a_{11}d_1 + a_{12}d_2 + \ldots + a_{1n}d_n &= -g_1 \\
a_{21}d_1 + a_{22}d_2 + \ldots + a_{2n}d_n &= -g_2 \\
a_{n1}d_1 + a_{n2}d_2 + \ldots + a_{nn}d_n &= -g_n
\end{aligned}
\qquad \text{(A.2-14)}
$$

If there are only two or three unknowns, you can solve these equations by simple algebraic techniques. For larger systems of equations, a matrix solution technique should be used. (See Carnahan, Luther, and Wilkes.)

4. Calculate $x_i^{(k+1)} = x_i^{(k)} + d_i$.
5. Test the values of d_i for convergence; for example, declare the procedure to have converged if

$$|d_i / x_i^{(k+1)}| < \varepsilon, \qquad i = 1, \ldots, n$$

If convergence has not been achieved, increase k by one and go back to step 2.

The Newton-Raphson method is based on a linearization of the functions g_1, \ldots, g_n about each estimated set of roots and a solution of the resulting linear equations to get the next estimate. (If you have no idea what all that means, don't worry about it.) It is an extremely efficient procedure to use when analytical derivatives of the functions g_1, \ldots, g_n are convenient to evaluate.

EXAMPLE A.2-2 Multivariable Root-finding

Find the roots of the following pair of functions:

$$
\begin{aligned}
g_1(x, y) &= 2x + y - (x + y)^{1/2} - 3 \\
g_2(x, y) &= 4 - y - 5/(x + y)
\end{aligned}
$$

1. By successive substitution.
2. By the bounded Wegstein algorithm.
3. By the Newton-Raphson method.

In each case, use a starting value $(x = 2, y = 2)$ and a relative convergence criterion $\varepsilon = 0.001$.

Solution

1. *Successive substitution*

The equations $g_1 = 0$ and $g_2 = 0$ must be rewritten to provide explicit expressions for x and y. One way of doing this is the following:

$$x_c = 0.5[3 - y_a + (x_a + y_a)^{1/2}]$$
$$y_c = 4 - 5/(x_a + y_a)$$

where the subscript "a" stands for assumed and "c" stands for calculated. We assume values of x and y, recalculate x and y using these expressions, and iterate until convergence is achieved. The calculation proceeds as follows.

Iteration	Assumed		Calculated	
	x	y	x	y
1	2.000	2.000	1.500	2.750
2	1.500	2.750	1.156	2.824
3	1.156	2.824	1.086	2.744
4	1.086	2.744	1.107	2.694
5	1.107	2.694	1.128	2.684
6	1.128	2.684	1.134	2.688
7	1.134	2.688	1.133	2.692
8	1.133	2.692	1.132	2.693
9	1.1320	2.6929	1.1314	2.6928

Since the relative changes in x and y in the last iteration are each less than 0.001, the calculation is terminated at this point, and the final values are accepted as the roots of the two given equations.

2. *Bounded Wegstein algorithm*

The same functions are used to generate calculated values of x and y from assumed values, only now the equations of Section A.2f are used to generate new assumed values after the first iteration. The results are as follows. (Check the first series of numbers with a hand calculator to make sure you know how to apply the formulas.)

Iteration	Assumed		Calculated	
	x	y	x	y
1	2.000	2.000	1.500	2.750
2	1.500	2.750	1.156	2.824
3	0.395	2.832	0.982	2.450
4	1.092	2.641	1.146	2.660
5	1.162	2.651	1.151	2.689
6	1.150	2.670	1.142	2.691
7	1.123	2.694	1.130	2.690
8	1.136	2.690	1.133	2.693
9	1.1320	2.6919	1.1318	2.6924

In this case, Wegstein's method did not accelerate the convergence. In fact, the large jump in the value of x in iteration 3 (when the Wegstein procedure was first used) could have been the first symptom of an instability, but the algorithm recovered well.

3. *Newton-Raphson method*

The necessary formulas are as follows.

$$g_1(x, y) = 2x + y - (x + y)^{1/2} - 3$$
$$g_2(x, y) = 4 - y - 5/(x + y)$$
$$a_{11}(x, y) = \partial g_1/\partial x = 2 - 0.5(x + y)^{-1/2}$$
$$a_{12}(x, y) = \partial g_1/\partial y = 1 - 0.5(x + y)^{-1/2}$$
$$a_{21}(x, y) = \partial g_2/\partial x = 5/(x + y)^2$$
$$a_{22}(x, y) = \partial g_2/\partial y = -1 + 5/(x + y)^2$$

Equations A.2-14 reduce for this two-dimensional problem to

$$a_{11}d_1 + a_{12}d_2 = -g_1$$
$$a_{21}d_1 + a_{22}d_2 = -g_2$$

The procedure is to assume values of x and y; calculate g_1, g_2, a_{11}, a_{12}, a_{21}, and a_{22} from the given formulas; solve the two preceding equations for d_1 and d_2, and calculate the new estimates of the roots as

$$x_c = x_a + d_1$$
$$y_c = y_a + d_2$$

The convergence test is then applied, and if the assumed and calculated values are not close enough together, the latter values are used to replace the former ones and the calculation is repeated. The results are shown here.

	Assumed		Calculated	
Iteration	x	y	x	y
1	2.000	2.000	1.130	2.696 (Verify!)
2	1.130	2.696	1.1315	2.6925
3	1.1315	2.6925	1.1315	2.6925

The superiority of the Newton-Raphson method to the others tested is clear in this example and is even more dramatic when more than two equations are to be solved simultaneously. Generally, when analytical derivatives are available, the Newton-Raphson method should be used for solving multiple nonlinear algebraic equations.

A.2i Programming Tips

Following are tips for carrying out trial-and-error solutions of algebraic equations by computer. All experienced programmers do these things routinely; most learned to do them the hard way.

1. Regardless of which method you use, you must first estimate the solution, and then use the method to proceed to (it is hoped) better and better estimates. The first guess can be critical in determining the success of the procedure. The closer the estimate is to the true solution, the more likely the method is to converge and to do so rapidly.

 You will often have a basis for a reasonable first guess—knowledge of the process system, physical constraints (an unknown mole fraction must be between 0 and 1, for instance), previous experience with similar problems, or common sense. If you have absolutely no basis, however, it is a good idea to calculate $f(x)$ for several randomly chosen values of x before beginning the procedure. The magnitude of f and the way it varies for the values of x you have chosen will often enable you to guess where a root is likely to be found. This guess then serves as the first estimate from which the systematic search procedure can proceed. The nature of the variation with x also may suggest an appropriate search procedure.

2. *Always impose an upper limit on the number of iterations performed.* If this limit is exceeded, terminate the procedure, print out an appropriate message, and stop the calculation.

 There is always a chance that a procedure will not converge, either because of an inherent difficulty with the problem or because the programmer made a mistake. If you fail to achieve convergence in the specified number of iterations, don't just raise the iteration limit and try again; instead, determine whether or not the program is functioning correctly and approaching a convergence limit. (The next step indicates how to accomplish this determination.)

3. At least the first time you run a program to perform a trial-and-error search, print out the values of the successive estimates $(x^{(1)}, x^{(2)}, \ldots)$ and the corresponding function values, and put a low limit on the number of iterations allowed—e.g., 10. Hand-calculate the value of $f(x^{(1)})$, and verify that your value is the same as the one calculated by the computer. If it is, you can have a reasonable degree of confidence in the validity of your program. Then verify that the successive estimates of x appear to be converging, if they did not in fact converge to within your specified tolerance. If they are converging, then either increase the number of allowed iterations or relax your convergence tolerance. If not, try a new initial guess. If the method still does not work, check your logic—something is probably wrong.

APPENDIX A.3 NUMERICAL INTEGRATION

In Chapter 8, we showed that the enthalpy change associated with the heating or cooling of a substance is evaluated by integrating the substance heat capacity $C_p(T)$ from the initial temperature to the final temperature. This is the first of many instances you will encounter in process analysis where an integration or the solution of a differential equation is required as part of a problem solution.

It often happens that required values of definite integrals or solutions of differential equations cannot be obtained using the methods of elementary calculus. If, for

example, you are called on to evaluate something such as

$$\int_0^{10} e^{-x^2}\, dx$$

you will not find help in a calculus book or a table of integrals—an analytical expression for the integral of exp $(-x^2)$ simply does not exist.

It is possible, however, to substitute for any mathematical operation such as differentiation or integration a series of arithmetic operations that yield approximately the same result. The arithmetic operations are usually simple but long and repetitious, and so are ideally suited to computers. The branch of mathematics that deals with techniques for solving mathematical problems on computers is called **numerical analysis**; a knowledge of the methodology of this field is becoming increasingly essential for engineers concerned with the design and analysis of chemical processes.

A.3.1 Quadrature

The general problem we will discuss is the evaluation of a definite integral:

$$I = \int_a^b y(x)\, dx \tag{A.3-1}$$

There are two possible reasons why you might not be able to evaluate I analytically: $y(x)$ may be a nonintegrable analytical function, such as exp $(-x^2)$, or it may not be an analytical function at all, but instead a series of tabulated (x, y) data points or a plot of y versus x.

A method of **numerical integration** (or **quadrature**, as it is also called) is required to evaluate I in either of these cases. The specific techniques we will present are algebraic, but the general approach to the problem is best visualized graphically. For the moment, we will suppose that all we have relating x and y is a table of data points:

x	$x_1(=a)$	x_2		x_{n-1}	$x_n(=b)$
y	y_1	y_2		y_{n-1}	y_n

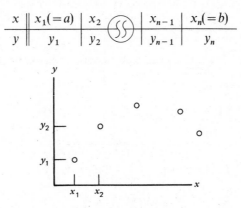

Let us graph the data points on a plot of y versus x. The integral we are trying to evaluate (I of Equation A.3-1) equals the area under the continuous curve of y versus x, but this curve is not available—we only know the function values at the discrete data points. The procedure generally followed is to fit approximating functions to the data points, and then to integrate these functions analytically.

The many existing quadrature formulas differ only in the choice of functions to fit to the data points. Two of the simplest approximations are to fit straight lines between successive points and sum the area under the lines, and to fit parabolas to successive triplets of points and sum the areas under the parabolas. These approximations lead to the quadrature formulas known respectively as the **trapezoidal rule** and **Simpson's rule.** We will discuss each in turn.

A.3.2 The Trapezoidal Rule

The area under a line through (x_1, y_1) and (x_2, y_2) on a plot of y versus x is easily calculated.

$$\Longrightarrow \quad I_o = \int_{x_2}^{x_1} y \, dx = (x_2 - x_1) \left(\frac{y_1 + y_2}{2} \right)$$

The area under a series of points from x_1 to x_n is obtained by a summation of such terms.

$$I = \tfrac{1}{2}[(x_2 - x_1)(y_1 + y_2) + (x_3 - x_2)(y_2 + y_3) + \ldots + (x_n - x_{n-1})(y_{n-1} + y_n)]$$

$$\Downarrow$$

Trapezoidal Rule:
$$\int_{x_1}^{x_n} y(x) \, dx \approx \tfrac{1}{2} \sum_{j=1}^{n-1} (x_{j+1} - x_j)(y_j + y_{j+1}) \qquad \text{(A.3-2)}$$

If the abscissa values of the data points are spaced at equal intervals, then the trapezoidal rule simplifies to

$$I = \frac{h}{2} [(y_1 + y_2) + (y_2 + y_3) + \ldots + (y_{n-1} + y_n)]$$

$$\Downarrow$$

Trapezoidal Rule—Equal Intervals:
$$\int_{x_1}^{x_n} y(x) \, dx \approx \frac{h}{2} \left[y_1 + y_n + 2 \sum_{j=2}^{n-1} y_j \right] \qquad \text{(A.3-3)}$$

where h is the distance between the x values of adjacent data points. Observe that to use the trapezoidal rule, you need not plot anything—simply substitute the tabulated data into Equation A.3-2 or (for equal spacing) Equation A.3-3.

The trapezoidal rule is an approximation, as are all quadrature formulas. Figure A.3-1 illustrates the nature of the error introduced by its use. The integral to be

FIGURE A3-1

Graphical illustration of the trapezoidal rule

evaluated

$$I = \int_a^b y(x)\,dx$$

is the area under the dashed curve of Figure A.3-1, while the trapezoidal rule, Equation A.3-2, would yield the area under the straight-line segments, which could be quite different. Note also, however, that if there were many more data points in the interval between a and b, the approximating series of lines would follow the dashed curve much more closely, and the estimate of the integral would accordingly be more accurate.

A.3.3 Simpson's Rule

A second and more accurate quadrature formula is the one most often used. It is applicable only to an odd number of equally spaced data points and is based on fitting and integrating parabolic functions between successive groups of three points.

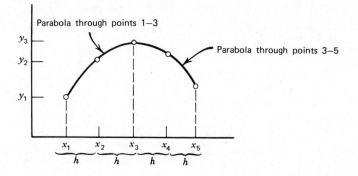

It can be shown after a fair amount of algebra that the area under a parabola through equally spaced points (x_1, y_1), (x_2, y_2), and (x_3, y_3) is

$$I_o = \frac{h}{3}(y_1 + 4y_2 + y_3)$$

where h is the interval between successive x values. Consequently, the area under a series of such parabolas fitted to n equally spaced points is

$$I = \frac{h}{3}[(y_1 + 4y_2 + y_3) + (y_3 + 4y_4 + y_5) + \ldots + (y_{n-2} + 4y_{n-1} + y_n)]$$

$$\Downarrow$$

Simpson's Rule:
$$\int_{x_1}^{x_n} y(x)\,dx \approx \frac{h}{3}\left[y_1 + y_n + 4\sum_{\substack{j=2, \\ 4,\ldots, \\ n-1}} y_j + 2\sum_{\substack{j=3, \\ 5,\ldots, \\ n-2}} y_i \right] \tag{A.3-4}$$

If you happen to have an even number of data points, you may integrate over all but the first or last subinterval (omit whichever subinterval contributes least to the integral) using Simpson's rule, and over the remaining subinterval using the trapezoidal rule.

EXAMPLE A.3-1 Simpson's Rule

The heat capacity of a gas is tabulated at a series of temperatures

$T(°C)$	20	50	80	110	140	170	200	230
$C_p(J/mol\cdot°C)$	28.95	29.13	29.30	29.48	29.65	29.82	29.99	30.16

Calculate the change in enthalpy for 3 mol of this gas going from $20°C$ to $230°C$.

Solution
$$\Delta H(J) = n \int_{20}^{230} C_p\,dT$$

The data points are evenly spaced in the independent variable (T), so that Simpson's rule can be applied to the integration, but since there are an even number of points the trapezoidal rule must be applied over the first or last temperature interval. Since C_p increases with temperature, we will apply the less accurate trapezoidal rule to the interval from $20°C$ to $50°C$, and use Simpson's rule between $50°C$ and $230°C$. If $\Delta T(=30°C)$ is the temperature interval between data points, Equations A.3-3 and A.3-4 yield

$$\int_{T_1}^{T_a} C_p\,dT \approx \frac{\Delta T}{2}(C_{p1} + C_{p2}) + \frac{\Delta T}{3}[C_{p2} + C_{p8} + 4(C_{p3} + C_{p5} + C_{p7})$$
$$+ 2(C_{p4} + C_{p6})]$$
$$= 6208\ J/mol$$
$$\Downarrow$$
$$\Delta H = (3\ mol)(6208\ J/mol) = \boxed{18\ 620\ J}$$

The trapezoidal rule is exact if the function to be integrated is in fact linear in each interval between data points, while Simpson's rule is exact if the real function is parabolic or cubic in each interval. (The latter point is far from obvious; for proof, see any reference on numerical analysis.)

A.3.4 Numerical Integration of Analytical Functions

When you have an analytical expression for an integrand $y(x)$ but you cannot perform the integration from $x = a$ to $x = b$ analytically, the procedure is to evaluate y at a series of values of x from a to b—that is, to generate a data table—and then to use a quadrature formula such as Simpson's rule to estimate the integral. Now, however, you have the choice of the number of evaluations of $y(x)$ to make.

As a rule, the accuracy of a quadrature formula increases with the number of points in the interval of integration, but so does the required computation time. Choosing the number of points to provide a suitable combination of accuracy and low computation time can be done using sophisticated numerical analysis techniques, but simple trial and error often suffices very well. A common procedure is to evaluate the integral using (say) 9 points, then 17, then 33, and so on ($n_{new} = 2n_{old} - 1$), until successively calculated values agree within a specified tolerance. The last value should be a good approximation to the exact value of the integral.

TEST YOURSELF

Suppose $f(x) = x^3 + 4$. Evaluate $\int_0^4 f(x)\, dx$:

1. Analytically.
2. Using the trapezoidal rule, with points at $x = 0, 1, 2, 3, 4$.
3. Using Simpson's rule, with points at $x = 0, 1, 2, 3, 4$. Explain the relationship between the answers to 1 and 3.

Appendix B
Physical Property Tables

TABLE B.1
Selected Physical Property Data[a]

Compound	Formula	Mol. Wt.	SG (20°/4°)	T_m(°C)[b]	$\Delta\hat{H}_m(T_m)$[c,j] kJ/mol	T_b(°C)[d]	$\Delta\hat{H}_v(T_b)$[e,j] kJ/mol	T_c(K)[f]	P_c(atm)[g]	$(\Delta\hat{H}°_f)$[h,j] kJ/mol	$(\Delta\hat{H}°_c)$[i,j] kJ/mol
Acetaldehyde	CH_3CHO	44.05	$0.783^{18°}$	−123.7	—	20.2	25.1	461.0	—	−166.2(g)	−1192.4(g)
Acetic acid	CH_3COOH	60.05	1.049	16.6	12.09	118.2	24.39	594.8	57.1	−486.18(l) −438.15(g)	−871.69(l) −919.73(g)
Acetone	C_3H_6O	58.08	0.791	−95.0	5.69	56.0	30.2	508.0	47.0	−248.2(l) −216.7(g)	−1785.7(l) −1821.4(g)
Acetylene	C_2H_2	26.04	—	—	—	−81.5	17.6	309.5	61.6	+226.75(g)	−1299.6(g)
Ammonia	NH_3	17.03	—	−77.8	5.653	−33.43	23.351	405.5	111.3	−67.20(l) −46.19(g)	−382.58(g)
Ammonium hydroxide	NH_4OH	35.03	—	—	—	—	—	—	—	−366.48(aq)	—
Ammonium nitrate	NH_4NO_3	80.05	$1.725^{25°}$	169.6	5.4	Decomposes at 210°C				−365.14(c) −339.36(aq)	—
Ammonium sulfate	$(NH_4)_2SO_4$	132.14	1.769	513	—	Decomposes at 513°C after melting				−1179.3(c) −1173.1(aq)	—
Aniline	C_6H_7N	93.12	1.022	−6.3	—	184.2	—	699	52.4	—	—
Benzaldehyde	C_6H_5CHO	106.12	1.046	−26.0	—	179.0	38.40	—	—	−88.83(l) −40.04(g)	−3520.0(l)
Benzene	C_6H_6	78.11	0.879	5.53	9.837	80.10	30.765	562.6	48.6	+48.66(l) +82.93(g)	−3267.6(l) −3301.5(g)

[a] Adapted in part from D. M. Himmelblau, *Basic Principles and Calculations in Chemical Engineering*, 3rd Edition, © 1974, Tables D.1 and F.1. Adapted by permission of Prentice-Hall, Inc., Englewood Cliffs, N. J.
[b] Melting point at 1 atm.
[c] Heat of fusion at T_m and 1 atm.
[d] Boiling point at 1 atm.
[e] Heat of vaporization at T_b and 1 atm.
[f] Critical temperature.
[g] Critical pressure.
[h] Heat of formation at 25°C and 1 atm.
[i] Heat of combustion at 25°C and 1 atm. Standard states of products are $CO_2(g)$, $H_2O(l)$, $SO_2(g)$, HCl(aq), and $N_2(g)$. To calculate $\Delta\hat{H}_c°$ with $H_2O(g)$ as a product, add $44.01n_w$ to the tabulated value, where n_w = moles H_2O formed/mole fuel burned.
[j] To convert $\Delta\hat{H}$ to kcal/mol, divide given value by 4.184; to convert to Btu/lb-mole, multiply by 430.28.

(continued)

TABLE B.1 (Continued)

Compound	Formula	Mol. Wt.	SG (20°/4°)	T_m(°C)[b]	$\Delta\hat{H}_m(T_m)$[c,j] kJ/mol	T_b(°C)[d]	$\Delta\hat{H}_v(T_b)$[c,j] kJ/mol	$T_c(K)$[f]	P_c(atm)[g]	$(\Delta\hat{H}_f°)$[a,j] kJ/mol	$(\Delta\hat{H}_c°)$[i,j] kJ/mol
Benzoic acid	$C_7H_6O_2$	122.12	1.266$^{15°}$	122.2	—	249.8	—	—	—	—	−3226.7(g)
Benzyl alcohol	C_7H_8O	108.13	1.045	−15.4	—	205.2	—	—	—	—	−3741.8(l)
Bromine	Br_2	159.83	3.119	−7.4	10.8	58.6	31.0	584	102	0(l)	—
1,2-Butadiene	C_4H_6	54.09	—	−136.5	—	10.1	—	446	—	—	—
1,3-Butadiene	C_4H_6	54.09	—	−109.1	—	−4.6	—	425	42.7	—	—
n-Butane	C_4H_{10}	58.12	—	−138.3	4.661	−0.6	22.305	425.17	37.47	−147.0(l) / −124.7(g)	−2855.6(l) / −2878.5(g)
Isobutane	C_4H_{10}	58.12	—	−159.6	4.540	−11.73	21.292	408.1	36.0	−158.4(l) / −134.5(g)	−2849.0(l) / −2868.8(g)
1-Butene	C_4H_8	56.10	—	−185.3	3.8480	−6.25	21.916	419.6	39.7	+1.17(g)	−2718.6(g)
Calcium carbide	CaC_2	64.10	2.22$^{18°}$	2300						−62.76(c)	—
Calcium carbonate	$CaCO_3$	100.09	2.93			Decomposes at 825°C				−1206.9(c)	—
Calcium chloride	$CaCl_2$	110.99	2.152$^{15°}$	782	28.37	>1600	—	—	—	−794.96(c)	—
Calcium hydroxide	$Ca(OH)_2$	74.10	2.24		(−H_2O at 580°C)					−986.59(c)	—
Calcium oxide	CaO	56.08	3.32	2570	50	2850	—	—	—	−635.6(c)	—
Calcium phosphate	$Ca_3(PO_4)_2$	310.19	3.14	1670	—	—	—	—	—	−4138(c)	—
Calcium silicate	$CaSiO_3$	116.17	2.915	1530	48.62	—	—	—	—	−1584(c)	—
Calcium sulfate	$CaSO_4$	136.15	2.96	—	—	—	—	—	—	−1432.7(c) / −1450.4(aq)	—
Calcium sulfate (gypsum)	$CaSO_4 \cdot 2H_2O$	172.18	2.32		(−1.5 H_2O at 128°C)					−2021(c)	—
Carbon (graphite)	C	12.010	2.26	3600	46.0	4200	—	—	—	0(c)	−393.51(c)
Carbon dioxide	CO_2	44.01	—	−56.6 at 5.2 atm	8.33	(Sublimes at −78°C)	—	304.2	72.9	−412.9(l) / −393.5(g)	—

Carbon disulfide	CS$_2$	76.14	$1.261^{22°/20°}$	-112.1	4.39	46.25	26.8	552.0	78.0	+87.9(l) +115.3(g)	-1075.2(l) -1102.6(g)
Carbon monoxide	CO	28.01	—	-205.1	0.837	-191.5	6.042	133.0	34.5	-110.52(g)	-282.99(g)
Carbon tetrachloride	CCl$_4$	153.84	1.595	-22.9	2.51	76.7	30.0	556.4	45.0	-139.5(l) -106.7(g)	-352.2(l) -385.0(g)
Chlorine	Cl$_2$	70.91	—	-101.00	6.406	-34.06	20.4	417.0	76.1	0(g)	—
Chlorobenzene	C$_6$H$_5$Cl	112.56	1.107	-45	—	132.10	36.5	632.4	44.6	—	—
Chloroethane	C$_2$H$_5$Cl	See ethyl chloride									
Chloroform	CHCl$_3$	119.39	1.489	-63.7	13.01	61.0	—	536.0	54.0	-131.8(l)	-373(l)
Copper	Cu	63.54	8.92	1083	—	2595	304.6	—	—	0(c)	—
Cupric sulfate	CuSO$_4$	159.61	$3.606^{15°}$	Decomposes > 600°C	—	—	—	—	—	-769.9(c) -843.1(aq)	—
Cyclohexane	C$_6$H$_{12}$	84.16	0.779	6.7	2.677	80.7	30.1	553.7	40.4	-156.2(l) -123.1(g)	-3919.9(l) -3953.0(g)
Cyclopentane	C$_5$H$_{10}$	70.13	0.745	-93.4	0.609	49.3	27.30	511.8	44.55	-105.9(l) -77.2(g)	-3290.9(l) -3319.5(g)
n-Decane	C$_{10}$H$_{22}$	142.28	0.730	-29.9	—	173.8	—	619.0	20.8	-249.7(l)	-6778.3(l) -6829.7(g)
Diethyl ether	(C$_2$H$_5$)$_2$O	74.12	$0.708^{25°}$	-116.3	7.30	34.6	26.05	467	35.6	-272.8(l)	-2726.7(l)
Ethane	C$_2$H$_6$	30.07	—	-183.3	2.859	-88.6	14.72	305.4	48.2	-84.67(g)	-1559.9(g)
Ethyl acetate	C$_4$H$_8$O$_2$	88.10	0.901	-83.8	—	77.0	—	523.1	37.8	-463.2(l)	-2246.4(l)
Ethyl alcohol (Ethanol)	C$_2$H$_5$OH	46.07	0.789	-114.6	5.021	78.5	38.58	516.3	63.0	-277.63(l) -235.31(g)	-1366.91(l) -1409.25(g)
Ethyl benzene	C$_8$H$_{10}$	106.16	0.867	-94.97	9.163	136.2	35.98	619.7	37.0	-12.46(l) +29.79(g)	-4564.9(l) -4607.1(g)
Ethyl bromide	C$_2$H$_5$Br	108.98	1.460	-119.1	—	38.2	24.7	504	61.5	-54.4(g)	—
Ethyl chloride	C$_2$H$_5$Cl	64.52	$0.903^{15°}$	-138.3	4.452	13.1	34.27	460.4	52.0	-105.0(g)	—
3-Ethyl hexane	C$_8$H$_{18}$	114.22	0.717	—	—	118.5	—	567.0	26.4	-250.5(l) -210.9(g)	-5407.1(l) -5509.8(g)
Ethylene	C$_2$H$_4$	28.05	—	-169.2	3.350	-103.7	13.54	283.1	50.5	+52.28(g)	-1410.99(g)
Ethylene glycol	C$_2$H$_6$O$_2$	62.07	$1.113^{19°}$	-13	11.23	197.2	56.9	—	—	-451.5(l) -387.1(g)	-1179.5(l)
Ferric oxide	Fe$_2$O$_3$	159.70	5.12	—	—	—	—	—	—	-822.2(c)	—
Ferrous oxide	FeO	71.85	5.7	—	—	Decomposes at 1560°C	—	—	—	-266.5(c)	—

(continued)

TABLE B.1 (Continued)

Compound	Formula	Mol. Wt.	SG (20°/4°)	T_m(°C)[b]	$\Delta\hat{H}_m(T_m)$[c,j] kJ/mol	T_b(°C)[d]	$\Delta\hat{H}_v(T_b)$[e,j] kJ/mol	T_c(K)[f]	P_c(atm)[g]	$(\Delta\hat{H}_f^\circ)$[h,j] kJ/mol	$(\Delta\hat{H}_c^\circ)$[i,j] kJ/mol
Ferrous sulfide	FeS	87.92	4.84	1193	—					-95.1(c)	—
Formaldehyde	H_2CO	30.03	0.815$^{-20°}$	-92	—	-19.3	24.48	—	—	-115.90(g)	-563.46(l)
Formic acid	CH_2O_2	46.03	1.220	8.30	12.68	100.5	22.25	—	—	-409.2(l) -362.6(g)	-262.8(l)
Glycerol	$C_3H_8O_3$	92.09	1.260$^{50°}$	18.20	18.30	290.0	—	—	—	-665.9(l)	-1661.1(l)
Helium	He	4.00	—	-269.7	0.02	-268.9	0.084	5.26	2.26	0(g)	—
n-Heptane	C_7H_{16}	100.20	0.684	-90.59	14.03	98.43	31.69	540.2	27.0	-224.4(l) -187.8(g)	-4816.9(l) -4853.5(g)
n-Hexane	C_6H_{14}	86.17	0.659	-95.32	13.03	68.74	28.85	507.9	29.9	-198.8(l) -167.2(g)	-4163.1(l) -4194.8(g)
Hydrogen	H_2	2.016	—	-259.19	0.12	-252.76	0.904	33.3	12.8	0(g)	-285.84(g)
Hydrogen bromide	HBr	80.92	—	-86	—	-67	—	—	—	-36.23(g)	—
Hydrogen chloride	HCl	36.47	—	-114.2	1.99	-85.0	16.1	324.6	81.5	-92.31(g)	—
Hydrogen cyanide	HCN	27.03	—	-14	—	-26	—	—	—	+130.54(g)	—
Hydrogen fluoride	HF	20.0	—	-83	—	20	—	503.2	—	-268.6(g) -316.9(aq, 200)	—
Hydrogen sulfide	H_2S	34.08	—	-85.5	2.38	-60.3	18.67	373.6	88.9	-19.96(g)	-562.59(g)
Iodine	I_2	253.8	4.93	113.3	15.1	184.2	—	826.0	—	0(c)	—
Iron	Fe	55.85	7.7	1535	15.1	2800	354.0	—	—	0(c)	—
Lead	Pb	207.21	11.337$^{20°/20°}$	327.4	5.10	1750	179.9	—	—	0(c)	—
Lead oxide	PbO	223.21	9.5	886	11.7	1472	213	—	—	-219.2(c)	—
Magnesium	Mg	24.32	1.74	650	9.2	1120	131.8	—	—	0(c)	—
Magnesium chloride	$MgCl_2$	95.23	2.325$^{25°}$	714	43.1	1418	136.8	—	—	-641.8(c)	—
Magnesium hydroxide	$Mg(OH)_2$	58.34	2.4	Decomposes at 350°C							—
Magnesium oxide	MgO	40.32	3.65	2900	77.4	3600	—	—	—	-601.8(c)	—

Mercury	Hg	200.61	13.546	−38.87	356.9	—	—	—	—	0(c)	—
Methane	CH_4	16.04	—	−182.5	−161.5	0.94	8.179	190.70	45.8	−74.85(g)	−890.36(g)
Methyl acetate	$C_3H_6O_2$	74.08	0.933	−98.9	57.1	—	—	506.7	46.30	−409.4(l)	−1595(l)
Methyl alcohol (Methanol)	CH_3OH	32.04	0.792	−97.9	64.7	3.167	35.27	513.20	78.50	−238.6(l) / −201.2(g)	−726.6(l) / −764.0(g)
Methyl amine	CH_5N	31.06	$0.699^{-11°}$	−92.7	−6.9	—	—	429.9	73.60	−28.0(g)	−1071.5(l)
Methyl chloride	CH_3Cl	50.49	—	−97.9	−24	—	—	416.1	65.80	−81.92(g)	—
Methyl ethyl ketone	C_4H_8O	72.10	0.805	−87.1	78.2	—	32.0	—	—	—	−2436(l)
Naphthalene	$C_{10}H_8$	128.16	1.145	80.0	217.8	—	—	—	—	—	−5157(g)
Nickel	Ni	58.69	8.90	1452	2900	—	—	—	—	0(c)	—
Nitric acid	HNO_3	63.02	1.502	−41.6	86	10.47	30.30	—	—	−173.23(l) / −206.57(aq)	—
Nitrobenzene	$C_6H_5O_2N$	123.11	1.203	5.5	210.7	—	—	—	—	—	−3092.8(l)
Nitrogen	N_2	28.02	—	−210.0	−195.8	0.720	5.577	126.20	33.5	0(g)	—
Nitrogen dioxide	NO_2	46.01	—	−9.3	21.3	7.335	14.73	431.0	100.0	+33.8(g)	—
Nitric oxide	NO	30.01	$1.63^{18°}$	−163.6	−151.8	2.301	13.78	179.20	65.0	+90.37(g)	—
Nitrogen pentoxide	N_2O_5	108.02	—	30	47	—	—	—	—	—	—
Nitrogen tetraoxide	N_2O_4	92.0	1.448	−9.5	21.1	—	—	431.0	99.0	+9.3(g)	—
Nitrous oxide	N_2O	44.02	$1.226^{-89°}$	−91.1	−88.8	—	—	309.5	71.70	+81.5(g)	—
n-Nonane	C_9H_{20}	128.25	0.718	−53.8	150.6	—	—	595	23.0	−229.0(l)	−6124.5(l) / −6171.0(g)
n-Octane	C_8H_{18}	114.22	0.703	−57.0	125.5	—	—	595.0	22.5	−249.9(l) / −208.4(g)	−5470.7(l) / −5512.2(g)
Oxalic acid	$C_2H_2O_4$	90.04	1.90	Decomposes at 186°C			—	—	—	−826.8(c)	−251.9(s)
Oxygen	O_2	32.00	—	−218.75	−182.97	0.444	6.82	154.4	49.7	0(g)	—
n-Pentane	C_5H_{12}	72.15	$0.63^{18°}$	−129.6	36.07	8.393	25.77	469.80	33.3	−173.0(l) / −146.4(g)	−3509.5(l) / −3536.1(g)

(continued)

TABLE B.1 (Continued)

Compound	Formula	Mol. Wt.	SG (20°/4°)[b]	T_m(°C)[b]	$\Delta \hat{H}_m(T_m)$[c,j] kJ/mol	T_b(°C)[d]	$\Delta \hat{H}_v(T_b)$[c,j] kJ/mol	T_c(K)[f]	P_c(atm)[g]	$(\Delta \hat{H}^\circ_f)$[h,j] kJ/mol	$(\Delta \hat{H}^\circ_c)$[i,j] kJ/mol
Isopentane	C_5H_{12}	72.15	0.62[19°]	−160.1	—	27.7	—	461.00	32.9	−179.3(l), −152.0(g)	−3507.5(l), −3529.2(g)
1-Pentene	C_5H_{10}	70.13	0.641	−165.2	4.94	29.97	—	474	39.9	−20.9(g)	−3375.8(g)
Phenol	C_6H_5OH	94.11	1.071[25°]	42.5	11.43	181.4	—	692.1	60.5	−158.1(l), −90.8(g)	−3063.5(s)
Phosphoric acid	H_3PO_4	98.00	1.834[18°]	42.3	10.54	($-\frac{1}{2}H_2O$ at 213°C)		—	—	−1281.1(c), −1278.6(aq, 1H₂O)	—
Phosphorus (red)	P_4	123.90	2.20	590[43 atm]	81.17	Ignites in air, 725°C		—	—	−17.6(c)	—
Phosphorus (white)	P_4	123.90	1.82	44.2	2.51	280	49.71	—	—	0(c)	—
Phosphorus pentoxide	P_2O_5	141.95	2.387			Sublimes at 250°C		—	—	−1506.2(c)	—
Propane	C_3H_8	44.09	—	−187.69	3.52	−42.07	18.77	369.9	42.0	−119.8(l), −103.8(g)	−2204.0(l), −2220.0(g)
Propylene	C_3H_6	42.08	—	−185.2	3.00	−47.70	18.42	365.1	45.4	+20.41(g)	−2058.4(g)
n-Propyl alcohol	C_3H_5OH	60.09	0.804	−127	—	97.04	—	536.7	49.95	−300.70(l), −255.2(g)	−2010.4(l), −2068.6(g)
Isopropyl alcohol	C_3H_8O	60.09	0.785	−89.7	—	82.24	—	508.8	53.0	−310.9(l)	−1986.6(l)
n-Propyl benzene	C_9H_{12}	120.19	0.862	−99.50	8.54	159.2	38.24	638.7	31.3	−38.40(l), +7.82(g)	−5218.2(l), −5264.48(g)
Silicon dioxide	SiO_2	60.09	2.25	1710	14.2	2230	—	—	—	−851.0(c)	—
Sodium bicarbonate	$NaHCO_3$	84.01	2.20	Decomposes at 270°C				—	—	−945.6(c)	—
Sodium bisulfate	$NaHSO_4$	120.07	2.742	—	—	—	—	—	—	−1126.3(c)	—
Sodium carbonate	Na_2CO_3	105.99	2.533	Decomposes at 854°C				—	—	−1130.9(c)	—
Sodium chloride	$NaCl$	58.45	2.163	808	28.5	1465	170.7	—	—	−411.0(c)	—

Sodium cyanide	NaCN	49.01	—	562	16.7	1497	155	—	—	-89.79(c)	—
Sodium hydroxide	NaOH	40.00	2.130	319	8.34	1390	—	—	—	-426.6(c) -469.4(aq) -466.7(c)	—
Sodium nitrate	NaNO$_3$	85.00	2.257	310	15.9	Decomposes at 380°C		—	—		—
Sodium nitrite	NaNO$_2$	69.00	2.168^0°	271	—	Decomposes at 320°C		—	—	-359.4(c)	—
Sodium sulfate	Na$_2$SO$_4$	142.05	2.698	890	24.3	—	—	—	—	-1384.5(c)	—
Sodium sulfide	Na$_2$S	78.05	1.856	950	6.7	—	—	—	—	-373.2(c)	—
Sodium sulfite	Na$_2$SO$_3$	126.05	2.63315°	Decomposes				—	—	-1090.3(c)	—
Sodium thiosulfate	Na$_2$S$_2$O$_3$	158.11	1.667	—	—	—	—	—	—	-1117.1(c)	—
Sulfur (rhombic)	S$_8$	256.53	2.07	113	10.04	444.6	83.7	—	—	0(c)	—
Sulfur (monoclinic)	S$_8$	256.53	1.96	119	14.17	444.6	83.7	—	—	+0.30(c)	—
Sulfur dioxide	SO$_2$	64.07	—	-75.48	7.402	-10.02	24.91	430.7	77.8	-296.90(g)	—
Sulfur trioxide	SO$_3$	80.07	—	16.84	25.48	43.3	41.80	491.4	83.8	-395.18(g)	—
Sulfuric acid	H$_2$SO$_4$	98.08	1.83418°	10.35	9.87	Decomposes at 340°C		—	—	-811.32(l) -907.51(aq)	—
Toluene	C$_7$H$_8$	92.13	0.866	-94.99	6.619	110.62	33.47	593.9	40.3	+12.00(l) +50.00(g)	-3909.9(l) -3947.9(g)
Water	H$_2$O	18.016	1.00^4°	0.00	6.0095	100.00	40.656	647.4	218.3	-285.84(l) -241.83(g)	—
m-Xylene	C$_8$H$_{10}$	106.16	0.864	-47.87	11.569	139.10	36.40	619	34.6	-25.42(l) +17.24(g)	-4551.9(l) -4594.5(g)
o-Xylene	C$_8$H$_{10}$	106.16	0.880	-25.18	13.598	144.42	36.82	631.5	35.7	-24.44(l) +18.99(g)	-4552.9(l) -4596.3(g)
p-Xylene	C$_8$H$_{10}$	106.16	0.861	13.26	17.11	138.35	36.07	618	33.9	-24.43(l) 17.95(g)	-4552.9(l) -4595.2(g)
Zinc	Zn	65.38	7.140	419.5	6.674	907	114.77	—	—	0(c)	—

TABLE B.2
Heat Capacitiesa

Form 1: $C_p(\text{J/mol}\cdot{}^\circ\text{C})$ or $(\text{J/mol}\cdot\text{K}) = a + bT + cT^2 + dT^3$

Form 2: $C_p(\text{J/mol}\cdot{}^\circ\text{C})$ or $(\text{J/mol}\cdot\text{K}) = a + bT + cT^{-2}$

Example: $(C_p)_{\text{acetone(g)}} = 71.96 + (20.10 \times 10^{-2})T - (12.78 \times 10^{-5})T^2 + (34.76 \times 10^{-9})T^3$, where T is in °C.

Note: The formulas for gases are strictly applicable at pressures low enough for the ideal gas law to apply.

Compound	Formula	Mol. Wt.	State	Form	Temp. Unit	a	$b \cdot 10^2$	$c \cdot 10^5$	$d \cdot 10^9$	Range (Units of T)
Acetone	CH$_3$COCH$_3$	58.08	l	1	°C	123.0	18.6			−30–60
			g	1	°C	71.96	20.10	−12.78	34.76	0–1200
Acetylene	C$_2$H$_2$	26.04	g	1	°C	42.43	6.053	−5.033	18.20	0–1200
Air		29.0	g	1	°C	28.94	0.4147	0.3191	−1.965	0–1500
			g	1	K	28.09	0.1965	0.4799	−1.965	273–1800
Ammonia	NH$_3$	17.03	g	1	°C	35.15	2.954	0.4421	−6.686	0–1200
Ammonium sulfate	(NH$_4$)$_2$SO$_4$	132.15	c	2	K	215.9				275–328
Benzene	C$_6$H$_6$	78.11	l	1	K	62.55	23.4			279–350
			g	1	°C	74.06	32.95	−25.20	77.57	0–1200
Isobutane	C$_4$H$_{10}$	58.12	g	1	°C	89.46	30.13	−18.91	49.87	0–1200
n-Butane	C$_4$H$_{10}$	58.12	g	1	°C	92.30	27.88	−15.47	34.98	0–1200
Isobutene	C$_4$H$_8$	56.10	g	1	°C	82.88	25.64	−17.27	50.50	0–1200
Calcium carbide	CaC$_2$	64.10	c	2	K	68.62	1.19	-8.66×10^{10}	—	298–720
Calcium carbonate	CaCO$_3$	100.09	c	2	K	82.34	4.975	-12.87×10^{10}	—	273–1033
Calcium hydroxide	Ca(OH)$_2$	74.10	c	1	K	89.5				276–373
Calcium oxide	CaO	56.08	c	2	K	41.84	2.03	-4.52×10^{10}		273–1173
Carbon	C	12.01	c	2	K	11.18	1.095	-4.891×10^{10}		273–1373
Carbon dioxide	CO$_2$	44.01	g	1	°C	36.11	4.233	−2.887	7.464	0–1500
Carbon monoxide	CO	28.01	g	1	°C	28.95	0.4110	0.3548	−2.220	0–1500
Carbon tetrachloride	CCl$_4$	153.84	l	1	K	93.39	12.98			273–343
Chlorine	Cl$_2$	70.91	g	1	°C	33.60	1.367	−1.607	6.473	0–1200
Copper	Cu	63.54	c	1	K	22.76	0.6117			273–1357

Cumene (Isopropyl benzene)	C_9H_{12}	120.19	g	1	°C	139.2	53.76	−39.79	120.5	0–1200
Cyclohexane	C_6H_{12}	84.16	g	1	°C	94.140	49.62	−31.90	80.63	0–1200
Cyclopentane	C_5H_{10}	70.13	g	1	°C	73.39	39.28	−25.54	68.66	0–1200
Ethane	C_2H_6	30.07	g	1	°C	49.37	13.92	−5.816	7.280	0–1200
Ethyl alcohol (Ethanol)	C_2H_5OH	46.07	l	1	°C	103.1				0
			l	1	°C	158.8				100
			g	1	°C	61.34	15.72	−8.749	19.83	0–1200
Ethylene	C_2H_4	28.05	g	1	°C	+40.75	11.47	−6.891	17.66	0–1200
Ferric oxide	Fe_2O_3	159.70	c	2	K	103.4	6.711	-17.72×10^{10}	—	273–1097
Formaldehyde	CH_2O	30.03	g	1	°C	34.28	4.268	0.0000	−8.694	0–1200
Helium	He	4.00	g	1	°C	20.8				All
n-Hexane	C_6H_{14}	86.17	l	1	°C	216.3				20–100
			g	1	°C	137.44	40.85	−23.92	57.66	0–1200
Hydrogen	H_2	2.016	g	1	°C	28.84	0.00765	0.3288	−0.8698	0–1500
Hydrogen bromide	HBr	80.92	g	1	°C	29.10	−0.0227	0.9887	−4.858	0–1200
Hydrogen chloride	HCl	36.47	g	1	°C	29.13	−0.1341	0.9715	−4.335	0–1200
Hydrogen cyanide	HCN	27.03	g	1	°C	35.3	2.908	1.092		0–1200
Hydrogen sulfide	H_2S	34.08	g	1	°C	33.51	1.547	0.3012	−3.292	0–1500
Magnesium chloride	$MgCl_2$	95.23	c	1	K	72.4	1.58			273–991
Magnesium oxide	MgO	40.32	c	2	K	45.44	0.5008	-8.732×10^{10}		273–2073
Methane	CH_4	16.04	g	1	°C	34.31	5.469	0.3661	−11.00	0–1200
			g	1	K	19.87	5.021	1.268	−11.00	273–1500
Methyl alcohol (Methanol)	CH_3OH	32.04	l	1	°C	75.86				0
			l	1	°C	82.59				40
			g	1	°C	42.93	8.301	−1.87	−8.03	0–700
Methyl cyclohexane	C_7H_{14}	98.18	g	1	°C	121.3	56.53	−37.72	100.8	0–1200
Methyl cyclopentane	C_6H_{12}	84.16	g	1	°C	98.83	45.857	−30.44	83.81	0–1200
Nitric acid	HNO_3	63.02	l	1	°C	110.0				25
Nitric oxide	NO	30.01	g	1	°C	29.50	0.8188	−0.2925	0.3652	0–3500

[a] Adapted in part from D. M. Himmelblau, *Basic Principles and Calculations in Chemical Engineering*, 3rd Edition, © 1974, Table E.1. Adapted by permission of Prentice-Hall, Inc., Englewood Cliffs, N.J.

(continued)

TABLE B.2 (Continued)

Compound	Formula	Mol. Wt.	State	Form	Temp. Unit	a	$b \cdot 10^2$	$c \cdot 10^5$	$d \cdot 10^9$	Range (Units of T)
Nitrogen	N_2	28.02	g	1	°C	29.00	0.2199	0.5723	−2.871	0–1500
Nitrogen dioxide	NO_2	46.01	g	1	°C	36.07	3.97	−2.88	7.87	0–1200
Nitrogen tetraoxide	N_2O_4	92.02	g	1	°C	75.7	12.5	−11.3		0–300
Nitrous oxide	N_2O	44.02	g	1	°C	37.66	4.151	−2.694	10.57	0–1200
Oxygen	O_2	32.00	g	1	°C	29.10	1.158	−0.6076	1.311	0–1500
n-Pentane	C_5H_{12}	72.15	l	1	°C	155.4	43.68			0–36
			g	1	°C	114.8	34.09	−18.99	42.26	0–1200
Propane	C_3H_8	44.09	g	1	°C	68.032	22.59	−13.11	31.71	0–1200
Propylene	C_3H_6	42.08	g	1	°C	59.580	17.71	−10.17	24.60	0–1200
Sodium carbonate	Na_2CO_3	105.99	c	1	K	121				288–371
Sodium carbonate	$Na_2CO_3 \cdot 10H_2O$	286.15	c	1	K	535.6				298
Sulfur	S	32.07	c (Rhombic)	1	K	15.2	2.68			273–368
			c (Monoclinic)	1	K	18.3	1.84			368–392
Sulfuric acid	H_2SO_4	98.08	l	1	°C	139.1	15.59			10–45
Sulfur dioxide	SO_2	64.07	g	1	°C	38.91	3.904	−3.104	8.606	0–1500
Sulfur trioxide	SO_3	80.07	g	1	°C	48.50	9.188	−8.540	32.40	0–1000
Toluene	C_7H_8	92.13	l	1	°C	148.8				0
			l	1	°C	181.2				100
			g	1	°C	94.18	38.00	−27.86	80.33	0–1200
Water	H_2O	18.016	l	1	°C	75.4				0–100
			g	1	°C	33.46	0.6880	0.7604	−3.593	0–1500

TABLE B.3
Vapor Pressure of Water[a]

p_v(mm Hg) versus $T(°C)$

Example: The vapor pressure of liquid water at 4.3°C is 6.230 mm Hg

$T(°C)$	0.0	0.1	0.2	0.3	0.4	0.5	0.6	0.7	0.8	0.9
Ice										
−14	1.361	1.348	1.336	1.324	1.312	1.300	1.288	1.276	1.264	1.253
−13	1.490	1.477	1.464	1.450	1.437	1.424	1.411	1.399	1.386	1.373
−12	1.632	1.617	1.602	1.588	1.574	1.559	1.546	1.532	1.518	1.504
−11	1.785	1.769	1.753	1.737	1.722	1.707	1.691	1.676	1.661	1.646
−10	1.950	1.934	1.916	1.899	1.883	1.866	1.849	1.833	1.817	1.800
−9	2.131	2.112	2.093	2.075	2.057	2.039	2.021	2.003	1.985	1.968
−8	2.326	2.306	2.285	2.266	2.246	2.226	2.207	2.187	2.168	2.149
−7	2.537	2.515	2.493	2.472	2.450	2.429	2.408	2.387	2.367	2.346
−6	2.765	2.742	2.718	2.695	2.672	2.649	2.626	2.603	2.581	2.559
−5	3.013	2.987	2.962	2.937	2.912	2.887	2.862	2.838	2.813	2.790
−4	3.280	3.252	3.225	3.198	3.171	3.144	3.117	3.091	3.065	3.039
−3	3.568	3.539	3.509	3.480	3.451	3.422	3.393	3.364	3.336	3.308
−2	3.880	3.848	3.816	3.785	3.753	3.722	3.691	3.660	3.630	3.599
−1	4.217	4.182	4.147	4.113	4.079	4.045	4.012	3.979	3.946	3.913
−0	4.579	4.542	4.504	4.467	4.431	4.395	4.359	4.323	4.287	4.252
Liquid water										
0	4.579	4.613	4.647	4.681	4.715	4.750	4.785	4.820	4.855	4.890
1	4.926	4.962	4.998	5.034	5.070	5.107	5.144	5.181	5.219	5.256
2	5.294	5.332	5.370	5.408	5.447	5.486	5.525	5.565	5.605	5.645
3	5.685	5.725	5.766	5.807	5.848	5.889	5.931	5.973	6.015	6.058
4	6.101	6.144	6.187	6.230	6.274	6.318	6.363	6.408	6.453	6.498
5	6.543	6.589	6.635	6.681	6.728	6.775	6.822	6.869	6.917	6.965
6	7.013	7.062	7.111	7.160	7.209	7.259	7.309	7.360	7.411	7.462
7	7.513	7.565	7.617	7.669	7.722	7.775	7.828	7.882	7.936	7.990
8	8.045	8.100	8.155	8.211	8.267	8.323	8.380	8.437	8.494	8.551
9	8.609	8.668	8.727	8.786	8.845	8.905	8.965	9.025	9.086	9.147

[a] From R. H. Perry and C. H. Chilton, Eds., *Chemical Engineers' Handbook*, 5th Edition, McGraw-Hill, New York, 1973, Tables 3-3 and 3-5. Reprinted by permission of McGraw-Hill Book Co.

(continued)

TABLE B.3 (Continued)

T(°C)	0.0	0.1	0.2	0.3	0.4	0.5	0.6	0.7	0.8	0.9
10	9.209	9.271	9.333	9.395	9.458	9.521	9.585	9.649	9.714	9.779
11	9.844	9.910	9.976	10.042	10.109	10.176	10.244	10.312	10.380	10.449
12	10.518	10.588	10.658	10.728	10.799	10.870	10.941	11.013	11.085	11.158
13	11.231	11.305	11.379	11.453	11.528	11.604	11.680	11.756	11.833	11.910
14	11.987	12.065	12.144	12.223	12.302	12.382	12.462	12.543	12.624	12.706
15	12.788	12.870	12.953	13.037	13.121	13.205	13.290	13.375	13.461	13.547
16	13.634	13.721	13.809	13.898	13.987	14.076	14.166	14.256	14.347	14.438
17	14.530	14.622	14.715	14.809	14.903	14.997	15.092	15.188	15.284	15.380
18	15.477	15.575	15.673	15.772	15.871	15.971	16.071	16.171	16.272	16.374
19	16.477	16.581	16.685	16.789	16.894	16.999	17.105	17.212	17.319	17.427
20	17.535	17.644	17.753	17.863	17.974	18.085	18.197	18.309	18.422	18.536
21	18.650	18.765	18.880	18.996	19.113	19.231	19.349	19.468	19.587	19.707
22	19.827	19.948	20.070	20.193	20.316	20.440	20.565	20.690	20.815	20.941
23	21.068	21.196	21.324	21.453	21.583	21.714	21.845	21.977	22.110	22.243
24	22.377	22.512	22.648	22.785	22.922	23.060	23.198	23.337	23.476	23.616
25	23.756	23.897	24.039	24.182	24.326	24.471	24.617	24.764	24.912	25.060
26	25.209	25.359	25.509	25.660	25.812	25.964	26.117	26.271	26.426	26.582
27	26.739	26.897	27.055	27.214	27.374	27.535	27.696	27.858	28.021	28.185
28	28.349	28.514	28.680	28.847	29.015	29.184	29.354	29.525	29.697	29.870
29	30.043	30.217	30.392	30.568	30.745	30.923	31.102	31.281	31.461	31.642
30	31.824	32.007	32.191	32.376	32.561	32.747	32.934	33.122	33.312	33.503
31	33.695	33.888	34.082	34.276	34.471	34.667	34.864	35.062	35.261	35.462
32	35.663	35.865	36.068	36.272	36.477	36.683	36.891	37.099	37.308	37.518
33	37.729	37.942	38.155	38.369	38.584	38.801	39.018	39.237	39.457	39.677
34	39.898	40.121	40.344	40.569	40.796	41.023	41.251	41.480	41.710	41.942
35	42.175	42.409	42.644	42.880	43.117	43.355	43.595	43.836	44.078	44.320
36	44.563	44.808	45.054	45.301	45.549	45.799	46.050	46.302	46.556	46.811
37	47.067	47.324	47.582	47.841	48.102	48.364	48.627	48.891	49.157	49.424
38	49.692	49.961	50.231	50.502	50.774	51.048	51.323	51.600	51.879	52.160

T(°C)	0.0	0.1	0.2	0.3	0.4	0.5	0.6	0.7	0.8	0.9
39	52.442	52.725	53.009	53.294	53.580	53.867	54.156	54.446	54.737	55.030
40	55.324	55.61	55.91	56.21	56.51	56.81	57.11	57.41	57.72	58.03
41	58.34	58.65	58.96	59.27	59.58	59.90	60.22	60.54	60.86	61.18
42	61.50	61.82	62.14	62.47	62.80	63.13	63.46	63.79	64.12	64.46
43	64.80	65.14	65.48	65.82	66.16	66.51	66.86	67.21	67.56	67.91
44	68.26	68.61	68.97	69.33	69.69	70.05	70.41	70.77	71.14	71.51
45	71.88	72.25	72.62	72.99	73.36	73.74	74.12	74.50	74.88	75.26
46	75.65	76.04	76.43	76.82	77.21	77.60	78.00	78.40	78.80	79.20
47	79.60	80.00	80.41	80.82	81.23	81.64	82.05	82.46	82.87	83.29
48	83.71	84.13	84.56	84.99	85.42	85.85	86.28	86.71	87.14	87.58
49	88.02	88.46	88.90	89.34	89.79	90.24	90.69	91.14	91.59	92.05

T(°C)	0.0	0.1	0.2	0.3	0.4	0.5	0.6	0.7	0.8	0.9
50	92.51	97.20	102.09	107.20	112.51	118.04	123.80	129.82	136.08	142.60
60	149.38	156.43	163.77	171.38	179.31	187.54	196.09	204.96	214.17	223.73
70	233.7	243.9	254.6	265.7	277.2	289.1	301.4	314.1	327.3	341.0
80	355.1	369.7	384.9	400.6	416.8	433.6	450.9	468.7	487.1	506.1

T(°C)	0.0	0.1	0.2	0.3	0.4	0.5	0.6	0.7	0.8	0.9
90	525.76	527.76	529.77	531.78	533.80	535.82	537.86	539.90	541.95	544.00
91	546.05	548.11	550.18	552.26	554.35	556.44	558.53	560.64	562.75	564.87
92	566.99	569.12	571.26	573.40	575.55	577.71	579.87	582.04	584.22	586.41
93	588.60	590.80	593.00	595.21	597.43	599.66	601.89	604.13	606.38	608.64
94	610.90	613.17	615.44	617.72	620.01	622.31	624.61	626.92	629.24	631.57
95	633.90	636.24	638.59	640.94	643.30	645.67	648.05	650.43	652.82	655.22
96	657.62	660.03	662.45	664.88	667.31	669.75	672.20	674.66	677.12	679.69
97	682.07	684.55	687.04	689.54	692.05	694.57	697.10	699.63	702.17	704.71
98	707.27	709.83	712.40	714.98	717.56	720.15	722.75	725.36	727.98	730.61
99	733.24	735.88	738.53	741.18	743.85	746.52	749.20	751.89	754.58	757.29
100	760.00	762.72	765.45	768.19	770.93	773.68	776.44	779.22	782.00	784.78
101	787.57	790.37	793.18	796.00	798.82	801.66	804.50	807.35	810.21	813.08

TABLE B.4
Properties of Saturated Steam: Temperature Table[a]

$T(°C)$	$P(bar)$	$\hat{V}(m^3/kg)$		$\hat{U}(kJ/kg)$		$\hat{H}(kJ/kg)$		
		Water	Steam	Water	Steam	Water	Evaporation	Steam
0.01	0.00611	0.001000	206.2	zero	2375.6	+0.0	2501.6	2501.6
2	0.00705	0.001000	179.9	8.4	2378.3	8.4	2496.8	2505.2
4	0.00813	0.001000	157.3	16.8	2381.1	16.8	2492.1	2508.9
6	0.00935	0.001000	137.8	25.2	2383.8	25.2	2487.4	2512.6
8	0.01072	0.001000	121.0	33.6	2386.6	33.6	2482.6	2516.2
10	0.01227	0.001000	106.4	42.0	2389.3	42.0	2477.9	2519.9
12	0.01401	0.001000	93.8	50.4	2392.1	50.4	2473.2	2523.6
14	0.01597	0.001001	82.9	58.8	2394.8	58.8	2468.5	2527.2
16	0.01817	0.001001	73.4	67.1	2397.6	67.1	2463.8	2530.9
18	0.02062	0.001001	65.1	75.5	2400.3	75.5	2459.0	2534.5
20	0.0234	0.001002	57.8	83.9	2403.0	83.9	2454.3	2538.2
22	0.0264	0.001002	51.5	92.2	2405.8	92.2	2449.6	2541.8
24	0.0298	0.001003	45.9	100.6	2408.5	100.6	2444.9	2545.5
25	0.0317	0.001003	43.4	104.8	2409.9	104.8	2442.5	2547.3
26	0.0336	0.001003	41.0	108.9	2411.2	108.9	2440.2	2549.1
28	0.0378	0.001004	36.7	117.3	2414.0	117.3	2435.4	2552.7
30	0.0424	0.001004	32.9	125.7	2416.7	125.7	2430.7	2556.4
32	0.0475	0.001005	29.6	134.0	2419.4	134.0	2425.9	2560.0
34	0.0532	0.001006	26.6	142.4	2422.1	142.4	2421.2	2563.6
36	0.0594	0.001006	24.0	150.7	2424.8	150.7	2416.4	2567.2
38	0.0662	0.001007	21.6	159.1	2427.5	159.1	2411.7	2570.8
40	0.0738	0.001008	19.55	167.4	2430.2	167.5	2406.9	2574.4
42	0.0820	0.001009	17.69	175.8	2432.9	175.8	2402.1	2577.9
44	0.0910	0.001009	16.04	184.2	2435.6	184.2	2397.3	2581.5
46	0.1009	0.001010	14.56	192.5	2438.3	192.5	2392.5	2585.1
48	0.1116	0.001011	13.23	200.9	2440.9	200.9	2387.7	2588.6
50	0.1234	0.001012	12.05	209.2	2443.6	209.3	2382.9	2592.2
52	0.1361	0.001013	10.98	217.7	2446	217.7	2377	2595
54	0.1500	0.001014	10.02	226.0	2449	226.0	2373	2599
56	0.1651	0.001015	9.158	234.4	2451	234.4	2368	2602
58	0.1815	0.001016	8.380	242.8	2454	242.8	2363	2606
60	0.1992	0.001017	7.678	251.1	2456	251.1	2358	2609
62	0.2184	0.001018	7.043	259.5	2459	259.5	2353	2613
64	0.2391	0.001019	6.468	267.9	2461	267.9	2348	2616
66	0.2615	0.001020	5.947	276.2	2464	276.2	2343	2619
68	0.2856	0.001022	5.475	284.6	2467	284.6	2338	2623

[a] From R. W. Haywood, *Thermodynamic Tables in SI (Metric) Units*, Cambridge University Press, London, 1968. \hat{V} = specific volume, \hat{U} = specific internal energy, and \hat{H} = specific enthalpy. *Note:* kJ/kg × 0.4303 = Btu/lb$_m$.

(continued)

TABLE B.4 (Continued)

T(°C)	P(bar)	$\hat{V}(\text{m}^3/\text{kg})$ Water	Steam	$\hat{U}(\text{kJ/kg})$ Water	Steam	$\hat{H}(\text{kJ/kg})$ Water	Evaporation	Steam
70	0.3117	0.001023	5.045	293.0	2469	293.0	2333	2626
72	0.3396	0.001024	4.655	301.4	2472	301.4	2329	2630
74	0.3696	0.001025	4.299	309.8	2474	309.8	2323	2633
76	0.4019	0.001026	3.975	318.2	2476	318.2	2318	2636
78	0.4365	0.001028	3.679	326.4	2479	326.4	2313	2639
80	0.4736	0.001029	3.408	334.8	2482	334.9	2308	2643
82	0.5133	0.001030	3.161	343.2	2484	343.3	2303	2646
84	0.5558	0.001032	2.934	351.6	2487	351.7	2298	2650
86	0.6011	0.001033	2.727	360.0	2489	360.1	2293	2653
88	0.6495	0.001034	2.536	368.4	2491	368.5	2288	2656
90	0.7011	0.001036	2.361	376.9	2493	377.0	2282	2659
92	0.7560	0.001037	2.200	385.3	2496	385.4	2277	2662
94	0.8145	0.001039	2.052	393.7	2499	393.8	2272	2666
96	0.8767	0.001040	1.915	402.1	2501	402.2	2267	2669
98	0.9429	0.001042	1.789	410.6	2504	410.7	2262	2673
100	1.0131	0.001044	1.673	419.0	2507	419.1	2257	2676
102	1.0876	0.001045	1.566	427.1	2509	427.5	2251	2679

TABLE B.5
Properties of Saturated Steam: Pressure Table[a]

P(bar)	T(°C)	\hat{V}(m³/kg) Water	\hat{V}(m³/kg) Steam	\hat{U}(kJ/kg) Water	\hat{U}(kJ/kg) Steam	\hat{H}(kJ/kg) Water	\hat{H}(kJ/kg) Evaporation	\hat{H}(kJ/kg) Steam
0.00611	0.01	0.001000	206.2	zero	2375.6	+0.0	2501.6	2501.6
0.008	3.8	0.001000	159.7	15.8	2380.7	15.8	2492.6	2508.5
0.010	7.0	0.001000	129.2	29.3	2385.2	29.3	2485.0	2514.4
0.012	9.7	0.001000	108.7	40.6	2388.9	40.6	2478.7	2519.3
0.014	12.0	0.001000	93.9	50.3	2392.0	50.3	2473.2	2523.5
0.016	14.0	0.001001	82.8	58.9	2394.8	58.9	2468.4	2527.3
0.018	15.9	0.001001	74.0	66.5	2397.4	66.5	2464.1	2530.6
0.020	17.5	0.001001	67.0	73.5	2399.6	73.5	2460.2	2533.6
0.022	19.0	0.001002	61.2	79.8	2401.7	79.8	2456.6	2536.4
0.024	20.4	0.001002	56.4	85.7	2403.6	85.7	2453.3	2539.0
0.026	21.7	0.001002	52.3	91.1	2405.4	91.1	2450.2	2541.3
0.028	23.0	0.001002	48.7	96.2	2407.1	96.2	2447.3	2543.6
0.030	24.1	0.001003	45.7	101.0	2408.6	101.0	2444.6	2545.6
0.035	26.7	0.001003	39.5	111.8	2412.2	111.8	2438.5	2550.4
0.040	29.0	0.001004	34.8	121.4	2415.3	121.4	2433.1	2554.5
0.045	31.0	0.001005	31.1	130.0	2418.1	130.0	2428.2	2558.2
0.050	32.9	0.001005	28.2	137.8	2420.6	137.8	2423.8	2561.6
0.060	36.2	0.001006	23.74	151.5	2425.1	151.5	2416.0	2567.5
0.070	39.0	0.001007	20.53	163.4	2428.9	163.4	2409.2	2572.6
0.080	41.5	0.001008	18.10	173.9	2432.3	173.9	2403.2	2577.1
0.090	43.8	0.001009	16.20	183.3	2435.3	183.3	2397.9	2581.1
0.10	45.8	0.001010	14.67	191.8	2438.0	191.8	2392.9	2584.8
0.11	47.7	0.001011	13.42	199.7	2440.5	199.7	2388.4	2588.1
0.12	49.4	0.001012	12.36	206.9	2442.8	206.9	2384.3	2591.2
0.13	51.1	0.001013	11.47	213.7	2445.0	213.7	2380.4	2594.0
0.14	52.6	0.001013	10.69	220.0	2447.0	220.0	2376.7	2596.7

P	T	\hat{V}	\hat{V}	\hat{U}	\hat{U}	\hat{H}	\hat{H}	\hat{H}
0.15	54.0	0.001014	10.02	226.0	2448.9	226.0	2373.2	2599.2
0.16	55.3	0.001015	9.43	231.6	2450.6	231.6	2370.0	2601.6
0.17	56.6	0.001015	8.91	236.9	2452.3	236.9	2366.9	2603.8
0.18	57.8	0.001016	8.45	242.0	2453.9	242.0	2363.9	2605.9
0.19	59.0	0.001017	8.03	246.8	2455.4	246.8	2361.1	2607.9
0.20	60.1	0.001017	7.65	251.5	2456.9	251.5	2358.4	2609.9
0.22	62.2	0.001018	7.00	260.1	2459.6	260.1	2353.3	2613.5
0.24	64.1	0.001019	6.45	268.2	2462.1	268.2	2348.6	2616.8
0.26	65.9	0.001020	5.98	275.7	2464.4	275.6	2344.2	2619.9
0.28	67.5	0.001021	5.58	282.7	2466.5	282.7	2340.0	2622.7
0.30	69.1	0.001022	5.23	289.3	2468.6	289.3	2336.1	2625.4
0.35	72.7	0.001025	4.53	304.3	2473.1	304.3	2327.2	2631.5
0.40	75.9	0.001027	3.99	317.7	2477.1	317.6	2319.2	2636.9
0.45	78.7	0.001028	3.58	329.6	2480.7	329.6	2312.0	2641.7
0.50	81.3	0.001030	3.24	340.6	2484.0	340.5	2305.4	2646.0
0.55	83.7	0.001032	2.96	350.6	2486.9	350.6	2299.3	2649.9
0.60	86.0	0.001033	2.73	359.9	2489.7	359.9	2293.6	2653.6
0.65	88.0	0.001035	2.53	368.6	2492.2	368.5	2288.3	2656.9
0.70	90.0	0.001036	2.36	376.8	2494.5	376.7	2283.3	2660.1
0.75	91.8	0.001037	2.22	384.5	2496.7	384.4	2278.6	2663.0
0.80	93.5	0.001039	2.087	391.7	2498.8	391.6	2274.1	2665.8
0.85	95.2	0.001040	1.972	398.6	2500.8	398.5	2269.8	2668.4
0.90	96.7	0.001041	1.869	405.2	2502.6	405.1	2265.6	2670.9
0.95	98.2	0.001042	1.777	411.5	2504.4	411.4	2261.7	2673.2
1.00	99.6	0.001043	1.694	417.5	2506.1	417.4	2257.9	2675.4
1.01325 (1 atm)	100.0	0.001044	1.673	419.1	2506.5	419.0	2256.9	2676.0

*From R. W. Haywood, *Thermodynamic Tables in SI (Metric) Units*, Cambridge University Press, London, 1968. \hat{V} = specific volume, \hat{U} = specific internal energy, and \hat{H} = specific enthalpy. *Note:* kJ/kg × 0.4303 = Btu/lb$_m$.

(continued)

TABLE B.5 (Continued)

P(bar)	T(°C)	V̂(m³/kg) Water	V̂(m³/kg) Steam	Û(kJ/kg) Water	Û(kJ/kg) Steam	Ĥ(kJ/kg) Water	Ĥ(kJ/kg) Evaporation	Ĥ(kJ/kg) Steam
1.1	102.3	0.001046	1.549	428.7	2509.2	428.8	2250.8	2679.6
1.2	104.8	0.001048	1.428	439.2	2512.1	439.4	2244.1	2683.4
1.3	107.1	0.001049	1.325	449.1	2514.7	449.2	2237.8	2687.0
1.4	109.3	0.001051	1.236	458.3	2517.2	458.4	2231.9	2690.3
1.5	111.4	0.001053	1.159	467.0	2519.5	467.1	2226.2	2693.4
1.6	113.3	0.001055	1.091	475.2	2521.7	475.4	2220.9	2696.2
1.7	115.2	0.001056	1.031	483.0	2523.7	483.2	2215.7	2699.0
1.8	116.9	0.001058	0.977	490.5	2525.6	490.7	2210.8	2701.5
1.9	118.6	0.001059	0.929	497.6	2527.5	497.8	2206.1	2704.0
2.0	120.2	0.001061	0.885	504.5	2529.2	504.7	2201.6	2706.3
2.2	123.3	0.001064	0.810	517.4	2532.4	517.6	2193.0	2710.6
2.4	126.1	0.001066	0.746	529.4	2535.4	529.6	2184.9	2714.5
2.6	128.7	0.001069	0.693	540.6	2538.1	540.9	2177.3	2718.2
2.8	131.2	0.001071	0.646	551.1	2540.6	551.4	2170.1	2721.5
3.0	133.5	0.001074	0.606	561.1	2543.0	561.4	2163.2	2724.7
3.2	135.8	0.001076	0.570	570.6	2545.2	570.9	2156.7	2727.6
3.4	137.9	0.001078	0.538	579.6	2547.2	579.9	2150.4	2730.3
3.6	139.9	0.001080	0.510	588.1	2549.2	588.5	2144.4	2732.9
3.8	141.8	0.001082	0.485	596.4	2551.0	596.8	2138.6	2735.3
4.0	143.6	0.001084	0.462	604.2	2552.7	604.7	2133.0	2737.6
4.2	145.4	0.001086	0.442	611.8	2554.4	612.3	2127.5	2739.8
4.4	147.1	0.001088	0.423	619.1	2555.9	619.6	2122.3	2741.9
4.6	148.7	0.001089	0.405	626.2	2557.4	626.7	2117.2	2743.9
4.8	150.3	0.001091	0.389	633.0	2558.8	633.5	2112.2	2745.7
5.0	151.8	0.001093	0.375	639.6	2560.2	640.1	2107.4	2747.5
5.5	155.5	0.001097	0.342	655.2	2563.3	655.8	2095.9	2751.7
6.0	158.8	0.001101	0.315	669.8	2566.2	670.4	2085.0	2755.5
6.5	162.0	0.001105	0.292	683.4	2568.7	684.1	2074.7	2758.9
7.0	165.0	0.001108	0.273	696.3	2571.1	697.1	2064.9	2762.0

7.5	167.8	0.001112	708.5	2573.3	709.3	2055.5	2764.8
8.0	170.4	0.001115	720.0	2575.5	720.9	2046.5	2767.5
8.5	172.9	0.001118	731.1	2577.1	732.0	2037.9	2769.9
9.0	175.4	0.001121	741.6	2578.8	742.6	2029.5	2772.1
9.5	177.7	0.001124	751.8	2580.4	752.8	2021.4	2774.2
10.0	179.9	0.001127	761.5	2581.9	762.6	2013.6	2776.2
10.5	182.0	0.001130	770.8	2583.3	772.0	2005.9	2778.0
11.0	184.1	0.001133	779.9	2584.5	781.1	1998.5	2779.7
11.5	186.0	0.001136	788.6	2585.8	789.9	1991.3	2781.3
12.0	188.0	0.001139	797.1	2586.9	798.4	1984.3	2782.7
12.5	189.8	0.001141	805.3	2588.0	806.7	1977.4	2784.1
13.0	191.6	0.001144	813.2	2589.0	814.7	1970.7	2785.4
14	195.0	0.001149	828.5	2590.8	830.1	1957.7	2787.8
15	198.3	0.001154	842.9	2592.4	844.7	1945.2	2789.9
16	201.4	0.001159	856.7	2593.8	858.6	1933.2	2791.7
17	204.3	0.001163	869.9	2595.1	871.8	1921.5	2793.4
18	207.1	0.001168	882.5	2596.3	884.6	1910.3	2794.8
19	209.8	0.001172	894.6	2597.3	896.8	1899.3	2796.1
20	212.4	0.001177	906.2	2598.2	908.6	1888.6	2797.2
21	214.9	0.001181	917.5	2598.9	920.0	1878.2	2798.2
22	217.2	0.001185	928.3	2599.6	931.0	1868.1	2799.1
23	219.6	0.001189	938.9	2600.2	941.6	1858.2	2799.8
24	221.8	0.001193	949.1	2600.7	951.9	1848.5	2800.4
25	223.9	0.001197	959.0	2601.2	962.0	1839.0	2800.9
26	226.0	0.001201	968.6	2601.5	971.7	1829.6	2801.4
27	228.1	0.001205	978.0	2601.8	981.2	1820.5	2801.7
28	230.0	0.001209	987.1	2602.1	990.5	1811.5	2802.0
29	232.0	0.001213	996.0	2602.3	999.5	1802.6	2802.2
30	233.8	0.001216	1004.7	2602.4	1008.4	1793.9	2802.3
32	237.4	0.001224	1021.5	2602.5	1025.4	1776.9	2802.3
34	240.9	0.001231	1037.6	2602.5	1041.8	1760.3	2802.1

(continued)

TABLE B.5 (Continued)

P(bar)	T(°C)	$\hat{V}(m^3/kg)$ Water	$\hat{V}(m^3/kg)$ Steam	$\hat{U}(kJ/kg)$ Water	$\hat{U}(kJ/kg)$ Steam	$\hat{H}(kJ/kg)$ Water	$\hat{H}(kJ/kg)$ Evaporation	$\hat{H}(kJ/kg)$ Steam
36	244.2	0.001238	0.0554	1053.1	2602.2	1057.6	1744.2	2801.7
38	247.3	0.001245	0.0524	1068.0	2601.9	1072.7	1728.4	2801.1
40	250.3	0.001252	0.0497	1082.4	2601.3	1087.4	1712.9	2800.3
42	253.2	0.001259	0.0473	1096.3	2600.7	1101.6	1697.8	2799.4
44	256.0	0.001266	0.0451	1109.8	2599.9	1115.4	1682.9	2798.3
46	258.8	0.001272	0.0430	1122.9	2599.1	1128.8	1668.3	2797.1
48	261.4	0.001279	0.0412	1135.6	2598.1	1141.8	1653.9	2795.7
50	263.9	0.001286	0.0394	1148.0	2597.0	1154.5	1639.7	2794.2
52	266.4	0.001292	0.0378	1160.1	2595.9	1166.8	1625.7	2792.6
54	268.8	0.001299	0.0363	1171.9	2594.6	1178.9	1611.9	2790.8
56	271.1	0.001306	0.0349	1183.5	2593.3	1190.8	1598.2	2789.0
58	273.3	0.001312	0.0337	1194.7	2591.9	1202.3	1584.7	2787.0
60	275.6	0.001319	0.0324	1205.8	2590.4	1213.7	1571.3	2785.0
62	277.7	0.001325	0.0313	1216.6	2588.8	1224.8	1558.0	2782.9
64	279.8	0.001332	0.0302	1227.2	2587.2	1235.7	1544.9	2780.6
66	281.8	0.001338	0.0292	1237.6	2585.5	1246.5	1531.9	2778.3
68	283.8	0.001345	0.0283	1247.9	2583.7	1257.0	1518.9	2775.9
70	285.8	0.001351	0.0274	1258.0	2581.8	1267.4	1506.0	2773.5
72	287.7	0.001358	0.0265	1267.9	2579.9	1277.6	1493.3	2770.9
74	289.6	0.001364	0.0257	1277.6	2578.0	1287.7	1480.5	2768.3
76	291.4	0.001371	0.0249	1287.2	2575.9	1297.6	1467.9	2765.5
78	293.2	0.001378	0.0242	1296.7	2573.8	1307.4	1455.3	2762.8
80	295.0	0.001384	0.0235	1306.0	2571.7	1317.1	1442.8	2759.9
82	296.7	0.001391	0.0229	1315.2	2569.5	1326.6	1430.3	2757.0
84	298.4	0.001398	0.0222	1324.3	2567.2	1336.1	1417.9	2754.0
86	300.1	0.001404	0.0216	1333.3	2564.9	1345.4	1405.5	2750.9
88	301.7	0.001411	0.0210	1342.2	2562.6	1354.6	1393.2	2747.8

90	303.3	0.001418	0.02050	1351.0	2560.1	1363.7	1380.9	2744.6
92	304.9	0.001425	0.01996	1359.7	2557.7	1372.8	1368.6	2741.4
94	306.4	0.001432	0.01945	1368.2	2555.2	1381.7	1356.3	2738.0
96	308.0	0.001439	0.01897	1376.7	2552.6	1390.6	1344.1	2734.7
98	309.5	0.001446	0.01849	1385.2	2550.0	1399.3	1331.9	2731.2
100	311.0	0.001453	0.01804	1393.5	2547.3	1408.0	1319.7	2727.7
105	314.6	0.001470	0.01698	1414.1	2540.4	1429.5	1289.2	2718.7
110	318.0	0.001489	0.01601	1434.2	2533.2	1450.6	1258.7	2709.3
115	321.4	0.001507	0.01511	1454.0	2525.7	1471.3	1228.2	2699.5
120	324.6	0.001527	0.01428	1473.4	2517.8	1491.8	1197.4	2689.2
125	327.8	0.001547	0.01351	1492.7	2509.4	1512.0	1166.4	2678.4
130	330.8	0.001567	0.01280	1511.6	2500.6	1532.0	1135.0	2667.0
135	333.8	0.001588	0.01213	1530.4	2491.3	1551.9	1103.1	2655.0
140	336.6	0.001611	0.01150	1549.1	2481.4	1571.6	1070.7	2642.4
145	339.4	0.001634	0.01090	1567.5	2471.0	1591.3	1037.7	2629.1
150	342.1	0.001658	0.01034	1586.1	2459.9	1611.0	1004.0	2615.0
155	344.8	0.001683	0.00981	1604.6	2448.2	1630.7	969.6	2600.3
160	347.3	0.001710	0.00931	1623.2	2436.0	1650.5	934.3	2584.9
165	349.8	0.001739	0.00883	1641.8	2423.1	1670.5	898.3	2568.8
170	352.3	0.001770	0.00837	1661.6	2409.3	1691.7	859.9	2551.6
175	354.6	0.001803	0.00793	1681.8	2394.6	1713.3	820.0	2533.3
180	357.0	0.001840	0.00750	1701.7	2378.9	1734.8	779.1	2513.9
185	359.2	0.001881	0.00708	1721.7	2362.1	1756.5	736.6	2493.1
190	361.4	0.001926	0.00668	1742.1	2343.8	1778.7	692.0	2470.6
195	363.6	0.001977	0.00628	1763.2	2323.6	1801.8	644.2	2446.0
200	365.7	0.00204	0.00588	1785.7	2300.8	1826.5	591.9	2418.4
205	367.8	0.00211	0.00546	1810.7	2274.4	1853.9	532.5	2386.4
210	369.8	0.00220	0.00502	1840.0	2242.1	1886.3	461.3	2347.6
215	371.8	0.00234	0.00451	1878.6	2198.1	1928.9	366.2	2295.2
220	373.7	0.00267	0.00373	1952	2114	2011	185	2196
221.2 (Critical point)	374.15	0.00317	0.00317	2038	2038	2108	0	2108

TABLE B.6

Properties of Superheated Steam[a]

P(bar) (T_sat,°C)		Sat'd Water	Sat'd Steam	Temperature (°C)→ 50	75	100	150	200	250	300	350
0.0	\hat{H}	—	—	2595	2642	2689	2784	2880	2978	3077	3177
(—)	\hat{U}	—	—	2446	2481	2517	2589	2662	2736	2812	2890
	\hat{V}			—	—	—	—	—	—	—	—
0.1	\hat{H}	191.8	2584.8	2593	2640	2688	2783	2880	2977	3077	3177
(45.8)	\hat{U}	191.8	2438.0	2444	2480	2516	2588	2661	2736	2812	2890
	\hat{V}	0.00101	14.7	14.8	16.0	17.2	19.5	21.8	24.2	26.5	28.7
0.5	\hat{H}	340.6	2446.0	209.3	313.9	2683	2780	2878	2979	3076	3177
(81.3)	\hat{U}	340.6	2484.0	209.2	313.9	2512	2586	2660	2735	2811	2889
	\hat{V}	0.00103	3.24	0.00101	0.00103	3.41	3.89	4.35	4.83	5.29	5.75
1.0	\hat{H}	417.5	2675.4	209.3	314.0	2676	2776	2875	2975	3074	3176
(99.6)	\hat{U}	417.5	2506.1	209.2	313.9	2507	2583	2658	2734	2811	2889
	\hat{V}	0.00104	1.69	0.00101	0.00103	1.69	1.94	2.17	2.40	2.64	2.87
5.0	\hat{H}	640.1	2747.5	209.7	314.3	419.4	632.2	2855	2961	3065	3168
(151.8)	\hat{U}	639.6	2560.2	209.2	313.8	418.8	631.6	2643	2724	2803	2883
	\hat{V}	0.00109	0.375	0.00101	0.00103	0.00104	0.00109	0.425	0.474	0.522	0.571
10	\hat{H}	762.6	2776.2	210.1	314.7	419.7	632.5	2827	2943	3052	3159
(179.9)	\hat{U}	761.5	2582	209.1	313.7	418.7	631.4	2621	2710	2794	2876
	\hat{V}	0.00113	0.194	0.00101	0.00103	0.00104	0.00109	0.206	0.233	0.258	0.282
20	\hat{H}	908.6	2797.2	211.0	315.5	420.5	633.1	852.6	2902	3025	3139
(212.4)	\hat{U}	906.2	2598.2	209.0	313.5	418.4	603.9	850.2	2679	2774	2862
	\hat{V}	0.00118	0.09950	0.00101	0.00102	0.00104	0.00109	0.00116	0.111	0.125	0.139
40	\hat{H}	1087.4	2800.3	212.7	317.1	422.0	634.3	853.4	1085.8	2962	3095
(250.3)	\hat{U}	1082.4	2601.3	208.6	313.0	417.8	630.0	848.8	1080.8	2727	2829
	\hat{V}	0.00125	0.04975	0.00101	0.00102	0.00104	0.00109	0.00115	0.00125	0.0588	0.0665
60	\hat{H}	1213.7	2785.0	214.4	318.7	423.5	635.6	854.2	1085.8	2885	3046
(275.6)	\hat{U}	1205.8	2590.4	208.3	312.6	417.3	629.1	847.3	1078.3	2668	2792
	\hat{V}	0.00132	0.0325	0.00101	0.00103	0.00104	0.00109	0.00115	0.00125	0.0361	0.0422
80	\hat{H}	1317.1	2759.9	216.1	320.3	425.0	636.8	855.1	1085.8	2787	2990
(295.0)	\hat{U}	1306.0	2571.7	208.1	312.3	416.7	628.2	845.9	1075.8	2593	2750
	\hat{V}	0.00139	0.0235	0.00101	0.00102	0.00104	0.00109	0.00115	0.00124	0.0243	0.0299
100	\hat{H}	1408.0	2727.7	217.8	322.9	426.5	638.1	855.9	1085.8	1343.4	2926
(311.0)	\hat{U}	1393.5	2547.3	207.8	311.7	416.1	627.3	844.4	1073.4	1329.4	2702
	\hat{V}	0.00145	0.0181	0.00101	0.00102	0.00104	0.00109	0.00115	0.00124	0.00140	0.0224
150	\hat{H}	1611.0	2615.0	222.1	326.0	430.3	641.3	858.1	1086.2	1338.2	2695
(342.1)	\hat{U}	1586.1	2459.9	207.0	310.7	414.7	625.0	841.0	1067.7	1317.6	2523
	\hat{V}	0.00166	0.0103	0.00101	0.00102	0.00104	0.00108	0.00114	0.00123	0.00138	0.0115
200	\hat{H}	1826.5	2418.4	226.4	330.0	434.0	644.5	860.4	1086.7	1334.3	1647.1
(365.7)	\hat{U}	1785.7	2300.8	206.3	309.7	413.2	622.9	837.7	1062.2	1307.1	1613.7
	\hat{V}	0.00204	0.005875	0.00100	0.00102	0.00103	0.00108	0.00114	0.00122	0.00136	0.00167
221.2(P_c)	\hat{H}	2108	2108	228.2	331.7	435.7	645.8	861.4	1087.0	1332.8	1635.2
(374.15)(T_c)	\hat{U}	2037.8	2037.8	206.0	309.2	412.8	622.0	836.3	1060.0	1302.9	1600.3
	\hat{V}	0.00317	0.00317	0.00100	0.00102	0.00103	0.00108	0.00114	0.00122	0.00135	0.00163
250	\hat{H}	—	—	230.7	334.0	437.8	647.7	862.8	1087.5	1331.1	1625.0
(—)	\hat{U}	—	—	205.7	308.7	412.1	620.8	834.4	1057.0	1297.5	1585.0
	\hat{V}	—	—	0.00100	0.00101	0.00103	0.00108	0.00113	0.00122	0.00135	0.00160
300	\hat{H}	—	—	235.0	338.1	441.6	650.9	865.2	1088.4	1328.7	1609.9
(—)	\hat{U}	—	—	205.0	307.7	410.8	618.7	831.3	1052.1	1288.7	1563.3
	\hat{V}	—	—	0.0009990	0.00101	0.00103	0.00107	0.00113	0.00121	0.00133	0.00155
500	\hat{H}	—	—	251.9	354.2	456.8	664.1	875.4	1093.6	1323.7	1576.3
(—)	\hat{U}	—	—	202.4	304.0	405.8	611.0	819.7	1034.3	1259.3	1504.1
	\hat{V}	—	—	0.0009911	0.00100	0.00102	0.00106	0.00111	0.00119	0.00129	0.00144
1000	\hat{H}	—	—	293.9	394.3	495.1	698.0	903.5	1113.0	1328.7	1550.5
(—)	\hat{U}	—	—	196.5	295.7	395.1	594.4	795.3	999.0	1207.1	1419.0
	\hat{V}	—	—	0.0009737	0.0009852	0.001000	0.00104	0.00108	0.00114	0.00122	0.00131

[a] Adapted from Haywood, *Thermodynamic Tables in SI (Metric) Units*, Cambridge University Press, London, 1968. Water is a liquid in the enclosed region between 50°C and 350°C. \hat{H} = specific enthalpy (kJ/kg), \hat{U} = specific internal energy (kJ/kg), \hat{V} = specific volume (m³/kg). *Note:* kJ/kg × 0.4303 = Btu/lb_m

P(bar) $(T_{sat.}°C)$		Temperature (°C)→ 400	450	500	550	600	650	700	750
0.0	\hat{H}	3280	3384	3497	3597	3706	3816	3929	4043
(—)	\hat{U}	2969	3050	3132	3217	3303	3390	3480	3591
	\hat{V}	—	—	—	—	—	—	—	—
0.1	\hat{H}	3280	3384	3489	3596	3706	3816	3929	4043
(45.8)	\hat{U}	2969	3050	3132	3217	3303	3390	3480	3571
	\hat{V}	21.1	33.3	35.7	38.0	40.3	42.6	44.8	47.2
0.5	\hat{H}	3279	3383	3489	3596	3705	3816	3929	4043
(81.3)	\hat{U}	2969	3049	3132	3216	3302	3390	3480	3571
	\hat{V}	6.21	6.67	7.14	7.58	8.06	8.55	9.01	9.43
1.0	\hat{H}	3278	3382	3488	3596	3705	3816	3928	4042
(99.6)	\hat{U}	2968	3049	3132	3216	3302	3390	3479	3570
	\hat{V}	3.11	3.33	3.57	3.80	4.03	4.26	4.48	4.72
5.0	\hat{H}	3272	3379	3484	3592	3702	3813	3926	4040
(151.8)	\hat{U}	2964	3045	3128	3213	3300	3388	3477	3569
	\hat{V}	0.617	0.664	0.711	0.758	0.804	0.850	0.897	0.943
10	\hat{H}	3264	3371	3478	3587	3697	3809	3923	4038
(179.9)	\hat{U}	2958	3041	3124	3210	3296	3385	3475	3567
	\hat{V}	0.307	0.330	0.353	0.377	0.402	0.424	0.448	0.472
20	\hat{H}	3249	3358	3467	3578	3689	3802	3916	4032
(212.4)	\hat{U}	2946	3031	3115	3202	3290	3379	3470	3562
	\hat{V}	0.151	0.163	0.175	0.188	0.200	0.211	0.223	0.235
40	\hat{H}	3216	3331	3445	3559	3673	3788	3904	4021
(250.3)	\hat{U}	2922	3011	3100	3188	3278	3368	3460	3554
	\hat{V}	0.0734	0.0799	0.0864	0.0926	0.0987	0.105	0.111	0.117
60	\hat{H}	3180	3303	3422	3539	3657	3774	3892	4011
(275.6)	\hat{U}	2896	2991	3083	3174	3265	3357	3451	3545
	\hat{V}	0.0474	0.0521	0.0566	0.0609	0.0652	0.0693	0.0735	0.0776
80	\hat{H}	3142	3274	3399	3520	3640	3759	3879	4000
(295.0)	\hat{U}	2867	2969	3065	3159	3252	3346	3441	3537
	\hat{V}	0.0344	0.0382	0.0417	0.0450	0.0483	0.0515	0.0547	0.0578
100	\hat{H}	3100	3244	3375	3500	3623	3745	3867	3989
(311.0)	\hat{U}	2836	2946	3047	3144	3240	3335	3431	3528
	\hat{V}	0.0264	0.0298	0.0328	0.0356	0.0383	0.0410	0.0435	0.0461
150	\hat{H}	2975	3160	3311	3448	3580	3708	3835	3962
(342.1)	\hat{U}	2744	2883	2999	3105	3207	3307	3407	3507
	\hat{V}	0.0157	0.0185	0.0208	0.0229	0.0249	0.0267	0.0286	0.0304
200	\hat{H}	2820	3064	3241	3394	3536	3671	3804	3935
(365.7)	\hat{U}	2622	2810	2946	3063	3172	3278	3382	3485
	\hat{V}	0.009950	0.0127	0.0148	0.0166	0.0182	0.0197	0.0211	0.0225
221.2(P_c)	\hat{H}	2733	3020	3210	3370	3516	3655	3790	3923
(374.15)(T_c)	\hat{U}	2553	2776	2922	3045	3157	3265	3371	3476
	\hat{V}	0.008157	0.0110	0.0130	0.0147	0.0162	0.0176	0.0190	0.0202
250	\hat{H}	2582	2954	3166	3337	3490	3633	3772	3908
(—)	\hat{U}	2432	2725	2888	3019	3137	3248	3356	3463
	\hat{V}	0.006013	0.009174	0.0111	0.0127	0.0141	0.0143	0.0166	0.0178
300	\hat{H}	2162	2826	3085	3277	3443	3595	3740	3880
(—)	\hat{U}	2077	2623	2825	2972	3100	3218	3330	3441
	\hat{V}	0.002830	0.006734	0.008680	0.0102	0.0114	0.0126	0.0136	0.0147
500	\hat{H}	1878	2293	2723	3021	3248	3439	3610	3771
(—)	\hat{U}	1791	2169	2529	2765	2946	3091	3224	3350
	\hat{V}	0.001726	0.002491	0.003882	0.005112	0.006112	0.007000	0.007722	0.008418
1000	\hat{H}	1798	2051	2316	2594	2857	3105	3324	3526
(—)	\hat{U}	1653	1888	2127	2369	2591	2795	2971	3131
	\hat{V}	0.001446	0.001628	0.001893	0.002246	0.002668	0.003106	0.003536	0.003953

ANSWERS TO TEST YOURSELVES

p. 11.

1. A ratio of equivalent values of a quantity expressed in different units.
2. (60 s)/(1 min)
3. $(1 \text{ min}^2)/(3600 \text{ s}^2)$
4. $(1 \text{ m}^3)/(10^6 \text{ cm}^3)$

p. 14.

1. **(a)** $(1000 \text{ mm})/(1 \text{ m})$; **(b)** $(10^{-9} \text{ s})/(1 \text{ ns})$; **(c)** $(1 \text{ m}^2)/(10^4 \text{ cm}^2)$;
 (d) $(1 \text{ m}^3)/(35.3145 \text{ ft}^3)$; **(e)** $(9.486 \times 10^{-4} \text{ Btu/s})/(1.341 \times 10^{-3} \text{ hp})$
2. m/s; cm/s; ft/s

p. 16.

1. Conversion factor (natural force unit/defined force unit). Values given by Equation 2.4-4.
2. 2 N; $(2/32.174) \text{ lb}_f$
3. No
4. 1 kg; same; less
5. 2 lb_m; same; less

p. 19.

1. All additive terms have the same dimensions. No. Yes.
2. $\text{m}^{-2} \cdot \text{s}^{-2}$
3. A multiplicative combination of factors with no net units; st^2/r or r/st^2.
4. $a(\text{lb}_f)$; Q is dimensionless.

p. 21.

1. **(a)** 1.22×10^4 (3 s.f.); **(b)** 1.22000×10^4 (6 s.f.); **(c)** 3.040×10^{-3} (4 s.f.)
2. **(a)** 134 000 (3 s.f.); **(b)** 0.01340 (4 s.f.); **(c)** 4200 (3 s.f.)
3. **(a)** 3 s.f.; **(b)** 2 s.f.; **(c)** 3 s.f., 11.2; **(d)** 2 s.f., 12
4. **(a)** 1460; **(b)** 13.4; **(c)** 1.76×10^{-7}
5. **(a)** 4.25–4.35; **(b)** 4.295–4.305; **(c)** 2.7775×10^{-3}–2.7785×10^{-3}; **(d)** 2450–2550;
 (e) 2499.5–2500.5

p. 23.

1. Substitute in Equation 2.7-1.
2. Correct; too high; too low; too low.

p. 28.

1. $y = a(x^2 - 2)$
2. **(b)** Plot $1/y$ versus $(x - 3)^2$: $a =$ slope, $b =$ intercept
 (c) Plot y^3 versus x^2: $a =$ slope, $b = -$intercept
 (d) Plot $1/\sqrt{\sin y}$ versus x: $a =$ slope, $b =$ intercept
 (e) Plot $\ln y$ versus x: $b = \ln (y_2/y_1)/(x_2 - x_1)$, $\ln a = \ln y_1 - bx_1$, $a = e^{\ln a}$
 (f) Plot $\ln y$ versus $\ln x$: $b = \ln(y_2/y_1)/\ln (x_2/x_1)$, $\ln a = \ln y_1 - b \ln x_1$, $a = e^{\ln a}$

p. 31.

1. **(a)** $P = at + b$; **(b)** $P = ae^{bt}$; **(c)** $P = at^b$; **(d)** $y^2 = 3 + a \exp (b/x^2)$;
 (e) $1/F = a(t^2 - 4)^b$
2. **(a)** P versus t on semilog paper; **(b)** P versus t on log paper;
 (c) P^2 versus t^3 on semilog paper; **(d)** $1/P$ (or P) versus $(t - 4)$ on log paper

p. 44.

1. Dimensionless.
2. 0.50 g/cm^3; 2.0 cm^3/g; 31 lb$_m$/ft^3; 1.5 g; 36 cm^3
3. Yes.
4. No—possibly different reference densities used for each.
5. Weigh the rock—determine m(g). Put V_1(ml) water in the cylinder; then put the rock in, and read the new cylinder liquid level V_2(ml). The volume of the rock is V(ml) $= V_2 - V_1$; $\rho = m/V$(g/ml); SG $= \rho/1.00$.

p. 46.

1. 10.0 cm^3/s
2. 159.5 g/min
3. Same; same; greater at outlet.

p. 47.

1. 100 ml/min; 100 g/min
2. Flowmeters—see Figure 3.2-1.
3. Too low (the gas is much less dense, so it must flow at a much higher rate than the liquid to raise the float to the same position).

p. 50.

1. **(a)** 6.02×10^{23} molecules; **(b)** M grams
2. The molecular weight of the species expressed in tons.
3. **(a)** 1 lb-mole, 2 lb$_m$; **(b)** 2 lb-moles, 2 lb$_m$

4. 2000

5. 50 kmol/h

p. 53.

1. **(a)** 80/81; **(b)** 0.5
2. 0.25 lb_m A/lb_m; 0.75 lb_m B/lb_m; 0.333 mole A/mole; 0.667 mole B/mole

p. 54.

1. n/V(mol/liter)
2. nM/V(g/liter)
3. $(20/C_A)$(liters)
4. $(120\,c_A)$g/h

p. 58.

1. See Figures 3.4-1 and 3.4-2, and Equation 3.4-2.
2. The fluid pressure is higher at the bottom than at the top (hydrostatic head effect). No. Maybe. Yes. (The answers of course depend on how large the tank is.)
3. No. Convert 1300 mm Hg to (say) dynes/cm^2, then multiply by 4 cm^2 to calculate F(dynes).
4. 79 mm Hg

p. 59.

1. No
2. Pressure relative to a vacuum; pressure relative to atmospheric pressure.
3. 735 mm Hg (absolute); 20 mm Hg of vacuum.
4. 4 in. Hg; 33.9 in. Hg

p. 62.

1. See Figure 3.4-3; 0–7000 atm; gauge.
2. See Figure 3.4-4.
3. **(a)** True; **(b)** true; **(c)** false
4. −14 mm Hg

p. 65.

1. Immerse in ice-water mixture, mark mercury level as 0°. Immerse in boiling water, mark level as 100°. Divide 0 to 100 interval into 100 equal subintervals, label appropriately.

2. $1°C$
3. $1°C$

p. 84.

1. Semibatch, transient
2. Batch, transient
3. Semibatch, transient
4. **(a)** Continuous, transient; **(b)** Continuous, steady state

p. 90.

3. Steady-state, either no reactions or no net mole change on reaction (e.g., $A \rightarrow B$ but not $A \rightarrow 2B$ or $A + B \rightarrow C$).
4. Steady-state, A is nonreactive.
5. Steady-state, no density change from inlet to outlet. (Good approximation for liquids and solids, effectively requires no reactions and constant temperature and pressure for gases.)

p. 95.

2. $W = 250(1 - x)/60$
3. $R = (75)(1.595)/(154)$
4. $B = 50 + Q$; $q = 0.25Q$; $x = 0.75Q/(50 + Q)$

p. 99.

1. $(\text{atoms})_{in} = (\text{atoms})_{out}$ for each atomic species; multiply all stream amounts by a constant factor; an assumed amount of an input or output stream.
2. **(a)** Flow rates are 1000, 20 000, and 21 000 (all kmol/h), mole fractions are unchanged.
 (b) Flow rates are 200, 100, and 100 (all lb_m/min), mass fractions are unchanged.

p. 102.

1. H_2: $(5 \, lb_m)_{in} = (1 \, lb_m + 4 \, lb_m)_{out}$; O_2: $(5 \, lb_m)_{in} = (4 \, lb_m + 1 \, lb_m)_{out}$; total mass: $(10 \, lb_m)_{in} = (10 \, lb_m)_{out}$
2. Balance on B, solve for Q_1; balance on C, solve for x; total mass balance, solve for M.

p. 120.

1. Yes
2. 4

3. (4 moles H_2O produced)/(6 moles O_2 consumed)
4. $(400)(6)/(4) = 600$
5. 200 mol/min

p. 122.

1. C_2H_4
2. 100%
3. 50 kmol O_2; 100 kmol C_2H_4O; 50 kmol
4. 50 kmol C_2H_4; 75 kmol O_2; 50 kmol C_2H_4O; 25 kmol
5. 0.80; 0.40; 40 kmol

p. 129.

1. 0.90
2. 80%
3. 16 mol B/mol C
4. 80 mol, 10 mol

p. 134.

1. 60 mol; 0.6
2. **(c)** $I = O + C$: $250 = 130 + (60 \text{ mol } CH_4)(2 \text{ mol } O_2/1 \text{ mol } CH_4)$
 (d) $G = O$: $(60 \text{ mol } CH_4)(2 \text{ mol } H_2O/1 \text{ mol } CH_4) = 120$
 (e) $I = O$: $250 = 130 + 60 + (120 \text{ mol } H_2O)(1 \text{ mol } O_2/2 \text{ mol } H_2O)$

p. 135.

Overall $= 100/110 = 0.909$; Single pass $= 100/200 = 0.50$

p. 144.

1. 21% O_2, 79% N_2; $79/21 = 3.76$ moles N_2/mole O_2
2. 25% H_2, 25% O_2, 50% H_2O; 50% H_2, 50% O_2
3. 20; 0.95; $5/95 = 0.0526$

p. 146.

1. 200 mol O_2/h
2. 200 mol O_2/h
3. (4.76×200) mol air/h
4. $(2 \times 4.76 \times 200)$ mol air/h
5. 50%

p. 184.

1. $200 \text{ cm}^3/\text{s}$; no; increase (assuming H_2O expands, which it does); look up $\rho_{H_2O}(75°C)$, $V = 200/\rho$

2. ρ_{Hg} varies with T, and $P = \rho g h$

3. $V_{tot} = \dfrac{m_1}{\rho_1} + \dfrac{m_2}{\rho_2} + \ldots + \dfrac{m_n}{\rho_n}$; $\quad \dfrac{1}{\rho} = \dfrac{V_{tot}}{m_{tot}} = \dfrac{m_1}{m_{tot}\rho_1} + \dfrac{m_2}{m_{tot}\rho_2} + \ldots = \dfrac{x_1}{\rho_1} + \dfrac{x_2}{\rho_2} + \ldots$

p. 187.

1. A relation between pressure, specific volume, and temperature of a substance; $P\hat{V} = RT$; a statement of a phenomenon that always occurs in the same manner under given conditions.

2. Same; $m_{CO_2} > m_{H_2}$

3. C_2H_4; double

4. $P\hat{V} = RT \quad 1/\hat{V} = P/RT$ (mol/volume)
 ρ (mass/volume) $= P/RT$ (mol/volume) $\times M$ (mass/mol) $= \boxed{PM/RT}$.

p. 189.

1. See Table 5.2-1.

2. $V_{new} = V_{old}/2$; $V_{new} = 2 V_{old}$

3. Decreases (n is unchanged, V increases); nothing

4. (a), (c), and (e)

p. 191.

1. (b), (d)

2. 5 bars; 50 m^3; p_{H_2} increases, v_{H_2} is unchanged

3. Greater than

p. 200.

1. a (pressure-volume2/mole2); b (volume/mole)

2. Critical temperature and pressure

p. 201.

1. Accurate, adaptable to computers; cumbersome

2. $P\hat{V} = zRT$; dimensionless

3. 81 liters/min

p. 207.

1. No; $T_r = (-190)/(T_c + 8)$, $P_r = 300/(P_c + 8)$, look up z on charts, calculate $V = znRT/P$.
2. Need a different chart for each species.
3. See p. 206.

p. 208.

See Example 5.3-4. Nonpolar compounds with similar critical properties.

p. 225.

1. Distillation. The naphthas come off from the top of the column, the lubricating oils from the bottom, and the heating oils from the middle.
2. Evaporation, filtration, centrifugation
3. Evaporation, reverse osmosis (membrane filtration at high pressure)
4. Condensation, absorption

p. 229.

1. $-5°C$, 3 mm Hg
2. $-56.6°C$, 5.112 atm
3. All CO_2 solidifies at 1 atm, solid melts at 9.9 atm and $-56°C$, liquid boils at 9.9 atm and $-40°C$
4. 1 atm; 9.9 atm
5. $-78.5°C$; $-56°C$, $-40°C$.

p. 235.

1. Cox chart (Figure 6.1-4); Antoine equation (Equation 6.1-4)
2. Plot $p*$ versus $1/T_{absolute}$ on semilog paper, draw line through points and extrapolate to $1/T_4$.
3. The plot is more likely to be linear on the Cox chart.

p. 237.

1. See p. 236.
2. **(a)** 0; **(b)** 2; **(c)** 2; **(d)** 3; **(e)** 3

p. 243.

1. Yes; yes.

2. 200 mm Hg; 600 mm Hg; 200/960; look up or calculate the temperature at which $p^*_{acetone} = 960$ mm Hg.

3. The temperature to which a gas must be cooled before any constituent condenses; superheated, saturated.

4. $y_{H_2O} = p^*_{H_2O}(T_0)/P_0$ **(a)** nothing; **(b)** condense; **(c)** condense; **(d)** nothing

5. $p^*_{H_2O}(T_{dp}) = y_{H_2O}P_0$. Look up the temperature at which the vapor pressure is $p^*_{H_2O}$.

6. $T_{dp} = T_0 - T_{sh}$; $y_{H_2O} = p^*_{H_2O}(T_{dp})/P_0$.

p. 244.

1. 82°C

2. 50%

3. $s_m = 0.111$, $s_p = 44.4\%$

p. 249.

1. Raoult's law: $p_A = x_A p^*_A$, where x_A is the mole fraction of A in the liquid phase; most likely to be valid as $x_A \rightarrow 1.0$.

2. Henry's law: $p_A = x_A H_A$; most likely to be valid when $x_A \rightarrow 0$.

3. The distribution coefficient is the ratio of the mole fraction of a component in one phase to the mole fraction of that component in a second phase, when the two phases are in equilibrium.

4. Henry's law for x_{CO_2} and p_{CO_2}, Raoult's law for x_{H_2O} and p_{H_2O}. Look up Henry's law constant for CO_2 (*Perry's Chemical Engineers' Handbook*, 6th Edition, p. 3-101) and vapor pressure of H_2O (Table B.3 of this book).

p. 256.

1. The temperature at which the first bubble of vapor forms if the liquid is heated at the given pressure. The temperature at which the first drop of liquid forms if the vapor mixture is cooled at the given pressure.

2. 92°C; 0.70 mol benzene/mol (from Figure 6.4-1a)

3. 99°C; 0.30 mol benzene/mol (from Figure 6.4-1a); decreases

4. Increase (see Equation 6.4-4); increase (see Equation 6.4-6)

5. The hydrostatic head of the liquid must be added to the pressure at the liquid surface. Convert 5 ft of water to atm, add it to 1 atm, and look up the boiling point of water at the corrected pressure. (The answer would still not be precisely correct since it neglects surface tension effects.)

p. 261.

1. *jk* indicates the composition of $MgSO_4 \cdot 7\,H_2O$ crystals in equilibrium with $MgSO_4 \cdot H_2O$ solutions whose compositions are given on line *cd*.

2. A hydrated salt has water present in its crystal lattice in a fixed stoichiometric ratio to the anhydrous species.

3. $136°F$; 52.7% $MgSO_4$–47.3% H_2O

4. Mass ratio of crystals to saturated solution $= 0.141$; mass fraction of crystals $= 0.123$.

p. 264.

1. A solution property that depends only on concentration, and not what the solute and solvent are. Vapor pressure, boiling point, and freezing point.

2. 850 mm Hg. Solute is nonvolatile, nondissociative, and nonreactive with the solvent; Raoult's law holds.

3. Greater. $p^* = (1000$ mm Hg$)/0.85 = 1176$ mm Hg.

4. It lowers the freezing point of water, so it keeps ice from forming at temperatures where it would otherwise form.

5. There would be more "particles" in solution, so the vapor pressure should be lowered by more than 20 mm Hg. If dissociation is complete, we might expect a lowering of 40 mm Hg.

p. 268.

1. See Answer to Test Yourself on p. 249, no. 3. Extraction is transference of a solute from one liquid solvent to another.

2. Less soluble; $m_{VA} \gg m_W$

p. 269.

1. A tie line connects compositions of two phases in equilibrium.

2. H_2O rich phase—95.0% H_2O, 2.5% acetone, 2.5% MIBK; MIBK-rich phase—92.5% MIBK, 2.5% H_2O, 5.0% acetone. Mass ratio of MIBK phase to H_2O phase $= \dfrac{0.950 - 0.450}{0.450 - 0.025} = 1.18$

p. 298.

1. Kinetic, potential, internal; heat, work

2. Heat is only defined in terms of energy being transferred.

3. $E + Q + W = E_f$

p. 299.

1. The mass flow rates are the same; $\rho_{outlet} < \rho_{inlet}$; $V_{outlet} > V_{inlet}$.

2. $\Delta E_p > 0$, $\Delta E_k > 0$

p. 301.

1. Closed system: no mass crosses system boundaries
 Open system: mass crosses system boundaries
 Adiabatic system: no heat transferred to or from system
2. $Q = 250$ J
 $W = -250$ J
3. $\Delta U = -50$ kcal
4. If the substance is a liquid or a solid, or a gas under nearly ideal conditions, it is reasonable to neglect the dependence of U on pressure.

p. 304.

1. $V_{in} = V_{out}$
2. $P_{in} > P_{out}$

p. 305.

1. 6000 cal
2. 1000 cal/min
3. The specific volume and pressure: $\hat{H} = \hat{U} + P\hat{V}$

p. 307.

1. $W_s = 0$
2. $Q = 0$
3. $\Delta E_k = 0$
4. $\Delta E_p = 0$

p. 311.

1. A property whose change of value in any process depends only on its initial and final states and not on the path between them.
2. (a) 0 (b) 5000 J/kg
 (c) $\Delta \hat{H} = \hat{H}_A(v, 0°C, 1\ atm) - \hat{H}_A(v, 30°C, 1\ atm)$
 $= (5000\ J/kg - 7500\ J/kg) = -2500$ J/kg
 (e) No—\hat{H} is a state property.

p. 321.

1. Incompressible fluid, negligible heat transfer, and no internal energy changes due to anything but friction.
2. Above, plus no friction or shaft work.

p. 342.

1. Any path for which data are available may be chosen for the evaluation of $\Delta\hat{H}$.
2. **(b)** Heat to the normal boiling point at 1 atm, vaporize the liquid, heat the vapor to 400°F at 1 atm.
 (c) Cool H_2O to 25°C, mix H_2O and NaOH at 25°C, heat mixture from 25°C to 50°C—all processes at 1 atm.
 (d) Cool gases to 25°C, react at 25°C [taking $H_2O(v)$ as a product], heat products from 25°C to 300°C.

p. 343.

1. **(a)** ii **(b)** i **(c)** iii
2. Determine z for C_2H_6 at each of the system conditions. If z is close to 1 for both states, it is reasonable to neglect $\Delta\hat{H}$.

p. 346.

1. $C_v = (\partial\hat{U}/\partial T)_v$, $C_p = (\partial\hat{H}/\partial T)_P$
2. **(a)** exact, **(b)** inaccurate, **(c)** good approximation
3. $\Delta H = (5\text{ g})(0.5\text{ cal/g·°C})(20°C) = 50$ cal

p. 352.

1. $C_v \approx 5$ cal/mol·K (ideal gas)
 $C_v \approx 7$ cal/mol·K (liquid)
2. **(a)** $\Delta\hat{H} = (1\text{ cal/g·°C})(25°C) = 25$ cal/g
 (b) $\Delta\hat{H} = 0 - (2\text{ cal/g·°C})(75°C) = -150$ cal/g
 (c) $\Delta\hat{H} = (2\text{ cal/g·°C})(75°C) - (1\text{ cal/g·°C})(25°C) = 125$ cal/g

p. 354.

1. $CaCO_3(s)$: $C_p = 26 + 7.5 + 3(17) = 84.5$ J/mol·°C
2. $\Delta H = [(2\text{ kg})(2.5\text{ kJ/kg·°C}) + (1\text{ kg})(1.8\text{ kJ/kg·°C})](10°C) = 68$ kJ
 $\Delta\hat{H} = 68$ kJ/3 kg $= 23$ kJ/kg
3. $\Delta\hat{H} = 25$ cal/g
4. $(C_p)_{mixture} = [(0.50)(1.00) + (0.50)(0.54)]$ cal/g $= 0.77$ cal/g

p. 364.

1. Yes; no
2. **(a)** 0; **(b)** 900 J/mol
 (c) Cool the vapor to 75°C; condense at 75°C; cool the liquid to 25°C.

3. $\Delta \hat{U}_f \approx 5500$ cal/mol
$\Delta \hat{U}_v = \Delta \hat{H}_v - RT = 26\,710$ cal/mol

4. Raising the temperature raises the average kinetic energy of the liquid molecules. More of these molecules therefore have enough energy to escape from the liquid, raising the vaporization rate. For the condensation rate to be raised by an equal amount (the condition for equilibrium), the frequency of collisions of vapor molecules with the liquid surface must be increased, which entails raising the pressure.

p. 366.

1. Equation 8.4-3
2. Equation 8.4-4
3. Slope of tangent to curve at $p^* = 1$ atm equals $- \Delta \hat{H}_v/R$
5. Equation 8.4-8
6. $\Delta \hat{H}_v(T_2) = \int_{T_2}^{T_1} C_{pl} \, dT + \Delta \hat{H}_v(T_1) + \int_{T_1}^{T_2} C_{pv} \, dT$

p. 375.

Air at 25°C, $h_r = 20\% \Rightarrow h_a = 0.0040$ kg H_2O/kg dry air;
$T_{wb} = 12.5°C$; $T_{dp} = 0.5°C$; $\hat{V}_H = 0.85$ m^3/kg dry air,
$\hat{H} = (35.00 - 0.27)$ kJ/kg dry air $= 34.73$ kJ/kg dry air

p. 378.

1. (a) $(C_p)_{gas}$, $(C_p)_{liquid}$, and $(\Delta \hat{H}_v)_{liquid}$ are independent of temperature, and the enthalpy change of unevaporated liquid is small compared to the heat of vaporization and the enthalpy of the gas.
 (b) Yes; (c) No
2. $T_{as} = 11°C$; $h_a = 0.0069$ kg H_2O/kg dry air;
 $h_r = 70\%$; $\hat{H} = (31.60 - 0.06)$ kJ/kg dry air $= 31.54$ kJ/kg dry air

p. 381.

1. (a) $\hat{H} = -40$ kJ/mol A (b) $\hat{H} = 20$ kJ/mol A
2. $Q = \Delta H = (5)(-40)$ kJ $= -200$ kJ (evolved)
3. $Q = 5(-60 + 40)$ kJ $= -100$ kJ (evolved)

p. 390.

1. 130°F, 0.15
2. 600 Btu/lb$_m$
3. Vapor fraction $\approx (0.50–0.18)/(0.88–0.18) = 0.46$

p. 425.

1. $H_{\text{products}} - H_{\text{reactants}}$ when stoichiometric quantities of reactants are fed, the reaction proceeds to completion, and reactants and products are at the same temperature and pressure. As above, with reactants and products at 25°C and 1 atm.

2. **(a)** -20 kJ/mol A reacted; **(b)** exothermic; **(c)** withdrawn; **(d)** higher

3. The latent heat of vaporization of C_6H_{14} at 25°C and 1 atm

4. $\Delta \hat{U}_r(T) = \Delta \hat{H}_r(T) + 2RT$

5. $\Delta \hat{H}_r = \sum_{\text{products}} v_i \hat{H}_i - \sum_{\text{reactants}} v_i \hat{H}_i$

 In general, $\hat{H}_i = \hat{U}_i + P\hat{V}_i$; for liquids and solids, $\hat{H}_i \approx \hat{U}_i$; and for ideal gases, $P\hat{V}_i = RT$, so that $\hat{H}_i = \hat{U}_i + RT$. To obtain the desired result, substitute for \hat{H}_i in the expression for $\Delta \hat{H}_r$, and recognize that

 $$\Delta \hat{U}_r = \sum_{\text{products}} v_i \hat{U}_i - \sum_{\text{reactants}} v_i \hat{U}_i$$

p. 427.

1. See last paragraph of Section 9.2.

2. Multiply equation 2 by -2 and add the result to equation 1

 $2A + B - 2A - 2D \rightarrow 2C - 2C - 6E$

 \Downarrow

 $B + 6E \rightarrow 2D, \quad \Delta \hat{H}_r = \Delta \hat{H}_{r1} - 2\Delta \hat{H}_{r2} = 3000$ kJ/mol

p. 429.

1. The formation reaction of CO is $C + \frac{1}{2}O_2 \rightarrow CO$, so the heat of formation is $-\frac{1}{2}$ times the heat of the given reaction, or -110.5 kJ/mol. (This checks with the value in Table B.1.)

2. The latent heat of vaporization of propane at 25°C.

3. $\Delta \hat{H}_r^\circ = (\Delta \hat{H}_f^\circ)_{CO_2} + 2(\Delta \hat{H}_f^\circ)_{H_2O(v)} - (\Delta \hat{H}_f^\circ)_{CH_4}$

p. 445.

1. **(a)** $\Delta \hat{H}_f^\circ(\text{soln}, n = 50) = (\Delta H_f)_A - 100 - 10 = -110$ kJ/mol A
 (b) $\Delta \hat{H}_f^\circ(\text{soln}, n = \infty) = -115$ kJ/mol A

2. **(a)** $\hat{H}(25°C) = -110$ kJ/mol A
 (b) $H(25°C) = -550$ kJ

p. 448.

1. $LHV = 2658$ kJ/mol
 $HHV = 2658$ kJ/mol $+ 5(44.013$ kJ/mol$)$
 $= 2878$ kJ/mol

2. $HHV = (0.40)(143 \text{ kJ/g}) + (0.60)(55 \text{ kJ/g}) = 90.2 \text{ kJ/g}$

3. For coal:

$$(\$150/\text{ton})(1 \text{ lb}_m/15\,000 \text{ Btu})(1 \text{ ton}/2000 \text{ lb}_m) = 5.0 \times 10^{-6} \text{ \$/Btu}$$

For natural gas:

$$x(\$/\text{ton})(1 \text{ ton}/2000 \text{ lb}_m)(1 \text{ lb}_m/23\,000 \text{ Btu}) = 5.0 \times 10^{-6} \text{ \$/Btu}$$
$$\Downarrow$$
$$x = \$230/\text{ton}$$

p. 452.

1. The product temperature when the fuel is burned completely in an adiabatic reactor.

2. Heat loss through reactor walls, incomplete reaction.

3. With an air feed, the heat released from combustion of a fixed amount (say, 1 mol) of fuel goes into heating the nitrogen in the air as well as the reaction products and excess oxygen, whereas only the latter must be heated for a pure O_2 feed. If a fixed quantity of heat is added to a greater amount of material, the resulting temperature rise must be less.

p. 455.

1. **(a)** Ignition—abrupt increase in the rate of an oxidation reaction.
 (b) Ignition temperature—the temperature at which ignition occurs when a combustible mixture is heated slowly.
 (c) Ignition lag—time between a mixture reaching its ignition temperature and the occurrence of ignition.
 (d) Flammability limits—composition limits outside of which ignition or explosion cannot occur.
 (e) Flash point of a liquid—temperature at which a liquid gives off enough vapor to form an ignitable mixture with the air above the liquid surface.

2. **(a)** An explosition or ignition; no reaction would occur.
 (b) Yes; the reaction would cease.
 (c) There is a region between the emerging jet (pure methane) and the bulk of the room air (almost no methane) within which the methane fraction falls between the flammability limits. The flame persists in this region.

p. 458.

1. A flame is a combustion zone in which various species present undergo transitions from states of high energy to state of lower energy; the energy lost in the transitions is given off as visible light.

2. In low-temperature flames, which occur when the fuel and oxygen are not well mixed, carbon particles are formed and remain unburned, but the heat in the

reaction zone raises them to a temperature at which they glow incandescently; the result is a yellow flame. At higher combustion temperatures, various intermediate species are formed and excited to high energy states, and emit blue light as they return to lower states; the result is a blue flame.

3. In a detonation, a pressure front (shock wave) forms and propagates at supersonic speed, rapidly compressing and igniting the flammable mixture, and giving the appearance of an instantaneous combustion. In the vicinity of the detonation, the force of the shock wave can demolish a building or propel a bullet. The noise is the vibration of the eardrums caused by the shock wave.

4. The rate of the reaction depends strongly on temperature; at room temperature it is immeasurably low.

p. 487.

1. (a) 1; (b) x_1
2. Ten variables (flows of A, B, C, and D at inlet, A, B, and C in vapor product, and B, C, and D in liquid product) − 4 balances (A, B, C, and D) − 2 relations (given split of B, equilibrium relation for C) = 4 degrees of freedom.

p. 489.

Four variables (inlet water, outlet hydrogen, oxygen, and water) − 3 balances (hydrogen, oxygen, water) + 1 reaction = 2 degrees of freedom. (We need a basis of calculation + one additional piece of information to be able to solve for the remaining variables.)

p. 493.

1. Mixer—3, separator—4
2. 5 degrees of freedom (7 local degrees of freedom − 2 ties)

p. 537.

1. (a) Accumulation, input, output (although the mass balance will yield the result that accumulation = 0); (b) All but generation; (c) Accumulation, output, consumption
2. $dm/dt = 10 \text{ kg/h} - 10 \text{ kg/h} = 0 \text{ kg/h}$

p. 549.

1. Neglect changes in E_k and E_p; no mass accumulation; \hat{U} and \hat{H} independent of P; no phase changes or chemical reactions; no spatial variations in T; constant C_v and C_p.

2. When the expression for U_{sys} is differentiated to yield Equation 11.3-9, if C_v varies with temperature (and hence with time) another term of the form $M(T_{sys} - T_r)(dC_v/dt)$ would have to be added.

p. 592.

1. $-1, +2$
2. The line for which the sum of the squares of the residuals is a minimum.
3. Yes. (Corollary—no.)
4. Positive and negative deviations from the line would cancel out, possibly making a terrible fit look good.

p. 594.

1. Linear
2. Nonlinear
3. Nonlinear
4. (a) Linear; (b) nonlinear
5. (a) Linear (multiply equation by z); (b) nonlinear; (c) linear

p. 595.

1. A straight line on a plot of f versus x can only intersect the x-axis at one point.
2. One root. (The intersection of a 45° line through the origin with a curve that begins at 1 when $x = 0$ and decreases, approaching 0 as x approaches infinity.)

p. 596.

1. (a) $x = 4$ and $x = 0$. (b) $x_p = 15/4$; $x_n = 5$. Converge to $x = 4$.
2. The equation for a line on an f·versus x plot through (x_n, f_n) and (x_p, f_p) is $f = f_n + [(f_p - f_n)/(x_p - x_n)](x - x_n)$. The intercept of this line on the x axis, which we will call x_{new}, is obtained by setting $f = 0$ and solving the resulting equation for x. The result is Equation A.2-1.

p. 599.

1. Yes. No. You can solve for the roots directly.
2. The value of $f'(x_0)$ would be 0, and the correction term f/f' would blow up.

p. 601.

1. (a) Modified successive substitution; (b) successive substitution; (c) acceleration
2. 13.6

p. 602.

1. 2.60
2. A line on an f versus x plot through (2, 2.3) and (2, 3.245) intersects the 45° line at $x = 2.60$.

p. 603.

1. Equation A.2-8, with $\varepsilon = 1.0 \times 10^{-6}$.
2. 0.001, 0.0001.
3. Equation A.2-8, since the denominator approaches zero as x approaches the root.

p. 613.

1. 80
2. 84
3. 80. The function is cubic, so Simpson's rule is exact.

ANSWERS TO SELECTED PROBLEMS

Note: Most of the given solutions are incomplete.

Chapter Two

2. (c) 5.37×10^3 kJ/min
6. 48 600 miles
8. (c) 10.0 lb$_f$
10. $W(N) = 3.45 \times 10^4 [(\pi/2) - (1 - h)\sqrt{1 - (1 - h)^2} - \sin^{-1}(1 - h)]$
16. Re $= 1.4 \times 10^3$, so the flow is laminar.
18. (b) $\Delta P = 5.37 [\mu^{0.15} N \rho^{0.85} \ldots]$
20. (b) $\rho = 70.5 \exp [8.27 \times 10^{-7} P]$
22. $Q = 6.69 \times 10^{-3} = 0.00669$
24. (b) $t(C = 0.1$ mol/liter$) = 1.12$ min (extrapolated), 1.70 min (exact)
26. (b) $y = -1 + [0.75 x^2 + 2.25]^{-1}$
28. (d) $y = \{1 + 2 \exp [3(x - 2)]\}^{1/2}$
30. $k = 9.3 \times 10^{-3}$ min^{-1}
32. $k = 1.567$
34. (b) $a = 31.5$ V \cdot (s/liter)$^{1/2}$ kPa
36. (b) $b = -0.525$
40. (c) $E = 7200$ cal/mol, $D_0 = 0.046$ cm^2/s

Chapter Three

2. (a) 3.00×10^2 kg
4. 3.32
6. Change in boat level $= (W_a/\rho_w A_b)[1 - (A_b/A_p)(1 - \rho_w/\rho_a)]$
·8. (f) 80.0 kmol H, (i) 80.64 kg H
10. (b) 2.60×10^2 kg/s
12. (c) 0.300 g $CaCO_3$/g
14. 0.431 g propane/g
16. 2010 mol air/h
18. Sample 1 contains 6.2% methane, 17.3% ethane, 35.3% propane, and 41.2% *n*-butane by mass.
20. (b) 0.095 g benzene/g
22. (c) $C_A(200\text{s}) = 5.30$ M
24. (b) 139 kPa
26. (b) 30.8 m
28. (b) 3880 mm
30. 8.9 mm Hg

32. **(b)** 23 mm Hg
34. **(c)** $v = 453$ ml/s
36. **(b)** 6.12 in^3
38. **(b)** 3.26 °C/s
40. **(b)** 918 lb$_m$ O$_2$
42. At $T = 94°C$, $k = 0.707$ $l/mol \cdot min$

Chapter Four

6. **(b)** Two unknowns must be specified.
8. 0.49 lb$_m$ strawberries
10. **(b)** 666 kg/h
12. $9.54/kg
14. 207 gal
16. **(b)** 1500 kg galena
18. **(c)** $400 000/day
20. **(b)** $R_A = 37.1$, $R_B = 33.3$, $R_x = 7.4$
22. **(c)** 3.4 h
24. 9.31 metric tons ore
26. 5.0 liters
28. **(c)** $x = 0.05$, $Q = 2.26$ m^3/h
30. 39.8%
32. **(b)** 206%
34. **(a)** 10.0% bypass
36. **(b)** 80.9 kmol/h
38. **(b)** 96.6% Whizzo
40. **(d)** 1305 lb$_m$ NO/ton N$_2$O$_4$
42. **(c)** 250% excess
44. **(d)** 3125 kg C$_2$H$_2$/h
46. 0.060 mol O$_2$/mol
48. **(b)** 138 ft^3, **(d)** 88%
50. **(b)** 742K
52. **(b)** 45 mol HCHO/mol CO$_2$
54. **(b)** 7.52 mol C$_2$H$_4$Cl$_2$/100 mol C$_2$H$_5$Cl
56. **(a)** Fractional yield = 0.865
58. **(b)** 20.0 kmol CH$_3$OH/h
60. **(c)** 124 kmol CH$_4$/h

62. 0.246 mol fresh feed/mol fed to reactor

64. 694 kmol/h i-C_4H_{10} recycled from still; 1480 kmol H_2SO_4/h, 799 kmol H_2O/h in acid recycle

66. 5.5% C_3H_8 (dry basis)

68. 50.0% excess air

70. **(b)** 178 kmol air/h

72. **(c)** 8.9% CO_2 (wet basis), 10.0% CO_2 (dry basis)

74. 30.1% excess air

76. **(b)** 46% excess air

78. 11.5% CO_2

80. **(b)** 372 mol air/metric ton coal

82. 30% bypass

84. **(a)** 10.4% CO_2, 0.18% SO_2, 7.3% O_2

Chapter Five

2. Bulk density $= 3.26$ g/cm^3

4. **(b)** 9.44 lb$_m$

6. $v = 151$ m/s

8. $M = 3.9$ g/mol

10. **(b)** 0.0046 g/s

12. **(a)** 2.75 m^3/h

14. 0.170 mol O_2/mol

16. **(b)** 58.5 g/liter

18. **(c)** 56.4 m/s^2

20. **(c)** 0.215 g/min

22. 22 510 SCMH

24. **(b)** 0.74

26. **(b)** 75%

28. **(d)** 1.46 g/liter

30. **(a)** 0.152 kg acetone condensed

32. 6.6 mole% HCl

34. **(b)** $C_{11}H_{20}O_{10}$

36. 383 m^3(STP)/100 kg ore

38. **(b)** 45 ppm SO_2

40. 590.2 kg CH_3COOH/h

42. **(b)** $K_p(350K) = 7.98$ atm

44. 25.8 mole% CO_2, 6.5% CO, . . .
46. $P(-50°C, 10 \text{ atm}) = 1.79$ liters/mol
48. For CO_2, $T_{max} = 464.7°C$
50. For CO_2 at 200K and 6.8 atm, $V = 2.11$ liter/mol
52. **(b)** 83.1 kg/m^3
54. 0.168 m^3/h
56. 0.082 m^3
58. 180 ft^3/min (answer depends on how chart is read)
60. 299 m^3/h

Chapter Six

2. **(b)** $P_f = 116$ mm Hg
4. $\Delta \hat{H}_v = 34.1$ kJ/mol
6. **(b)** $\Delta \hat{H}_v = 36.7$ kJ/mol
8. **(c)** 0.797 g MEK(v)/liter
10. **(b)** 83.7% condensation
12. The air contains 4.38 mole% water
14. 0.310 m^3 outlet air/m^3 feed air
16. 80.6 min
18. $P_{out} = 2015$ mm Hg
20. **(a)** 89% recovery if cooling tower water is used
22. 2.1 gal condensate/min
24. **(b)** Cool to 7.8°C
26. **(a)** 0.0142 kmol methanol/min
28. 7.5% H_2O in the exiting pulp
30. 0.872 kg condensed/kg entering
32. 0.791 mol CH_4/mol
34. **(b)** 3.35×10^{-3} mol C_3H_8/mol
36. **(b)** 138 m^3 H_2O(l) fed to the absorber
38. **(b)** 0.134 m^3/kg liquid feed
40. 5.45×10^{-3} mol H_2O/mol
42. **(b)** 1.98×10^{-4} mol/liter
44. **(a)** 0.768 mol C_6H_6/mol in the vapor phase
46. 0.038 mol C_6H_6/mol in the liquid feed to the reboiler
48. $T_{dp} = 63.5°C$
50. From Raoult's law, $T_{bp} = 150°F$
52. **(d)** At 50°C, 5 stages are required

54. **(c)** $x_B = 0.30$, $y_B = 0.51$
56. **(d)** $T = 71°C$
58. $T_b = 72.2°C$
60. 5820 mol solvent/h
62. 88 kg H_2O evaporate/min
64. **(c)** 23.0 kg A
66. 44% precipitates
68. $p^*(25°C) = 89.8$ mm Hg
70. $M = 73.5$
72. 36 g in the benzene phase
74. 1.67% unextracted
76. Phase A contains 27% acetone, 5% MIBK, and 68% water
78. 81% extracted

Chapter Seven

2. **(a)** 715 Btu
4. 113 J/s
6. 0.0660 J/s, -44.1 J/s
10. **(b)** $Q = -675$ J
12. **(b)** 2338 J/mol
14. **(c)** $Q = 0.485$ kJ
16. **(b)** $\hat{U}(45°C) = 440$ J/mol
18. 0.425 kW
20. **(b)** 2.96×10^5 cal/min
22. **(a)** 650 kJ/min
24. **(i)** 3.98×10^7 J/s
26. $Q = 84\,570$ kW
28. **(a)** 1.90×10^5 Btu/min
30. **(b)** 13.4°C for a clothed person
32. **(a)** 3.00×10^5 kJ
34. **(a)** Every 13 sec
36. $m_B = 0.601$ kg
38. **(c)** 11.8 hp
40. 2.00×10^2 kJ/SCM
42. **(a)** 26 300 cal
44. 0.994 kg vapor/kg
46. 0.0828 lb_m

48. (a) $T_{\text{oven}} = 122°C$
50. $T_f = 210°C$, $P_f = 19$ bars
52. (b) 4.6 cm
54. (a) 27.7 m/s
56. 82.5 m³/min
58. 0.55 kW

Chapter Eight

2. If three terms are used, $Q = 11\ 518$ J
4. (d) 25 960 J/mol
8. (b) 13 500 kJ
12. 2137 kW
14. 1890 Btu/h
16. (b) 37.21 J/mol·°C
18. -8.6% error
20. 433 kJ/kg
22. 1810 m²
24. 3560 kJ
26. 10.2 g/min
28. 0.186 kg steam/m³ gas
30. 1.39×10^6 J/day
32. 42.5°C
34. 693°C
36. (b) 3.44×10^9 J/min
38. 2670 kW
40. 43.8 kJ/mol
42. 10.4 kJ/min
44. (b) 2.55×10^8 J
46. 100 kJ/mol
48. (b) 34.3 kJ/mol
50. 112 tons
52. (b) 3080 Btu/min
54. (c) $(T_{f2} - 35) = -16.8°C$
56. 339 kJ/kg feed
58. (b) 99.6 kg/h additional steam
60. (c) 3733 kg/h fresh water
62. (c) 0.51 g evaporate

64. 1760 kW

66. (b) 764 lb_m coolant/h

68. (b) $T_F = 52.2°C$

70. (a) 0.826 kg steam generated in the boiler; 973 kJ removed in the condenser

72. $T_{wb} = 80°F$

74. (d) 23 J

76. (a) 79.6 tons

80. (b) 0.113 kg H_2O/min

82. (b) 0.137 lb_m H_2O/min

84. (b) $-14\,000$ Btu/h

86. (b) 13.94 kJ/mol HCl

88. 51\,700 J/mol H_2SO_4

90. 497 kJ/liter

92. $Q = +1890$ kJ/min

94. (d) 3290 lb_m/h

96. (b) $T_{max} = 130°C$

98. (b) $\hat{H}_f = 84$ Btu/lb_m

100. (b) $\hat{H} = 44$ Btu/lb_m

102. 86\,000 Btu/h

Chapter Nine

2. (a) $Q = -18\,370$ kW

4. (b) 97.4 kJ

6. 122.7 kJ/mol

8. (c) $Q = -126\,240$ kJ/h

10. (b) $Q = -165$ kJ/m³(STP)

12. 4.99×10^6 Btu/ton Fe

14. 1290 kg/h

16. -0.048 kW (reactor), -0.807 kW (condenser)

18. (a) $Q = -1.3$ kJ/mol feed

20. (b) $Q = -279$ kW (process), -265 kW (reactor)

22. (b) $Q_A = 3.67 \times 10^5$ kJ/h, $Q_F = 1.34 \times 10^6$ kJ/h, $Q_C = 5.65 \times 10^4$ kJ/h

24. (c) 55.9 kmol recycled/h

26. (d) $f = 0.360$ mol ethane reacted/mol fed

28. 80.1% conversion

30. 0.417 mol C_2H_2/mol CH_4 consumed

32. $H = -4424$ kJ

34. **(b)** $Q = -69.3$ kJ
36. **(c)** $50.0°C$
38. **(a)** 177 kJ/mol $(NH_4)_2SO_4$
40. $94.5°C$
42. **(b)** 1080 kJ/kg C_6H_5Cl
44. HHV = 2400 kJ/mol
46. 985.6 kJ/mol
48. **(b)** 1.01×10^{-6} kg SO_2/kJ
50. -6.2×10^5 kJ/h
52. **(b)** 83.1 MW
54. **(c)** 12 100 kg steam/h
56. **(a)** 99% excess air
58. $2040°C$
60. **(a)** LHV = 42.3 kJ/g
62. $V = 174$ m^3/h
64. **(a)** 1155 SCMH CH_4, 706 SCMH $O_2 (+N_2)$

Chapter Ten

4. **(a)** $n_{df} = 7$
6. **(b)** 1.8 mol recycle/mol fresh feed
8. **(c)** $Q = 19.5$ kW
10. **(d)** $Q = -24.3$ kW
12. **(c)** $T_{out} = 1222$ K

Chapter Eleven

4. **(b)** 0.24 lb-mole O_2
6. **(c)** $t = 26.5$ min
8. **(b)** $C(2 \text{ min}) = 3.82 \times 10^{-7}$ mol SO_2/liter
10. **(c)** $V = 20.1$ m^3
12. **(b)** 17.2 h
14. **(c)** $k = 0.582$ liter/mol·s **(e)** 70.4 min
16. **(d)** For $n_p = 51$, $t = 90.4$ min
20. **(b)** 2860 W
22. **(c)** 100 min
24. **(b)** 200 g benzene remain
26. **(b)** $54.8°C$

Index

ATOMIC WEIGHTS AND NUMBERS

Atomic weights apply to naturally occurring isotopic compositions and are based on an atomic mass of $^{12}C = 12$

Element	Symbol	Atomic number	Atomic weight	Element	Symbol	Atomic number	Atomic weight
Actinium	Ac	89	—	Iridium	Ir	77	192.2
Aluminum	Al	13	26.9815	Iron	Fe	26	55.847
Americium	Am	95	—	Krypton	Kr	36	83.80
Antimony	Sb	51	121.75	Lanthanum	La	57	138.91
Argon	Ar	18	39.948	Lawrencium	Lr	103	—
Arsenic	As	33	74.9216	Lead	Pb	82	207.19
Astatine	At	85	—	Lithium	Li	3	6.939
Barium	Ba	56	137.34	Lutetium	Lu	71	174.97
Berkelium	Bk	97	—	Magnesium	Mg	12	24.312
Beryllium	Be	4	9.0122	Manganese	Mn	25	54.9380
Bismuth	Bi	83	208.980	Mendelevium	Md	101	—
Boron	B	5	10.811	Mercury	Hg	80	200.59
Bromine	Br	35	79.904	Molybdenum	Mo	42	95.94
Cadmium	Cd	48	112.40	Neodymium	Nd	60	144.24
Calcium	Ca	20	40.08	Neon	Ne	10	20.183
Californium	Cf	98	—	Neptunium	Np	93	—
Carbon	C	6	12.01115	Nickel	Ni	28	58.71
Cerium	Ce	58	140.12	Niobium	Nb	41	92.906
Cesium	Cs	55	132.905	Nitrogen	N	7	14.0067
Chlorine	Cl	17	35.453	Nobelium	No	102	—
Chromium	Cr	24	51.996	Osmium	Os	75	190.2
Cobalt	Co	27	58.9332	Oxygen	O	8	15.9994
Copper	Cu	29	63.546	Palladium	Pd	46	106.4
Curium	Cm	96	—	Phosphorus	P	15	30.9738
Dysprosium	Dy	66	162.50	Platinum	Pt	78	195.09
Einsteinium	Es	99	—	Plutonium	Pu	94	—
Erbium	Er	68	167.26	Polonium	Po	84	—
Europium	Eu	63	151.96	Potassium	K	19	39.102
Fermium	Fm	100	—	Praseodymium	Pr	59	140.907
Fluorine	F	9	18.9984	Promethium	Pm	61	—
Francium	Fr	87	—	Protactinium	Pa	91	—
Gadolinium	Gd	64	157.25	Radium	Ra	88	—
Gallium	Ga	31	69.72	Radon	Rn	86	—
Germanium	Ge	32	72.59	Rhenium	Re	75	186.2
Gold	Au	79	196.967	Rhodium	Rh	45	102.905
Hafnium	Hf	72	178.49	Rubidium	Rb	37	84.57
Helium	He	2	4.0026	Ruthenium	Ru	44	101.07
Holmium	Ho	67	164.930	Samarium	Sm	62	150.35
Hydrogen	H	1	1.00797	Scandium	Sc	21	44.956
Indium	In	49	114.82	Selenium	Se	34	78.96
Iodine	I	53	126.9044	Silicon	Si	14	28.086